AF863036

Lecture Notes in Electrical Engineering

Volume 882

Series Editors

Leopoldo Angrisani, Department of Electrical and Information Technologies Engineering, University of Napoli Federico II, Naples, Italy
Marco Arteaga, Departament de Control y Robótica, Universidad Nacional Autónoma de México, Coyoacán, Mexico
Bijaya Ketan Panigrahi, Electrical Engineering, Indian Institute of Technology Delhi, New Delhi, Delhi, India
Samarjit Chakraborty, Fakultät für Elektrotechnik und Informationstechnik, TU München, Munich, Germany
Jiming Chen, Zhejiang University, Hangzhou, Zhejiang, China
Shanben Chen, Materials Science and Engineering, Shanghai Jiao Tong University, Shanghai, China
Tan Kay Chen, Department of Electrical and Computer Engineering, National University of Singapore, Singapore, Singapore
Rüdiger Dillmann, Humanoids and Intelligent Systems Laboratory, Karlsruhe Institute for Technology, Karlsruhe, Germany
Haibin Duan, Beijing University of Aeronautics and Astronautics, Beijing, China
Gianluigi Ferrari, Università di Parma, Parma, Italy
Manuel Ferre, Centre for Automation and Robotics CAR (UPM-CSIC), Universidad Politécnica de Madrid, Madrid, Spain
Sandra Hirche, Department of Electrical Engineering and Information Science, Technische Universität München, Munich, Germany
Faryar Jabbari, Department of Mechanical and Aerospace Engineering, University of California, Irvine, CA, USA
Limin Jia, State Key Laboratory of Rail Traffic Control and Safety, Beijing Jiaotong University, Beijing, China
Janusz Kacprzyk, Systems Research Institute, Polish Academy of Sciences, Warsaw, Poland
Alaa Khamis, German University in Egypt El Tagamoa El Khames, New Cairo City, Egypt
Torsten Kroeger, Stanford University, Stanford, CA, USA
Yong Li, Hunan University, Changsha, Hunan, China
Qilian Liang, Department of Electrical Engineering, University of Texas at Arlington, Arlington, TX, USA
Ferran Martín, Departament d'Enginyeria Electrònica, Universitat Autònoma de Barcelona, Bellaterra, Barcelona, Spain
Tan Cher Ming, College of Engineering, Nanyang Technological University, Singapore, Singapore
Wolfgang Minker, Institute of Information Technology, University of Ulm, Ulm, Germany
Pradeep Misra, Department of Electrical Engineering, Wright State University, Dayton, OH, USA
Sebastian Möller, Quality and Usability Laboratory, TU Berlin, Berlin, Germany
Subhas Mukhopadhyay, School of Engineering & Advanced Technology, Massey University, Palmerston North, Manawatu-Wanganui, New Zealand
Cun-Zheng Ning, Electrical Engineering, Arizona State University, Tempe, AZ, USA
Toyoaki Nishida, Graduate School of Informatics, Kyoto University, Kyoto, Japan
Luca Oneto, Department of Informatics, Bioengineering., Robotics, University of Genova, Genova, Genova, Italy
Federica Pascucci, Dipartimento di Ingegneria, Università degli Studi "Roma Tre", Rome, Italy
Yong Qin, State Key Laboratory of Rail Traffic Control and Safety, Beijing Jiaotong University, Beijing, China
Gan Woon Seng, School of Electrical & Electronic Engineering, Nanyang Technological University, Singapore, Singapore
Joachim Speidel, Institute of Telecommunications, Universität Stuttgart, Stuttgart, Germany
Germano Veiga, Campus da FEUP, INESC Porto, Porto, Portugal
Haitao Wu, Academy of Opto-electronics, Chinese Academy of Sciences, Beijing, China
Walter Zamboni, DIEM - Università degli studi di Salerno, Fisciano, Salerno, Italy
Junjie James Zhang, Charlotte, NC, USA

The book series *Lecture Notes in Electrical Engineering* (LNEE) publishes the latest developments in Electrical Engineering - quickly, informally and in high quality. While original research reported in proceedings and monographs has traditionally formed the core of LNEE, we also encourage authors to submit books devoted to supporting student education and professional training in the various fields and applications areas of electrical engineering. The series cover classical and emerging topics concerning:

- Communication Engineering, Information Theory and Networks
- Electronics Engineering and Microelectronics
- Signal, Image and Speech Processing
- Wireless and Mobile Communication
- Circuits and Systems
- Energy Systems, Power Electronics and Electrical Machines
- Electro-optical Engineering
- Instrumentation Engineering
- Avionics Engineering
- Control Systems
- Internet-of-Things and Cybersecurity
- Biomedical Devices, MEMS and NEMS

For general information about this book series, comments or suggestions, please contact leontina.dicecco@springer.com.

To submit a proposal or request further information, please contact the Publishing Editor in your country:

China

Jasmine Dou, Editor (jasmine.dou@springer.com)

India, Japan, Rest of Asia

Swati Meherishi, Editorial Director (Swati.Meherishi@springer.com)

Southeast Asia, Australia, New Zealand

Ramesh Nath Premnath, Editor (ramesh.premnath@springernature.com)

USA, Canada:

Michael Luby, Senior Editor (michael.luby@springer.com)

All other Countries:

Leontina Di Cecco, Senior Editor (leontina.dicecco@springer.com)

**** This series is indexed by EI Compendex and Scopus databases. ****

More information about this series at https://link.springer.com/bookseries/7818

Mohd Hasnun Arif Hassan ·
Mohd Hafizi Zohari · Kumaran Kadirgama ·
Nik Abdullah Nik Mohamed ·
Amir Aziz
Editors

Technological Advancement in Instrumentation & Human Engineering

Selected papers from ICMER 2021

Springer

Editors
Mohd Hasnun Arif Hassan
Faculty of Mechanical and Automotive Engineering Technology
Universiti Malaysia Pahang
Pekan, Pahang, Malaysia

Mohd Hafizi Zohari
Faculty of Mechanical and Automotive Engineering Technology
Universiti Malaysia Pahang
Pekan, Malaysia

Kumaran Kadirgama
Faculty of Mechanical and Automotive Engineering Technology
Universiti Malaysia Pahang
Pekan, Malaysia

Nik Abdullah Nik Mohamed
Faculty of Mechanical and Automotive Engineering Technology
Universiti Malaysia Pahang
Pekan, Malaysia

Amir Aziz
Faculty of Mechanical and Automotive Engineering Technology
Universiti Malaysia Pahang
Pekan, Malaysia

ISSN 1876-1100 ISSN 1876-1119 (electronic)
Lecture Notes in Electrical Engineering
ISBN 978-981-19-1576-5 ISBN 978-981-19-1577-2 (eBook)
https://doi.org/10.1007/978-981-19-1577-2

© The Editor(s) (if applicable) and The Author(s), under exclusive license
to Springer Nature Singapore Pte Ltd. 2023
This work is subject to copyright. All rights are solely and exclusively licensed by the Publisher, whether the whole or part of the material is concerned, specifically the rights of translation, reprinting, reuse of illustrations, recitation, broadcasting, reproduction on microfilms or in any other physical way, and transmission or information storage and retrieval, electronic adaptation, computer software, or by similar or dissimilar methodology now known or hereafter developed.
The use of general descriptive names, registered names, trademarks, service marks, etc. in this publication does not imply, even in the absence of a specific statement, that such names are exempt from the relevant protective laws and regulations and therefore free for general use.
The publisher, the authors, and the editors are safe to assume that the advice and information in this book are believed to be true and accurate at the date of publication. Neither the publisher nor the authors or the editors give a warranty, expressed or implied, with respect to the material contained herein or for any errors or omissions that may have been made. The publisher remains neutral with regard to jurisdictional claims in published maps and institutional affiliations.

This Springer imprint is published by the registered company Springer Nature Singapore Pte Ltd.
The registered company address is: 152 Beach Road, #21-01/04 Gateway East, Singapore 189721, Singapore

To SARS-CoV-2 that has been rocking the world since 2019, on your bike and give us a break!

Foreword

Research in the Mechanical Engineering field has evolved into becoming a multidisciplinary field that is not only focusing on Mechanical Engineering per se, but also integrating knowledge in electronics, instrumentation, programming, signal processing, materials science, and so on. This is due to the fact that the solution to a problem nowadays does not only rely on a single discipline approach, but often requires an integration with other disciplines. Take a car for example, a purely mechanical system in the past has now been highly integrated with electronics, control system, artificial intelligence, and so on that results in only 50% or less components in a car being purely mechanical. This is just one of many examples of what was then a purely mechanical system, now has evolved into an integrated system that involves other disciplines.

Technological Advancement in Instrumentation & Human Engineering gathers selected papers presented in the International Conference on Mechanical Engineering Research 2021 (ICMER 2021) that covers topics related to human engineering, instrumentation, materials, modelling and simulation, and signal processing. This proceeding book symbolizes the multidisciplinary Mechanical Engineering research has evolved into. It is hoped that this book will promote multidisciplinary research in the Mechanical Engineering field towards a more efficient and comprehensive solution to a problem at hand. As a wise man who goes by the name of Isaac Newton once said, "If I have seen further, it is by standing on the shoulders of giants".

Regards,

Mohd Hasnun Arif Hassan
Corresponding Editor

Contents

Human Engineering

Driving Data Analysis for the Development of Kuala Terengganu Driving Cycle . . . 3
I. N. Anida, J. S. Norbakyah, W. N. J. H. W. Yussof, P. Walker, and A. R. Salisa

Investigation on Lane Changing Manoeuvres of an Autonomous Vehicle by Using Model Predictive Control (MPC) and Proportional-Integral-Derivative Control (PID) for Steer-by-Wire (SBW) System . . . 15
P. M. Heerwan, N. E. Z. Mohamad, M. A. Zakaria, M. I. Ishak, and S. M. Asyraf

Improvement of Brake Response Time on Trailing Vehicle: A Review . . . 29
Norrizal Mustaffa, Syabillah Sulaiman, Fathul Hakim Zulkifli, Khairu Kamarudin, and Shaiful Rizal Masrol

The Validity of Football Skills Test for 12-Year-Old Male Players . . . 41
Adjah Naqkiah Mazlan, Mohd Hairul Azam Mesnan, Mohamad Rasidi Pairan, and Mohd Hizwan Mohd Hisham

Factors Contributing to the Pedal Error or Pedal Misplacement Among Malaysian Car Drivers: A Survey . . . 51
Mohamad Zairi Baharom, Zulkifli Ahmad, Nursya Mimie Ayuny Ismail, Mohd Hasnun Arif Hassan, Juffrizal Karjanto, and Khairil Anwar Abu Kassim

Preliminary Study on the Influence of Boot Studs on Rugby Players' Sprinting Performance . . . 61
Sharul Hadi Turiman, Zulkifli Ahmad, and Nasrul Hadi Johari

Ergonomic Chair Design in Minimizing MSD Chassis Assembly Workers Complaints Using Ergonomic Function Deployment (EFD) . . . 73
Nelfiyanti, Nik Mohamed, M. F. F. A. Rashid, and Chon Chin Seik

A Rack and Pinion Integrated Mechanism for Gym Forearm Machine Power Generation . . . 89
Ammar A. M. Al-Talib, Ain Atiqa, and Lih Jing Soo

Drowsiness Detection for Safe Driving Using PERCLOS and YOLOv2 Method . . . 101
Ngo Kah Lock, Weng Mun Ng, Norrabiatul Adawiyah Jusoh, Nazhatul Hafizah Kamarudin, Rizauddin Ramli, and Zuliani Zulkoffli

A Study of the Effect of Industrial 4.0 on Improving the Manufacturing Performance: A Case Study in Miri and Bintulu . . . 113
Mohd Adzrie, Brandon Chagat, Radhika Lahei Arechinan, Sri Tharunan Naidu, and Umar Abdul Karim

Instrumentation

Design and Fabrication of Nutrient Film Technique (NFT) Hydroponic System . . . 123
Mohd Rafiuddin Rozilan, Ana Syahidah Mohd Rodzi, Ahmad Faiz Zubair, Abdul Rahman Hemdi, Rasdi Deraman, and Nor Diyana Md Sin

Performance of Food Dehydrator Using Different Type of Distributor Base Plate . . . 145
Muhamad Hakimi Mokhtar, Muhamad Helmi Ashraf Termizi, A. S. M. Yudin, and A. Alias

Comparing Industry Training Using Virtual Reality Against Conventional Training: A Case Study . . . 161
Christian Stark, Salome Wiener, and Jeng Feng Chin

IoT Enabled Piezoelectric Energy Harvesting Floormat . . . 177
Chong Lye Lim, Hashwinni Rajaretnam, Sarina Tajudin, and Mohammed W. Muhieldeen

Dynamic Modelling of Drone Systems Using Data-Driven Identification Methods . . . 191
Osama Yaser Osman Suliman, Ir. Fahri Heltha, Ts. Muhamad Faliq, and Aulia Rahman

Design of Intelligent Mixer Machine for Food Applications . . . 213
A. S. F. Mahamude, D. Ramasamy, W. S. W. Harun, K. Kadirgama, J. Mogan, Suhelmy Azhar Bin Said, and Kaniz Farhana

Temperature Profile of Mixed Mode Solar Cabinet Coconut Dryer . . . 231
Keith Yvonne B. Diez, Jao Philip A. Yap, and Cresencio P. Genobiagon Jr.

Performance of Solar Cabinet Dryer Utilizing Thermosyphon Water Heater 243
Cresencio P. Genobiagon Jr.

Development of Small-Scale PETE Plastic Bottle Shredder with Electronic Sensor Controls for Cap Separation: Basis for Reverse Vending Machine 253
Alec Christian M. Cervantes, Ace Vann Cardiff T. Aleria, Christian Carlo B. Taer, and Cresencio Jr. P. Genobiagon

Design of Flora Inspired Savonius Wind Turbine 265
Apple Jee P. Ladao and Delan S. Bacus

Arduino-Based Electronic Bicycle Transmission Switching System 275
Lance Gabriel M. Atencio, John Stephen E. Cena, Ian U. Opalla, and Randy E. Angelia

A Review on the Bolted Flange Looseness Detection Method 287
Mohd Padzly Radzi and Mohd Hafizi Zohari

Pedestrian Lane Control System with Alert System 299
Emmanuel V. Galang, Noel R. Portillo, and Jetron J. Adtoon

Design and Analysis of 6 Degree of Freedom Robotic Whiteboard Cleaner 313
Mark Ryan M. Estrera and Cresencio P. Genobiagon

Materials

Study on Mangrove Barks Activated Carbon (MBAC) for Fibre Reinforce Plastic (FRP) Rehabilitation Pile Structure 327
Z. Salleh and T. M. I. A. T. Mazlan

Preliminary Tensile Investigation of FDM Printed PLA/Coconut Wood Composite 339
J. Kananathan, K. Rajan, M. Samykano, K. Kadirgama, K. Moorthy, and M. M. Rahman

Preliminary Tensile Investigation of FDM Printed PLA/Copper Composite 351
A. Kottasamy, K. Rajan, M. Samykano, K. Kadirgama, K. Moorthy, and M. M. Rahman

A Narrative Review: Bamboo Fiber as an Alternative Source for Pulp and Paper 363
Mohd Faizal Esa, Nor Mazlana Main, Mohd Nazrul Roslan, Noraini Marsi, Khairu Kamarudin, and Latifah Jasmani

Investigation of Curing Process of Silver Conductive Ink on Polymer Substrates Using Halogen Lamp and Oven 387
Muhammad Najmi Zainal, Rd Khairilhijra Khirotdin, Siti Nur Elida Eraman, Muhamad Fahrul Nizam Suhaimi, and Nurhafizzah Hassan

Review: Two-Dimensional Layered Material Based Electrodes for Lithium Ion and Sodium Ion Batteries 399
Omama Javed and Radhiyah Binti Abd Aziz

Design of an Experimental Test Rig to Determine the Battery Internal Resistance Beyond Normal Operating Condition 419
Abdullah Jubair, Zul Hilmi Che Daud, Izhari Izmi Mazali, Zainab Asus, and Mohd Kameil Abdul Hamid

Mechanical Performance of Cornstalk Fibre/Fibreglass/Polyphenylene Sulfide (PPS) Composites Bicycle Frame Using Finite Element Analysis 431
Humeshvaren Maganathan, Cik Suhana Hassan, Muhamad Faliq Mohamad Nazer, and Nor Fazilah Abdullah

Study on Low Frequency Vibration Isolation Characteristics of Transformer by Phononic Crystal 445
Zhaokai Sun, Bo Zhang, Yudong Pan, Shilong Lu, and Yuyang Mao

Study on the Band Gap Characteristics of Two-Dimensional Local Resonant Phononic Crystals 459
Yuan Xing, Bo Zhang, Yao Zhang, Jiaxing Song, and Meng Wang

Modelling and Simulation

Numerical Study of Encased RTP Behavior Under Internal Pressure 473
M. S. Noorazizi and N. A. H. Jasni

Airflow Analysis of Contra-Rotating Fans Performance by Numerical Simulation 483
Teo Ting He, Muhammed Abdelfattah Sayed Abdelaal, Izzuddin Zaman, Djamal Hissein Didane, and Bukhari Manshoor

Scouring Around Rigs-to-Reefs Jacket Platform with Different Sitting Configurations on Seabed 497
Mohd Asamudin A. Rahman, Mohd Hairil Mohd, Ahmad Fitriadhy, Muhammad Nadzrin Nazri, Erwan Hafizi Kasiman, and Fatin Alias

Conceptualizing an Industry 4.0's Predictive Maintenance System in a Medical Devices Manufacturing Enterprise 513
Christian Stark and Jeng Feng Chin

Optimization of Temperature Rise in Turning Using Single Objective Genetic Algorithm 527
Mimi Muzlina Mukri and Nor Atiqah binti Zolpakar

Alternative Railway Tools and Sustainability in RAMS: A Review 541
M. A. Muhammed Nor, A. F. Yusop, M. A. Hamidi, M. N. Omar, N. A. Abdul Hamid, and W. M. Wan Mohamed

Brief Review on Recent Advancement of Computational Analysis on Hemodynamics in Peripheral Artery Disease 555
U. Z. Shahrulakmar, M. N. Omar, and N. H. Johari

Using Microsoft Project as Planning, Monitoring and Controlling Tool for Project Success – A Case Study of Construction Projects in Malaysia 573
Kanesan Muthusamy, Nagesparan Ainarappan, Elango Natarajan, and Batumalay Kaliannan

Numerical Studies for Small-Scale Solar Chimney Power Plants with Various Geometric Configurations 587
Mohd Noor Asril Saadun, Nor Azwadi Che Sidik, Teng Meng Xian, and Mohd Afzanizam Mohd Rosli

Design and Simulation of Diffuser Augmented Wind Turbine (DAWT) for Urban Areas 605
Delan S. Bacus and Cresencio P. Genobiagon Jr.

Modeling the Effect of Different Locations of Carotid Atherosclerosis on Hemodynamics Parameters 615
A. Fahmi Huwaidi M. Noor and Nasrul Hadi Johari

Preliminary Study of Offshore Wind Turbine Foundation in Malaysia 629
M. A. N. Mutasim, M. H. Mansor, and H. A. Salaam

Study on the Recognition of Driver's Starting Intentions Based on Fuzzy Inference-SVM Cascade Algorithm 643
Hongtao Hao and Chao Zhang

Signal Processing

A Review on the Application of Fiber Bragg Grating Sensors in Bolted Joints Health Monitoring 655
M. S. N. A. Adhreena and Z. M. Hafizi

High-Speed Camera Analysis of Tool Eccentricity During Friction Stir Welding of Thick Plate Aluminium Alloys 669
L. H. Ahmad Shah, S. Walbridge, and A. Gerlich

Lamb Wave Actuation Techniques for SHM System-A Review 677
N. Ismail, Z. M. Hafizi, Kok-Sing Lim, and Harith Ahmad

The Development of Euler Solver Based on Flux Vector Splitting and Modified TVD Schemes 687
Iman Fitri Ismail, Bambang Basuno, Akmal Nizam Mohammed, Farzad Ismail, and Nurul Farhana Mohd Yusof

Applicability of Peak Detection Methods for Composite Fatigue FBG Wavelength 703
M. Loman, M. H. Zohari, and F. Lamin

Controllability Analysis of Convergence on Water Distribution System Architecture with Fault Using Cellular Automata Sequence 711
Nurul Hannah Mohd Yusof, Nurul Adilla Mohd Subha, Norikhwan Hamzah, Fatimah Sham Ismail, Mohd Ariffanan Mohd Basri, and Anita Ahmad

Fault Diagnosis of Spark Plug in a Spark Ignition Engine by Using Wavelet Power Spectrum 723
A. A. Azrin, I. M. Yusri, A. Aziz, M. F. Jamlos, and R. Mamat

Image Reconstruction Technique Using Radon Transform 735
Teh Chia Ai, Wan Zailah binti Wan Said, Norsuzlin Mohd Sahar, and Mohammad Tariqul Islam

Human Engineering

Driving Data Analysis for the Development of Kuala Terengganu Driving Cycle

I. N. Anida, J. S. Norbakyah, W. N. J. H. W. Yussof, P. Walker, and A. R. Salisa

Abstract Driving cycle plays a vital role in the production and evaluating the performance of the vehicle. Driving cycle is a speed-time data set and as an important input for vehicle emission models. A problem coming with the second-by-second speed driving data is to analyze the big driving data. To analyze this big data, it is necessary to choose best big data analysis methods, which give opportunity to store, preprocess, detect outlier and apply classification or clustering algorithms. In this study, a set of driving data is stored, managed and analyzed using Tall Arrays (TA) and *k*-means clustering algorithms in MATLAB for the development of Kuala Terengganu Driving Cycle (KTDC). The objectives of this paper are; to store and manage driving data using TA in MATLAB, to develop a KTDC by using *k*-means clustering, and lastly to analyze the energy consumption and emissions of KTDC. Firstly, the driving data is collected in five different routes in Kuala Terengganu city at go-to-work times. Then the data is stored and analyzed in the MATLAB. The development of KTDC is by using *k*-means clustering approach. Finally, the energy consumption and emissions of KTDC is analyzed by using AUTONOMIE software. KTDC is successfully developed with 35.15 km/h in average speed and 12 micro-trips.

Keywords Driving cycles · Hybrid electric vehicles · Fuel economy · Emissions · Tall arrays · *k*-means clustering · Micro-trips

---

I. N. Anida · J. S. Norbakyah · W. N. J. H. W. Yussof · A. R. Salisa (✉)
Faculty of Ocean Engineering Technology and Informatics, Universiti Malaysia Terengganu, 21030 Kuala Nerus, Terengganu, Malaysia
e-mail: salisa@umt.edu.my

J. S. Norbakyah · A. R. Salisa
Energy Storage Research Group (ESRG), Universiti Malaysia Terengganu, 21030 Kuala Nerus, Terengganu, Malaysia

Renewable Energy and Power Research Interest Group (REPRIG), Universiti Malaysia Terengganu, 21030 Kuala Nerus, Terengganu, Malaysia

P. Walker
School of Mechanical and Mechatronic Engineering, Faculty of Engineering and Information Technology, University of Technology Sydney, Ultimo, NSW 2007, Australia

© The Author(s), under exclusive license to Springer Nature Singapore Pte Ltd. 2023
M. H. A. Hassan et al. (eds.), *Technological Advancement in Instrumentation & Human Engineering*, Lecture Notes in Electrical Engineering 882,
https://doi.org/10.1007/978-981-19-1577-2_1

1 Introduction

Driving cycle is a representative speed-time profile of driving conduct of a particular locale or city [1–3]. The driving cycle characterizes the conduct of the vehicle on the road and has a broad range of users, from designing activity control frameworks to deciding the execution of vehicles. It is additionally utilized within the emission testing of vehicles for certification of emission standards [4]. It is widely used of application for vehicle manufacturers, environmentalists and traffic engineers.

Generally, the well-known drive cycles such as the WLTC (Worldwide harmonized Light vehicle Test Cycle), NEDC (New European Drive Cycle), FTP (Federal Test Procedure) and Japanese modes were seen originated from Europe, United States, Japan, China and India which are significant in the global automotive industry. Nevertheless, there are also other areas that developed local drive cycles such as Toronto, Canada [5], Iran [6], Singapore [7], Hong Kong [8] and others.

However, each one of the constructed driving cycles is not representing the actual situation in Kuala Terengganu (KT), the capital city of the state of Terengganu, Malaysia. According to [9], driving cycles play a critical role in vehicle design, and if vehicle manufacturers focus solely on a single driving cycle during vehicle construction and design, there is a risk that the design is optimized for this specific driving cycle, and the result for another driving cycle may be non-robust and suboptimal. Therefore, an actual KT city work route driving cycle needs to be constructed and characterized in order to help other researchers to continue the studies related to emissions and fuel consumption in KT city.

Vehicle driving cycle is a series of point for speed of vehicle versus time which is mainly used to evaluate the performance of either the vehicle or engine. However, tremendous and big parallel data for thousands of seconds' data will cram the system and will cause various problems such as data storage efficiency, security issues, services quality, data governance policies and many more. Big Data (BD) is a new popular term used to describe the remarkably rapid increase in the volume of structured and unstructured data. Big data accuracy can lead to more assured decision making, and better decisions can lead to increased operational efficiency, cost savings, and risk reduction [10].

In the MATLAB R2020a version, users are presented with a data array format called Tall Arrays (TA) to make BD operations easy [11]. With this type of data, it is now possible to process BDs that are quite long, normally unprocessed and scattered. Moreover, with this new data type, users continue to use the MATLAB code and functions they are accustomed to.

In this study, a sample of TA application was performed using the actual driving data in Kuala Terengganu city. The second-by-second speed data are collected at Go-to-Work (GTW) time which at 7.30, 8.00 and 8.30 a.m. with 10 runs of data along five different routes. Then the data is analyzed to develop a Kuala Terengganu Driving Cycle (KTDC).

The structure of the work is organized as follows: in the second chapter, a brief description of the methodology involving data collection, TA in MATLAB, *k*-means clustering, and KTDC development was provided. Finally, in the third chapter, the results and analysis are presented.

2 Methodology

Figure 1 shows the flow chart to develop a driving cycle in Kuala Terengganu (KT) along five different routes which are Route A, B, C, D and E in Kuala Terengganu city at 7.30, 8.00, and 8.30 a.m. The KT driving cycle's inputs are second-by-second speed which collected at GTW times of 7.30, 8.00, and 8.30 a.m., with 10 data runs. All five routes were chosen for their traffic volume in KT city. In this research, the on-board measurement method will be used using a Global Positioning System (GPS). The data gathered then will be stored and managed using TA in MATLAB. Then the features from each micro-trip such as average speed and percentage of idle will be extracted. The clustering of the micro-trips using *k*-means method will be took place in order to develop the final driving cycle of Kuala Terengganu along Route A, B, C, D and E at GTW times.

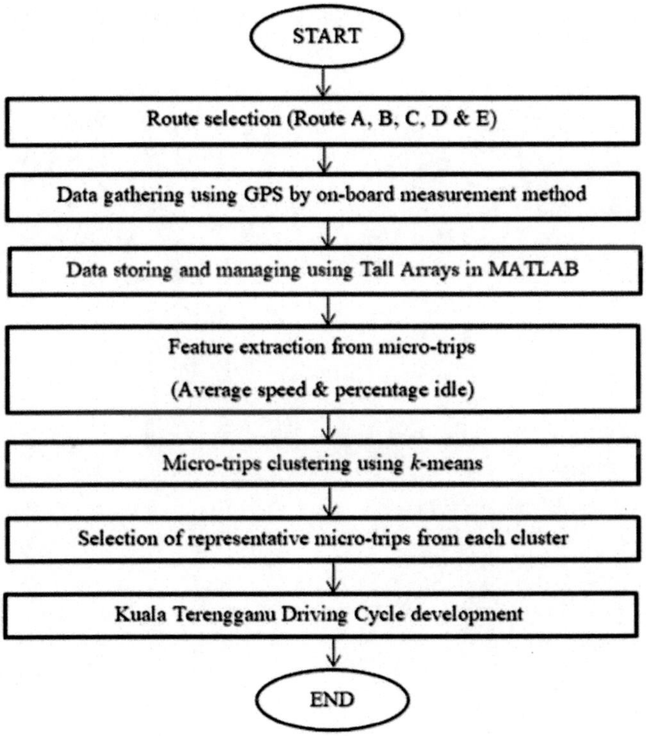

Fig. 1 Flow chart of the study

2.1 Data Collection

Route A, Route B, Route C, Route D, and Route E are the selected routes for KT driving cycle from Kampung Wakaf Tembesu to Wisma Persekutuan as shown in Fig. 2. According to the Malaysian Ministry of Works, these five routes are the most frequently used by Kuala Terengganu residents as 'route-to-work' routes [12]. In this study, GPS-based on-board measurement method speed-time data are collected along the selected route from Kampung Wakaf Tembesu to Wisma Persekutuan. Because of its population, Kampung Wakaf Tembesu was chosen as the starting point. On the other hand, Wisma Persekutuan was chosen as the end point because most government sectors are located in Wisma Persekutuan and it is nearby.

With 10 runs, data was collected at GTW times along the chosen road. There are three types of data collection techniques or methods: chase car technique, on-board measurement technique, and combination of on-board measurement and circulation driving. A chase car technique is a type of data collection that involves following a target vehicle while recording second-by-second speed data. On-board measurement is when speed-time data is collected using a real-time logging system installed on a specific vehicle along a predetermined route. Finally, the combination of two techniques of on-board measurement and chase car is also known as the hybrid method [13]. For the KT driving cycle, the on-board measurement technique will be used for

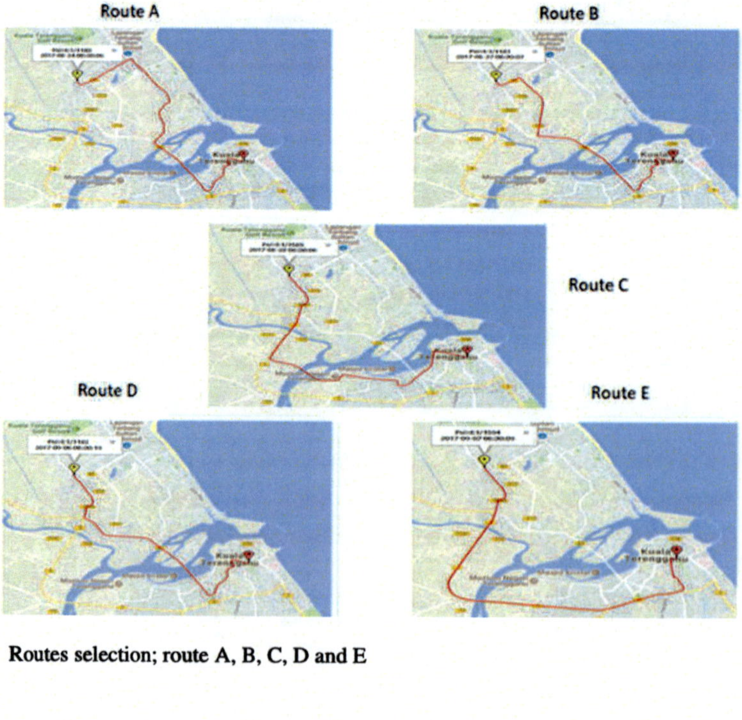

Fig. 2 Routes selection; route A, B, C, D and E

data collection because it is more appropriate for KT drivers' irregular behaviour to avoid risks such as accidents and sudden loss of control.

2.2 *MATLAB Tall Arrays*

Presently, the amount of data observed in business, science, and even individual applications has increased to such an extent where storing and processing data on personal computers is nearly impossible to produce significant results. In point of fact, an analytical system is made up of data collected from any scientific instrument and its associated systems. When working with a BD that is complex to tackle using standard methods on multiple devices, the following problems occur where [14]; Conventional methods do not work, accessing and analyzing data is difficult, learning new tools and programs for BD analysis is complicated, and managing data in physical environments is quite impossible.

Converting huge data sets into smaller bits is one technique to solve these issues. It will also be easy for users to use recognized or recognizable programming language codes (such as MATLAB codes). TA which is firstly introduced with MATLAB R2016b is a new filing scheme that treats different types of data in many different files as a single array/table [15]. TA supports many data types such as numeric, table, date time, categorical and character format. The data obtained can be used in conjunction with statistics and machine learning toolboxes. Figure 3 shows the stages of BD analysis in MATLAB [15].

In this work, MATLAB TA will be used to process a driving dataset which has the size of 60,000 × 6 and also to understand analyzing big data sets that cannot be processed with traditional functions and individual computers. Regardless of the fact that our dataset is not notably big, the functions and applications of the study can be used to examine any non-memory data, despite the number of rows. MATLAB TA analyses large volumes of data in the same way that it analyses small volumes of data. The most significant distinction is that tall arrays do not utilize memory to assess all rows of the dataset unless requested by the user.

A data store must be setup before processing using TA. Users will be able to access a data collection as a result of this. The data in the data collection can be found in

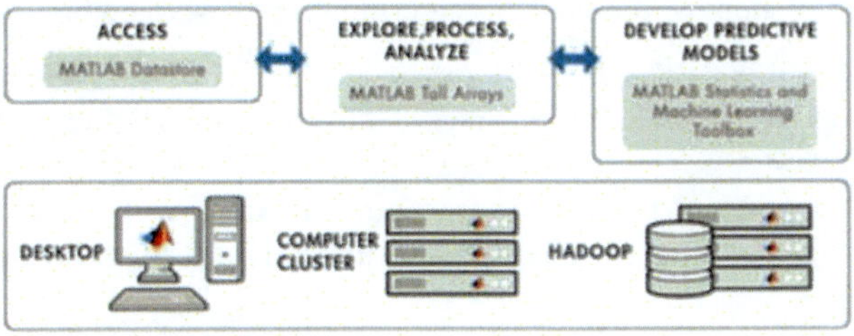

Fig. 3 Analyzing big data using MATLAB

one huge file or multiple smaller files. The data in TA could be numerical, logical, date, time category, or character structured. Only the "tall" function can be used to convert data in the data store into TA format. One of the most important features of MATLAB TA is that it does not immediately evaluate the operations performed, and it should not be evaluated until the user wants to see the results. For this study, only average speed and percentage idle will be calculated in order to develop a driving cycle.

2.3 *K-means Clustering*

Micro trips are used to develop a drive cycle. A micro-trip is a trip between two consecutive time points with zero vehicle velocity [16]. Each micro trip begins with an idle phase and concludes with a decelerating phase that approaches zero. The entire data set must be divided into a number of micro-trips. Following this process, a large number of micro-trips can be obtained for all collected data. The micro-trips are then divided into groups based on traffic conditions, such as congested traffic flow, medium traffic flow, and clear traffic flow. The *k*-means algorithm will be used to cluster the micro-trips.

K-means is one of the most basic unsupervised learning algorithms for clustering. The procedure follows a simple and easy method for classifying a given data set using a fixed number of clusters (assume *k* clusters). Clustering in N-dimensional Euclidean space, R_N, is the process of portioning a given set of n points into a number, say *K*, of groups, based on some similarity or dissimilarity metric. Let the set of n points $\{x_1, x_2, \ldots, x_n\}$ be presented by set *S*, and *K* clusters be represented by $C_1, C_2, \ldots, C_k$. Then:

$$C_i \neq \emptyset \text{ for } i = 1, \ldots, K$$
$$C_i \cap C_j = \emptyset \text{ for } i = 1, \ldots, K, j = 1, \ldots, K \text{ and } i \neq j$$
and:
$$U_{i=1}^{K} C_i = S \quad (1)$$

The *k*-means algorithm endeavours to solve the clustering issue by optimizing a given metric. The steps of the *k*-means algorithms are described briefly below (17):

Step 1: Determine a value for *k*. The value of *k* in this study is determined by the traffic condition, which includes congested traffic flow, medium traffic flow, and clear traffic flow.
Step 2: Set up the *k* cluster centres (randomly, if necessary).
Step 3: Assign the class memberships of the total data, N, to the nearest cluster centre.
Step 4: Re-estimate the *k*-cluster centres based on the memberships found above.

Step 5: Exit if none of the N data changed memberships in the previous iteration. Otherwise, proceed to Step 3.

2.4 KTDC Development

The proposed method for developing the driving cycle is to cluster micro-trips. To cluster the micro-trips, driving features must first be extracted. The micro-trips can be used to extract a variety of driving characteristics. However, only two features will be used for this purpose: average speed and percentage of idle. These two characteristics were chosen because they will have the greatest impact on emission [18]. The micro-trips are clustered into three groups as shown in Fig. 4 using the *k*-means clustering method. Each group has its own characteristics and represents a different traffic condition: clear traffic, medium traffic, and congested traffic.

The representatives of micro-trips are then identified in order to generate the driving cycle for each cluster. The representative micro-trips will be those that are closest to the cluster centre. The micro-trips chosen for each group are depicted in Figs. 5, 6 and 7. The micro-trips will then be combined to produce the final driving cycle of Kuala Terengganu along Routes A, B, C, D, and E at GTW times.

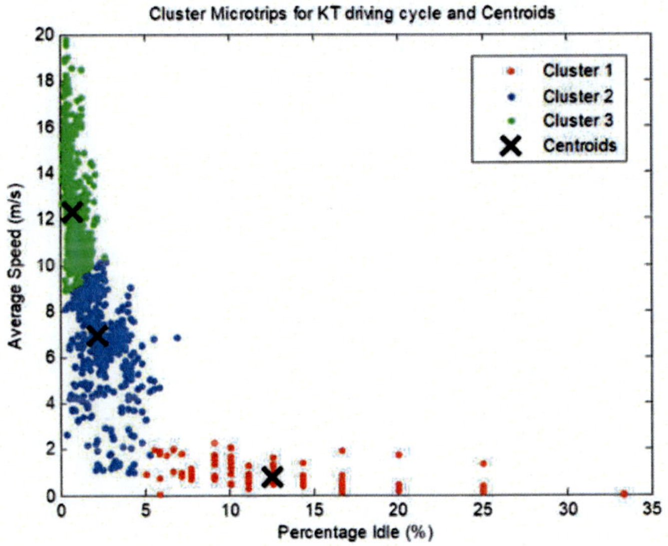

Fig. 4 Clustering of micro-trips

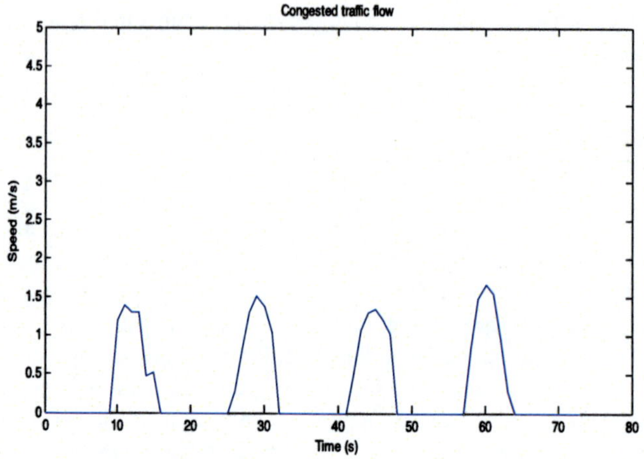

Fig. 5 Cluster one; congested traffic flow

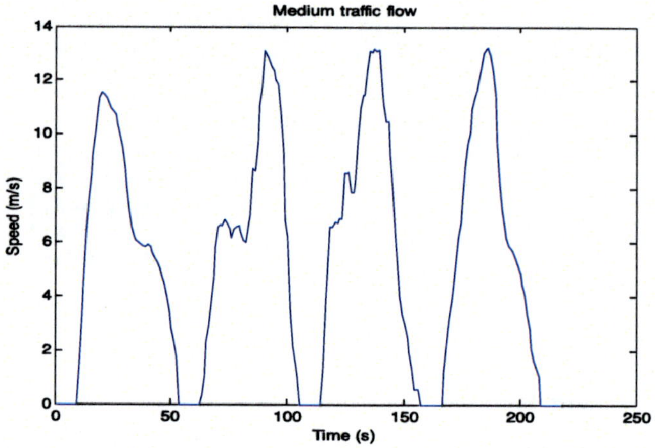

Fig. 6 Cluster two; medium traffic flow

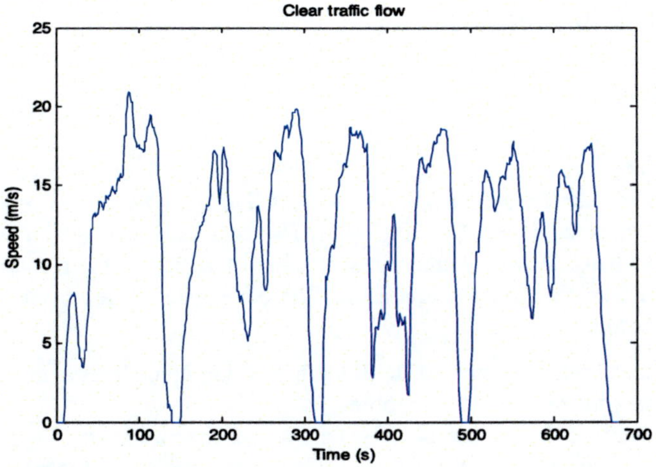

Fig. 7 Cluster three; clear traffic flow

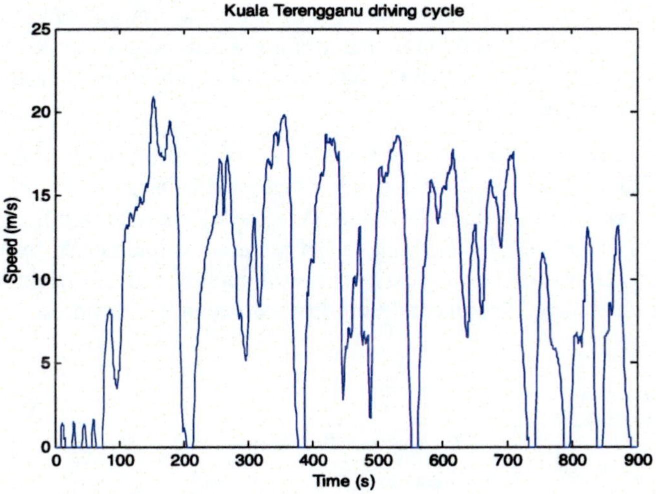

Fig. 8 Kuala Terengganu driving cycle

3 Results and Analysis

3.1 KTDC

After collecting all of the driving cycle data along five different routes at 7.30, 8.00, and 8.30 a.m., the final development of the KT driving cycle can be completed. Figure 8 depicts the final KT driving cycle. The total distance is 8.8 km, with a total of 12 micro-trips. Table 1 summarises the characteristics of the KT driving cycle in terms of nine assessment parameters. The developed driving cycle yielded the following results:

1) The speed range of more than 10 km/h was dominant. This is due to the high volume of traffic on KT city routes.
2) Micro-trips at higher speed ranges are longer than micro-trips at lower speed ranges. This is because a vehicle in free flow travels at a higher speed range with fewer stops due to less traffic congestion.
3) A 35.15 km/h average speed was recorded in the developed KT driving cycle, indicating that the vehicles are moving at a slower speed and that there are more micro-trips found below the average speed. As a result of the frequent stops along the road, more fuel is consumed and emissions are produced during that time period.

Table 2 listed the comparison of the KTDC and existing standard driving cycle which is NEDC (New European Driving Cycle) and UDDS (Urban Dynamometer Driving Schedule). From the table, it shows that there are not much different between KTDC, NEDC and also UDDS since all of them are driving cycle for the city. In Malaysia, NEDC is still being used by the local authorities for legislation purposes and by the local manufacturers and suppliers for evaluation purposes.

Table 1 Assessment parameters of KTDC

Parameters	KTDC
Distance travelled (km)	8.8
Total time (s)	953
Average speed (km/h)	35.15
Average running speed (km/h)	40.55
Average acceleration (m/s^2)	0.59
Average deceleration (m/s^2)	0.56
RMS (m/s^2)	0.73
Percentage idle (%)	12.10
Percentage cruise (%)	5.11
Percentage acceleration (%)	40.51
Percentage deceleration (%)	42.29

Table 2 Comparison between KTDC, NEDC and UDDS

Parameters	KTDC	NEDC	UDDS
Distance travelled (km)	8.8	11.02	12.89
Total time (s)	953	1180	600
Average speed (km/h)	35.15	33.6	31.51
Average running speed (km/h)	40.55	42.24	38.85
Average acceleration (m/s^2)	0.59	0.53	0.50
Average deceleration (m/s^2)	0.56	0.72	0.58
RMS (m/s^2)	0.73	0.14	0.68
Percentage idle (%)	12.10	16.95	17.66
Percentage cruise (%)	5.11	38.81	7.96
Percentage acceleration (%)	40.51	23.56	39.71
Percentage deceleration (%)	42.29	17.29	34.67

Table 3 Comparison of PHEV, HEV and conventional vehicle engine

Parameters	PHEV	HEV	Conventional
Fuel economy (mile/gallon)	212.84	53.78	32.2
Fuel consumption (l/100 km)	1.11	4.37	7.3
CO_2 emissions (g/mile)	26.1	103.3	172.82

3.2 Energy Consumption and Emissions Analysis

After developing the driving cycle, the fuel rate, such as fuel consumption and fuel economy, and emission can be calculated using AUTONOMIE software version v1210. AUTONOMIE is a design, simulation, and analysis tool for automotive control systems. It is forward simulation software that is mathematically based and it is based on MATLAB, with MATLAB data and configuration files and models built in Simulink. Table 3 displays the KTDC fuel rate and emissions for conventional engine vehicles, hybrid electric vehicles (HEV), and plug-in hybrid electric vehicles (PHEV). A vehicle's emissions will produce carbon monoxide (CO_2) gas.

The table clearly shows that the PHEV is the best powertrain among conventional engines and HEVs, with the lowest fuel consumption and emission values and the highest fuel economy value. This is due to the power split type PHEV's design, which incorporates one engine, two motor-generators (MGs), and multiple planetary gears [19]. As a result, it will reduce emissions and energy consumption.

4 Conclusion

The development of KT driving cycle is successfully done using micro-trips and *k*-means clustering method. The data are collected from predetermined initial location to final location along Route A, B, C, D and E at Go-to-Work times which are 7.30, 8.00, and 8.30 a.m. All the driving data is successfully stored and managed in the Tall Arrays in MATLAB for the development of the driving cycle. For the energy consumption and emission analysis, it is proven that PHEV uses less fuel consumption and emits less emission compared to HEV and conventional vehicle powertrain. For the future work, in order to develop an accurate Malaysian Driving Cycle, every states and cities in Malaysia should be considered.

References

1. Zhao X, Zhao X, Yu Q, Ye Y, Yu M (2020) Development of a representative urban driving cycle construction methodology for electric vehicles: a case study in Xi'an. Transp Res D 81:102279
2. Borlaug B, Holden J, Wood E, Lee B, Fink J, Agnew S, Lustbader J (2020) Estimating region-specific fuel economy in the United States from real-world driving cycles. Transp Res D 86:102448
3. Kaymaz H, Korkmaz H, Erdal H (2019) Development of a driving cycle for Istanbul bus rapid transit based on real-world data using stratified sampling method. Transp Res D 75:123–135
4. Arun NH, Mahesh S, Ramadurai G, Shiva SM (2017) Development of driving cycles for passenger cars and motorcycles in Chennai, India. In: Sustainable cities and society
5. Amirjamshidi G, Roonda MJ (2015) Development of simulated driving cycles: case study of the Toronto Waterfront Area. Transp Res D 34:255–266
6. Fotouhi A, Montazeri-Gh M (2013) Tehran driving cycle development using the *k*-means clustering method. Scientia Iranica Trans A Civil Eng 20(2):286–293
7. Ho S-H, Wong Y-D, Chang VW-C (2014) Developing Singapore driving cycle for passenger cars to estimate fuel consumption and vehicular emissions. Atmos Environ 97:353–362
8. Hung WT, Tong HY, Lee CP, Ha K, Pao LY (2007) Development of a practical driving cycle construction methodology: a case study in Hong Kong. Transp Res D 12:115–128
9. Nyberg P, Frisk E, Nielsen L (2016) Driving cycle equilance and transformation. IEEE Trans Veh Technol
10. Stergiou C, Psannis KE, Gupta BB, Ishibashi Y (2018) Security, privacy, efficiency of sustainable cloud computing for big data and IoT. Sustain Comput
11. Tordoff B (2016) Big data and tall arrays. In: MATLAB EXPO 2016 United Kingdom. http://www.matlabexpo.com/uk/2016/. Accessed 15 Apr 2018
12. Ministry of Works Malaysia (2015) 2014 Road Traffic Volume Malaysia (RTVM). Ministry of Works Malaysia, Highway Planning Division
13. Galgamuwa U, Perera L, Bandara S (2015) Developing a general methodology for driving cycle construction comparison of various established driving cycles in the world to propose a general approach. J Transp Technol 5:191–203
14. Ekici S, Unal F, Akbulut Y, Sengur A (2018) Electricity consumption analysis with Matlab Tall Arrays. In: IETS conference, May 2018
15. MathWorks (2018) Big Data with MATLAB. https://ch.mathworks.com/solutions/bigdata-matlab.html. Accessed 15 Apr 2018
16. Wang Q, Huo H, Yao H, Zhang Z (2008) Characterization of vehicle driving patterns and development of driving cycles in Chinese cities. Transp Res D 13:289–297
17. Maulik U, Bandyopadhyay S (2000) Genetic algorithm based clustering technique. Pattern Recogn 33(9):1455–1465
18. Fotouhi A, Montazeri-Gh M (2013) Tehran driving cycle development using the *k*-means clustering method. Scientia Iranica Trans A Civil Eng 20(2):286–293
19. Son H, Park K, Hwang S, Kim H (2017) Design methodology of a power split type plug-in hybrid electric vehicle considering drivetrain losses. Energies 10:437

Investigation on Lane Changing Manoeuvres of an Autonomous Vehicle by Using Model Predictive Control (MPC) and Proportional-Integral-Derivative Control (PID) for Steer-by-Wire (SBW) System

P. M. Heerwan, N. E. Z. Mohamad, M. A. Zakaria, M. I. Ishak, and S. M. Asyraf

Abstract Lane changing is one of the crucial tasks for an autonomous vehicle to avoid an obstacle. This task can be performed by controlling the throttle, brake, and steering actuators appropriately based on the analysis of the vehicle's surroundings. The problem with lane changing is that the control strategy is too complex and needs a high processor for real-time data analysis. In addition, lane changing involves high-level control for vehicle trajectory and low-level control for controlling the steering actuator. This study proposed a well-known method, namely Model Predictive Control (MPC), to determine the vehicle's lateral position and yaw angle during lane changing maneuver. The optimum steering angle command can control the steer-by-wire (SBW) system from the lateral position and yaw angle in MPC. The Proportional-Integral-Derivative (PID) controller is implemented to control the steering wheel angle in the SBW's system. Then, the SBW system will turn the wheel of the vehicle plant. From the simulation result, the PID controller can converge the error although the vehicle's speed is increasing. The result shows that the mean absolute error (MAE) of the SBW system decreases slightly from 0.0115 to 0.0079 as the speed increase from 16 to 41 km/h. From this study, it can be concluded that the MPC and PID controllers can control the vehicle's trajectory during lane changing by calculating an optimum lateral motion and yaw angle to provide an optimum steering angle for the vehicle to change lanes successfully.

P. M. Heerwan (✉) · S. M. Asyraf
Faculty of Mechanical and Automotive Engineering Technology, Universiti Malaysia Pahang, 26600 Pekan, Pahang, Malaysia
e-mail: mheerwan@ump.edu.my

N. E. Z. Mohamad · M. A. Zakaria
Faculty of Manufacturing and Mechatronic Engineering Technology, Universiti Malaysia Pahang, 26600 Pekan, Pahang, Malaysia

M. I. Ishak
College of Engineering, Universiti Malaysia Pahang, 26600 Pekan, Pahang, Malaysia

© The Author(s), under exclusive license to Springer Nature Singapore Pte Ltd. 2023
M. H. A. Hassan et al. (eds.), *Technological Advancement in Instrumentation & Human Engineering*, Lecture Notes in Electrical Engineering 882,
https://doi.org/10.1007/978-981-19-1577-2_2

Keywords Steer-by-wire (SBW) · Model Predictive Control (MPC) · PID control · Lane changing · Autonomous vehicle

1 Introduction

According to a Malaysian Institute of Road Safety Research (MIROS) report, frontal collisions are the most frequent type of accident in Malaysia [1]. The fatality rate is also alarming, which is approximately 7000 road users have died due to road accidents every year, while thousands have experienced other incapacities [2]. Furthermore, the number of pedestrians killed in traffic accidents is also worrying as it accounts for about 25% of the total number of accidents every year [3].

On the other hand, research on autonomous vehicles has given significant attention due to the rapid development of sensors and computing technology. These developments have given rise to the possibility of autonomous vehicle technology, which reduces vehicle accidents that cause by human beings, such as distraction [5] and limits of attention and perception [6]. Since safety is the primary aspect of the autonomous vehicle, the controller must be robust to uncertain conditions [4]. One example of uncertain conditions is when a vehicle try to avoid an unforeseen obstacle. The steering system is the primary system to avoid obstacles and manoeuvre the vehicle to a safe trajectory. In the steering system of an autonomous vehicle, the controllers and actuators replace the mechanical linkage of the steering system. This system is also known as the steer-by-wire (SBW) system, where the control algorithm for vehicle trajectory and optimum steering wheel angle input is compiled in the controllers. Then, an electrical signal will be transferred to the actuator, connected to the steering rack and pinion to steer the wheel [3].

The autonomous system has many control techniques to track the vehicle's motion based on the reference trajectory. Trajectory tracking is crucial to force a vehicle to follow the parameterised references, such as the geometrical path against time. There are two critical variables in trajectory tracking: the reference trajectory and the actual vehicle position [6]. The main purpose of the controller is to decrease the error between the reference trajectory with exact vehicle position by correcting the position errors.

In [8], trajectory tracking development can be classified into conventional control and intelligent control method. The conventional control is an established control method such as MPC, PID and neural network. However, intelligent control is an improvement of the established control method. This improvement usually tackles the conventional control method problems and increases the controller's robustness and aggressiveness in various driving conditions.

Pauca et al. proposed MPC for the vehicle's longitudinal and lateral dynamics during the overpassing manoeuvre, where both the vehicle's longitudinal velocity and trajectory satisfy the reference and constraints [7]. The authors used a nonlinear vehicle model to simulate an actual vehicle model. Then, the vehicle model is linearised before being simulated into the MPC algorithm.

M. Ali et al. proposed the Imperialist Competitive Algorithm (ICA) for optimising PID control parameters of the SBW system. The 10 degrees of freedom vehicle model has been simulated, and the result shows that the lateral motion and vehicle yaw angle errors can be reduced. Furthermore, their proposed method can also tune the vehicle's plant output to the desired trajectory to maintain its stability.

An improved MPC method for a vehicle's lateral control is recently introduced in [8], where the vehicle's stability is improved during high-speed cases. Li et al. used the steady-state response as the controller input instead of the proportional effect of the error. The authors also used saturation limits to constraint the input. The improved MPC that can steer the vehicle does not exceed the side slip limitation as the controller's input is the steady-state error instead of heading error.

In [9], the authors introduce an adaptive and predictive controller that focuses on improving the autonomous vehicle's tracking performance and lateral stability by considering the actual control such as torque input to the steering actuator. The researchers used a regression model based on previous data from input–output where the recursive least squares (RLS) algorithm identifies the parameters. Therefore, a control model is developed to adapt to varying road conditions with faster settling time and minimal overshoot. However, as the road gets more slippery, the controller has less aggressive manoeuvres to maintain lateral stability.

Furthermore, Pratama et al. [10] introduced an adaptive trajectory controller that can follow the desired trajectory reference while estimating the unknown slip parameter. The authors used the backstepping algorithm, which can determine the unknown slip parameter of the vehicle based upon the Lyapunov criterion. The simulation and experimental results of this research show that the angular velocity error vector of the dynamic tracking wheel is around 0.06 rad/s.

This paper investigates the performance of lane changing manoeuvre of an autonomous vehicle with SBW when integrated with MPC and PID control system. This study has two main objectives; the first objective is to establish a reference value for lateral and steer angle by conducting several lanes changing experiments at various velocities using an actual vehicle without SBW. The second objective is to evaluate the lane changing manoeuvre when integrated with MPC and PID control system using simulation based on the experimental reference value. The simulation result shows that the MPC and PID controllers can control the vehicle's trajectory during lane changing while maintaining the vehicle's stability even though the speed is increasing.

2 Experimental Setup

Figure 1 shows the experimental vehicle model, namely UMP Test Car, that equipped with a gyroscopic sensor and encoder. By using a gyroscope and rotary encoder, the lateral position and yaw angular velocity can be adopted. This collected data will be used in the MPC-PID control system as a trajectory reference. This ensures the

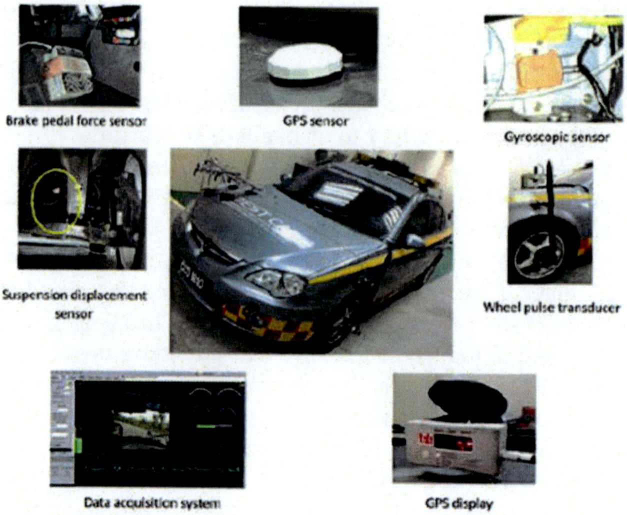

Fig. 1 Experimental setup

MPC-PID controller can control the vehicle's trajectory with a minimal steady-state error during lane changing manoeuvres at a different constant speed.

In the experiment, the driver will drive a vehicle with a constant speed of 16 km/h. When a vehicle is steady at 16 km/h, the driver will turn the steering wheel to change the lane, as shown in Fig. 2. Concurrently, the assistant at the rear seat will record the trajectory data, yaw angle and linear velocity of the vehicle by using a gyroscopic sensor and rotary encoder, respectively. The data is saved in the ROS bag, where the message data from ROS topics is stored. ROS bag and ROS topic are the tools inside the Robot Operating System (ROS). The experiment is repeated for 20, 29 and 41 km/h, respectively.

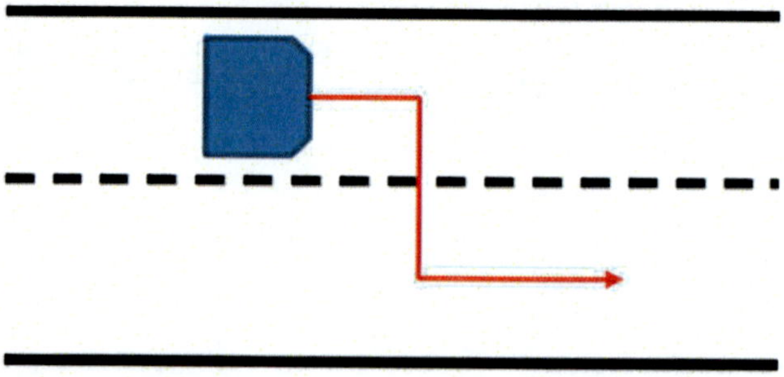

Fig. 2 Lane changing manoeuvre

3 Mathematical Modelling

In this study, the mathematical model of the SBW system is developed in the Matlab Simulink. Then, the MPC and PID control system is designed to analyse the lane changing manoeuvre of the vehicle.

3.1 Mathematical Modelling of SBW

Figure 3 shows the bicycle model of the SBW system that consists of electric motor, gear, steering shaft and wheel. Based on Fig. 3, the motor modeling can be described as follows:

$$J_m \ddot{\delta}_m + B_m \dot{\delta}_m + T_{wm} = T_m \tag{1}$$

where J_m is the moment of inertia of the motor, B_m is the viscous friction of the motor, $\ddot{\delta}_m$ is the angular acceleration of the motor, $\dot{\delta}_m$ is the angular velocity of the motor, T_{wm} is torque wheel to motor and T_m is the torque motor. In this study, the first step is to determine the T_m since it can influence the steering performance. From Fig. 3, the motor in SBW modelling is directly connected to the steering shaft to turn the wheel. Then, the turning wheel modelling is described by considering the moment of inertia and torque of the wheel, as shown in Fig. 1.

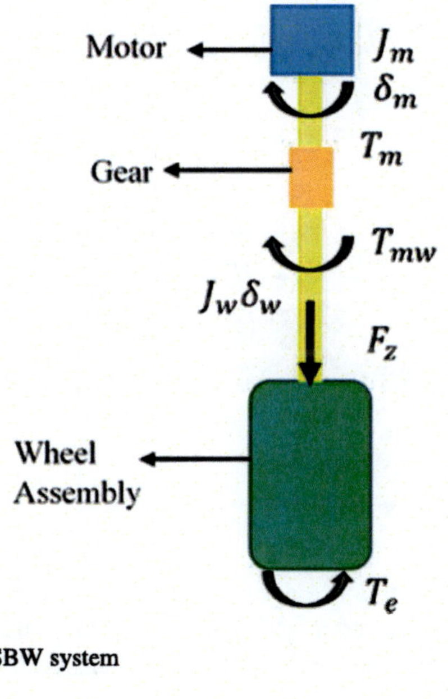

Fig. 3 Bicycle model of SBW system

Table 1 SBW system parameters

Parameters	Symbol	Value	Unit
Moment of inertia equivalent	J_{eq}	2	$kg.\ m^2$
Vicious friction equivalent	B_{eq}	20	$Nms.\ rad$
Torque load	T_1	7	Nm
Gear ratio	k	60	–

$$J_w\,\ddot{\delta}_w + B_w\,\dot{\delta}_w + T_F + T_e = T_{mw} \tag{2}$$

The J_w is the moment of inertia of the wheel, B_w is the vicious friction of the wheel, $\ddot{\delta}_w$ is the angular acceleration of the wheel, $\dot{\delta}_w$ is the angular velocity of the wheel, T_F is the coulomb friction, T_e is the self-aligning torque and T_{mw} is the torque motor to wheel. The gear ratio is defined as k, where it is the ratio of the torque of the motor and the wheel.

$$\frac{\ddot{\delta}_m}{\ddot{\delta}_w} = \frac{\dot{\delta}_m}{\dot{\delta}_w} = \frac{\delta_m}{\delta_w} = \frac{T_m}{T_{mw}} = k \tag{3}$$

By substituting Eqs. 2 and 3 into Eq. 4, the relationship between motor and wheel can be determined as follows:

$$J_w\ddot{\delta}_w + B_w\,\dot{\delta}_w + T_F + T_e = k(T_m - J_m\ddot{\delta}_m - B_m\dot{\delta}_m) \tag{4}$$

Then, Eq. 4 can be simplified as follows:

$$J_{eq}\ddot{\delta}_w + B_{eq}\dot{\delta}_w + T_L = T_{eq} \tag{5}$$

where,

$$J_{eq} = J_w + k^2 J_m$$
$$B_{eq} = B_w + k^2 B_m$$
$$T_{eq} = kT_m$$
$$T_L = T_e + T_F$$

The parameters of the SBW system are defined in Table 1 shown below.

3.2 MPC Control System

Figure 4 shows the MPC block diagram where the model inputs and differences between the process outputs and model outputs, also known as residuals, are used to predict the output, which is the vehicle trajectory. The model operates parallel with

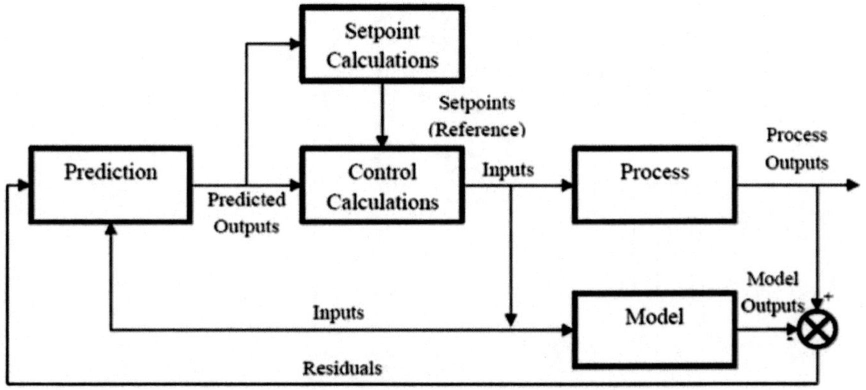

Fig. 4 Block diagram of MPC control system for vehicle trajectory

the process while the residuals act as the feedback signal to the Prediction block. The predicted output is used in the MPC and the setpoint calculation and control calculation. These two calculations can include the upper and lower limit for the input and output variables, also known as constraints.

The MPC computed the current control action online. At each time step, MPC uses the plant's current input and output values, the plant's current state, and the plant's model to measure the future control sequence that can optimise the plant. These calculations are done over the finite horizon. The MPC can produce a good response performance while satisfying the constraint of the control action. Besides that, the plant's input can be obtained from the first control in sequence.

3.3 PID Controller

The purpose of the PID controller in the SBW system is to modify the steering angle error by tuning the proportional, integral, and derivative (PID) gain. The required steering angle, which is the input to the steering wheel angle, can be obtained from PID tuning. The setpoint is the desired steering angle that the SBW system needs to achieve. The PID algorithm uses the difference between the setpoint and current steering angle to calculate the required torque motor for the BLDC motor. Hence, the closed-loop system is produced as constant feedback is provided. The PID equation can be found in Eq. 6 as follows:

$$
\begin{aligned}
error &= \delta_f \\
\Delta\, error &= \delta_{curr} - \delta_{cmd} \\
p &= k_p \times error
\end{aligned}
$$

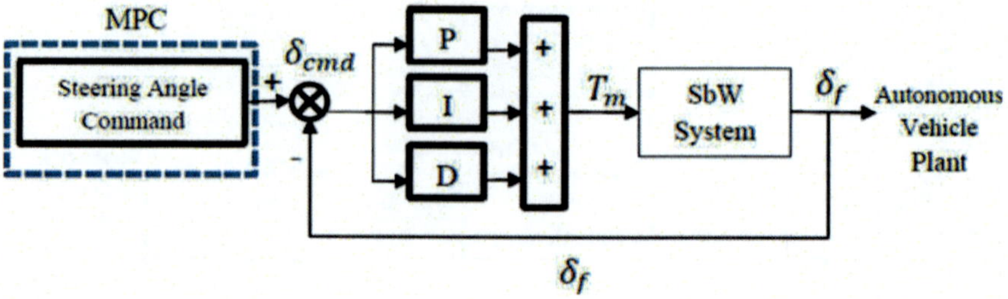

Fig. 5 Block diagram of the SBW system with PID controller

$$
\begin{aligned}
i &= k_i \times \sum_n error \\
d &= k_d \times \frac{\Delta error}{\Delta time} \\
e_{Total} &= p + i + d
\end{aligned} \tag{6}
$$

where k_p is the proportional gain, k_i is the integral gain, k_d is the derivative gain, δ is the front steering angle, δ_{curr} is the current steering angle and δ_{cmd} is the steering command. The motor torque of the SBW system is controlled by the feedback from e_{Total}. The value of e_{Total} is inserted in Eq. 6, which the steering angle is the output. Figure 4 shows the control system of the SBW system using PID controller.

From Fig. 5, the PID in the SBW system will get the optimum steering angle command from the MPC. The controller will monitor the steering wheel inputs and send the corrected torque motor to the SBW system. Then, the SBW system will give feedback to the PID regarding the current steering angle. Next, the PID controller compares the current and the desired steering angle, which is the difference that also known as error. The PID will decide how much correction should apply to make the current steering angle closer to the desired angle. This ensures the desired steering angle of the wheel is attained. Thus, a closed-loop system is developed to enhance the positioning of the system.

The primary function of the MPC in this vehicle control system is to make predictions about future steering outputs based on the vehicle model. At each time step, the MPC can solve the optimisation problem where the optimal control can be obtained. The optimal control can drive the predicted steering angle output to the desired reference as close as possible. It also tries to minimise the steering wheel angle change from one-time step to another. The optimised steering angle is sent to the PID controller to control the BLDC motor in the SBW system.

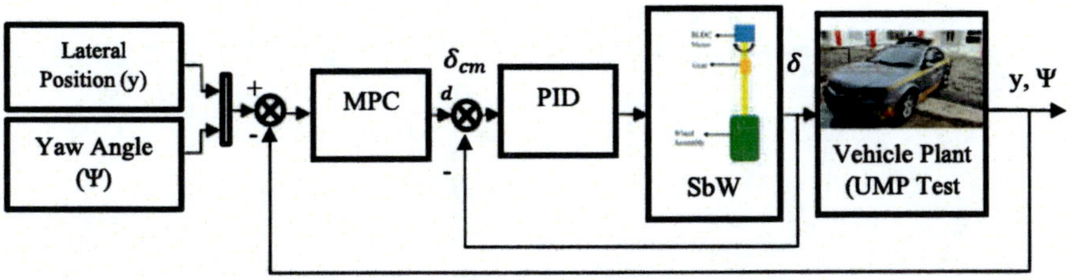

Fig. 6 Vehicle control system with integration of MPC-PID controller

3.4 MPC-PID Control System

Figure 6 shows the control system for steer control of the autonomous vehicle. The MPC will receive two kinds of input, which are the lateral position and yaw angle. Then the MPC will send the output to the SBW system as the steering angle command. In the MPC, the wheels' angle constraint is specified so that the vehicle's trajectory able to track the reference trajectory while satisfying the constraint. The SBW system will send the steering angle to the autonomous vehicle plant, where the wheel's angle will change according to the desired angle. Next, the autonomous vehicle plant will send back the current lateral position and yaw angle to the MPC for comparison. This ensures that the lateral position and yaw angle of the vehicle closely tracks with the reference value obtained. However, the SBW needs another controller to control the steering angle. Therefore, PID is used to control the BLDC motor of the SBW system.

4 Results and Discussion

4.1 Steady-State Cornering

In this study, the first step is to validate the vehicle modelling by comparing the simulation data with the experimental results. The validation is conducted when a vehicle is cornering at a steady- state condition. The vehicle needs to corner at two different constant steering angles, 180° and 360°. Simultaneously, the vehicle's speed also needs to be in constant motion, which approximate 8 km/h. Figure 7(a) and (b) shows the trajectory of the vehicle during experiment and simulation with steering angle 180° and 360° respectively.

Based on Fig. 7, the vehicle's mathematical modelling in the simulation is able to simulate the same behaviour as real vehicle such as the vehicle rotates at a specific circumference when constant steering is applied. Nevertheless, the circle's circumference between the initial parameters response and experiment response is not the

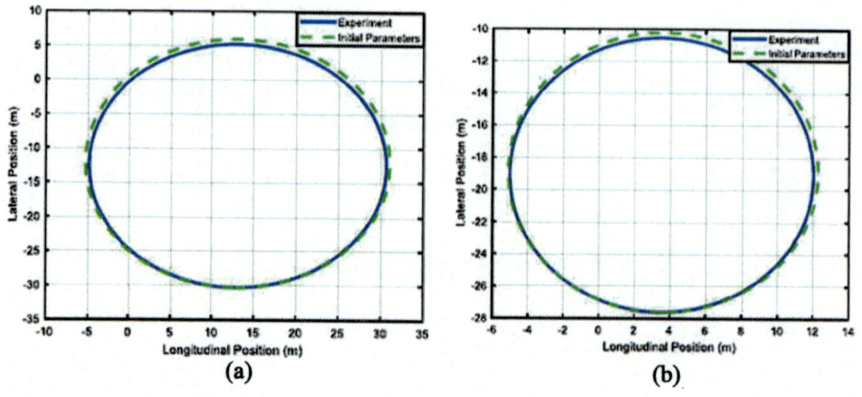

Fig. 7 Trajectory of the vehicle during steady-state cornering

same for both the steering angle. Hence, the vehicle's parameters need to be calibrated until the simulated circle's circumference is almost the same as the experiment circle's circumference. The initial parameters need to be modified are inertial moment of the vehicle, I_z, cornering stiffness of front and rear tires, C_f and C_r.

Figure 8 shows the trajectory of the vehicle after parameters calibrated. It can be seen that after calibrated, the trajectory response of the vehicle model is corresponding to the experiment. The initial and calibrated parameters are shown in Table 2. These verified mathematical modelling and parameters are used in the vehicle's trajectory control.

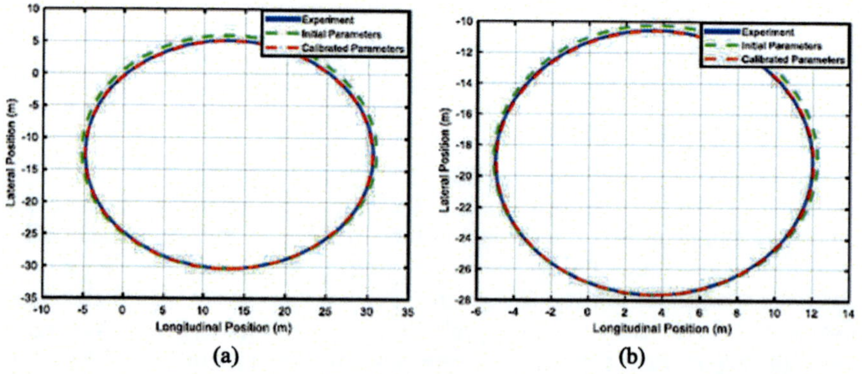

Fig. 8 Comparison the trajectory of the vehicle during steady-state cornering after parameters calibrated

Table 2 Initial and calibrated vehicle's parameters

Parameters	I_z	C_f	C_r
Initial	1,413	15,000	33,000
Calibrated	2,000	35,000	35,000

4.2 Vehicle Trajectory Control

In the vehicle trajectory analysis, the mean absolute error (MAE) is used as the performance index for the MPC-PID controller at a different constant vehicle's speed. MAE calculate the absolute error, which is the absolute value of the difference between the vehicle's trajectory and reference trajectory. The vehicle's trajectory response is simulated using the vehicle's mathematical modelling and parameters that have been validated in the steady-state corner experiment.

To analyse the lane changing manoeuvre, the trajectory reference used in this simulation is obtained from the lane changing manoeuvre experiment. There are four different constant speeds of the trajectory reference used in this simulation, such as 16, 20, 29 and 41 km/h. Figure 9(a)–(d) shows the desired lateral position from the experiment and the lateral position that has been simulated in the SBW system with MPC-PID controller. Figure 9(a)–(d) shows that the MPC-PID controller possesses a fast error converge rate. The vehicle's trajectory can track the reference trajectory during the lane change manoeuvre with minimal error even though the constant vehicle's speed increase. The controller also can quickly track the reference trajectory due to the controller's robustness. Even though the controller exhibits small oscillation before reaching a steady-state, the controller can reduce the oscillation and steady-state error.

Table 3 shows the tracking performance at different constant vehicle's speeds. From Table 3, the MAE decreases slightly as the vehicle's speed increases. This indicates that the controller has satisfactory tracking performance even though the vehicle's speed increases. It also approved that the proposed controller follow the trajectory reference during the lane change maneuver at various constant vehicle speeds. The controller also has a short response time, slight overshoot and steady precision. Furthermore, the MPC-PID controller can produce good response performance while satisfying the wheels' angle constraint.

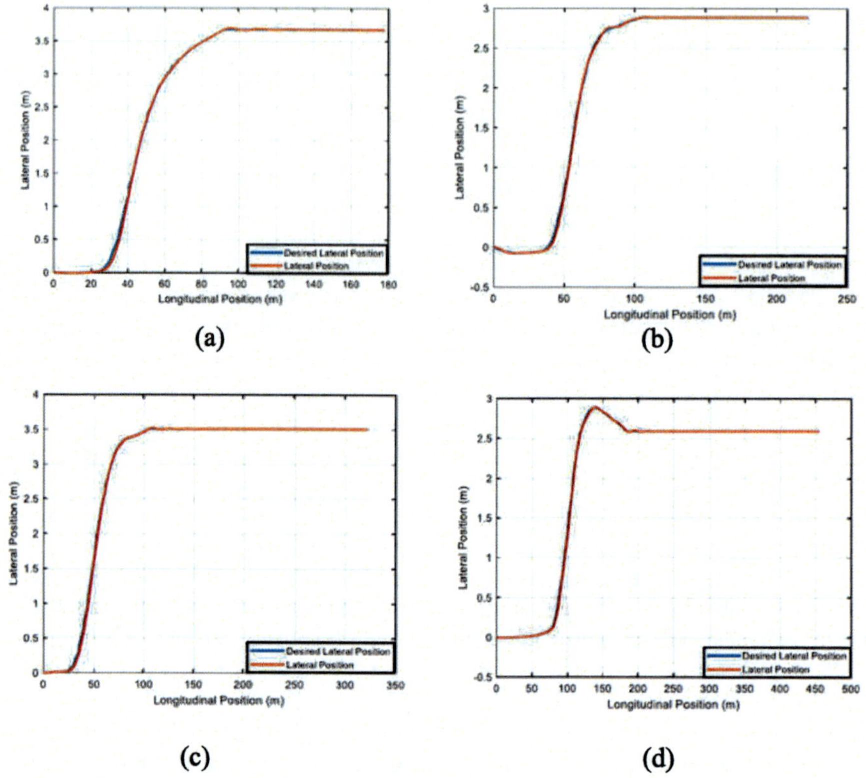

Fig. 9 Vehicle trajectory response at **a** 16 km/h, **b** 20 km/h, **c** 29 km/h, **d** 41 km/h

Table 3 Tracking error with various vehicle's speed

Vehicle's speed (km/h)	MAE
16	0.0115
20	0.0111
29	0.0084
41	0.0079

5 Conclusions

In this study, two controllers are used to track the trajectory of the autonomous vehicle, namely the MPC and PID controller. The simulation result proves that implementing MPC in the autonomous vehicle improves the steering control of the SBW system by optimising the steering angle for the PID controller. In addition, the PID helps the SBW system controls the steering angle of the wheel by controlling the torque motor in BLDC motor. The future scope of this study is to validate the performance of MPC and PID for SBW system by experiment.

Acknowledgements The authors would like to be obliged to Universiti Malaysia Pahang for providing laboratory facilities and financial assistance under the Flagship Grant no. RDU192219. Special thanks to Automotive Engineering Center, Universiti Malaysia Pahang (www.ump.edy.my) for providing test car and technical support.

References

1. Baharuddin M, Khamis N, Kassim KA, Mansor M (2019) Autonomous emergency brake (AEB) for pedestrian for ASEAN NCAP safety rating consideration: a review. J Soc Automot Eng Malaysia 3(1)
2. Raol JR, Gopal AK (2016) Mobile intelligent autonomous systems. CRC Press, Boca Raton
3. Hojjati-Emami K, Dhillon B, Jenab K (2012) Reliability prediction for the vehicles equipped with advanced driver assistance systems (ADAS) and passive safety systems (PSS). Int J Ind Eng Comput 3(5):731–742
4. Schram R, Williams A, van Ratingen M (2013) Implementation of Autonomous Emergency Braking (AEB), the next step in Euro NCAP'S safety assessment. In: ESV, Seoul
5. Hamid UZA et al. (2018) Autonomous emergency braking system with potential field risk assessment for frontalcollision mitigation. In: 2017 IEEE conference on systems, process and control, ICSPC 2017, January 2018. Institute of Electrical and Electronics Engineers Inc., pp 71–76
6. Zakaria MA et al. (2015) Dynamic curvature steering control for autonomous vehicle: performance analysis. In: IOP conference series: materials science and engineering, 2nd international manufacturing engineering conference and 3rd Asia-Pacific conference on manufacturing systems (iMEC-APCOMS 2015), Kuala Lumpur, Malaysia, 12–14 November 2015, vol 114
7. Pauca O, Caruntu CF, Lazar C (2019) Predictive control for the lateral and longitudinal dynamics in automated vehicles. In: 2019 23rd international conference on system theory, control and computing (ICSTCC)
8. Li Y, Chai S, Chai R, Liu X (2020) An improved model predictive control method for vehicle lateral control. In: 2020 39th Chinese control conference (CCC)
9. Ercan Z, Gokasan M, Borrelli F (2017) An adaptive and predictive controller design for lateral control of an autonomous vehicle. In: 2017 IEEE international conference on vehicular electronics and safety (ICVES)
10. Pratama PS, Jeong JH, Jeong SK, Kim HK, Kim HS, Yeu TK, Hong S, Kim SB (2016) Adaptive backstepping control design for trajectory tracking of automatic guided vehicles. In: AETA 2015: recent advances in electrical engineering and related sciences, pp 589–602

Improvement of Brake Response Time on Trailing Vehicle: A Review

Norrizal Mustaffa, Syabillah Sulaiman, Fathul Hakim Zulkifli, Khairu Kamarudin, and Shaiful Rizal Masrol

Abstract This present study discussed the improvement of brake response time in reducing the rear end collision. Lost of attention and fail to response on the leading vehicle brake light were the reasons of this occurrence. A brief review on the current improvement on rear end collision has been studied. Center High Mounted Stop Lamp (CHMSL) is the earlier method to reduce the brake response time, however this method required an improvement as the revolution of technology improved. Lighting technology also plays an important role to cater on this matter as the bulb has changed from incandescent to LED bulb. Recent study more focused on the flashing brake light and this method showed promising reduction of brake response time that capable to capture the driver attention of following vehicle. The combination between several method also been discussed.

Keywords Brake light · Response time · Flashing light

1 Introduction

World Health Organization (WHO) has reported that more 1.2 million people die each year as the impact of traffic accidents at all over the world and the number is forecasted to increase yearly [1]. National Highway Traffic Safety Administration (NHTSA) has outlined that in the year of 2018, 7.2% of total fatal crashes is accounted from rear end collisions. Rear end collisions may involve all types of vehicle cruising on the highway either motorcycles, cars, busses, and lorries [2]. The main factor that contributed to this type of collisions may be categorized into two types which were human and hardware. In general, the occurrence of rear end collisions is due to the trailing driver fails to response sufficiently on time as the lead vehicle press the brake pedal.

Human factor such as age of driver, driving position and gender give different response time as the lead vehicle decelerate [3–5]. Following lead vehicle too closely,

N. Mustaffa (✉) · S. Sulaiman · F. H. Zulkifli · K. Kamarudin · S. R. Masrol
Centre of Automotive and Powertrain Technology (CAPT), Faculty of Engineering Technology, Universiti Tun Hussein Onn Malaysia, 86400 Panchor, Johor, Malaysia
e-mail: norrizal@uthm.edu.my

© The Author(s), under exclusive license to Springer Nature Singapore Pte Ltd. 2023
M. H. A. Hassan et al. (eds.), *Technological Advancement in Instrumentation & Human Engineering*, Lecture Notes in Electrical Engineering 882,
https://doi.org/10.1007/978-981-19-1577-2_3

looking at the wrong place during driving, failure to estimate the front car slowing, stopping or stopped, uses phone while driving and under the influences of alcohol are some of the example of driving behavior that contributed to rear end collisions [6–8]. In view of hardware factor, it is directly related to the brake light itself. The installed brake light may fail to notice the trailing driver as the brake pedal is pressed due to several reason such as size of brake light, light intensity, visibility of the brake light, brake light position, bulb malfunction and others [9]. All these two main factors are closely related to the brake response time of the trailing vehicle. There are lots of research has been conducted to improve the alertness of trailing driver especially that focused on hardware factor.

Theoretically, the effective brake light will reduce the response time of the trailing vehicle to slow down. According to the SAE J2944, response time is the time interval from the brake pedal being pressed or activation of brake light of lead vehicle until the first foot contact with the brake pedal of trailing vehicle [10]. Generally, the brake response time is measured in seconds or milliseconds depending on the situation. The brake response time is the most common term used in many studies or literature, however the definition the brake response time may slightly differ between the studies. Therefore, it is very crucial to determine and clearly identify the starting point and ending point of the measured brake response time.

Figure 1 shows the sequence of braking operation defined by Ecker et al. [3] in their study. From the figure, total braking time is the combination of several times called latency time, movement time, brake response time, pressure build-up time and full braking time. Focused on the brake response time, the definition is almost identical with SAE J2944 where the starting point is when the brake light switch is

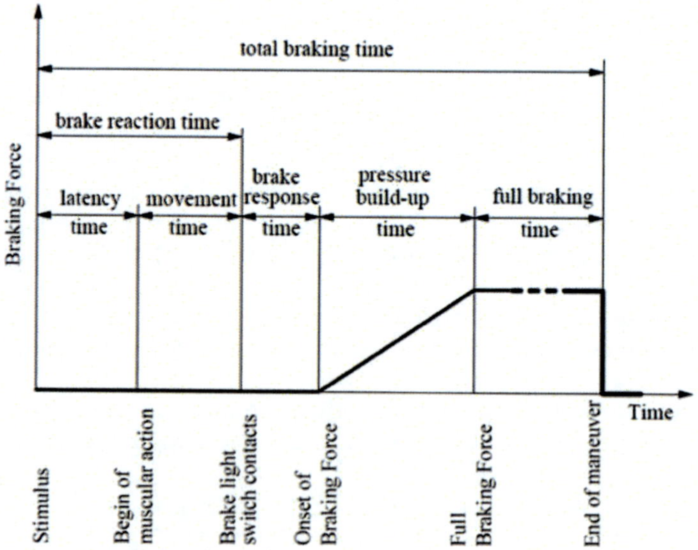

Fig. 1 Sequence of braking operation [3]

in contact meanwhile the ending point is when the braking force starts to apply. The time interval between these two defined points present the effectiveness of trailing vehicle driver react based on the lead vehicle. The measurement of brake response time is very critical and required high precision of measurement as the time normally less than 1 s. The shorter the time the better the brake response time.

2 Review of Brake Light and Its Response Time

2.1 Brake Light

Lighting on a vehicle is highly focused about the safety and communications. Headlight and taillight installed are to notice others highway user about the availability of the vehicle and to assist the driver to drive the vehicle in most limited visibility especially at night. The vehicle lighting standard is mandated by the National Highway Traffic Safety Administration (NHTSA) under Title 49 of the United States Code, Chapter 301, Part 571, Federal Motor Vehicle Safety Standards (FMVSS), Standard No. 108, Lamps, Reflective Devices, and Associated Equipment. According to the Standard 108, the functional purpose of the taillight is to indicate the vehicle's presence and width meanwhile the brake light is to indicate braking [11].

Switch of brake light is linked to the brake pedal. As the driver of the vehicle press the brake pedal regardless the distance or force of the brake pedal experienced, brake light switch will trigger, and the brake light is activated. The brake light will turn off as the driver release the brake pedal to its normal position. This is clearly expressed that the vehicle is under braking situation and the driver of the trailing vehicle should take an appropriate action as they saw the brake light turn on. However, the brake light does not indicate the braking vehicle velocity and vehicle deceleration information [10]. This information might be valuable if the following vehicle may evaluate what is happening. In this situation, the effectiveness of the brake light plays very important role.

Nowadays, most of the vehicle manufacturer has employed light emitting diode (LED) for headlight, taillight, and brake light as the replacement of incandescent bulb. The widely used of LED in automotive industry is attributed to the advantages of LED over incandescent such as longer useful life, larger mechanic resistance to vibration, high level of luminous efficiency, lower in power consumption, less heating effect and quick response time [12]. Research conducted by Lin et al. [12] has summarized that the usage of LED as brake light also may enhanced the warning effect by adjusting the brightness and the circuit design structure is much simpler only by using few elements that leads to lower construction cost as compared to incandescent bulb.

2.2 Analysis in Reducing Brake Response Time

There are lots of study related to the improvement of brake light effectiveness which aims to increase the reaction or response of the trailing vehicle driver by capturing their attention. Center High Mounted Stop Lamp (CHMSL) is the earlier study that focused on this matter by locating or mounting the brake light closer to the line of sight of following vehicle [13–17]. Some of the field study conducted found CHMSL is only effective in simpler crash [16]. The placement of the brake light that visible through the back glass of the leading vehicle gives trailing driver an earlier indication of what is currently happening about the traffic condition. However, another study found that CHMSL did not gives major impact on the brake response time at medium speed of vehicle [13]. Even so, most of the vehicle manufacturer starts to install CHMSL in the new vehicle model in the 1990s. Study conducted by Kahane and Hertz [18] summarize that at initial, the usage of CHMSL has reduced about 50% of rear-end collisions but the effectiveness has reduced to 4.3% after few years later. As the brake light and taillight used the same color, the CHMSL and luminance was applied to compensate between these two lights. Despite that, several factors have given high impact and resulted on the reduction of the CHMSL efficiency. Wickens et al. [19] stated ambient lighting condition during the event, number of bulbs used, various size of light, and shapes of light were among the most influential factors.

Red color has been gazette as a standard color of brake light and taillight worldwide. In the study conducted by McIntyre [10], combination of yellow taillight with red brake light was experimentally compared to the default condition of taillight and brake light which is red. It is proved that the combination of yellow color used as taillight has increased the capturing attention to the brake light of trailing driver. Brake response time for the combination color was lower than default color as shown in Fig. 2. The introduction of yellow color as taillight has increased the capturing attention of trailing driver as the color different makes the trailing driver easy to detect and react to the brake light.

Another method of reducing the brake response time is effective warning method. It is working by giving a warning information such as warning form, warning lead time and warning content and these have significantly affected the warning attributes. The warning method might be categorized into three types: visual warning, tactile warning, and auditory warning. Among these three types of warning, auditory warning is observed better than the others as the auditory warning may capture the attention of driver regardless any situations and no matter the driver focused on, the sound will hit the ear of driver [20, 21]. Cheng et al. [22] has conducted a study on a driver response to a leading vehicle collision warning through driving simulator. Audible means of collision warning were used, and the audible sound varies depending on three dangerous setup scenes. Based on the study, audible warning sound resulted on the better action by trailing driver as the sound notify the dangerous situation in the setup situations. The statistical analysis proved that there is a significant effect between warning sound and the brake response time. Other study related to the effect of auditory on driver response time during collision avoidance process has

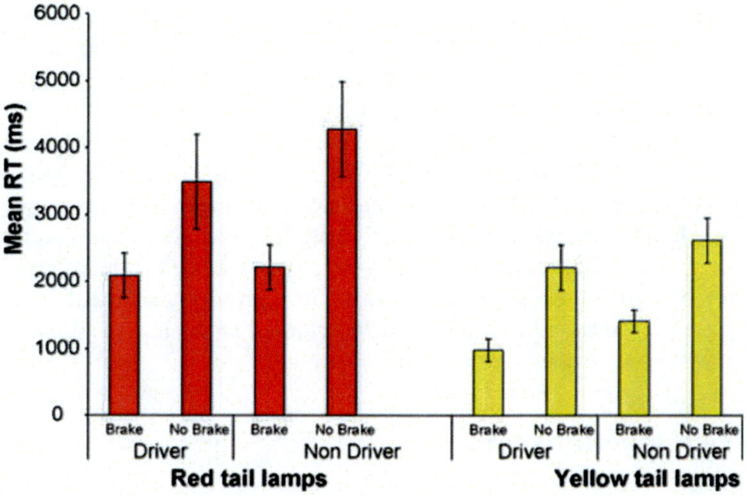

Fig. 2 Response time of red and yellow taillight [10]

been performed by Xiang et al. [23]. The study was conducted based on a driving simulator. Several audible warning methods has been compared that focused on warning lead time and warning content. The findings of the study revealed that auditory warning capable to reduce the brake response time and collision occurrence rate effectively. In addition, 5 s early warning is more effective than 3 s late warning, but late warning also observed gives higher impact on lowering the collision rate by generating 37% improvement. Wege et al. [24] has carried out the study related to the collision warning and found the findings is similar with the earlier study conducted by Refs. [22, 23].

The revolution of lighting technology also plays an important role in reducing the brake response time. Recent study related to the lighting technology was conducted by Palaniappan et al. [25]. The aim is to study the different effects of driver reaction time based on different types of brake light. 10 sets of brake light with 8 from the light are LED and remaining light is incandescent bulb were employed from various car manufacturers. 22 participants were involved, and they were classified into experienced and inexperienced participants. The experimental methodology relied on custom built hardware and software to present random brake light to the participants. Real physical brake light was used in the mean to setup the experiment. Based on the results, they found that experienced driver may react faster than inexperienced driver when brake light was activated. In view of lighting types used, incandescent bulb has resulted on slower driver response to the brake light than LED. This is attributed to the incandescent bulb take longer time to illuminate than LED, generally 50 ms after switch is activated. The finding about the LED has better reaction time was observed in agreement with previous study conducted by Jilek et al. [26] and Bullough et al. [27]. Lin et al. [11] has modified the incandescent bulb of a car brake light into LED and the arrangement of the light is in the form of "X" alphabet. The study focused on to get an attention from leading vehicle and the result indicated that by introducing the "X" alphabet arrangement as brake light may improve the road safety especially in critical traffic condition.

There are various more improvement on brake light system that has been proposed and studied in reducing the brake response time including increase the luminance of the brake light when the vehicle decelerate, resize the brake light into bigger size, brake light illuminated when the driver released the accelerator pedal and provided optical loom cues of the front vehicle [28–31]. However, the continuity and implementation of some of the study seems does not received lots of attention even the results have proved an improvement may be achieved on the brake response time.

2.3 Flashing Brake Light

In other study, flashing brake lamp was also has been evaluated using various types of vehicle including motorcycle. Lots of researcher agreed that by introducing flashing light, visibility of the light may be improved, and it helps to attract the attention and alertness of following vehicle. Early study related to this flashing light was conducted by Tang [32] using motorcycle. The turn signal was integrated to the brake light and when the brake pedal is applied, brake light is activated together with the flashing turn signal. In the event of brake pedal is released, turn signal retains to its actual function. Testing was conducted in actual urban and suburban traffic that covered various types of illuminations including noontime, afternoon, and nighttime. During daytime, author found that direct sunlight illuminates the brake light as the sunlight hit the brake light cover. The illumination and reflection make the contrast between the activated and deactivated brake light, therefore it creates difficulty in analyzing the brake light. The addition of flashing turn signal helped the trailing driver to identify the activation of brake light. Experimental data revealed that with the addition of flashing turn signal on the brake light activation has shorten 80 ms in overall of the brake response time as compared to the default brake light for all testing conditions.

Another study of modified brake light involving turn signal has been conducted by Tang et al. [33]. In their study, both turn signal will flashing with the activation of brake light. Due to single bulb of motorcycle taillight, it is crucial to provide fast and sufficient information to the trailing vehicle as the brake pedal is applied. The motivation of integrated with the turn signal is due to the previous study that has highlighted blue and yellow were the recommended color for better recognition and may reduce the error at any level of light condition. The experiment was conducted based on computer driving simulator equipped with motorcycle taillight. With the addition of both flashing turn signal as aided brake light, brake response time was found significantly reduced by 200 ms on average and it is more effective especially during nighttime.

Alfendinck [34] performed an experimental analysis of brake response time on five types of brake light with the combination of fog lamp, stop light, and hazard signal. Each types of combination have flashing light with different frequency and the details arrangement is shown in Fig. 3. 21 participants have involved during the experiments that consists of 11 male and 10 female. The experiments were conducted using driving simulator software at the speed of 50 km/h. Two condition were tested, normal driving task and distraction driving task. The result found that brake response time of normal brake light was better than the integration of flashing light as the addition to the brake light at both conditions. In view of distraction driving task, brake light with flashing stop light at 5 Hz was the worse brake reaction time as compared to the normal brake light with the reaction time was 80 ms longer than the mean reaction time of other brake light condition. Averagely, distraction driving task has increased 240 ms of reaction time than the normal driving task.

In European, Mercedes also conducted an experiment on flashing brake light. The finding has proved that the usage of flashing brake light greatly reduced the driver reaction to hit the brake pedal by 0.2 s faster than normal brake light. In addition, the study also discovered that at 80.5 km/h of vehicle speed, flashing brake light has reduced the stopping distance by 4.4 m meanwhile at the speed of 105 km/h by approximately 6.1 m. This clearly explained when a vehicle equipped with flashing brake light, visibility, and alertness of trailing vehicle in reducing rear end collision might be reduce [35]. Wang et al. [36] also has performed a study on effectiveness of flashing brake light and flashing hazard in avoiding rear end collision. An advanced simulator with 6 degree of freedom and moving base was used in providing the

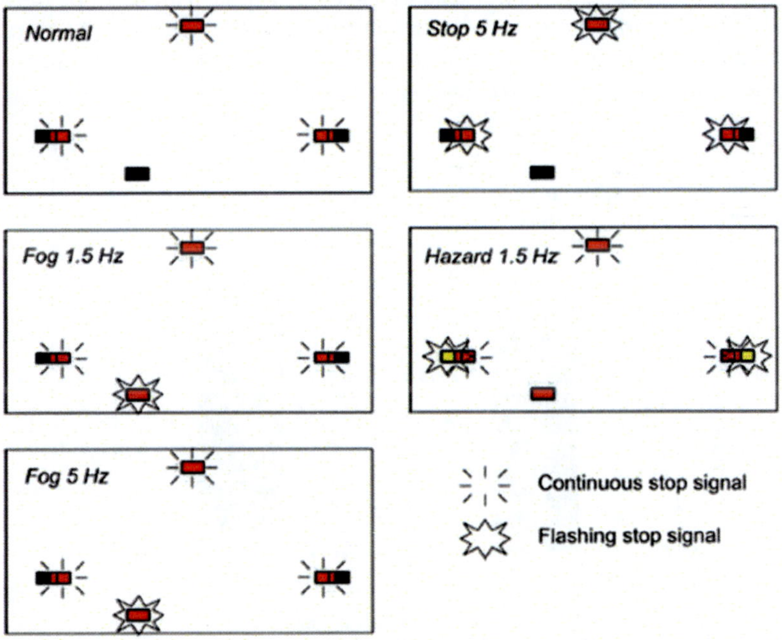

Fig. 3 Arrangement of brake light, fog lamp, taillight, and turn signal of the test condition [34]

actual and realistic driving experience. The driving situations were projected on a front screen and integrated with audio system to provide simulation of engine, traffic, and road condition. Various combination of scenario, speed, deceleration, acceleration, and steering angle were simulated. 21 male participants were involved in completing the study. The analysis showed if the brake response time may be reduced by 0.2 s, the stopping distance may be reduced by 4.4 m provided the initial vehicle speed is 80 km/h with the deceleration is 0.8 g. This agreed with previous study performed by Ref [35]. Study conducted by Wierwille et al. [37] also revealed almost identical findings, but the time response reduce is 0.25 s. Wang et al. [36] summarized and concluded with the addition of flashing brake light and hazard light, the brake response time had reduced by 0.14–0.62 s and 0.03–0.95 s, respectively as compared to the normal brake light. Extension to their study, brake light color also plays an important parameter that may contribute to the reduction of brake response time. Figure 4 shows the comparison of red and amber color with respect to the brake response time. It presented that amber color has reduced 10% or brake response time over red color. However, brake light size seems does not gives a significant impact on the brake response time.

Recent research related to the flashing brake light has been conducted by Sohrabi [38]. Analysis conducted not limited to the brake light response time, but effects of age, gender, and driving license were also considered in the study. The study was performed on 46 participants consist of 10 males and 36 females. Driving simulator was employed to simulate the driving condition with flashing brake light with a frequency of 7 Hz. All the required data were recorded through the installed software. The increased of driver's age resulted in reduction of brake response time averagely by 11.58 ms with the age range is between 20 to 46 years. Men has gives significant brake response time (75.52 ms) over women and this presented men attention during

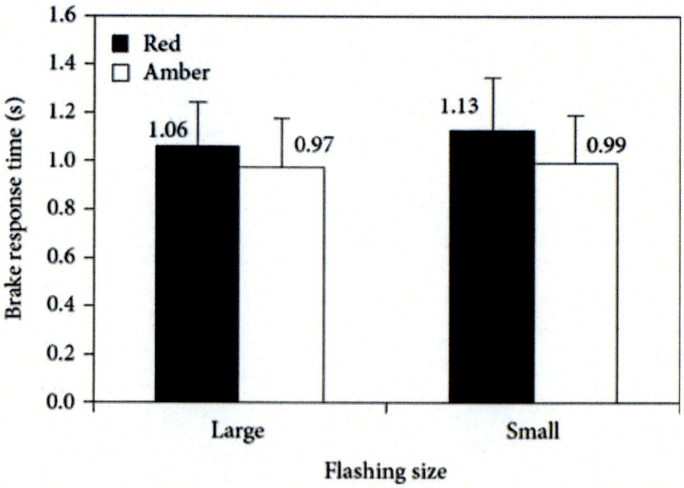

Fig. 4 Effects of color and size of brake light on brake response time [36]

driving was better than women. Driving license also plays an important criterion as the driver with driving license has improved the brake response time as compared to driver without driving license. The climate of the study revealed that with the usage of flashing brake light, brake response time has significantly improved by 323.42 ms. The improvement was equivalent to 7.19 and 10.78 m with the vehicle speed of 80 and 120 km/h, respectively.

3 Conclusions

A brief review of brake light improvement in reducing brake response time that leads to rear end collision has been discussed in this paper. Several methods have been studied and tested involving CHMSL, yellow color of brake light, warning method, revolution from incandescent bulb to LED, integration of brake light with other light and hazard light, and implementation of flashing brake light. CHMSL has been applied in most of the vehicle nowadays meanwhile flashing brake light is another potential method that may be applied, as flashing brake light has resulted in significantly reduced the brake response time and greatly attract the trailing driver attention. Further study that covers all types of vehicle related to this this matter is highly required to analyze the overall effectiveness in reducing rear end collision. The gathered information of flashing brake light will be used for future study by applying the findings on motorcycle application.

Acknowledgements The authors would like to thank the Universiti Tun Hussein Onn Malaysia for supporting this research under the Tier 1 Grant H805.

References

1. World Health Organization (2018) Global status report on road safety 2018: summary, World Health Org., Washington, DC, USA, Technical report. WHO/NMH/NVI/18.20
2. National Center for Statistics and Analysis, Traffic safety facts 2018 annual report (2020) A compilation of motor vehicle crash data, Report No. DOT HS 812 981, National Highway Traffic Safety Administration
3. Ecker H, Wassermann J, Ruspekhofer R, Hauer G (2001) Brake reaction times of motorcycle riders. Int Mot 1–11
4. Woods DL, Wyma JM, William Yund E, Herron TJ, Reed B (2015) Age-related slowing of response selection and production in a visual choice reaction time task. Front Hum Neurosci 9(APR):1–12
5. Jeong H, Green P (2012) Forward collision warning modality and content: a summary of human factors studies, Univ. Michigan, Ann Arbor, MI, USA, Technical report. UMTRI-2012-35
6. Isler RB, Starkey NJ (2010) Evaluation of a sudden brake warning system: effect on the response time of the following driver. Appl Ergon 41(4):569–576
7. McKnight AJ, McKnight AS (1993) The effect of cellular phone use upon driver attention. Accid Anal Prev 25(3):259–265

8. Drunk Driving | NHTSA. https://www.nhtsa.gov/risky-driving/drunk-driving. Accessed 27 May 2021
9. Tewari A (2010) Method and apparatus for vehicle brake light control, U.S. Patent No. 7, 834, 751, Washington, DC, USA, U.S. Patent and Trademark Office
10. Operational Definitions of Driving Performance Measures and Statistics (Recommended Practice J2944), Warrendale, Pennsylvania: Society of Automotive Engineers (2013)
11. McIntyre SE (2007) Capturing attention to brake lamps. Accid Anal Prev 40(2):691–696
12. Lin YC, Wang HY, Cheng WJ (2009) A novel vehicle brake light system. In: IIH-MSP 2009 - 2009 5th international conference on intelligent information hiding and multimedia signal processing, pp 398–401
13. Sivak M, Olson PL, Farmer KM (1981) Effects of highmounted brake lights on the behavior of following drivers, Technical report. UM-HSRI-81-1, University of Michigan Highway Safety Research Institute, Ann Arbor, Mich, USA
14. Allen K, McCormick IA, NZ Accident Compensation Corporation (N.Z.) (1987) The evaluation of high mounted auxiliary stop-lights, A New Zealand Field Test, Final report, The Corporation, Wellington, New Zealand
15. McKnight AJ, Shinar D (1992) Brake reaction time to center high-mounted stop lamps on vans and trucks. Hum Factors 34(2):205–213
16. Kahane CJ, Hertz E (1998) The long-term effectiveness of Center High Mounted Stop Lamps in passenger cars and light trucks, Technical report. DOT HS 808 696, U.S. Department of Transportation, National Highway Traffic Safety Administration, Washington, DC, USA
17. Office of the Federal Register, FMVSS (Federal Motor Vehicle Safety Standard) 108 (Lamps, reflective devices, and associated equipment), in 49 Code of Federal Regulations (Part 571. 108), U.S. Government Printing Office, Washington, DC, USA (1985)
18. Kahane CJ, Hertz E (1998) The long-term effectiveness of center mounted stop lamps in passenger cars and light trucks, NHTSA. Technical report Dot HS 808 696
19. Wickens CD, Gordon SE, Liu Y (1998) An introduction to human factors engineering. Longman, New York, pp 107, 108
20. McCallum M (2006) Integrated vehicle-based safety system heavy truck driver vehicle interface (DVI) literature review, Center for Human Performance and Safety
21. Wege C, Will S, Victor T (2013) Eye movement and brake reactions to real world brake-capacity forward collision warnings – a naturalistic driving study. Accid Anal Prev 58:259–270
22. Cheng B, Hashimoto M, Suetomi T (2002) Analysis of driver response to collision warning during car following. JSAE Rev 23(2):231–237
23. Xiang W, Yan X, Weng J, Li X (2016) Effect of auditory in-vehicle warning information on drivers' brake response time to red-light running vehicles during collision avoidance. Transp Res Part F Traffic Psychol Behav 40:56–67
24. Wege C, Will S, Victor T (2013) Eye movement and brake reactions to real world brake-capacity forward collision warnings - a naturalistic driving study. Accid Anal Prev 58:259–270
25. Palaniappan R, Mouli S, Fringi E, Bowman H, McLoughlin I (2021) Incandescent bulb and LED brake lights: novel analysis of reaction times. IEEE Access 9:29143–29152
26. Jilek P, Vrábel L (2020) Change of driver's response time depending on light source and brake light technology used. Sci J Silesian Univ Technol Ser Transp 109:45–53
27. Bullough JD, Van Derlofske J, Yan H (2001) Evaluation of automotive stop lamps using incandescent and sweeping neon and LED light sources, SAE technical paper, no. 724
28. Gail J, Lorig M, Gelau C, Heuzeroth D, Sievert W (2001) Optimisation of rear signal pattern for reduction of rear-end accidents during emergency braking Manoeuvres, Federal Highway Research Institute
29. Li Z, Milgram P (2008) An empirical investigation of a dynamic brake light concept for reduction of rear-end collisions through manipulation of optical looming. Int J Hum Comput Stud 66:158–172
30. Shinar D (1995) Field evaluation of an advance break warning system. Hum Factors 37:746–751
31. Shinar D (2000) Fleet study evaluation of an advance brake warning system. Hum Factors 42:482–489

32. Tang KH (2003) A field study on validation of supplemental brake lamp with flashing turn signals for motorcycles. Int J Ind Ergon 31(5):295–302
33. Tang KH, Tsai LC, Lee YH (2006) A human factors study on a modified stop lamp for motorcycles. Int J Ind Ergon 36(6):533–540
34. Alfendinck JWAM (2004) Evaluation of emergency brake light display (EBLD) systems, October, vol 2, no May, pp 4–8
35. Kim PP (2010) Automatic flashing brake lights and associated method. Patent Application Publication, United States
36. Li G, Wang W, Li SE, Cheng B, Green P (2014) Effectiveness of flashing brake and hazard systems in avoiding rear-end crashes. Adv Mech Eng 6:792670 2014
37. Wierwille WW, Lee SE, DeHart MC, Perel M (2006) Test road experiment on imminent warning rear lighting and signaling. Hum Factors 48(3):615–626
38. Sohrabi MS (2019) Effects of flashing brake lights on drivers' brake reaction time and releasing accelerator gas pedal time. Heal Emerg Disas Q 4(4):209–216

The Validity of Football Skills Test for 12-Year-Old Male Players

Adjah Naqkiah Mazlan, Mohd Hairul Azam Mesnan, Mohamad Rasidi Pairan, and Mohd Hizwan Mohd Hisham

Abstract This study aims to determine the validity of the Football Skills Test conducted on 12-year-old male players through three test: Agility, 50-m running with a ball and Zigzag with a ball. The test was conducted on two football teams, the elite players group and the non-elite players group. The elite group consists of Akademi Tunas Johor (ATJ) players, while the non-elite players are from Perdana United Fc. A total of 20 players were selected as the study sample, with 10 players from each group. The Independent Sample t-Test was used to determine any significant difference between the test scores of players in the elite and the non-elite group. The analysis results show a difference between the Agility Test scores of players in the elite group (14.882 ± 0.863) and the non-elite group (16.105 ± 1.608). The analysis also showed a difference between the players' scores in the 50-m running with a ball test, with the elite group (12.664 ± 0.448) and the non-elite group (13.103 ± 1.208). Furthermore, there is a difference in the elite and non-elite group scores in the zigzag with a ball test, with elite players (17.878 ± 2.120) and non-elite players (20.263 ± 4.028). In this regard, the study's findings show a significant difference between the score of players in the elite group and the non-elite group in the Agility Test conducted. The study's findings confirm that the Agility Test is suitable for measuring the skills of 12-year-old football players.

Keywords Football skills test battery · 12-year-old male players · Test validity · Elite and non-elite players

1 Introduction

Football is a game that requires endurance strength, an important component of fitness in the recruitment and selection of young football talents [1]. Therefore,

Supplementary Information The online version contains supplementary material available at https://doi.org/10.1007/978-981-19-1577-2_4.

A. N. Mazlan (✉) · M. H. A. Mesnan · M. R. Pairan · M. H. M. Hisham
Universiti Teknologi Malaysia, 81310 Johor Bahru, Malaysia
e-mail: adjahnaqkiah@utm.my

© The Author(s), under exclusive license to Springer Nature Singapore Pte Ltd. 2023
M. H. A. Hassan et al. (eds.), *Technological Advancement in Instrumentation & Human Engineering*, Lecture Notes in Electrical Engineering 882,
https://doi.org/10.1007/978-981-19-1577-2_4

young players should be trained to show high performance in football from an early age to achieve optimal performance in football [2]. Programmes like the young talent development programme have supported grassroots-level talent development in many football clubs. In this regard, clubs have started to recruit players as young as 6 to 12-year-old to follow specialised football training programmes. During this late childhood, transitional skills improve and most children are able to master complex motor skills. Children are ready to learn strategy and more complex play combinations with an increasing emphasis on strategy and tactics [3]. Test batteries have been proven a viable tool for assessing physical performance and football skills [4]. In this light, UTM Sports Science students have designed a football skills test battery as part of a development programme for a club academy. The skills battery test comprises an agility test, a 50-m running with a ball test and a Zigzag with a ball test to measure the performance levels of 12-years-old football players. However, the validity of the skill tests conducted on the players has been yet to be reported. Validity indicates whether an instrument could test what it supposes to measure or whether the instrument used could measure what the researcher wants to measure accurately [5]. In other words, a test is deemed valid if it could accurately measure all aspects to be measured. In this study, a football skills test battery for adult players was adopted and modified to suit 12-year-old players. There is a need to determine the validity of the skill test battery conducted because it has been modified. The test's validity in assessing young players' football skills has yet to be tested.

Thus, as an objective in this study, a comparison will be made to determine whether there is a difference between the scores of 12-year-old players from an elite player group and those from non-elite player group. A significant difference between these groups will confirm that the appropriate football skills test is used to measure the football skills of 12-year-old players.

2 Research Objective

The objectives of this study are:

(a) To study the validity of the football skills test battery for 12-year-old male players.
(b) To determine the significant differences between the scores of the players in the elite group and non-elite group in each skill test.

3 Methodology

This study focuses on conducting a validity test on football skills test battery conducted on 12-year-old male elite and non-elite players. The method used in this study is a quantitative data collection method. This study used 3 football skill tests conducted on 12-year-old male players. The players were divided into 2 groups:

elite player and non-elite player groups with 10 players in each group. Each test was conducted separately on different dates and implemented at the same time at 4.00 pm. The tests were arranged according to stations where each player would take turns to perform the test. The data were collected using a scoring sheet: test score form where each player's scores in each skill test were recorded.

3.1 Sampling

The sampling method used in this study is the purposive sampling method. The researchers chose this sampling method as it allows data and information to be collected from a sample representative of the target population. The study's sample consists of 20 12-years-old male football players; 10 football players are from the elite player group, and 10 are from the non-elite player group. The appropriate minimum sample size in the experimental study was 15 and for the experimental study that having a high control group only 8–10 sample were required [6]. The classification for elite player is those who are in the national football development program that is Akademi Tunas Johor (ATJ) while for non-elite player in this study are those who participate in trained football clubs only. The data obtained from the sample were used to examine the validity of the skills tests conducted. Two sets of forms were used to collect their personal information and their test scores, which were used as indicators to assess the validity of the skills test.

3.2 Instrument

The instruments used in this study include the form used to collect the participants' personal information and the form to collect their skills test scores. The test score form used records the participants' scores in each test conducted. The personal information form was used to obtain the participants' personal information. The test battery comprises three main components representing the required skills in football, namely agility test, 50-m running with a ball test test and zigzag a ball test (Fig. 1).

3.3 Procedure

The procedure implemented involved a study sample that performed three tests of football skills. Permission requested from the coaches for each team to conduct tests on their players of 10 people for each group and followed the prescribed procedures. Before conducting the test on the sample, the researcher explained the procedure of performing each test as well as safety measures to the subject so that unwanted things do not happen such as the risk of injury. The study sample consisted of 12-year-old

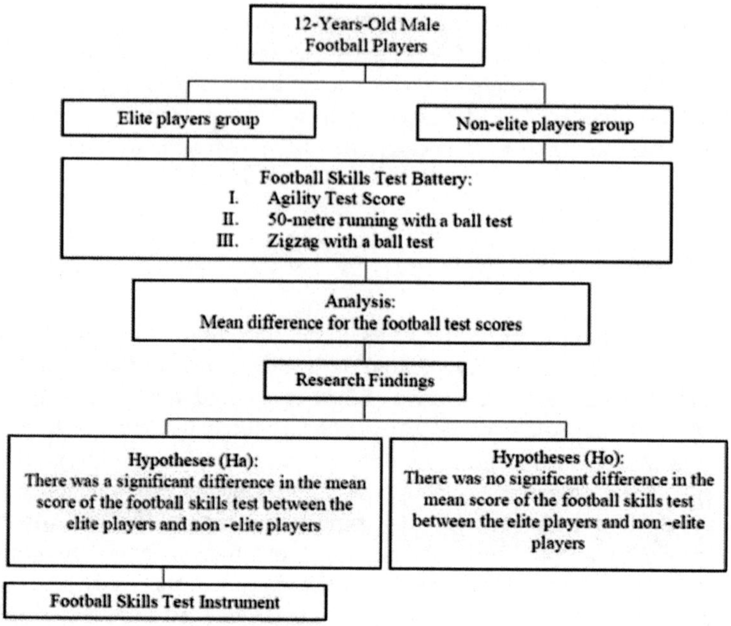

Fig. 1 Conceptual framework for the test

children who needed to be briefed before performing a test. A pre-test warm-up session and a post-test cool-down session should be performed to avoid the risk of injury. The procedure is shown in the Table 1.

The agility test is to measure an individual's speed in changing direction or body position. For 50-m running with a ball, the test was to measure the speed of the running subject while carrying and controlling the ball at the foot. While Zigzag with a ball test was to assess the subject's ability to change direction rapidly with the ball. In overall, method of measurement for these three tests will be assessed through the fastest time defect a participants can perform. The design diagram of the test conducted as shown in Fig. 2.

3.4 Data Analysis

Researchers use quantitative data through skills testing form which involves descriptive and inferential data. The descriptive analysis will use units such as mean, standard deviation, weight (kg) and height (cm). Inferential analysis used the Independent Sample t-Test method to look at the differences of 12-year-old male soccer players between the elite group of players and the non-elite group of players. The significant value set by the researcher in this study for comparative analysis was p value =

Table 1 The procedure on the three skills test

Skills test battery	Procedure	Measurement
Agility test	a) Participant must be in a state of readiness and have done a warm-up b) Participant has to run according to the marked cone: Step 1: Start by running forward from cone A to cone B where at the same time the time record is started, then change direction to the right running to cone C Step 2: Run from cone C to cone A with a zigzag run past the three cones in the middle and break back to cone C with the same zigzag run Step 3: Run from cone C to cone D, then change direction to the right run from cone D to cone A and finish c) Time will be stopped after the participant passes cone A	– The assessment will be done through the fastest time defect that the participants can do
50-m running with a ball test	a) A marker cone is placed at a distance of 25 m and the starting line b) The participant has to stand behind the starting line and has to run with the ball when the 'start' instruction is given c) The participant should run with the ball controlled using the foot from the start line up to the 25 m line and then make a 'turning' run back to the start line again d) Time will be stopped once the participant reaches the starting line again	– Time is used as a measurement where the shorter the time taken to reach the finish line, the better the athlete's performance

(continued)

Table 1 (continued)

Skills test battery	Procedure	Measurement
Zigzag with a ball test	a) The participant stands at the starting line and waits for the "ready" signal to be given b) When the whistle signal is sounded, the participant should run while controlling the ball around the cone marked in a zigzag pattern c) Time will be taken after the whistle is blown and terminated when the participant reaches the finish line d) The participant should start the run from the starting cone and circle the stacked cone and return to the starting cone e) The scorer will take time and make observations of the ball control behavior performed by the participant	– The assessment will be done through the calculation of the fastest time while carrying the ball to the end and control over the ball is given attention

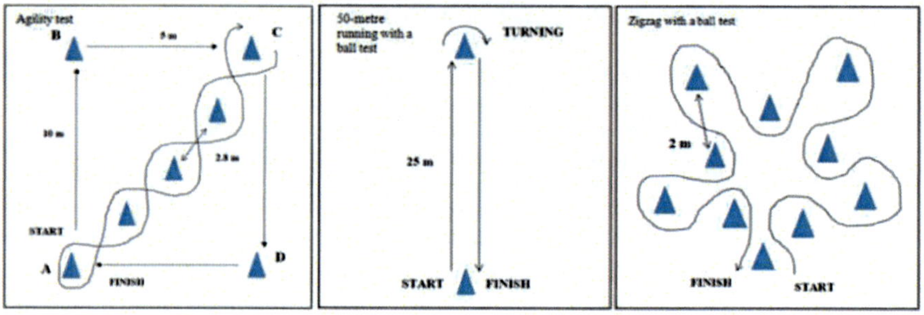

Fig. 2 The design diagram of the test

0.05. T-test suitable to be implemented when the groups have 20–30 samples and usually used in cases where the experimental subjects are divided into two independent groups. Beside that, it is one of the most widely type of statistical test that used to compare the means of two groups [7].

The test results obtained were recorded and analyzed into Microsoft Office Excel 2010. All data are presented as mean ± standard deviation (SD).

4 Findings

The participants' scores in each skill test conducted, namely agility test, 50-m running with a ball test and the zigzag with a ball test, were analysed to obtain the mean ± standard deviation (SD).

4.1 *Descriptive Statistics Analysis*

Descriptive statistics were used to analyse the raw score data on the participants' score in the skill test. For participants in the elite group, the mean score and standard deviation for the agility test are (14.882 ± 0.863); for the 50-m running with a ball test (12.664 ± 0.448), and for the ball zigzag test (17.878 ± 2.120). Meanwhile, participants in the non-elite group, the mean score and standard deviation for the agility test are (16.105 ± 1.608); for the 50-m running with a ball test (13.103 ± 1.208), and for the ball zigzag test (20.263 ± 4.028).

4.2 *Inferential Statistics Analysis*

The results of the t-test analysis obtained are shown in the table above. The Independent Sample t-Test analysis for the Agility Test scores of players in the elite and non-elite teams is significant at $t\ (18) = 2.119, p = 0.048$. Moreover, the study found a difference in the Agility Test scores between 12-year- old players in the elite group (14.882 ± 0.863) and players in the non-elite group (16.105 ± 1.608). Based on this result, hypothesis Ha1 was accepted, proving a significant difference between the agility test score for 12-year-old male players in the elite and non-elite groups. Meanwhile, hypothesis Ho1 stating there is no significant difference between the agility test score for 12-year-old male players in the elite and the non-elite group were rejected (Table 2).

The t-Test analysis for the scores of the 50-m running with a ball test is not significant at $t\ (18) = 1.077, p = 0.295$. This result indicates no difference between the scores of the 50-m running with a ball test for 12-year-old players in the elite group (12.664 ± 0.448) and the non-elite group (13.103 ± 1.208). Therefore, this

Table 2 Results of the independent sample T-test analysis on the 3 tests score for participants in the elite player group and non-elite player group

Skills test battery	Sig.	t	Df	Sig. (2-t)
Agility test	0.024	2.119	18	0.048
50-m running with a ball test	0.148	1.077	18	0.295
Zigzag with a ball test	0.057	1.657	18	0.115

study accepted hypothesis Ho2, stating no significant difference between the scores of the 50-m running with a ball test for players in the elite and non-elite players. In contrast, hypothesis Ha2, which states a significant difference between the scores, was rejected.

The t-Test analysis for the scores of the Zigzag with a ball test is not significant at $t(18) = 1.657$, $p = 0.115$. This result indicates no difference between the scores of the Zigzag with a ball test for the players in the elite group (17.878 ± 2.120) and the non-elite group (20.263 ± 4.028). Therefore, this study accepted hypothesis Ho3, stating no significant difference between the scores of the zigzag with a ball test for the 12-year-old male players in the elite and non-elite players. In contrast, hypothesis Ha3, which states there is a significant difference between the scores, was rejected.

5 Discussion

It can be concluded that there is a significant difference between the agility test scores of players in the elite and the non-elite group. Hachana et al., supported this finding, which showed that elite players performed significantly better than non-elite players in the Modified Illinois Agility test [8]. Speed, agility and agility training is a progressive training system and which aims to develop basic motor skills to improve the ability of athletes to be more proficient at faster speeds and with better accuracy [9]. It is important to look for opportunities to incorporate an element of speed in each training session for young players. To achieve the optimal level of agility, several factors must be considered in order for the desired level of fitness to be achieved. In addition, a player's level of physical fitness will affect the performance and mastery of skills in the game [10]. Based on the results, the study fulfills the research objective. It found that the agility test conducted on the participants is valid an this finding confirms that the appropriate agility test is used to measure the performance of 12-year-old players in the elite and non-elite groups.

Meanwhile, this study found no significant difference between the scores of the 50-m running with a ball test conducted on players in elite and non-elite players. Therefore, regarding the difference between the elite players and non-elite players, it could be observed that players in elite and non-elite players groups have a similar ability in acceleration and running with a ball. Based on the findings from this observer, the researcher argued that players can still improve their achievement through more difficult assessments with repeated training. This is supported by the opinion of others researcher who argued that movement skills should be learned and trained in stages and repetitively so it can improve movement efficiency [11]. A football player who has a good level of muscular endurance will have a good level of cardiovascular endurance. While running, muscular endurance during the run is required in maximizing the run [12]. However, the data indicating the tests are not valid and suitable for measuring 12-year-old players' performance. Thus, for the improvement, the researcher proposed to produce an appropriate and accurate reference norm in terms of testing, measurement and evaluation procedures based on the sprint test with the ball. This is supported by others study stating that the evaluation process becomes more effective if the evaluation procedure has validity and reliability values on the sample [13].

Similarly, there is no significant difference between the players' scores in the zigzag with a ball test between elite and non-elite players. Therefore, it can be concluded that the elite players and non-elite players in this study showed a balanced performance in ball control and agility, as shown in the zigzag with a ball test. Based on the findings from this observer, the researcher argued that players need to improve in the repeated sprint training with multi-directional change-of-direction movement because based on others researcher, change of direction movements are performed in response to a visual stimulus, which may result in specific adaptations that improve speed and reactive agility in young highly trained football players [14]. Thus, it can be concluded that the deficiency in mastery of multi-directional movements with speed resulted in insignificant findings. Indicating the tests are not valid and suitable for measuring 12-year-old players' performance.

However, these findings may result from a number of factors that can be considered in the limitations and recommendations of the study. An increase in the number of samples should be made and some controllable factors should be taken into account. Increase the number of test batteries, diversify test variations, study test reliability, take into account the physiological conditions of the sample such as sleep duration, heart rate, food intake, previous exercise activities and others. From a practical point of view, in the present study researcher assume that skilled players can balance physically and mentally aspect. As well, highly skilled players might be able to conserve physical effort due to their technically and tactically advantages [14].

6 Contributions and Conclusion

In conclusion, there is a significant difference between the agility test scores of 12-year-old football players in the elite and the non-elite group. This finding proves that the agility test conducted distinguished between elite and non-elite players in the selected test sample group. Meanwhile, there is no difference between the players' scores in the 50-m running with a ball test and the zigzag with a ball test. These findings indicate that both tests could not distinguish between elite and non-elite players in the test sample group. However, these test can be improved in terms of study and modification procedures through distance and load because re-testing is able to improve the quality of test. This is supported by other studies stating that the evaluation process becomes more effective if the evaluation procedure has validity and reliability values on the sample [13]. Validity could be proven when there is a significant difference between the two groups [12]. Therefore, the agility test conducted could measure what needs to be measured. Moreover, this study proved that this test is suitable for measuring football skills among 12-year-old male players in the selected test sample group.

References

1. Deprez D, Coutts AJ, Lenoir M, Fransen J, Pion J, Philippaerts R, Vaeyens R (2014) Reliability and validity of the Yo-Yo intermittent recovery test level 1 in young football players. J Sports Sci 32(10):903–910. https://doi.org/10.1080/02640414.2013.876088
2. Francisco ARL (2013) Analisis Kemampuan Teknik Dasar Bermain Sepakbola Pada Pemain Usia 16 Tahun. J Chem Inf Model 53(9):1689–1699. https://doi.org/10.1017/CBO9781107415324.004
3. Purcell L (2005) Sport readiness in children and youth. Paediatr Child Health 10(6):343–344 https://doi.org/10.1093/pch/10.6.343
4. Rösch D, Hodgson R, Peterson L, Graf-Baumann T, Junge A, Chomiak J, Dvorak J (2000) Assessment and evaluation of football performance. Am J Sports Med 28(5 SUPPL):29–39. https://doi.org/10.1177/28.suppl_5.s-29
5. Validity T, Design F, Instrument C (2012) Kesahan dan kebolehpercayaan instrumen kompetensi Rekaan Fesyen Pakaian (RFP). Jurnal Pendidikan Malaysia 37(2):11–19–19. https://doi.org/10.17576/JPEN-2012
6. McMillan JH, Schumacher S (2010) Research in Education: Evidence Based Inquiry, 7th edn. Pearson Education Incorporated, New Jersey
7. Kim TK (2015) T test as a parametric statistic. Korean J Anesthesiol 68:220–223
8. Hachana Y, Chaabène H, Ben Rajeb G, Khlifa R, Aouadi R, Chamari K, Gabbett TJ (2014) Validity and reliability of new agility test among elite and subelite under 14-football players. PLoS ONE 9(4):1–6. https://doi.org/10.1371/journal.pone.0095773
9. Azmi K, Kusnanik NW (2018) Effect of exercise program speed, agility, and quickness (SAQ) in improving speed, agility, and acceleration. J Phys Conf Ser 947(1). https://doi.org/10.1088/1742-6596/947/1/012043
10. Lovell TWJ, Bocking CJ, Fransen J, Kempton T, Coutts AJ (2018) Factors affecting physical match activity and skill involvement in youth soccer. Sci Med Football 2(1):58–65. https://doi.org/10.1080/24733938.2017.1395062
11. Chen W, Zhu W, Mason S, Hammond-Bennett A, Colombo-Dougovito A (2016) Effectiveness of quality physical education in improving student's manipulative skill competency. J Sport Health Sci 5:231–238
12. Hafiz Ismail M, hashim A, Mohd Rashid N, Kunci K, Pasti Bakat Bola Sepak M, Spesifik Sukan Bola Sepak U, … Diri dan Orientasi Ego K (2013) Mengenal Pasti Bakat Sukan Bola Sepak Dalam Kalangan Pelajar Lelaki Berumur 14 dan 15 Tahun. Nd International Seminar on Quality and Affordable Education (Isqae), pp 468–477 (2013)
13. Harkimi M, Jani J, Razak SMA (2018) Tahap Kelajuan Pemain Bola Sepak Bawah 13 Tahun Pusat Latihan Daerah (PLD) SMK Anderson. e-Jurnal Penyelidikan dan Inovasi 5(2):188–201
14. Born DP, Zinner C, Duking P, Billy S (2016) Multi-directional sprint training improves change-of-direction speed and reactive agility in young highly trained soccer players. J Sports Sci Med 15:314–319

Factors Contributing to the Pedal Error or Pedal Misplacement Among Malaysian Car Drivers: A Survey

Mohamad Zairi Baharom, Zulkifli Ahmad, Nursya Mimie Ayuny Ismail, Mohd Hasnun Arif Hassan, Juffrizal Karjanto, and Khairil Anwar Abu Kassim

Abstract The driver's foot movement is unpredictable, which could lead to an error in foot placement. The farther the foot is from the target pedal, the more likely it is to make an error while pressing the pedal. The purpose of this study is to identify the mistakes and factors that contribute to incorrect pedal placement among Malaysian drivers. In addition, to learn more about the elements which cause road accidents in Malaysia. An online survey was carried out to collect the data for the study using the Google Form platform. According to the survey, the study was successful in determining foot placement on the pedal by Malaysian drivers when driving, as well as the factors, which contribute to incorrect foot placement on the pedal and road accidents in Malaysia. A total of 321 responses with age range between 20 and 60 years old participated in the survey. Based on the survey, 10% of the respondents are using both legs while driving and use their left foot to press the brake pedal in an automatic transmission car. The study also discovered various sources and causes that contribute to Malaysian drivers' wrong pedal placement, including being surprised, interrupted, and confused.

Keywords Pedal error · Pedal misapplication · Sudden Unintended Acceleration (SUA) · Foot pedal

M. Z. Baharom (✉) · Z. Ahmad · N. M. A. Ismail · M. H. A. Hassan
Faculty of Mechanical and Automotive Engineering Technology, Universiti Malaysia Pahang, 26600 Pekan, Pahang, Malaysia
e-mail: mohamadzairi@ump.edu.my

J. Karjanto
Faculty of Mechanical Engineering, Universiti Teknikal Malaysia Melaka, Hang Tuah Jaya, Durian Tunggal, 76100 Melaka, Malaysia

K. A. Abu Kassim
Malaysian Institute of Road Safety Research (MIROS), 43000 Kajang, Selangor, Malaysia

© The Author(s), under exclusive license to Springer Nature Singapore Pte Ltd. 2023
M. H. A. Hassan et al. (eds.), *Technological Advancement in Instrumentation & Human Engineering*, Lecture Notes in Electrical Engineering 882,
https://doi.org/10.1007/978-981-19-1577-2_5

1 Introduction

Motor vehicle accidents are one of the most dominant causes of deaths, disabilities, and hospitalization. A total of 9,355 deaths from road accidents occurred in Malaysia is recorded from January 2018 to December 2019 [1]. Motorcycle and passenger vehicle users contribute to the majority of deaths, which is 84% of all deaths. According to a research by Malaysian Institute of Road Safety (MIROS), human error causes most of the road accidents (80.6%), 13.2% due to road condition, and 6.2% of the accidents happened due to the vehicles [1].

Some of the common causes that lead to road accidents in Malaysia are including road constraints. The designs of urban infrastructure can be considered as deemed barriers when there are too many construction projects underway at the same time [2]. It could confuse some of the drivers to drive along the road, especially for drivers who are unfamiliar with the route. It might be difficult for them to drive smoothly. Other than that, distracted driving is another factor that contributes to road accidents. It can occur in a variety of ways, including using a mobile phone while driving, which can induce mental distraction [3]. Overtaking, driving in the opposite lane, and tailgating are additional contributing causes in Malaysian road accidents [4]. In Malaysia, speeding, running red lights, and driving under the influence are all factors that contribute to road accidents [5].

Despite the fact that the causes of accidents in Malaysia are quite diverse, sudden unintentional acceleration (SUA) is one of the most important difficulties that results in injuries and deaths on the road every year. SUA is defined as an unintentional, unexpected, and high-powered acceleration from a stationary or moving position that is accompanied by a loss of apparent braking effectiveness [6]. The unintended acceleration can also occur when a driver attempts to step on the brake pedal with his or her right foot while shifting from Park (P) to a drive gear (drive (D) or reverse (R), but instead steps on the accelerator pedal (full-throttle acceleration and cause a crash) [7]. A number of authors have considered that pedal error is the major factor of these kind of accidents [7, 8]. For decades, pedal misapplication occurrences that result in collisions have been a topic of discussion [9]. Over the years, designers have attempted to construct various mixed brake-accelerator pedals in order to reduce the chance of the driver pressing the wrong pedal [10].

The target group or participants who participated in several studies related to the pedal error are among the young and senior drivers. Meanwhile, most of the other studies recruited participants aged between 22 and 65 years old [6, 11–15]. However, Wu et al. (2017) believes that the questionnaire is critical in determining the elements that may contribute to pedal errors [16]. It is made up of a series of questions about age, gender, education, driving history, and daily driving habits.

The goal of this study is to assess how Malaysian drivers place their feet on the pedal while driving and to identify the sources of foot placement mistakes and factors that contribute to incorrect pedal placement among Malaysian drivers.

2 Methodology

Since the pandemic of COVID-19 is limiting the research team, the survey is conducted online. The platform of Google Form is used to collect the data with 21 questions given to the respondents. All the questions are divided into three different sections which are the demographic and driving history (Part 1), driving experience (Part 2), and factors of road accidents (Part 3). The questions are constructed based on several research conducted before [17, 18]. The main topic of the survey is on the pedal error and pedal misapplication among the Malaysian drivers.

2.1 Participants

The study focuses on the Malaysian drivers aged between 20 and 65 years old who are able to drive an automatic transmission car with a valid driving license with at least 2 years driving experience. A total of 321 respondents (175 male and 146 female) participated in the online survey. Most of the respondents are from Malaysia, aged between 20 to 65 years old.

2.2 Questionnaire

The first part of the survey is the respondent's demographic and driving history. In this case, the characteristics that are included in the survey are gender, age, driving experience (years), and driving duration in a day.

Table 1 shows the descriptive statistics of the respondents participated in the survey. It can be observed that most of the respondents are male. The respondents were grouped into a few age groups and most of them are aged between 21 and 40 years old. The survey has the least participants who aged less than 20 years old. Majority of the respondents (36.1%) are experienced drivers with 11–20 years of driving experience. It can be said that most of them are experienced drivers based on their years of driving. Almost 98% of the respondents drive for less than 5 h in a day. Only a few participants drive for 6–15 h per day.

3 Results and Discussion

3.1 Driving Experience

The next part of the questionnaire in the survey is to obtain the respondent's response regarding the pedal error experienced. It is aimed to collect the data whether they

Table 1 Descriptive statistics of the respondents (N = 321)

Characteristics	Category	N	%
Gender	Male	175	54.5
	Female	146	45.5
Age	Less than 20 years old	20	6.2
	21–30 years old	87	27.1
	31–40 years old	97	30.2
	41–50 years old	56	17.4
	51–60 years old	61	19
Driving experience	Less than 2 years	27	8.4
	3–10 years	80	24.9
	11–20 years	116	36.1
	21–30 years	77	24
	More than 31 years	21	6.5
Driving period in a day	Less than 5 h	314	97.8
	6–10 h	2	0.6
	11–15 h	5	1.6

have experienced pedal misapplication during their driving years or not. Another additional question is about the driving behavior, either they use both feet while driving or only one foot to drive.

Figure 1 shows the histogram for driving experience of the participants regarding the pedal error issue and their foot behavior on the pedal. From the data, it can be observed that from all of 321 respondents, 36.76% of them have experienced pedal error during their driving years. Some of the respondents (60.4%) have wrongly

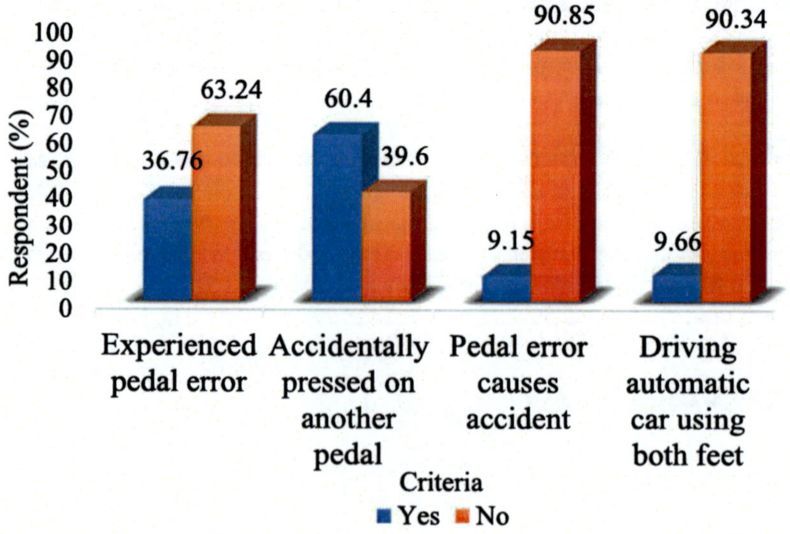

Fig. 1 Percentage of respondent response

pressed on the accelerator pedal instead of to press on the brake pedal. However, 9.15% of the respondents who experienced pedal error have caused an accident due to the pedal misapplication. Among the respondents, 9.66% (31 participants) drive the automatic transmission car using both feet.

3.2 Pedal Error Conditions

Other than that, pedal error conditions were also asked in the survey. It is to observe the conditions where the respondents have accidentally experienced pedal error while driving. As we know there are many possible conditions of when the pedal error occurred. Hence, these questions are aimed to gain more accurate information regarding the pedal error conditions during driving.

Figure 2 shows the pedal error conditions that the respondents have answered in the survey. There is a total of 118 respondents who had experienced pedal error during their driving years. The most agreed pedal error conditions are during parking, driving and hard braking. The highest percentage of the pedal error condition experienced by the respondents is during parking (37%). Other conditions are including during driving the vehicle and when applying hard braking, which contribute 24% of the responses each. Besides that, the response recorded on the pedal error conditions are including while they are sleepy, while shifting the gear, and reversing. One of them has experienced the pedal error after switching from manual transmission car to automatic transmission car. It might be because of the gear differences and they might forget that automatic car only has 2 pedals instead of 3 pedals like a manual car.

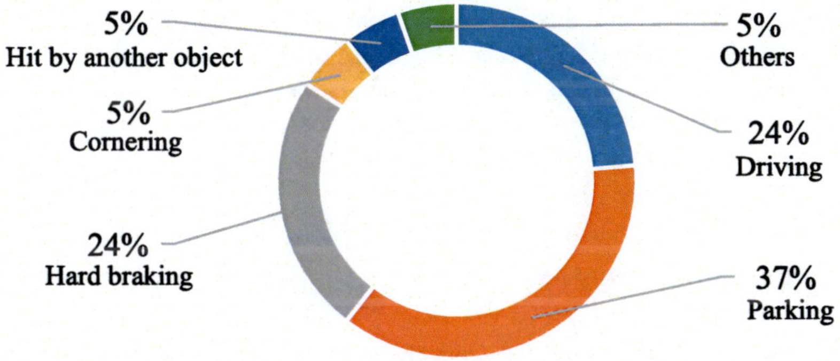

Fig. 2 Pedal error conditions

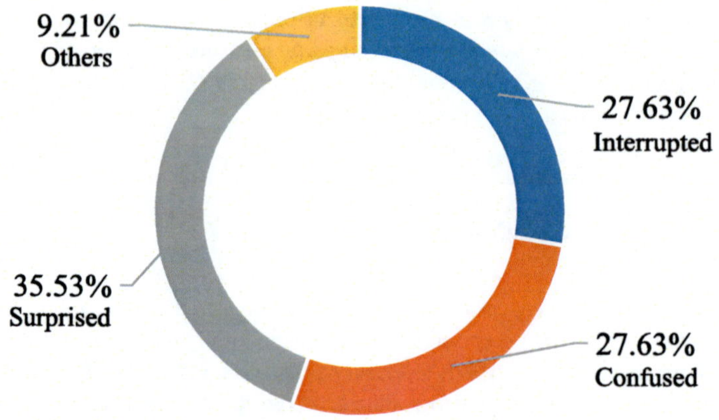

Fig. 3 Factors of pedal error while driving

3.3 Factors of Pedal Error While Driving

Other characteristics that are considered crucial in this study include the causes of pedal error while driving. The data for the factors of pedal error to occur while driving have been analyzed. It is shown that there are a few factors which contribute to the pedal misapplication during driving.

Figure 3 shows the factors of pedal error while driving by the respondents of the online survey. The data was obtained from the response of 118 respondents who have experienced pedal error while driving. Based on the data, 35.53% of the respondents experienced the pedal error because they were surprised and then wrongly pressed the pedal. There are two drivers mentioned on how their foot stuck on the carpet during driving which lead them to press the wrong pedal while driving. There are some cases where an accident occurred because of the car floor mat [19, 20]. The driver's foot was stuck during driving and caused pedal error to occur. Other than that, one of the respondents faced the pedal misapplication because of the thought that the gear was already in 'Neutral (N)' gear.

3.4 Factors of Road Accidents

The last section of the survey is about the factors causing the road accidents in Malaysia. A few factors are stated in the survey and the respondents have to answer based on the 5-points Likert scale score, which begins with strongly disagree (low score) until strongly agree (high score). Then, each of the factors are analyzed based on the given score.

Figure 4 shows the boxplot for the factors of road accidents based on the online survey data recorded. Based on the analyzed data, driving under influence has the

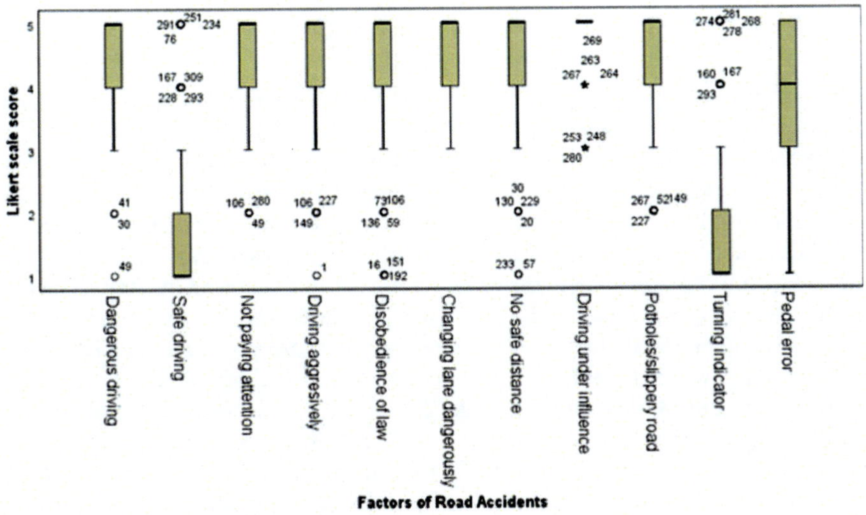

Fig. 4 Boxplot of factors of road accidents

highest mean of Likert scale score which is 4.71 [21–27], followed by changing lane dangerously which scores 4.58 [28–30]. These variables have resulted in many deaths and injuries, and it is a significant issue that everyone should be aware of. It can be observed that driving safely and using the turning indicator correctly are the least agreed as factors of road accidents, which demonstrates that the respondents did not simply answer the questions without thinking about their options. The survey result can be stated to be free of response bias based on the Likert scale scores.

Based on the data obtained from the online survey, we observed that almost 10% of the respondents drive an automatic transmission car using both feet. In this study, we can name it as a two-legged driving behavior. From the data, we analyzed the respondents with two-legged driving behavior by two categories, which are the age group and gender. According to the data collected for two-legged driving behavior, female drivers make up 58% and males make up 42% of the participants who drive using both feet. For male drivers who use both feet to drive the automatic transmission car, most of them are younger drivers aged 20–30 years old (46%) and 31–40 years old (31%). In comparison to male drivers, many female drivers use both feet to drive an automatic transmission car; 77% of female drivers are senior drivers aged 41–50 years old (44%) and 51–60 years old (33%). There is also limitation while conducting the survey. For instance, there are a few respondents who answered the survey multiple times. However, the duplicated feedback from the same respondent has been screened and removed.

4 Conclusions

Based on the survey conducted, we are able to observe the two types of driving behavior, including one-legged and two-legged driving behavior. According to the survey data, only 37% of those polled had experienced pedal mistake during their driving years. The vast majority of them erroneously pressed the accelerator pedal rather than the brake pedal. Some of those who have had a pedal error (9%) have caused an accident as a result of the misapplication of the pedal. Parking (37%) is the most common condition where the pedal error occurred, followed by during driving and hard braking which contribute 24% for each of the condition. According to this study, 36% of the respondents made a pedal error because they were astonished and then pressed the pedal incorrectly, followed by confusion and interruption, which make up 28%, respectively. It is clear that driving while drunken and changing lanes dangerously are significant contributors to traffic accidents according to high scores given by the respondents. In this study, 10% of the respondents use both of their feet to drive the automatic transmission car. It means that their left foot might be always on the brake pedal and their right foot is on the accelerator pedal. Most of them are female senior drivers aged over 40 years old which contribute 45% of the whole respondents who use both feet to drive the automatic transmission car. It can be stated that even when they are accelerating, their left foot is always placed on the brake pedal. However, this case might cause confusion to the other drivers because the car brake lights might turn on even when they do not have intention to press on the brake pedal. It is because of the force applied by their left foot on the brake pedal. Hence, an accident might occur due to the confusion while driving and it is very dangerous to the other drivers.

Acknowledgements The author would like to express the gratitude and special acknowledgements to ASEAN NCAP for the ASEAN NCAP Collaborative Holistic Research (ANCHOR III) research grant which funded this project (UIC201510). Also, to our collaborators from Universiti Teknikal Malaysia Melaka (UTeM) and Malaysian Institute of Road Safety Research (MIROS) for the continuous support to conduct this project.

References

1. JKJR (2019) "Buku Statistik Keselamatan Jalan Raya"
2. Noh NC (2021) "223 kemalangan akibat fizikal jalan raya,". https://www.bharian.com.my/berita/kes/2021/01/773051/223-kemalangan-akibat-fizikal-jalan-raya. Last Accessed 02 Aug 2021
3. Bernama (2016) "Penggunaan telefon bimbit antara punca utama kemalangan jalan rayalAstro Awani,". https://www.astroawani.com/berita-malaysia/penggunaan-telefon-bimbit-antara-punca-utama-kemalangan-jalan-raya-109199. Last Accessed 02 Aug 2021
4. Ibrahim MF (2018) "6 kesalahan utama pengguna jalan raya," Harian Metro. https://www.hmetro.com.my/utama/2018/03/322201/6-kesalahan-utama-pengguna-jalan-raya. Last Accessed 02 Aug 2021

5. Farid A (2020) "Kemalangan Di Jalan Raya: 5 Punca Ia Terjadi Di Malaysia!," Hello Doktor. https://hellodoktor.com/kesihatan/fakta-menarik/kemalangan-di-jalan-raya/. Last Accessed 02 Aug 2021
6. Wu J, Yang J, Yoshitake M (2014) Pedal errors among younger and older individuals during different pedal operating conditions. Hum Factors 56(4):621–630. https://doi.org/10.1177/0018720813487200
7. Schmidt RA, Young DE (2010) "Cars gone wild: the major contributor to unintended acceleration in automobiles is pedal error,". Front Psychol 1(Nov):1–4. https://doi.org/10.3389/fpsyg.2010.00209
8. Jonas R, Crump C, Brinkerhoff R, Krake A, Watson H, Young D (2018) "Variability in circumstances underlying pedal errors: an investigation using the national motor vehicle crash causation survey,". SAE Tech Pap 2018(April):1–8. https://doi.org/10.4271/2018-01-0493
9. Padmanaban J, Fitzgerald M, Marsh J (2013) "Pedal misapplication: Crash characteristics and contributing factors,." SAE Int J Passeng Cars—Mech Syst 6(2):601–607. https://doi.org/10.4271/2013-01-0446
10. Nilsson R (2002) Evaluation of a combined brake-accelerator pedal. Accid Anal Prev 34(2):175–183. https://doi.org/10.1016/S0001-4575(01)00011-2
11. Deng TM, Hou FJ, Shao YM, Peng JS, Xu J (2019) "Pedal operation characteristics and driving workload on slopes of mountainous road based on naturalistic driving tests." Saf Sci 119(December 7):40–49. https://doi.org/10.1016/j.ssci.2018.10.011
12. Jammes Y, Behr M, Llari M, Bonicel S, Weber JP, Berdah S (2017) Emergency braking is affected by the use of cruise control. Traffic Inj Prev 18(6):636–641. https://doi.org/10.1080/15389588.2016.1274978
13. McGehee DV, et al (2016) "The wagging foot of uncertainty: data collection and reduction methods for examining foot pedal behavior in naturalistic driving,". SAE Int J Transp Saf 4(2). https://doi.org/10.4271/2016-01-1526
14. Mcgehee DV, Mazzae EN, Baldwin GHS (1998) "Driver reaction time in crash avoidance research: validation of a driving simulator study on a test track,"
15. Tran C, Doshi A, Trivedi MM (2012) "Investigating pedal errors and multi-modal effects: driving testbed development and experimental analysis,". IEEE Conf Intell Transp Syst Proc ITSC 1137–1142. https://doi.org/10.1109/ITSC.2012.6338908
16. Wu Y, Boyle LN, McGehee D, Roe CA, Ebe K, Foley J (2017) Foot placement during error and pedal applications in naturalistic driving. Accid Anal Prev 99:102–109. https://doi.org/10.1016/j.aap.2016.10.019
17. Karali S, Gyi DE, Mansfield NJ (2017) Driving a better driving experience: a questionnaire survey of older compared with younger drivers. Ergonomics 60(4):533–540. https://doi.org/10.1080/00140139.2016.1182648
18. Wu Y (2015) "Quantifying drivers foot movements and pedal misapplication errors,"
19. Hard (2009) "Misaligned floor mat may have caused calamity,". https://www.consumerreports.org/cro/news/2009/09/misaligned-floor-mat-may-have-caused-calamity/index.htm. Last Accessed 08 Aug 2021
20. Jalil S (2021) "Alas kaki getah terlipat punca kemalangan rempuh tiga gerai makanan dan kereta|Careta,". https://careta.my/article/alas-kaki-getah-terlipat-punca-kemalangan-rempuh-tiga-gerai-makanan-dan-kereta. Last Accessed 08 Aug 2021
21. Aminnuraliff M (2021) "Penunggang motosikal maut dirempuh pemandu mabuk," Karangkraf. https://www.sinarharian.com.my/article/122620/BERITA/Semasa/Penunggang-motosikal-maut-dirempuh-pemandu-mabuk. Last Accessed 10 Aug 2021
22. Bahaudin NH (2021) "Pemandu mabuk punca kemalangan lima kenderaan,". Harian Metro, New Straits Time Press (M) Bhd. https://www.hmetro.com.my/mutakhir/2021/03/687030/pemandu-mabuk-punca-kemalangan-lima-kenderaan. Last Accessed 10 Aug 2021
23. Fuad F (2020) "Penunggang maut, dilanggar pemandu mabuk dari belakang,". BH, New Straits Time Press (M) Bhd. https://www.bharian.com.my/berita/kes/2020/11/759067/penunggang-maut-dilanggar-pemandu-mabuk-dari-belakang. Last Accessed 10 Aug 2021

24. Mokhtar NS (2020) "Pelajar maut angkara pemandu mabuk,". BH, New Straits Time Press (M) Bhd. https://www.bharian.com.my/berita/nasional/2020/09/734009/pelajar-maut-angkara-pemandu-mabuk. Last Accessed 10 Aug 2021
25. Nordin R (2020) "Driving while high: six years' jail, RM11K fine for woman who killed Navy officer in accident|The Star,". https://www.thestar.com.my/news/nation/2020/10/15/driving-while-high-six-years039-jail-rm11k-fine-for-woman-who-killed-navy-officer-in-accident. Last Accessed 10 Aug 2021
26. Raja Rahim RN (2020) "9 kemalangan maut angkara pemandu mabuk,". Harian Metro, New Straits Time Press (M) Bhd. https://www.hmetro.com.my/mutakhir/2020/06/584929/9-kemalangan-maut-angkara-pemandu-mabuk. Last Accessed 10 Aug 2021
27. Ramli NS (2021) "Pemuda maut motosikal dilanggar pemandu mabuk - Utusan Digital,". Media Mulia Sdn. Bhd. https://www.utusan.com.my/berita/2021/03/pemuda-maut-motosikal-dilanggar-pemandu-mabuk/. Last Accessed 10 Aug 2021
28. Awang Chik H (2020) "Tukar lorong punca penunggang motosikal kemalangan, Bung Moktar anggap pemandu kurang siuman|Berita|ProjekMM. https://www.projekmm.com/news/berita/2020/08/05/tukar-lorong-punca-penunggang-motosikal-kemalangan-bung-moktar-anggap-peman/1891154. Last Accessed 10 Aug 2021
29. Sabri N (2021) "Kesan pemandu kereta tukar lorong sebabkan kemalangan,". Harian Metro, New Straits Time Press (M) Bhd. https://www.hmetro.com.my/mutakhir/2021/04/695230/kesan-pemandu-kereta-tukar-lorong-sebabkan-kemalangan. Last Accessed 10 Aug 2021
30. Zulkiffli Z (2020) "Tukar lorong tiba-tiba punca kemalangan,". Harian Metro, New Straits Time Press (M) Bhd. https://www.hmetro.com.my/mutakhir/2020/03/550829/tukar-lorong-tiba-tiba-punca-kemalangan. Last Accessed 10 Aug 2021

Preliminary Study on the Influence of Boot Studs on Rugby Players' Sprinting Performance

Sharul Hadi Turiman, Zulkifli Ahmad, and Nasrul Hadi Johari

Abstract Sprinting ability in rugby players is an essential skill for the purpose of chasing and tackling opponents, and to run fast while carrying the ball to the try line. One of the key requirements in sprinting performance is the right boots selection. This study aims to investigate the influence of different boot studs towards sprinting performance of amateur rugby players. A total of ten rugby players took part in the 40 m sprinting test on the natural grass rugby pitch. Each player performed the sprinting using three different shapes of boot studs i.e., cone, blade and triangle stud shapes. Fully automated timing gates were placed at the 0 m and 40 m mark of the sprint track. Data obtained from the experiments were recorded according to the studs' shapes for statistical analysis. The results showed that majority of the players were sprinting faster using blade studs compared to other two studs with mean time recorded of 6.18 ± 0.33 s. The statistical analysis revealed moderate differences in sprint performance between cone and blade with (0.99 ± 0.25) while differences between cone and triangle was trivial (0.05 ± 0.18). However, the differences between blade and triangle were moderate (−0.86 ± 0.28). The analysis showed that players experienced similar sprinting ability when they changed their boots from cone to blade, and blade to triangle. The findings demonstrate that amateur rugby players should use boots with blade studs to establish better sprinting ability for natural grass pitch.

Keywords Boot studs · Rugby · Sprint

Supplementary Information The online version contains supplementary material available at https://doi.org/10.1007/978-981-19-1577-2_6.

S. H. Turiman · Z. Ahmad · N. H. Johari (✉)
Faculty of Mechanical and Automotive Engineering Technology, Universiti Malaysia Pahang, 26600 Pekan, Pahang, Malaysia
e-mail: nhadi@ump.edu.my

Z. Ahmad · N. H. Johari
Centre for Human Technology, Universiti Malaysia Pahang, 26600 Pekan, Pahang, Malaysia

© The Author(s), under exclusive license to Springer Nature Singapore Pte Ltd. 2023

M. H. A. Hassan et al. (eds.), *Technological Advancement in Instrumentation & Human Engineering*, Lecture Notes in Electrical Engineering 882,
https://doi.org/10.1007/978-981-19-1577-2_6

1 Introduction

No matter what kind of sports and the level of the tournament, athletes' performance largely depend on the skill and the equipment they used. In rugby, physical ability plays a crucial part as this modern game is now becoming faster and the players are getting bigger. Therefore, it is important for the players to wear the correct boots so that they can reach their maximum potential on the field especially during sprinting. If the shoes worn by the player are not able to supply enough traction, the player will easily fall or slip, and it can cause disadvantage to the team. The main factor that determines whether the player can move freely while playing is the type of shoes they worn [1].

To reduce the risk of injury to the feet, the interaction factor between the pitch surface and the surface of sports boots should be considered [2]. Most athletes are more concerned with performance characteristics such as traction and stability when choosing sports boots [3]. There are many studded boots on the market such as round studded and bladed studded, but it should be noted that certain boots are not designed to be played on some surfaces. Rounded stud on sports boots makes it easier for players to spin while playing. Rounded studs are longer compared to bladed stud, and this allows for a larger contact surface area between the shoe surface and the pitch surface which is ultimately able to maximize the player's traction and torque. Due to the rounded shape of the stud, it is recommended to wear round studded boots when playing on natural grass pitch and it is suggested that round studded boots to be used as they are more suitable on all pitch playing surfaces[4].

Stud characteristics such as number of studs, measurement of stud length and distance between studs are important factors influencing player movement. Demonstration from [5] study has shown that friction between player and field surface is influenced by changes in force exerted by player, distance between studs and measurements stud. The study shows that the design of the stud not only affects the peak attraction, but it also affects the amount of movement that the player has to do to produce maximum resistance during sports. Players need good traction, stability and suitable shoes to perform fast movements during tournament. The design of the sport boots must take into account the movement of the player when accelerating and changing direction. This is so that the sport boots produced are able to supply the appropriate tractional force to the player. During fast-moving movements, the interaction between the boots surface and the pitch surface will produce different tractional forces depending on the shape, length and arrangement of the studs. Some important information has been identified from the results of a study conducted by [3]. The study found that the time taken to run was 26% slower when using a sports boot without studs. A time difference of 3% was recorded when running using a sports boot with different types of geometry studs on the surface of the artificial pitch. However, a significant time difference was recorded where the running time was 20% longer when running on a field surface that had a different surface other than artificial turf. This study concludes that the traction and stability characteristics of the boots can guarantee the player to perform fast movements such as accelerating and

Table 1 Anthropometric characteristics and playing experience of the participants

Variable	Rugby player n = 10
Weight (kg)	72.81 ± 13.20
Height (m)	1.76 ± 0.12
Age (years)	18.6 ± 0.70
Playing experience (years)	3.6 ± 0.70

changing direction. Study by [6] stated that the production of a good boot design can be optimized if the interaction between the shoe stud and the pitch surface is considered during the design process. Studies related to different stud shapes have been conducted by [7]. Their study used seven types of studs' shape. The results showed that rhombus and triangular studs shape gave a higher negative rate of acceleration when compared to other stud shapes. This study concludes that the penetration effect of stud by sports boots on the pitch surface is lower when using studs that have a larger perimeter length.

The main focus of this study is to investigate the influence of different boot studs on rugby players sprinting performance. It is hoped that this study will provide technical information to the rugby community and assist the rugby player and coach in choosing the appropriate boots that suit them well before playing.

2 Methodology

2.1 Subjects

Ten male junior rugby players from local college team participated in the study (Table 1). The players competed as backs at state and national levels. Seven players had been trained in rugby for 4 years, whilst the others were between 3 to 2 years. However, they were in the same training regime, supervised by the coaching staff. The players were ranging from 16 to 19 years and categorized as junior as supported by [8] and [9]. Before participating in the sprinting test, all players were informed of the aims, benefits, risks, and procedures of the study. The researcher used participants who are still studying in high school for this study. All players had signed an informed-consent form prior to the experiment.

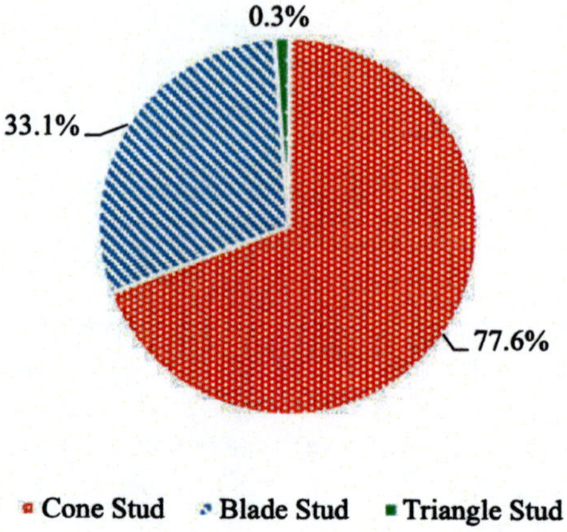

Fig. 1 Studs type selection by rugby player

2.2 *Equipment*

Before actual experiments were initiated, the researcher first conducted a pilot study using a questionnaire. This questionnaire was conducted on the respondents of the study to obtain data and views of respondents' involvement in rugby and information related to equipment use by respondent. This questionnaire was built using Google Forms application and distributed through WhatsApp and Facebook applications to the entire rugby community in Malaysia such as Malaysian Rugby Union (KRM), KV & SMT Rugby Community and Pahang Rugby. The respondents were given 1 week to answer this questionnaire until it no longer received any response. It was found that a total of 350 respondents had answered the questionnaire. Based on the results of the questionnaire, it was found that 77.6% of rugby players chose shoes with cone shaped studs, 33.1% chose blade shaped studs and 0.3% chose triangular shaped studs as referred to Fig. 1.

Although the percentage of rugby players who use boots with triangular studs is very low, researchers still use boots with triangular studs in this study because this type of boot is widely sold in the Malaysian market. Figure 2 shows the stud shape used in the sprinting test and Table 2 shows the characteristic for each type of boots used for the sprinting test.

Fig. 2 Different stud shapes of rugby boots used in sprinting test. (From left to right) conical stud, bladed stud and triangular stud

Table 2 Boots characteristic

Boots model	Stud material	Stud shape	Stud length (mm)
Puma rapido II	Hard rubber	Cone	9
Nike phantom VNM club	Hard rubber	Blade	9
Adidas goletto	Hard rubber	Triangle	9

Recording the time of the sprint test is very important because it analyzes the player's acceleration. Therefore, the equipment used to measure and record the time must be accurate. It is necessary to ensure that errors during time taking can be reduced. Most studies conducted for sprinting tests such as [10] use timing gates equipment as the primary equipment for recording time. This timing gate equipment, however, is not widely used by local rugby teams due to its high price and difficulty of installing the equipment on the sport's pitch. The researcher has conducted a search related to the appropriate tool to record the time of the sprint test. The researcher has decided to use a Photo Finish-fully automatic timing application (Photo Finish 2020) as the main equipment to record time. This application uses smart phone as a medium to record time. The advantages of using this application are that it is easy to use, easy to install from the Google Play application and it records the time automatically (Fig. 3).

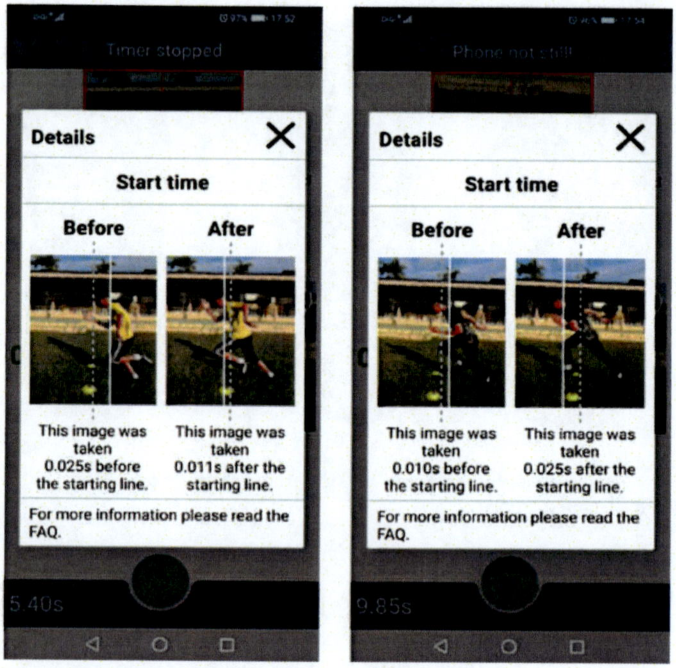

Fig. 3 Picture and time taken by Photo Finish 2020 application.

2.3 *Procedure*

Sprint speed was evaluated using the 40 m dash (Fig. 4). Marker poles were positioned at starting point and 40 m. Participants were instructed to run as quickly as possible along the 40 m distance from a standing start. All testing trials were recorded using two smart phones installed with "Photo Finish – Fully Automatic Timing App" application. All the smart phones were connected to the application via Bluetooth.

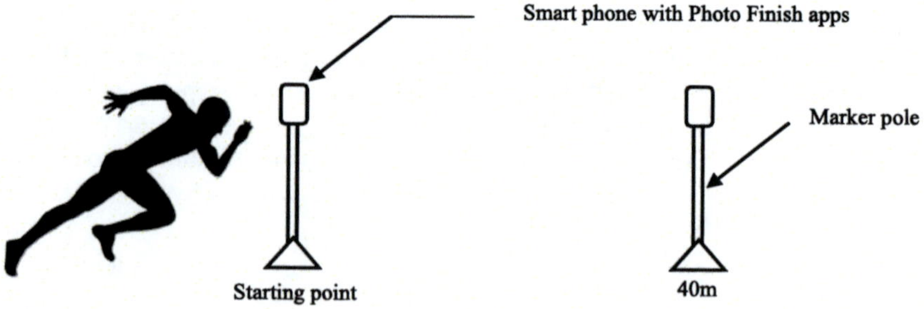

Fig. 4 Sprinting assessment of rugby players

The criteria used to determine the completion of the sprinting segment was the first body part to cross the marker pole. Time was measured to the nearest 0.01 s. Each participant performed three attempts. The time for each attempt was recorded for analysis. All sprints were performed on normal surface pitch. All participants performed the sprint test in rugby boots with different stud shapes made with hard plastic. Figure 4 shows the sprinting assessment conducted by the researcher.

2.4 *Statistical Analysis*

All data obtained from the experiments conducted were reported as mean ± standard deviation (SD) after being analysed using SPSS software. To measure the paired samples t-tests, analysed differences between the players sprinting time of cone stud, blade stud and triangle stud, the calculation used was based on Cohen's *d* and 90% confidence limit effect size principle. The probability of different values of the effect size is described as trivial effect: 0–0.19, small effect: 0.2–0.59 and moderate effect: 0.6–0.79 refer to [11]. Results for sprinting performance of all stud types were also analysed using repeated-measure ANOVA. The results were also compared with other studies using the one sample t-test analysis method.

3 Results and Discussion

The average 40 m sprinting performance for each type of boot stud was recorded in Table 3. Based on the tests conducted, the results showed that players who sprinted using boot with cone stud recorded mean of 6.43 ± 0.14 s, 6.18 ± 0.33 s for boot with blade stud and 6.42 ± 0.22 s for boot with triangle stud. Based on the results of the 40 m sprinting test conducted, the researcher conducted an analysis through Cohen's *d* method to identify the most suitable stud to be used for the purpose of sprinting on the surface of natural grass sports pitch. The formula for Cohen's d is as follows:

Table 3 40 m sprint performance with different stud type

	Stud type (Mean ± SD)		
	Cone stud	Blade stud	Triangle stud
40 m sprint (s)	6.43 ± 0.14	6.18 ± 0.33	6.42 ± 0.22

SD = Standard Deviation

Table 4 Sprint performance difference between stud type

	Cohen's *d* (90% CI)	Qualitative outcome
Cone vs Blade	0.73 (0.13 to 1.31)	Moderate
Blade vs triangle	−0.71 (−1.28 to −0.11)	Moderate
Cone vs Triangle	0.03 (−0.49 to 0.55)	Trivial

CI = Confidence Interval

$$d = \frac{M_E - M_C}{Sample\ SD\ pooled} \times \left(\frac{N-3}{N-2.25}\right) \times \sqrt{\frac{N-2}{N}}$$

Correction factor for small samples <50

where M_E and M_C are the mean of group 1 and 2, respectively, and N is the sample size.

Referring to Table 4, the comparison between cone stud and blade stud showed a moderate result of 0.73. Blade stud and triangle stud also showed a moderate result of −0.71, however the comparison between cone stud and triangle stud showed a trivial result of 0.03. Sprint performance difference between stud type can also be seen through the descriptive plot in Figs. 5. The results of the experiments and analysis conducted show that rugby players sprint faster by wearing boots with blade studs compared to wearing boots with cone studs and triangle studs.

However, the analysis using repeated-measure ANOVA showed that the sprinting performance of rugby players using boot stud type cone, triangle and blade did not show significant difference. Wilks'lambda = 0.617, f (2,8) = 2.486, $p = 0.145$, η^2

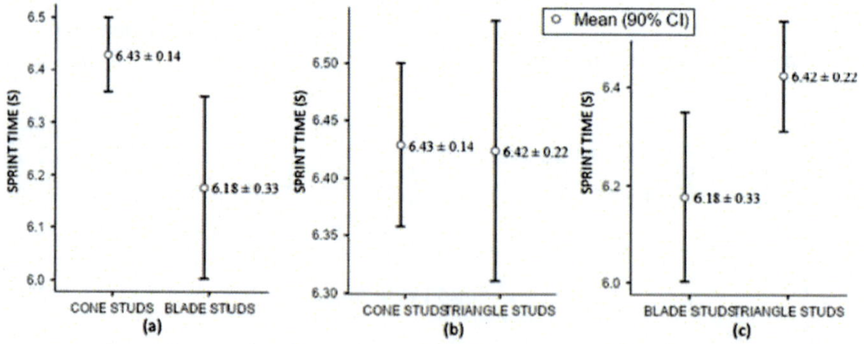

Fig. 5 Descriptive plot of sprint performance between **a** cone and blade studs, **b** cone and triangle studs, and **c** blade and triangle studs

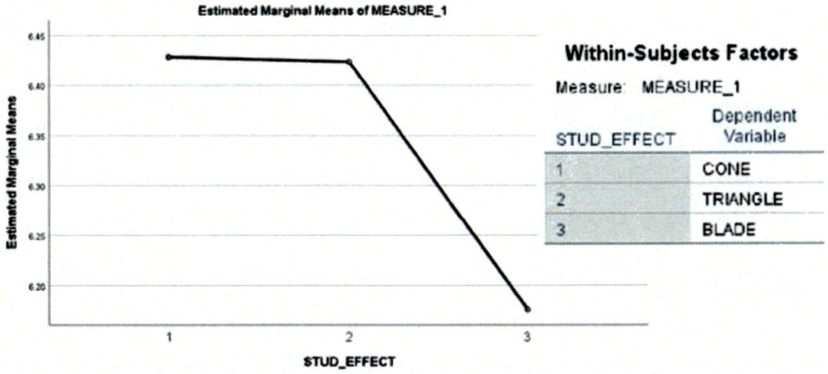

Fig. 6 Estimated marginal means of the cone, triangle, and blade stud effect

= 0.383. Figure 6 show the estimated marginal means of the stud effect to the rugby players sprinting performance.

This study was conducted with the aim of differentiating the sprint performance of rugby players when using boots with different studs on a normal turf surface. Significant differences were obtained in player sprint performance tests based on the type of studs used. The experimental results show that the use of boots with blade type studs makes that rugby players sprint faster compared to the use of boots with cone and triangle type studs. However, the results of this study contradict the statement from [4] which stated that rounded shaped studs are more suitable for use on the surface of natural grass sports pitch.

The results from this study are also supported by a study from [7] which stated that studs with a large surface area have a low penetration rate on the pitch surface. This means that if penetration to the outsole was to be achieved, the increase in skin friction may have a beneficial effect when running on softer surfaces and the greater surface area involved in the boot-surface interaction could provide increased friction and improved traction. The study by [12] stated that traction is produced when the stud on a sports boot uses a horizontal shear force on the boot surface and initially it will be resisted by the shear resistance resulting from the compression of the soil layer.

Referring to Table 5, when comparing with 40 m sprinting performance from other studies using one sample t-test, it is found that the sprinting performance of

Table 5 Comparison of sprint assessment of rugby player

Reference	*N*	Playing level	Assessment (s)
			40 m Sprint
Darrall-Jones et al. [13]	28	Junior	5.57 ± 0.18
Seitz et al.[10]	24	Junior	5.39 ± 0.23
Present study	10	Junior	6.18 ± 0.33

rugby players from this study is low. Studies from [10] show junior rugby player are significantly faster (5.39 ± 0.23) than present study, $t(9) = 7.454$, $p \leq 0.05$. Studies from [13] also show junior rugby player are significantly faster (5.57 ± 0.18) than present study, $t(9) = 5.747$, $p \leq 0.05$. Taken into account the best sprinting performance for this study with 6.18 ± 0.33 s, this shows that that foreign players for the junior category are faster than the local junior rugby players.

4 Conclusion

The ability of a player to sprint well is a very important aspect in the sport of rugby. Therefore, the use of appropriate sports boots is important to ensure that rugby players are able to sprint at their best during the games. This study was conducted to find out the most suitable sports boots stud to be worn by rugby players for sprinting purpose. The results found that boots with blade studs gave the best results compared to boots with cone and triangle studs. However, when comparing the sprinting performance between foreign and local players, the results showed that for junior category, foreign rugby players were faster than the local junior rugby players in this study. It is hoped that in the future, researchers will be able to conduct study on the influence of studs toward sprinting performance of rugby players on different types of pitch surfaces such as puddle and muddy pitch surface. It is hoped that the results of this study can assist the players and coaches in choosing the appropriate boots to be used in training or rugby games.

Acknowledgements The authors would like to thank the management of Kuantan Vocational College for providing facilities and the rugby players, and to Universiti Malaysia Pahang for financial assistance under grant no. PGRS210345.

References

1. Van Groningen D (2016) Effects of outsole shoe patterns on athletic performance. ProQuest Diss. Theses, p 63
2. Iacovelli JN, Yang J, Thomas G, Wu H, Schiltz T, Foster DT (2013) The effect of field condition and shoe type on lower extremity injuries in American Football. Br J Sports Med 47(12):789–793
3. Hennig EM, Sterzing T (2010) The influence of soccer shoe design on playing performance: a series of biomechanical studies. Footwear Sci. 2(1):3–11
4. Mansfield MM, Bucinell RB (2016) Effects of playing surface and shoe type on ACL tears in soccer players. Am J Eng Appl Sci 9(4):1150–1157
5. Severn KA, Fleming PR, Clarke JD, Carre MJ (2011) Science of synthetic turf surfaces: investigating traction behaviour. Proc Inst Mech Eng Part P J Sport Eng Technol 225(3):147–158
6. Ranson C, George J, Rafferty J, Miles J, Moore I (2018) Playing surface and UK professional rugby union injury risk. J Sports Sci 36(21):2393–2398
7. Dé R, James D (2014) The effect of stud shape on penetration characteristics through synthesized natural turf in football. Procedia Eng. 72:648–653

8. Till K, Jones B, Darrall J, Emmonds S, Cooke C (2015) Longitudinal development of anthropometric and physical characteristics within academy rugby league players. J Strength Cond Res 29(6):1713–1722
9. Darrall J, Jones B, Till K (2016) Anthropometric, sprint, and high-intensity running profiles of English academy rugby union players by position. J Strength Cond Res 30(5):1348–1358
10. Seitz LB, Barr M, Haff GG (2015) Effects of sprint training with or without ball carry in elite rugby players. Int J Sports Physiol Perform 10(6):761–766
11. Cohen J (1988) Statistical power analysis for the behavioral science
12. Forrester S, Fleming P (2019) Traction forces generated during studded boot-surface interactions on third-generation artificial turf: a novel mechanistic perspective. Eng. Reports 1(5):1–21
13. Darrall JD, Jones B, Roe G, Till K (2015) Reliability and usefulness of linear sprint testing in adolescent rugby union and league players. J Strength Cond Res 30(5):1359–1364

Ergonomic Chair Design in Minimizing MSD Chassis Assembly Workers Complaints Using Ergonomic Function Deployment (EFD)

Nelfiyanti, Nik Mohamed, M. F. F. A. Rashid, and Chon Chin Seik

Abstract The assembly line is the most critical part of the automotive industry that produces car units manually. Chassis assembly is one of the existing processes in the assembly line with a higher level of difficulty than other processes because it involves under body process. MSD complaints of the actual condition felt by workers using the RULA and REBA methods obtained a total score of "7 and 11 points" with a risk level of "Very High" which could have pain in the spine and other limbs. This study designed an ergonomic chair to minimize MSD complaints felt by chassis workers by using the Ergonomic Function Deployment (EFD) method. The results obtained after simulating the application of ergonomic chair design using RULA and REBA methods gained a total score of "2 and 3 points", which shows a risk level of "Low". It shows a significant improvement to the scores, thus the impact of ergonomic chairs can minimize workers' MSD complaints.

Keywords Design · Ergonomics · Musculoskeletal disorders · REBA · RULA

1 Introduction

The most popular business sector in the global market is the automotive industry [1]. One of the rapidly growing automotive industries is car production. An influential line in automotive products that can adapt to varying consumer demands is the assembly line [2]. An Assembly line is an activity that assembles all the components needed into a new production unit that is the most important part of the production of the

Supplementary Information The online version contains supplementary material available at https://doi.org/10.1007/978-981-19-1577-2_7.

Nelfiyanti (✉) · N. Mohamed · M. F. F. A. Rashid · C. Chin Seik
College of Engineering, Universiti Malaysia Pahang, 26300 Kuantan, Pahang, Malaysia
e-mail: nelfiyanti@ftumj.ac.id

Nelfiyanti
Department of Industrial Engineering, Faculty of Engineering, Universitas Muhammadiyah Jakarta, 10510 Jakarta, Indonesia

© The Author(s), under exclusive license to Springer Nature Singapore Pte Ltd. 2023
M. H. A. Hassan et al. (eds.), *Technological Advancement in Instrumentation & Human Engineering*, Lecture Notes in Electrical Engineering 882,
https://doi.org/10.1007/978-981-19-1577-2_7

automotive industry [3]. The assembly process is done manually and monotonously using humans as the main source [4].

One of the assembly processes that have a high level of work accident hazard compared to the others is the chassis line assembly process activity [5]. The Chassis Assembly Line is made up of a series of workstations where humans conduct assembly jobs. Vehicle types of chassis frames were constructed as they progressed down the line. A computer is installed at each workstation [6]. Each Chassis frame received a share of the entire effort. Musculoskeletal Disorder (MSD) is a complaint that is often complained by chassis assembly workers during the work process.

MSD itself is a pain felt in the skeletal muscles that are mild and even very painful [7, 8]. The work process performed monotonously and continuously is a major factor in the occurrence of MSD complaints in workers [9, 10]. To know and analyze the working system of interaction between humans, machines and the environment can use an ergonomic approach [11].

RULA and REBA are ergonomic approaches that can be used for postural risk assessment [12]. There are 8 types of chassis line assembly work that have a risk level of MSD complaints that are in the risk category "Very High" with a final score of REBA 11 points and RULA 7 points which means that it needs investigation and repair as soon as possible. The work is done manually with the help of several work aids such as a hand drill.

This work is done continuously during working hours by using body limb bent forward, bent sideways, neck up and tilted and wrists twist. The eight types of work consist of installing fuel tank, install the propeller, fill the transmission oil, install pipe air cleaner, install pipe exhaust tail, install the shift level cable clamp, install the mudguard left fender and install the mudguard right fender.

Improvements in minimizing the complaints of MSD workers chassis assembly line by designing an ergonomic work chair design. This design uses the Ergonomic Function Deployment (EFD) method. EFD itself is a method that simplifies the design process, decision -making that is organized in the matrix so that it can be examined and modified [13]. EFD is a development of Quality Function Deployment (QFD) in product design by considering the needs and desires of consumers, as well as ergonomic aspects when used [14].

EFD is widely used in designing and developing product designs such as designing slimmer bicycle bags that make it easier to carry [15], designing transports to increase productivity [16], designing train passenger chairs to provide comfort for passengers [17] and chair design economy bus to provide a sense of security. Most EFDs are used to design other manufacturing products, and none are focused on assembly workers especially the chassis process. Therefore, this research aims to design an ergonomic chair design for chassis workers in minimizing the perceived MSD complaints.

2 Methodology

This research has several stages that must be done to produce an ergonomic chair design for chassis assembly line workers. Figure 1 provides an explanation of the stages that must be performed in this study.

In Fig. 1 it can be clearly seen that the research process begins by conducting direct observations on the work environment of chassis line assembly workers. The stages consist of:

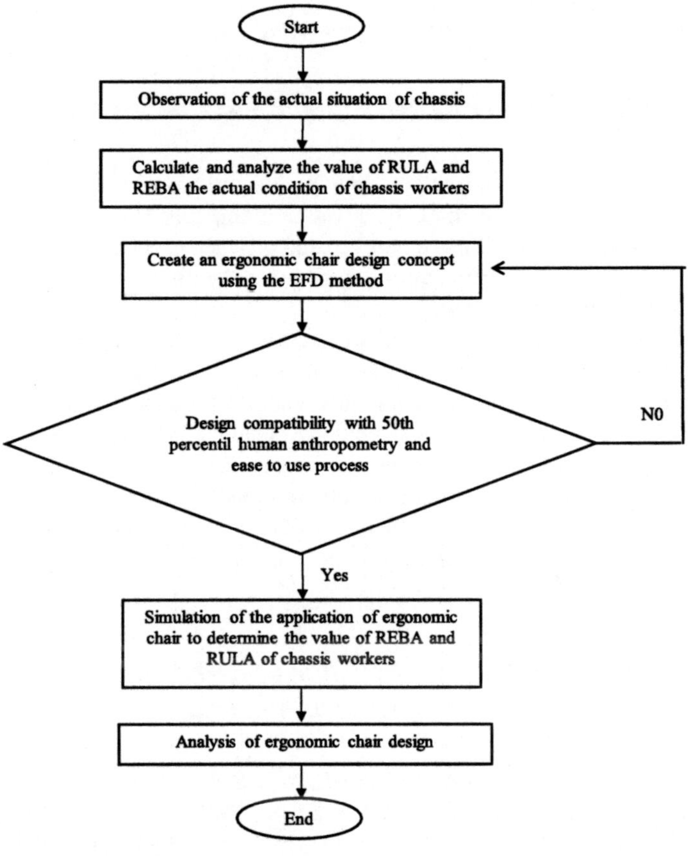

Fig. 1 Flow chart of research methodology

Fig. 2 The manner and posture of workers when working on chassis line assembly

2.1 Observation of the Actual Situation of Chassis Workers

The work and the method of functioning with the work postures utilized during the assembly process necessitate direct observation in this case study. Working manner and posture, location of facilities employed and distance necessary from retrieval of facilities, as well as components to be assembled to the underside of the automotive units are among the descriptions acquired.

To analyze ergonomic risks in the workplace, observational approaches might be utilized [18]. A method that can be paired with observation is the interview. Conducting interviews with the people involved can help identify workplace risks [19]. Figure 2 shows the process and method of chassis line assembly workers.

It can be seen in Fig. 2 the process and the way the work is done by the chassis worker during the work process that is done repeatedly. There are some limbs that are used not in accordance with the standards that give the impact of pain. Such as Neck, head, arms, back, waist and legs.

2.2 Calculate and Analyze the Value of REBA and RULA, the Actual Condition of Chassis Workers

The ergonomic assessment instruments REBA and RULA were used to measure MSD complaint evaluation experienced by chassis line assembly employees in this study. RULA (Rapid Upper Limb Assessment) is a method for evaluating the posture, style, and movements of upper-limb occupational activities [20]. RULA calculations and analysis performed on chassis assembly workers using the help of CATIA V5 software in the Ergonomic design section and analysis of the Human measurements' editor sub section.

The REBA (Rapid Entire Body Assessment) method is an ergonomic analysis method that includes all limb movements that are thought to be hazardous during workstation operations [12]. REBA form from Microsoft excel is an application used to calculate and analyze complaints felt by chassis assembly workers.

2.3 *Create an Ergonomic Chair Design Concept Using the EFD Concept*

2.3.1 Identification of Product Requirement Attributes

Attribute identification was performed to determine the needs and desires of chassis assembly line workers related to the ergonomic aspects of ENASE (Effective, Comfortable, Safe, Healthy and Efficient) [13]. These attributes have been obtained that will be translated as employee needs in the product planning to be made [12].

2.3.2 Preparation House of Ergonomic Matrix (HOE)

The making of the House of Quality matrix is based on consumer desires in accordance with ergonomic principles that will be the attributes used [14]. HOE is a development of the House of Quality (HOQ) matrix [15].

2.4 *Work Chair Chassis Assembly Line Design to Minimize MSD*

Once all product specifications are determined, then the next step is to create an ergonomic chair drawing design tailored to the needs of the chassis assembly line workers using the SolidWorks software. SolidWorks is a complete computer-aided design software with the ability to design, simulate, communicate, manage data and can see the impact on the environment [21].

2.5 *Simulation of the Application of Ergonomic Chair Design to Determine the Value of RULA and REBA of Chassis Workers*

The application of ergonomic chair design of workers chassis assembly line can be done by the simulation to determine the RULA and REBA scores obtained whether there is a reduced risk of MSD complaints about workers. The parameters used in the RULA analysis are the working position and posture, the load lifted, the force used (static/dynamic) and the amount of work. The posture used in the assessment is the palm of the hand, the posture that is assessed consists of the palm, upper arm, lower arm, back and neck. Posture in the RULA analysis is divided into 2 groups, namely groups A and B. Group A consists of the upper arm, lower arm and wrist. While group B consists of the neck, spine and legs [23, 24].

REBA calculation and analysis using working position or posture parameters during the work process. The assessed work posture consists of the neck, back, arms, wrists and legs. In addition to workers' posture, REBA is also influenced by coupling factors, external loads felt by the body and activities performed by workers. REBA has more value than other methods because it analyzes all parts of the worker's body through a focus on the overall posture, which is expected to minimize MSD complaints [22].

Analysis of RULA and REBA in this study using CATIA V5 and Microsoft excel software in performing simulations to determine the value and category of risk felt by workers with posture and workload parameters. The simulation method itself is software used to determine the results of the implementation of improvements[25].

3 Results and Discussion

3.1 Analysis of RULA Score Calculation and REBA Actual Condition of Chassis Workers

Analysis of the actual condition of the process and the way chassis workers during the work process is done to determine the total score that is included in the categories and actions that need to be done. Figure 3, 4 is an analysis of MSD complaints felt by chassis workers using RULA and REBA methods for current conditions.

Fig. 3 RULA analysis of the current condition of chassis assembly line workers

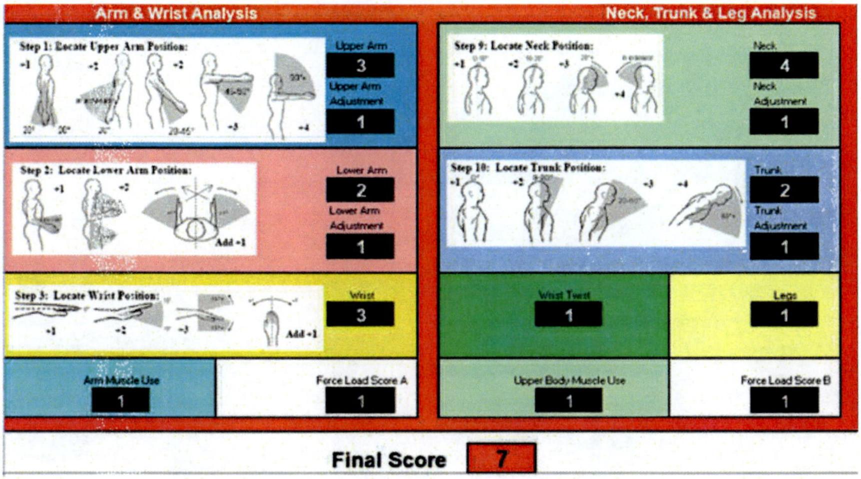

Fig. 4 Chassis relevance worker RULA evaluator manual

Figure 3 is the result of RULA calculation of chassis assembly workers using CATIA V5 software with the help of mannequins whose body posture is adjusted to the actual conditions. The total score of the RULA calculation result was obtained as 7 points with the category "very high" which immediately requires corrective action. Point 7 is obtained by providing an assessment of the limbs used in working from the arms, wrists, neck, body and limbs in accordance with the working condition by following the assessment rules described in Fig. 4.

Figure 4 provides clear information on how to give points to each limb that is needed in the RULA so that a total score of 7 points is obtained. Each assessment is given in accordance with the situation and condition of the worker which is adjusted to the existing weight.

Figure 5 describes the assessment given in each Movement made by workers. Obtained a total score of REBA Analysis of 11 points which are included in risk level 4 with a "very high"-risk category with the necessary actions Necessary national action possible.

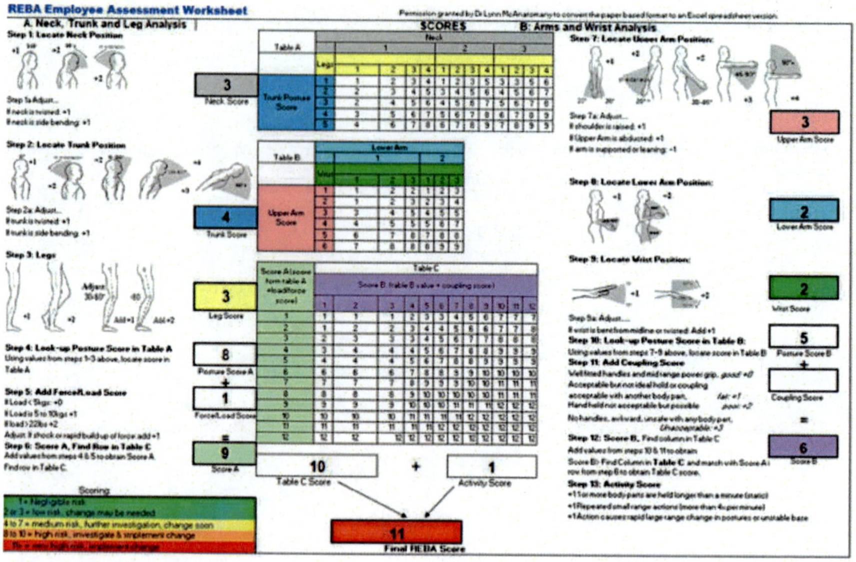

Fig. 5 REBA analysis the current condition of chassis assembly line workers

3.2 *Ergonomic Chair Design Using the Concept of EFD*

After the analysis stage of the actual condition of the chassis assembly workers using the RULA and REBA approach with the results obtained included in the category of "Very High". Therefore, corrective action is needed as soon as possible to minimize the MSD complaints. So, the next step that must be done is to make an ergonomic chair design that can provide comfort for workers during the work process. The ergonomic chair was designed using the EFD approach. The EFD stage in the design of ergonomic chairs for chassis assembly workers consists of:

3.2.1 Identify the Needs of Ergonomic Workers

Needs attributes were obtained from observations and direct interviews with chassis assembly line workers. Interviews conducted on the problems felt during work which is described in the aspects of ENASE (Effective, Comfortable, Safe, Healthy and Efficient) which can be seen in Table 1.

Table 1 Identification of product requirements attributes

No	Ergonomic aspects	Product requirements attributes	Product attribute details
1	Effective	Effective in the design of ergonomic chairs is to provide an interior that is easy to use	a. The interior consists of a precise design of sitting, wrists, feet connected with ergonomic tools b. Customized design of workers' anthropometric measurements
2	Comfortable	Provide comfort for workers when using it	The chair design corresponds to the body size of the chassis worker a. The dimensions of the chair are not narrow and not too roomy b. The design of the device connector with the foot that does not provide pain and lightness
3	Safe	Does not harm workers when using it such as loose and falling	a. Ergonomic backrest, comfortable armrests, and leg connectors b. The wheels on the chair can walk according to the movement of the workers' legs
4	Healthy	Can minimize employee MSD complaints	The design of the chair can reduce pain complaints especially in the neck, back and legs a. There is a comfortable sling connected to the armrest
5	Efficient	Lighter energy consumption and product usage time	a. Ergonomic chairs are easy to use according to the working position b. The interior used is in a position that suits the user from the sides of the legs, back, arms and neck

3.2.2 Construction of House of Ergonomics (HOE)

House of Ergonomics (HOE) is made based on the needs attributes that have been obtained from the identification results. HOE serves to see how the various aspects of the product need attributes relate. Figure 6 shows the relationship of various aspects of ergonomic chair in minimizing the complaints of MSD workers chassis assembly line.

Figure 6 shows the relationship between consumer desires and technical requirements in the design of ergonomic chairs that will be made. Obtained the section "dimensions of limb and ease of use" got the highest point of 7.2 which means that this section is significant as the primary reference used in the design. Furthermore, the technical part considered in the design, which is also included in providing ease of use, is "foot connector to tool and wheelchair legs" for comfort for workers while using the chair without giving side effects apart from the chair. Connecting the chair

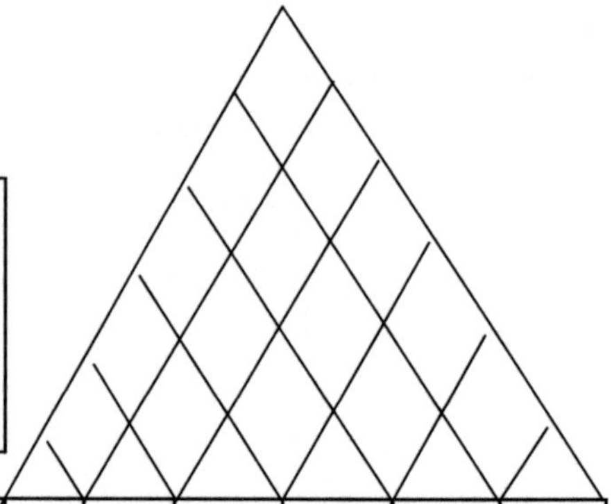

(+9)	●	Strong
(+3)	○	Currently
(+1)	Δ	Weak

Product requirements attributes	Technical requirements	Worker chair shape	Dimensions of limbs	Ease of use	Foot connector to tool	Wheelchair legs	Chair accessories
Able to reduce the workload of workers	The design of the chair is adapted to the anthropometric needs of workers in carrying out assembly activities.	○	●	●		○	Δ
The chair is easy to use with comfortable dimensions	The seat design corresponds to the body size of the chassis worker	●	●		●		○
Does not harm workers when using it	The design of the chair can provide comfort for workers during the work process	○	●	●		●	
Can minimize MSD employee complaints	The design of the chair can reduce pain complaints especially in the neck, back and legs		●	●	○	●	
Uses less energy and is easy to use	Desain chairs have a connector to the worker, so it does not give a detachable impact.	○		●	●		Δ
	Average value	3,4	7,2	7,2	4,2	4,2	1

Fig. 6 Matrix of the relationship of consumer desire with technical requirement (HOE)

to the worker's body is made on the legs and body so that the chair can follow the worker's movement.

3.3 *Work Chair Chassis Assembly Line Design to Minimize MSD*

Chassis assembly strip worker chair design is made and planned based on the results of employee needs and desires found in the HOE contact matrix. The technical

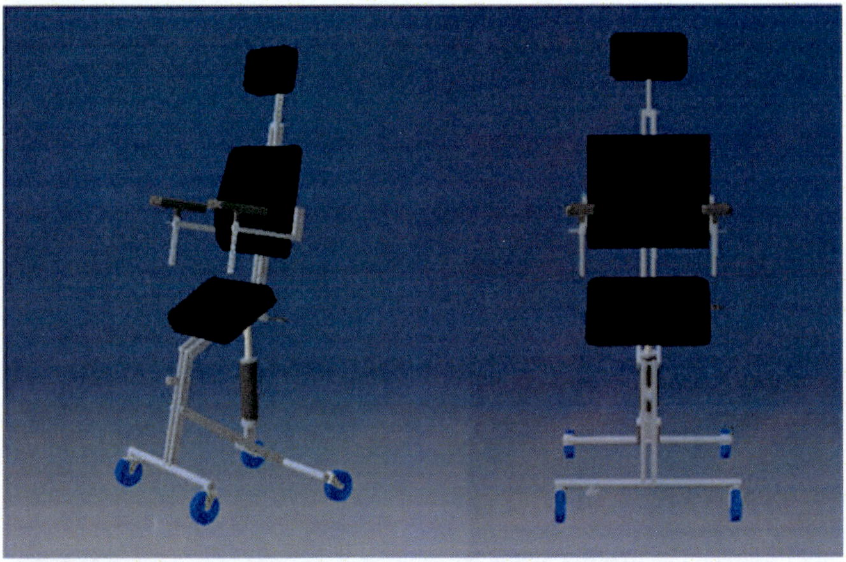

Fig. 7 Results of ergonomic chair drawing of workers chassis assembly line

requirements selected based on the HOE results are "dimensions of limb and ease of use" so that the ergonomic chair design will be adapted to the technical requirements as the main reference. At the same time, other technical requirements are supported to complement the function of the ergonomic chair. After obtaining the technical requirements of the main ergonomic chair design and support, then the next step is to design the image of the ergonomic chassis worker chair using the help of SolidWork software. The design of the ergonomic chair can be seen in Fig. 7.

As seen in Fig. 7, the ergonomic chair has a connector with the worker's legs to facilitate the process of use, which follows the worker's movement during the work process. In addition, the design of the chair has back and arm straps that can be adjusted according to the comfort of workers.

Figure 7 explains the ergonomic chair design that chassis assembly workers will use to facilitate and minimize MSD complaints. The ergonomic chair design uses the high average grip of Asian workers. Average anthropometric data used from age 17–50 years for all sexes. The anthropometry data used is in the 50th percentile with an average body height of 163.37 cm.

Ergonomic chairs are designed to be flexible so that they provide comfort for workers when using them. Like the backrest that can be adjusted back and forth, the high and low are adapted to the needs of workers when using it. Adjustable seat and backrest connectors use pneumatic cylinders. The ergonomic chair has wheels for movement, brakes that act as a stopper when the chair is stopped or moved. In addition, the central part that connects workers with ergonomic chairs is the presence of hooks on the chair and legs. This hook facilitates access to the use of ergonomic chairs that provide a sense of comfort for workers.

3.4 Simulate the Application of Ergonomic Chair to Determine the Value of RULA and REBA Chassis Workers

From the results of the chassis assembly line workers chair design, the next step is to perform a simulation using CATIA to determine the RULA value and excel form to determine the REBA value of the workers if using this chair design. Figures 7 and 8 are simulation results of the application of ergonomic chairs to determine the values of RULA and REBA.

Figure 8 is the result of a simulation of RULA calculation in the use of ergonomic chair design for chassis assembly workers using CATIA V5. The total RULA score obtained is 2 points which means it is included in the category of "Low" at level 0 so it does not require corrective action but requires development in a better direction.

A total score of 2 was obtained by simulating workers (using mannequins in CATIA V5 software) using an ergonomic design chair. The posture of the mannequin is adjusted according to the working position. After that, perform RULA analysis automatically utilizing the application that produces a total score of 2. The process and results of RULA can be seen in Fig. 8.

Figure 9 is the result of REBA Analysis based on simulation using chair design for chassis workers with a total score of 3 which is included in risk level 1 with "Low" risk category and Action may be required.

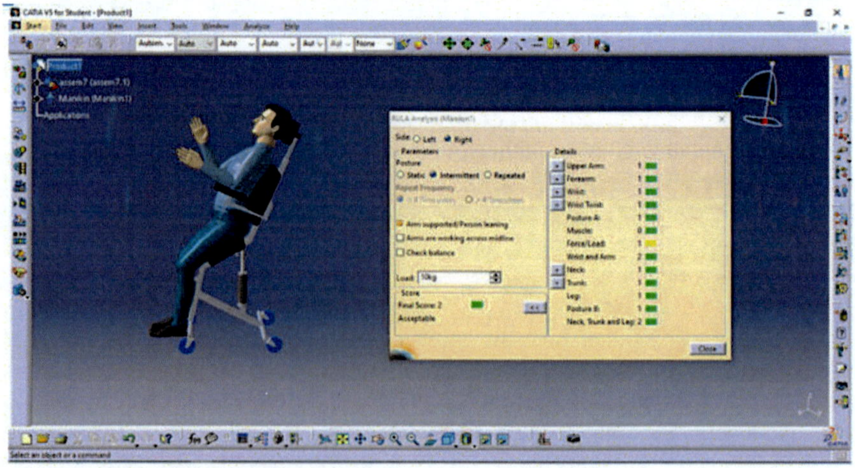

Fig. 8 Simulation of RULA analysis after using the chair design

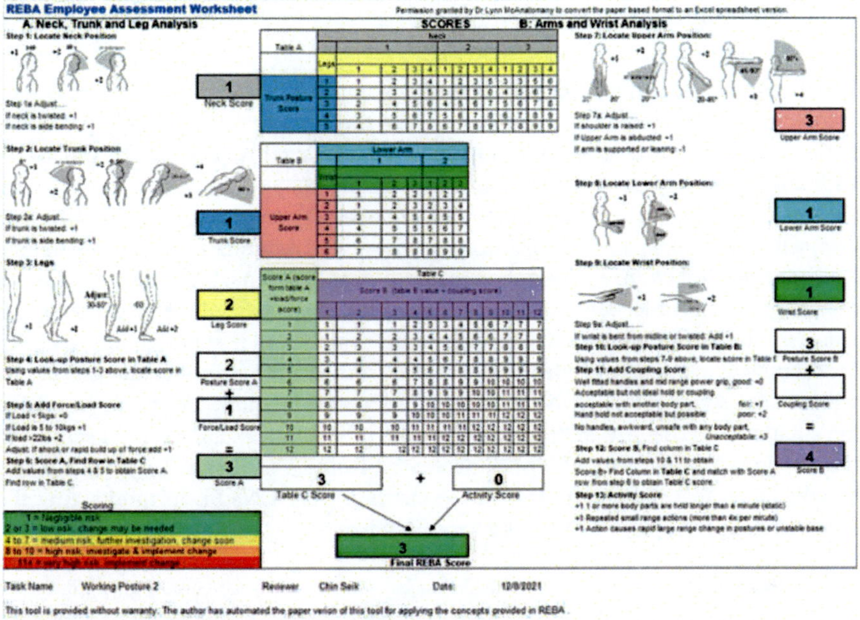

Fig. 9 Simulation of REBA analysis using an ergonomic chair

3.5 *Analyze the Results of Ergonomic Chair Design Simulation*

After the simulation using ergonomic chair design, the total score value of RULA and REBA decreased. RULA score in the actual condition is obtained as much as 7 points with the "Very high" risk category that requires Investigation and repairs are needed as soon as possible. RULA analysis based on simulation using ergonomic chair design found a decrease in total score to 2 points with "Low" risk category and Action No problem with body posture.

Meanwhile, the REBA analysis in the actual condition obtained a total score of 11 which is included in the category of "Very High" risk with Necessary national action possible. After simulating the use of ergonomic chairs, a total score of 3 points was obtained which was included in the "Low" category with Action may be required.

The use of ergonomic chairs is very ENASE for workers in carrying out their work in the assembly line. The change occurred from the "Very high" to "Low" category which can minimize MSD workers' complaints so that the production process runs smoothly.

4 Conclusion

The assembly line is the process of combining the components needed to produce a complete car unit that is ready for use. One of the existing line assembly processes with a high level of complexity and risk is chassis assembly. The chassis process is an assembly activity that is carried out under the car body that is done manually with the full use of manpower that has an impact on MSD complaints. Ergonomic chair design was done using the EFD method with the aim of minimizing MSD complaints.

Simulation of the use of ergonomic chairs was performed by using CATIA to obtain the results of RULA and excel form for REBA. Obtained the result of RULA Analysis for the simulation result of 2 points which was previously in the actual condition was at point 7. While the initial analysis of REBA was obtained as 3 points with a total score value of the actual condition of 11 points. From the results of the simulation of the use of ergonomic chairs, it is obtained that the total score and risk category for RULA and REBA decreased from "Very High" to "Low". It can be concluded that the use of ergonomic chairs is very ENASE in minimizing the complaints of MSD workers chassis assembly line.

Acknowledgements The authors would like to be obliged to Ministry of Higher Education, Malaysia and Universiti Malaysia Pahang for providing financial assistance under project no. FRGS/1/2019/TK03/UMP/02/20.

References

1. Rosa C, Silva FJG, Ferreira LP (2017) Improving the quality and productivity of steel wire-rope assembly lines for the automotive industry. Procedia Manuf 11:1035–1042
2. Wilson JM (2014) Henry Ford vs. assembly line balancing. Int J Prod Res 52(3):757–765
3. Mohamed NMZ, Khan MK (2012) Decomposition of manufacturing processes: a review
4. Nelfiyanti N, Mohamed N, Azhar NAJ (2021) Identification of ergonomic issues among malaysian automotive assembly workers by using the nordic body map method. In: Recent Trends in Manufacturing and Materials Towards Industry 4.0: Selected Articles from iM3F 2020. Springer, Malaysia
5. Raja SJ et al (2013) Hazard identification & risk assesment in new chassis assembly line. Appl Mech Mater 376:468–475
6. Mhatre H, Thorat HT, Desai TN (2019) Work content reduction of chassis assembly line using MOST: a case study
7. Wahyu S, Gunawan G (2018) Design of standard operating procedure (SOP) based at ergonomic working attitude through musculoskeletal disorders (Msd's) complaints. In: MATEC web of conferences, vol. 218. EDP Sciences
8. Mohamed NMZN (2020) Quick response manufacturing and ergonomic consequences in manufacturing environment. In: IOP conference series: materials science and engineering, vol 788, no 1
9. Ferguson SA et al (2011) Musculoskeletal disorder risk as a function of vehicle rotation angle during assembly tasks. Appl Ergon 42(5):699–709
10. Yazdanirad S et al (2018) Comparing the effectiveness of three ergonomic risk assessment methods—RULA, LUBA, and NERPA—to predict the upper extremity musculoskeletal disorders. Ind J Occup Environ Med 22(1):17

11. Ray PK, Tewari VK, Saha E (2018) Ergonomic performance and evaluation of worksystem: a few applications.In: Ergonomic design of products and worksystems-21st century perspectives of Asia. Springer, Singapore, pp 1–12
12. Micheletti CM et al (2019) Risk assessment for musculoskeletal disorders in forestry: A comparison between RULA and REBA in the manual feeding of a wood-chipper. Int J Environ Res Public Health 16(5):793
13. Liansari GP, Novirani D, Subagja RN (2016) Rancangan blueprint alat cetak kue balok yang ergonomis dengan metode ergonomic function deployment (EFD). Jurnal Rekayasa Sistem Industri 5(2):106–117
14. Liansari GP, Asterina F, Putra ATGT (2018) Usulan rancangan house of ergonomic (HOE) produk interior toilet gerbong kereta penumpang kelas ekonomi menggunakan metode ergonomic function deployment (EFD). Penelitian dan Aplikasi Sistem dan Teknik Industri 12(1):328395
15. Firmansyah RA, Desrianty A, Herni F (2014) Usulan rancangan tas sepeda trail menggunakan metode ergonomic function deploymeNT (EFD). Reka Integra 2(2):353–363
16. Pradani WR, et al (2019) Design of wood pellets carrier using ergonomic function deployment (EFD) approach to increase productivity of work: a research at PTPN VIII ciater. In: IOP conference series: materials science and engineering, vol 528, no 1. IOP Publishing
17. Surya RZ, Rusdi B, Gasali M (2016) Aplikasi ergonomic function deployment (EFD) pada Redesign Alat Parut Kelapa untuk Ibu Rumah Tangga. Jurnal Optimasi Sistem Industri 13(2):771–780
18. Grooten WJA, Johansson E (2018) Observational methods for assessing ergonomic risks for work-related musculoskeletal disorders. a scoping review. Revista Ciencias de la Salud 16:8–38
19. Geraldo Z, dos Santos L, Vieira GB (2015) Lean Manufacturing and ergonomic working conditions in the automotive industry. Procedia Manuf 3:5947–5954
20. Jadhav GS, et al (2014) Ergonomic evaluation tools RULA and REBA analysis: case study. In: National conference on industrial engineering and technology management (NCIETM) NITIE, Mumbai
21. Benotsmane R, Dudás L, Kovács G (2020) Simulation and trajectory optimization of collaborating robots by application of solidworks and matlab software in industry 4.0. Acad J Manuf Eng 18(4):191–197
22. Sulaiman F, Sari YP (2018) Analisis postur kerja pekerja proses pengeasahan batu akik dengan menggunakan metode reba. Jurnal Optimalisasi 1(1):1–11
23. Tungga RD, Herwanto D, Nugraha B (2021) Analisis postur kerja pegawai pada line packing refrigerator dengan metode rapid upper limb assessment (Rula) Di Pt. Xyz. Ina J Ind Qual Eng 9(1):35–47
24. Siboro BAH, Purbasari A (2018) Pengembangan perangkat lunak metode rula secara digital untuk memudahkan penilaian ergonomi resiko kerja. Proficiensi J Ind Eng Study Program 6(1):16–24
25. Saffar S, Jamaludin Z, Jafar FA (2017) Improving material handling system performance in automotive assembly line using Delmia quest simulation. In: Asian Simulation Conference. Springer, Singapore

A Rack and Pinion Integrated Mechanism for Gym Forearm Machine Power Generation

Ammar A. M. Al-Talib, Ain Atiqa, and Lih Jing Soo

Abstract The gym forearm machine has been integrated with a DC generator to harness wasted energy from the exercising equipment and convert it into electrical power. The system is including a rack and pinion arrangement and a motor to generate the power as the forearm machine pull up and pull down. The power generated can be used in some advantageous application like charging a handphone or a laptop computer. The power is stored in lithium batteries. This generator can generate up to 2.56 kW after encountered 500 raises of the forearm exercise, which is sufficient to charge a smart phone with battery capacity of 2750 mAh. The time required for a full battery charge is 130 min by using 5 V USB port. The output has given a positive impact towards green and clean energy generation by utilizing the wasted kinetic energy from humans exercising when using gymnastic machines.

Keywords Energy harvesting · Renewable energy · Fitness gymnasium equipment

1 Introduction

In the era of rapidly evolving world, researchers have discovered the way of transferring the energy produced from humans' activities to generate useful energy. This green technology goal is to fully utilize wasted energy comes from human activities into beneficial energy [1–3]. One of those applications can be the power generation from forearms gym machine. It can convert the kinetic energy from forearms machine into electrical power. Piezoelectric is believed to be the most stable method to provide an alternative to traditional rechargeable batteries. This is because most human activities are based on mechanical movement which is irrespective to the environment [4]. According to Vineesh [5] and Halim and Park [6], Piezoelectric

A. A. M. Al-Talib (✉) · A. Atiqa · L. J. Soo
Mechanical Engineering Department, Faculty of Engineering, Technology and Built Environment, UCSI University, 56000 Kuala Lumpur, Malaysia
e-mail: ammart@ucsiuniversity.edu.my

A. Atiqa
e-mail: AinAtiqa@ucsiuniversity.edu.my

© The Author(s), under exclusive license to Springer Nature Singapore Pte Ltd. 2023

M. H. A. Hassan et al. (eds.), *Technological Advancement in Instrumentation & Human Engineering*, Lecture Notes in Electrical Engineering 882,
https://doi.org/10.1007/978-981-19-1577-2_8

discs are also used in converting the mechanical energy into electrical energy. The piezoelectric effect is basically defined into two categories, which are direct effect and the indirect effect.

The principle of operation for the system proposed in this study is to connect a DC generator to a rack and pinion gear mechanism in the forearm machine, which is going to convert the mechanical up and down exercising motion into rotational energy. This can generate a nonconventional electrical energy. When using the gym forearm fitness machine, the frequent lifting and lowering of the handle rod, will make the rack moves vertically through the guided plate and it will cause the matched pinion to rotate. This rotation will be transferred to the motor via a shaft and proper coupling, to convert the mechanical energy into electrical energy.

2 Working Principle

The proposed design in this study for a power generator forearm machine, is by adding a rack and pinion arrangement to produce a rotational mechanical energy which can be coupled to a DC generator to produce electrical energy [7, 8], also [9]. The rack and pinion pair will produce the rotation of the pinion through the up and down rack frequent exercising motion.

In this study, an energy harvesting system that utilize the movement of the forearm machine in the gym is constructed for converting the kinetic energy into electrical energy as shown in Fig. 1. The system aims is to convert the vertical movement of human forearm to rotational motion. The rotational motion hence will rotate the DC generator and as a result will produce the useful electrical power.

This harvester is using a DC motor arrangement. When a person is lifting the handle rod during the normal exercise, the vertical force will be transferred to the

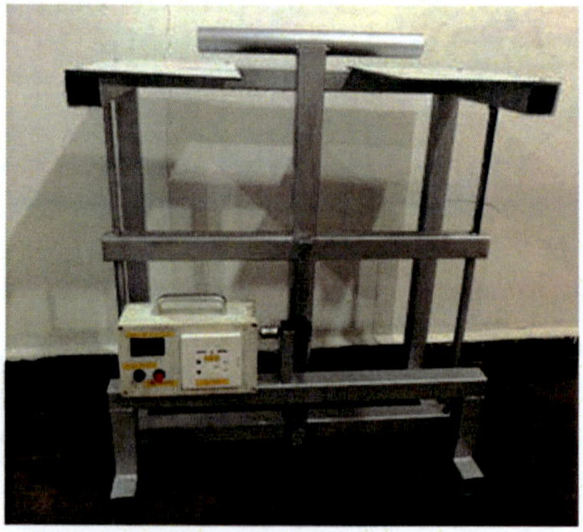

Fig. 1 Prototype of forearm machine energy harvester

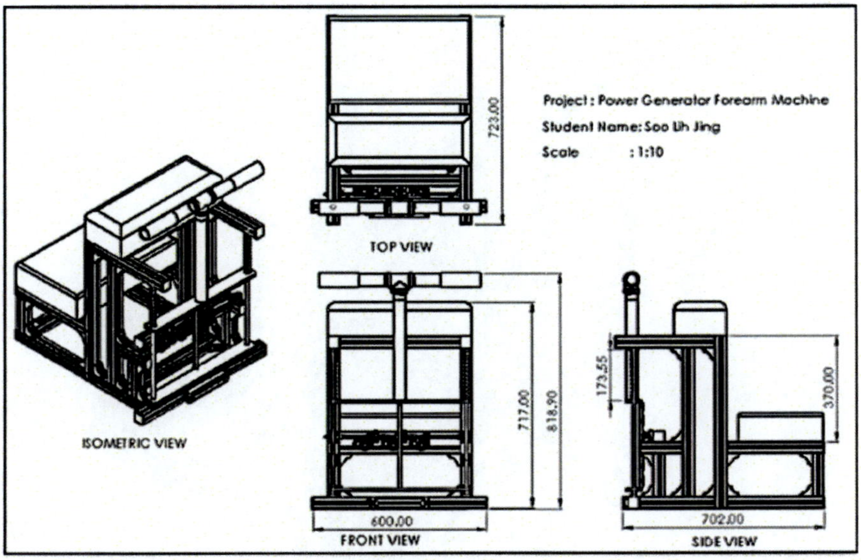

Fig. 2 Prototype illustration using SOLIDWORK

rack gear, and it will move upward together with the guide plate. The vertical force from human motion will result in the pinion to be rotated and the vertical force will be converted into rotational motion which will power the DC generator. The conversion of the exercising mechanical energy into electrical energy can be saved in a battery and can power up the utility panel.

SOLIDWORK software has been chosen for the modelling purpose and to illustrate the details of the prototype. The detailed components and assembly of the prototype is shown in Fig. 2 as an orthographic projection. The total height of the prototype is 818.90 mm and the total width x depth is 702 × 600 mm. The details of the prototype with its components list are shown in Fig. 3 and tabulated in Table 1.

The use of materials has been prudently selected to ensure the robustness and rigidness of the fabricated machine. Steel hollow square tubes has been selected. The required gear rack length had been cut and welded on the structure properly. Arc welding has been used to join all the structure bars. Furthermore, to assure a long lifespan of the machine and to restrain from rust or oxidizing, anti-rust paint has been used to spray all the structure parts.

For the DC generator section as shown in Fig. 4, the design has been established by using 12 V DC generator, 6600 mA dry cell, transformer, and LED lamp. These components and equipment were installed inside an insulator box and fastened with screws to ensure the safety and to minimize the risk. The design is intended for the purpose of AC appliances application; thus, an inverter has been employed to convert the DC power which was generated by the motor into standard 240 V AC power.

The rack and pinion are playing the main role in the performance of the proposed energy harvesting system. Proper design calculations have been implemented to find

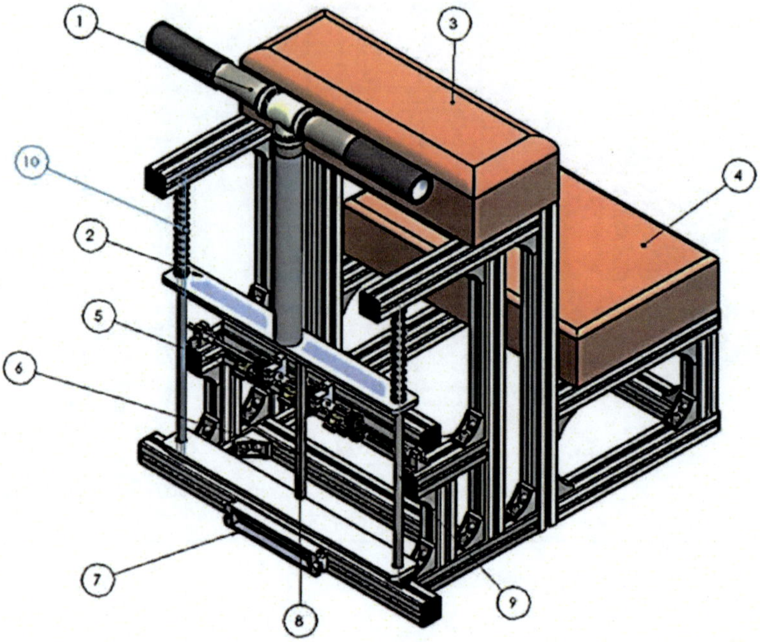

Fig. 3 Prototype assembled parts

Table 1 Materials for power generator forearm machine

Item no	Component name	Quantity	Materials
1	Pulling handle rod	1	Mild steel
2	Guide plate	1	Mild steel
3	Cushion base for arm rest	1	Fabrics
4	Seat cushion base	1	Fabrics
5	DC generator arrangement	1	/
6	Rack	1	Aluminium profile
7	LED straight bar	1	/
8	Rack and pinion arrangement	1	Mild steel
9	Guide rod	1	Mild steel
10	Spring	1	Stainless steel

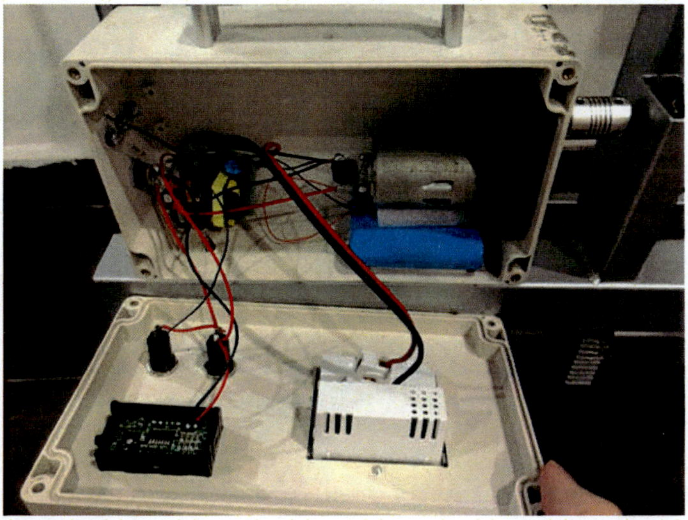

Fig. 4 DC Generator Circuit

the right module of pinion, the pitch diameter of gear, the number of gear teeth, and the depth of teeth, the pressure / force angle and outer diameter of pinion.

The following values are found important to be mentioned, as they are affecting the performance of the whole process,

$$\text{Revolutions of gear per lift, R} = 4.55 \text{ revolutions}$$
$$\text{Revolutions of gear per cycle (lift and lower)}, R^c = R \times 2$$
$$= 9.1\text{revolutions}$$

$$\text{Revolution per minute of gear, RPM} = 30 cycles \times 9.1$$

$$= 273 \text{ revolutions.}$$

Real exercising on the proposed machine has shown that one cycle of arm lifting and lowering can be completed within 2 s. This implies that in 1 min, 30 exercising cycles can be accomplished and the DC generator will develop 273 revolutions knowing that the rack working distance is 200 mm.

Simulations and real exercising have shown that the DC generator can successfully power up the 5 V USB port and 240 V power point on the utility panel. In addition, the generated electrical power is stored in the lithium rechargeable cell which can be used for future other applications. Since the forearm machine is not a continuous use, then, the use of the rechargeable cell will make it more convenient for continuous power supply.

Fig. 5 Gear and rack combination

Based on the readings recorded by a dynamic voltage meter, the lithium battery has been capable of constantly providing a voltage at 9.53 V which is sufficient for the intended applications.

Figure 5 is showing the power generation arrangement in the forearm machine. A flexible shaft coupling has been used between the gear shaft and the DC motor shaft to ensure efficient rotational motion transfer and generating the electricity.

To validate the output experimental data, participants falling into the average weight category for Asians have been asked to participate in the study. According to Sarah Catherine Walpole (2012), the average weight relied on combined statistics for Asian men and women is 127.2 pounds which can be converted into 57.7 kg. These values have been obtained from data collected in years 2005 until 2012.

The experiments have involved five categories of people shown in Table 2. There are five participants in each category, and they have gone through 20 trials and the average output for the currents and voltage generated have been recorded.

While carrying out the test, the multimeter has been set at the range of 20 V and 20 mA to measure the output of the DC generator. The readings from the multimeter have been recorded for each arm raise from time to time and the average values are considered for evaluation and analysis.

Table 3 is showing the details of the participants.

Table 2 Experimental group participants

Group	Gender	Weight (Kg)
1	Male	41–50
2	Male	51–60
3	Male	61–70
4	Female	41–50
5	Female	51–60

Table 3 Age, weight, and height of the participants

Group	Subject	Age	Weight (Kg)	Height (cm)
1	1	21	50.00	167
	2	19	49.90	170
	3	18	48.45	169
	4	17	43.40	162
	5	15	41.00	158
2	1	23	59	173
	2	23	58.56	175
	3	22	54	167
	4	22	52.10	166
	5	19	54	160
3	1	25	70	182
	2	24	68	178
	3	23	67.6	177
	4	22	65	177
	5	18	62	174
4	1	23	48.3	163
	2	23	43.20	165
	3	23	40.2	156
	4	23	41	160
	5	21	40	162
5	1	23	52.03	165
	2	24	54	157
	3	25	51	165
	4	30	50.3	163
	5	29	52	166

3 Results and Discussion

The output results for the different experiments conducted are shown in this section. Figures 6, 7 and 8 are showing the output voltage, current and power accordingly.

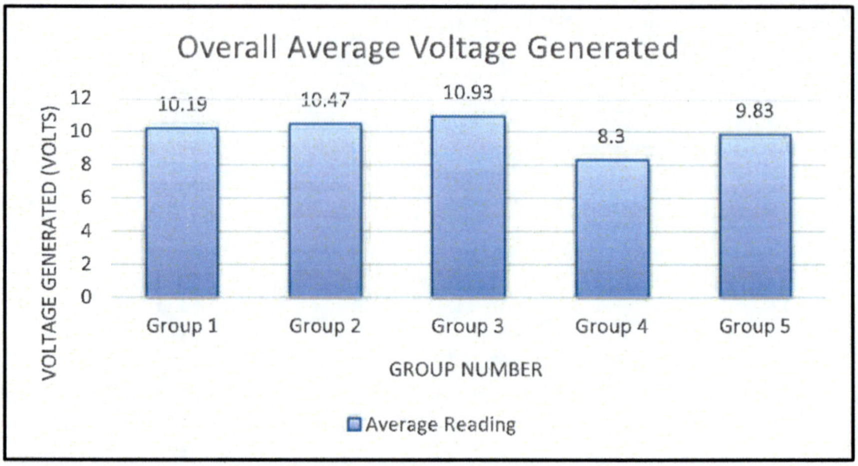

Fig. 6 Overall average voltage generated by Group 1–5

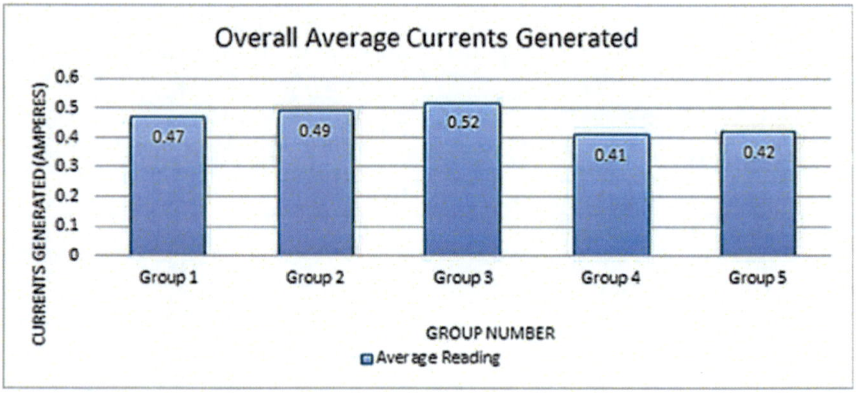

Fig. 7 Overall average currents generated by Group 1–5

It's found that the power generated from the power generator forearm machine is 2.56 kW after encountered 500 raises of the forearm exercise. The power generated has been stored in the lithium battery. The targeted objective is to charge a smart phone with battery capacity of 2750 mAh using a charging cable connected to 5 V USB port. Figure 9 is showing the connection from the generator to the smart phone. The procedure has been started with zero charge of the smart phone to ensure the effectiveness of the experiment. The phone battery charge has been recorded at 15 min intervals.

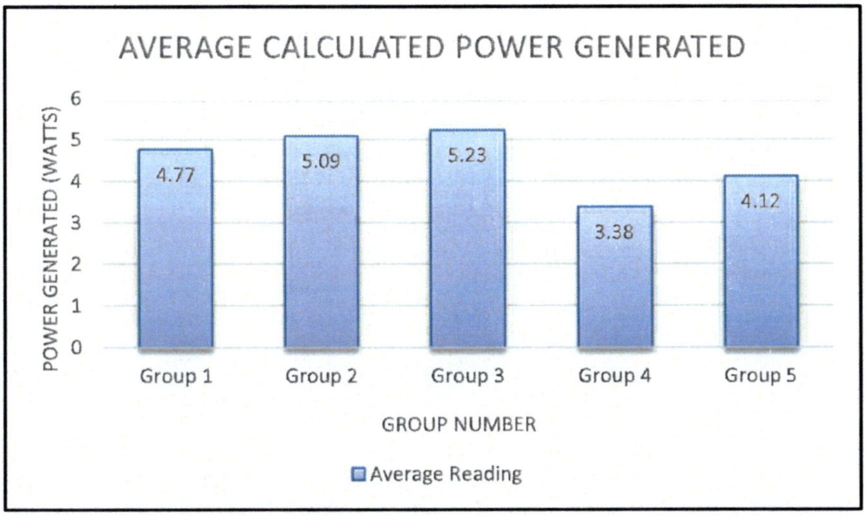

Fig. 8 Average power generated by Group 1–5

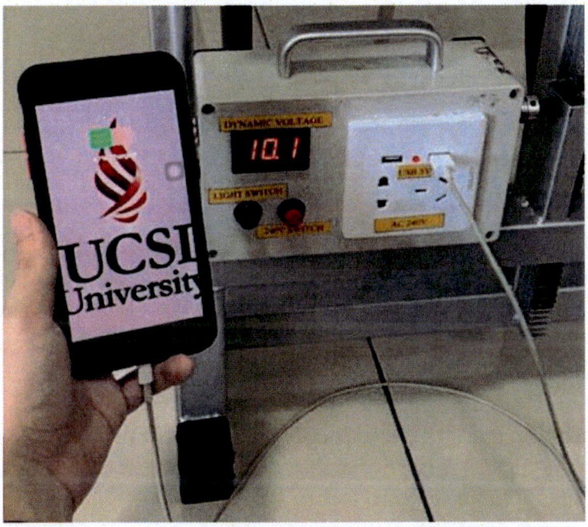

Fig. 9 A phone charging using cable connected to power generator forearm machine

The charging results are shown in Table 4.

It can be observed From Table 4 that charging a smart phone with a battery capacity of 2750 mAh, will require 130 min to complete a 100% charge by using 5 V USB port. In comparison with normal wall socket, the speed is considered lower by the factor of battery efficiency, wiring resistance and generator itself. Nevertheless,

Table 4 Phone battery charge percentage during the charging process using the utility panel

Time (minutes)	Phone battery charge (%)
0	0
15	12
30	30
45	42
60	53
75	70
90	88
105	92
120	95
130	100

the output give positive impact towards green and clean energy objective which is utilized the waste kinetic energy from human gymnastic activities.

The capacity rating of the lithium battery used in the prototype is 6600 mA which can theoretically being able to fully charge the phone model with 2750 mAh battery for 2.4 cycles at an ideal working condition.

4 Conclusion

The objective of design to utilize wasted energy comes from exercise equipment to generate electrical power is achieved throughout the modifications implemented in this study. The power generated from the power generator forearm machine is 2.56 kW after encountered 500 raises of the forearm exercise. It's found that charging a smart phone with a battery capacity of 2750 mAh, will require 130 min to complete a 100% charge by using 5 V USB port. The urge to establish renewable energy to perform the stability of the ecosystem is seen crucial by improving the reliability of the system. The simplicity of the design for this rack and pinion of power generator forearm machine also confers interesting features than other mechanisms. The power generator forearm machine is very useful and much pertinent for the peoples that have the daily habits to go to gym. It provides affordable and economical energy to power nearby simple applications. Other than that, it is one of the electrical power generations that does not pollute the environment due to the zero-fuel input and zero greenhouse gas emissions. It can be suggested to do further improvements by adding bearings for a smooth running of the rotating shaft to reduce the friction. Also, springs with proper stiffness can be added to the up and down moving shafts to ease the operation and increase the working cycles efficiency.

Acknowledgements The authors wanted to express their thanks and appreciation to the Center of Excellence for Research, Value Innovation and Entrepreneurship (CERVIE) at UCSI University for funding the conference participation and the ever support during the research time.

References

1. Zou H-X, Zhao L-C, Gao Q-H, Zuo L, Liu F-R, Tan T, Wei K-X, Zhang W-M (2019) Mechanical Modulations for Enhancing Energy Harvesting: Principles. Methods and Applications. Elsevier Applied Energy 255:113871
2. Gammaitoni L (2012) There's Plenty of Energy at the Bottom (Micro and Nano Scale Nonlinear Noise Harvesting). Contemp Phys 53(2):119–135
3. Zhu, D. and Beeby, S., 2011. Kinetic energy harvesting. In Energy harvesting systems (pp. 1–77). Springer, New York, NY.
4. Soin, N., Anand, S.C. and Shah, T.H., 2016. Energy harvesting and storage textiles. In Handbook of Technical Textiles (pp. 357–396). Woodhead Publishing.
5. Vineesh K, Amarnath KT, Lavanya M, Priya RA (2015) Review on footstep power generation by piezoelectric transducer. Int J Appl Eng Res 10:3814–3816
6. Halim MA, Park JY (2014) Theoretical modeling and analysis of mechanical impact driven and frequency up-converted piezoelectric energy harvester for low-frequency and wide-bandwidth operation. Sens Actuators, A 208:56–65
7. Mr.Shingare Pratik Vinod, Mr. Somwanshi Vikas Balkrushna, Mr. Sonawane Vijay Kailas, & Mr.Tore Tushar Pandharinath. (2018) Gym Power Generation Mechanism. International Journal of Advance Research And Innovative Ideas In Education 4(6):192–195
8. Kumar M, Mundada GS (2017) Energy Harvesting from Gym Equipments. International Journal of Innovative Research in Electrical, Electronics, Instrumentation and Control Engineering 5(07):127–131
9. Shahosseini I, Najafi K (2014) Mechanical amplifier for translational kinetic energy harvesters. J Phys: Conf Series, IOP Publish 557(1):012135

Drowsiness Detection for Safe Driving Using PERCLOS and YOLOv2 Method

Ngo Kah Lock, Weng Mun Ng, Norrabiatul Adawiyah Jusoh, Nazhatul Hafizah Kamarudin, Rizauddin Ramli, and Zuliani Zulkoffli

Abstract Driver drowsiness is considered as a significant risk factor that impacted to large number of road accidents. This paper presents a vehicle safety system developed using computer vision technique to detect driver drowsiness while driving. Deep learning method, YOLOv2 has been chosen as the artificial intelligence model to detect the percentage of eyes closure of the driver. A threshold value for percentage eye closure for the driver, where is when the set threshold value of 20% has been reached, response from a connected buzzer will alert the driver. This system development involves a few stages of dataset preparation and training. An accuracy of 86.58% has been reached to detect eyes closure in this training model. The trained model had been simulated using NVIDIA Jetson Nano platform to simulate real world application and find out a few areas of improvement such as the frequency of yawning additional feature parameter to detect drowsiness and self-training mode for face recognition and eyes dataset of the driver was recommended to personalize the development application and minimizing inaccurate output.

Keywords Computer vision · Autonomous vehicle technology · Drowsiness detection · Jetson nano · PERCLOS and YOLO

1 Introduction

Fatigue and drowsiness driving has been identified as primary reasons for road accidents, especially when driving in the long-distance and monotonous motorway [1–5].

Supplementary Information The online version contains supplementary material available at https://doi.org/10.1007/978-981-19-1577-2_9.

N. K. Lock · W. M. Ng · N. A. Jusoh · N. H. Kamarudin · Z. Zulkoffli (✉)
Faculty of Engineering, Technology and Built Environment, UCSI University, 56000 Kuala Lumpur, Malaysia
e-mail: Zuliani@ucsiuniversity.edu.my

R. Ramli
Department of Mechanical and Manufacturing Engineering, Universiti Kebangsaan Malaysia, UKM, 43600 Bangi, Selangor Darul Ehsan, Malaysia

© The Author(s), under exclusive license to Springer Nature Singapore Pte Ltd. 2023
M. H. A. Hassan et al. (eds.), *Technological Advancement in Instrumentation & Human Engineering*, Lecture Notes in Electrical Engineering 882,
https://doi.org/10.1007/978-981-19-1577-2_9

Under these unwanted situations, it could become very dangerous for an individual to drive on the road, especially when the driver is not aware of his or her consciousness at that particular moment. However, although most individuals are aware of these significant issues, under such circumstances where drivers cannot keep themselves awake due to tiredness from daily activities, it is hard for drivers to stay focus and not fall asleep while driving on the road.

There are several approaches that were proposed to measure the driver drowsiness while driving. These approaches can be categorized into three main categories which included physiological approach, vehicle-based approach and behavioral metrics [6, 7]. The physiological and vehicle-based approach were a contact-based method that needs to attach sensors to the vehicle or the driver while the behavioral based approach was a contactless based method [8–10]. In addition, behavioral based approach can be divided into three sub-categories, which included appearance-based, template-based, and feature-based methods [11, 12]. Drowsiness detection using physiological approach is highly relying on the instrument, and it is not applicable for real-time detection and the vehicle-based approach is not reliable as it tends to get different results in a different environment. As a result, the behavioral approach will be used in the application to detect the drowsiness level of the driver. Percentage of eyelid closure (PERCLOS) over the pupil over time is the most often utilized method or technology in terms of measuring the drowsiness of the driver under the behavioral approach. To solve the problems of real time detection, low detection accuracy, and slow processing speed, a new PERCLOS detection model using YOLOv2 is proposed in this paper [13–16].

2 Methodology

This system development consists of two stages of programming part and testing part which is embedded in Jetson Nano. For programming part, the flowchart development for this drowsiness measurement was shown in Fig. 1. The system starts when the camera automatically turns on and captures images with a maximum rate of 30 frames per second. After that image acquisition process, the deep learning object detector, YOL0v2 will detect the regional proposals and classify the interest object of eyes simultaneously. It is set the toggle will be "1" if the eyes are detected in a closing state, and toggle will be "0" if the eyes are detected in an opening state.

The internal system detects the vector results of 5 s with 30 frames per seconds. The system will be undergone preprocessing stage of deleting the first array, which is zero array vector if the array size is bigger than 150 and save the toggle value into the last array. PERCLOS percentage was obtained using the sum of the toggle value in the array divided by 150, which equals the total number of the array and multiply by 100. If the PERCLOS is larger than 20%, then the driver is considered as "sleepy" whereas if the "PERCLOS" is lower or equal to 20%, then the driver is considered as "awake". The system will shut down once the "Q" button is pressed, or else the system will loop again to detect regional proposal and classify the object.

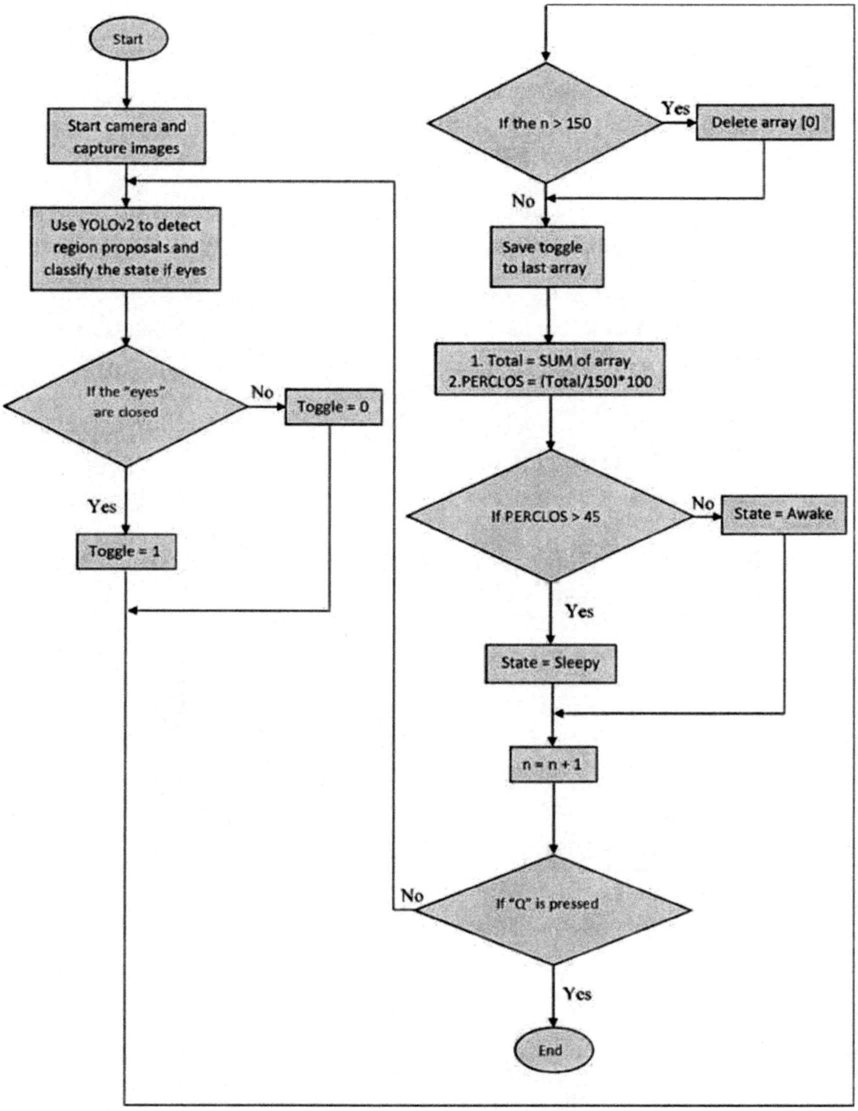

Fig. 1 Flowchart of PERCLOS detection

2.1 Hardware Setup

The device used to measure the drowsiness of the driver included a NVIDIA Jetson Nano microprocessor to process the input and generate output, a camera to capture images as input to NVIDIA Jetson Nano and a buzzer to provide response alert to the driver. Figure 2(a) shows the technical specifications of NVIDIA Jetson Nano used in this development. Figure 2(b) shows the schematic hardware setup of Jetson Nano with triple connections. Below is the procedure of hardware connections:

i) The connection of 5 V input voltage parallel with the car radio,
ii) Pi camera connect to the camera socket of the NVIDIA Jetson Nano using a ribbon cable
iii) Buzzer to alert the driver will connect to GPIO pins of the Jetson Nano, which positive terminal will connect to pin 12 of the GPIO pins and negative terminal connected to the ground pin.

As this device will be installed in car, devices housings for protection to this device is needed and made it able to fix in car. The devices housing has been designed using Solidworks software and printed using 3D Printer. Figure 2(c) shows the illustration of devices housing using Solidworks software.

GPU	128-core Maxwell
CPU	Quad-core ARM A57 @ 1.43 GHz
Memory	4GB 64-bit LPDDR4 25.6 GB/s
Storage	microSD
Video Encode	4K @ 30 | 4 X 1080p @ 30 | 9 X 720p @ 30 (H.264/H.265)
Video Decode	4K @ 60 | 2 X 4K @ 30 | 8 X 1080p @ 30 | 18 X 720p @ 30 (H.264/H.265)
Camera	2 X MIPI CSI-2 DPHY lanes
Connectivity	Gigabit Ethernet, M.2 Key E
Display	HDMI and display port
USB	4 X USB 3.0, USB 2.0 Micro-B
Others	GPIO, I²C, I²S, SPI, UART
Mechanical	69 mm X 45 mm, 20-pin edge connector

(a)

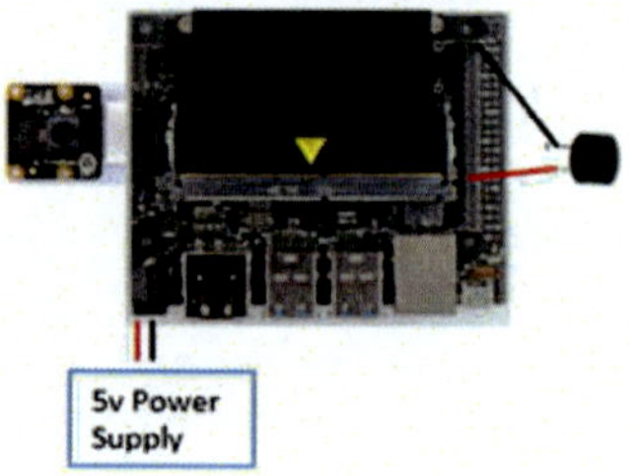

(b)

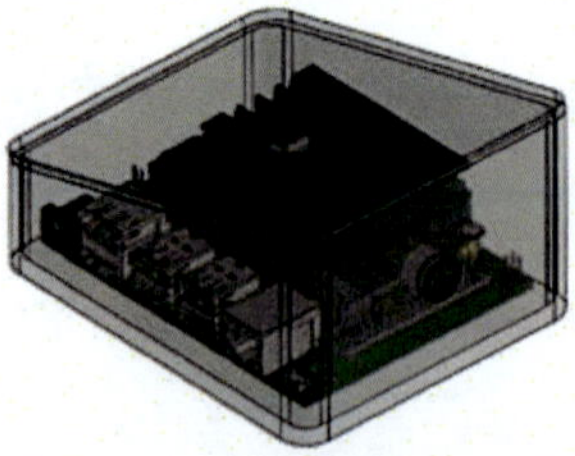

(c)

Fig. 2 **a** Jetson Nano specifications **b** Schematic diagram for Jetson Nano hardware setup and **c** Illustrations of device's housing sketched using solidworks software

2.2 *Algorithm Development of Eye Closure Detection Using Deep Learning Technique*

YOL0v2 was chosen for this system development using deep learning model pre-trained on an architecture which is called Darknet-19. The structure of Darknet-19 is consisting of 19 convolutional layers and five maxpooling layers. Each Convolution layer has the BatchNorm normalization and then Leaky ReLu activation except for the last Convolution block. The Darknet-19 was initially trained with 1000 classes of classification dataset which is at 224 × 224 pixel^2 from standard ImageNet.

2.3 *Data Acquisition for Deep Learning's Model Training*

In this development to detect the drowsiness level of the driver, closed eyes in the wild (CEW) dataset and KomNet dataset were combined and used in deep learning model training. CEW dataset contains 2423 images in total, including 1192 images with close eyes and 1231 images with open eyes, collected directly over internet. Meanwhile, the KomNet dataset consists of 50 persons and 24 images each, a total of 1200 images. The collected files were in the format of.jpg. Face images were taken with a frontal face facing the camera.

2.4 *Data Pre-processing*

Datasets were prepared and generated the annotation files for the computer to recognize before performing training on deep learning's model. Beforehand, a python script code to rename the images from 0 to the total number of dataset images. This code allows the other python script to read the images more easily to generate annotate file. Next, a "data labelling" python script code is sketched to display the images, where it allows the user to use the rectangular selector to select the object. Then, the key binding is used to classify the classes of objects and the "generate.xml" python script is sketched in order to generate annotation files for training purpose. In this application, the images need to be classified into two classes, which are the opened eyes class and the close eyes class. Figure 3 shows the flowchart of the "data labelling" and "generate.xml" in python script.

Heretofore, there is no complete dataset for open eyes and close eyes, all the images downloaded from CEW dataset or KomNet dataset need to go through pre-processing stage before training stage. Pre-processing stage essential to align the dataset information and maximize the performance of the model. Example of open eyes data is shown in Fig. 4(a), which is circled with a green ellipse while close eyes data in Fig. 4(b), which is circled with a red ellipse.

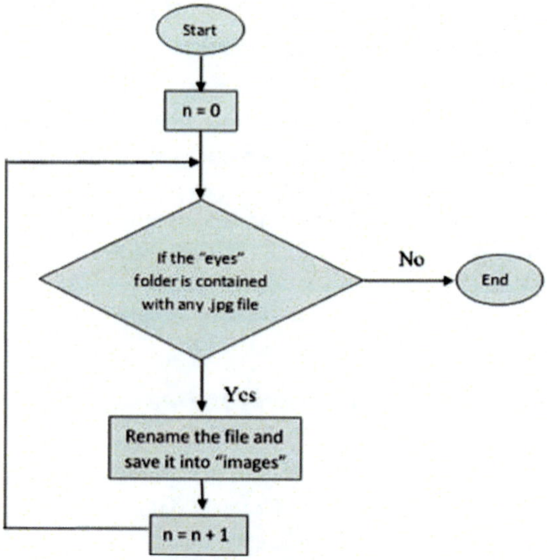

Fig. 3 Flowchart for "data labelling" and "generate.xml files"

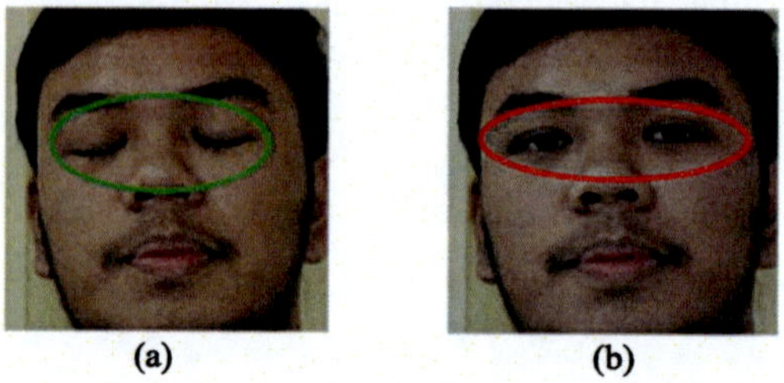

Fig. 4 Dataset examples **a** closed eyes and **b** open eyes

2.5 Deep Learning's Model's Training and Fine-Tuning

To train the model detecting open eyes and closed eyes, the pre-trained model must be obtained. It is to get the pre trained image feature extractor process to infer its bounding boxes. The pre-trained weights of the Darknet 19 was downloaded from the original YOL0v2 project website [13, 14]. At this point, with the complete pre-trained model, images dataset and annotations files, the model is ready to train.

Multi-scale training using YOL0v2 that is robust to run on different input images sizes, since the model uses only convolutional and pooling layers was chosen in this project development. YOL0v2 architecture changes the network every few iterations and randomly choose a new image dimension size from the dimension set. This makes the network able to predict the object at different resolutions. Neural networks use optimizing strategies like stochastic gradient descent to minimize the error in the algorithm and loss function is used to compute this error, to quantify how good or bad the model is performing [14]. YOL0v2 is composed of 3 losses, including confidence loss, localization loss and classification loss. Therefore, the losses are calculated and combined [16] as shown in Eq. 1.

Confidence Loss

$$\sum_{i=0}^{S^2}\sum_{j=0}^{B} 1_{ij}^{obj}(C_i - \hat{C}_j)^B + \lambda_{noobj}\sum_{i=0}^{S^2}\sum_{j=0}^{B} 1_{ij}^{no\,obj}(C_i - \hat{C}_j)^2$$

$$\lambda_{coord}\sum_{i=0}^{S^2}\sum_{j=0}^{B} 1_{ij}^{obj}\left[(x_i - \hat{x}_j)^2 + (y_i - \hat{y}_j)^2\right]$$

Localization Loss

$$+\lambda_{coord}\sum_{i=0}^{S^2}\sum_{j=0}^{B} 1_{ij}^{obj}\left[(\sqrt{w_i} - \sqrt{\hat{w}_i})^2 + (\sqrt{h_i} - \sqrt{\hat{h}_i})^2\right]$$

Classification Loss

$$\sum_{i=0}^{S^2} 1_{ij}^{obj} \sum_{c \in lasses} (pi(c) - \hat{p}i(c))^2$$

$$\text{Losses} = \text{Confidence loss} + \text{Localization loss} + \text{Classification loss} \quad (1)$$

Which:
$\sum_{i=0}^{S^2}$ is for each grid cell
$\sum_{i=0}^{S^2}$ is for each box
1_{ij}^{obj} is one if the object appears in cell i and j-th box detect it, otherwise 0
C_i is the ground truth confidence score
$\widehat{C_i}$ is predicted confidence score
$\widehat{p_i}(c)$ denotes the conditional class probability for class c in cell i
λ_{coord} increase the weight for the loss in the boundary box coordinates
λ_{noorbj} set to 0.5 to decrease the loss for empty boxes
1_{ij}^{noobj} is one if there is no object in the i-th cell, otherwise 0

2.6 *Evaluation of Deep learning's Model*

A confusion matrix is a technique to summarize the performance of the model being trained used to detect the status of the eyes of the driver. The number of correct and incorrect predictions are summarized with count value and break down by each class. The confusion matrix output shows the errors being made by the model, but most importantly the types of error being made by the model to detect. In this application

Table 1 Confusion matrix

	Open eyes	Close eyes
Open eyes	True positive	False positive
Close eyes	True negative	False negative

to detect the status of the eyes of the driver, the model is trained to detect two classes, which are open eyes and closed eyes. As a result, two-row by two-column confusion matrices will be applied to this application. The event row will be assigned to "positive" while the no event row will be assigned to "negative". Meanwhile, the event column of the prediction is "true" and no event is "false". In this application to detect the status of the eyes of the driver, "True positive" is assigned to predict open eyes correctly, "False positive" is assigned to mispredict open eyes. In contrast, "True negative" is assigned to predict close eyes correctly while "False Negative" is assigned to predict close eyes wrongly, as shown in Table 1.

With these parameters from the confusion matrix, the accuracy, precision, recall and Fl score can be determined. Accuracy is the most intuitive performance measure, and it is simply a ratio of correctly predicted, which included correctly predict open eyes and correctly predicted close eyes over the total test dataset, and the formula is as Eq. 2. Precision is the ratio of the correctly predicted open eyes over the total number of open eyes and represented in Eq. 3. Recall or called sensitivity, is the ratio of correctly predicted open eyes to the correct total test dataset, and the mathematical representation shown in Eq. 4. F1 score is the weighted average of Precision and Recall. As a result, the F1 score considers both false positive and false negative. Fl score represents the harmonic score evaluation of precision and sensitivity. It represents an unbiased accuracy since it does not affect by imbalanced data, the mathematical representation shown in Eqs. 2 to 5.

$$Accuracy = \frac{True\ Positive + True\ Negative}{Total\ Prediction} \tag{2}$$

$$Precision = \frac{True\ Positive}{True\ Positive + False\ Positive} \tag{3}$$

$$Recall = \frac{True\ Positive}{True\ Positive + False\ Negative} \tag{4}$$

$$F1score = \frac{2 \times Precision \times Sensitivity}{Precision + Sensitivity} \tag{5}$$

3 Results and Discussion

Performance of the developed model is being evaluated by using accuracy, precision, recall and Fl score. An improvement version of the confusion matrix was used, which is the two-by-three matrix used instead of the two-by-two matric due to some images that cannot be detected. As a result, the third column is added to the confusion matrix for the images that cannot be detected to provide more apparent evaluation results.

The tested data which are used to evaluate the performance of the model to detect the status of the eyes of the driver consists of ten percent of the dataset, which is not included in training model. The evaluated model's performance using Fl score below 60% was set as threshold of not applicable to real life. After that, the combined dataset had been processed by filtering out the defective data and the model is retrained by using new dataset. The deep learning model to detect the status of the driver's eyes reached 86.58% accuracy in Fl score the processed dataset. Transfer learning approach was implemented in this application to transfer the weightage of the neural network that has learned from previous tasks and apply that weightage to detect the status of the driver's eyes.

3.1 Eye Closure Detection

The trained model to detect the status of the eyes of the driver was tested with a personal computer and a webcam. The trained model is applied with the PERCLOS method to measure the drowsiness level of the driver. As mentioned previously, with the average frame rate of 30 frames per second, 450 frames will be captured and will take into consideration for PERCLOS, if the PERCLOS value is larger than 20%, the state of the driver will become "sleepy". Figure 5(a) shows the status of the eyes of the driver which is detected by using the trained model when the status of the eyes of the driver is in open eyes state. From the figure, the driver is in open eyes state and the model is also able to detect that the status of the eyes of the driver is in open eyes state. Figure 5(b) shows the status of the eyes of the driver which is detected

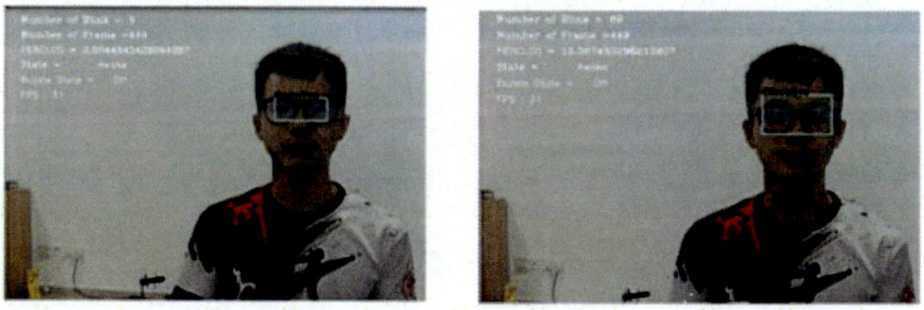

(a) (b)

Fig. 5 **a** Status of the eyes of the driver when in open eyes and **b** Status of the eyes of the driver when in closed eyes

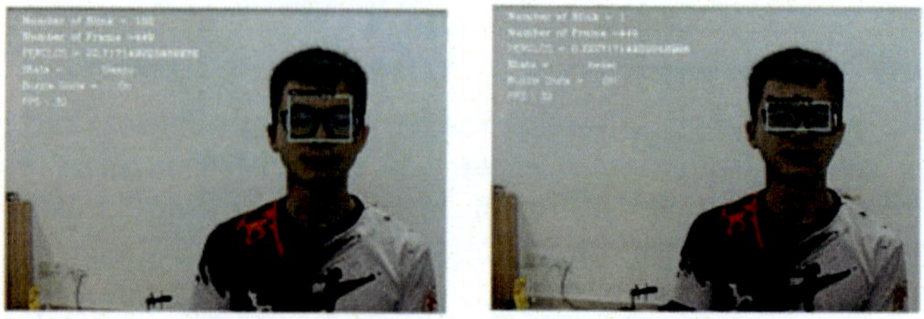

(a) (b)

Fig. 6 **a** Drowsiness level of the driver when the PERCLOS is higher than 20% and **b** Drowsiness level of the driver when the PERCLOS is lower than 20%

by using the trained model when the status of the eyes of the driver is in close eyes state. From the figure, the driver is in open eyes state and the model is also able to detect that the status of the eyes of the driver is in open eyes state. Figure 6(a) shows the drowsiness state of the driver when the PERCLOS value is higher than 20%, the state of the driver is changed to drowsy. Figure 6(b) shows the drowsiness state of the driver when the PERCLOS value is lower than 20%, the state of the driver is changed to awake.

3.2 *Deep Learning Model Tests with PERCLOS on NVIDIA Jetson Nano*

A monitor display was connected to the NVIDIA Jetson Nano, to view the PERCLOS value and the buzzer state in a terminal. In this development, the state of the driver will be sleepy when the PERCLOS value is higher than 20%, explaining when the state of the driver will become sleepy if his eyes are closed more than 20% over a period of time.

Figure 7(a) is showing that the deep learning model is detecting that the driver is in open eyes state, the PERCLOS value does not add as there is no closing eye.

(a) (b)

Fig. 7 **a** The PERCLOS value is not adding when the driver is on open eyes state and **b** The PERLOS value is adding when the driver is in close eyes state

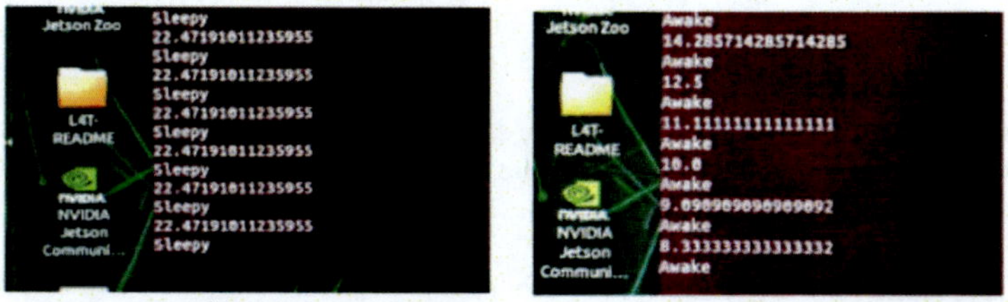

(a) (b)

Fig. 8 **a** The state of the driver is changed to sleepy when the PERCLOS is higher than 20% and **b** The state of the driver is changing back to awake when the PERCLOS is lower than 20%

Figure 7(b) is when the driver is closing his eyes, the figure shows that the PERCLOS value is adding as the number of close eyes is adding over the specific amount of time. However, the state of the driver is awake as the PERCLOS value is below 20%.

When the PERCLOS value is higher than 20%, the state of the driver changed to sleepy. It is shown in Fig. 8(a) when the PERCLOS value is higher than 20%, the state of the driver is becoming sleepy. While the PERCLOS value is dropping to below 20%, the state of the driver will become awake as shown in Fig. 8(b). The buzzer will be triggered as well when the state of the driver is becoming sleepy.

4 Conclusion

This paper reported algorithm development and testing to provides a solution to the problem of detecting the state of drowsiness using arithmetic based method. This system uses percentage of eye closure method to detect sleepiness. Deep learning method was chosen and carried out to the combined datasets of from CEW dataset and KomNet. The algorithm development was based on the concept of eye-tracking. To obtain finer results, three thousand six hundred and thirty-two images from these datasets have been used. The developed algorithm was partially tested and found to be effective and able to provide response to the system from the buzzer. This developed method provides a non-intrusive approach for drowsiness detection. In future research, a few areas of improvement were suggested such as the frequency of yawning can also be used as a feature parameter to detect drowsiness and self-training mode for face recognition and eyes dataset of the driver was recommended to personalize the development application and minimizing inaccurate output.

References

1. Rajkumarsingh B, Totah D (2021) Drowsiness detection using android application and mobile vision face API. R & D J South Afr Inst Mech Eng 37(1):26–34

2. Zhipeng M, Shuwan Y, Jian Z, Jingjiu Q, Jun S, Jingshuang D (2018) Research on drowsy-driving monitoring and warning system based on multi-feature comprehensive evaluation. IFAC Papers Line 5131(1):784–789
3. Karthikeyan S, Subha R, Elakkiya G, Kowshika R, Dhanvarshni S (2021) Driver-drowsiness detection system. Turk J Comput Math Educ 12(9):1892–1898
4. Rukhsar K, Shruti M, Shivraj P, Suraj A, Saritha LR (2019) Human drowsiness detection system. Int J Eng Adv Technol (IJEAT) 8(4):2249–8958
5. Jabbar R, Shinoy M, Kharbeche M, Al-Khalifa K, Krichen M, Barkaoui K (2020) Driver drowsiness detection model using convolutional neural networks techniques for android application. In: 2020 IEEE international conference on informatics, IoT, and enabling technologies (ICIoT), vol 2, no 1, pp 237–242
6. Sahayadhas A, Sundaraj K, Murugappan M (2012) Detecting driver drowsiness based on sensors: a review. Sensors (Basel) 12(12):16937–16953
7. Prakash C, Rahul S, Gautam S, Smarjeet DA (2016) Survey paper on drowsiness detection & alarm system for drivers. Int Res J Eng Technol (IRJET) 3(12):1433–1437
8. Messaoud D, Abdelmadjid B, Veronique C (2018) A light on physiological sensors for efficient driver drowsiness detection system. Sensors Transd J 224(8):39–50
9. Kiran VBN, Raksha R, Anisoor R, Varsha KN, Nagamani NP (2020) Driver drowsiness detection. Int J Eng Res Technol (IJERT) 8(15):33–35
10. Maryam H, Alireza M, Aliasghar BS (2020) Driver safety development: real-time driver drowsiness detection system based on convolutional neural network. SN Comput Sci 1(289):1–10
11. Mohsen P, Adel M, Gebraeil NS, Mohammad MB, Alireza K, Mohammad HE (2018) Using image processing in the proposed drowsiness detection system design. Iran J Public Health 47(9):1371–1378
12. Walid M, Belhassen A, Roobaea A, Abdulmajeed A (2019) Automated drowsiness detection through facial features analysis. Computación y Sistemas 23(2):511–521
13. Redmon AJ, Farhadi A (2016) YOLO: real-time object detection. In: Proceedings of the IEEE conference on computer vision and pattern recognition, vol 1, no 2, pp 779–788
14. Redmon J, Farhadi A (2017) YOL09000: Better, faster, stronger. In: Proceedings of the IEEE conference on computer vision and pattern recognition, vol 1, no 2, pp 1–9
15. Zhao ZQ, Zheng P, Xu ST, Wu X (2019) Object detection with deep leaming: a review. IEEE Trans Neural Netw Learn Syst 1(2):3212–3232
16. Feiping N, Zhanxuan H, Xuelong L (2018) An investigation for loss functions widely used in machine learning. Commun Inf Syst 18(1):37–52

A Study of the Effect of Industrial 4.0 on Improving the Manufacturing Performance: A Case Study in Miri and Bintulu

Mohd Adzrie, Brandon Chagat, Radhika Lahei Arechinan, Sri Tharunan Naidu, and Umar Abdul Karim

Abstract Manufacturing industry nowadays have become more onto digitalization to keep the competitiveness. The correlation of digitalization with industry revolution 4.0 (IR4.0) or "Industry 4.0" has brought many new technologies to the world as people are compelled to use high technologies gadgets which are faster and more efficient. The aim of this paper is to determine the level of impact of IR4.0 on the performance of manufacturing industry in Miri and Bintulu and assess the implication of IR4.0 on the manufacturing sustainability. The method used to conduct this study is qualitative method through questionnaire. Data analysis was done using descriptive statistics tool to evaluate the mean value, and Cronbach's alpha test to measure its reliability for the five-point Likert scale. The findings indicate that 46% of the respondents have awareness of the IR4.0, meanwhile 71% of the respondents have awareness of the manufacturing sustainability. Regarding the first objective, mean value of 3.27 out of 5.00 was obtained which indicates that IR4.0 gave moderate impact on the manufacturing performance. For the second objective, mean value of 3.28 out of 5.00 was obtained. This indicates that the implication of IR4.0 help moderately on the sustainability of manufacturing industry within Miri and Bintulu.

Keywords Industry 4.0 · Manufacturing · Sustainability

1 Introduction

Currently, the event towards Industry 4.0 has been observed and provides significant influence on the manufacturing industry [1]. On regards of sustainability, IR4.0 is believed to be ready to improve the present industrial system sustainability [2]. This research paper will study on how the event of IR4.0 assistance on improving the manufacturing industries performances as well as its impact on the manufacturing sustainability.

M. Adzrie (✉) · B. Chagat · R. Lahei Arechinan · S. Tharunan Naidu · U. Abdul Karim
Mechanical Engineering Program, Faculty of Engineering, Universiti Malaysia Sabah, Jalan UMS, 88400 Kota Kinabalu, Sabah, Malaysia
e-mail: mohdadzrie@ums.edu.my

© The Author(s), under exclusive license to Springer Nature Singapore Pte Ltd. 2023

M. H. A. Hassan et al. (eds.), *Technological Advancement in Instrumentation & Human Engineering*, Lecture Notes in Electrical Engineering 882,
https://doi.org/10.1007/978-981-19-1577-2_10

The manufacturing industry went through an exceptional timeline of rapid changes which results in the evolution of management strategies, technological implementation, customer demands and nature in competition [3]. This rapid change forcing the modern industries to include the cost of training their employees for them to obtain knowledge on handling the modern machines. Besides that, the state-of-the-art of mathematical programming and computer science by the cognitive robots used in manufacturing industry has been developed and evolved. Implementation of Cyber-Physical Systems (CPS), Internet of Things (IoT) and Internet of Services (IoS) in autonomous production has influenced the current industrial revolution [4]. Mainly there are two objectives of this project. The objectives are as follow:

i. To determine the level of impact of IR 4.0 on the performance of modern manufacturing industry.
ii. To assess the implication of IR 4.0 on the manufacturing sustainability.

2 Research Methodology

The location of this study are two cities in Sarawak, which are Bintulu and Miri as both cities are well known for its rapid development. The focus of this study will more on the manufacturing industry. The research method is through qualitative method which is through survey-based questionnaire. Conducted surveys on the employers and employees can be the influenced factors for successful data obtained [5]. The reason behind this is due to movement control order (MCO) that is still ongoing in our country of Malaysia.

2.1 Development of Questionnaire

The questionnaire was developed according to the objectives of this study. The questionnaire was divided into three parts according to the objectives of the study, which are as follows:

i. The respondents background—name, age, job position, company's background
ii. The impact of implementation of IR4.0 on manufacturing company performance
iii. The implication of IR4.0 on the manufacturing sustainability which focus on economy element and environment element

Table 1 Questions in Part (i) on correlation of IR4.0 with performance factors

No.	Questions
1	Industry 4.0 increase the efficiency [6]
2	Industry 4.0 helps increased productivity [6]
3	Industry 4.0 help in monitoring the quality control of the products [7]
4	Industry 4.0 help in reducing the cost required to make a product [6, 8]
5	Focused and narrowed to the specific area of maintenance activities supported by Industry 4.0 [9]

Table 2 Questions in Part (ii) on impacts of IR4.0 on manufacturing sustainability—economy element

No.	Questions
6	Implementation of Internet of Thing (IoT) have impact on the sustainability [10, 11]
7	Implementation of Artificial Intelligence (A.I) can help on the sustainability of manufacturing industry [10, 12]
8	Industry 4.0 on manufacturing sustainability can increase profitability [9]

Table 3 Questions in Part (ii) on impacts of IR4.0 on manufacturing sustainability—environment element

No.	Questions
9	Smart manufacturing (industry 4.0) helps in reducing the energy used [13]
10	Sustainable manufacturing industry can reduce waste emission [13]

Tables 1, 2, 3 shows the number and type of questions for asked each part which were based on relevant references. The questions were provided with filling in the blank space and single choice question for Part (i), and five-point Likert scale for both Part (ii) and Part (iii). The five-scale Likert point range from 1 to 5 where the number indicates:

1. Very Low
2. Low
3. Moderate
4. High
5. Very High

2.2 Data Collection

Data obtained from this research were collected by using qualitative method which by distribution of the questionnaire to the respondents. The questionnaires were distributed to the respondents which were the employers and the employees of the company. The targeted company for the distribution of questionnaire was that related to manufacturing industry, which based in Miri and Bintulu. The collection of data took approximately four weeks to give enough time for the respondents from the company to respond to the questionnaire. Email and phone number were attached in the questionnaire description for any enquiries from the respondents.

2.3 Data Analysis

Data were transferred to the software called Statistical Package for the Social Science (SPSS) for further analysis to generate results. Two methods were used to analyze the data which is descriptive analysis and Cronbach's Alpha Reliability test analysis. As the SPSS software was used, two commands were used to make descriptive analysis of the data, which is "descriptive" and "frequencies". Through SPSS, it can simplify the task interpreting information by creating graphs and charts.

Reliability test is measured using the common method, which is the Cronbach's alpha. This method is widely used for the purpose of measure of reliability in social and organizational sciences [14]. There were certain rules of thumb in consideration of reliability test by Cronbach's alpha method which is >0.9—excellent, >0.8—good, >0.7—acceptable, >0.6—questionable, >0.5—poor, and <0.5—unacceptable [15].

3 Results and Discussion

The results and discussions are presented in three subsections according to IR4.0 on the manufacturing performance and sustainability, and its Cronbach's Alpha reliability test analysis.

3.1 IR4.0 on Manufacturing Performance

There were five questions asked that related to the first objective. The data obtained from the 100 respondents after they answered the questionnaire, the average mean of the data is 3.27. The average mean value indicates the level of impact of industry 4.0 on the manufacturing industry performance. It shows that the implementation of IR4.0 gave a moderate impact. Table 4 below shows the summary of the data.

Table 4 Descriptive statistics for correlation of IR4.0 with performance (N = 100)

No.	Questions	Mean
1	How much does it increase the works efficiency?	3.29
2	How much does it helps increased productivity?	3.32
3	How much does it help in monitoring the quality control of the products?	3.32
4	How much does it help in reducing the cost required to make a product?	3.30
5	How often does it take to do maintenance in a year?	3.14

Table 5 Descriptive statistics for correlation of IR4.0 with manufacturing sustainability on economy (N = 100)

No.	Questions	Mean
6	How much does your company implement the using of Internet of Thing (IoT)?	3.44
7	How much does your company implement the using of Artificial Intelligence (A.I)?	3.07
8	How much do you think the ratio of production sales to production cost?	3.25

Table 6 Descriptive statistics for correlation of IR4.0 with manufacturing sustainability on environment (N = 100)

No.	Questions	Mean
9	How often do you think your company tracking, reporting, and reducing energy use?	3.30
10	How often do you think your company tracking, reporting, and reducing waste and emission?	3.29

3.2 The Implications of IR4.0 on Manufacturing Sustainability

Meanwhile, the question related to the second objective divided into two parts, which is on the company's economy and the benefits of implementation of IR4.0 in a company on the environment. For the first part which is on the economy, there were three questions. The average mean value obtained from the three questions is 3.25. Finally, there were only two questions that was asked for the second part. From those two questions, average mean value that was obtained is 3.30. In term of manufacturing sustainability, implication of industry 4.0 helps moderately as the average mean value from the two parts was 3.28. Tables 5 and 6 shows the summary of data that was analyzed using descriptive analysis.

Table 7 Reliability Test by using Cronbach's Alpha (N = 100)

No.	Question	Cronbach's Alpha	Cronbach's Alpha if item deleted
Factors correlation of IR4.0 with performance			
1	How much does it increase the works efficiency?	0.936	0.911
2	How much does it helps increased productivity?		0.900
3	How much does it help in monitoring the quality control of the products?		0.903
4	How much does it help in reducing the cost required to make a product?		0.915
5	How often does it take to do maintenance in a year?		0.955
Factors correlation of IR4.0 with manufacturing sustainability—Economy			
6	How much does your company implement the using of Internet Of Thing (IoT)?	0.863	0.793
7	How much does your company implement the using of Artificial Intelligence (A.I)?		0.811
8	How much do you think the ratio of production sales to production cost?		0.813
Factors correlation of IR4.0 with manufacturing sustainability—Environment			
9	How often do you think your company tracking, reporting, and reducing energy use?	0.933	0.000
10	How often do you think your company tracking, reporting, and reducing waste and emission?		0.000

3.3 *The Implications of IR4.0 on Manufacturing Sustainability*

Data collected were analyzed by using Cronbach's Alpha Reliability Test from IBM SPSS software shown in Table 7.

From the table above, factors correlation of IR4.0 with performance shows the Cronbach's Alpha value of 0.936. Meanwhile, factors correlation of IR4.0 with manufacturing sustainability on economy shows the Cronbach's Alpha value of 0.836. Then, factors correlation of IR4.0 with manufacturing sustainability on environment shows the Cronbach's Alpha value of 0.933.

On the rightest column of the table shows the value of Cronbach's Alpha value if the item is deleted. This is a very important column as this will estimates the Cronbach's Alpha value if the particular item being removed. Higher Cronbach's Alpha value indicates more reliability. If the item deleted shows lower Alpha value than the current Alpha value, this means that the item suppose cannot be removed.

As observed from the table, 9 out of 10 questions cannot be removed as it has lower Cronbach's Alpha value if item deleted than its Alpha value, only question 5 have higher which is 0.955. That means the Cronbach's alpha of this scale would rise from 0.936 to 0.955 if you got rid of that item.

3.4 Impacts of IR4.0 in Oil and Gas Industry

In oil and gas industry, implementation of IR4.0 help to reduce the cost for maintenance activities, by using method such as Predictive Maintenance (PM). Through PM, sensing technologies are being employed where sensors have been used on the site to constantly monitor the condition of the machines and equipment by acquiring the information by detecting physical, chemical, or biological property quantities and convert them into readable signal. The data constantly been collected and recorded to predict next possible maintenance to avoid unplanned downtime of the machine or equipment, thus helps reducing resources and energy used.

4 Conclusion

From research findings, it gave a clear insight that the implementation of fourth industrial revolution (IR4.0), which also known as Industry 4.0, on the manufacturing industry can help its business to be more competitive. Industry 4.0 is a new era of digitalization that can help improve the performance of an industry in term of works efficiency, productivity, and profitability.

As stated in the first objective of this study which is to determine the level of impact of industrial revolution 4.0 on the performance of modern manufacturing industry, it can be conclude that its bring benefits to modern manufacturing industry. Hence, the first objective of this study is achieved. Industry 4.0 bring benefits on the performance of the manufacturing industry. It also helps the sustainability of a company on the economy and environment. Through the application of Internet of Thing (IoT) and Artificial Intelligence on a company, it brings great benefits on the company's economy. In term of environment, industry 4.0 helps in reducing waste and emission from a company which could harm the environment through its manufacturing process and effective recycling of materials. It also helps in reducing the consumption of energy. Thus, the second objective of this study which is to assess the implication of IR 4.0 on the manufacturing sustainability is achieved.

Acknowledgements The authors would like to express gratitude to the University Malaysia Sabah for the support given throughout the study.

References

1. Stock T, Seliger G (2016) Opportunities of sustainable manufacturing in industry 4.0. Procedia CIRP 40, 536–541
2. Bai C, Dallasega P, Orzes G, Sarkis J (2020) Industry 4.0 technologies assessment: a sustainability perspective. Int J Prod Econ 229:1–15
3. Raut R, Naoufel C, Kharat M (2017) Sustainability in the banking industry: a strategic multi-criteria analysis. Bus Strateg Environ 26(4):550–568
4. Yuan XM (2019) Impact of Industry 4.0 on inventory systems and optimization. Industry 4.0-impact on intelligent logistics and manufacturing. InTechOpen, London
5. Sundar R, Balaji AN, Satheesh Kumar RM (2014) A review on lean manufacturing implementation techniques. Procedia Eng 97:1875–1885
6. Kumar N, Kumar J (2019) Efficiency 4.0 for industry 4.0. Human Technol 15(1):55–78
7. Aldag MC, Eker B (2018) What is quality 4.0 in the Era of industry 4.0?. 3rd International conference on quality of life, pp. 31–33. University of Kragujevac, Serbia
8. Salimbeni S (2020) Industry 4.0, quality 4.0 & new product development. ResearchGate, Buenos Aires
9. Ejsmont K (2021) The impact of industry 4.0 on employees—insights from Australia. Sustainability 13(6):1–31
10. Ejsmont K, Gladysz B, Kluczek A (2020) Impact of industry 4.0 on sustainability—bibliometric literature review. Sustainability 12(14):1–29
11. Beier G, Niehoff S, Xue B (2018) More sustainability in industry through industrial internet of things? Appl Sci 8(2):1–12
12. Khakurel J, Penzenstadler B, Porras J, Knutas A, Zhang W (2018) The rise of artificial intelligence under the lens of sustainability. Technologies 6(4):1–18
13. Posinasetti N, Sustainable manufacturing: principles, applications and directions. Industr.com, https://www.industr.com/en/sustainable-manufacturing-principles-applications-and-directions-2333598. Accessed 20 August 2021
14. Bonett D, Wright TA (2014) Cronbach's Alpha reliability: interval estimation, hypothesis testing, and sample size planning. J Organ Behav 36(1):3–15
15. Wadkar S, Singh K, Chakravarty R, Argade SD (2016) Assessing the reliability of attitude scale by Cronbach's Alpha. J Global Commun 9(2):113–117

Instrumentation

Design and Fabrication of Nutrient Film Technique (NFT) Hydroponic System

Mohd Rafiuddin Rozilan, Ana Syahidah Mohd Rodzi, Ahmad Faiz Zubair, Abdul Rahman Hemdi, Rasdi Deraman, and Nor Diyana Md Sin

Abstract Hydroponic is an effective solution introduced at urban houses as it conserves water and space. The purpose of this project is to design and fabricate a Nutrient Film Technique (NFT) hydroponic system for urban domestic residents to grow green leafy vegetables. The best concept design of the NFT hydroponic system was modelled in solid three-dimensional (3D) by using SolidWorks software. The structural and flow simulation were analyzed to determine the outcome predictions for such design. The project fabrication method mainly uses UPVC pipe and other local material for the ease of assembly. The project prototype occupied an area of 0.48 m^2 with the dimension of 1.3 m height, 0.6 m width, 0.7 m length, and with crop capacity of 18 plants. The temperatures, EC, pH, relative humidity, plant length and weight are among the various parameter that been analyze in the project. Plant growth and other parameters showed that the NFT hydroponic system was a success in terms of the design for urban houses and the output on plant productivity.

Keywords Hydroponic system · Nutrient Film Technique (NFT) · SolidWorks

1 Introduction

1.1 Hydroponic History

Soilless culture is defined as "any method of growing plants without using soil as a rooting medium, in which the inorganic nutrients taken by the roots are provided by irrigation water". The mineral nutrients required by the plants are dissolved in a suitable concentration in irrigation water, resulting in the "nutrient solution". The

M. R. Rozilan · A. S. Mohd Rodzi (✉) · A. Faiz Zubair · A. R. Hemdi · R. Deraman
School of Mechanical Engineering, College of Engineering, Cawangan Pulau Pinang, Universiti Teknologi MARA, Kampus Permatang Pauh, 13500 Permatang Pauh, Pulau Pinang, Malaysia
e-mail: anasyahidah@uitm.edu.my

N. D. Md Sin
School of Electrical Engineering, College of Engineering, Universiti Teknologi MARA, Cawangan Johor, Kampus Pasir Gudang, 81750 Masai, Johor, Malaysia

© The Author(s), under exclusive license to Springer Nature Singapore Pte Ltd. 2023
M. H. A. Hassan et al. (eds.), *Technological Advancement in Instrumentation & Human Engineering*, Lecture Notes in Electrical Engineering 882,
https://doi.org/10.1007/978-981-19-1577-2_11

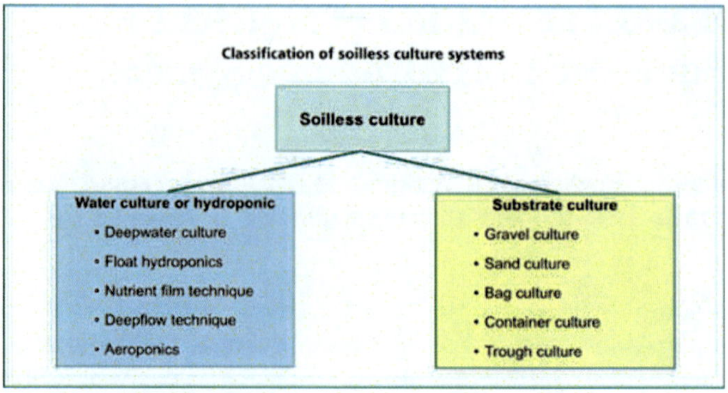

Fig. 1 The classification of the soilless culture system

method of soilless culture can be divided into two types, which are water culture or hydroponic and substrate culture. Figure 1 shows the classification of soilless culture. While hydroponic is a recent term used for growing plants without soil, the concept idea origins back much earlier. The term was created from Greek words hydro ("water") and ponos ("labor"), which translated directly meant "water working" [13].

According to [26], in 1600, the first experiment showing that plants do obtain substances from the water was recorded by a Belgian scientist. The first serious discussions and analyses of hydroponic emerged during the year 1925 when the greenhouse industry was facing problems. Greenhouse soils need to be replaced often to maintain good soil structure and fertility. This makes researchers became aware of the soilless culture potential to replace conventional soil agricultural methods. Since that, in early 1930, he grew vegetables and flowers on a commercial scale and the result was successful until it been recognized as the discovery of the century [12].

According to [11], 70% of the population will live in urban area in an estimation to year 2050. At that year, a smart way to ensure healthy population such as fresh food production without losing quality and their nutrition values. Hydroponics system are among the best choices [8]. But it is highlighted the perfect choice is the Aquaponics system. The advantages in this method are while generating fish and plants, it can also producing organics food. To optimize the sustainability of aquaponic production, water must be recirculated between both subsystems to take use of the metabolic waste of fish that serves as a nutrient for plants [30].

1.2 Nutrient Film Technique (NFT)

NFT is the type of hydroponic system that most people imagine when they hear the term "hydroponics". NFT system works with continuous water and nutrient to the plants by flow and water depth control. The nutrient solution is pumped into the growth tray and flows over the roots of the plants, and then drains back into the reservoir. This makes oxygen-rich conditions available for the plants. Usually, the plant is supported with a sponge or other growing media in a small basket and the roots will be dangling into the nutrient solution. Many green veggies may be produced readily in this method. As seen in Fig. 2, the most utilised commercially is for lettuce production [28] and [14].

The Nutrient Film Technique (NFT) is a water-culture technique that involves growing plants with their roots enclosed in a plastic film trough or a rigid tube through which nutrient solution is continually circulated [5], who pioneered the NFT approach, found the most crucial result, and Fig. 3 depicts the layout set up

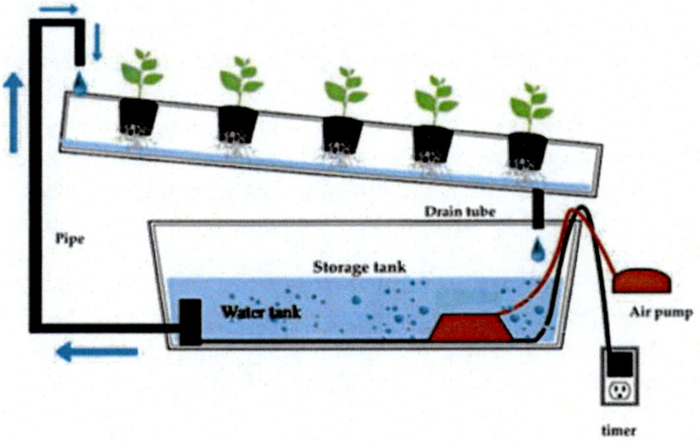

Fig. 2 Structure of the NFT system

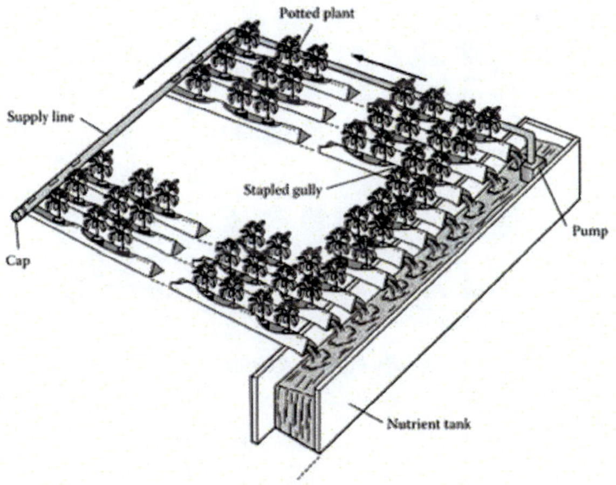

Fig. 3 A set of NFT gullies and a solution tank are built in

NFT. He has claimed that the level of liquid flowing through the roots of the plants should be very shallow in order for the plant roots to receive enough oxygen.

[27] experimented with several methods to increase the number of plants that may be cultivated in each space in a greenhouse. Among the tasks performed with lettuce are vertical pipes through which nutritional solution is dripped, moistening and feeding the plants. Then, using the NFT system, he develops and implements methods to increase the use of vertical space in the greenhouse for low-growing crops such as lettuce. He designed the cascade system, which is presented in Fig. 4, by cutting Polymer Vinyl Chloride (PVC) plastic tubing with a 3-in. (7.6 cm) diameter and stacking one over the other up to eight troughs high [26].

[15] developed a low-cost recirculating aquaculture hydroponic system for the use in science or mathematics as laboratory and teaching material. The setup includes Fish Culture Tank where fish stocks are kept, water treatment tank that is built from sediment basin, biofilter and clean water basin to treat water before use. Next, Packed Column Aerator, and lastly Hydroponic Unit where the crop is grown. For less than $600, the authors designed an aquaculture-hydroponic unit which was ideal for use in laboratory settings. The aquaculture-hydroponic recirculating device is an ideal hands-on system that can be used to learn the applications of science and math in agriculture.

[20] evaluated if alternate hydroponic gravel bed, floating, and NFT systems provided advantages in nutrient stripping, water usage, plant productivity, or aquaponic fish production. The experimental aquaponic system was made up of 12 individual and identical aquaponic units, allowing for the repetition of experimental methods. Each aquaponic unit included a fish holding tank, a biofilter, and a hydroponic subsystem. The first is controlled, followed by gravel, floating, and finally NFT. The regulated system will serve as the reference point for the other three techniques.

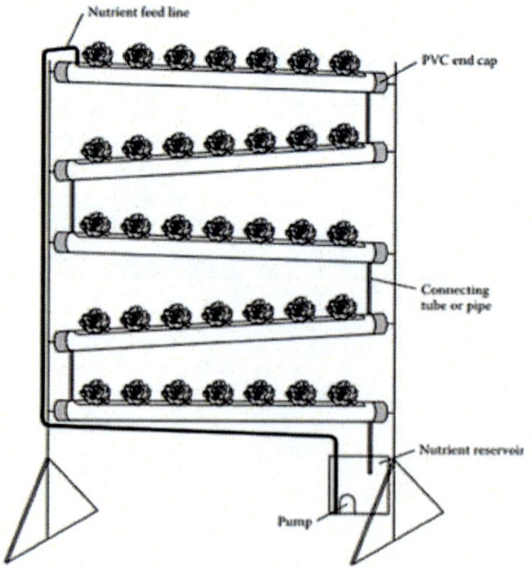

Fig. 4 Type of a cascade NFT system

Researchers found that NFT hydroponic subsystems were substantially less effective than either gravel or floating culture technologies for lettuce production, nitrate removal, and phosphate removal, as well as water utilisation and buffering efficiency. Due to little root interaction with water in comparison to gravel and floating systems, NFT may have a lower result than other systems.

[3] conducted research to determine whether hydroponic lettuce cultivation in Arizona is an acceptable and more sustainable alternative to conventional lettuce farming. Crop budgets and government agricultural statistics provided information on traditional lettuce production. Hydroponic lettuce development data was generated using engineering equations and literature values that were then used to an 815-square-meter hydroponic greenhouse as a case study. There were temperature controls, artificial lighting, and water circulation pumps. That's the moment in time when hydroponic lettuce farming can't be deemed a more effective alternative to traditional lettuce farming approaches. Despite the fact that, it offers interesting suggestions that could lead to more sustainable food production. However, land and water are better utilised in hydroponic lettuce gardening than in traditional farming. [31] built a project from the design, producing, and testing of a prototype indoor hydroponic tower coupled with an automated monitoring system. The study includes an analysis of various parameters such as plant length, height of the leaf, fresh weight, water temperature, ambient temperature, relative humidity, pH, water level, and total dissolved solids. For better calibration in data collection, an Arduino-based monitoring system was built from scratch and used to collect the data required along with physical instruments.

[22] research on discovering the optimum design and the least cost of hydroponic manufacturing. It was intended to target all urban populations to make hydroponic leafy vegetables and to reuse cooking waste. This paper built a vertical hydroponic (VHFS) to control the irrigation with sensor and Arduino hardware and solar panel. Arduino software used to maintain nutrient levels. For a small nucleus family of size four, VHFS will enable the daily vegetable requirement (1.2 kg leafy vegetable). Such a system has a life span of more than 5 years. Figure 5 shows the design NFT used from [21]. Hydroponically produced lettuce was compared to soil-grown lettuce under protected settings using the nutrient film technique.

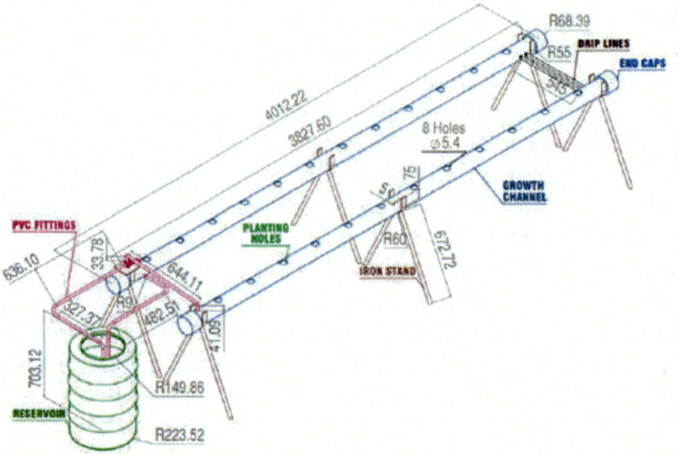

Fig. 5 Isometric view of NFT system

1.3 Automation

Hydroponic system automation also allows the owner to maintain their crop at a healthy level with a hydroponic system control assist. Nutrient element monitoring can be divided into two sections, which is crop growth monitoring and nutrient solution quality monitoring.

The development and production of an automated hydroponic growing crop for domestic and especially those without farming knowledge were done [10]. This paper uses artificial growing light which is suitable for cultivation. The automation system for the hydroponic is based on PLC and HMI, and a cabinet for growing crops with 3 shelves was designed. The goal of the project is to combine technology available with modern clean farming techniques for people who lack the knowledge to grow but need green crops. This project was designed to accommodate a small group of people to serve fresh herbs/vegetables, which can be put within a kitchen for a family of five.

[29] conducted study to determine the power, electrical energy, and actual efficiency of photovoltaics when employed as a source of the DC water pump activator at the hydroponic water spinach facility under the impact of sunshine intensity. A simple hydroponic construction consisting of 4 PVC pipes, which were fitted with a pipe elbow with 13 holes for the plant and the gap between the holes was 5 cm with plastic hoses at the top of the pipe and a bucket was put under the pipe as water and nutrients container. The plastic roof was designed, and the solar array was mounted on top of a pole with the location on the side next to the hydroponic installation, then the solar array was connected through wires to the DC stepdown on the poles and would continuously pump water. DC water pump with average sunlight intensity of

373.71–310.65 W/m^2 with an average temperature of 30–33 °C will operate optimally at 11.00–15.00 h. [17] also monitoring and controlling their aquaponics system using Internet of thing (IoT). On top of it, they developed a web-based interface for monitoring and controlling the aquaponic system. The pH value, the temperature, and the water level are all measured in this study. In the proposed work, the Web Socket protocol is used to transfer information, establish secure connections to the server, and maintain the system in real-time. As a result, it is enhanced vertically, which reduces space requirements with a fully automated platform utilising IoT and big data analysis [33]. They also create a standard VF system that is similar to [18].

NFT hydroponic farming for urban households to grow vegetables is the objective of this study. By arranging the plants in a vertical array, we may make the most of the available space. For the hydroponic system, UPVC pipes were used to ensure the project's success. UPVC pipes make it easier to assemble and disassemble the frame and hydroponic tray for storage. To monitor the growth conditions of the plants, an Arduino-based monitoring system will be installed.

2 Design and Fabrication of Hydroponic System

2.1 Conceptual Design

Three conceptual designs were developed by using Solidworks software and analyzed by using Pugh Matrix [24]. The best selection of hydroponics system was fabricated and monitoring by Arduino-based monitoring system. Figure 6 shows the three conceptual designs that were developed by Solidworks. The Pugh selection method was applied to evaluate each of the propose concept design. Pugh concept selection method is used to compare and select the best design from a set of conceptual design. Based on this method, the design criteria are listed, and the weightage of each criterion are determined according to how important it is in the system. Then, each design

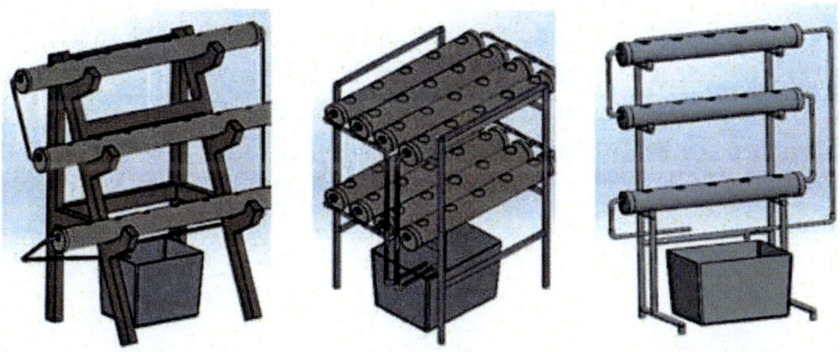

(a) Design 1 (b) Design 2 (c) Design 3

Fig. 6 Three of conceptual design in hydroponic systems

Table 1 Pugh concept selection

Design Criteria	Weight age	Design 1	Design 2	Design 3
Plant capac ity	1	0	+1	−1
Overall weight	2	−1	0	0
Frame design	1	+1	+1	1
Nutrient volume	2	0	−1	0
Cost	2	−1	−1	+1
Area usage	2	0	0	+1
	+0 −	1 5 4	2 4 4	5 5 1
	Total	−3	−2	4

was rated based on how well each design versus the criteria relative to the baseline. The highest total number of rating is the ones that should be considered for the final design selection.

The detail analysis of design concept selection of proposed hydroponic system was presented in the Table 1. A list of important criteria in the hydroponic system are obtained from the previous studies such as area usage, plant capacity, nutrient solution volume, frame design, cost, and overall weight. Each of design concepts will be rated according to the selected design criteria. According to the Table 1, design 3 scores highest cumulative rating which is 4 points and selected as the best concept design for hydroponic system.

2.2 *Final Design*

Based on the analysis of concept design as discussed in previous section, the design 3 was further improved and detail final design was produced as shown in Fig. 7 Compared to the other design that receives a rating of not more than 1 point. This final design has benefits in terms of space area that it will occupy which suitable for urban home, and lowest cost for fabrication. The detail specification for the final design is tabulated ~~shows~~ in Table 2.

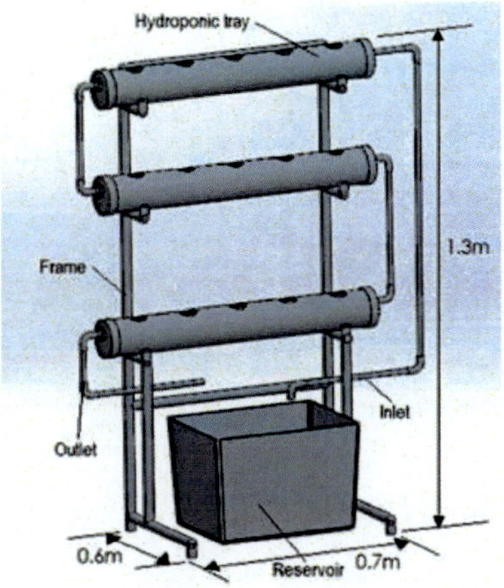

Fig. 7 Final design of NFT hydroponics system

Table 2 Final design specification

Criteria	Specification
Dimension	Length = 0.6 m, Width = 0.7 m, Height = 1.3 m
Weight	Minimum weight = 0.8 kg Maximum weight = 1.0 kg
Area covered	0.42 m^2
Plants capacity	18
Nutrient volume	15 L
Material	UPVC

2.3 *Hydroponic Pump Losses*

The length and diameter of the cylindrical hydroponic tray are 0.8 and 0.11 m illustrated shown in Fig. 3 It was assumed that the nutrient solution will occupy half of the cylindrical hydroponic tray volume for the safety factor, thus the following equation was used to calculate the half-cylindrical volume of the hydroponic tray (Fig. 8):

$$V_{cyl} = \frac{1}{2}\left(\pi r^2 h\right) = \frac{1}{2}\left(\pi (0.055)^2 (0.8)\right) = 0.003801\,\text{m}^3 \tag{1}$$

The, the volume can be converted into weight and litre as in Eq. (2) and (3) respectively.

$$0.003801\,\text{m}^3 \times \left(\frac{100\,\text{cm}^3}{1\,\text{m}^3}\right) \times \left(\frac{0.917\,\text{g}}{1\,\text{cm}^3}\right) \times \left(\frac{1\,\text{kg}}{1000\,\text{g}}\right) = 3.48\,\text{kg} \tag{2}$$

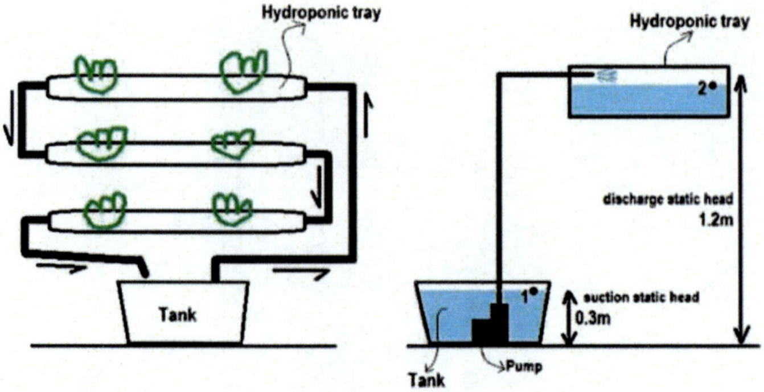

Fig. 8 Position static pump head of the system

$$0.003801\,\mathrm{m}^3 \times \left(\frac{100\,\mathrm{cm}^3}{1\,\mathrm{m}^3}\right) \times \left(\frac{1\,\mathrm{ml}}{1\,\mathrm{cm}^3}\right) \times \left(\frac{11}{1000\,\mathrm{ml}}\right) = 3.8011 \tag{3}$$

The total weight and volume for the system which consists of three cylindrical hydroponic trays are 10.44 kg and 11403 L, respectively. Next, with the total volume, the desired flow rate at the hydroponic tray, Q_2 can be determined. To get the hydroponic tray to be filled with 11.403 L within a minute, it equals 0.019 L/s or 1.1403 L/min for every 10 min. Thus, with the piping diameter of 15 mm, the calculated velocity at the hydroponic tray, v_2 can be determined by Eq. (4).

$$v_2 = 1273.23\left(\frac{Q_2}{D^2}\right) \tag{4}$$

where, Q_2 is the flow rate at hydroponic tray (L/s) and D is the pipe diameter (mm)

$$v_2 = 1273.23\left(\frac{0.019}{15^2}\right) = 0.1075\,\mathrm{m/s}$$

In this project, the assumption of 30 °C average water temperature during operation and the viscosity, v of water is 801 *CSt*, then the Reynold number, *Re* can be determined by Eq. (5).

$$\mathrm{Re} = 1000 \times \left(\frac{v \times D}{v'}\right) \tag{5}$$

where, v is - the velocity (m/s), D is the pipe diameter (mm), and v is the viscosity (CSt)

$$\mathrm{Re} = 1000 \times \left(\frac{0.1075 \times 15}{801}\right) = 2.01$$

The resulted Reynold number suggests that the system operates in laminar condition. Hence, the following Eq. (6) was used to determine the friction parameter, f:

$$f = \frac{64}{\text{Re}} = \frac{64}{2.01} = 31.84 \tag{6}$$

From Eqs. (4) and (6), the friction factor, $\Delta H_{FF}/L$ can be determined using Darcy-Weisbach equation as in Eq. (7).

$$\frac{\Delta H_{FF}}{L} = 10^5 f\left(\frac{v^2}{D \times 2g}\right) \tag{7}$$

where, $\Delta H_{FF}/L$ is the- friction factor per 100 m pipe (m/100 m pipe),

f is the - friction parameter, v is the - velocity (m/s),
D is the - pipe diameter (mm), and
g is the -acceleration due to gravity (9.81 m/s^2).

$$\frac{\Delta H_{FF}}{L} = 10^5(31.84)\left(\frac{0.1075^2}{15 \times 19.62}\right)125.025\,\text{m}/100\,\text{mpipe}$$

Estimated total length of piping used was 2 m, calculation of the pipe friction loss, H_{FP} with the following equation was used:

$$\Delta H_{FP} = \frac{\Delta H_{FF}}{L} \times \frac{L}{100} = 125.025 \times \frac{2}{100} = 2.5\,\text{m} \tag{8}$$

To find the pump total head for the system, the following equation was used:

$$Pump\,Total\,Head,\ H_{PT} = H_{TS} + H_{FP} \tag{9}$$

$$Total\,Static\,Head,\ H_{TS} = H_{DS} - H_{SS} \tag{10}$$

where, H_{DS} - is the Discharge static head (m)H_{SS} - is the Suction static head (m), and H_{FP} - is the Pipe friction loss head (m).

Based on the Fig. 3.7, the value for H_{DS} and H_{SS} are 1.2 m and 0.3 m respectively and the value for H_{FP} calculated from Eq. (8) will be used.

Hence, results *Pump Total Head* as below:

$$Pump\,Total\,Head,\ H_{PT} = (1.2 - 0.3) + 2.5 = 3.4\,\text{m}$$

Next, to find the required velocity at supply, v_1, the following Bernoulli energy equation will be used. The pressure head is atmospheric, so P_1 and $P_2 = 0$.

$$H_{TS} + Z_1 + \frac{P_1}{\gamma} + \frac{v_1^2}{2g} = H_{FP} + Z_2 + \frac{P_2}{\gamma} + \frac{v_2^2}{2g} \quad (11)$$

where, Z_1 & Z_2 is the Suction and discharge tank level elevation (m), H_{FP} is the Pipe friction loss head (m), H_{TS} is the Total static head (m), P_1 & P_2 are the Inlet and outlet pressure (Pa), v_1 & v_2 are the Inlet and outlet velocity (m/s), γ specific weight (N/kg), and g is the acceleration due to gravity (9.81 m/s^2).

$$0.9 + 0.3 + \frac{v_1^2}{19.62} = 3.4 + 1.2 + \frac{0.1075^2}{19.62} = 8.16\,\text{m/s}$$

The cross-section area for 15 mm diameter pipe is 1.76×10^{-4} m^2. The flow rate at supply, Q_1 can be find with the following equation:

$$Q_1 = Av_1 = 1.76 \times 10^{-4}(8.16) = 1.4 \times 10^{-3}\,\text{m}^3/\text{s} \quad (12)$$

Lastly, the following equation was used to find the power for the pump:

$$Power,\, P = (\gamma Q_1 H_{PT}) = (1000 \times 9.81)(1.4 \times 10^{-3})(3.4) = 46.6\ \text{W} \quad (13)$$

Based on the calculated parameters, the suggested specification for the pump is 47 W and 1.2 m head.

2.4 Arduino for Monitoring Hydroponic System

The monitoring system designed for the hydroponic system was an Arduino-based microcontroller system. The microcontroller acquired all data from the inputs and processed it in sequence as programmed and then sends it to the output to perform the tasks. This project used an Arduino Mega because it has sufficient memory and many digital/analog input and output pins. The arrangement between input, microcontroller, and output of the system is shown in Fig. 9. There are five components as inputs to the system, which are pH sensor, water temperature sensor, ambient temperature, and humidity sensor, EC sensor, and real-time clock (RTC) module. Every sensor will measure its corresponding value every second and send it to the microcontroller. The output of the system consists of two components, which are the SD card module and OLED display. The SD card will store the data inside the memory card when being activated by the timer. The OLED display was included to easily monitor the current condition of the system for every minute. After the hardware components are constructed, the execution command for the tasks was developed using Arduino programming. The overall process of the system is shown in Fig. 10.

The system starts with the RTC module as a real-time clock and a timer. At the same time, all sensors are secure at their own designated places such as in the nutrient

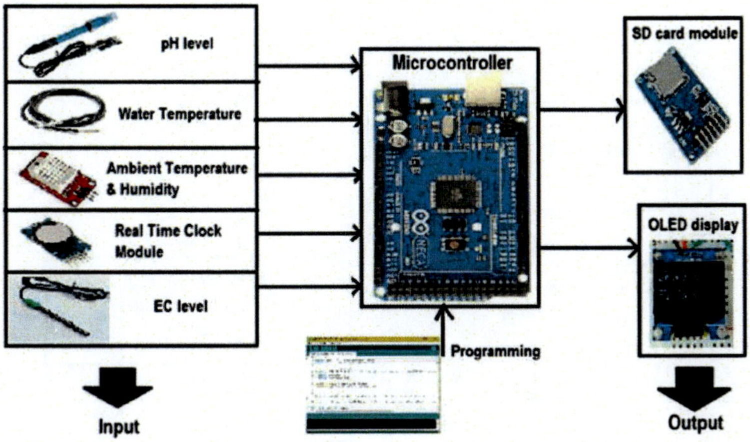

Fig. 9 Arduino Mega types use in the NFT system

solution reservoir and hydroponic frame. Thus, the sensors will operate and display individual value to the OLED display for every second. When the real-time clock matches the set timer, the system will trigger all the sensor values to be written and saved automatically with.txt format inside the SD card module. The data acquisition system was programmed to collect data every 9 a.m., 1, and 9 p.m. for 28 days to further analyzed the performance of the hydroponic system.

3 Results and Discussion

3.1 Simulation Analysis

A model of the piping system has been built based on the selected design for the flow simulation analysis. Based on the analysis of the pumping system, the parameters were used to run the flow simulation. From Fig. 11, the simulation results in the highest and the lowest velocity within the pipe were about 8.54 m/s at point 1 and 1.067 m/s at point 2. At point 1, the pipe diameter is 15 mm which results with a flow rate of 1.5×10^{-3} m^3/s or 90 L/min, while a steady flow rate of 100 L/min was measured. The differences show an error of 10% from simulation and actual data.

The frame being selected with the hydroponic tray weight is about 34.14 N load filled with water was considered as the load, a vertical downward gravity force was provided. The simulation results as shown in Fig. 12 indicates that the hydroponic frame tends to bend in the middle towards outward and experience concentrated stress of 2 MPa. The yield strength for UPVC is about 50 MPa, the designed frame passes the failure test.

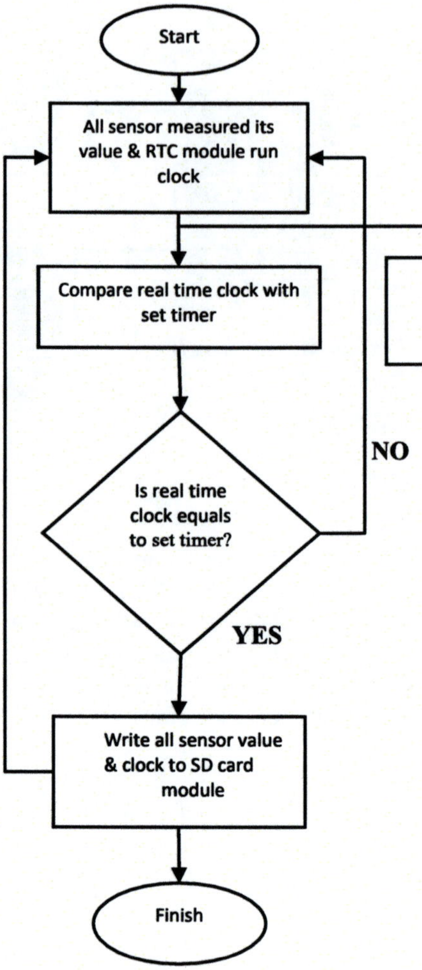

Fig. 10 Flow diagram of the monitoring system

Thus, the design was reconsidered and redesigned with improvement to ensure better withstand stress deformation with the same load acted upon it. Figure 14 shows the improved design for better support to eliminate the outward bend failure. The simulation analysis shown in Fig. 13 notices a little frame inclination but still can maintain staying upright which makes it better improvement with the lower concentrated of 0.735 MPa.

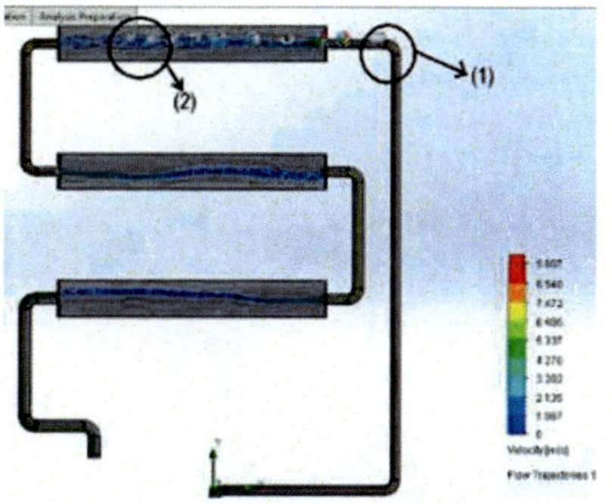

Fig. 11 Simulation of velocity within piping system

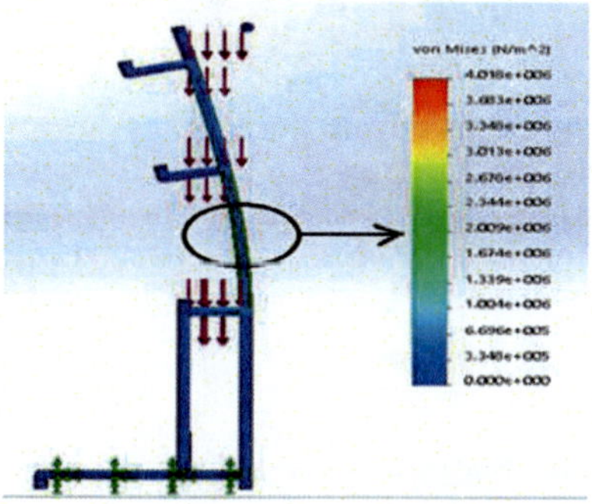

Fig. 12 Simulation of stress deformation at 2 MPa upon the frame

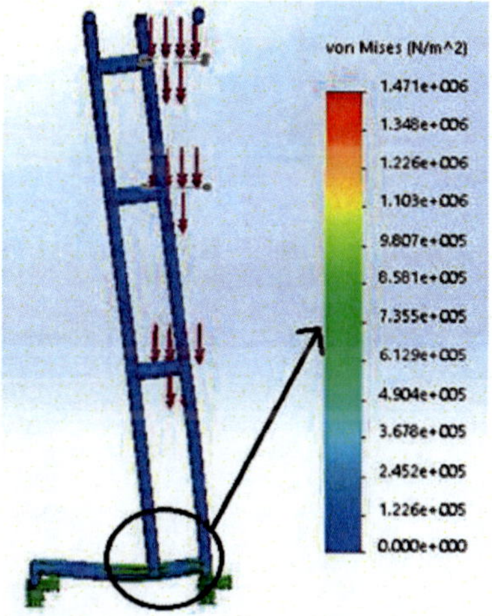

Fig. 13 Simulation of stress deformation is about 0.735 MPa after improving the design

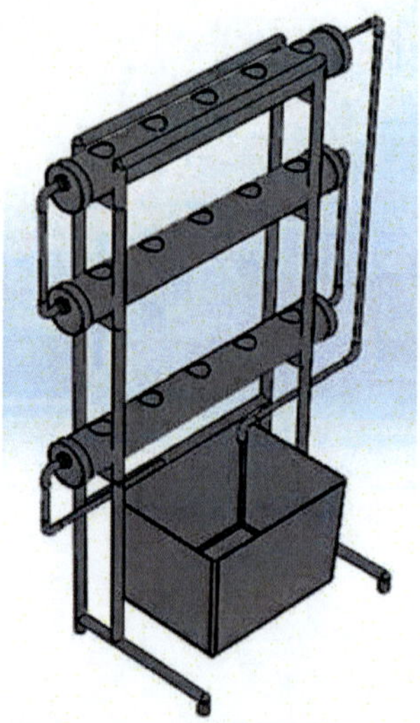

Fig. 14 Improved design NFT hydroponic system

3.2 NFT Hydroponic System

The prototype of the NFT hydroponic system that has been built with UPVC as the main material is shown in Fig. 15. It comprises three levels of hydroponic tray arranged vertically on the frame. The hydroponic tray was made from a 110 mm

Fig. 15 NFT fabrication of hydroponic system with dimension

Table 3 Contains of nutrient use in NFT hydroponics system [16]

Nutrient A (Macro)	Nutrient B (Micro)
Monoammonium phosphate Calcium nitrate Potassium nitrate	Magnesium sulfate Manganese sulfate Boric acid Copper sulfate Zinc sulfate Ammonium molybdate Iron chelate

diameter UPVC pipe so it can fit the net pots. The urban home NFT hydroponic system consists of 18 planting slots with 16 cm spacing for crops to be grown in the net pots. The hydroponic system has a base covering an area of 0.48 m^2 and a total height of 130, 70 cm width, and 60 cm length. Every hydroponic tray weighs about 2.5 kg, which cumulated to the total weight of 8 kg. The planting kit has been built to fit perfectly in the balcony.

As compared to conventional farming, the designed NFT hydroponic system has a significant improvement in terms of the yield produced per unit area [26]. By using garden pot in conventional farming, the maximum numbers of plants that can be planted in a 0.48 m^2 space with 16 cm apart were 11 plants. But with the designed hydroponic system, it can hold 18 plants in the same square area. This is an almost 63% increase in the maximum yield per unit area with the use of the NFT hydroponic system. The circular-shaped hydroponic tray of the hydroponic system was designed to store nutrient solution. AB Mix is a mixture of fertilizer A and fertilizer B. Fertilizer A contains potassium while fertilizer B contains sulfate and phosphate as shown in Table 3 for details contents of AB Mix [7]. This feature allowed the plants to have a continuous supply of nutrient solution even during power interruptions and the pump does not work.

The choy sum vegetable growth was very successful in the NFT hydroponic system as shown in Fig. 16. The plants survived growing in the balcony area of high-rise houses, indicate that they received enough sunlight for their photosynthesis process. The leaves observed were green and thick. Also, the roots developed were white and long. Plant length, leaf count, and weight of the plant are shown in Table 4.

3.3 Monitoring System for NFT Hydroponic Using Arduino

The microcontroller used in this project is Arduino Mega 2560 that processes the sensor data and triggers the output. Most of the components were connected to the microcontroller via jumper cables and wires that were soldered to secure them from getting loose. The data acquisition system was able to gather data of different parameters from the sensors with a total of 84 data points. The parameter such as temperature, humidity, pH and EC level of hydroponic system were closely monitored as shown in Fig. 17(a), (b), (c), (d) and (e). All five parameters were monitored in

Fig. 16 NFT hydroponic plant growth progress in days

Table 4 The development of characteristics of leaf length, leaf count and plants weight

	Leaf length (cm)	Leaf count	Plant weight (g)
Minimum	15	7	13
Maximum	20	10	30
Average	17.75	8.5	21

28 days. Data showed that some conditions were slightly out of range, but still a tolerable range and plant growth in healthy condition. According to Fig. 17(a) and (b), it shows that the ambient and reservoir temperature with an average mean of 28.9 and 27.8 °C, respectively. Besides that, the relative humidity data pattern shows an average mean of 72.0% as shown in Fig. 17(c). According to Fig. 17(d), a trend of rapid change was detected because of manual refill nutrient solution was done and the average mean for pH level is 6.9. For hydroponic plants, controlling the optimal acidic pH range is important since it influences solubility, availability, and uptake of several essential nutrient ions, according to [21].

The conductivity values of the solution were linearly related to the equivalent salt concentration in the solution. The conductivity of the full-strength fresh and unused solution registered on the conductivity meter was 3.5 mS/cm. As the plants absorbed nutrients, the solution's conductivity value decreased due to a decrease in salt concentration. Based on Fig. 17(e), it can be seen that then nutrient level

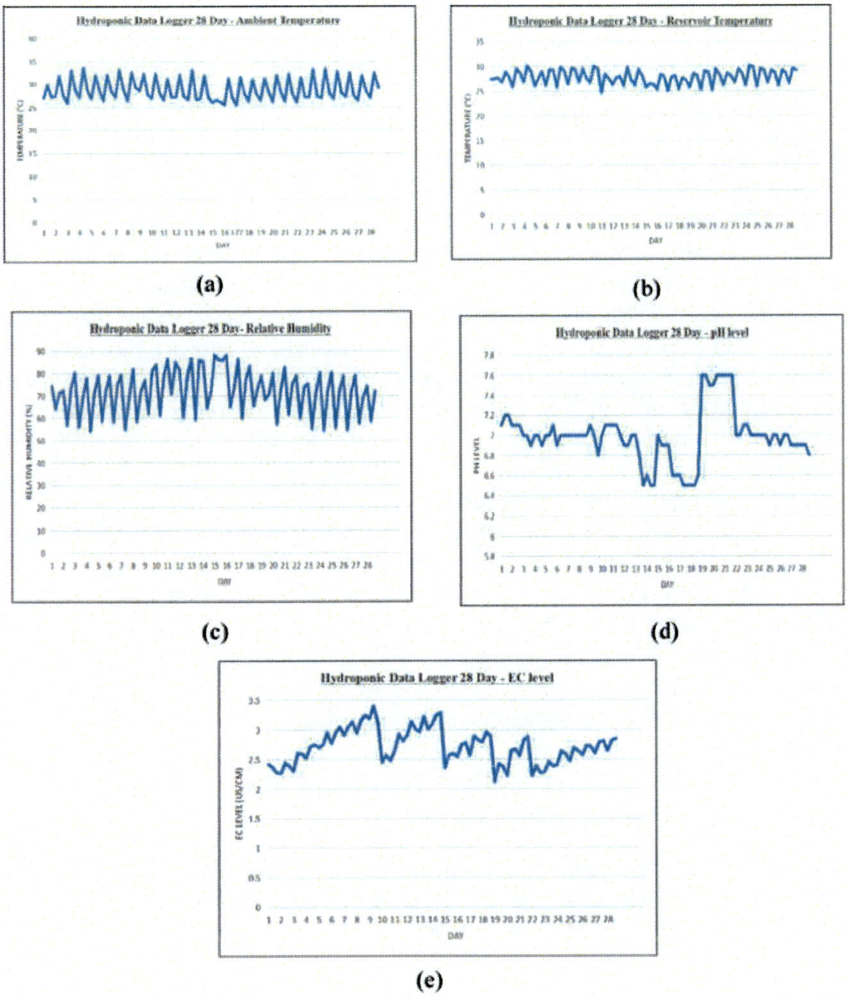

Fig. 17 Data acquisitions system recorded on 28 day of plant growth, **a** Ambient Temperature, **b** Reservoir Temperature, **c** Relative Humidity, **d** pH Level, and **e** EC Level

gradually increases as water evaporates leaving salt concentration higher, followed by a sudden drop that indicates the manual refill of nutrient solution. The electrical conductivity of the nutrient solution was in the range 2–3.5 mS/cm throughout the experiment. The average mean for the EC level is 2.71 mS/cm. The results EC were similarly controlled by Kwon et al. (2021), with an electrical conductivity (EC) of 2–4 mS/cm applied to all farming systems.

4 Conclusion

The design of suitable hydroponic system for urban use has been proposed. The utilization of space limitation and the load stress concentration on structural frame was fully considered in the proposed design. The Pugh selection method was applied to select the best design concept for urban purpose. According to the structural analysis simulation, a stable design has been achieved by improving the prior design by redesigning the frame to support the hydroponic tray. The initial design simulation outcome shows 2 MPa of concentrated stress while the improved design simulation outcome shows 0.735 MPa of concentrated stress only. This has increased the strength of the frame to withstand the load. The flow simulation also results in a 10% difference between simulated and measured flow rate of the system. The pumping parameters have also been identified 47 W as the pump power rating and the pump head must achieve a minimum of 1.2 m as it is the highest height needed to be pumped in the piping system. The completion of this NFT hydroponic system in designing, fabrication, and operation was successfully done without any major problem. The NFT hydroponic system for urban houses only takes a space of 0.48 m^2 but with crop capacity of 18 plants, making it suitable for the urban landscape due to its lean structure. Also, it is designed to be modular, so it can be moved to another location or kept in storage with ease. This benefits the user so they can produce locally grown leafy vegetables. The monitoring system for NFT hydroponic system in urban area was also successfully developed. The system can save data automatically according to the preset timer and also displaying the current data. Various kinds of sensors are used along with the Arduino microcontroller for measuring and monitoring the plant growing parameters. The time duration observed in this project until the crops are harvested is 28 days.

Acknowledgements This research is supported by the Ministry of Higher Education Malaysia and Universiti Teknologi MARA, Cawangan Pulau Pinang, Kampus Permatang Pauh and Universiti Teknologi MARA, Cawangan Johor, Kampus Pasir Gudang.

References

1. Ahamed T, Takigawa T, Noguchi R, Tian L (2014) Bioproduction engineering: a road map for sustainable agricultural practices. Nova Science, Japan. https://novapublishers.com/shop/bioproduction-engineering-a-road-map-for-sustainable-agricultural-practices/
2. Asao T (2012) Hydroponics- a standard methodology for plant biological research. InTech Design Tem, Crotia. https://www.intechopen.com/books/1781
3. Barbosa GL, Gadelha FD, Kublik N, Proctor A, Reichelm L, Weissinger E, Wohlleb G (2015) Comparison of land, water, and energy requirements of lettuce grown using hydroponic vs. conventional agricultural methods. Int J Environ Res Public Health 12:6879–6891. https://doi.org/10.3390/ijerph120606879
4. Bevier G (2012) Global food system: feeding the world. Reprod Domest Anim 47:77–79. https://doi.org/10.1111/j.1439-0531.2012.02058.x

5. Copper A (1978) Methods of establishing young plants in a nutrient-film tomato crop. J Horticult Sci 53:189–193. https://doi.org/10.1080/00221589.1978.11514818
6. Despommier D (2010) The vertical farm: feeding the world in the 21st century. Scientific American. https://www.semanticscholar.org/paper/The-Vertical-Farm%3A-Feeding-the-World-in-the-21st-Despommier/babed74420c8f706f218014570c6b12c6e020d1d
7. Dewi SK, Rahayu YS, Bashri D (2019) The effectiveness of nutrient variation to hydroponic Caisim (Brassica juncea L.) growth. J Phys Conf Ser 1417(1):012038. https://doi.org/10.1088/1742-6596/1417/1/012038
8. Ezzahoui I, Abdelouahid RA, Taji K, Marzak A (2021) Hydroponic and aquaponic farming: comparative study based on internet of things IoT technologies. Procedia Comput Sci **191**:499–504. https://www.sciencedirect.com/science/article/pii/S1877050921014642?via%3Dihub
9. Fahmi Zaim, Samah Bahaman Abu, Abdullah Haslinda (2013) Paddy industry and paddy farmers well-being: a success recipe for agriculture industry in Malaysia. Asian Social Sci 9(3):177–181. https://doi.org/10.5539/ass.v9n3p177
10. Francy R, Soori PK, Chacko S (2016) Intelligent in-house mini-automated farming. J Adv Agric Technol 3:286–291
11. Gentry M (2019) Local heat, local food: integrating vertical hydroponic farming with district heating in Sweden. Energy 174:191–197
12. Gericke W (1930) Excessive tax on soil fertility by crop production on poor land. Science 72:202–203
13. Giorgio G, Munoz P, Pardossi A, Ramazzoti S, Savvas D (2013) Soil fertility and plant nutrition. In: Gianquinto G (ed.) Good Agricultural Practices for greenhouse vegetable crops: Principle for Mediterranean climate area. FOA Plant Production and Protection Paper, Rome, pp 205–269. https://cris.unibo.it/handle/11585/173273#.YT1NQp0zbiA
14. Ike C, Orakwe C, Ezeagu C (2019) Hydroponic water requirement estimation for cucumber using FAO-CROPWAT Model in Awka, Anambra State, Nigeria. J Eng Appl Sci 15:118–129
15. Johnson D, Wardow G (1997) A prototype recirculating aquaculture-hydroponics system. J Agric Mech
16. Jones J (2016) Hydroponics: a practical guide for the soilless. CRC Press Library of Congress
17. Khaoula T, Abdelouahid R, Ezzahoui I, Marzak A (2021) Architecture design of monitoring and controlling of IoT-based. Procedia Comput Sci 191:193–498
18. Kozai T, Niu G, Takagaki M. (2016) An indoor vertical farming system for efficient quality food production. Plant Fact. https://doi.org/10.1017/CBO9781107415324.004
19. Lambin Eric F, Meyfroidt Patrick (2011) Global land use change, economic globalization, and the looming land scarcity. Proc Natl Acad Sci 108(9):3465–3472
20. Lennard W, Leonard B (2006) A comparison of three different hydroponic sub-systems (gravel bed, floating and nutrient film technique) in an Aquaponic test system. Aquacult Int 14:539–550
21. Majid M, Khan J, Ahmad Shah Q, Masoodi K, Afroza B, Parvaze S (2021) Evaluation of hydroponic system for the cultivation of Lettuce (Lactuca sativa L., var. Longifolia) and comparison with protected soil-based cultivation. Agric Water Manag J 245:1–13. https://doi.org/10.1016/j.agwat.2020.106572Received
22. Mishra B, Rath B, Mishra A (2017) Simulation of vertical hydroponics farming setup for growinh vegetables to address nutritional security. Int J Eng Sci Math, 1–6
23. Pardev P (2010) Food: the growing problem. Nature 466(1):546–547. https://www.nature.com/articles/466546a
24. Pugh S (1981) Concept selection: a method that works. In: Hubka V (ed.) Review of design methodology. Proceedings international conference on engineering design. Zurich, pp 497–506
25. Rabu MR, Muhammad RM (2015) The potential of urban fanning technology in Malaysia: policy intervention. In: Food and Fertilizer Technology Center for Asian and Pacific Region, pp 1–5
26. Resh H (2013) Hydroponic Food Production: A Definitive Guidebook for the Advanced Home Gardener and the Commercial Hydroponic Grower, 7th edn. CRC Press Taylor & Francis Group, New York

27. Schippers P (1977) The rate of respiration of potato tubers during storage 1. Review of literature. Springer Link, pp 173–188
28. Sharma N, Kumar K, Acharya S, Chaurasia O (2018) Hydroponics as an advanced technique for vegetableproduction: an overview. J Soil Water Conserv 17:364–371. https://doi.org/10.5958/2455-7145.2018.00056.5
29. Sigalingging R, Honora P (2019) Utilization of solar power as DC water pump movement in hydroponic plants. IOP Conf Ser Earth Environ Sci 260(1):012038. https://doi.org/10.1088/1755-1315/260/1/012038
30. Somerville C, Cohen M, Pantanella E, Stankus A, Lovatelli A (2014) Small-scale aquaponic food production Integrated fish and plant farming, vol. 589. FAO Fisheries and Aquaculture Technical Paper, Rome
31. Tagle S, Benoza H, Pena R, Aleeza O (2018) Development of an indoor hydroponic tower for urban farming.In: 6th DLSU Innovation and Technology Fair 2018. DSLU, Manila, pp 1–7
32. Zezza A, Tasciotti L (2010) Urban agriculture, poverty, and food security: empirical evidence from a sample of developing countries. Food Policy 35(4):265–273. https://doi.org/10.1016/j.foodpol.2010.04.007
33. Zhang Z, Rod M, Hosseinian F (2021) A Comprehensive review on sustainable industrial vertical farming using film farming technology Zitian. Sustain Agric Res 10(1):46–53. https://doi.org/10.5539/sar.v10n1p46

Performance of Food Dehydrator Using Different Type of Distributor Base Plate

Muhamad Hakimi Mokhtar, Muhamad Helmi Ashraf Termizi, A. S. M. Yudin, and A. Alias

Abstract Dehydrating food is the oldest method in preserving food in the history of mankind. Food dehydration is the process of decreasing the moisture content of food to low levels in order to extend its shelf life [1]. As a result of scientific advancements, food dehydration became a global sector capable of supplying convenient and healthy meals [2]. Today, the technology in dehydrating food keeps growing and variation of food dehydrators has been produced by engineers. Reducing the number of product failures by inventing a way to detect production faults at their earliest potential stage along with the full value-added stream is one way to minimize the manufacturing costs [3]. One of the most common types of dryers for food is the food dryer. This method is typically utilized on a smaller scale. The aim of the food dehydrator is to prevent the growth of bacteria and fungus on food through the removal of moisture. In this experiment one of the food dehydrators is Morgan Food Dehydrator which is used in this project. The main objective of this project is to analyze the rate of dehydration of the existing dehydrator in market and design the best pattern to increase the dehydration rate. It will come with 3 different designs that follow the concept of design distributor of fluidized bed dryer. In this study, the temperature and air velocity are maintained and the only concern at the end of the experiment is the time taken for the food to dry depending on the suggested moisture content left. From these experiments, the result shows that the different type of distributor base plate have a better drying rate than the current existing product.

Keywords Drying process · Loss on drying · Flow · Thermal

Supplementary Information The online version contains supplementary material available at https://doi.org/10.1007/978-981-19-1577-2_12.

M. H. Mokhtar (✉) · A. Alias
Faculty of Mechanical and Automotive Engineering Technology, Universiti Malaysia Pahang, 26600 Pekan, Pahang, Malaysia
e-mail: hakimimokhtar@gmail.com

A. Alias
e-mail: azribalias@ump.edu.my

M. H. A. Termizi · A. S. M. Yudin
College of Engineering, Universiti Malaysia Pahang, 26600 Pekan, Pahang, Malaysia

© The Author(s), under exclusive license to Springer Nature Singapore Pte Ltd. 2023
M. H. A. Hassan et al. (eds.), *Technological Advancement in Instrumentation & Human Engineering*, Lecture Notes in Electrical Engineering 882,
https://doi.org/10.1007/978-981-19-1577-2_12

1 Introduction

The fast-growing human population's food demand has increased. Knowledge of food quality and hygiene, food protection and market are one of the major issues to be tackled. Minimal prepared foods including fresh fruit or vegetables, normally consumed raw, may be infected by different micro-organisms like Salmonella spp. Escherichia coli O157:H7 and Listeria Monocytogenes. This can lead to food loss and foodborne diseases. Consumers want minimum packaged foods with a pure and pleasing taste, free from preservatives and microbiologically healthy [4]. Fruits, vegetables, and even meats can be preserved in order to extend their shelf life. Since ancient times, there have been a variety of methods for preserving food. Canning [5], dry salting [6], smoking [7] or even freezing [8] has been used to increase the food lifetime. One of the most typical and commonly method in preserving food is drying.

In terms of food preservation, drying is one of the oldest methods. Dried items have a low moisture content, which makes them safe for storage and transportation. This reduces the risk of microbial growth, which is why drying is an energy-intensive technique in food processing. It's common to find fresh food goods that are overly wet. If these products aren't dried properly in time, they'll rot or mould [9]. Several research studies of the drying of food products have been observed. There is many type of methods used in drying food such as fluidized-bed [10], ultra sound drying [11, 12], microwave vacuum dryer [13], infrared radiation [14], spray-dried [15]. Different fruit was used such as blackberry, grape [16], watermelon [17], apricot [18] and mango [19] were dried.

With an increasing number of industrial drying equipment for mass production available, the food dehydrator is one of the machines that is easy and cost-effective. Food dehydrator is one of the most common electrical kitchen appliances used nowadays to dry practically all vegetables, fruits, and meat. In a circular or square container, it circulates hot air in the same way. It is compact and widely used at home [20–22]. Existing food dehydrators suffer performance issues such as drying time and cost. Due to the flat base of the dehydrators, hot air could not circulate properly. No matter how much hot air is directed to the dehydrator via the angled hole on flat plate surface, the flow circulation is still poor. As a result of the improved plate surface design, hot air flow circulation can be greatly improved as a result. In order to get the greatest results from the dehydrator's hot air circulation, it is vital to examine the numerous base pattern possibilities available to you.

2 Experimental Set Up

2.1 Design of Experiment

Food Dehydrator
A food dehydrator is a kitchen appliance used to remove moisture from a variety of foods so that it can continue to last longer. It is an economical way to make and keep fresh meals at your home comfort. Food dehydrators are classified into two types: shelf dehydrators and stacking dehydrators. A shelf dehydrator has multiple food trays placed in a very similar position to the racks, allowing easy access and simultaneous airflow. On the other hand, a dehydration stacking unit comprises of four to ten trays each with a source of hot air either above or below the unit [23].

The Specification of Morgan Food Dehydrator
The present works aims to study the design of the base of the food dehydrator affected to the time taken for drying the specimen and rather it can keep the nutrient. The current existing food dehydrator in the market was used in this experiment. The chosen food dehydrator to be developed in this project was Morgan Food Dehydrator-A8 (MFD-A8). Matching with the company motto, 'For A Quality Lifestyle', this product is one of the household electrical appliances invented by the company that is of high quality, affordable and easy-to-use. The dehydrator consists of six main parts which is the cover, five interchangeable and height adjustable trays, under pan, base, temperature knob and power switch with indicator light as shown in the Fig. 1 below.

This model's measurements are 315 mm broad, 315 mm long, and 308–395 mm high. In addition, the rated voltage ranges from 220 to 240 V, with a frequency of 50 or 60 Hz. The MFD-A8 dehydrator uses between 200 and 240 watts of power and weighs 2.85 kg when empty. Transparent trays and covers allow the user to see the food's condition. The temperature can be changed to accommodate a range of food

Fig. 1 Morgan food dehydrator

thicknesses. With this dehydrator's eight temperature settings, you may dry food at any temperature between 35 and 70 °C [2].

2.2 Loss-on Drying

Loss-on drying (LOD) is the percentage of weight loss of any substances or volatile matter after it undergo any process of removing the moisture or water content [24]. The LOD approach is straightforward. In the first step, the sample's initial weight is recorded. It is then allowed to dry for a set period of time using any drying equipment. Afterwards, the sample's weight is recorded once more. Difference between before and after drying represents the amount of moisture lost in the sample. People utilize this strategy a lot. You can determine your weight reduction using a simple procedure and even a low-cost apparatus [26]. The calculation to find the percentage of weight loss is shown in below.

$$LOD = \frac{Initial\ weight\ of\ sample - final\ weight\ after\ drying}{Initial\ weight\ of\ sample} \times 100\% \quad (1)$$

2.3 The Different Method of Karl Fischer and Loss-On Drying

While both Karl Fischer and LOD are used to quantify the water content of a product, Karl Fischer Titration is limited to the measurement of water content in the sample. Whereas LOD measures the weight shift after drying, including other chemical compounds such as alcohol or fat [26].

Karl Fischer's method involves a chemical process. In order to make the water non-conductive, a reagent must be applied to the sample. Loss-on-Drying, on the other hand, compares the weight of the food after it has been dried. The difference in weight is used to calculate the product's moisture content [26]. To decide which method to utilize in the experiment, you must first consider the outcome or response variable. Karl Fischer method can be use when counting the amount of volatiles such as alcohol or fat. Whereas Loss-on-Drying is sufficient for determining water content in food if we are solely concerned with its water content.

Fig. 2 Weighing slice of apple before experiment

3 Experimental Procedure

3.1 Preparing the Specimen

The green apple was decided to be used as specimen in this experiment. Apples is one of the most widely consumed fruits throughout seasons around the world. Apple consumption and production worldwide are roughly 81 million tonnes according to the Food and Agriculture Organization Statistic Databases [27]. Because of its strong flavor, green apples before and after drying have a similar sour taste. The green apple should be dried in a regular dehydrator for around 10 h, with slices between 3 and 5 mm thick. Before the drying process, the weight of each apple will be measured after it has been sliced into thin slices in the 3–5 mm range. Each apple slice will be weighed on a miniature scale, as shown in the Fig. 2.

3.2 Drying the Specimen

After weighing and taking the data for the starting weight of each piece, the application slices are placed on the tray and a uniform distance is established so every piece of air gets the same airflow. The specimen is dried at a temperature of 45 °C. When the apple chunks are heated, the weighted scale measures their weight every two hours to observe how much they have shrunk. Apples must have a water content of at least 20% if they are to keep their flavor and nutritional value (Fig. 3).

Fig. 3 Dehydrating apples in the Morgan food dehydrator

3.3 *Original Base of the Dehydrator*

SolidWorks is used to create the dehydrator's original base, which is used as a guide when creating new patterns. The original base is represented in isometric perspective in the Fig. 4 below and the dimension in Fig. 5.

3.4 *Design Proposed for Base of Morgan Food Dehydrator*

To achieve the best dehydration rate on the food dehydrator, new design has been proposed for the base of the Morgan food dehydrator where the existing base did not have any pattern on it. Three designs have been proposed for the design pattern on base for the food dehydrator. The idea in designing the pattern is taken from a few

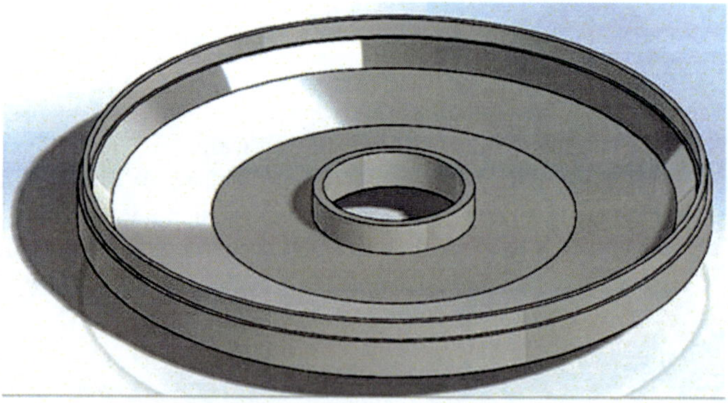

Fig. 4 Original base Morgan food dehydrator

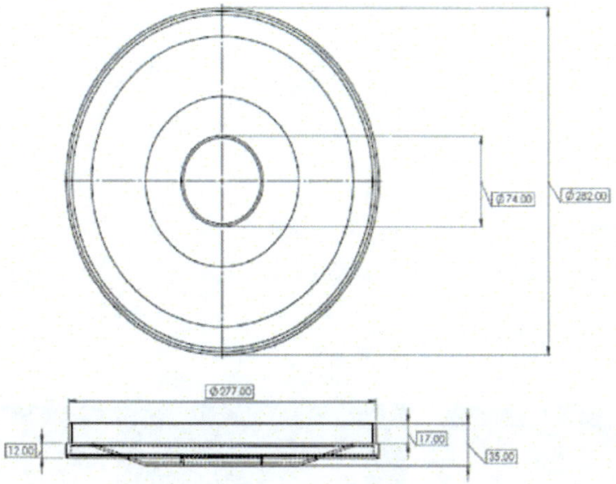

Fig. 5 Dimension of the original base Morgan food distributor

concepts of air flow such as the distributor of fluidized bed dryer, the shape of fan's blade and the curve shape of centrifugal blade from water pump.

Three designs have been proposed for the base of the food dehydrator which is straight pattern, fan blade pattern and curve pattern. The fan blade pattern is taken from the concept of fan-blade where it has some angle to smooth the airflow. Theoretically, this pattern has less resistance, and the air can flow upward and in circular due to the angle projected. The angle use in the Fan-blade design is 45°. The third design is based on the concept of impeller blade of centrifugal pump where it has curve along the plate. The all three design was drawn in SolidWorks and the dimension for each blade pattern have been shown as Fig. 6 and Fig. 7 below.

(a) Straight Pattern (b) Curve Pattern (c) Fan-Blade Pattern

Fig. 6 Blade pattern design

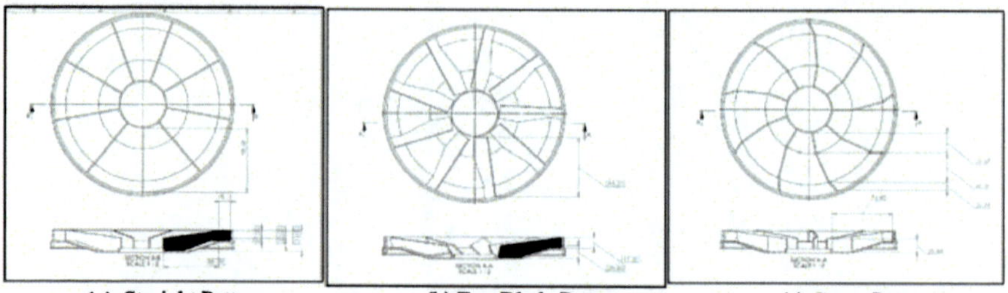

(a) Straight Pattern (b) Fan-Blade Pattern (c) Curve Pattern

Fig. 7 Detail drawing of blade design

(a) Straight Pattern (b) Fan-Blade Pattern (c) Curve Pattern

Fig. 8 Fabricated design of blade pattern

3.5 Fabrication Method

The blade for each design was designed by using the thin aluminium sheet. Every pattern is cut out of the sheet according to its shape. The blade pattern will be set-up on the dehydrator base before use. As for the experimental purpose only, plasticine was used to hold the pattern in place so that it wouldn't fall over. The fabrication is shown in Fig. 8.

Table 1 Initial weight of specimen in each experiment

Experiment	Weight of specimen (g)									
	1	2	3	4	5	6	7	8	9	10
Original	10.89	8.95	8.77	9.27	9.91	13.51	8.00	8.92	9.82	8.62
Straight	7.46	8.55	9.61	12.54	8.27	10.16	8.70	6.24	11.47	8.55
Fan	7.81	10.09	6.18	8.46	11.88	9.28	5.90	8.55	10.82	7.12
Curve	10.92	9.67	11.46	10.15	12.07	9.34	13.44	10.35	10.00	10.27

4 Results and Discussion

4.1 Initial Weight Data Before Drying

Table 1 shows the initial weight of specimen in each experiment. Based on the table below, the range of the specimen is from 7 to 14 g depends on the thickness of the apple slices.

4.2 Result After the Experiment

Table 2 shows the data of weight loss of the specimen when using the original base of the food dehydrator within 10 h.

The graph of the weight of the specimen against time for experiment on the original base is shown in Fig. 9. The graph shows that the relation between the weight and the time is inversely proportional. Each specimen has shown a decreased of weight in the graph from 0 to 10 h of drying.

From the Table 3, it shows the weight loss of the specimen when using the straight pattern base of the food dehydrator from 0 to 10 h.

Table 2 Weight loss of specimens in experiment on the original base

Time (hrs)	Weight of Specimen (g)									
	1	2	3	4	5	6	7	8	9	10
0	10.89	8.95	8.77	9.27	9.91	13.51	8	8.92	9.82	8.62
2	7.69	5.73	5.54	6.23	6.66	9.59	5.07	5.83	6.53	6.06
4	5.65	3.88	3.67	4.41	4.66	7.35	3.42	4.09	4.62	4.43
6	4.15	2.71	2.38	3.14	3.2	5.52	2.27	2.94	3.17	3.18
8	3.01	1.96	1.67	2.28	2.32	4.2	1.6	2.23	2.34	2.38
10	2.44	1.59	1.32	1.85	1.88	3.54	1.3	1.88	1.93	1.98

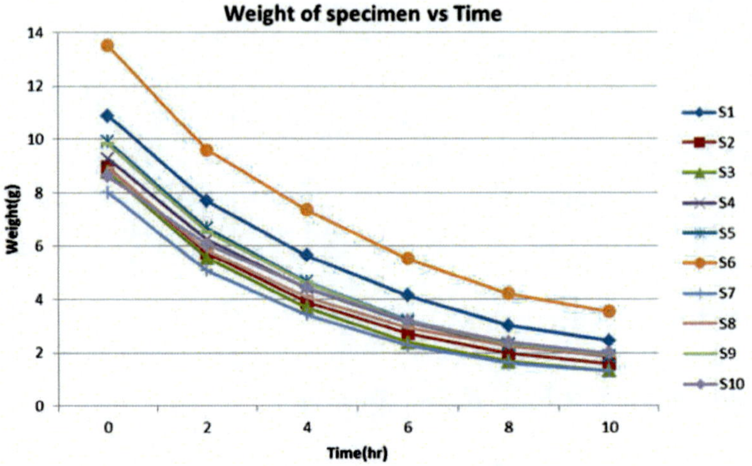

Fig. 9 Graph of weight against time for experiment on original base

Table 3 Weight loss of specimens in experiment using straight pattern

Time (hrs)	Weight of specimen (g)									
	1	2	3	4	5	6	7	8	9	10
0	7.46	8.55	9.61	12.54	8.27	10.16	8.7	6.24	11.47	8.55
2	5.18	6.2	7.33	9.18	5.84	7.43	6.21	4.42	8.77	5.61
4	3.65	4.42	5.26	6.86	4.25	5.28	4.55	3.05	6.44	3.83
6	2.12	2.64	3.18	4.54	2.66	3.13	2.88	1.67	4.11	2.04
8	1.74	2.15	2.55	3.83	2.23	2.56	2.38	1.37	3.34	1.65
10	1.34	1.67	1.89	2.89	1.67	1.94	1.78	1.08	2.41	1.38

The graph for the weight loss of the specimen when using the straight pattern base are shown in the Fig. 10 below. From the graphs, we can see that weight loss shows a significant drop from 0 to 6 h of drying process.

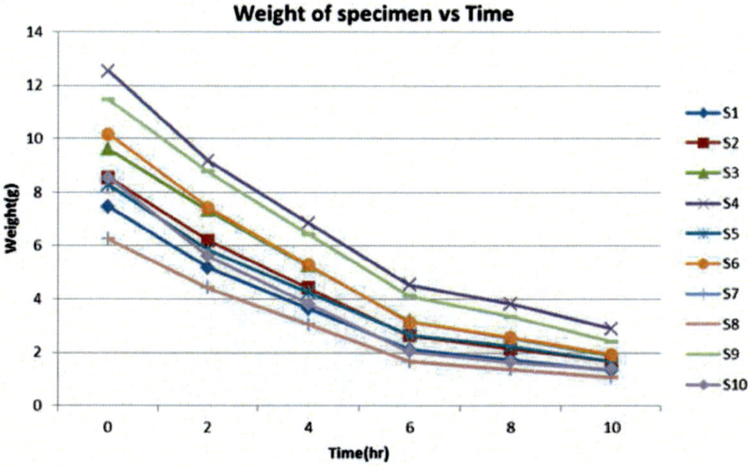

Fig. 10 Graph of weight against time for experiment of straight pattern

Table 4 Weight loss of specimens in experiment using fan-blade pattern

Time (hrs)	Weight of specimen (g)									
	1	2	3	4	5	6	7	8	9	10
0	7.81	10.09	6.18	8.46	11.88	9.28	5.90	8.55	10.82	7.12
2	4.56	6.34	3.30	5.19	7.92	5.89	3.21	4.90	6.83	3.69
4	3.47	5.01	2.36	4.13	6.59	4.82	2.42	3.68	5.35	2.63
6	1.84	3.23	1.24	2.56	4.48	2.91	1.23	2.01	3.30	1.37
8	1.36	2.46	0.94	1.97	3.51	2.23	0.94	1.45	2.42	1.09
10	1.10	1.78	0.88	1.47	2.69	1.68	0.81	1.23	1.80	1.01

From the Table 4, it shows the weight loss of the specimen when using the Fan-Blade pattern base of the food dehydrator from 0 to 10 h.

The graph for the weight loss of the specimen when using the Fan-Blade pattern base are shown in the Fig. 11 below. From the graphs, we can see that weight loss shows a significant drop from 0 to 2 h of drying process before it is steadily decreased until 10 h of drying.

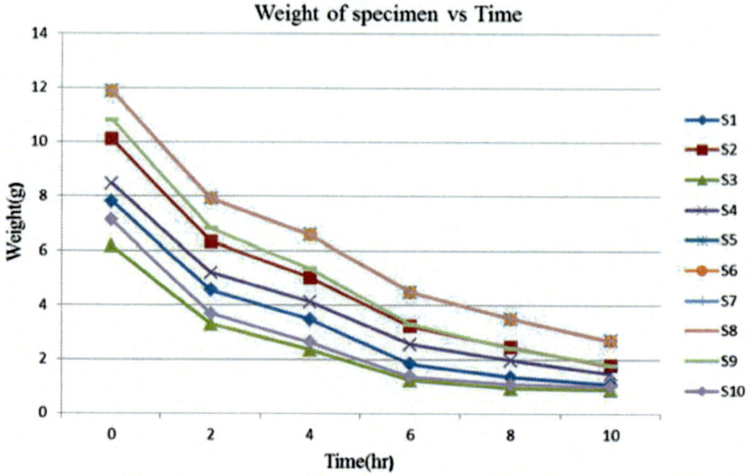

Fig. 11 Graph of weight against time for experiment of fan-blade pattern

Table 5 Weight loss of specimens in experiment using curve pattern

Time (hrs)	Weight of specimen (g)									
	1	2	3	4	5	6	7	8	9	10
0	10.92	9.67	11.46	10.15	12.07	9.34	13.44	10.35	10.00	10.27
2	7.48	6.73	7.59	5.90	7.80	5.61	9.15	6.75	6.74	6.73
4	3.70	3.69	3.92	2.45	3.94	2.62	5.05	2.98	3.26	3.32
6	2.89	2.98	3.13	2.10	3.19	2.09	3.37	2.37	2.49	2.66
8	2.07	2.26	2.34	1.75	2.43	1.56	2.95	1.75	1.73	1.99
10	1.73	1.82	2.00	1.70	2.09	1.42	2.46	1.61	1.60	1.69

From the Table 5, it shows the weight loss of the specimen when using the Curve Pattern base of the food dehydrator from 0 to 10 h.

The graph for the weight loss of the specimen when using the curve pattern base are shown in the Fig. 12 below. From the graphs, we can see that weight loss shows a significant drop from 0 to 2 h of drying process before it is steadily decreased until 10 h of drying, and this pattern of dehydration can be analyzed from the Curve Pattern.

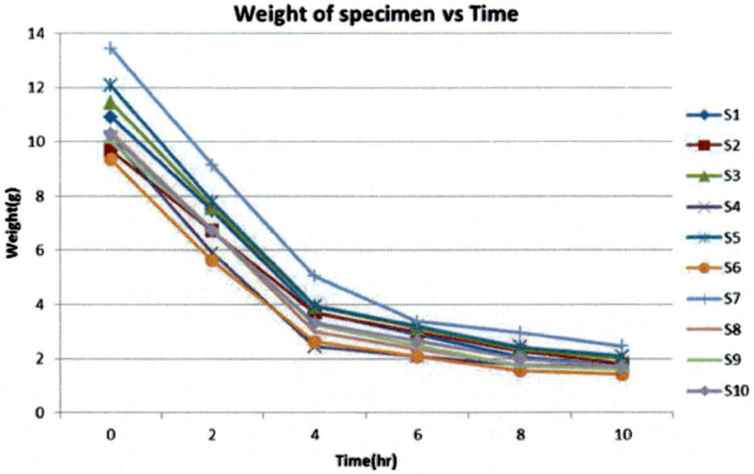

Fig. 12 Graph of weight against time for experiment of curve pattern

4.3 *Loss on Drying (LOD) Calculation*

The loss on drying (LOD) is calculated by finding the difference between the initial and final weight of the specimen and the result is divided by initial weight of specimen before it is multiple by 100%. Hence, the percentage of weight loss for each design pattern of base can be determined. Table of loss of drying for each design pattern show below (Tables 6, 7, 8 and 9).

From the table of LOD above, when using the original base, the specimen failed to achieve the 80% loss of drying. Only 60% of the specimens can achieved the required

Table 6 Loss of drying of specimens in experiment on the original base

Time (hrs)	Weight of specimen (g)									
	1	2	3	4	5	6	7	8	9	10
0	10.89	8.95	8.77	9.27	9.91	13.51	8.00	8.92	9.82	8.62
10	2.44	1.59	1.32	1.85	1.88	3.54	1.30	1.88	1.93	1.98
LOD	77.59	82.23	84.95	80.04	81.03	73.80	83.75	78.92	80.35	77.03

Table 7 Loss of drying of specimens in experiment on the straight pattern base

Time (hrs)	Weight of specimen (g)									
	1	2	3	4	5	6	7	8	9	10
0	7.46	8.55	9.61	12.54	8.27	10.16	8.7	6.24	11.47	8.55
10	1.34	1.67	1.89	2.89	1.67	1.94	1.78	1.08	2.41	1.38
LOD	82.04	80.47	80.33	76.95	79.81	80.91	79.54	82.69	78.99	83.86

Table 8 Loss of drying of specimens in experiment using fan-blade pattern base

Time (hrs)	Weight of specimen (g)									
	1	2	3	4	5	6	7	8	9	10
0	7.81	10.09	6.18	8.46	11.88	9.28	5.9	8.55	10.82	7.12
10	1.1	1.78	0.88	1.47	2.69	1.68	0.81	1.23	1.8	1.01
LOD	85.92	82.36	85.76	82.62	77.36	81.90	86.27	85.61	83.36	85.81

Table 9 Loss of drying of specimens in experiment using curved pattern base

Time (hrs)	Weight of specimen (g)									
	1	2	3	4	5	6	7	8	9	10
0	10.92	9.67	11.46	10.15	12.07	9.34	13.44	10.35	10.00	10.27
10	1.73	1.82	2.00	1.70	2.09	1.42	2.46	1.61	1.60	1.69
LOD	84.16	81.18	82.55	83.25	82.68	84.80	81.70	84.44	84.00	83.54

moisture content left and the others need more time to achieve it. For the straight pattern, almost of the specimen succeed to reach 80% of weight loss. There is 60% of specimen that achieved the targeted percent of weight loss. However, because of some of specimen that did not reach the target, this type of pattern still cannot be considered as the best pattern yet.

Next, the Fan-Based pattern show the LOD of the specimen that 90% are achieved the target percent of weight loss. This shows a huge difference compared to the two others pattern. The angle of pattern helps to increase the dehydration rate by projecting the airflow to the specimen faster compare the two pattern which has no angle plus, the straight pattern has a 90° angle which act as resistance to the flow of air. For the Curved pattern the calculated result shows that all the specimens reach the 80% of weight loss after 10 h of drying. Specimen 4 reached the 80% of weight loss at the early of 6 h of drying while 50% of the specimens reached the target at 8 h of drying.

This result shows that high efficiency of airflow occurs when using this impeller-type pattern. The pattern's curve helps the flow of the hot air to swirl around the base and up to the specimen. By having the swirling air, the rate of dehydration become faster hence reduce the time taken to reach the target of 20% moisture content left.

5 Conclusions

In this paper, the designed pattern for the base of the food dehydrator for making an improvement on drying process were examine. From this experiment some error can detected because of the pattern of the blade is using the aluminum sheet and it just use plasticine to hold the aluminum sheet. This can effected the pattern of the airflow because of the pattern blade is not rigid. 3D printing suggested to fabricate the blade for a better airflow. However, it can be concluded that the objective of the experiment was achieved successfully within the time of drying. From the data collected, we can clearly the original base of Morgan Food Dehydrator cannot reach the target of 20% moisture content left due to the flat surface of the base.

For straight pattern base, we can see the result is almost the same as the original base. The 90° angle of the pattern increase the resistance and it disturb the flow of the hot air. Plus, a dead zone area is created when using this straight pattern which contributes more to the resistance of air flow. Fan-blade and curve pattern shows almost the same result by having only 0.5 percent of different of LOD. The fan-blade pattern has the upper hand due to the type of air flow created by the design. The fan-blade pattern creates an airflow that hit the specimen at angle of 45° while the curve pattern produces a swirl hot air flow that continuously flow parallel to the surface of the specimen. As observed, the fan-blade pattern helps the hot air to hit the specimen faster by projecting the air at 45° to the specimen, this helps the moisture in the specimen to evaporate faster than other patterns.

However, the swirling air that produce by the curve pattern also helps to increase the evaporation process. The pattern slightly less in difference because of the time taken for the hot air to hit the specimen is longer than the fan-blade pattern. In conclusion, the fan-blade pattern design gives the best result in achieving the highest target by having 83.70% of weight loss in drying the apples. This can be approved when the drying of some specimens was finished at the early of 6 h instead of 10 h of drying with the same temperature of 45 °C and all the specimens reached the target at 8 h compared to the straight pattern and the curve pattern.

Acknowledgements The authors would like to thank the Ministry of Higher Education for providing financial support under Fundamental Research Grant Scheme (FRGS) No. FRGS/1/2019/STG02/UMP/02/4 (University reference RDU1901184) and Universiti Malaysia Pahang for laboratory facilities as well as additional financial support under Internal Research grants PGRS210342, and PGRS190306.

References

1. Jayas DS (2016) Food dehydration. In: Reference module in food science. Elsevier. https://doi.org/10.1016/B978-0-08-100596-5.02913-9
2. Barbosa-Cánovas GV, Vega-Mercado H (1996) Dehydration of foods. Springer Science & Business Media, New York. https://doi.org/10.1007/978-1-4757-2456-1
3. Sulaiman SHB, Ramli HB, Fadzil I, Kbin K (2020) The food dehydrator
4. Hyun J-E, Lee S-Y (2020) Blue light-emitting diodes as eco-friendly non-thermal technology in food preservation. Trends Food Sci Technol 105:284–295. https://doi.org/10.1016/j.tifs.2020.09.008
5. Naczk M, Artyukhova AS (2020) Canning. In: Seafood: resources, nutritional composition, and preservation. CRC Press, pp 181–198
6. Wu T, Guo H, Lu Z et al (2020) Reliability of LipidSearch software identification and its application to assess the effect of dry salting on the long-chain free fatty acid profile of tilapia muscles. Food Res Int 138:109791. https://doi.org/10.1016/j.foodres.2020.109791

7. Ruiz-Alonso SA, Girón-Hernández LJ, López-Vargas JH, Muñoz-Ramírez AP, Simal-Gandara J (2021) Optimizing salting and smoking conditions for the production and preservation of smoked-flavoured tilapia fillets. LWT 138:110733. https://doi.org/10.1016/j.lwt.2020.110733
8. Ojha KS, Kerry JP, Tiwari BK, O'Donnell C (2016) CBT-RM in FS. Freezing for Food Preservation. Elsevier. https://doi.org/10.1016/B978-0-08-100596-5.03108-5
9. Huang D, Yang P, Tang X, Luo L, Sunden B (2021) Application of infrared radiation in the drying of food products. Trends Food Sci Technol 110:765–777. https://doi.org/10.1016/j.tifs.2021.02.039
10. Sozzi A, Zambon M, Mazza G, Salvatori D (2021) Fluidized bed drying of blackberry wastes: drying kinetics, particle characterization and nutritional value of the obtained granular solids. Powder Technol 385:37–49. https://doi.org/10.1016/j.powtec.2021.02.058
11. Huang D, Men K, Li D et al (2020) Application of ultrasound technology in the drying of food products. Ultrason Sonochem 63:104950. https://doi.org/10.1016/j.ultsonch.2019.104950
12. Musielak G, Mierzwa D, Kroehnke J (2016) Food drying enhancement by ultrasound – A review. Trends Food Sci Technol 56:126–141. https://doi.org/10.1016/j.tifs.2016.08.003
13. Monteiro RL, Gomide AI, Link JV, Carciofi BAM, Laurindo JB (2020) Microwave vacuum drying of foods with temperature control by power modulation. Innov Food Sci Emerg Technol 65:102473. https://doi.org/10.1016/j.ifset.2020.102473
14. Nowak D, Lewicki PP (2004) Infrared drying of apple slices. Innov Food Sci Emerg Technol 5(3):353–360. https://doi.org/10.1016/j.ifset.2004.03.003
15. Kuriakose R, Anandharamakrishnan C (2010) Computational fluid dynamics (CFD) applications in spray drying of food products. Trends Food Sci Technol 21(8):383–398. https://doi.org/10.1016/j.tifs.2010.04.009
16. Essalhi H, Benchrifa M, Tadili R, Bargach MN (2018) Experimental and theoretical analysis of drying grapes under an indirect solar dryer and in open sun. Innov Food Sci Emerg Technol 49:58–64. https://doi.org/10.1016/j.ifset.2018.08.002
17. Lingayat A, Chandramohan VP, Raju VRK, Kumar A (2020) Development of indirect type solar dryer and experiments for estimation of drying parameters of apple and watermelon. Therm Sci Eng Prog 16:100477. https://doi.org/10.1016/j.tsep.2020.100477
18. Cârlescu P-M, Arsenoaia V, Roşca R, Tenu I (2017) CFD simulation of heat and mass transfer during apricots drying. LWT Food Sci Technol 85:479–486. https://doi.org/10.1016/j.lwt.2017.03.015
19. Mugodo K, Workneh TS (2021) The kinetics of thin-layer drying and modelling for mango slices and the influence of differing hot-air drying methods on quality. Heliyon 7(6):e07182. https://doi.org/10.1016/j.heliyon.2021.e07182
20. Xia E, Tong W, Hou Y et al (2020) The reference genome of tea plant and resequencing of 81 diverse accessions provide insights into its genome evolution and adaptation. Mol Plant 13(7):1013–1026. https://doi.org/10.1016/j.molp.2020.04.010
21. Sabarez H. Advanced Drying Technologies of Relevance in the Food Industry. 2021.
22. Mousakhani-Ganjeh A, Amiri A, Nasrollahzadeh F et al (2021) Electro-based technologies in food drying - A comprehensive review. LWT 145:111315. https://doi.org/10.1016/j.lwt.2021.111315
23. Manual U (2017) Food dehydrator reviews 8(Im):3–6
24. Chaudhary A What is loss on drying? and determination of loss on drying : pharmaceutical guidelines
25. Gatenby A (2019) How Many Ways to Measure Moisture? Published 2019. https://www.cscscientific.com/csc-scientific-blog/ramblings-on-how-many-ways-to-measure-moisture
26. Ranowsky A. Karl Fischer vs. Loss-On Drying - Which Method is the Best? CSC Scientific Blog
27. Kahraman O, Malvandi A, Vargas L, Feng H (2021) Drying characteristics and quality attributes of apple slices dried by a non-thermal ultrasonic contact drying method. Ultrason Sonochem 73:105510. https://doi.org/10.1016/j.ultsonch.2021.105510

Comparing Industry Training Using Virtual Reality Against Conventional Training: A Case Study

Christian Stark, Salome Wiener, and Jeng Feng Chin

Abstract Virtual reality (VR) for industry training has recently received much attention. In VR, a simulation environment of the actual process and situation is provided. In this study, the effectiveness of the VR-based industry training for operators to acquire a process skill was examined. Selected operators from a medical manufacturing plant were divided into two groups and separately undertook the training of the same process either via VR or real-life demonstration (conventional training). Their related skill levels were measured and compared in terms of knowledge retention and knowledge interpretation. Results showed that no significant difference exists between the two groups. Theoretical knowledge should be enhanced in VR demonstration. The survey and observations also showed several advantages of VR over real-life demonstrations.

Keywords Virtual reality · Industry training · Knowledge retention · Knowledge interpretation

1 Introduction

Virtual reality (VR) is defined as a virtual recreation of the real world [1] through simulation environments for actual processes and situations [2]. The technology allows humans to build a virtual world using people's imagination and interaction and human senses of seeing, hearing, and touching to experience the situations [3]. VR is regarded by Digital Twin Consortium [4] as one of the important key technologies in industrial training.

The objective of the study is to examine the effectiveness of VR-based industry training for operators in acquiring a process skill. Selected employees (direct operators) from a medical manufacturing plant located in Malaysia were divided into two groups and separately undertook training of the same process either via VR or

C. Stark · S. Wiener · J. F. Chin (✉)
School of Mechanical Engineering, Universiti Sains Malaysia (USM), Engineering Campus, 14300 Nibong Tebal, Penang, Malaysia
e-mail: chinjengfeng@usm.my

© The Author(s), under exclusive license to Springer Nature Singapore Pte Ltd. 2023

M. H. A. Hassan et al. (eds.), *Technological Advancement in Instrumentation & Human Engineering*, Lecture Notes in Electrical Engineering 882,
https://doi.org/10.1007/978-981-19-1577-2_13

real-life demonstration. The related skill levels of the two groups were measured and compared.

This paper is organized as follows. The following section provides a brief review of VR usage in training. Section 3 explains the methodology in performing the experiment. Sections 4 and 5 present the results and the discussion, respectively. Finally, the conclusion is presented in Sect. 6.

2 VR in Training

The progress of industry 4.0 technologies makes the VR application commonly available in training for industrial and nonindustrial processes, including production, assembly plant, medicine, military, education, and tourism [5]. Mathur [6] proposed VR devices, such as Oculus Rift and Razer Hydra, for training and instruction purposes. Unit 3D is considered one of the most widely used VR software [2, 7, 8].

Andaluz et al. [7] presented the use of VR in the industrial training process to train new operators without exposing them to chemical, biological, ergonomic, and psychological risks that are common in the industry. The training is designed using graphic engine Unity 3D and Oculus Rift head-mounted display. Another chemical plant also uses VR training modules to conduct training by virtually exposing chemical operators to unsafe situations [9]. VR training also provides an innovative approach for safety training in a precast concrete company [10] to ensure that the safety training is conducted in a cost-effective and repeatable manner. Similar VR devices with haptic control interfaces and joysticks were also adopted by Lustosa et al. [11] to train workers to drive a forklift. Herrera et al. [12] detailed the use of Unity 3D graphic engine to develop a training system for electric welding applied to vehicle assembly industry. Muszynska et al. [13] discussed using an interactive VR system in the training for the servicing and maintenance of robots and robotic stations without damaging expensive equipment. Li et al. [14] proposed an interactive VR training for the hydroelectric generator maintenance. The generators can be broken down into interactive and standardized virtual elements. Abdul Aziz et al. [15] discussed the implementation of VR training for the maintenance of an onshore pipeline platform for oil and gas industry. The VR tool designed by Mirauda et al. [16] aimed to provide a virtual laboratory for the training of their technical and nontechnical workers monitoring the water discharge in the open channel flows. The system was developed using Unity 3D graphics engine with Oculus Rift head-mounted device. A VR system is developed by Lacko [17] to improve the capabilities of industrial workers and allow them to make correct decisions at times of high stress within the industry by training and evaluating these workers in terms of safety and health at work.

In the medical field, Mathur [6] used a low-cost VR system for daily medical training, such as identifying certain organs when making complex incisions. VR training is also adopted to simulate an anatomy system in medical student training

[18]. The system consists of surgical scene construction, result determination, and interface development. Rajeswaren et al. [19] proposed the design considerations of a VR simulation for the intubation training of beginners and experts training for complex and special cases. VR was also adopted in developing a portable, low-cost VR system to conduct classroom management skill courses [20]. The VR application was designed to provide users with situations where they encounter problems in public speaking and a platform where they can practice their speech and improve their linguistic skills [21].

VR provides users with interactive and innovative learning processes that are in contrast with the slightly appealing conventional training [22]. VR allows companies to conduct training in various scenarios in an effective and realistic environment before the workers handle the actual job [23]. It also allows workers to experience certain simulated situations without risks while learning how to react to the situation in a safe environment [24].

VR is proposed as a present and future alternative for the efficient training of human resources in the industrial field, thereby improving the capabilities of its users [1]. VR can be used in testing new configurations of production chains to shorten and speed up the start-up of new assembly lines and even entire factories working directly on their virtual models [25]. In this way, the correct operation of all production systems can be ensured before they actually begin, thereby avoiding production delays, planning errors, and the inefficient scheduling of automated systems.

Gavish et al. [26] and Roldán et al. [27] studied the effectiveness and efficiency of VR training in an electrical assembly plant and compared them with conventional training methods. Matsas and Vosniakos [28] developed a gamelike interactive VR training system to simulate the cooperation between industrial robotic manipulators and operators. Vergnano et al. [29] reported a successful industrial case study of a simulation-based training workbench in a steel plant.

3 Methodology

3.1 Bending Process

The participants of this study were trained for the bending process of surgical instruments. This process is required for surgical products with curved ends. First, the bending tool must be set up according to the article number to achieve eventually the correct bend during the actual bending process. After the setup, a corresponding straight product is inserted into the bending tool (see Fig. 1). The tip of the product is manually bent using two levers according to the drawing. This step is followed by the measurement of relevant masses and checking of the shape before proceeding to the next instrument.

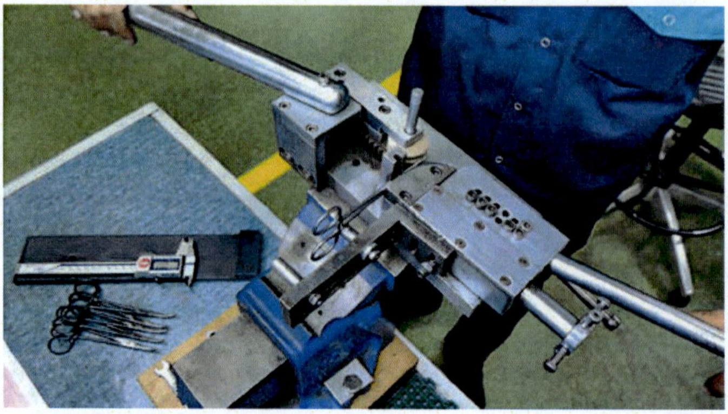

Fig. 1 Bending tool with straight product

3.2 VR Setup

For the VR training, a mobile VR device from Facebook called Oculus Quest was used (Fig. 2). The device works autonomously without any separate computer. Two controllers were used to navigate in the VR. The head and body movements were tracked for recording and transmitted into the VR with six degrees of freedom. The organic light-emitting diode (OLED) display has a resolution of 1440 × 1600 and a refresh rate of 72 Hz.

A movie that includes all of the relevant movements from the trainer's point of view was created to set up the VR glasses. Descriptions highlighting the critical parameters of the process and warning information were also added. The video was edited using a standard video editing software (Camtasia Studio 8), and the content was uploaded on the VR glasses.

Fig. 2 Oculus Quest

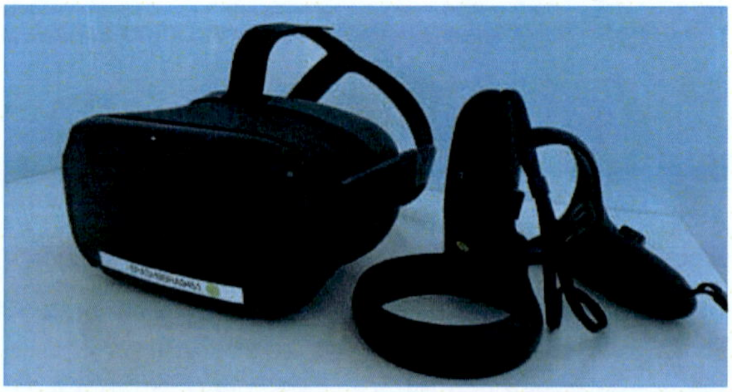

3.3 Participants

A total of 40 volunteers (all female) with an average age of 24 years (standard deviation [SD] = 5.5) participated in the case study. All employees participated voluntarily, and they were recruited from the company. The employees were randomly selected. The only selection criterion indicated that the employees were either newly hired and therefore had no experience in the production of surgical instruments or, in the case of previously hired employees (maximum duration of being employed: 12 months), had not been trained yet in the bending process. Out of the 40 volunteers, six employees were newcomers.

3.4 Design/Procedure

The staff attended a training session on the bending process in the production of surgical instruments. As shown in Fig. 3, half of the participants underwent the conventional training, during which the trainer explained the process to five participants at the same time and demonstrated it on the bending machine. The participants could ask questions during the training and ask the trainer to explain or demonstrate the process again. The training was finished after the participants ensured that they had no further questions and that they understood the process. The training was conducted for 2 consecutive days, once in the morning and once in the afternoon, with five participants in each group. The required time for the training was measured with a stopwatch. The second half of the participants underwent a VR training for the

Fig. 3 Conventional training

same process. This training was also conducted for 2 consecutive days, once in the morning and once in the afternoon, with five persons at a time. First, a short introduction about the handling of the VR glasses was given. Then, the employees (Fig. 4) watched the video of the bending process (Fig. 5) in the virtual room. This video could be repeated within 30 min (up to seven repetitions) until the participants were convinced that they had internalized the process. The time that the trainer needed to explain the VR glasses and the time for watching the videos were measured separately. All trainees spent the full 30 min watching the videos. The details provided during the conventional training and the VR training were comparable, and both followed the method in accordance with Training Within Industry.

The participants were randomly assigned to one of the two training setups. After receiving the respective training, the trainees were asked to complete two tests and one evaluation form to compare the conventional training and the VR training. The two tests were based on the user study by Gonzalez-Franco et al. [30]. First, a so-called knowledge retention test was performed, in which the participants had to complete

Fig. 4 VR training

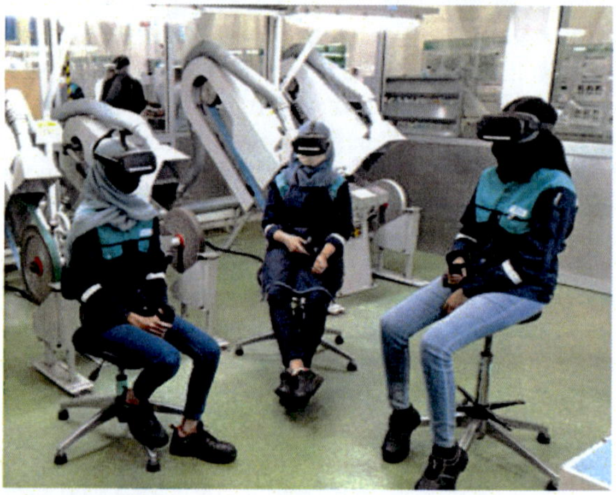

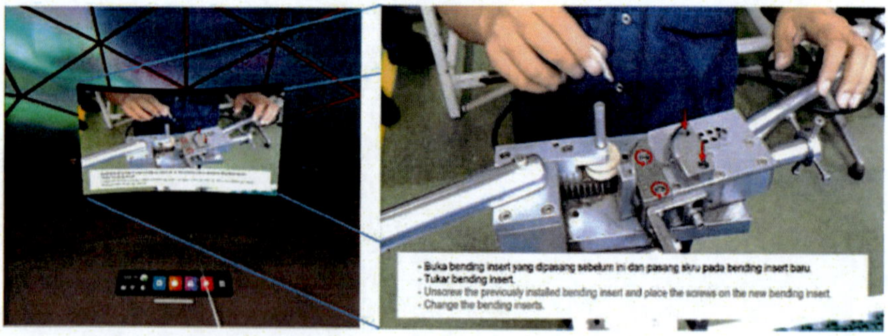

Fig. 5 Training video in VR

seven multiple-choice questions on the theoretical understanding of the process. The questionnaire is presented in Appendix 1. In the knowledge interpretation test (details presented in Appendix 2), the participants were tested on how well they performed in the practical execution of the process, i.e., how well they understood the procedure. The participants were asked to perform the process step by step until the end. If a step was forgotten or done incorrectly or if the participant had to ask questions before continuing, one point was deducted. In total, 24 points (one point per process step) could be achieved.

In the questionnaire, the training itself (Appendix 3) was evaluated with two questions ("How much would you like to participate in another training like this?" and "How do you evaluate the overall training on a scale from 1 to 6?"). Intrinsic motivation and self-estimated competence were also asked. For the last two questions, a shortened scale according to Wilde et al. [31] was developed. The individual questions were recorded using a five-point Likert scale (0 = *not true at all*; 1 = *a little true*; 2 = *partly true*; 3 = *very true*; 4 = *completely true*).

4 Results

4.1 Knowledge Capture

No statistically significant differences were found for knowledge retention between the conventional training and the VR training ($p = 0.408$). The average achieved scores for the VR training and for the conventional training were $M = 4.35$ ($SD = 1.42$) and $M = 4.00$ ($SD = 1.21$), respectively. Both training methods did not seem to provide sufficient theoretical knowledge, as the scores for both methods were not very high and far below the maximum of 7 (see Fig. 6).

No statistically significant differences were found for knowledge interpretation between the conventional training and the VR training ($p = 0.826$). The average achieved scores for the VR training and for the conventional training were $M = 20.60$ ($SD = 2.48$) and $M = 20.75$ ($SD = 1.74$), respectively. With regard to the practical knowledge of the employees after the training, both training methods can be considered successful based on the high scores (see Fig. 7).

For both training methods, no relation was found between both kinds of knowledge capture (knowledge retention and knowledge interpretation). This finding indicated that no correlation existed between the top performers for the knowledge retention test and the top performers of the knowledge interpretation test. Although a small positive correlation was determined, it was not statistically significant (VR training: $r = 0.250$ [$p = 0.287$]; conventional training: $r = 0.199$ [$p = 0.400$]).

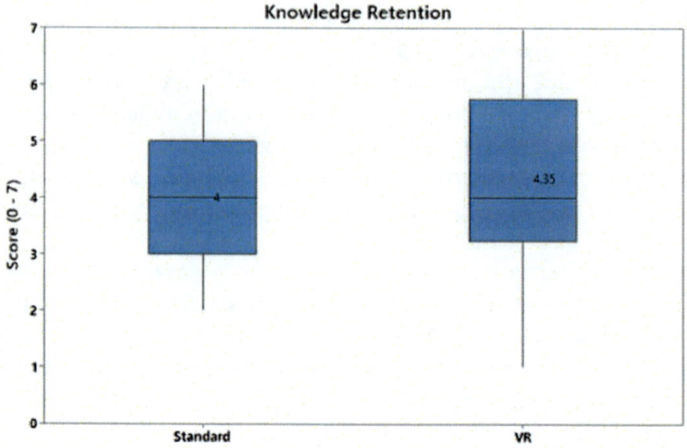

Fig. 6 Scores of knowledge retention test for conventional training and VR training

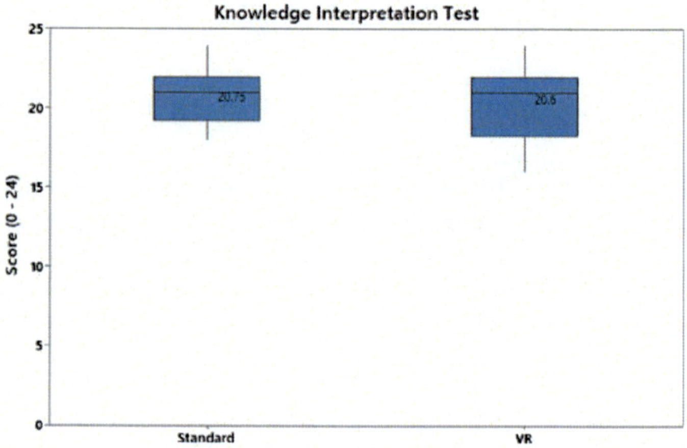

Fig. 7 Scores of knowledge interpretation test for conventional training and VR training

4.2 Time

The time needed for the trainer to complete the training (e.g., explain the content for the conventional training and explain the usage of the VR glasses for the VR training) for the conventional training was significantly higher ($M = 17.40$, $SD = 5.28$) than that for the VR training ($M = 12.00$, $SD = 5.17$, $p = 0.002$; see Table 1).

Table 1 Training duration for standard and VR training

	Standard training (min)	VR training (min)
Training duration	17.40	12.00
Self-study duration for participants	0	30.00

4.3 Evaluation

Cronbach's alpha was calculated to prove the reliability of the subscales for intrinsic motivation and for self-estimated competence. Both subscales revealed sufficient reliability with $\alpha = 0.7568$ for intrinsic motivation and $\alpha = 0.7947$ for self-estimated competence. The intrinsic motivation for the conventional training group was slightly higher (M = 3.77, SD = 0.35) than that for the VR training group (M = 3.60, SD = 0.59). However, these differences had no statistical significance ($p = 0.465$). The same pattern could be observed for the self-estimated competence with an average of M = 3.23 (SD = 0.60) for the conventional training and an average of M = 2.78 (SD = 0.79) for the VR training. No statistical significance was also found ($p = 0.083$).

The answer to the question whether the participants would repeat the training was very similar for both training methods (conventional training: M = 4.5, SD = 0.5; VR training: M = 4.4, SD = 1.0) and showed no statistical significance ($p = 1.00$). The results showed that participants in both groups would be likely to repeat the respective training. The overall rating (from 1 = *very good* to 6 = *not satisfying*) of the VR training was slightly better (M = 2.55, SD = 2.06) than that of the conventional training (M = 2.90, SD = 2.02). However, no statistical significance could be determined ($p = 0.591$).

For the conventional training, no statistically significant correlation could be found between voluntary repetition of the training and intrinsic motivation or self-estimated competence nor between the overall rating and intrinsic motivation or self-estimated competence ($p > 0.05$). The same applies to the correlation between the knowledge retention or knowledge interpretation and intrinsic motivation and self-estimated competence. No statistically significant correlation could be found ($p > 0.05$; see Table 2).

Table 2 Correlations for standard training (*p < 0.05)

Standard training	Intrinsic motivation	Self-estimated competence
Repetition	0.138	0.363
Overall rating	−0.369	−0.384
Knowledge retention	0.176	−0.088
Knowledge interpretation	0.249	0.005

Table 3 Correlations for VR training (*p < 0.05)

VR training	Intrinsic motivation	Self-estimated competence
Repetition	0.804*	0.319
Overall rating	−0.131	0.315
Knowledge retention	0.119	0.224
Knowledge interpretation	0.456*	0.218

However, the VR training showed a statistically significant positive correlation between the voluntary repetition of the training and the intrinsic motivation ($p<0.05$). A significant positive correlation also existed between the knowledge interpretation test and the intrinsic motivation ($p < 0.05$). For the other correlations, no statistical significance could be found ($p > 0.05$; see Table 3).

In Tables 2 and 3, a negative correlation coefficient in the case of the overall rating would indicate a high motivation/self-estimated competence.

5 Discussion

According to the study, no significant differences were found between the VR training and the conventional training. Both methods were comparable as they achieved relatively high scores for knowledge interpretation test and low scores for knowledge retention test. The overall time for the VR training was longer than that for the conventional training because the trainees in the VR training watched the videos for up to 30 min. However, the time for the trainer in the VR training was shorter than that in the conventional training, and the trainees in the former can conduct the training by themselves unsupervised after the first introduction. The results for the VR training showed a significant positive correlation between intrinsic motivation and a voluntary repetition of the training and between intrinsic motivation and knowledge interpretation test. Therefore, the motivation of the trainees in the VR training was higher than that of the trainees in the conventional training when they scored high in the practical test.

In general, practical performance is more important for operational tasks than theoretical knowledge. Nevertheless, the observation reveals mistakes by trainees in measuring the articles. Theoretical knowledge should be enforced in future training to understand the process comprehensively. The employees must be familiarized with the technical terminologies and understand the questions fully to have a proper theoretical test. Their level of comprehension dictates the design of the questions.

According to the observations and interviews with the trainees and trainers, the VR training has several advantages. First, the preparation of the video content takes several hours. Trainers must also familiarize themselves with the content and emphasize the critical elements in the video, such as safety, quality, and efficiency of the process. Therefore, the training content carries much detail and clarity in the form of specifically captured video. Second, the video was edited and replayed multiple times, allowing opportunities to improve the video progressively. Video editing and the processing software are user-friendly and require minimum training and experience. Third, the preparation time and cost can be justified and offset when the video content is reused in the future training. Fourth, VR training is highly duplicative and reproducible. The training quality becomes consistent when VR training video is used as the conventional training material.

Fifth, VR training permits training individualization. For example, the trainees were observed adjusting the video running speed during the training or replaying the videos multiple times until they were confident about the process. Sixth, compared with conventional training requiring physical parts, VR training serves as a cost-saving option. Seventh, VR can be used in safety-related training, such as evacuation training. Eighth, VR allows training off-site, an element that has become increasingly important during the recent pandemic, to provide process knowledge as realistic as possible.

Nevertheless, online training sometimes affects the stability of the processes and potential loss of productivity, efficiency, quality issues, and safety [32, 33]. It is important because the training aims to ensure consistency and accuracy of the process retained. On the basis of the conducted study and its results, a combination of VR training and conventional training is recommended to provide the highest possible training efficiency.

6 Conclusion

The study presented a comparison of VR training and conventional training in an actual manufacturing setting. The effectiveness of both training methods was measured based on knowledge retention test and knowledge interpretation test. The results showed that both training methods are comparable with no significant difference. The observation also outlined the advantages of a VR training. In the future, VR training will be able to cover other training contents. In particular, identifying individualized training needs and providing guidance throughout a process training will be much interactive using artificial intelligence (AI) in combination with VR. Furthermore, extended reality in combination with AI will open up opportunities to provide guidance for an efficient, standardized, qualitative effective, and individualized training that can even cover the personal aspect of training by using holograms instead of actual trainers.

Appendix 1: Questionnaire for Knowledge Retention Test

1. Where can you find the article number?
   (a) On the instrument
   (b) On the drawing
   (c) Ask your supervisor
2. Which side of the instrument needs to be on top when you insert it into the bending tool?
   (a) The leaf part
   (b) The box part
   (c) It doesn't matter, each side can be on top
3. How does the instrument need to be aligned on the bending tool?
   (a) The slant edge of the jaw is in line with the marking on the bending insert
   (b) The tip of the instrument is in line with the marking on the bending insert
   (c) The edge of the ring is in line with the marking on the bending insert
4. Where does the stopper need to be positioned?
   (a) At the middle of the instrument
   (b) At the tip of the instrument
   (c) At the edge of the ring
5. Which values need to be measured after bending the first instrument?
   (a) Bending height and total length
   (b) Jaw height and total length
   (c) Bending length and bending height
6. After bending the first instrument, you detect that one of the measurements is not correct. What do you need to do?
   (a) Report to your supervisor
   (b) Proceed with bending the other instruments
   (c) Adjust the position of the stopper
7. After bending some instruments, you recognize form issues. What do you need to do?
   (a) Proceed with bending the other instruments
   (b) Report to your supervisor
   (c) Adjust the position of the stopper

Appendix 2: Evaluation of the Training

1. How much would you like to participate in another training like this?
   1 = not at all; 2 = hardly; 3 = No comment; 4 = Likely; 5 = Extremely like
2. How do you evaluate the overall training on a scale from 1 to 6
   1 = very good; 6 = not satisfying
3. Please rate the following statements on a scale from 0 to 4 according to your opinion:
   0 = not true at all; 1 = a little true; 2 = partly true;3 = likely true; 4 = completely true

    1. I enjoyed the training
    2. I found the training interesting
    3. The training was entertaining
    4. I am satisfied with my performance in training
    5. During training I am clever
    6. I think I was pretty good at the training

Appendix 3: Knowledge Interpretation Test

Setup

1. Check article number on the drawing
2. Select the correct bending insert according to the article number
3. Unscrew the previously installed bending insert
4. Place the screws on the new bending insert
5. Change the bending inserts
6. Screw tighten the new bending insert
7. Place one instrument on the bending tool (leaf part is on top and slant edge of the jaw is in line with the marking on the bending insert)
8. Tighten the bending insert by turning the spindle
9. Loosen the nut of the stopper with a spanner
10. Move the stopper until the edge of the ring
11. Tighten the stopper
12. Move the roller up until the jaw is in line with the groove
13. Fix the roller by turning the rod
14. Bend the first instrument
15. Measure the bending height regarding to the drawing
16. Calibrate the height gauge (move the caliper down and calibrate it by pressing the "Zero" button)
17. Measure the total length of the instrument regarding to the drawing (place the instrument below the caliper and move it down to measure)

18. If measurements are correct: proceed; if measurements are not correct: change position of stopper.

Bending Process

19. Place the instrument on the bending tool (edge of the ring is placed precisely on the stopper, jaw is laying on the pin)
20. Hold the instrument and tighten the bending insert
21. Hold the bending tool on both handles. Pull the right handle and push the left handle at the same time until the instrument is in a straight position
22. Pull both handles at the same time towards yourself to bend the end of the jaw
23. Remove the instrument from the bending insert
24. Check the instruments for misalignment of the form or jaw displacement (if there is an issue, report to supervisor).

References

1. Naranjo JE, Sanchez DG, Robalino-Lopez A, Robalino-Lopez P, Alarcon-Ortiz A, Garcia MV (2020) A scoping review on virtual reality-based industrial training. Appl Sci 10(22):1–31
2. Dixken M, Diers D, Wingert B, Hatzipanayioti A, Mohler BJ, Riedel O, Bues M (2019) Distributed, collaborative virtual reality application for product development with simple avatar calibration method. In: Proceedings of the 2019 IEEE conference on virtual reality and 3d user interfaces (VR), Osaka, Japan, 23–27 March, pp 1299–1300
3. Han TW (2017) A study on the development direction of virtual reality games. Treatise Plast 20:287–294
4. Bécue A, Maia E, Feeken L, Borchers P, Praça I (2020) A new concept of digital twin supporting optimization and resilience of factories of the future. Appl Sci 10:1–32
5. Cýrus J, Krčmařík D, Petrů M, Kočí J (2020) Cooperation of virtual reality and real objects with HoloLens. Adv Comput Vis 94–106
6. Mathur AS (2015) Low-cost virtual reality for medical training. In: Proceedings of the 2015 IEEE virtual reality conference, VR 2015 proceedings, Arles, France, 23–27 March, pp 345–346
7. Andaluz VH, Pazmiño A, Perez J, Carvajal C (2017) Training of tannery processes through virtual reality. In: Lecture notes in computer science (Including Subseries Lecture notes in artificial intelligence and Lecture notes in bioinformatics), vol 2. Springer International Publishing, Cham, Switzerland, pp 75–93
8. Caiza G, Garcia CA, Naranjo JE, Garcia MV (2020) Flexible robotic teleoperation architecture for intelligent oil fields. Heliyon 6:1–13
9. Garcia Fracaro S, Chan P, Gallangher T, Tehreem Y, Toyoda R, Bernaerts K, Glassey J, Pfeiffer T, Slot B, Wachsmuth S, Wilk M (2021) Towards design guidelines for virtual reality training for the chemical industry. Educ Chem Eng 36c:12–23
10. Joshi S, Hamilto M, Warren R, Faucett D, Tian WM, Wang Y, Ma JF (2021) Implementing Virtual Reality technology for safety training in the precast/prestressed concrete industry. Appl Ergon 90:3–11
11. Lustosa EBS, De MacEdo DV, Rodrigues MAF (2018) Virtual simulator for forklift training. In: Proceedings of the 2018 20th symposium on virtual and augmented reality, SVR 2018, Foz do Iguacu, Brazil, 28–30, pp 18–26

12. Herrera DF, Bolívar Acosta S, Quevedo WX, Balseca JA, Andaluz VH (2018) Training for bus bodywork in virtual reality environments. In: Lecture notes in computer science (Including Subseries Lecture notes in artificial intelligence and Lecture notes in bioinformatics), vol 10850. Springer International Publishing, Cham, Switzerland, pp 67–85
13. Muszynska M, Szybicki D, Gierlak P, Kurc K, Burghardt A, Uliasz M (2019) Application of virtual reality in the training of operators and servicing of robotic stations. IFIP advances in information and communication technology, vol 568. Springer, International Publishing, Cham, Switzerland, pp 594–603
14. Li B, Bi Y, He Q, Ren J, Li Z (2018) A low-complexity method for authoring an interactive virtual maintenance training system of hydroelectric generating equipment. Comput Ind 100:159–172
15. Abdul Aziz F, Alsaeed ASMA, Sulaiman S, Ariffin MKAM, Al-Arhabi ARY (2020) Virtual reality training platform in onshore pipeline. In: Advances in material sciences and engineering. Lecture notes in mechanical engineering. Springer, Singapore, pp 207–216
16. Mirauda D, Capece N, Erra U (2020) Sustainable water management: virtual reality training for open-channel flow monitoring. Sustainability 12:757
17. Lacko J (2020) Health safety training for industry in virtual reality. In Proceedings of the 30th international conference on cybernetics and informatics, K and I 2020, Velke Karlovice, Czech Republic, pp 1–5
18. Wang X, Wang X (2018) Virtual reality training system for surgical anatomy. In: AIVR 2018: proceedings of the 2018 international conference on artificial intelligence and virtual reality, pp 30–34
19. Rajeswaran P, Kesavadas T, Jani P, Kumar P (2019) AirwayVR: virtual reality trainer for endotracheal intubation-design considerations and challenges. In: Proceedings of the 26th IEEE conference on virtual reality and 3D user interfaces, Osaka, Japan, 23–27 March, pp 1130–1131
20. Latoschik ME, Lugrin JL, Habel M, Roth D, Seufert C, Grafe S (2016) Breaking bad behavior: immersive training of class room management. In: Proceedings of the ACM symposium on virtual reality software and technology, VRST, New York, NY, USA, 2 April, pp 317–318
21. Pellett K, Zaidi SFM (2019) A framework for virtual reality training to improve public speaking. In: VRST'19: 25th ACM symposium on virtual reality software and technology, pp 1–2
22. Naranjo JE, Ayala P, Altamirano S, Brito G, Garcia MV (2018) Intelligent oil field approach using virtual reality and mobile anthropomorphic robots. In: Lecture notes in computer science (Including Subseries Lecture notes in artificial intelligence and Lecture notes in bioinformatics), vol 10851. Springer International Publishing, Cham, Switzerland, pp 467–478
23. Karagiannis P, Theodoros T, Michalos G, Makris S (2021) Operators training using simulation and VR technology. Procedia CIRP 96:290–294
24. Zhang G, Hansen JP, Minakata K (2019) Hand- and gaze-control of telepresence robots. In: ETRA'19: Proceedings of the 11th ACM symposium on eye tracking research and applications, Denver, CO, USA, pp 1–8
25. Oliveira DM, Cao SC, Hermida XF, Rodriguez FM (2007) Virtual reality system for industrial training. In: 2007 IEEE international symposium on industrial electronics, Vigo Spain, pp 1715–1720
26. Gavish N, Gutiérrez T, Webel S, Rodríguez J, Peveri M, Bockholt U, Tecchia F (2015) Evaluating virtual reality and augmented reality training for industrial maintenance and assembly tasks. Interact Learn Environ 23(6):778–798
27. Roldán JJ, Crespo E, Martín-Barrio A, Peña-Tapia E, Barrientos A (2019) A training system for Industry 4.0 operators in complex assemblies based on virtual reality and process mining. Robot Comput Integr Manuf 59:305–316
28. Matsas E, Vosniakos GC (2017) Design of a virtual reality training system for human–robot collaboration in manufacturing tasks. Int J Interact Des Manuf 11:139–153
29. Vergnano A, Berselli G, Pellicciari M (2017) Interactive simulation-based-training tools for manufacturing systems operators: an industrial case study. Int J Interact Des Manuf 11:785–797

30. Gonzalez-Franco M, Lanier J (2017) Model of illusions and virtual reality. Front Psychol 8(1125):1–8
31. Wilde M, Baetz K, Kovaleva A, Urhahne D (2019) Testing a short scale of intrinsic motivation
32. Segura Á, Diez HV, Barandiaran I, Arbelaiz A, Álvarez H, Simões B, Posada J, García-Alonso A, Ugarte R (2020) Visual computing technologies to support the Operator 4.0. Comput Indus Eng 139
33. Romero D, Stahre J, Taisch M (2020) The Operator 4.0: towards socially sustainable factories of the future. Comput Indus Eng 139

IoT Enabled Piezoelectric Energy Harvesting Floormat

Chong Lye Lim, Hashwinni Rajaretnam, Sarina Tajudin, and Mohammed W. Muhieldeen

Abstract Energy harvesting through clean energy sources has become a demand nowadays. Internet of Things (IoT) is recommended in an automation system that provides high efficiency and immediate data response. The present study presents a hardware design and software platform used for an IoT enabled piezoelectric energy harvesting floormat. The harvested energy from 16 diaphragm type piezoelectric transducers was stored in a supercapacitor. As for the IoT, the HC-05 Bluetooth module is employed. The prototype has been designed, such that when the user presses the push button, which acts as the switch of the doorbell, a buzzer will sound, and a customized LED lighting pattern will be displayed. With the IoT enabled, the output will be displayed on the PC and cell phone. From the design analysis, the average weight was calculated to be 75 kg while the average Voltage (DC) was 18.5 V and the average current was at 8.5 μA. On a bigger scale, it is projected to be able to generate 42.5 μA per m^2. A typical Built-To-Order (BTO) 4-rooms flat in Singapore which is about 92 m^2, will be able to generate 3.9 mA with 20 steps of a person that weighs 75 kg. Analytical prediction of 29.4 thousand steps required to charge a 250 mAh battery. Considering the average minimum steps produced by an individual is 3.5 thousand steps, it takes around 4.5 to 8.5 days to charge the battery fully.

Supplementary Information The online version contains supplementary material available at https://doi.org/10.1007/978-981-19-1577-2_14.

C. L. Lim (✉) · H. Rajaretnam · S. Tajudin
School of Engineering and Technology, PSB Academy, Singapore 319579, Singapore
e-mail: chonglye.lim@psb-academy.edu.sg

H. Rajaretnam
CU Scarborough, Coventry University, Scarborough YO11 2JW, England

M. W. Muhieldeen (✉)
Mechanical Engineering Department, Faculty of Engineering, Technology and Built Environment, UCSI University, 56000 Kuala Lumpur, Cheras, Malaysia
e-mail: al-gailani@ucsiuniversity.edu.my

UCSI-Cheras Low Carbon Innovation Hub Research Consortium, 56000 Kuala Lumpur, Cheras, Malaysia

© The Author(s), under exclusive license to Springer Nature Singapore Pte Ltd. 2023

M. H. A. Hassan et al. (eds.), *Technological Advancement in Instrumentation & Human Engineering*, Lecture Notes in Electrical Engineering 882,
https://doi.org/10.1007/978-981-19-1577-2_14

Keywords Piezoelectric · Renewable energy · IoT · Green energy

1 Introduction

The world has evolved over 3 revolutions in the past centuries and is in a transition towards the fourth industrial revolution (Industry 4.0) today. Industry 4.0 encourages the use of automation which in turn emphasizes the use of cyber-physical systems. With the involvement of the Internet of Things (IoT), systems can be monitored using digital sensors, and information can be transmitted and received in real-time in addition to storing data over cloud servers and networks [1]. With such accelerated advancements in technology, the use of electricity is inevitable in modern society [2]. Hence, a huge emphasis has been placed on alternate methods of energy harvesting. Energy harvesting through renewable resources refers to harvesting energy that is clean and constantly replenished through natural resources. The most commonly known energy harvesting sources are wind energy, tidal energy, geothermal energy, and solar energy [3]. However, energy harvesting can also be implemented small scale by converting mechanical energy in the surroundings to usable energy [4]. Energy can be harvested from the environment by capturing small-scaled wasted energy and using it to supply power to systems that require low power, ranging from μA – mA [5]. Wasted energy is the energy that has not been fully utilized or lost to the ambient after its primary source. Wasted energy exist in forms such as heat, force, light, and vibrations [6]. One such way of harvesting force as wasted energy is by utilizing human movement activities as the source of input [7]. The average number of steps per day ranges from about 3.5 thousand to 6.9 thousand for an individual across the world [8]. For every step taken, useful energy is dissipated for movement in a desired direction. However, the force applied to enable this movement is gone to waste. Such wasted energy is targeted to be harvested by using a piezoelectric energy harvesting floormat.

Piezoelectric crystals are made of atoms that are positively charged and negatively charged. By compressing the quartz crystals in the transducer, the centre of charge between the positively charged atoms and the negatively charged atoms changes in an opposing direction. As a result, when a force is applied, at the faces of the crystals, it is observed that on one end it is positively charged while the is negatively charged. By wiring both ends, there will be a transfer of electrons thus, electricity is generated [9]. The piezo pavement was installed on an actual walkway and used to power walkway signal indicators. The data collected on the amount of electrical energy is monitored by a mobile application and website information by the means of wireless communication. The maximum output power of this harvester (with a resistance of 10 kΩ) was measured as 148.3 mW, as a person weighing 100 kg walked over the pavement [10]. The idea of piezo pavements had been developed by Pavegen and installed in a few countries such as Hong Kong, London, United Arab Emirates [11]. Pavegen harnesses the power of every footstep to generate instant clean electricity for applications such as ambient lighting or data screens and display monitors. The piezoelectric energy harvester shown by Jeong et al. (2019) are designed to power LED strips that are fitted at the base of the shoes [12]. The shoes consist of a piezoelectric device (PZT ceramic), which is placed at the insoles of the shoes. With every step taken, the force input by the motion of the user is converted into electrical energy. The highest output of the piezoelectric energy harvester is measured to be 800 μW with a resistance of 400 kΩ. It is able to control the LED switching circuit thereby blinking upon the user's movement. The rectified voltage can be connected to the supercapacitor to store energy, or a load can be added at the DC out to power other components for the piezoelectric energy harvesting system [13].

IoT is a digitalized network communication system that enables real-time interactions between humans and machines. This is made possible through wireless communications between networks and sensors [14]. IoT network communication encompasses a wide spectrum of communication platforms. IoT devices include cell phones, routers, smart wearables such as smartwatches, and also Global Positioning System (GPS) that is installed in vehicle navigation systems. The communication technologies involve Bluetooth, LoRa, ZigBee, Wi-Fi, LTE, and 5G network. It can be further developed into applications and network services for various functions that include monitoring and data analysis as well [15].

Bluetooth is widely used for smart home automation purposes. It is highly favourable due to contributing factors such as low power consumption, affordable cost, and compact within devices [16]. A smart switching motor control by Karthick et al. (2021) was introduced in an industrial setting where typically a distance of 500 m is travelled to turn on/off motors [17]. The HC-12 Bluetooth module was utilized to transmit data and thereby enabling control of the ON/OFF functions of motors. This is done by the user providing the input to either ON or OFF the motor. This information is transmitted to the receiver over a 500 m range during which the LED will illuminate. The PIC microcontroller will then regulate the motor according to the user's input. A home automated control system with IoT using Bluetooth had been designed by Amoran et al. (2021) [18]. The HC-06 Bluetooth module was programmed with an Arduino microcontroller to enable the user to control the ON/OFF functions of home-based electrical appliances. It supports the Google assistant voice-prompt feature that is incorporated within the android operating system thereby functions well with smartphones too. It has an additional 12-h timer function which can be set by the user via smartphone. The working principle is such that the user connects to the Bluetooth module and sets the preferred setting which will be transmitted to the appliances that are connected to the Bluetooth. The appliances then receive the data and execute the command accordingly.

In this study, it is identified that the prototype for energy harvesting can be done by using piezoelectric transducers with PZT ceramics such as the diaphragm type or cantilever type. The circuitry for energy harvesting can be done by connecting an array of piezoelectric transducers input to the bridge rectifier and the output can be stored in a supercapacitor. IoT can be incorporated via Bluetooth module. Human movement can be perceived in various forms such as walking, jogging, jumping, or many more. However, the main focus in this study is human movement through footsteps-related activities, specifically walking on a floormat.

2 Methodology

The methodology flowchart for the present study shown in Fig. 1. The present study begins with the identification of the required list of components from the literature study, followed by the design brainstorming and design selection. Then the hardware and software design for an IoT enabled piezoelectric energy harvesting floormat will be discussed.

2.1 List of Components

In this study, there are several components have been to build the prototype and display the harvesting energy. All the components have been set accordingly to collect the proper and accurate data. These components are functioning by sending the signal to the PC or the phone to indicate the number of the footsteps, temperature, as well as the buzzer sound to notify the owner of the unit if there is any visitor has been pressed the doorbell. All these components have set by LPC 1768 microcontroller. The list of components used for the prototype is listed in Table 1.

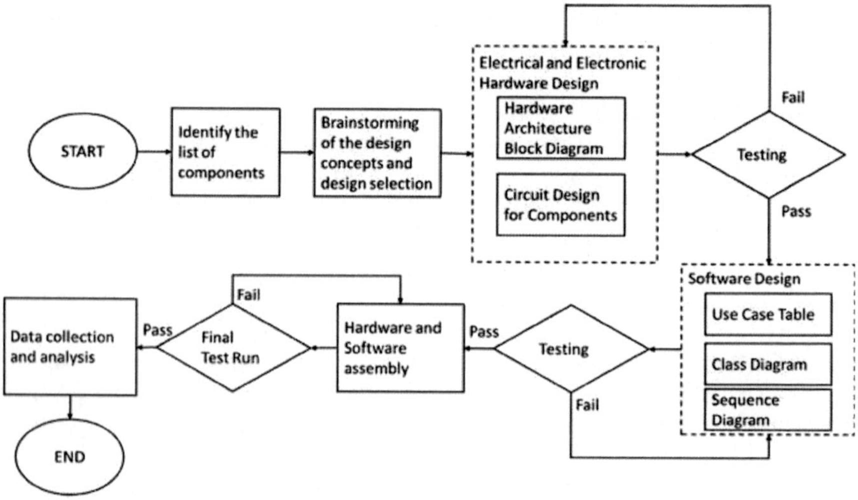

Fig. 1 Methodology flowchart

Table 1 List of components

Component	Function
LPC 1768	Microcontroller (MCU) used to support the monitoring system and execute command for the output functions
Piezoelectric Transducers	To harvest energy from user 1 footsteps
Buzzer	For audio alert to notify the owner (user 2) that a visitor (user 1) has pressed the doorbell
Temperature Sensor	This is an additional feature that measures surrounding temperature and provides data to user 2
Push Button	For the start-up of the device and the doorbell switch function
Bluetooth module	Sends the data to user 2 mobile phone and PC

2.2 *Selected Design Concept*

The selected design concept is the piezoelectric energy harvesting floormat. The floormat concept allows a low cost and small-scale prototype to be developed. For better visualization of the expected outcome of the design concept, a "SketchUp" model was done. SketchUp is a 3D design software that allows users to perform 3D modelling online [26]. It has templates that are ready for use and can be put together to bring ideas to a realistic modelling concept as shown in Fig. 2. A prototype, 20 by 20 cm, was built based on the finalized idea as shown in Fig. 3.

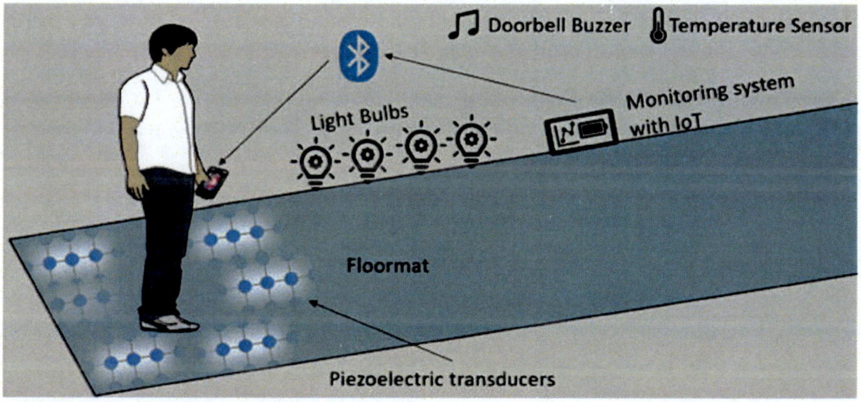

Fig. 2 Design visualization

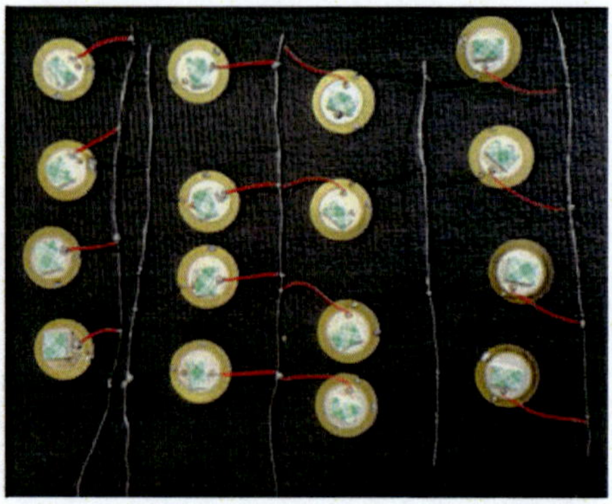

Fig. 3 Prototype of piezoelectric floormat

2.3 *Electrical and Electronic Hardware Design*

Hardware Architecture Block Diagram
The hardware architecture block diagram maps the connections between the electronic components and the MCU as shown in Fig. 4.

Circuit Design for Components
The National Instruments MultiSim 14.2 software was used to design the complete circuitry of the prototype. In the energy harvesting circuit, the voltage generated from the piezoelectric transducer is in AC. Therefore, it needs to go through a bridge rectifier to convert the AC voltage to a DC voltage. 4 diodes are connected in a

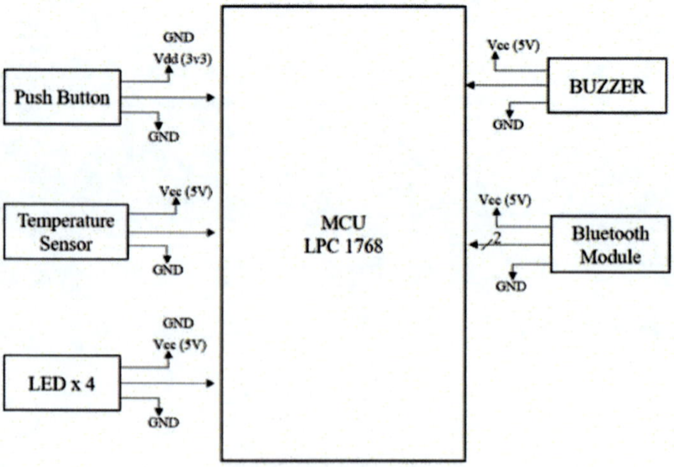

Fig. 4 Electronic hardware architecture block diagram

full wave rectifier configuration. DC voltage across the rectifier output terminals is connected in parallel to the supercapacitor to be charged.

There are 4 LEDs used in this prototype and connected to the MCU board. The HC-05 Bluetooth module communicates via the transmit and receive pins to exchange information with the MCU. The HC-05 module is powered up by 5 V Vcc. Since it is a bi-directional communication, the transmit pin of the HC-05 should be connected to receive pin of the MCU, vice versa. Since the push button is designed to execute an interrupt event in this prototype, it is connected to Pin 13 of the MCU. As the buzzer draws a high current therefore a resistor should be placed before connecting it to the MCU to protect the component.

In this prototype, the LMT 84 is connected directly to the MCU and read as an analog signal. This analog signal is manipulated in the program code and output a voltage within the allowable limit of the MCU. The schematics of the individual components of the prototype were all combined together to form the Project's Schematic Capture as shown in Fig. 5.

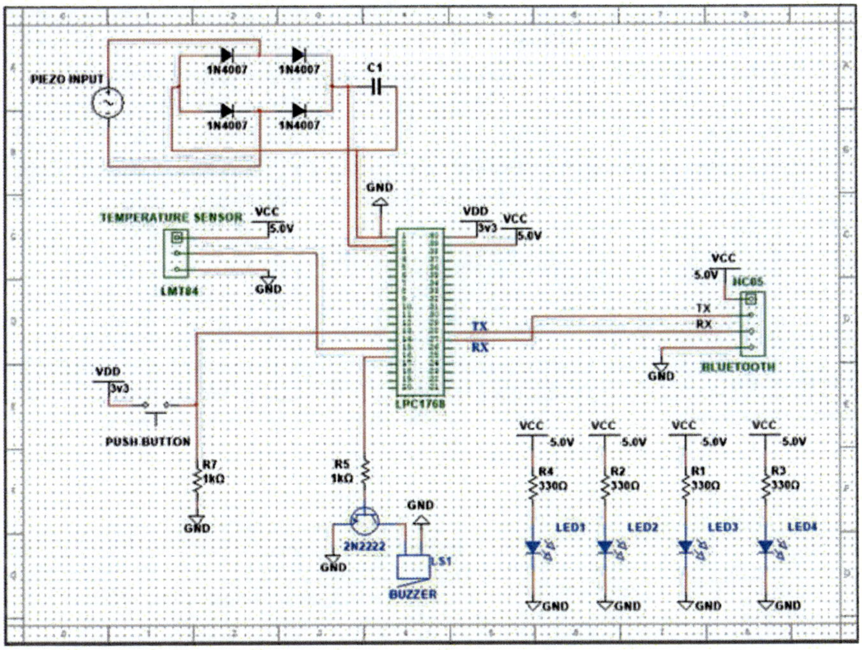

Fig. 5 Project's schematic capture

2.4 Software Design

The software design was modelled by an online program, "Visual Paradigm" while the coding was done by an online compiler, "Mbed Online Compiler". The "Mbed Online Compiler" can be downloaded or used as an online tool to perform "C" or "C++" coding and build the program. By accessing it online, the software updates and libraries are automatically stored up to date and each program file can be accessed anywhere over the web. Upon completion of coding, it builds a successfully coded program in a ".bin" binary file output. Visual Paradigm is an online tool that is used by professionals to model through the Unified Modelling Language (UML) diagrams. The UML segregates the software into 2 aspects, structural and behavioural. The UML modelling in software development is crucial in Object-Oriented programming, as it can be used to build the program based on its Use Case, Classes, and Sequence of execution, each of which will be presented in the sub sections.

Use Case Table

The Use Case can exist in the form of a table or in the form of a diagram representation. In this study, it will be represented in the form of a table. It consists of information on the system's response to the Human Action Input as shown in Table 2.

Class Diagram

A Class diagram is a representation of the classes, the "attributes" and the "operations" as shown in Fig. 6. The "attributes" denoted with the "-" in the beginning, indicates the variables that are used in the coding while the "operations", denoted with "+" in the beginning, indicated the method of the execution.

Table 2 Use case table

Human action	System response
Turn on the device	Initialization of device and components Main Scenario: • MCU initializes the components • Buzzer turns off • Temperature sensor initializes • Bluetooth initializes Optional Features: • Bluetooth displays "Bluetooth Ready"
Visitor detection (visitor press pushbutton)	• MCU sends a signal to initialize Buzzer • MCU transmit information to Bluetooth to notify the Owner • Buzzer sounds for 5 s • LED light display for 5 s (customized pattern)
Temperature detection (additional feature)	• MCU initialize Temperature sensor • Temperature is measured and data is sent to MCU • MCU transmits information to Bluetooth to transmit to the Owner

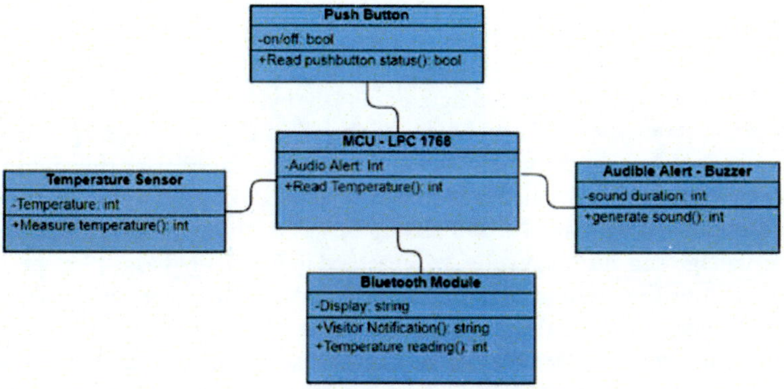

Fig. 6 Class diagram

Sequence Diagram

As shown in Fig. 7, the sequence of events is depicted through the Sequence Diagram. User 1 refers to the Visitor at the doorstep, while User 2 refers to the Owner. As visitors' step on the Energy harvesting floormat, it will store the energy in a rechargeable battery which can power up the MCU. The MCU will then begin to send temperature feedback to User 2. When the push button, doorbell switch, is pressed the Buzzer will give an audible output to both Users. In addition to that, User 2 receives a notification via Bluetooth that there is a visitor while User 1 gets a customized LED output lighting, which could display a "Welcome" sign.

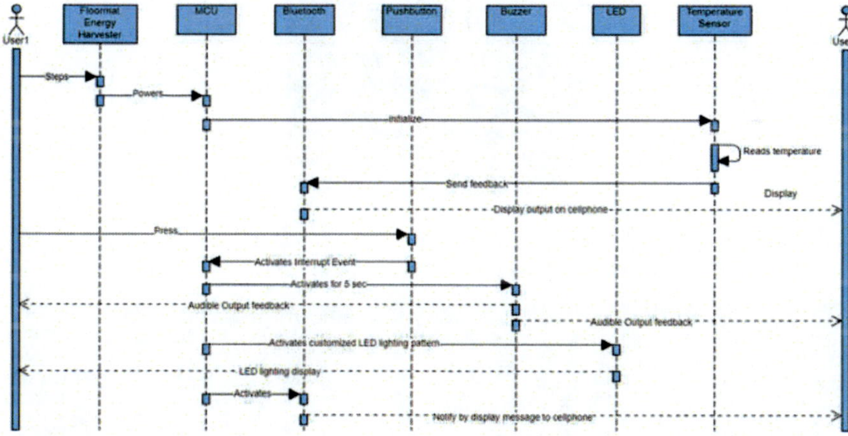

Fig. 7 Sequence diagram

3 Results and Discussion

The hardware and software were integrated, and the evaluation results were recorded. The floormat prototype was stepped on, as shown in Fig. 8. The energy generated was monitored by using the multimeter and the microcontroller as shown in Fig. 9. The prototype has been evaluated and 20 data has been collected from 3 load tests, which each loaded with 44, 77, and 105 kg, respectively (family member weight). The DC voltage and current results are presented in Fig. 10 and Fig. 11 respectively.

As shown in Fig. 10 and Fig. 11, the different weights have an impact on the amount of energy harvested. It is observed that both voltage and current produced varies with the weights. Heavier weights cause higher mechanical pressure on the piezo transducer, generating higher amounts of electrical energy. Taking the average readings, the average weight was calculated to be 75 kg while the average Voltage (DC) was 18.5 V and the average current was at 8.5 μA.

Since a 20 cm-by-20 cm floormat that fits 16 piezoelectric transducers can generate about 8.5 μA, in a bigger scale, it is projected to be able to generate 42.5 μA per m^2. A typical BTO 4-room Flat in Singapore which is about 92 m^2 will be able to generate 3.9 mA with 20 steps of a person that weighs 75 kg.

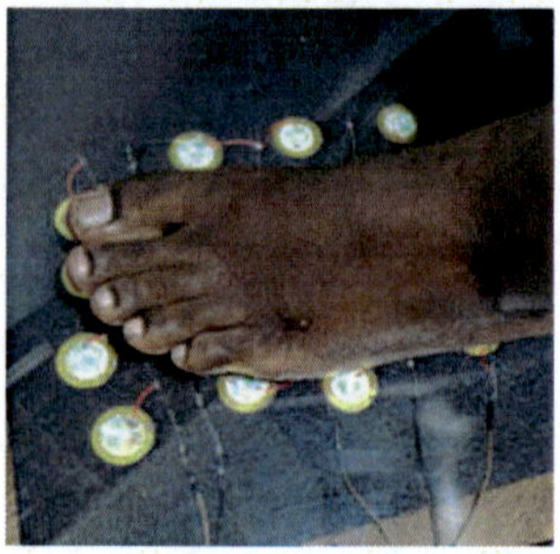

Fig. 8 Stepping on Piezo floormat

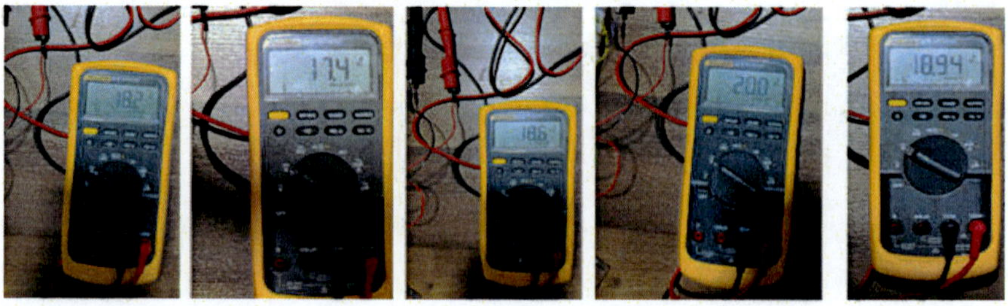

Fig. 9 DC Voltage reading

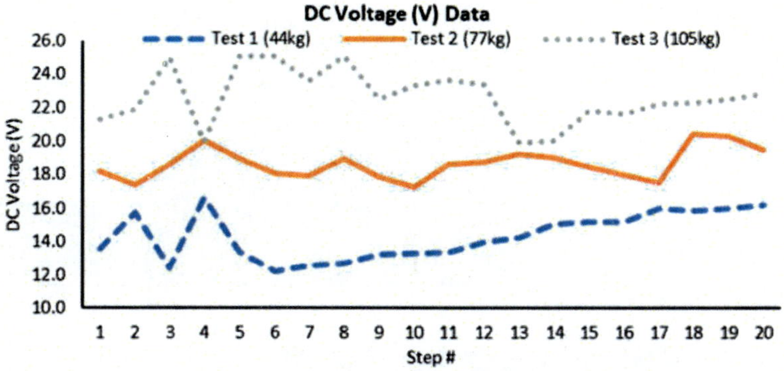

Fig. 10 DC Voltage (V) data

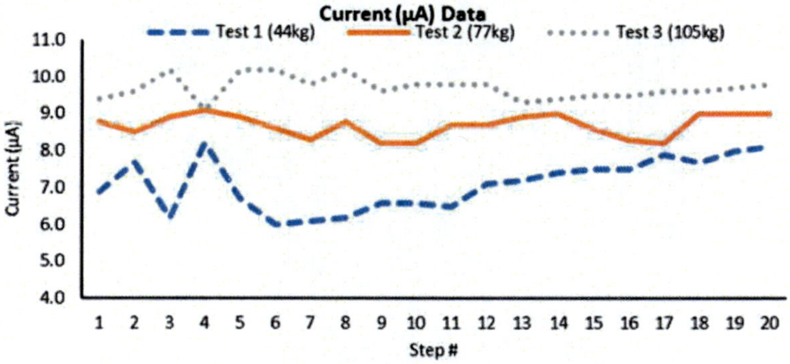

Fig. 11 Current (μA) data

With the average current, a 250 mAh battery will take about 29.4 thousand steps to be charged fully as calculated based on Eq. (1).

$$\begin{aligned} \textit{Number of Step Required} &= (\textit{Battery Capacity})/(\textit{Average Current}) \\ \textit{Number of Step Required} &= (250\,\text{mA})/(8.5\,\mu\text{A}) \approx 29412 \end{aligned} \tag{1}$$

Based on the average number of steps taken for an individual which ranges from 3.5 thousand steps to 6.9 thousand steps [8], it is projected to take approximately 4.5 to 8.5 days to completely charge the battery fully.

The software was tested to see if the system can be triggered, initialized, and display deliverables. When the MCU is turned on, the system was able to read the temperature, charged capacity of the battery and its status. When the push button is pressed, the system was able to detect and notify the user. The expected readings were captured in the Tera Term terminal as shown in Fig. 12.

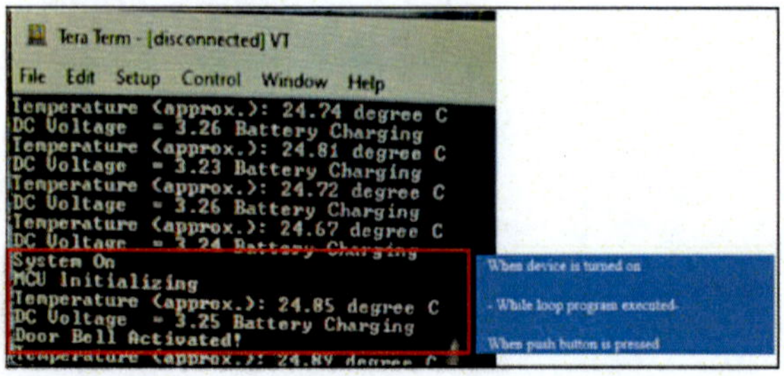

Fig. 12 Capture of PC output

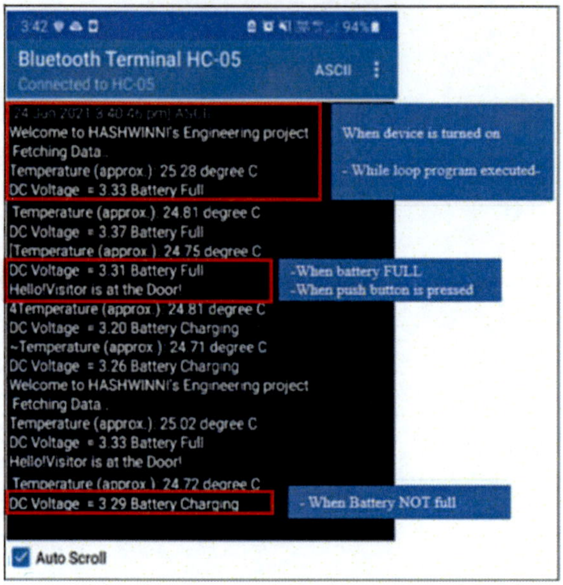

Fig. 13 Capture of cell phone output

The readings were also programmed to be displayed on the cell phone by accessing through a mobile application, "Bluetooth Terminal HC-05", that can be downloaded from *Android Playstore* as shown in Fig. 13 as the capture of cell phone output.

4 Conclusions

In conclusion, this study presented the design of an IoT enabled piezoelectric energy harvesting floormat, the small-scale prototype has been designed and fabricated for evaluation. The prototype consists of diaphragm type piezoelectric transducers powered by the LPC 1768 MCU and the HC-05 Bluetooth module for data transmission. Analytical calculation results based on the data collected from the floormat estimated the average output of 18.5 V (DC) and 8.5 μA with 20 steps of a person that weighs 75 kg. It is also projected with 42.5 μA output per m^2 and 29.4 thousand steps to fully charge a 250 mAh battery which takes approximately 4.5 to 8.5 days.

References

1. Devos M, Masek P (2017) Battery monitoring within Industry 4.0 landscape: solution as a service (SaaS) for industrial power unit systems. In: Galinina O, Andreev S, Balandin S, Koucheryavy Y (eds) Internet of Things, smart spaces, and next generation networks and systems. ruSMART NsCC NEW2AN 2017 2017 2017. Lecture notes in computer science, vol 10531, pp 40–52. Springer, Cham. https://doi.org/10.1007/978-3-319-67380-6_4
2. U.S. Energy Information Administration - EIA - Independent Statistics and Analysis. https://www.eia.gov/todayinenergy/detail.php?id=41433. Accessed 02 Apr 2021
3. Energy Market Authority, Renewable Energy – Overview. https://www.ema.gov.sg/Renewable_Energy_Overview.aspx. Accessed 04 June 2021
4. Covaci C, Gontean A (2020) Piezoelectric energy harvesting solutions: a review. Sensors 20:3512
5. Pan H, Qi L, Zhang Z, Yan J (2021) Kinetic energy harvesting technologies for applications in land transportation: a comprehensive review. Appl Energy 286:116518
6. Soin N (2018) Magnetic Nanoparticles—Piezoelectric Polymer Nanocomposites for Energy Harvesting. Magnetic Nanostructured Materials, pp 295–322
7. Mariello M, Fachechi L, Guido F, De Vittorio M (2021) Multifunctional sub-100 μm thickness flexible piezo/triboelectric hybrid water energy harvester based on biocompatible AlN and soft parylene C-PDMS-Ecoflex™. Nano Energy 83:105811
8. Healthline, Average Steps Per Day by Age, Gender, Occupation, and Country. https://www.healthline.com/health/average-steps-per-day#country. Accessed 07 May 2021
9. Sezer N, Koç M (2021) A comprehensive review on the state-of-the-art of piezoelectric energy harvesting. Nano Energy 80:105567
10. Song GJ et al (2019) Development of a pavement block piezoelectric energy harvester for self-powered walkway applications. Appl Energy 256:113916
11. Smart Cities. https://pavegen.com/smart-cities/. Accessed 14 Mar 2021
12. Jeong SY et al (2019) Piezoelectric device operating as sensor and harvester to drive switching circuit in LED shoes. Energy 177:87–93
13. Yamuna MB, Sundar KS (2014) Design of piezoelectric energy harvesting and storage devices. Int J Adv Res Electr Electron Instrum Eng 03:10945–10953
14. Wang J, Lim MK, Wang C, Tseng M-L (2021) The evolution of the Internet of Things (IoT) over the past 20 years. Comput Ind Eng 155:107174
15. Aboubakar M, Kellil M, Roux P (2021) A review of IoT network management: current status and perspectives. J King Saud Univ Comput Inf Sci
16. Kothandaraman D, Harshavardhan A, Manoj Kumar V, Sunitha D, Korra SN (2021) BLE in IoT: improved link stability and energy conservation using fuzzy approach for smart homes automation. Mater Today Proc
17. Karthick S et al (2021) Realization of industrial automation using Bluetooth technologies. Mater Today Proc
18. Amoran AE, Oluwole AS, Fagorola EO, Diarah RS (2021) Home automated system using Bluetooth and an android application. Sci Afr 11:e00711
19. mbed LPC1768. https://os.mbed.com/platforms/mbed-LPC1768/. Accessed 01 June 2021

20. NXP Semiconductors. https://www.nxp.com/docs/en/data-sheet/LPC1769_68_67_66_65_64_63.pdf. Accessed 01 Apr 2021
21. Element 14, Piezo element. https://sg.element14.com/multicomp/mcft-27t-4-2al-127/piezo-element/dp/1801061. Accessed 10 Apr 2021
22. Element 14, Buzzer, Piezo. https://sg.element14.com/projects-unlimited/ai-1440-twt-12v-r/buzzer-piezo-12vdc-14mm-pcb-mnt/dp/1838978. Accessed 10 Apr 2021
23. Texas Instruments, LMT84. https://www.ti.com/product/LMT84. Accessed 12 Apr 2021
24. Hokuriku Electric Industry Co., Ltd, KSMC612A Datasheet (PDF). https://www.alldatasheet.com/datasheet-pdf/pdf/484325/HOKURIKU/KSMC612A.html. Accessed 14 Apr 2021
25. GM Electronic (2021) HC-05 Bluetooth module user's manual v1.0. https://www.gme.cz/data/attachments/dsh.772-148.1.pdf. Accessed 18 Apr 2021
26. SketchUp. https://www.sketchup.com. Accessed 31 May 2021

Dynamic Modelling of Drone Systems Using Data-Driven Identification Methods

Osama Yaser Osman Suliman, Ir. Fahri Heltha, Ts. Muhamad Faliq, and Aulia Rahman

Abstract Dynamic modeling to understand and predict the behaviors of complex physical systems are still a challenge in system identification and control theory. Especially quadcopters as they are considered an unstable MIMO dynamics system. In this Paper, system identification is performed using the black box approach. A designed quadcopter is utilized as a black box system to estimate models for the quadcopter orientation. Using the quadcopter onboard Blackbox chip data records that have been collected during flights, and MATLAB identification toolbox to obtain an estimated nonlinear ARX and ARIMAX models that best represent the physical system. Simulation and tuning are conducted to investigate the model and fit it to the targeted PID controller's design. The approach shows good potential in modeling quadcopters using Blackbox records with higher accuracy when compared to previous studies on the approach.

Keywords Dynamic modeling · UAV modeling · Data-driven · Blackbox

1 Introduction

Dynamic modeling of complex physical systems in this case drones, break down the complex behaviors into a simplified equation concerning time. That gives the ability to mimic and predict its behaviors. Since they explain how system properties change over time, the models are referred to as dynamic models. The monitored inputs are linked to the drone response using a dynamic model. This approximate model can be as basic as an input-output graph or as complex as a series of differential equations of motion. They are especially needed for the analysis of aircraft stability, flight controller design, and aircraft handling characteristics [1]. A drone is

O. Y. O. Suliman (✉) · Ir. F. Heltha · Ts. M. Faliq
Faculty of Engineering, UCSI University, Kuala Lumpur, Malaysia
e-mail: eng.osamayaser@gmail.com

Ir. F. Heltha · A. Rahman
JTEK, USK, Banda Aceh, Indonesia

© The Author(s), under exclusive license to Springer Nature Singapore Pte Ltd. 2023
M. H. A. Hassan et al. (eds.), *Technological Advancement in Instrumentation & Human Engineering*, Lecture Notes in Electrical Engineering 882,
https://doi.org/10.1007/978-981-19-1577-2_15

a nonlinear dynamical system with four inputs and six degrees of freedom (DOFs)—three cartesian coordinates and three orientation angles—that is naturally unstable [2]. A dynamic drone model with six (6) degrees of freedom can be estimated using data analysis and identification since functional data analysis leads inevitably to dynamic systems [3]. To accurately model such complex systems, system identification methods are proposed using the black box approach. Black Box models have no physical insight of the observed system, but they showed good edibility and have been successful in the past [4]. This black box identification method is much more difficult. The main reason for that is it's fully dependent on the data of the flight tests, therefore exposed to all possibilities and limitations of the experiment and easily effected by internal noises and oscillations, and also the environmental disturbances such as wind [5]. The design is modeled in a computer-aided framework such as MATLAB and is based on basic stability and control derivative definitions and transfer function of the airframe dynamics as obtained from the first-principles aerodynamic theory [6]. A linear model structure, such as the ARX model structure, may be used to approximate the device model [7]. The art and science of identifying mathematical models of dynamical processes from observable input and output signals are known as identification. Identification model creates representations of dynamical systems and their processes by comparing previous knowledge about the mechanism to information gathered from observable input-output data to create a system capable of mimicking and reproducing the behaviors of the pross [8]. The black box model is not understood physically or is not used, but the structure of the specific model is considered universal and belongs to a family that was effective in the past [9]. Models with a black box, unlike the grey-box model, do not require any previous device modelling. This is a strictly statistical construct that can be estimated using a variety of methods. This research would only cover black-box models created with MATLAB. Glad and Ljung are the founders for the approaches used in this study for device recognition [10]. Experimental data in the form of datasets are required to approximate a device model. This paper presents a new approach using onboard Blackbox data collection from a 5 in. FPV quadcopter and simulated evaluation of the resulted nonlinear ARX and ARIMAX models for three orientation angles (Roll, Pitch, Yaw). Using FPV mini quadcopters flight controllers onboard Blackbox came to the picture with a suitable solution for drone's data collection since they offer controlled logging frequencies and eliminate the need for internet connection during flight. FPV mini quadcopter is a new technology stole the spotlight in recent years with companies racing to deliver new hardware with better capabilities to establish a name in this new market. FPV (First Person View) is not a new concept. It has his fame in the game developing field, yet in drone's world where most companies focus on aerial photography. This is a fairly new approach.

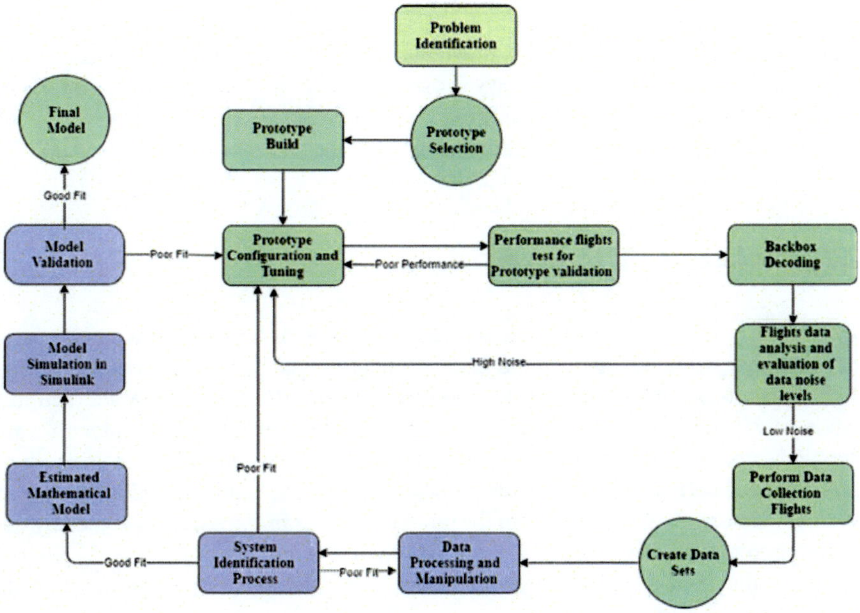

Fig. 1 Workflow

2 Methodology Approach

Dynamic Black box modelling using data driven methods gives a higher accuracy yet very agile and risky approach, where the focus will be to obtain the best data sets possible, this shifts our perception of the process to prioritize the data collection process and the verified prototype used in the data collection to obtain the best representation for the system. The reason for choosing this costume mini quadcopter is the ability for some modern flight controllers to log flight data using embedded flash memory chips called Blackbox. The key aim of the prototype is to be able to capture data inputs and outputs since most drones on the market are completely integrated and cannot be accessed nor send its data to platforms—Laptops or Servers—to hold a record for further analysis. Using FPV Mini quadcopter will resolve these issues by saving all the records inside the flight controller and access it uses Blackbox decoders afterword. Building such systems for research not only gives an enhance data collection method but it opens new possibilities for innovation and researchers experiment to fully explore aerial vehicles abilities. Moreover, modern FPV quadcopters break the limitation in performance and control been a fully experimental and open-source platform. Dynamic Blackbox modelling for UAV's has limited experimental research due to the high level of complexity and high cost for such agile robots. Developing a closed—loop (HIL—Hardware-In-Loop) experiment to collect the data with high sampling time that give a correct representation for the model in the study during flites. Using MATLAB identification tool to estimate a model and verify the model results by analyzing the accuracy of the result against the actual data sets collected from the system in the study. This flow chart gives a general visual explanation for the approach of the study, the arrows show the workflow starting from the problem identification and ending with the final model with labelled indicators in case of decision making and undesired results (Fig. 1).

2.1 Building FPV Quadcopter

Assembling FPV quadcopters is a delicate process with high risk since most parts doesn't have protection circuits and fuses. The build must be checked multiple times. All connections and solder points must be correct and solid to avoid any short circuit when plug in the battery or during flight. When building the mini quadcopter all parts should be connected correctly but in order for them to communicate with each other, protocols and software managing the process is needed. Specialty in this fully custom system where every single aspect of the build must be configured and designed by the designer to get the desired performance. To do that, a number of software is needed to upload the firmware to the flight controller board and assign the correct protocols of communication.

Software's used in this process:

1. Betaflight: Betaflight is an open-source flight configurator software used to upload firmware and configure. Multi-rotor drones and fixed wing craft. This software leads the development of UAV's and connects developers and the community delivering supports and ideas sharing.
2. BLHeli Configurator: BLHeli Configurator is a chrome app, similar to Betaflight configurators. Created by Andrey Mironov and considered the first free cross-platform ESC programming/flashing tool for BLHeli. Used to configure the motor's electronic speed controllers (ESC) (Fig. 2).

Fig. 2 Prototype

2.2 *Data Collection Approach*

The approach for the data collection method was set to use the flight controller Blackbox flash memory chip, to do so first the flash memory should be enabled and its logging rate set to different rates. The collected data was observed to determine the best logging rate that fully describes the quadcopter motion and results in less data manipulation later (Fig. 3).

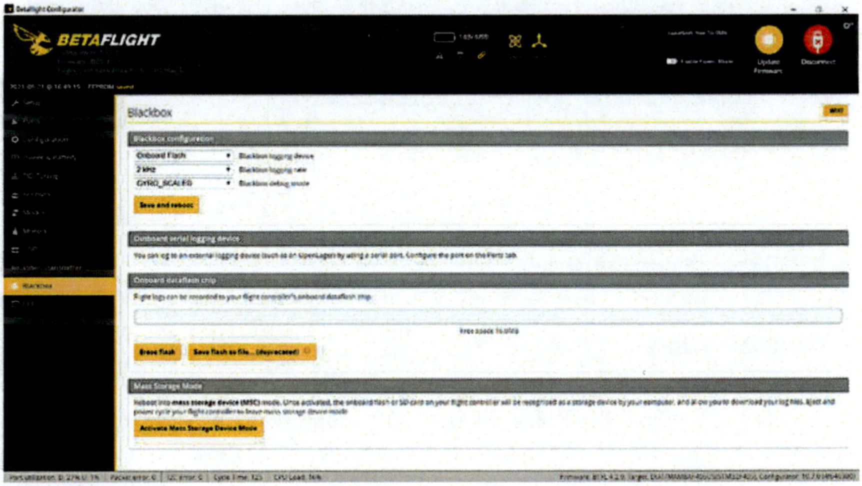

Fig. 3 Flight controller Blackbox configuration

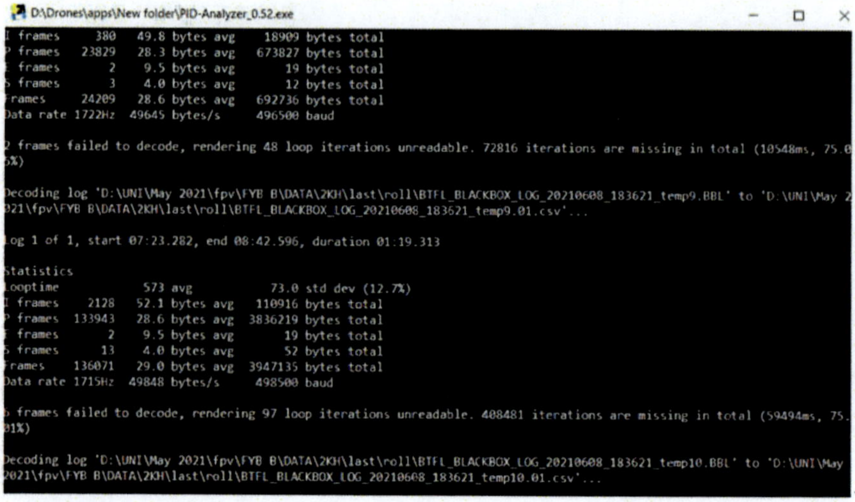

Fig. 4 PID analyzer software

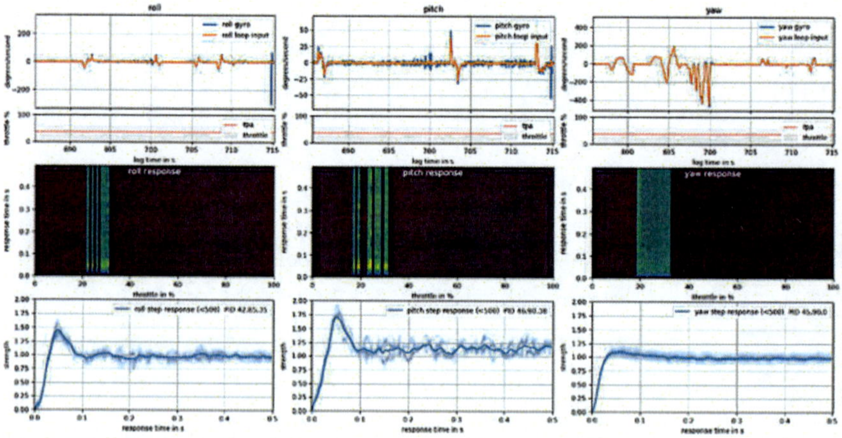

Fig. 5 PID analyzer decoding results

3 Visualization

To be able to visualize the Blackbox data PID Analyzer is used, this software is designed to decode the Blackbox data and shows graphs of the noise levels and the effects of the PID tuning in the performance. PID Analyzer isn't designed to export the Blackbox data, but it is done as a middle process and the observation in Fig. 4. Shows the process where it separates each flight in different files the orange line in Fig. 5 shows the set point developed by the inner loop of the flight controller and the blue line shows the recorded response from the drone.

4 Data Collection and Manipulation

First, a suitable area to conduct the flight test was identified. The area chosen was a closed area in order to minimize wind interference. A total of 5 tests were conducted, each taking approximately 30 s for a total of 2.5 min. The test consisted of several small maneuvers which required a competent control on behalf of the pilot to avoid any unwanted accidents or crashes that might affect the reliability of the data. Afterward, the Blackbox log was collected from the drone using Betaflight software and decoded using PID Analyzer software into spreadsheets. Then the required inputs and outputs were identified and transferred to a new spreadsheet to prepare the data sets in order to import them to MATLAB. The Inputs selected are the roll, pitch, yaw commanded angles against the roll, pitch, yaw gyro readings during the flight to describe the drone response orientation. Working with high logging rates and highly unstable systems noise is inevitable since the drone is one of its greater sources. The data collected had to undergo filtration to exclude the high frequency noise recorded by the gyro as observed in Figures 6 and 7 which is the leading factor for poor fit in the estimation process.

Filtration must be done precisely and accurately till we reach a desirable result without effecting the actual output signal which can be resulted due to high filtration and manipulation. Moreover, it was observed that taking off and landing results in high noise and oscillation due to ground effect if recorded, so the logging started

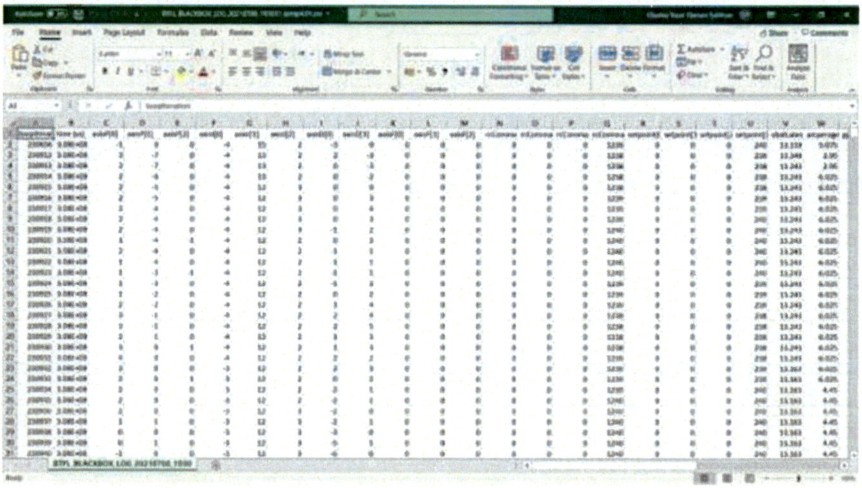

Fig. 6 The decoded flights record

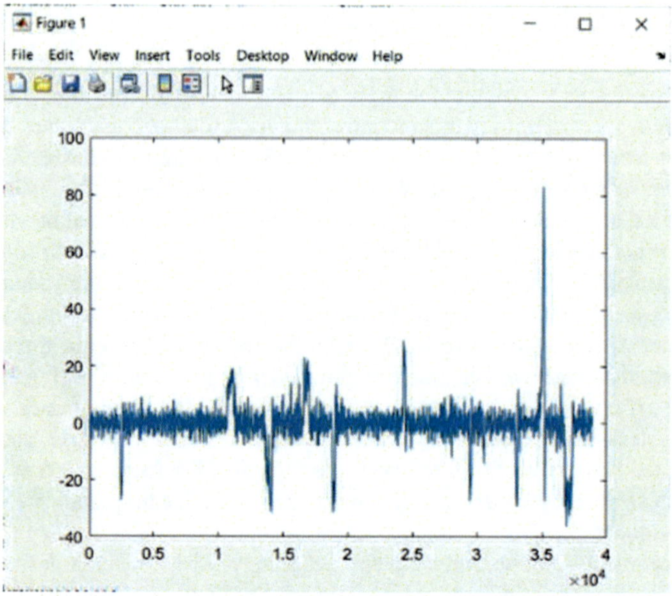

Fig. 7 Gyro raw data for the Roll output before filtering

mid-flight using a designated switch and ended before approaching landing. The approach taken was the moving average technique that calculates the average of multiple set points along with the data set, this technique is highly effective against oscillation and lowers the high frequency with maintaining the signal characteristic, hence this approach was selected for data filtering. The main variable to control the filtration using this technique is the number of subpoints that the algorithm calculates the average from, giving a higher number will result in intensive filtration that will lead the signal to lose its characteristic and the ability to represent the system in study. The filtration results are shown in Fig. 8.

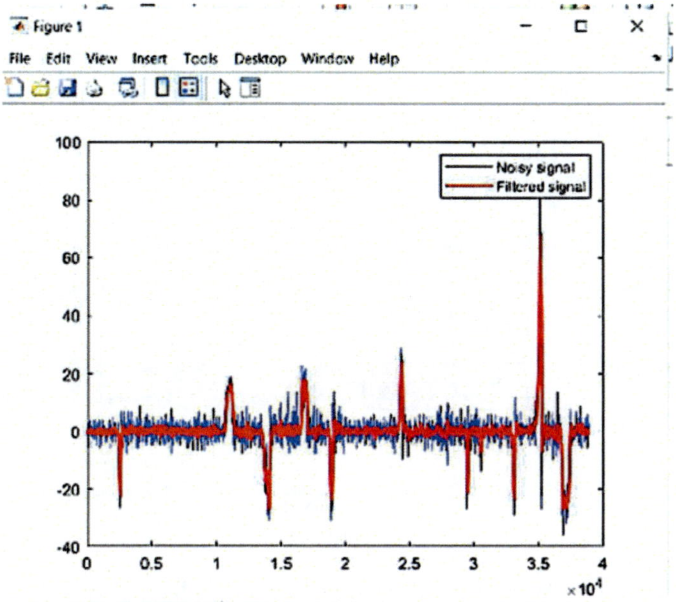

Fig. 8 Implementing moving average filter

5 The Estimated Models

The model estimation process showed a high fit to the collected data using the data sets obtained from the manipulation process. Number of ARX and ARIMAX were estimated for each set of data to select the best two fits of each data sets. The models selected were tested for fit against all data sets to select the model that best represent the angle response for all data sets. The following Fig. 9 shows the estimation process for the roll angle model.

The fitting test was repeated for the three sets of data and the results showed in the following Table 1 and Fig. 10.

The ARIMAX model 1 shows the highest fit to all the data sets and fit to estimation data: 100% (prediction focus). so, it is selected to represent the pitch angle. The model 3 details is shown below

$$\text{Roll Model orders} = \text{na} = 4 \;\; \text{nb} = 4 \;\; \text{nc} = 4 \;\; \text{nk} = 1$$

Discrete-time ARIMAX model:

$$A(z)y(t) = B(z)u(t) + \left[\frac{C(z)}{1 - z^{-1}}\right]e(t) \tag{1}$$

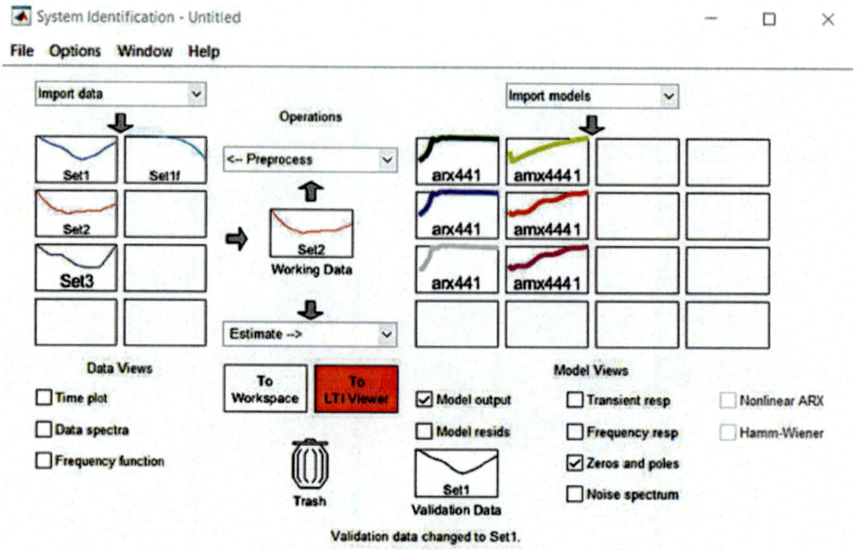

Fig. 9 Roll model estimation shows the developed ARX and ARIMAX models

Table 1 Fitting results for ARIMAX roll models against all data sets

Data sets	Model 1	Model 2	Model 3
Set 1	84.85	60.29	60.29
Set 2	89.29	61.4	61.4
Set 3	84.35	62.41	62.41

Fig. 10 Idmodel parameter interface in Simulink

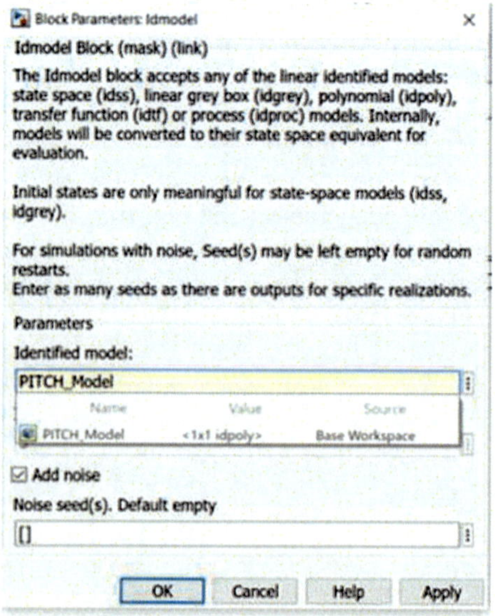

Table 2 Fitting results for pitch models

Data sets	Model 1	Model 2	Model 3
Set 1	83.99	84.5	84.94
Set 2	80.43	84.61	84.23
Set 3	83.11	85.33	85.43

where;

$$A(z) = 1 - 3.948(+/- 0.0004265)z^{(-1)} + 5.848(+/- 0.001273)z^{(-2)} - 3.852(+/- 0.001267)z^{(-3)} + 0.9519(+/- 0.0004206)z^{(-4)}.$$

$$B(z) = -0.001949(+/- 5.114e - 05)z^{(-1)} + 0.006098(+/- 0.0001544)z^{(-2)} - 0.006352(+/ - 0.0001554)z^{(-3)} + 0.002204(+/- 5.221e - 05)z^{(-4)}.$$

$$C(z) = 1 + 2.044(+/- 0.1441)z^{(-1)} + 1.508(+/- 0.1468)z^{(-2)} + 0.439(+/- 0.1501)z^{(-3)} + 0.04979(+/- 0.148)z^{(-4)}.$$

The proses repeated for the pitch angle. Number of ARX and ARIMAX were estimated for each set of pitch data to select the best two fits of each data sets. The models selected were tested for against all the data sets to select the model that best represent the pitch angle. The fitting test was repeated for the three sets of data and the results showed in the following Table 2.

The ARIMAX model 3 shows the highest fit to all the data sets and fit to estimation data: 99.89% (prediction focus). so, it is selected to represent the pitch angle. The model 3 details is shown below

$$\text{Pitch Model orders} = \text{na} = 3 \;\; \text{nb} = 3 \;\; \text{nc} = 3 \;\; \text{nk} = 0$$

Discrete-time ARIMAX model:

$$A(z)y(t) = B(z)u(t) + C(z)e(t) \tag{2}$$

where;

$$A(z) = 1 - 1.532(+/ - 0.7059)z^{(-1)} + 0.1034(+/ - 0.1394)z^{(-2)} + 0.4293(+/ - 0.06888)z^{(-3)}.$$

$$B(z) = -0.0004586(+/(-0.0001725)) + 0.0001421(+/(-0.0002327))z^{(-1)} + 0.0005056(+/(-0.0001783))z^{(-2).}$$

$$C(z) = 1 + 0.1397(+/ - 0.07037)z^{(-1)} - 0.07031(+/ - 0.02163)z^{(-2)} + 0.09186(+/ - 0.00651)z^{(-3)}.$$

Table 3 Fitting results for yaw models

Data sets	Model 1	Model 2	Model 3
Set 1	90.4	88.23	81.03
Set 2	85.843	90.08	85.09
Set 3	75.93	84.65	86.83

The prosses repeated for the yaw angle. Number of ARX and ARIMAX were estimated for yaw of each data set to select the best two fits of each data set. The models selected were the one which tested the best for all the data sets to represent the yaw angle. The fitting test was repeated for the three sets of data and the results showed in the following Table 3.

For yaw angle the ARX model 2 shows the highest fit to all data sets with a fit to estimation data of: 99.97% (prediction focus). so, it is selected to represent the yaw angle. The model 3 details is shown below

$$\text{Yaw Model orders} = \text{na} = 3 \;\; \text{nb} = 1 \;\; \text{nk} = 1$$

Discrete-time ARX model:

$$A(z)y(t) = B(z)u(t) + e(t) \tag{3}$$

where;

$$\begin{aligned} A(z) &= 1 - 1.776(+/- \; 0.004987)z^{(-1)} + 0.5838(+/- \; 0.009846)z^{(-2)} \\ &\quad + 0.1925(+/(-0.004866))z^{(-3)}. \\ B(z) &= 0.0002361(+/- \; 2.437e - 06)z^{(-1)}. \end{aligned}$$

6 Selected Models Simulation

The three models are simulated with a feedback loop using a PID block and Idmodel block Simulink. Idmodel block is an element in system identification toolbox in Simulink that mimics the output of a is a previously specified linear model estimated or constructed state-space (idss), linear grey-box (idgrey), polynomial (idpoly), transfer function (idtf), or process (idproc) model using time-domain input data, the block parameters are shown in Fig. 10.

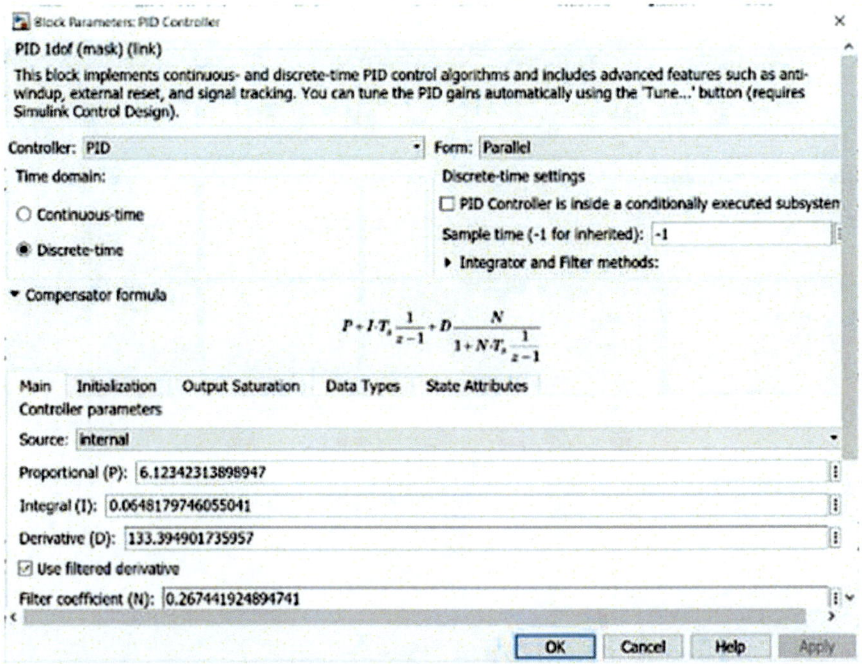

Fig. 11 PID controller parameter interface in Simulink

On the other hand, the PID (proportional, integral, and derivative control) controller block is the most used control technique in the industry hence is used to control the plant. The block parameters are shown in Fig. 11.

To simulate the models an idmodel block is constructed and connected to the selected model to be considered as the plant, along with a PID controller tuned using the PID tuner app that allows importing a specific model to tune according to it. Then upload the resulted tunning to the controller. The Figs. 12, 13 and 14 show this process for the roll model in detail.

The complete drone model is designed as showed in Fig. 15.

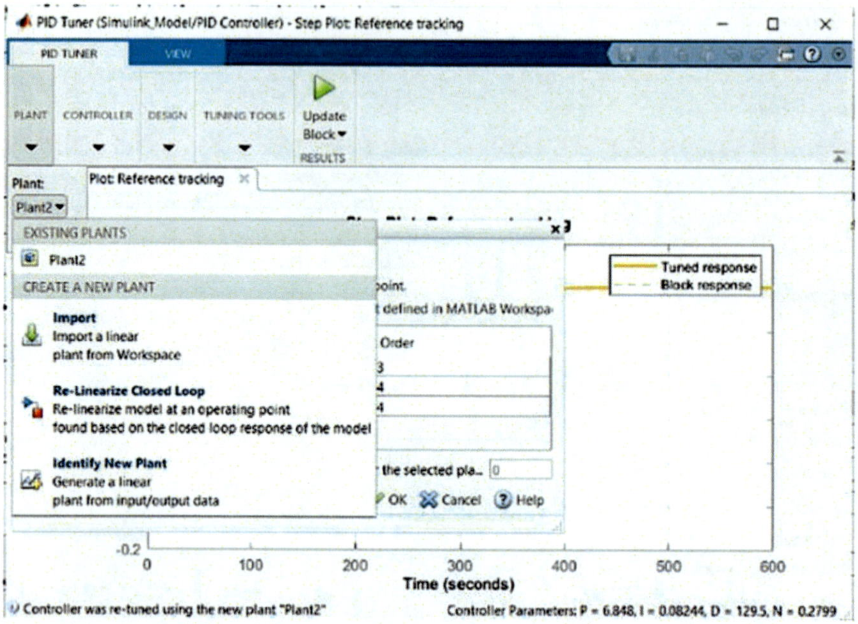

Fig. 12 Importing roll model to PID tuner

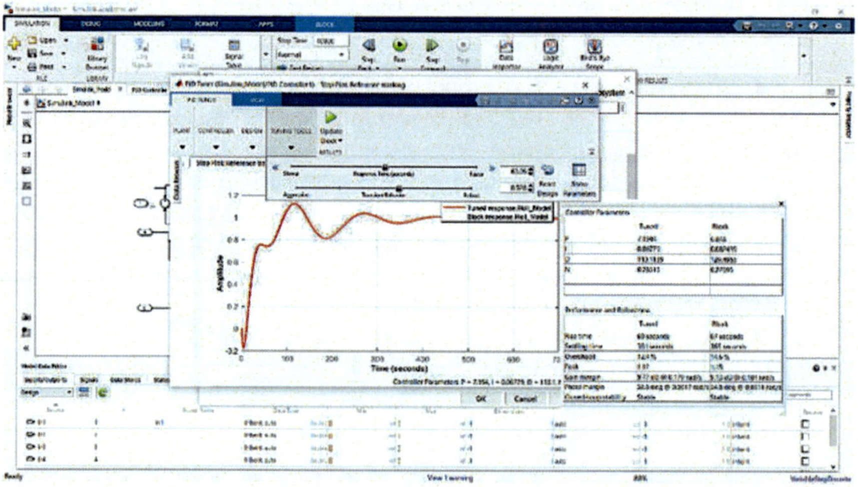

Fig. 13 Roll model tunning result

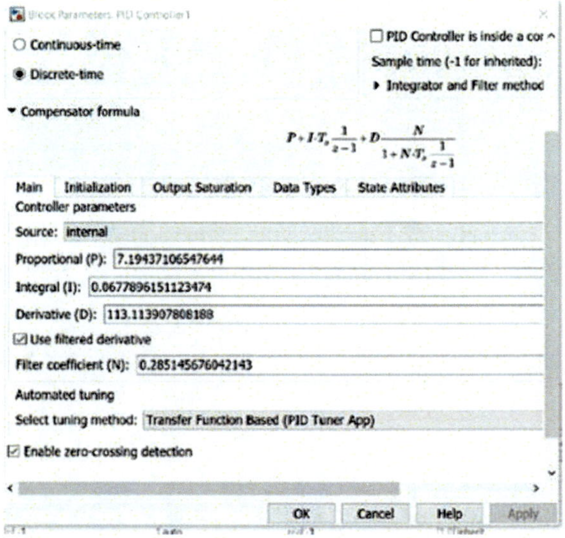

Fig. 14 Final roll model PID block parameters

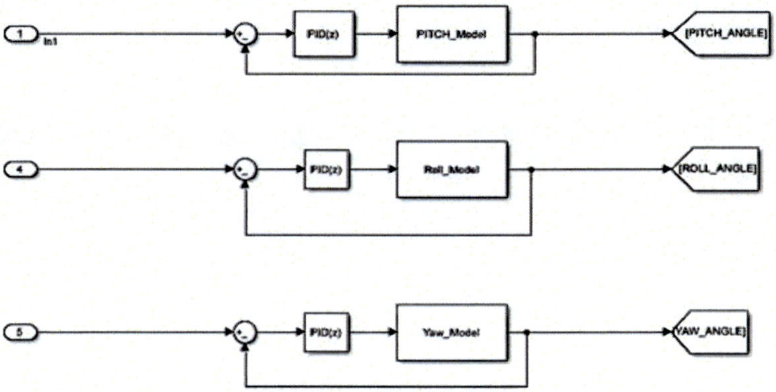

Fig. 15 The complete drone control models in Simulink

7 Validation

The generated models are designed simulated for validation and demonstrating the application feasibility for the study. The data sets that have been gathered are supplied as an input signal to the model, then compared with the output generated. This offers an overview of the model behavior and if it matches the data set output utilized in this investigation. The method is then duplicated for validating all models (Fig. 16).

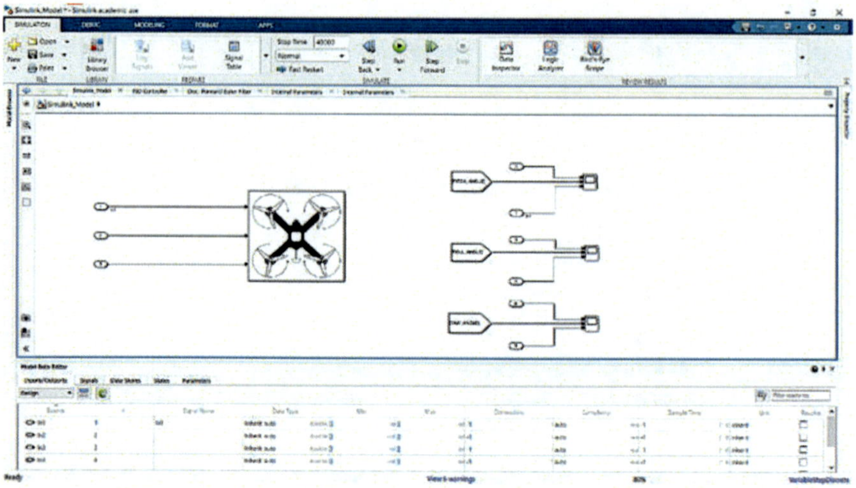

Fig. 16 The complete drone model in Simulink

Data set 2 is used for the validation scenario in Simulink. The data set inputs were used as inputs for the simulation and the data outputs are fed to the scope along with the generated outputs for comparison (Fig. 17).

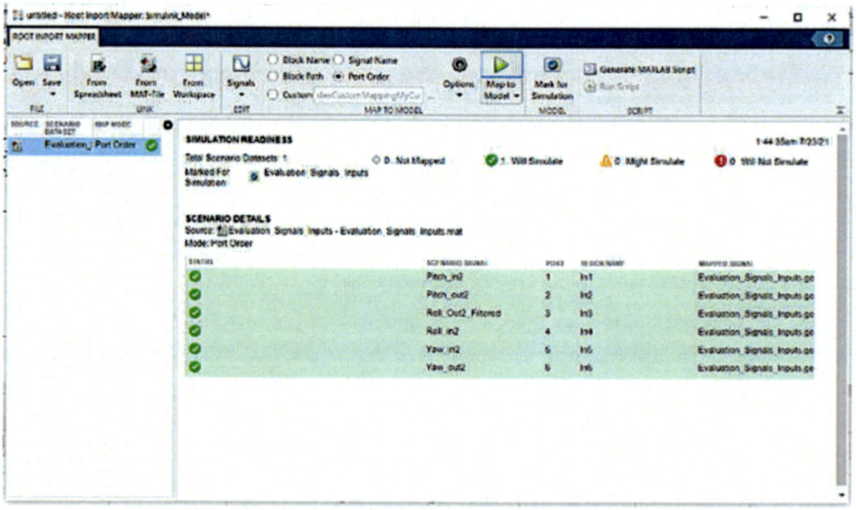

Fig. 17 The simulation scenario mapping for all inputs and outputs in Simulink

7.1 Roll Model Validation

The selected model for roll angle was model 1 with an overall fit of 86.163 to all data sets. Using this complete drone model, the simulation result for the validation data set is shown in Fig. 18. In the following figures the blue signal represents the simulated output, green signal is the recorded output and red signal is the reference control input signal for both the prototype and simulated models.

In this simulation, As seen, the model output followed the control input signal very well. The model output had enough gain to accurately follow the control input signal, which is higher than the recorded output signal, the assumption for that is the neglection of the drone weight in the simulation which is around 500 g. However, the model output had a slightly longer settling time than the recorded output signal which showed small oscillation as shown in Fig. 19.

This oscillation is mainly the result of the aggressive tunning for the PID controller which can be enhanced with more robust tunning. In the following figures the blue signal represents the simulated output, the green signal is the recorded output and the red signal is the reference control input signal for both the prototype and simulated models.

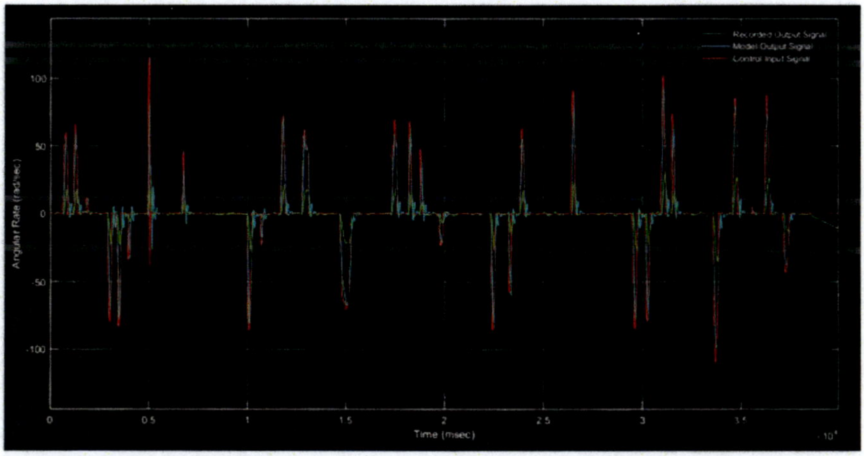

Fig. 18 A plot of the behavior of the roll model output and the recorded output signal against the control input signal

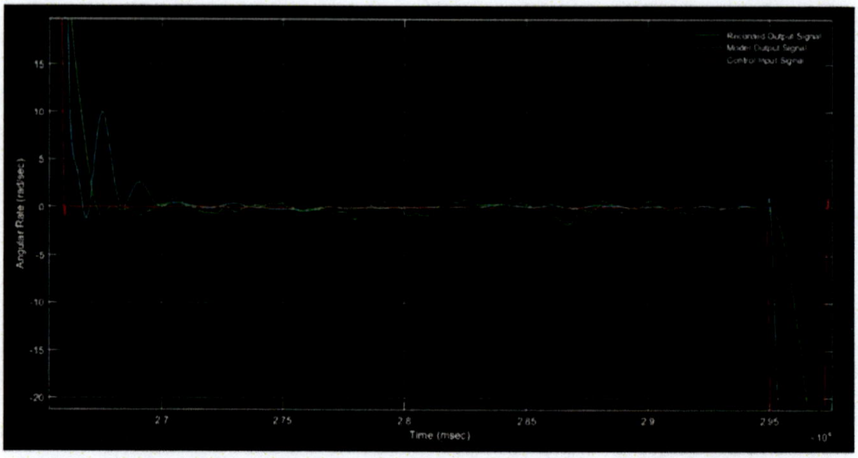

Fig. 19 A closer plot of the small oscillation between the model output and the control signal

7.2 Pitch Model Validation

The selected model for Pitch angle was model 3 with an overall fit of 84.867 to all data sets. The Pitch model simulation for the validation data set in Fig. 20 shows a better fit. In the following figures the blue signal represents the model output signal, the yellow signal is the recorded output and the red signal is the reference control input signal for both the prototype and simulated models.

As seen, the pitch model output follow the control input signal very well without oscillation and had enough gain to accurately follow the control input signal correctly.

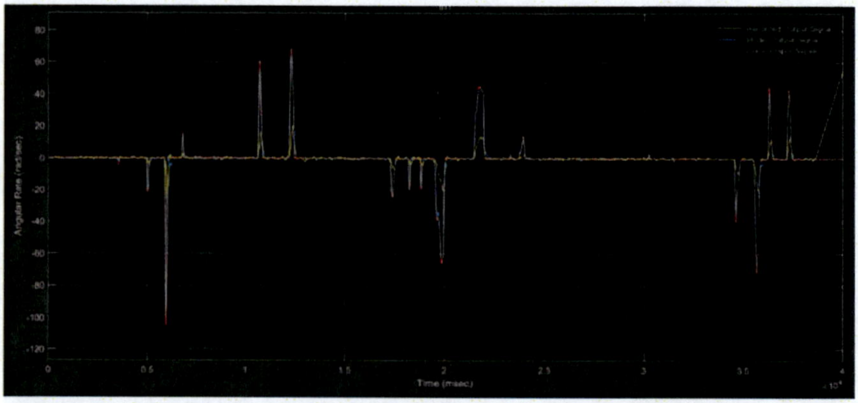

Fig. 20 A plot of the behavior of the pitch model output and the recorded output signal against the control input signal

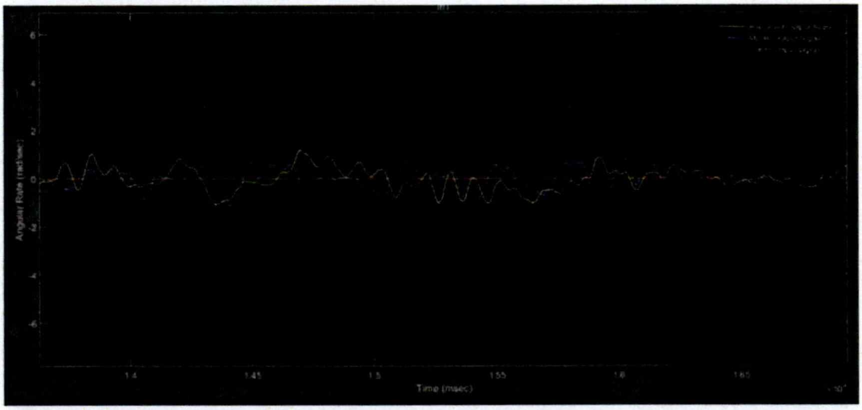

Fig. 21 A closer plot of the similar oscillation between the model output and the recorded output signal

This similar behavior to the roll model strength the assumption of the cause to be the weight of the drone been neglected in the simulation.

However, the model showed similar oscillation to the measured output signal around zero degrees as in Fig. 21. This error is mainly the result of disturbance and noise present during the data collection that still present after the filtration which effected the estimated model.

7.3 Yaw Model Validation

The selected model for Pitch angle was model 2 with an overall fit of 87.65 to all data sets. The Yaw model simulation for the validation data set in Fig. 22 shows the best results among the estimated models. The model output shows similar behaviors having enough gain to accurately follow the reference correctly In the following figures the blue signal represents the yaw model output, the yellow signal is the recorded output and the red signal is the reference control input signal for both the prototype and simulated models.

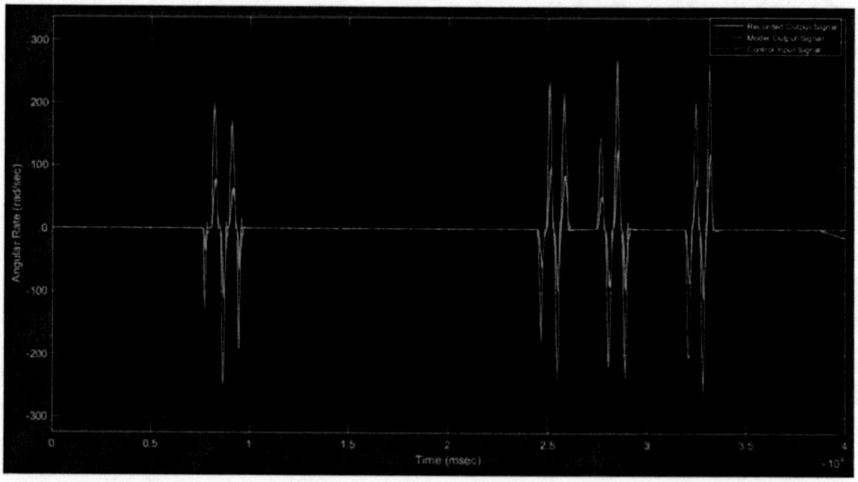

Fig. 22 A plot of the behavior of the yaw model estimated and the recorded output signals against the control input signal

As seen, the simulated output follows the measured output very well, without oscillation. However, it shows a small oscillation to the measured output signal around zero degrees as shown in Fig. 23.

This error is mainly the result of disturbance and noise present in the data set that can be solved with more filtration and enhancement in the tunning process.

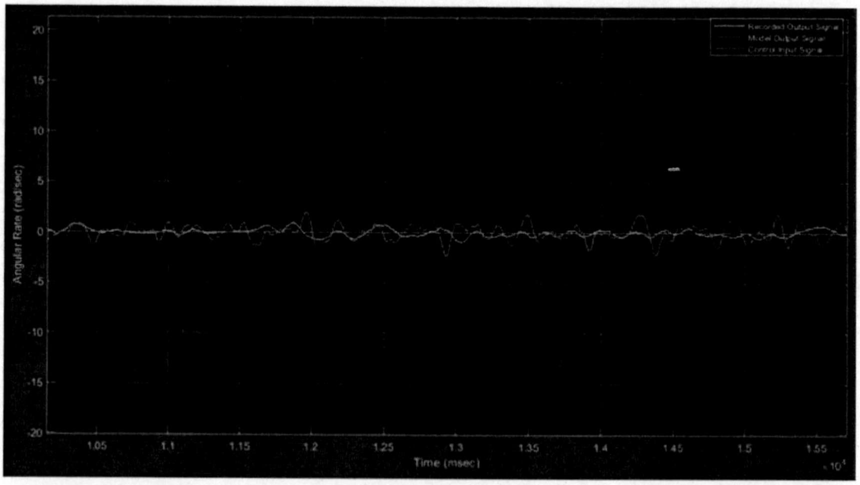

Fig. 23 A closer plot of the error between the simulated data using the estimated pitch model and the measured signals

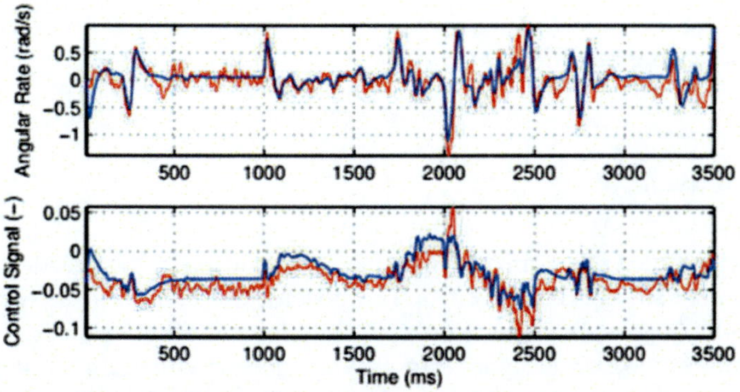

Fig. 24 A plot shows the results of the estimated roll and pitch model against the recorded data in previous studies [7]

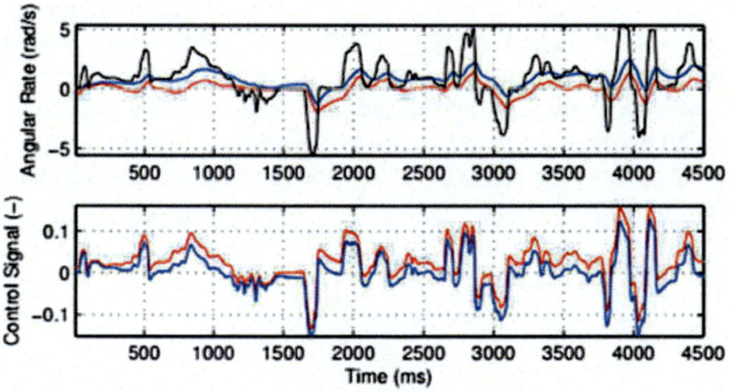

(a) No added system disturbance

Fig. 25 A plot shows the results of the estimated yaw model against the recorded data in previous studies [7]

8 Significance

The paper presents a method to enhance the data collection and noise levels in Blackbox modeling that gave an overall fit of more than 80% to all models. When comparing the obtained results to previous studies we noticed that the developed method gives a higher fit and less noise, taking the similar approach using Raspberry pi to log data from an inertial measurement unit (IMU) by Kugelberg [7] where the estimated fit for the Pitch and Roll to data set was 48.5% shown in the Fig. 24. The red signal is the recorded measured signal against the model simulated signal in blue.

Similarly, for the yaw model, the fit was around 65.5%. The enhancement comes from the use of the black box chip since it gives rich data for its ability to give high and controlled logging rates. moreover, the use of moving average filters gave higher fits than downsampling (Fig. 25).

9 Conclusion

The study presented a data-driven (Blackbox) dynamic model for a quadcopter using system identification of Blackbox data. The model describes the orientation of a quadcopter (Euler Angles) in space using gyro readings and control signals. The developed approach utilizes a designed 5 in. FPV quadcopter with a closed-loop i/o data collection system (Blackbox) for data collection. The use of BlackBox chip onboard provide to give reliable and controlled logging rates which enhanced the quality of the data and enhanced the overall estimation process. The system identification process for the collected and analyzed experimental data to obtain a mathematical model was conducted successfully and resulted in models that mimic the system in the study with high accuracy and a fit of more than 80% to all data sets. Moreover, a designed PID control models was developed in Simulink for validation against the measured data. The developed controllers performed very well and followed the control signal accurately with a difference in amplitude to the recorded signal which is a result of the physical features of the drone been neglected such as weight and environmental disturbance.

References

1. Ellner SP, Guckenheimer J (2006) Dynamic Models in Biology. https://doi.org/10.1515/9781400840960
2. Beltramini F, Bergamasco M, Lovera M (2011) Experiment design for MIMO model identification, with application to rotorcraft dynamics. In: IFAC proceedings volumes (IFAC-PapersOnline), vol 44, no 1, Part 1. IFAC. https://doi.org/10.3182/20110828-6-IT-1002.03463
3. Bickel P, Diggle P, Fienberg S, Gather U, Olkin I, Zeger S (n.d.) Springer series in statistics
4. Sjoberg J, Zhang Q, Ljung L, Benveniste A, Deylon B, Glorennec P (1995) Nonlinear black-box modeling in system identification : a unified overview. https://doi.org/10.1016/0005-1098(95)00120-8
5. Zhang X, Li X, Wang K, Lu Y (2014) A survey of modelling and identification of quadrotor robot. Abst Appl Anal. https://doi.org/10.1155/2014/320526
6. Tischler MB, Directorate A, Atcom USA (1995) System identification methods for aircraft flight control development and validation system. https://doi.org/10.21949/1403895
7. Kugelberg I (2016) Black-box modeling and attitude control of a quadcopter. http://urn.kb.se/resolve?urn=urn%3Anbn%3Ase%3Aliu%3Adiva-125649
8. Ljung L (2003) System Identification and Simple Process Models. Linköping University Electronic Press. https://www.diva-portal.org/smash/record.jsf?pid=diva2:316777
9. Isermann R, Münchhof M (2011) Identification of dynamic systems. In: Identification of dynamic systems. https://doi.org/10.1007/978-3-540-78879-9
10. Ljung L, Glad T (1994) Modeling of dynamic systems. Prentice-Hall, Inc. https://www.diva-portal.org/smash/record.jsf?pid=diva2%3A318401&dswid=-4217

Design of Intelligent Mixer Machine for Food Applications

A. S. F. Mahamude, D. Ramasamy, W. S. W. Harun, K. Kadirgama, J. Mogan, Suhelmy Azhar Bin Said, and Kaniz Farhana

Abstract As numerous innovative technologies are taking place in food manufacturing industries, the design of mixer machine for food application is also in concern. Food mixer is primarily used for mixing, blending and grinding food components all together. The main focus of this literature is to design an intelligent mixer machine for food application namely "keropok lekor" for consumption. The idea is brainstorming and the final design is selected based on three concepts. Criteria were set on the basis of ease of machining, stability, strength, withstand force, cost effectiveness and design simplicity. Various processes and machines such as milling machine, lathe machine, drilling machine and bending machine were used to design the mixer. There is a calculation of the torque and power of the shaft which is further used in simulations to evaluate the strength of the design. The goal of the paper will be achieved if the mixer machine is successfully produced and works as planned. In conclusion, the machine that has been designed can be used to process keropok lekor successfully.

Keywords FEA · Design · Factor of Safety

Supplementary Information The online version contains supplementary material available at https://doi.org/10.1007/978-981-19-1577-2_16.

A. S. F. Mahamude · D. Ramasamy (✉) · W. S. W. Harun · J. Mogan
Department of Mechanical Engineering, College of Engineering, Universiti Malaysia Pahang, 26300 Gambang, Kuantan, Pahang, Malaysia
e-mail: deva@ump.edu.my

D. Ramasamy
Automotive Excellence Center, Universiti Malaysia Pahang, 26600 Pekan, Pahang, Malaysia

K. Kadirgama
Faculty of Mechanical and Automotive Engineering Technology, Universiti Malaysia Pahang, 26600 Pekan, Pahang, Malaysia

S. A. B. Said
Keropok Lekor Cenderawasih, No. 57, Taman Geliga, Jalan Kuala Kemaman, 24000 Kemaman, Terengganu, Malaysia

K. Farhana
Department of Apparel Engineering, Bangladesh University of Textiles, Dhaka 1208, Bangladesh

© The Author(s), under exclusive license to Springer Nature Singapore Pte Ltd. 2023

M. H. A. Hassan et al. (eds.), *Technological Advancement in Instrumentation & Human Engineering*, Lecture Notes in Electrical Engineering 882,
https://doi.org/10.1007/978-981-19-1577-2_16

1 Introduction

Keropok is a conventional wafer item in Southeast Asian nations. It is called "kaogrieb" in Thailand, "keropok" in Malaysia and "krupuk", "kerupuk" or "kroepoek" in Indonesia and "bánh ph`ông tôm" in Vietnam. In Malaysia, ish wafer is devoured as new (Keropok Lekor) and puffed (Keropok Keping). This is like the three southern boundary territories of adjoining Thailand, which have in excess of 200 limited scope makers. Likewise, in the territories of Kelantan, Terengganu and a few pieces of Pahang Malaysia, there are in excess of 100 limited scope makers of this item. This demonstrates the significance of keropok to these networks. Keropok is a dried sea food item generally well known in South-East Asian nations, and presently its assembling is for the most part drilled in limited scope. It appears to have market potential for development. In any case, the nature of keropok fabricated in limited scope appears to be not to be reliable. Additionally, the progress from little producers to bigger scope fabricating requires a designing and scientific understanding of the crude material and handling alternatives Practically speaking, wafer extension is just grounded quantitative trait of the initial item that is known for firmly relating with the taste inclinations of a human board. It is seasoned with salt and sugar and is made from fish and sago flour. The colour is mildly greyish and gives a fishy flavour [1]. It is produced using minced fish tissue blended in with starch, salt, squashed ice and monosodium glutamate (MSG) to further develop surface and taste. The preparing and creation of keropok lekor may go from lawn to present day little or medium-scale industrial facility creation [2, 3]. It is a challenge, to improve the quality of product crackers tastier, hygienic and interesting. Also the product should be made within target price [4, 5].

Nowadays, the driving demand for keropok lekor has risen more than previous years, not only in the local market but also international ones. Thus, it needs a higher production rate of keropok lekor to satisfy the demand of the market. Besides, the conventional mixer machine typically used in the food industry is costly, especially in terms of maintenance. Besides, the conventional mixer machine is immobile, which is quite inconvenient for the user to move it from one place to another [6].The main component of this mixer machine is aluminium alloy rods, flat aluminium bar, flat aluminium sheet, mild steel angle bar, nuts and bolts, wheels, classical 'V' timing belts, hex washer head screws, push-button switch, mild steel for the main body frame, stainless steel mixing tank, motor and speed reducer. The objective is to simulate the strength of the mixer design [7]. The method of making new productive arrangements blending machines is conceivable by decreasing their energy power, energy costs for the creation of the feed blend [8].

Once the necessary information was gathered, the selected designs were screened and given scores to analyse the required criteria of ease of machining, Stability, Strength, Withstand force, cost effectiveness and design simplicity. SolidWorks software was used to sketch the parts involved from the design criteria, and this is used for the selection process. The materials chosen were then simulated to test their sturdiness and endurance. Finally, the product design is sent for fabrication and tested in real conditions to ensure that it was functioning.

2 Methodology

In food ventures, blending of flour to shape mixture has been a significant activity in its creation interaction. Indeed, even in many homes, blending of flour for preparing food variations has got viral; thus the requirement for a reasonable flour blending machine is on the increase. The challenge of creating blenders with minimal expense had prompted the advancement blending machines which acquired ubiquity in profoundly serious bread kitchens [9]. The execution of exercises is centred around suitable mechanical advancement that meets the prerequisites naming less innovative touch, and venture, and attraction as a saltine massaging machine. The examination technique utilised in this investigation is the designing strategy. The designing technique is to do an arrangement movement, plan, development, applied which isn't standard, so that there are changes and new commitments, both in cycles and items [10].

The primary point of food blending is to guarantee the homogeneity of the items. The gear utilized and the blending power connections are critical to know to appropriately comprehend blending applications in the food business and the intricacies looked by each. Furthermore, the issue of pollution/neatness of the gear can impact the whole creation (clump), which can cause changes in the properties of the item or even lead to the total debasement of the item [11]. After finishing the process of collecting idea and data on how to design the intelligent mixer machine, we come out with several designs of the machine, which are design 1, design 2 and design 3. Furthermore, we had done the research on each design on its advantages and disadvantages as the additional information for choosing the best outcome. Thus, to select the best design, all the three design will go through the screening and scoring process to determine which design is better.

3 Design Selection

In this design selection, there are three sketches of the proposed design before the final design of the mixer machine is being selected for the fabrication.

Figure 1 and 2 shows the first and second design of the project, respectively. Meanwhile, Figs. 3 and 4 show the third and fourth design for our project. The design uses motor and speed reducer to move the blade to rotate the mixture. For the mixing tank, circle shape is being used to increase the performance of the mixer machine.

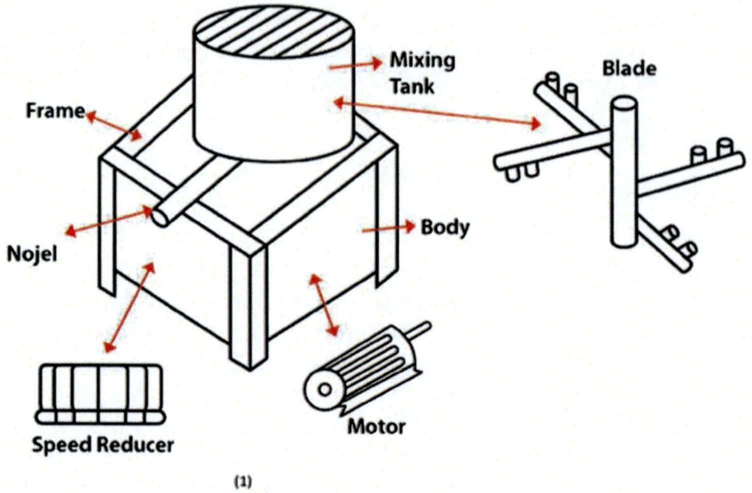

Fig. 1 First design of the mixer

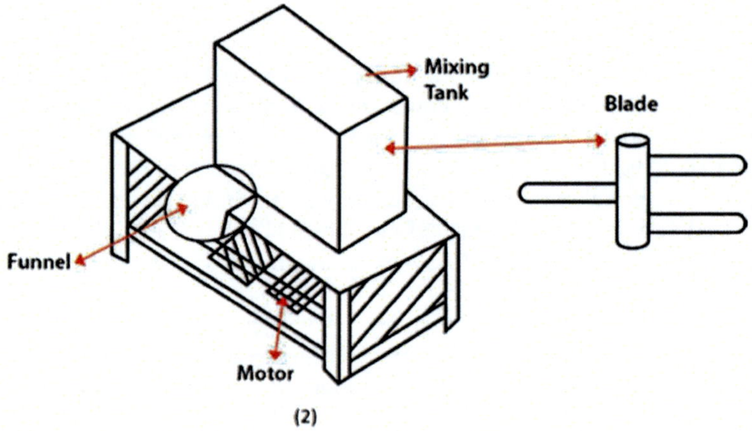

Fig. 2 Second design of the mixer

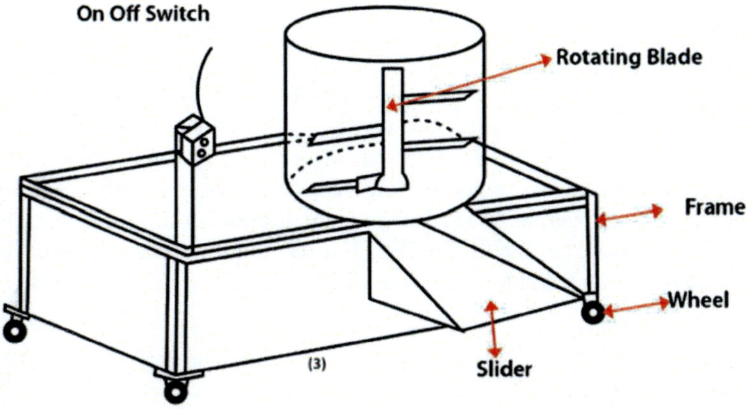

Fig. 3 Third design of the mixer

Fig. 4 Forth design of the mixer

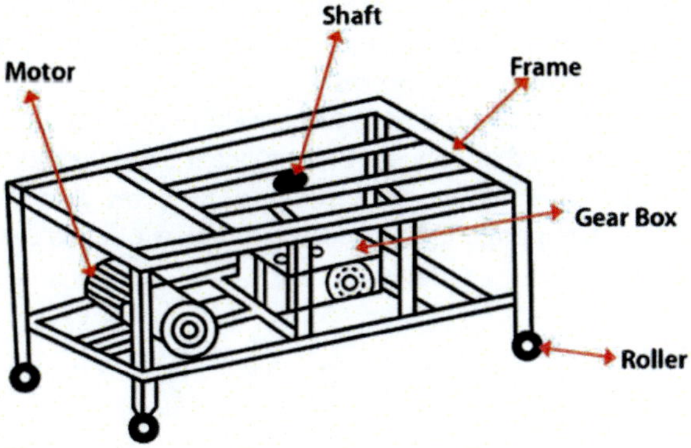

4 Concept Selection

Concept selection is the process of deciding which design is the best and able to cope with the reference concept. There are two types of selection which are screening concept and scoring concept. This process of selection is done by evaluating and comparing the concept of design, strength, and weakness of each concept criterion with the current reference criteria and after comparing and scoring method, the chosen design will be selected as the final design. For this project, design 3 has the highest rating and is being selected for the fabrication.

5 Screening Concept

This concept of screening method is the quickest and inexpensive method to evaluate designs as in Table 1. Screening concept process is enlisted to identify and compare design being able to complete with the reference criteria and which design has the highest promising concepts.

As it can be seen from the screening concept in Table 2, concept 3 is in the first rank followed by concept 1 and concept 2 and the total score for the concept 3 are five which supersede the other design. This scoring method is more specific compared to the screening method because the parts of each design are being evaluated at this process (Tables 3, 4 and 5).

Table 1 Screening concept for design evaluation

	Reference	Concept 1	Concept 2	Concept 3
Safety	0	0	–	0
Manufacturing cost	0	0	+	+
Part availability	0	+	0	+
Stability	0	+	–	+
Easy to assemble	0	+	+	0
Simplicity	0	0	+	0
Ease of machining	0	0	0	0
Durability	0	+	0	+
Appearance	0	–	0	0
Portability	0	–	0	+
Total	0	2	1	5
Rank		**2**	**3**	**1**

Table 2 The scoring concept for body design

	REFERENCE	CONCEPT 1	CONCEPT 2	CONCEPT 3
Description		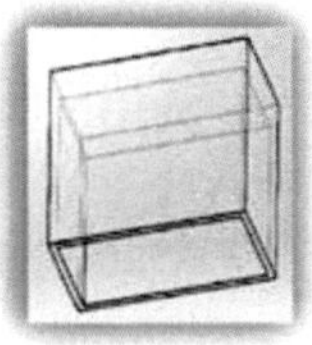		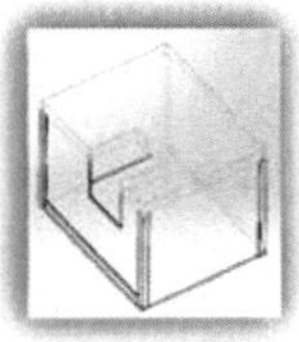
Ease of machining	0	+	0	0
Stability	0	0	-	+
Strength	0	0	-	+
Withstand force	0	-	-	+
Low cost	0	+	+	+
Design simplicity	0	+	+	0
Total	0	2	-1	4
Rank		**2**	**3**	**1**

Table 3 The scoring concept for frame design

	REFERENCE	CONCEPT 1	CONCEPT 2	CONCEPT 3
Description				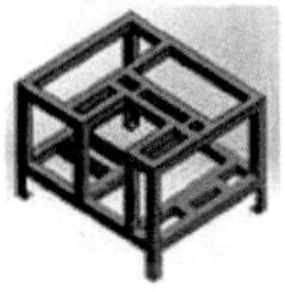
Ease of machining	0	+	0	0
Stability	0	-	0	+
Strength	0	0	0	+
Withstand force	0	-	0	+
Low cost	0	+	+	+
Design simplicity	0	+	+	-
Total	0	1	2	3
Rank		**3**	**2**	**1**

Table 4 Scoring concept for mixing tank design

	REFERENCE	CONCEPT 1	CONCEPT 2	CONCEPT 3
			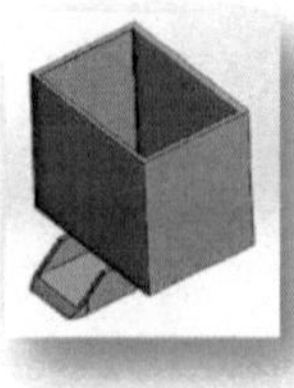	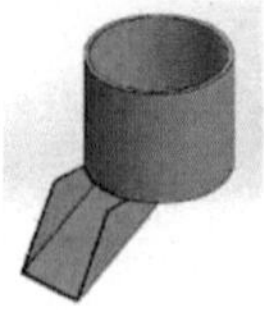
Ease of machining	0	+	0	0
Ease of assemble	0	-	0	+
Ease of use	0	-	+	+
Withstand force	0	-	-	+
Low cost	0	0	+	+
Design simplicity	0	+	0	-
Total	0	-1	1	3
Rank		**3**	**2**	**1**

From the result in the table above, it is concluded that, concept 3 wins against the criteria with the highest score. Therefore, concept 3 is being chosen for the betterment and continuation of the final project. (design 3).

Table 5 The scoring concept for blade design

	REFERENCE	CONCEPT 1	CONCEPT 2	CONCEPT 3
		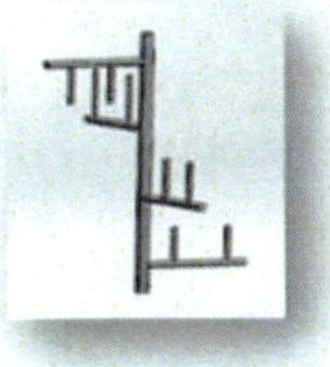	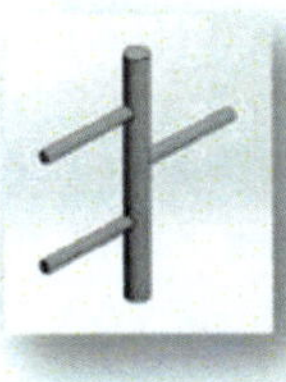	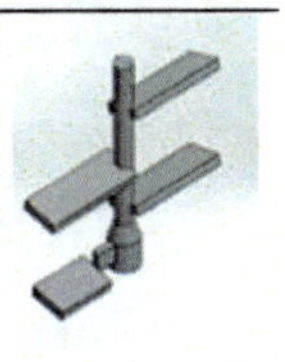
Ease of machining	0	0	+	0
Ease of assemble	0	0	+	0
Strength	0	+	-	+
Design simplicity	0	-	0	0
Low cost	0	+	+	+
Time is taken to mix	0	+	-	+
Total	0	2	1	3
Rank		2	3	1

6 Advantages and Disadvantages for Each Design

Quality review of food and farming production are troublesome. All the while, with expanded assumptions for food results of superior grade and security principles, the requirement for exact, quick and target quality assurance of these attributes in food items keeps on developing [12]. In general, expected functions from a vision machine are the exploitation and imposition of the environmental constraint of a scene, the capturing of the images, analysis of those captured images, recognition of certain objects and features within each image, and the initiation of subsequent actions in order to accept or reject the corresponding objects. After a vision system performs all these stages, the task in hand is almost completed. Here, the sequence and proper functioning of each system and sub-system in terms of high-quality images is explained. In operation, there is a scene with some constraint, first step for the machine is the image acquisition, pre-processing of image, segmentation, feature extraction, classification, inspection, and finally actuation, which is an interaction with the scene under study. At the end of this report, industrial image vision applications are explained in detail [13].

Table 6 Advantages and disadvantages of designs

Design 1	Advantages:
	i. All the parts are easy to design except the blade part ii. The mixing tank has higher durability to withstand high force iii. The cover has a simple design which is easy for the assembling process
	Disadvantages:
	i. The design of the blade is very complex and it needs higher experience to design the blades ii. The frame can't withstand high force and has low stability iii. The machine only can be fixed in one place
Design 2	Advantages:
	i. All the parts have a simple design ii. The cost to make the machine is cheaper because it uses fewer materials iii. The machine has high stability and can withstand high force
	Disadvantages:
	i. The machine has low efficiency because its shape cannot mix all the mixture to the surface of the mixing tank ii. The body cover of the machine isn't fully covered to prevent from the foreign objects iii. The machine only can be fixed in one place
Design 3	Advantages:
	i. The mixer machine can hold extreme weight and has high stability ii. The machine is portable because it has wheels to move the machine iii. The blade and the mixing tank have high efficiency and performance
	Disadvantages:
	i. The machine is very heavy

Table below shows the advantages and disadvantages of each design (Table 6).

7 Final Design

The final design was selected from the concept selection, and it was chosen based on the specification characteristic in the concept screening. Hence, it was confirmed that the design improvement is finalised on the basis of concept design C. The 2D design is later converted to the 3D design by using the SolidWorks software. This software is suitable for more details in the 3D modelling to see the image of the project like the real product. Besides, the designer can eventually identify the wrong position occur in the 3D modelling as it can easily correct it because the software can provide a clear image of the design. Lastly, this software is also equipped with assembling mode whereby the designer can make part by part and assemble them. Figure 5 shows the 3-D sketching, while Fig. 5 shows the 3-D drawing of the final design.

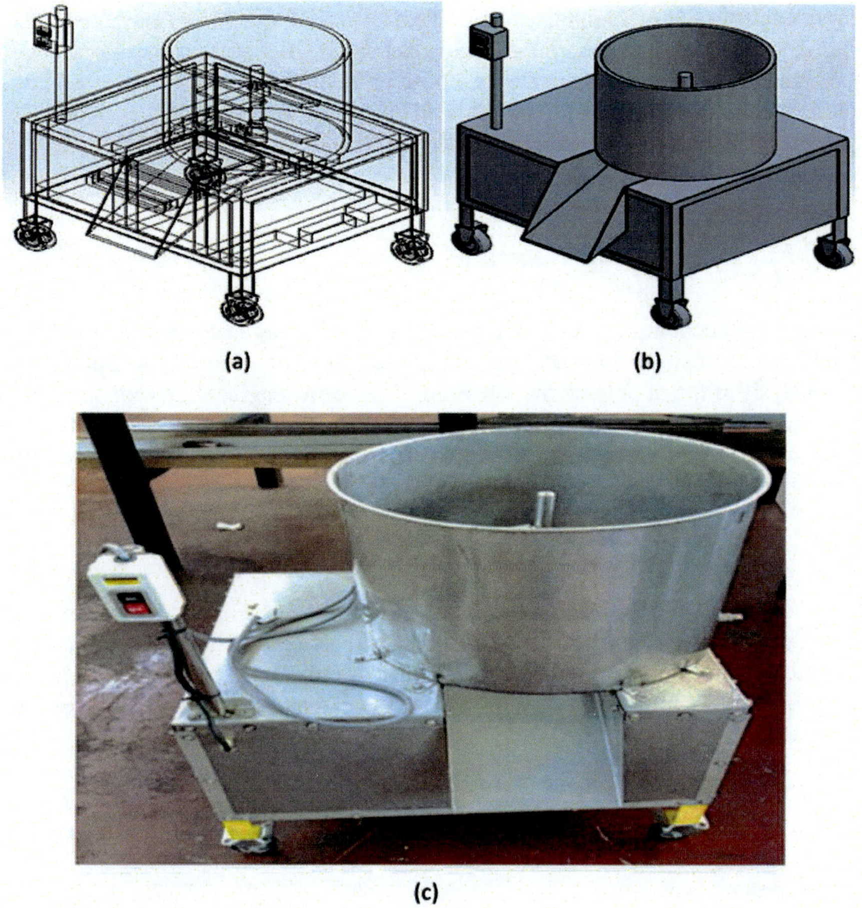

Fig. 5 3-D sketching

8 Results and Discussion

Several improvements can be made for this mixer machine to make it more efficient. The first one is to change the blade design to the hydrofoil blade. Hydrofoil blade will produce better vertical circulation. Advantages of hydrofoils design include higher pumping for a given power and shaft speed, as well as more axial flow pattern and better control of shear-thinning fluids. The next one is to consider the timer rather than the push-button switch for the mixer machine. It will help the owner to decrease the manual labour as the machine only need to be on for the timer, and it will stop automatically. So, the owner does not need to monitor the mixing of the mixture as the timer will be set by the time when the mixing already is done.

8.1 Engineering Economics

Engineering economics, previously known as engineering economy, is a subset of economics concerned with the use and application of the economic principle in the analysis of the of engineering decisions. The term "engineering economic decision" refers to all investment decisions relating to engineering projects. The most interesting facet of an economic decision, from an engineer's point of view, is the evaluation of costs and benefits associated with making a capital investment.

8.1.1 Break-Even Analysis

Break-even analysis in economics, business and cost accounting refers to the point in which total cost and total revenue are equal. A break-even point analysis is used to determine the number of units or revenue needed to cover total costs (fixed and variable costs).

The formula for break-even analysis is as follows:

$$\text{Break}-\text{even quantity} = \text{Fixed costs}/(\text{Sales price per unit}-\text{Variable cost per unit})$$

i. **Variable Cost**

Table 7 Variable cost

Item	Parts	Total cost (RM)
1	Mild steel angle	1600.00
2	Aluminium sheet metal	150.00
3	Aluminium flat bar	200.00
4	304 stainless steel sheet metal	200.00
5	Aluminium alloy rod	30.00
6	Aluminium alloy rod	50.00
7	Speed reducer	800.00
8	Motor drive	850.00
9	Pulley	400.00
10	Belting	200.00
11	Wheels	60.00
12	Switch	10.00
13	Connecting wires	11.00
14	3 pin plug	3.00
15	Miscellaneous	16.50
16	Labor cost	300.00
17	Manufacturing cost	500.00
	Total	RM 5380.50

(See Table 7).

ii. **Fixed Cost**

(See Table 8).

$$Fixed\ Cost, C_f = \text{RM } 1{,}915.00$$

Table 8 Fixed cost

Item	Parts	Total cost (RM)
1	Wire cutter	5.00
2	Wire stripper	10.00
3	Cutter machine	300.00
4	Wielding machine	500.00
5	Drilling machine	300.00
6	CNC machine	200.00
7	Shear machine	100.00
8	Electricity bill	500.00
	Total	1,915.00

$$Variable\,Cost,\,C_v = \text{RM5, 380.50}\;Price,\,p = \text{RM7, 00.00}$$

$$\begin{aligned} v &= \frac{Fixed\,Cost}{Price - Variable\,Cost} \\ &= \frac{Fixed\,Cost}{Price - Variable\,Cost} \\ &= \frac{1915.00}{7000 - 5380.50} \\ &= 1.18\,\text{units} = 2\,\text{units} \end{aligned}$$

$$\begin{aligned} Total\,Cost,\,C_t &= Fixed\,Cost - Variable\,Cost \\ &= \text{RM 1, 915.00} + \text{RM 5, 380.50} \\ &= \text{RM 7295.50} \end{aligned}$$

$$\begin{aligned} Total\,Revenue,\,R_v &= 2 \times \text{RM 7, 295.50} \\ &= \text{RM 14, 591.00} \end{aligned}$$

Figure 6 shows the break-event diagram. We concluded that 2 units of the mixer machine need to be sell to cover the fixed cost.

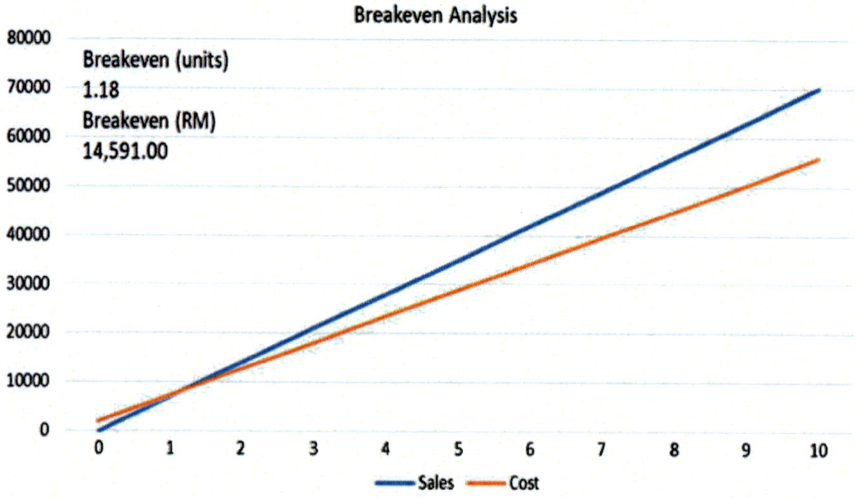

Fig. 6 Break-even analysis graph

8.1.2 Return of Investment Analysis

From the calculation here, the target design to have a profit of RM 100,000. Thus, a need to sell 73 units of this product with the cost of investment of RM 394,691.00. The gain from the investment made is RM511, 000.00 with 30% return of investment.

$$\text{Target profit} = \text{RM } 100,000.00$$

$$\text{Unit to reach target profit} = \text{Target profit} \times v$$
$$\text{Price} - \text{Variable cost}$$
$$= \frac{1000}{70000 - 5380.50} \times 1.18$$
$$= 73 \text{ units}$$

$$\text{Cost of investment} = \text{Fixed cost} + (\text{Variable cost} \times \text{No. of unit})$$
$$= 1915 + (5380.50 \times 73)$$
$$= \text{RM } 394\,691.50$$

$$\text{Cost of investment} = \text{Fixed cost} + (\text{Variable cost} \times \text{No. of unit})$$
$$= 1915 + (5380.50 \times 73)$$
$$= \text{RM } 394\,691.50$$

$$\text{Gain of investment} = \text{No. of unit} \times \text{Price}$$
$$= 73 \times \text{RM } 7000$$
$$= \text{RM } 511,000.00$$

$$\text{Return of Investment (ROI)} = \frac{\text{Gain of Investment} - \text{Cost of Investment}}{\text{Cost of Investment}}$$
$$= \frac{511,000 - 394,691}{394,691}$$
$$= 0.295$$
$$= 30\%$$

9 Improvement and Recommendation

Several improvements can be made for this mixer machine to make it more efficient. The first one is to change the blade design to the hydrofoil blade. Hydrofoil blade will produce better vertical circulation. Advantages of hydrofoils design include higher pumping for a given power and shaft speed, as well as a more axial flow pattern and better control of shear-thinning fluids. The next one is to consider the timer rather than the push-button switch for the mixer machine. It will help the owner to decrease the manual labour as the machine only need to be on for the timer, and it will stop automatically. The owner does not need to monitor the mixing of the mixture as the timer will be set to the time when the mixing already is done. To improve the heating process by low cost energy is a challenge for making the chips. So it has to prepare a safe and cheap boiler for the process control of fish boil and mixing.

10 Conclusion

The device has been designed smoothly and verified by using the SolidWorks software and a simple calculation of torque and power. In brief, throughout completing the fabrication of the mixer machine, all the objectives have been achieved which will comfort the end user for processing keropok lekor. The system is successfully designed to ease the user conveniently. It is fully functional, follows the scale given by the project coordinator as expected from the project and able to cut, slice and mix frozen fish to process keropok lekor into pieces by pre-cutting into a smaller size as well as the system designs have been studied regarding the safety measures.

Acknowledgements The authors would like to thank for financial UIC190816 and RDU192808 assistance for the project.

References:

1. Taewee TJ (2011) Cracker "'Keropok"': a review on factors influencing expansion. Int Food Res J 18(3):855–866
2. Khaizura MN, et al (2020) Quantification of coliform and Escherichia coli in Keropok lekor (Malaysian Fish Product) during processing. J Appl Sci Res. (November):1651–1655
3. Omar M et al (2011) Sustaining traditional food: consumers' perceptions on physical characteristics of Keropok Lekor or fish snack. Int Food Res J 18(1):117–124
4. Mujanah S, Ratnawat T (2016) The improvement of product quality through the appropriate technology for crackers in small scale entrepreneur in kenjeran district of Surabaya
5. Yu S, Tan LJ (1990) Acceptability of crackers ('Keropok') with fish protein hydrolysate. Int J Food Sci Technol 25(2):204–208
6. Murad NS et al (2017) The effect of mixing time and mixing sequence during processing on the physicochemical and sensory properties of keropok lekor. J Sci Technol 9(4):88–95

7. Khidir TJ (2018) Designing, remodeling and analyzing the blades of portable concrete mixture. Int. J. Mech. Eng. Rob. Res. 7(6):674
8. Kushnir V, Gavrilov N, Kim SJ (2016) Justification of the design of the two-shaft mixer of forages. Procedia Eng. 150:1168–1175
9. Okafor BEJ (2015) Design of power driven dough mixing machine. Int. J. Eng. Technol. 5(2):76–79
10. Hendrawan Y et al (2020) The design and application of cracker dough kneading machine for increasing productivity and quality of mackerel crackers in Tambakasri village Malang. IOP Conf. Ser. Earth Environ. Sci. 524(1):012022. https://doi.org/10.1088/1755-1315/524/1/012022
11. Elia B, et al (2011) Food mixing in the industrial processes
12. Patel KK et al (2012) Machine vision system: a tool for quality inspection of food and agricultural products. J Food Sci Technol 49(2):123–141
13. Golnabi H, Asadpour AJ (2007) Design and application of industrial machine vision systems. Rob. Comput.-Integr. Manuf. 23(6):630–637

Temperature Profile of Mixed Mode Solar Cabinet Coconut Dryer

Keith Yvonne B. Diez, Jao Philip A. Yap, and Cresencio P. Genobiagon Jr.

Abstract A low-cost mixed-mode natural convection dryer was designed to dry coconut kernels. The dryer was made of low-cost materials readily available in local areas. The dryer consists of two major parts, the heat collector and the chamber. The study aim to compare the performance of solar dryer with sun drying. The experiment was conducted in the open field at the University of Mindanao, Davao City, Davao del Sur, Philippines. Coconut cups were placed in the drying chamber and dried until the ideal moisture content is achieved. The temperature, and moisture content were observed during the experiment, as well as the weight of coconut. The drying chamber has a maximum temperature of 78.4 °C and a minimum temperature of 64.6 °C under a clear sky sunny weather. The dryer has reduced the initial moisture content of 54.86 to 6.82% in 21 h of drying.

Keywords Copra · Solar dryer · Open sun drying · Heat collector · Natural convection

1 Introduction

The world's largest importers of coconut oil include the European Union, the United States, Malaysia, China, and Korea. In 2018, the European Union alone imported 475,000 metric tons of coconut oil. Dried coconut meat (copra) is squeezed to yield coconut oil, with the byproducts themselves finding many new uses. Fresh coconut kernel contains 32–35% oil, corresponding to a moisture content of 54–49%. Copra is the processed, dried kernel of coconut utilized in the extraction of coconut oil. Copra is utilized for a variety of purposes. It is used as food as an important source of nutrients, including proteins, vitamins, and minerals [1]. It is commonly produced by small coconut farmers using open sun-drying or smoke-kiln methods to reduce its moisture content from 50 to 6%. A process that significantly influences the quality

K. Y. B. Diez · J. P. A. Yap · C. P. Genobiagon Jr. (✉)
Mechanical Engineering Program, College of Engineering Education, University of Mindanao, Matina, Davao City, Philippines
e-mail: cresencio_genobiagon@umindanao.edu.ph

© The Author(s), under exclusive license to Springer Nature Singapore Pte Ltd. 2023
M. H. A. Hassan et al. (eds.), *Technological Advancement in Instrumentation & Human Engineering*, Lecture Notes in Electrical Engineering 882,
https://doi.org/10.1007/978-981-19-1577-2_17

of the product [2]. The long drying process is known to give rise to various problems, including the growth of molds that jeopardizes the quality of the product.

In the Philippine context, the traditional method of copra processing is still widely used. In the coconut sun-drying method, the nut is split and dried under the sun for five days to reach the desired moisture content of 7%. While the process is cheap and straightforward, it exposes the coconut meat to various risks, including uncontrolled drying, insect infestation, and animal attack [3]. It also exposes the product to bacterial damage and growth of molds like the common Aspergillus flavus and other aflatoxin related molds. This has the potential to accumulate and spread around. Aflatoxins are the most potent carcinogens in animal and human populations and interfere with the functioning of the immune system [4, 5]. Aflatoxins in copra cake, fed to animals, can be passed on in milk or meat, leading to human illnesses. In sun drying, the temperature ranges from 25 to 40 °C, with a drying cycle of about 5–7 days. This brings the moisture level of the copra to 10%, making it a slow drying process [6].

A "Tapahan" or a direct smoke dryer is commonly used by Philippine coconut farmers and similarly popular in other copra producing countries. This dryer is made of split bamboo, grill platform where nut halves in shell are placed. The resulting copra from this method is usually dark, and smells of smoke with scorches at times and considerably lower grades. The kiln method also exposes the copra to considerable quality risk including contamination by polycyclic aromatic hydrocarbon (PAH) caused by incomplete combustion, some of which are highly carcinogenic including benzo[a]pyrene, and dibenzo[a,h]anthracene that is present in smoke-cured foods [7, 8]. Although the efficiency of drying is considerably higher than that of natural drying, this method leads to poor quality of dried copra i.e., lousy odor and discoloration. The drying cycle time is about 48 h. In indirect hot-air driers, the coconut meat does not come into contact with combustion gases and smoke from the fuel and hence the name of this drying method. The copra thus obtained is usually of very high quality. However, unless the method of heat exchange is efficient, there is a considerable loss of heat [9].

The drying process of food matter is dependent on various factors. The conditions set by these factors affect the speed and efficiency of the whole drying procedure. These factors include drying temperature, ambient humidity level, and the properties of the coconut being dried. The first one is the drying temperature. In a research in the International Journal of Agricultural Science and Research (IJASR), a 90 °C drying temperature took only 12 h to reach the desired moisture content of 6.1%, while an 80, 60 and 70 °C took 15, 42, and 24 h respectively. However, copra dried at 80 to 90 °C produce low-quality results [10]. Another factor affecting the drying process of the matter is ambient humidity. An increase in air temperature allows it to absorb more liquid, which means humidity is decreased. Low relative humidity helps in the faster drying of solid matter [11]. Cashew nuts were tested for drying on various dryers, including a designed and fabricated convection drying, and showed that moisture of cashew nuts in the convection dryer was brought down to 5% in 6 h, which is 8 h faster than in open-sun drying [12]. The same experiment was also conducted on black turmeric using the same design principle and have shown satisfactory results in achieving the desired product state [13]. The properties of the coconuts are also factors that have a considerable effect on the process of drying. Nuts vary a lot, and there is no constant size or moisture content. This affects the drying process; similarly, the age of nuts also affects the drying procedure [14].

Mechanical dryers were employed using available fuel such as coconut husk and shell. The use of solar dryers for coconut were not evident in rural areas due to lack of expertise and knowledge of this methods. The use of a solar coconut dryer working on the principle of natural convection drying holds the potential of preventing these problems. The objectives of this study is to design and fabricate a mixed mode solar dryer for coconuts. It aims to determine the performance of the dryer in terms of its temperature profile and its drying rate.

The design, fabrication, and development of a practical and efficient direct heat solar, natural convection coconut dryer holds the potential of speeding up the copra production among small and medium-sized coconut farmers in the country without additional cost on energy. The Philippines is one of the four countries, which are main producers of coconut, together with India, Indonesia, and Sri Lanka [19]. It will benefit future studies that aim to develop a more sophisticated or similar but improved design on direct solar heat natural convection drying.

This study will utilize locally available and cheap construction materials without any variation on the materials' types, dimensions, and specifications. It will also be conducted in a definite range of time with no regard to changes in atmospheric temperature, given that the testing will be done under the availability of the heat source (the sun). It will also be tested on a particular set of coconuts for drying with no variation on coconut variety, size, and specification. Moreover, the testing will only be conducted within the City of Davao and not on areas that have a potentially higher or lower atmospheric temperatures.

2 Materials and Methods

2.1 Conceptual Framework

The conceptual framework of the study (see Fig. 1), the inputs include the properties of coconut meat, the drying temperature and the average solar insolation. The process will involve the design and fabrication of the solar coconut dryer and the performance of evaluation of the solar dryer. The output is the mixed mode solar dry dryer. The intervening factors of the process are the variation in heat input from the heat source (solar), variation in the internal air temperature of the dryer, and the span of time.

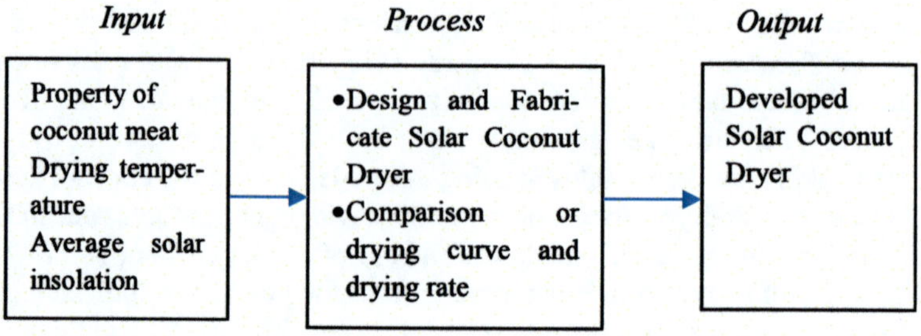

Fig. 1 Conceptual framework

2.2 Materials and Resources

Various materials are to conduct this study, such as glass cover, since it is commonly used in dryers for more heat absorption [18, 19]. Also, plywood was purchased from a local lumber store. It would serve as the frame of the dryer. The plastic screen mesh was bought by one of the researchers from a local hardware store, together with the black paint, paintbrush, common nails, wood glue, and hinges. Measuring materials such as steel rule, Steel Square, and tape measure were borrowed from the school shop. The coconuts used were purchased from a local fruit market and weighed for about 6 kg. All materials were purchased from local hardware stores. The researchers thoroughly canvassed every local hardware, and made sure that all materials were low-cost but with good quality.

2.3 Design

During the designing stage, the proponents made sure that all factors, which may affect the performance of this study, were considered. There are previous journals such as [15–17], which provided the researchers with ideas that can help them on how to make a design for the machine. The researchers came up with two concepts based on the references. The first design concept was an indirect dryer with a separated heat collector, while the second design is a combined direct and indirect dryer wherein the chamber is in direct contact with the heat collector.

Design Concepts. The concept made by the researchers (see Fig. 2) is an indirect type of dryer, which has a separating air dryer or also knows as a heat collector. This type of dryer uses the forced convection mode, which requires a fan that exhausts the air from the inlet to the outlet, which could be a chimney [15, 16]. The heat collector is connected to the drying chamber through a pipe [17]. The second design concept (see Fig. 3), which is a combination of a direct and an indirect type of dryer.

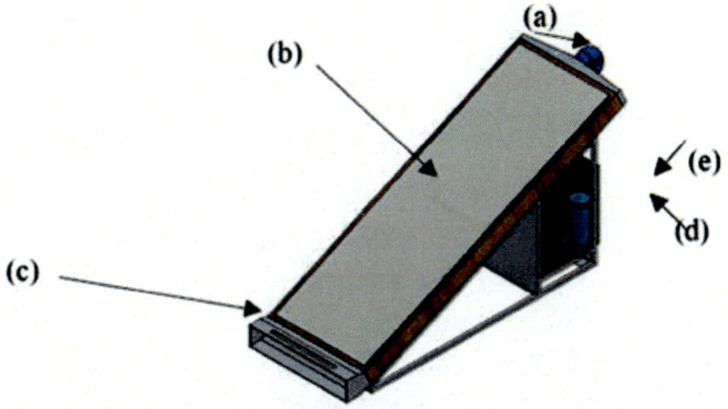

Fig. 2 Concept 1 Assembly **a** a connecting pipe installed with fan, **b** Solar Collector, **c** Air Inlet, **d** Air Outlet (chimney), **e** Chamber

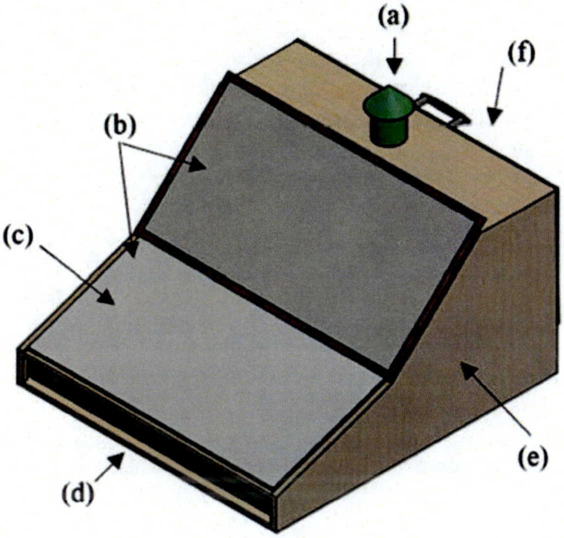

Fig. 3 Concept 2 Assembly **a** Chimney (air outlet), **b** Glass Cover, **c** Heat Collector, **d** Air Inlet, **e** Wooden Frame, **f** Door

2.4 Fabrication

The frame used for the dryer was fabricated by the researchers at the school shop where they were able to borrow tools, which significantly supported the fabrication process. Nevertheless, the researchers still had to ask for guidance and approval from experts from the field in food drying, thus requiring assistance for the overall process of fabrication.

2.5 Testing

The testing was done until the coconut inside the dryer achieve the ideal moisture weight. The set-up was using a digital Hygrometer as a measuring device. It would measure both the temperature and relative humidity inside and outside the surface of the dryer. The researchers bought two batches of coconuts from a local fruit market, and each weighed about 6 kg. The first batch would be used for the dryer, and second is for open sun drying. The researchers used a moisture meter for measuring the moisture content of the coconut. The mean initial moisture content of the coconut was 54.86%. A digital weighing scale is also used for measuring the weight of the coconuts. The two batches of coconuts were cut into half. The first batch was then placed onto the tray made of the plastic screen inside the dryer. The second batch was laid on the field for open sun drying. The drying time is at 9:00 AM–4:00 PM. The weight and moisture content of the coconut would be measured every hour as well as the chamber and surroundings' temperature during drying time. The testing was conducted under sunny weather and cloudy weather.

From the studies [17, 18], proponents were able to calculate the moisture ratio of the product. The amount of the moisture ratio of the product can be calculated using the formula presented in Eq. (1):

$$\text{Moisture ratio (MR)} = \frac{M}{Mo} \tag{1}$$

where M is the moisture content at any time, while Mo is the initial moisture content of the product, the average drying rate can also be computed using Eq. (2):

$$\text{Average drying rate, } M_{ave} = \frac{m}{t} \tag{2}$$

M is equivalent to the amount of moisture removed while t is the number of daily sunshine hours.

3 Results and Discussions

3.1 Design and Fabrication

The second concept was selected because it followed most of the desired objectives for the study, which also includes the economic aspect. The selected concept is a combined direct and indirect type of dryer, which uses the natural convection mode to use airflow by buoyancy force [15, 16]. The actual dryer (see Fig. 4), which was successfully fabricated. It also comes with a door on the rear side of the dryer where products to be dried pass through. The drying cabinet consists of 2 drying trays. The dimension of each tray is 12 × 46 in. The trays are fitted with a wooden frame and

Fig. 4 Actual dryer

uses plastic screen as its container. The frame of the dryer is made of marine plywood with ¾ in. thickness, as well as the heat collector. The heat collector and the overall frame of the dryer are coated with flat black acrylic paint for more emissivity [13]. The heat collector, and the drying cabinet is covered with two 24 × 46 in. glass cover with ¾ in. thickness. To achieve natural air circulation, three air inlet rectangular holes were put on the bottom front of the dryer, and a chimney as the air outlet was placed at the top of the dryer. Each air inlet has a dimension of 3.5 × 1.5 in. The air outlet or chimney was bought in a local hardware store.

3.2 The Functionality of the Equipment

Based on the studies of previous journals, the proponents have found a method for analyzing the essential data in this study. Both inside, and outside the temperature of the dryer for each hour of the whole duration of the drying process.

The maximum inside surface temperature of the dryer was 78.4 °C at 2:00 PM under a clear sky condition on the first day with load testing with a corresponding outside temperature of 41.7 °C (see Fig. 5). The mean inside chamber temperature during the first day was 74.5 °C, while the mean outside surface temperature was 44 °C.

The relative humidity for each hour of the drying process (see Fig. 6). On the first day with load testing under clear sky weather, the minimum inside relative humidity

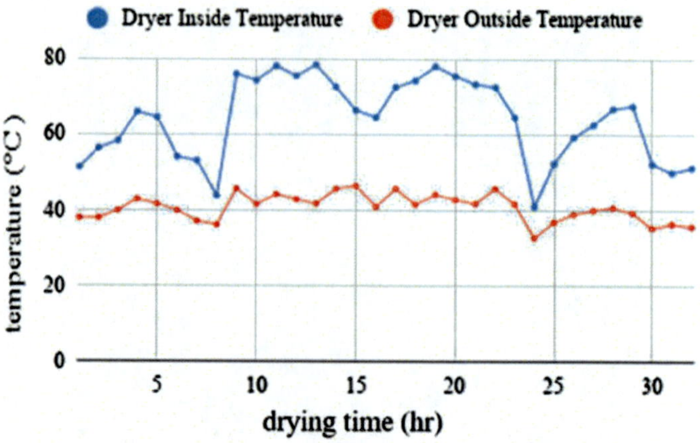

Fig. 5 Temperature profile of the solar cabinet dryer

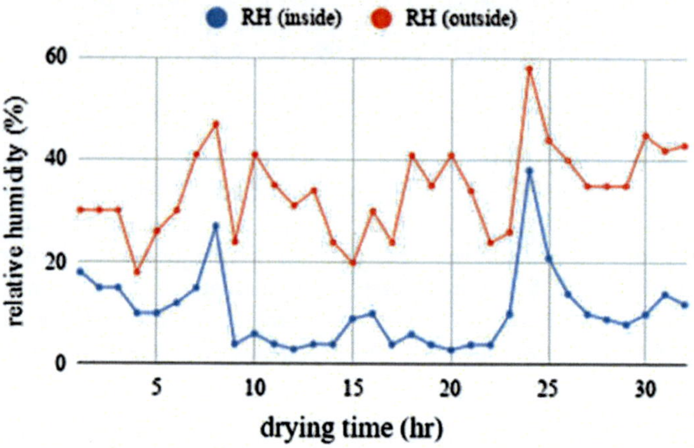

Fig. 6 Hourly Relative Humidity of the solar dryer

was 4%, accompanied by 35% outside ambient relative humidity. On the second day, the minimum inside relative humidity ranged to 3% while having an ambient relative humidity of 31%. When the dryer attained the minimum inside temperature and outside temperature of 41.1 and 32.8 °C, respectively, among all recorded data, it also came with a maximum inside relative humidity of 38 and 58% ambient relative humidity on the third day of the drying process.

The physical appearance comparison between the use of the dryer and open sun-drying method is also presented in this section [10]. Figure 7 shows the physical difference between the use of the dryer and open sun drying. Figure 7a shows the result of the use of the dryer while Fig. 7b shows the result of the open sun drying.

Fig. 7 **a** Dryer result **b** Open sun-drying result

The use of the dryer produced some burnt kernels; this was due to exposure of high temperature. The open sun drying method produced dried kernels with molds present.

Figure 8 shows that moisture ratio decreases continuously with drying time for both machine drying and open sun drying The continuous decrease of the moisture ratio with time indicates the effectiveness of the dryer. The machine dried samples took less time to achieve the moisture ratio of 0.124, which came with 6.82% of moisture content.

Table 1 shows the weight of each batch of coconuts dried over time using the dryer and open sun drying method. Each batch had an initial weight of about 6 kg. In 21 h of drying, the machine dried coconuts have lost much higher than the open sun drying. The moisture loss of the copra from the dryer resulted in 48.044% in 21 h while in open sun drying, the copra had a moisture loss of 22.95%. The average drying rate of the dryer is equivalent to 0.02288 kg/kg-hr for three days of drying, which is higher in capacity than of the open sun-drying that results in 0.01093 kg/kg-hr.

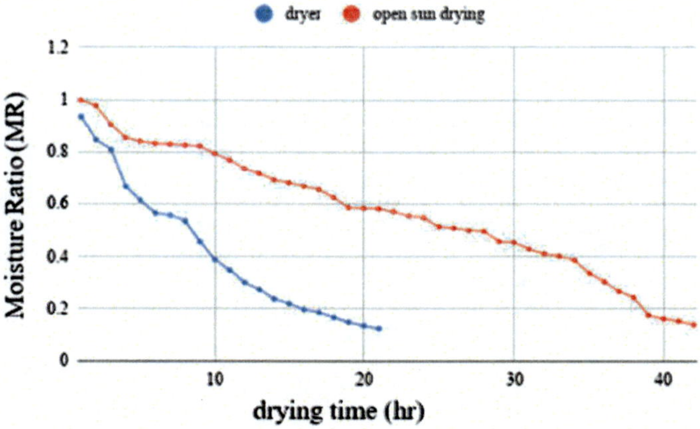

Fig. 8 Moisture Ratio (MR) of coconut during solar dryer and sun drying

Table 1 Weight analysis

Hour	Dryer (kg)	Open sun drying (kg)	Hour	Dryer (kg)	Open sun drying (kg)
1	6.442	6.458	12	4.139	5.635
2	6.209	6.451	13	3.976	5.521
3	5.901	6.378	14	3.886	5.454
4	5.765	6.12	15	3.754	5.367
5	5.263	5.946	16	3.689	5.326
6	5.078	5.895	17	3.608	5.28
7	4.902	5.862	18	3.567	5.236
8	4.875	5.852	19	3.496	5.129
9	4.804	5.839	20	3.432	4.992
10	4.528	5.826	21	3.385	4.983
11	4.284	5.725	22	3.347	4.976

4 Conclusions and Future Works

Drying equipment was designed and developed, the device used solar heat of the sun. In this study, the unit was tested in the open field inside the University of Mindanao, Maa, Davao City. It is located at latitude 7.0635°N and longitude 125.6082°E where enough sunlight is available. The equipment produced copra with a moisture content of 6.82% in a 21 h of drying and have shown better performance than the previous study where the moisture content of 6.74% was attained in 28 h [20].

The drying process can be made faster by augmenting the heating capacity of the dryer by using other heating materials and increasing the surface area of the equipment.

References

1. Shcheglov S (2018) What is copra and how is it made? WorldAtlas
2. Kerala Agricultural University (2011) Package of practices recommendations: crops, 14th edn. Kerala Agricultural University, Thrissur, p 360
3. Kulanthaisami S, Subramanian P, Mahendiran R, Venkatachalam P, Sampathrajan A (2009) Drying characteristics of coconut in solar tunnel dryer. Madras Agric J 98:265–269
4. Karunarathna NB, Fernando CJ, Munasinghe DMS, Fernando R (2019) Occurrence of aflatoxins in edible vegetable oils in Sri Lanka. Food Control 101:97–103
5. Padmanaban G, Palani PK, Thilak VMM (2017) Grey relation analysis of solar drying process parameter on copra. Ital J Food Sci 29:434–442
6. Singh L, Varshney JG, Agarwal T (2016) Polycyclic aromatic hydrocarbons' formation and occurrence in processed food. Food Chem 199:768–781
7. Kumar P, Mahato DK, Kamle M, Mohanta TK, Kang SG (2017) Aflatoxins: a global concern for food safety, human health and their management. Front Microbiol 7:2170
8. ILO (1992) Small-scale oil extraction from groundnuts and copra. https://www.appropedia.org/Welcome_to_Appropedia

9. IDMP& DpGI (2014) Solids drying: basics and applications. Chemengonline
10. Deepa J, Rajkumar P, Arumuganathan T (2015) Quality analysis of copra dried at different drying air temperatures. Int J Agric Sci Res 5:1–5
11. Sasongko SB, Hadiyanto H, Djaeni M, Perdanianti AM, Utari FD (2020) Effects of drying temperature and relative humidity on the quality of dried onion slice. Heliyon 6:e04338
12. Dhanushkodi S, Wilson VH, Sudhakar K (2017) Mathematical modeling of drying behavior of cashew in a solar biomass hybrid dryer. Res Efficient Technol 3:359–364
13. Kondareddy R, Sivakumaran N, Nayak PK (2019) Drying kinetics of black turmeric (Curcuma caesia) with optimal controller assisted low cost solar dryer. Food Res 3:373–379
14. Chaudhari AD, Salve SP (2014) A review of solar dryer technologies. Int J Res Advent Technol 2:218–232
15. Sharma K, Wadhawan N (2018) Effect of natural and forced convection solar dryers in retention of proximate nutrients in tomato. Int J Curr Microbiol Appl Sci 7:1175–1186
16. Jiskani SA, Ahmed Chandio I, Mehdi G, Memon AH, Raqeeb Bhutto A, Sandilo UG (2020) Fabrication & performance analysis of direct type passive solar dryer for chilies and grapes drying. In: 2020 3rd international conference on computing, mathematics and engineering technologies: idea to innovation for building the knowledge economy, iCoMET 2020
17. Tabassum S, Bashar M, Islam M, Sharmin A, Debnath S, Parveen S et al (2019) Design and development of solar dryer for food preservation. Bangladesh J Sci Ind Res 54:155–160
18. Genobiagon CP Jr, Alagao FB (2019) Performance of low-cost dual circuit solar assisted cabinet dryer for green banana. J Mech Eng Res Dev 42(1):42–45
19. Bakti AI, Gareso PL (2018) Characterization of active carbon prepared from coconuts shells using FTIR, XRD and SEM techniques. J Ilm Pendidik Fis Al-Biruni 7:33–39
20. Krishna VS, Mathew G (2018) Development of a solar copra dryer incorporated with evacuated tubes. Int J Curr Microbiol Appl Sci 3:2655–2658

Performance of Solar Cabinet Dryer Utilizing Thermosyphon Water Heater

Cresencio P. Genobiagon Jr.

Abstract The model of optimal design of solar assisted cabinet dryer was formulated for green banana. The simulation model is composed of three systems: one for the water collector, one for the air collector and other for the drying cabinet. Economic and manufacturability factors were considered in the optimal design. A solar cabinet banana dryer utilizing a thermosyphon solar water heater as the main source of heat and a solar air heater as auxiliary source was tested with green Cavendish variety of banana as the drying specimen. From period April to June the experiment was conducted in the rooftop to ensure good solar irradiance. It was found out the solar water collector can generate a maximum water temperature of 60 °C and a drying chamber temperature of 50 °C during noon and at clear sky condition. Also the air heater generates about 40 °C. The temperature profile of the drying chamber resembles a sine curve. The temperature profile was lower than that of the mathematical model. These were contributed from the construction of the solar heater but overall performance.

Keywords Air collector · Banana solar dryer · Cabinet dryer · Thermosiphon · Water heater

1 Introduction

1.1 Background of the Study

In the Philippines, Region XI is the leading producer of Cavendish Banana. The majority of these plantations are independent small-scale banana growers. Banana produce is exported to the Middle East, Japan and China.

Low quality bananas are peeled, sliced and sun dried. These dried bananas are the raw materials for the growing banana flour industry and banana feed meal for the livestock industry. The drying process is done chiefly through sun drying mostly on

C. P. Genobiagon Jr. (✉)
Mechanical Engineering Program, University of Mindanao, Davao City, Philippines
e-mail: cresencio_genobiagon@umindanao.edu.ph

© The Author(s), under exclusive license to Springer Nature Singapore Pte Ltd. 2023

M. H. A. Hassan et al. (eds.), *Technological Advancement in Instrumentation & Human Engineering*, Lecture Notes in Electrical Engineering 882,
https://doi.org/10.1007/978-981-19-1577-2_18

roadsides, pavement, GI sheets, stone slabs and concrete basketball courts [1]. Dried products are subject to contamination by extraneous materials such as sand, stones, soils, leaves and incursion by rats, insects and animal excreta [2]. During the rainy season, discoloration and spoilage are inevitable [3]. Flour mills can acquire large scale drying technologies but banana growers and farmers don't have this alternative and only use sun drying.

There are three categories of drying methods including sun drying, solar drying and mechanical drying. Sun drying and solar drying utilizes sun's energy. Solar dryers, compared to sun drying can generate higher air temperature and lower relative humidity. Mechanical dryer uses energy resources such as biomass, fossil fuel and electricity [4].

Solar dryers are generally classified based on whether the product is directly exposed to insolation, the mode of air flow through the dryer, and the air temperature circulated to the drying chamber. In direct dryers, the crop is exposed to the sun but, in indirect dryers, the crop is placed in an enclosed drying chamber and thus shielded from insolation [5]. Indirect dryers can be either passive (natural convection) or active (forced convection) dryer. Solar collectors are directly coupled with the dryer. A greenhouse dryer is a type of direct solar dryer. These dryers are either dome or flat roof structures [6]. Dome type can have maximized the global solar radiation while the flat roof structure take advantage with good mixing of hot air. Solar dryers can utilized a thermal storage system and water ha a good thermal storage properties [7]. Water heaters were employed as heat source for indirect dryers [8]. It consist of a drying cabinet, heat exchanger, water heater solar collector and a heat storage tank.

Čipliene, Novošinskas, Raila, and Zvicevičius [9] made an experimental dryer that utilized two types of solar collectors. The air heater used for direct heating of the drying agent and the flat plate solar collector accumulated the converted energy and stored it in the tank. Daghigh and Shafieian, [10] utilized a heat-pipe evacuated tube solar dryer with heat recovery system and water as the heating medium.

The alternative drying technique to be described in this study is an indirect solar dryer. The dryer will utilize two types of collectors, and a thermosyphon solar water heater was used to supply heat in the drying chamber by a heat exchanger. Supply air was preheated in solar air heaters. However, available water heaters such as the water-in-glass evacuated tube solar heaters and the heat pipe solar heaters were expensive and were not appropriate for small-scale farmers. Thus study investigates the actual performance of the solar dryer and compares it to the simulated result.

1.2 *Design of the Solar Cabinet Dryer*

Indirect solar dryers are those in which the crops are placed in an enclosed drying cabinet, thereby being shielded from direct exposure to solar radiation. An indirect solar dryer basically consists of two major components: an air heater, which is used to raise the temperature of the drying air and a drying chamber which is the enclosure that accommodates the crops. Kwendakwema [11] designed and tested an indirect forced air solar food dryer consisting of flat-plate solar collector, drying cabinet and a brick heat storage unit. This solar cabinet dryer consist of three major parts: the solar water heater; the solar air heater; and the drying cabinet.

The dryer utilized a thermosyphon solar water heater. The 0.84 m^3 flat plate solar collector is made of six-1.4–10 mm copper tubes and 20 mm-riser and header. A 0.29 mm aluminum plate acts as the solar absorber, painted black, molded to the copper tube with a silicon adhesive to eliminate air pockets. The collector is covered with 2 layer 5 mm glass panel and insulated with 25 mm thick mineral wool. The solar air heaters consist of two solar collectors with a total area of 1.5 m^3. The force convection collector utilized a used car air conditioning blower.

The drying cabinet consists of 4 drying trays. The dimension of the tray is 40 × 50 cm. A heat exchanger made of an old car radiator with the radiator fan serves as a circulating fan.

2 Theoretical Mathematical Model

A mathematical model was developed to simulate the performance of the solar water heater, the air collector and the drying cabinet.

2.1 *Solar Collector*

The mathematical model of the solar heater was the energy balance of the collector taking into consideration a steady-state condition. Under steady-state conditions, the useful energy output of the collector [12]

$$Q_u = A_c\left[S - U_L\left(T_{pm} - T_a\right)\right] \tag{1}$$

and can be expressed in terms of fluid inlet temperature and the collector removal factor

$$Q_u = A_c F_R[S - U_L(T_i - T_a)] \tag{2}$$

the collector removal factor can be expressed as

$$F_R = \frac{\dot{m}C_p}{A_C U_L}\left[1 - exp\left(-\frac{A_c U_L F'}{\dot{m}C_p}\right)\right] \tag{3}$$

The collector efficiency factor F’ is the ratio of actual useful energy gain to the useful gain of the collector absorbing surface at the local fluid temperature. This collector efficiency factor is constant at a particular collector design. This research collector design, the collector efficiency of the water heater collector is,

$$F' = \frac{1}{\frac{WU_L}{\pi Dh} + \frac{WU_L}{C_{bond}} + \frac{W}{D+(W-D)F}} \tag{4}$$

$$F = \frac{\text{Tanh}\, m(W-D)/2}{m(W-D)/2} \tag{5}$$

and for the air heater collector, is

$$F' = \frac{1}{1 + \frac{U_L}{h_1 + \frac{1}{\frac{1}{h_2} + \frac{1}{h_r}}}} \tag{6}$$

where

$$h_r = \frac{\sigma(T_1^2 + T_2^2)(T_1 + T_2)}{\frac{1}{\varepsilon_1} + \frac{1}{\varepsilon_2} - 1} \tag{7}$$

the mean fluid temperature and the mean plate temperature are expressed as

$$T_{fm} = T_{fi} + \frac{Q_u/A_c}{F_R U_L}(1 - F") \tag{8}$$

and

$$T_{pm} = T_{fi} + \frac{Q_u/A_c}{F_R U_L}(1 - F_R) \tag{9}$$

The collector's overall heat transfer coefficient can be determined by adding the top loss coefficient and the bottom coefficient. The bottom coefficient is easily determined by conduction heat transfer analysis and in some cases of air heaters, convection analysis is also considered. The top loss heat transfer coefficient empirical equation develop by Klien [4] is useful for hand and computer computations.

$$U_t = \left(\frac{N}{\frac{C}{T_{pm}}\left[\frac{(T_{pm}-T_a)}{(N+f)}\right]^e} + \frac{1}{h_w} \right) + \frac{\sigma(T_{pm} + T_a)(T_{pm}^2 + T_a^2)}{\frac{1}{\varepsilon_p + 0.00591 N h_w} + \frac{2N+f-1+0.133\varepsilon_p}{\varepsilon_g} - N} \tag{10}$$

where

$$f = (1 + 0.089h_w - 0.1166 h_w \varepsilon_p)(1 + 0.078866N)$$
$$C = 520(1 - 0.000051\beta^2) \text{ for } 0 < \beta < 70: \text{ for } 70 < \beta < 90. \text{ use } \beta = 70$$
$$e = 0.4330(1 - 100/T_{pm})$$

The fluid motion of the thermosyphon developed from the momentum balance of the collector tube[13].

$$\frac{d\dot{V}}{dt} = -B - F\dot{V} \tag{11}$$

the "bouyancy" term B is expressed as

$$B = \frac{\frac{\pi g}{4\rho}\sum_{k=1}^{n}\rho_k L_k sin\theta_k}{\sum_{k=1}^{n}\frac{L_k}{D_k^2}} \tag{12}$$

the "frictional" term F is expressed as

$$F = \frac{\frac{128\mu}{4\rho}\sum_{k=1}^{n}\frac{(L_{eq})_k L_k}{D_k^3}}{\sum_{k=1}^{n}\frac{L_k}{D_k^2}} \tag{13}$$

3 Experimentatal Setup

3.1 Solar Cabinet Dryer

The solar dryer consists of three heating systems integrated into one; a water heater collector, an air heater, and a drying chamber. The cabinet dryer was set up on the rooftop to minimize the shades of trees. An open ground was set up was not possible because of the restriction of data logging equipment. The collector was facing north. Both collectors are inclined 30° with the horizontal (See Fig. 1).

3.2 Data Collection

Data collection was done with DATATAKER 80 series. The temperature was taken by a type k thermocouple placed at entrance and exit of each collector and heat exchanger Calibration of these thermocouples were made using the boiling and ice points of water. Instrument accuracy (±1%)was determined. Solar radiation was measured by a pyranometer(Kipp and Zonen model CP3 accuracy ±5%). The weight of the banana was measured by a load cell. Measurements signals from the thermocouple, pyranometer and load cell (accuracy ±5%) were recorded every 15 s by a 15-channel data logger (DATATAKER DT80 series 3). The air velocity was measured manually using a hot wire anemometer (Lutron AM4204 accuracy ±5%).

The experiment runs from 7:00 AM to 6:00 PM at clear sky conditions and cloudy conditions. Green bananas were peeled, washed and sliced into 2 mm thick slices. A 5 kg sample were evenly distributed to the four drying trays. The tank of the water heater serves as energy storage.

Fig. 1 The solar cabinet Dryer

3.3 Simulation

The solution of the mathematical model is simulated in MS Excel. An Excel Macro was created using the Visual Basic features of Excel to solve the performance and characteristics of the collector. A temperature profile was created using the actual isolation data from the data logger. Using the excel spreadsheets, the numerical calculation of Eq. 10 determines the overall heat transfer coefficient.

4 Results

The experiment was performed during summer from April 2012-June 2012. Experiment was taken out during a clear sky and cloudy sky conditions. The insolation during the clear sky and cloud sky, respectively. These insolation values were used to determine the projected collector mean plate temperature and the collector mean fluid temperature (see Fig. 2 and Fig. 3).

Results of the simulation and the actual data collected was compared. Using the Normalized root-mean-square deviation (NRMSD), the two data sets were compared [13]. During the clear sky simulation and experimental setup (Fig. 4), the performance of the collector was compared to the simulation results and the actual data from the logger. The figure clearly show that the simulated condition were higher than the

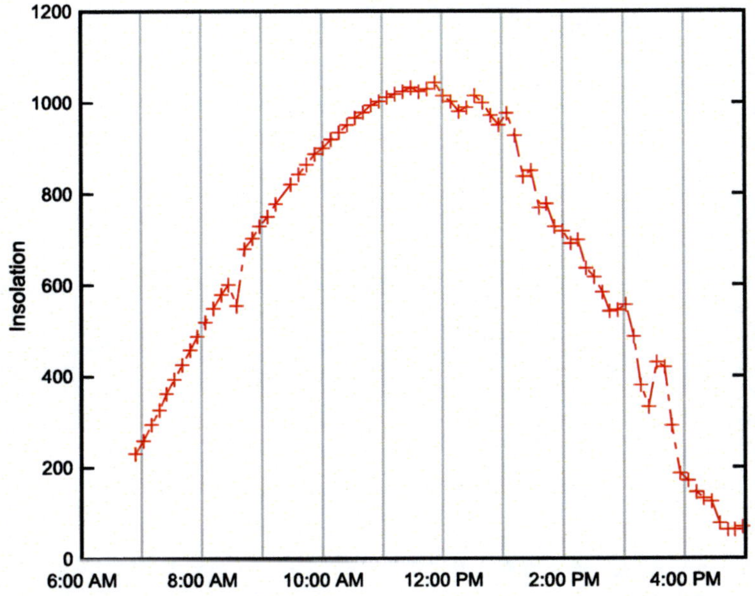

Fig. 2 Clear sky solar insolation

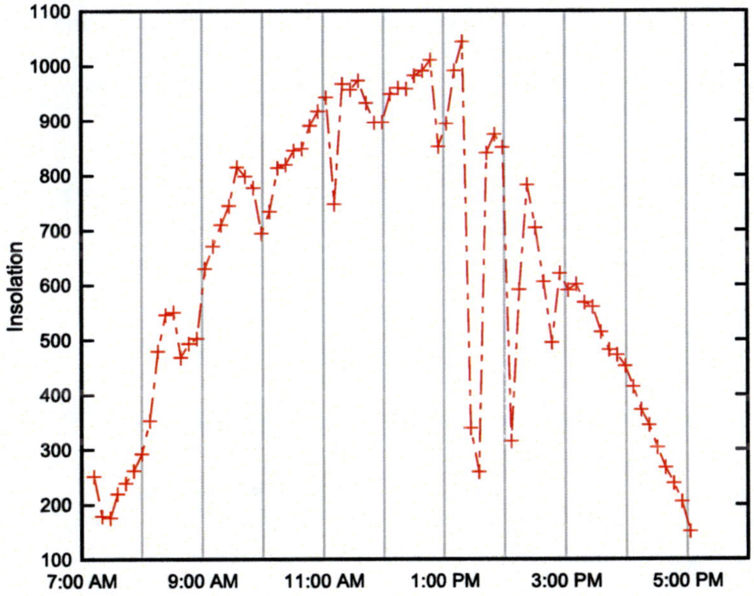

Fig. 3 Cloudy sky insolation

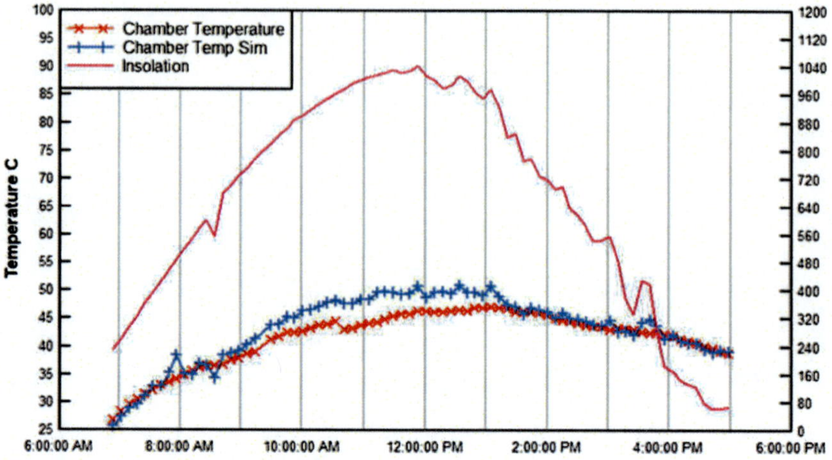

Fig. 4 Chamber temperature vs. simulation result during clear sky condition

experimental value. Moreover, shown in Fig. 5 is the cloudy sky simulation and experimental setup.

The temperature profile of the water heater at clear sky condition was compared to the simulated temperature (see Fig. 6). Both inlet and outlet temperature shows an NRMSD value of 8 and 8% respectively. Moreover, temperature during the intermetent cloudy condition was compared to the simulated temperature and shows both inlet and outlet temperature shows NRMSD value of 30 and 18% respectively (see Fig. 7).

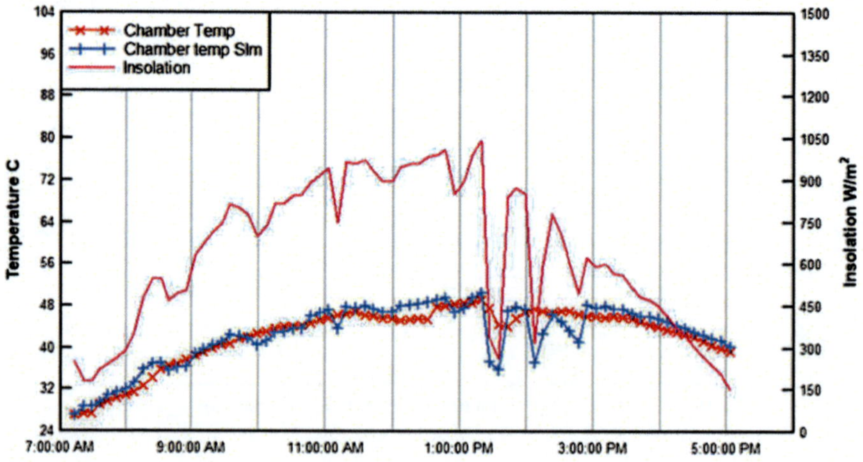

Fig. 5 Chamber temperature vs. simulation result during cloudy condition

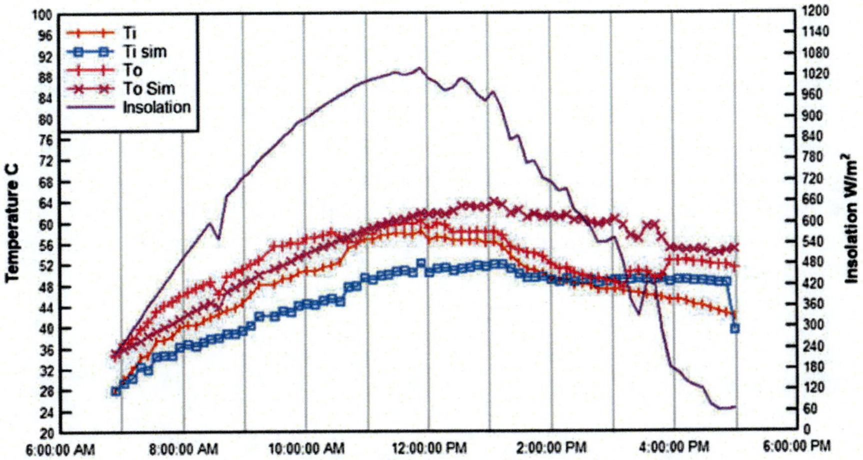

Fig. 6 Clear sky temperature profile of water heater vs simulation results

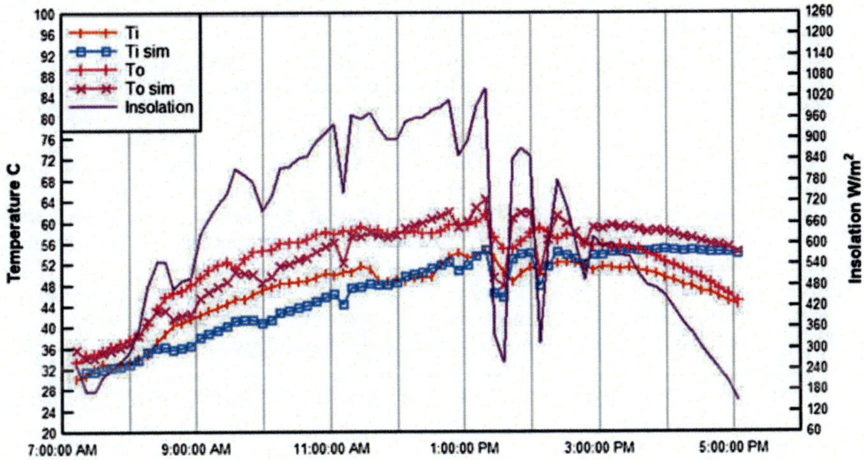

Fig. 7 Cloudy sky temperature profile of water heater vs simulation results

5 Conclusion and Recommendation

The experimental solar dryer performance was evaluated. Solar insolation and chamber temperature was monitored during testing. Actual temperatures were below the simulated values may points to several factors in the construction of the dryer. The maximum water temperature and the chamber temperatures were 56 and 48 °C. The results can be applied in the performance of the water heater and the cabinet dryer when considering the temperature generated by the collectors and the heat exchanger in general. These temperatures are sufficient enough for low temperature drying of food products. The solar dryer efficiency when intermittent cloud cover was observed was comparable with the efficiency on clear sky conditions. This is an indication that the heat storage is dual collector system that address the problem of variability of the solar radiation.

An opportunity on the design of the collector, a study may have been conducted to investigate the behavior of the solar water heater during high solar radiation and the temperature between the inlet and outlet on the water collector. A study on the thermal storage capacity of the liquid water and considered as a secondary heat source during nighttime extended drying hours.

References

1. Murali S, Amulya PR, Alfiya PV, Delfiya DSA, Samuel MP (2020) Design and performance evaluation of solar - LPG hybrid dryer for drying of shrimps. Renew Energy 147:2417–2428
2. Kaddumukasa P, Kyamuhangire W, Muyonga J, Muranga FI (2005) The effect of drying methods on the quality of green banana flour. In: African crop science conference proceedings
3. Tomar V, Tiwari GN, Norton B (2017) Solar dryers for tropical food preservation: thermophysics of crops, systems and components. Sol Energy 154:2–13
4. Sandeep TN, Channabasamma BB, Gopinandhan TN, Nagaraja JS (2021) The effect of drying temperature on cup quality of coffee subjected to mechanical drying. J Plant Crop 49:35–41
5. El-Ghobashy El-Hagar MM (2020) Design, building and performance evaluation of a mixed-mode solar dryer for agricultural products. WSEAS Trans HEAT MASS Transf
6. Amarulzaman A, Hasanuzzaman M, Rahim NA (2021) Global advancement of solar drying technologies and its future prospects: a review. Solar Energy 221:559–582
7. Dina SF, Ambarita H, Napitupulu FH, Kawai H (2014) Study on effectiveness of continuous solar dryer integrated with desiccant thermal storage for drying cocoa beans. Case Stud Therm Eng 5:32–40
8. Čipliene A, Novošinskas H, Raila A, Zvicevičius E (2015) Usage of hybrid solar collector system in drying technologies of medical plants. Energy Convers Manag 93:399–405
9. Daghigh R, Shafieian A (2016) An experimental study of a heat pipe evacuated tube solar dryer with heat recovery system. Renew Energy 96:872–880
10. Nabnean S, Janjai S, Thepa S, Sudaprasert K, Songprakorp R, Bala BK (2016) Experimental performance of a new design of solar dryer for drying osmotically dehydrated cherry tomatoes. Renew Energy 94:147–156
11. Silva MA, Katekawa ME (2007) Drying rates in shrinking medium: case study of banana. Braz J Chem Eng 24:561–569
12. Duffie JA, Beckman WA (2013) Solar Engineering of Thermal Processes, 4 edn
13. Dejchanchaiwong R, Arkasuwan A, Kumar A, Tekasakul P (2016) Mathematical modeling and performance investigation of mixed-mode and indirect solar dryers for natural rubber sheet drying. Energy Sustain Dev 34:44–53

Development of Small-Scale PETE Plastic Bottle Shredder with Electronic Sensor Controls for Cap Separation: Basis for Reverse Vending Machine

Alec Christian M. Cervantes, Ace Vann Cardiff T. Aleria, Christian Carlo B. Taer, and Cresencio Jr. P. Genobiagon

Abstract Plastic waste management has long been a problem in the Philippines; recent studies show that plastic wastes in the country end up polluting water bodies due to improper handling and disposal. Plastic bottles, which are commonly made of Polyethylene Terephthalate (PETE) plastic, are among the main constituents of these plastic wastes. Seeing this existing problem, the proponents conducted this study to provide a new approach to managing plastic bottle waste by using a shredder, which not only reduces the volume of solid waste to maximize storage space but also introduces new methods for plastic recycling. The shredder was paired with electronic control features to separate the caps from its bottles before shredding since they have different plastic compositions. Different sizes of PETE plastic bottles were tested, and results show that 300 to 500 mL bottles have achieved average volume reduction percentages of within 50–65% after shredding. 300–400 mL bottle sizes had high shredding success rate at around 92%, and the cap detection feature has acquired an average accuracy rating of 88%. The results concluded that this project has successfully developed a small-scale PETE plastic bottle shredder applicable for reverse vending machine application.

Keywords PETE bottles · Plastic shredder · Reverse vending machine · Arduino uno · Plastic waste

1 Introduction

The use of plastic is unarguably in demand by many. In the Philippines, plastic use has widely been welcomed not only by industries but also by educational institutions. Asian countries contribute a significant share to the total amount of plastic waste floating in the world's oceans, estimated to be around 12 million tons in 2010 [1].

A. C. M. Cervantes · A. V. C. T. Aleria · C. C. B. Taer · C. Jr. P. Genobiagon (✉)
University of Mindanao, Matina, Davao, Philippines
e-mail: cresencio_genobiagon@umindanao.edu.ph

A. C. M. Cervantes
e-mail: acervantes@umindanao.edu.ph

© The Author(s), under exclusive license to Springer Nature Singapore Pte Ltd. 2023

M. H. A. Hassan et al. (eds.), *Technological Advancement in Instrumentation & Human Engineering*, Lecture Notes in Electrical Engineering 882,
https://doi.org/10.1007/978-981-19-1577-2_19

This info is supported by a similar study [2] stating that the Philippines generates about 3 million tons of plastic waste annually, and 20 percent of that leaks into the oceans. With this, it could be said that the bulk of plastic waste in the country calls for minimization. In this case, a machine needs to be designed to decrease the plastic waste produced in the country, most specifically in academic institutions, where the school community also produces a lot of plastic wastes.

Multiple pieces of literature regarding the design and testing of small-scale polyethylene terephthalate (PETE) plastic bottle shredders are available on the web. Previous studies more favor the "rotary blade against a fixed blade" design due to its simplicity and effectiveness [3, 4]. However, the materials used for the blades are varying from carbon steel to alloy steel, depending on different cost considerations by previous researchers [3–5]. Plastic recycling facilities who purchase and use shredded plastic for mixing and re-melting are more commonly found in progressive foreign countries [6]. The plastic wastes are turned into plastic lumber composites made of a mixture of plastic, plant fiber, wood flour, and binders [7].

Despite the presence of already-existing small-scale plastic bottle shredders, only a few, if none, have electronic sensor controls that facilitate the separation of the cap from its bottle. The implementation of plastic shredders in reverse vending machines for plastic bottles is also not yet familiar. If such features are integrated into small-scale plastic shredders, people, especially students, could be more encouraged to dispose of plastic bottles in school properly. These opportunities for innovation are the inspiration for this study.

The researchers target to achieve the following objectives: to develop and fabricate a small-scale PETE plastic shredder for a reverse-vending machine; to integrate electronic sensor-control features in the machine to provide a machine process that separates bottle caps from the PETE plastic bottles before disposal; and test the machine's functionality.

2 Materials and Methods

2.1 Conceptual Framework

Figure 1 below shows the conceptual framework which the researchers followed in conducting this study. The input consists of the following: design constraints, material selection, and design concepts. The proponents must process these inputs to successfully design and fabricate the machine according to the specified requirements. Lastly, the output is to a fully working small-scale PETE plastic bottle shredder applicable for reverse vending machine applications.

For the machine itself, the process flow for its operation is shown in Fig. 2. First, the sensor detects if a legitimate bottle cap is placed in the cap passage. Next, the microcontroller assesses the conditions and then triggers relay switches to accept or reject the inserted cap. Then, it turns on the actuator and the shredder for a specific

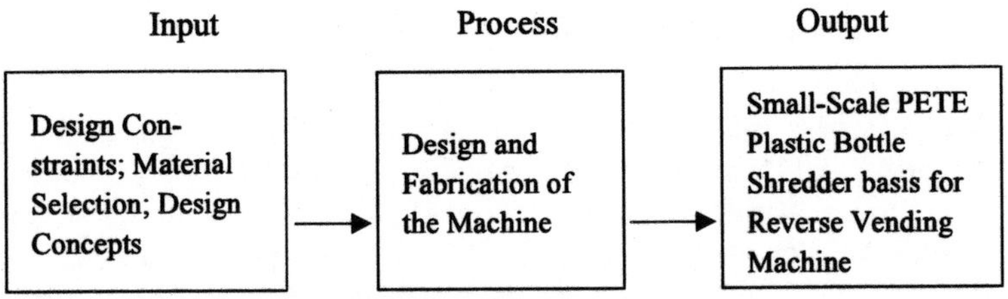

Fig. 1 Conceptual framework

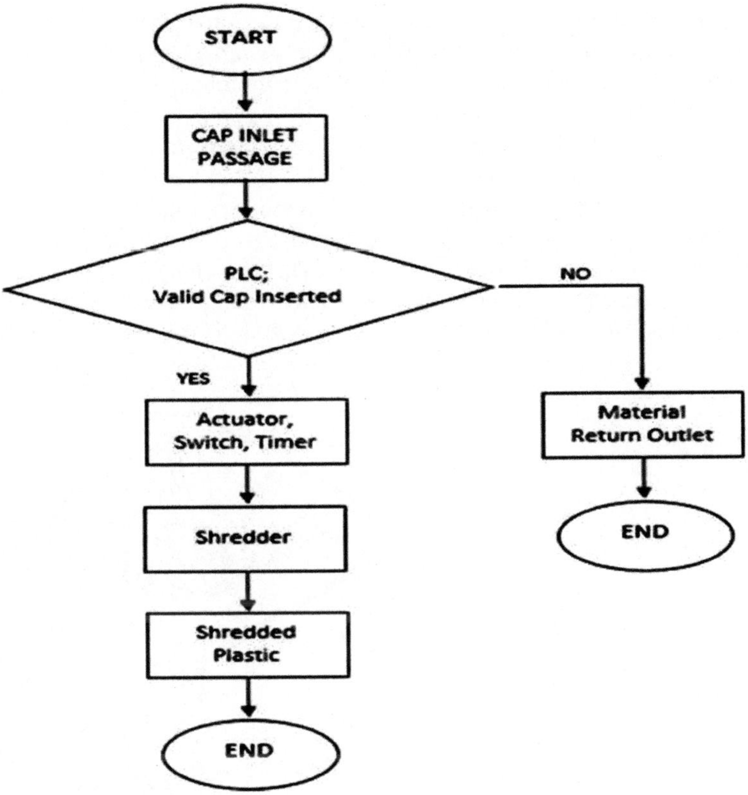

Fig. 2 Process flowchart

amount of time. After the process, bits of shredded PETE plastic will be collected and stored in a container.

2.2 Design Procedure and Material Selection

In the design process, the proponents planned to adopt design from existing plastic bottle shredders and then test it if it is suitable for shredding PETE bottles. This verification process was done through Solidworks stress simulation for the shredder components. The proponents specifically planned to adopt the rotor–stator design [3] as shown in Fig. 3 below, where rotating toothed blades shred against stationary knives. This design was chosen due to its simplicity since it involves only a few parts.

Motor Power. The proponents were able to acquire a 1.5 HP single-phase AC motor and a plastic shredder with the desired rotor–stator design from a former undergraduate thesis in the University with consent from its proponents [8]. The acquired shredder was innovated by adding suitable drivetrain, frame, and electronic control system to fit the design constraints determined by the proponents. The acquired AC motor already has a 1.5 HP rating, and the proponents had to determine if this rating is suitable for the specific application in this study. To do so, the proponents calculated the thickness of PETE plastic that a 1.5 HP power rating can shred and compared this calculated data to the gathered data on the dimensions of plastic bottles commonly found in the University as shown in Table 1. The Maximum Principal and Shear Stress Theory equations shown below, as adapted from [9], was used to facilitate the calculation. This calculation method was based upon the nature of the blades'

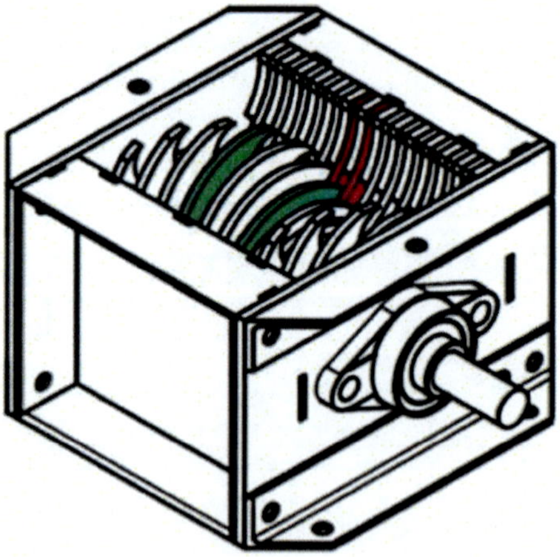

Fig. 3 Rotor–stator shredder design

Table 1 Dimensions of sample PETE bottles (mm)

Beverage (ml)	Max. diameter	Height	Max. thickness	Min. thickness
Nature's spring (500)	60	210	2	0.3
Coca-cola (300)	55	200	1.4	0.7
Real leaf (500)	66.2	211	2.1	0.8

shredding action, as calculated by a similar study [5].

$$\sigma_{max} = \frac{\sigma}{2} + \sqrt{\left(\frac{\sigma}{2}\right)^2 + \tau^2} \quad (1)$$

$$\tau_{max} = \sqrt{\left(\frac{\sigma}{2}\right)^2 + \tau^2} \quad (2)$$

Drivetrain. To achieve low output speeds, the proponents needed to add a speed reduction component. The proponents opted using a gear speed reducer because it saves space, and it reduces noise levels.

Flexible power transmission elements were used to connect the reducer input to the motor and the reducer output to the shredder. Belt-drive was chosen for the input side of the reducer since belt-drives are suitable for high-speed applications [9]. For the reducer's output side, a coupling was used to connect the reducer to the shredder. A coupling was used since it is compact, so it takes less space and produces minimal noise, and it does not introduce further power losses to the drivetrain [9].

Framework. Small-sized tubular steel was considered a primary choice for the machine frame. Steel tubes provide the necessary stiffness for the machine frame at a lower weight and cost than the conventional angle iron [10]. For the machine covering, the proponents planned to use plastic material.

Electronic Controls. An Arduino microcontroller was used to implement the electronic control functions of the machine, for it can provide a convenient platform for managing inputs and controlling outputs of multiple electronic devices [11]. The Arduino would be combined with the color sensor, the sorting mechanism, the actuator, and the motor. The color sensor and sorting mechanism are part of the cap detection and recognition feature of the machine. An RGB color sensor was used for this since it is a cost-effective option [12], and a servo motor was used for the sorting mechanism. The actuator was included as a mechanism to push the bottle down into the shredder blades during shredding. A DC linear actuator was used for this purpose.

2.3 Testing and Statistical Methods

Shredding Performance. The machine's shredding capability was used as a measure of its performance. The machine's shredding capability was tested by repeated trials of putting bottles in the shredding process. The proponents counted the number of instances that successful shredding occurs. Successful shredding is recorded when the entire bottle was consumed and shredded by the machine without any problems, and unsuccessful shredding was recorded otherwise. The success rate of shredding was then recorded by dividing the number of successful shredding instances to the total number of trials.

Sensor Accuracy. This was evaluated by testing the caps for validity using the sensor in repeated trials. The accuracy of the sensor was mathematically represented by dividing the number of the sensor's correct detections to the total number of trials. Similar methods of testing as highlighted above was found in similar study of reverse vending machine shredder [13]. For the interpretation of these data, simple comparative and inferential statistics were used [14].

3 Results and Discussions

3.1 Machine Specifications

A summary of the major mechanical components of this project, including their specifications or descriptions, is shown in Table 2. These data were the results of the thorough design and material selection process, further discussed in the following sections.

Motor Power. From thorough research, the proponents found that the average ultimate tensile strength of PETE plastic was at 38.8 MPa [15] and that the average efficiency of worm gear reducers at high-speed ratios is at 70% [16]. Considering these gathered data, it was calculated that with a set-up of 1.5 HP motor coupled with worm gear speed reducer, the resulting effective output force at the shredder blades was at 4348.8 N. Consequently, the calculated thickness of PETE plastic that this force can shred was at 2.263 mm. Looking back on the data presented in Table 1, it was determined that this calculated thickness is greater than the maximum body thickness of locally found plastic bottles. Thus, it was confirmed that the acquired 1.5 HP motor provided enough power and was applicable for this study.

Drivetrain. A worm gear speed reducer with a 60:1 speed ratio was selected and purchased to meet the shredder's low-speed requirement, as was desired by the proponents. For the belt-drive connecting the motor to the speed reducer input, based on

Table 2 Summary of major mechanical components

Part	Qty.	Specs./description
AC motor	1	1.5 HP rating
Plastic shredder assembly	1	Single shaft; Rotor–stator design
Worm gear reducer	1	WPA-60
Roller chain coupling	1	RC 4016
V-belt pulley	2	3.4", Double groove
V-belt	2	A-37

Fig. 4 Drivetrain configuration

mechanical engineering design standards [9], it was determined that two Type-A V-belts must be used on two double-grooved 3.4-in pulleys mounted on the motor and the worm, respectively. A 4016 roller chain coupling was used connecting the worm gear and the shredder, which was the best option based on local availability. The resulting configuration of the drivetrain from the motor to the shredder is shown in Fig. 4.

Machine Frame and Covering. The materials selected for the framework are 2 " × 1" rectangular and 1" × 1" square tubes, 1.5 mm thick. Outside dimensions of the frame were at 30 × 29 × 68 in. These outside measurements were based on standard dimensions of existing vending machines. The stresses on the machine frame were no longer critically analyzed.

For the covering on the machine frame, the researchers used Sintra PVC Board. An actual image of the shredder machine with its covers and labels is shown in Fig. 5.

3.2 Stress Analysis

The proponents utilized the help of Solidworks software to simulate and show the von Mises stresses induced on the machine members during machine operation. For the simulation, AISI 1023 carbon steel sheet was set as the material for the components, with a yield strength of 2.827×10^8 Pa., and the ultimate tensile strength of 4.25×10^8 Pa. based on the Solidworks database. The calculated effective force for shredding PETE plastic was used as input force. The simulation was done on the blades, which are the most critical parts that experience most stress during operation.

Fig. 5 Actual front image of shredder machine

Figure 6 shows the von Mises stress distribution one of the rotating blades during shredding action. The highest stress recorded in this simulation is at 132 MPa.

Figure 7 shows the von Mises stress distribution on a set of stationary knives during shredding action. The highest stress recorded in this simulation is at 30.49 kPa, which occurs on the inside corners, which are stress concentration areas.

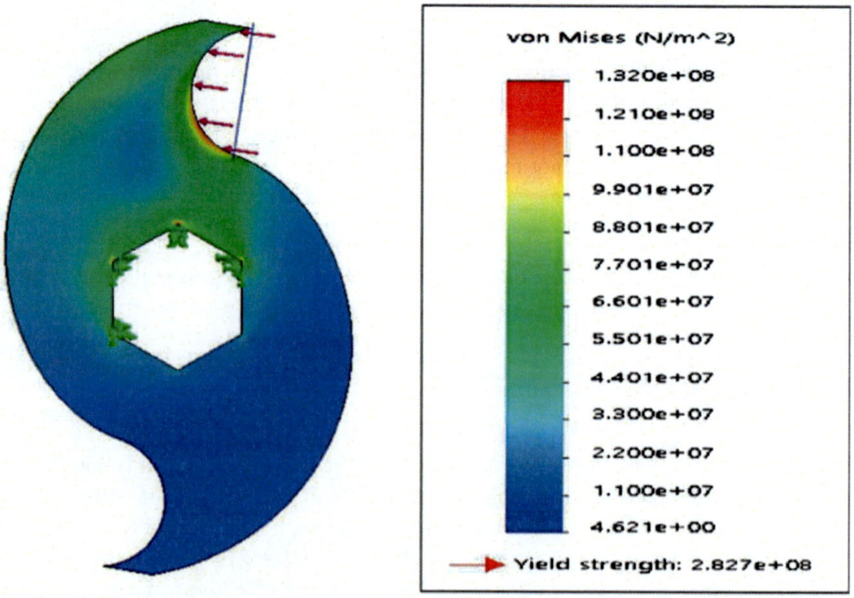

Fig. 6 Stress simulation results on rotating blades

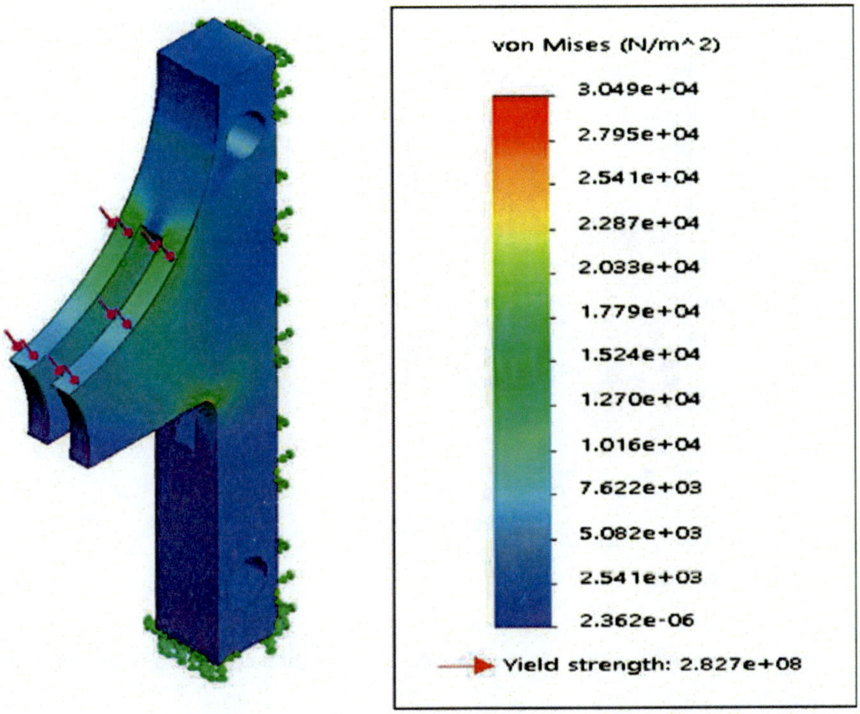

Fig. 7 Stress simulation results on stationary knives

It can be seen from these simulation results that the von Mises stresses on the critical components are well below the yield strength of the material. The subsequent safety factors were equal to or greater than 2, making it a sufficiently safe design for shredding PETE bottles. Similar studies on plastic shredder designs recorded higher von Mises stress values [17] compared to the results of this study. The lower stress values can be attributed to this study's simpler design.

3.3 Shredding Performance

A total of 97 trials were done for testing the shredding capability using PET bottles of various volume capacity. The summary of results for the testing on shredding capability is shown in Fig. 8.

In the results shown in the figure, a lower shredding success rate is recorded for 200 mL bottles, only at 60%. A slightly lower success rate is also recorded on 500 mL bottles. But overall, a high shredding success rate was achieved, especially in bottles within 300 to 400 mL capacity, reaching up to 92%. This shredding capability is at

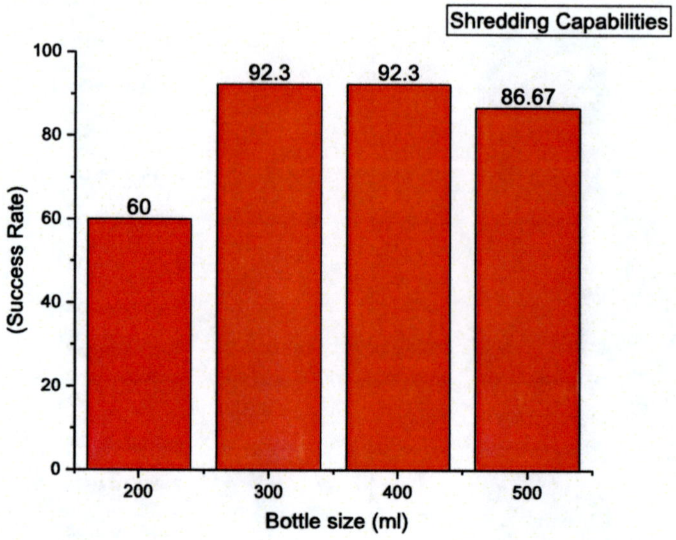

Fig. 8 Shredding success rate graph

par with other similar study on reverse vending machine shredders employing the same design [18].

3.4 Sensor Accuracy

The accuracy test results show that the sensor and its set-up exhibit high accuracy, reaching up to 100% on some cap variants, with 70% as the lowest accuracy percentage on a particular beverage brand. These high accuracy values signify that the cap separation process of the shredder is reliable and effective (Table 3).

Table 3 Sensor accuracy

Bottle cap	Total trials	Correct detection	Incorrect detection	Accuracy (%)
Sprite	10	9	1	90
Real leaf	10	7	3	70
Nature's spring	10	10	0	100
Coke	10	9	1	90
Royal	10	9	1	90

4 Conclusion

From the results, it can be said that the proponents have successfully developed a small-scale PETE plastic bottle shredder that is applicable for reverse vending machine applications. The proponents have successfully fabricated an aesthetic shredder machine that can effectively shred medium-sized bottles. The Results and Discussion have proven that the shredder's design allows it to resist material stresses during operation and is safe. The tabulated data show that the shredder can effectively shred PETE plastic bottles, reaching a 92% shredding success rate on 300–400 mL sizes. The sensor set-up for cap detection has garnered high accuracy rate of 88% across multiple cap variants after multiple trials, making it a reliable machine for unsupervised operation.

References

1. Sample I (2015) Coastal communities dumping 8 m tonnes of plastic in oceans every year. The Guardian
2. Vila A (2018) Philippines plastic pollution: why so much waste ends up in oceans. South China Morn Post
3. Reddy S, Raju T (2018) Design and Development of mini plastic shredder machine. In: IOP conference series: materials science and engineering
4. Jadhav N, Patil A, Lokhande H, Turambe D (2018) Development of plastic bottle shredding machine. Int J Waste Res 8:1000336
5. Ikpe AE, Ikechukwu O (2017) Design of used PET bottles crushing machine for small scale industrial applications. Int J Eng Technol IJET 3:157–168
6. Hopewell J, Dvorak R, Kosior E (2009) Plastics recycling: challenges and opportunities. Phil Trans Roy Soc B Biol Sci 364:2115–2126
7. dos Santos FA, Canto LB, da Silva ALN, Visconte LLY, Pacheco EBV (2020) Processing and properties of plastic lumber. In: Thermosoftening plastics
8. Bh C (2018) Development of recycled plastic shredder and injector: interlocking plastics brick maker. University of Mindanao
9. Budynas R, Nisbett JK (2015) Shigley's mechanical engineering design. Mechanical engineering
10. Nora RR, Masud UD, Ravi GM (2015) Comparison between conventional (angular) steel section and tubular steel section. Int J Eng Res 10:539–541
11. Louis L (2016) Working principle of arduino and using it as a tool for study and research. Int J Control Autom Commun Syst 1:21–29
12. Kaushik AA (2013) RGB color sensing technique. Int J Adv Res Sci Eng. 2(9):78–87
13. Rahim NHA, Khatib ANHM (2021) Development of pet bottle shredder reverse vending machine. Int J Adv Technol Eng Explor 8:24–33
14. Ali Z, Bhaskar SB (2016) Basic statistical tools in research and data analysis. Ind J Anaesth 60:662
15. Bamberger Polymers. Overview of materials for Polyethylene Terephthalate (PET), Unreinforced. http://www.matweb.com/
16. Skulić A, Krsmanovic D, Radosavljevic S, Ivanovic L, Stojanovic B (2017) Power losses of worm gear pairs. Acta Technica Corviniensis Bull Eng 10(3):39–45

17. Nasr M, Yehia K (2019) Stress analysis of a shredder blade for cutting waste plastics. J Int Soc Sci Eng 1:9–12
18. Ekman R (2018) Development of a Plastic Shredder. Master Thesis, Department of Design Sciences Faculty of Engineering LTH, Lund University

Design of Flora Inspired Savonius Wind Turbine

Apple Jee P. Ladao and Delan S. Bacus

Abstract Wind energy is a promising source of renewable energy. However, integrating energy harvester in an urban landscape poses a significant challenge. This study developed a wind turbine, a bio-inspired savonius that will blend with the environment. Savonius wind turbine is a vertical-axis wind turbine type used to convert wind force into torque on a rotating shaft. Using SolidWorks flow simulation, the different configuration was considered such as twist angle and guide vanes. Simulation shows a promising potential of this turbine. A scaled prototype was made and tested. Results showed that a 135° angle of twist with guide vanes is ideal and has the highest torque during its running condition in theoretical data. When it is made into a prototype, its actual torque proves to be lower than its theoretical torque. The actual data presents a promising power efficiency of 32.45%, higher than the usual average range for savonious wind turbines.

Keywords Bioinspired savonius · Flora inspired savonius · Savonius wind turbine · Wind energy · Wind turbine · Savonius type

1 Introduction

One way of preserving the world is by increasing renewable energies, which are quickly replenished. The wind is one of its promising sources because it is one of the cleanest and most sustainable ways to generate electricity [1]. A wind turbine is a device used to harness power from the wind's kinetic energy. The blades convert it to mechanical energy, and the generator produces the electrical energy [2].

Many researchers are trying to develop and innovate a range of different designs of wind turbine to make it more convenient, efficient, and effective [3]. Advance

A. J. P. Ladao (✉) · D. S. Bacus (✉)
Mechanical Engineering Program, College of Engineering Education, University of Mindanao, Matina, Davao, Philippines
e-mail: ajladao@gmail.com

D. S. Bacus
e-mail: dbacus@umindanao.edu.ph

© The Author(s), under exclusive license to Springer Nature Singapore Pte Ltd. 2023
M. H. A. Hassan et al. (eds.), *Technological Advancement in Instrumentation & Human Engineering*, Lecture Notes in Electrical Engineering 882,
https://doi.org/10.1007/978-981-19-1577-2_20

application software like SolidWorks tests the designs and sees their performance before proving it by making a prototype [4].

Researchers make efforts to invent and acting in preserving nature. That is why wind turbines are a big help because of their clean way of generating electricity. It is less to no emission significant in minimizing the effects of global warming [5]. Many wind turbine designs are inspired by things that exist, correlate, and complement it in harmony with nature.

The tree-shaped wind turbine work of Jérôme Michaud- Larivière in Place de la Concorde in Paris, France, is an example of a wind turbine that is inspired by nature. He hopes that it will eventually be installed in people's yards in an urban center. It can generate electricity at a wind speed average of 3.4 m/s [6, 7].

Thus, this latest bioinspired savonius wind turbine is developed. A flower-shaped that is compared to the tree-shaped will generate electricity in a wind speed average of 2 m/s. For its being stylish and functional, it can be installed and used anywhere without damaging anything in nature and making it ecofriendly.

A Prototype wind turbine also inspires this study; blades have been conceived to be used for a straight type VAWT; in other words, straight cylindrical elements parallel to the vertical axis for each section have been considered. As a side effect of being a time-saver, prototyping will make the development process more cost-efficient since the overall development cycle becomes shorter. Additionally, it only needs fewer resources when prototyping compared to a complete design [8].

To design, develop and fabricate a floral-inspired savonius wind turbine that is efficient and ecofriendly is the aim of this study. Using SolidWorks application software, the other objectives will be achievable: to easily simulate and collect the theoretical data [9]. Comparisons will be made for the multiple designs considering its twist angle, guide vanes, and torque and velocity output to produce a power coefficient value more than the average of 5 to 30% for savonius wind turbine [10]. Develop and test a scaled prototype of the design that generates the most torque among all the others using a 3D printer, ensuring that the output meets the input.

This study would focus on developing a design of savonius wind that inspired flowers. The electrical system shall not be covered in this study. Also, this study is limited only to determining the variations of wind pressure and velocity of the design concepts. Wind conditions in Davao City shall only be the scope of this study.

2 Materials and Methods

The researcher's study is about applied science which aims to design a floral-inspired savonius wind turbine. Comparisons will be made for the multiple designs considering its twist angle, guide vanes, and torque output. This chapter will explain the materials, resources, methods, and procedures used to conduct the study.

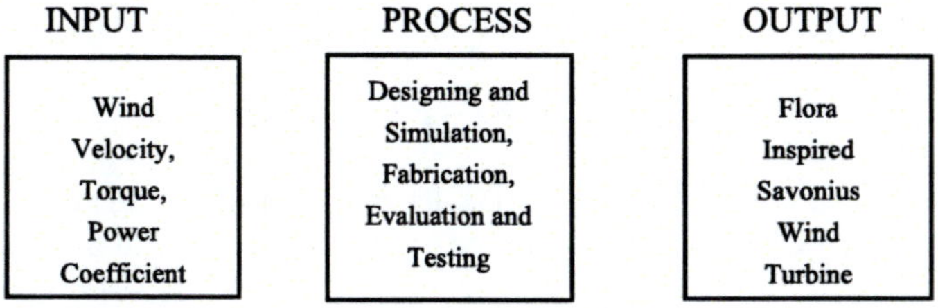

Fig. 1 Conceptual framework

2.1 *Conceptual Framework*

The input, process, and output concepts of the system are shown in Fig. 1. The idea of acquiring this savonius wind turbine is to design a flower-shaped blade considering its twist angle and guide vanes. It will utilize and make use of the wind that moves from a different direction. The wind's kinetic energy that spins the blade will generate mechanical energy, and the DC motor will convert it to electrical power. After this study, a promising wind turbine is developed.

2.2 *Materials and Resources*

To accomplish this study, various materials and resources are needed. SolidWorks is used to draw the project, and collecting the theoretical data by simulation enables deciding what design will be used for fabrication. An ABS filament will be used as the material for the 3D printing of the prototype. Ceramic Bearing will maintain proper alignment of turning parts and enable high-speed rotation of making a layer by layer build-up [12]. It is formed like puzzle pieces so that it can be connected and assembled easily.

2.3 *Testing and Evaluation*

The prototype was then tested to determine the power output and its coefficient of performance. The output power is determined by subjecting the rotor shaft of the turbine with a radius of 4 mm to a belt pulley system whereby a belt is subjected to the rotating shaft, and two spring balance is attached to the ends of the belt. Angular speed (f) is determined using a tachometer. Using the data that was gathered, power transmitted (P_{out}) by the shaft would be then computed using the formula presented in [10]

$$P_{out} = 2\pi f T \quad (1)$$

whereby torque (T) is determined by net belt pull (F_1 and F_2) of the spring balance

$$Torque = (F_2 - F_1)x\ \ radius \tag{2}$$

The coefficient of performance (Cp) is determined using the blade's formula as presented in [9]. A Tachometer is a device used to measure the turbine's rotational speed.

$$Cp = \frac{P_{out}}{0.5x\rho_{air}xAxv^3} \tag{3}$$

Equation (3) sets the density (ρ_{air}) as 1.2 kg/m3 and velocity (v) as 1.5 m/s following the environmental conditions of the research locale. The swept area (A) is computed using the blade diameter.

2.4 *Flow Simulations*

This study utilized the Solidworks Flow Simulation application in investigating the pressure and velocity variation needed in this study. Solidworks Flow Simulation is an intuitive Computational Fluid Dynamics (CFD) solution embedded within SolidWorks 3D CAD that enables the researcher to simulate liquid and gas flows quickly and easily through and around the designs [11]. The simulation was set to be an incompressible flow, and the designs were assumed to be adiabatic. Initial parameters were set to have a 1.5 m/s, the average wind velocity in Davao city. Pressures at atmospheric pressures of 101.325 kPa. The simulation also used a single reference frames (SRF) model to simulate the incompressible and steady-state flow field.

2.5 *Fabrication*

After gathering and analyzing the collected theoretical data, the chosen design was then fabricated as a prototype to further test and gather the actual data. Fabrication of the physical structure of the prototype was made in several trials, where it underwent several modifications until the final design was achieved. Using a 3D printer to enable the production of the complex structure of the flora inspires savonius wind turbine easily and quickly with guide vanes design. There will be less material to be used compared to the traditional method of manufacturing it. ABS filament is fed to the printer, and it will Whereby the density is taken as 1.2 kg/m^{-3} and velocity as 2.3 m/s^{-2} by the research locale's environmental conditions. The swept area is computed using the length multiplied by the diameter of the blade. [13].

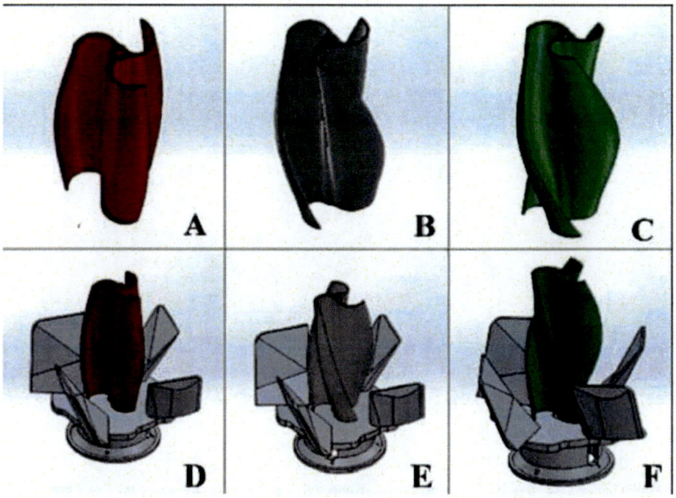

Fig. 2 Design of the project (**a** No Angle of Twist, **b** 90° Angle of Twist, **c** 135° Angle of Twist, **d** No Angle of Twist with Guide Vanes, **e** 90° Angle of Twist with Guide Vanes, **f** 135° Angle of Twist with Guide Vanes)

3 Results and Discussions

This chapter will compare the blade designs with the help of its theoretical data gathered from the simulation in SolidWorks. Actual data are collected in the printed prototype.

3.1 Flora Inspired Savonius Blade Design

As a result of the intense research and experimenting with the angle of twist and guide vanes in the blades for the possible design of the project, the researcher finally came up with six designs that will be compared and considered using the data gathered during simulation. In Fig. 2 below, the first three designs are savonius blades with a different angle of twist, and the last three have the same angle of twist as the first, but each has guide vanes.

3.2 Wind Velocity

The figure below shows the flow of each design's wind velocity. The colors indicate its contrast. It starts with zero velocity, the dark blue, and an average velocity of 2 m/s red. It is visible that there is a difference in wind velocity around the surface area of

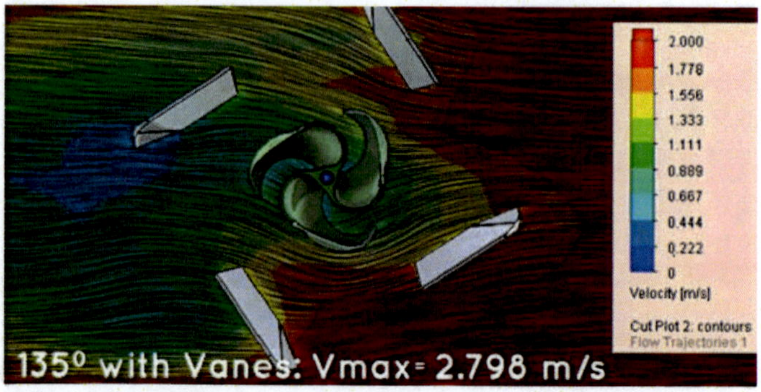

Fig. 3 Wind velocity flow trajectory (Top View)

the blades. This indicates that the design spins successfully. Since blades only spin when there are differences in wind velocity that passes through it.

As observed in Fig. 3, the design with a 135° angle of twist with guide vanes has a maximum velocity of 2.798 m/s during its simulation in SolidWorks. It is the lowest maximum velocity compared to the other five designs. Ideally, a wind turbine can generate power even if the blades spin at the lowest possible wind velocity.

3.3 Torque Analysis

In this project, torque is one essential piece of data to identify the best among the six designs. It is a force measurement that enables the turbine to rotate. To generate high electricity, it is essential to have high torque and low speed. Therefore, the higher the toque, the better. Shown in Fig. 4 below is the torque line graph for 135° angle of twist with guide vanes.

The simulation for torque shows at the graph in design F (135° Angle of Twist with Guide Vanes) that it has the highest torque during its running condition compared to the other five designs that mostly have negative torque. This design has a torque of 0.002054347 J.

3.4 Design Fabrication

Thru the results in the simulation, the most efficient and optimal among the six designs are identified. It was 3D printed as a prototype with dimensions of 8 cm diameter, 19.5 cm length, 135° angle of twist, and four guide vanes. Guide vanes in wind turbines help effective direct angles arranged to accelerate wind and serve as a wind booster for better efficiency [14]. ABS filament was used in printing the

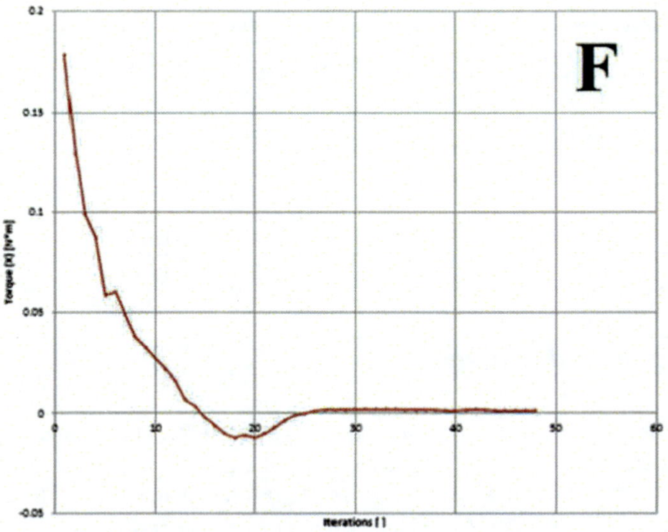

Fig. 4 Torque graph (135° Angle of Twist with Guide Vanes)

parts of the project because of its better mechanical and thermal properties. Figure 5, shown below, is the printed design.

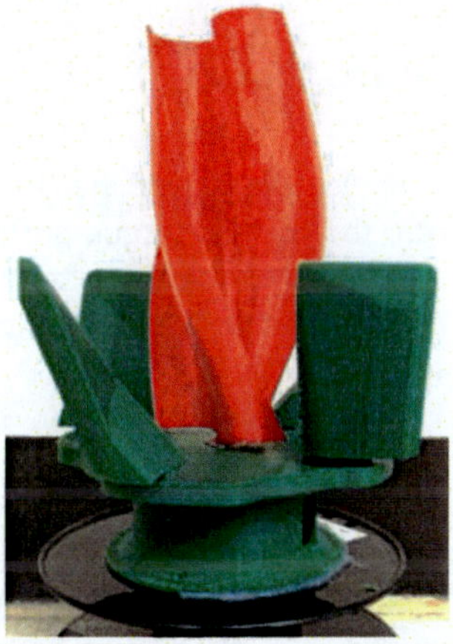

Fig. 5 135° Angle of Twist with Guide Vanes

Table 1 I actual data (135 ° Angle of Twist with Guide Vanes)

F1 (Newton)	F2 (Newton)	Torque (Joules)	RPM	Power (mWatts)	Cp
0.6865	0.3923	1.1768×10^{-3}	149.84	18.5	0.324
0.6865	0.3923	1.1768×10^{-3}	159.67	18.4	0.324
0.6865	0.3923	1.1768×10^{-3}	150.1	18.5	0.325
0.6865	0.3923	1.1768×10^{-3}	150.25	18.5	0.325
0.6865	0.3923	1.1768×10^{-3}	149.98	18.5	0.325
Average				18.5	0.3245

3.5 Testing and Evaluation

The table presents the actual operational data gathered from the prototype (Table 1).

The actual operational data gathered from the prototype presented in the table above shows that the average torque is 0.0011768 J. It is lower than the theoretical, which is equal to 0.002054347 J. This also has an average power output of 18.5 mW. It uses the area of the blade 19.5 cm length times 4 cm radius and velocity of 2.3 m/s to get the power efficiency of 32.45%. This efficiency is higher than the usual average range for typical savonius wind turbines as presented in [10], the average range of 5–30% efficiency. This result is also lower than Beltz's limit of 59.3%, the maximum theoretical efficiency of any wind turbine [15].

4 Conclusions and Future Works

Among the six-blade designs made, it is proved that 135° angle of twist with guide vanes is the most ideal having the highest torque during its running condition in theoretical data. When it is made into a prototype, its actual torque is proved lower than its theoretical torque. This design has the highest torque and lowest maximum wind velocity among the designs, both in theoretical and actual data. Its power efficiency is also higher than the usual power average of a savonius wind turbine and lower than the maximum theoretical efficiency of any wind turbine (Beltz's Limit). Compared to the tree-shaped wind turbine with a wind speed average of 3.4 m/s, this only has an average of 2.3 m/s. Therefore, this project is commendable.

The research also recommends that an electrical system be included in the design of this device, as well as including the application or location in the optimization analysis for the turbine and considering losses.

References

1. Jaber S (2014) Environmental impacts of wind energy. J Clean Energy Technol 251–254

2. Ploumakis E, Bierbooms W (2018) Enhanced kinetic energy entrainment in wind farm wakes: large eddy simulation study of a wind turbine array with kites. Airborne Wind Energy Green Energy Technol 165–185
3. Pérez JMP, Márquez FPG, Tobias A, Papaelias M (2013) Wind turbine reliability analysis. Renew Sustain Energy Rev 23:463–472
4. Rosly N, Mohd S, Zulkafli MF, Ghafir MFA, Shamsuddin SS, Muhammad WNAW (2017) Flow simulation of modified duct system wind turbines installed on vehicle. J Phys: Conf Ser 914:012006
5. Igba J, Alemzadeh K, Anyanwu-Ebo I, Gibbons P, Friis J (2013) A systems approach towards reliability- centred maintenance (RCM) of wind turbines. Procedia Comput Sci 16:814–823
6. Oh SJ, Han HJ, Han SB, Lee JY, Chun WG (2009) Development of a tree-shaped wind power system using piezoelectric materials. Int J Energy Res 34(5):431–437
7. Barber M, Urban 'Wind Trees' generate electricity from breezes, curbed. https://www.curbed.com/2017/3/14/14914302/wind-tree-turbine-for-sale. Last Accessed 14 Mar 2017. (Online)
8. Söderlund T, 6 reasons why you should be prototyping more, medium. https://medium.com/ux-prototyping/6-reasons-why-you-should-be-prototyping-more. Last Accessed 06 Feb 2015. (Online)
9. Akwa JV, Vielmo HA, Petry AP (2012) A review on the performance of Savonius wind turbines. Renew Sustain Energy Rev 16(5):3054–3064
10. Beer FP, Johnston ER, Dewolf JT, Mazurek DF (2012) Mechanics of materials, 6th ed. Mcgraw-Hill, New York, USA
11. The Solidworks website (2002). https://www.solidworks.com/product/solidworks-flow-simulation. (Online])
12. Blok LG, Longana ML, Yu H, Woods BKS (2018) An investigation into 3D printing of fibre reinforced thermoplastic composites. Addit Manuf 22:176–186
13. Heikal HA, Abu-Elyazeed OS, Nawar MA, Attai YA, Mohamed MM (2018) On the actual power coefficient by theoretical developing of the diffuser flange of wind lens turbine. Renew Energy 125:295–305
14. Korprasertsak N, Leephakpreeda T (2015) Optimal design of wind boosters for low speed vertical axis wind turbines. Appl Mech Mater 798:195–199
15. J. Bukala K, Damaziak K, Kroszczynski M, Krzeszowiec, Investigation of parameters influencing the efficiency of small wind turbines. J Wind Eng Ind Aerodyn 146:29–38

Arduino-Based Electronic Bicycle Transmission Switching System

Lance Gabriel M. Atencio, John Stephen E. Cena, Ian U. Opalla, and Randy E. Angelia

Abstract Drivers can classify today's driving into two main types of experiences, manual and automatic. A manual vehicle gives the driver power to choose whatever gear they want, while an automatic vehicle chooses the bicycle gears for the driver's convenience. It applies not only to automobiles but also to bicycles as well. This study aims to create an electronic transmission system for bikes using an Arduino as its brain and a motor to control its transmission. It will give the riders the freedom to choose the gears they want to use while automatically selecting bicycle gears for their convenience. In this study, the researchers have created an electronic transmission system for the bicycle with both manual and automatic modes. The transmission accuracy was measured and compared to calculated data by conducting actual ride tests on the device. Overall, the system produced outstanding results with 98% accuracy using a simple statistical analysis method. The device works just like it was programmed using Arduino, and it should improve the quality and safety of riding the bike compared to conventional bike shifters.

Keywords Arduino · Automatic transmission · Bike shifter · Drivetrain · Electronic shifter

1 Introduction

The number of bicycles has quintupled for the last five decades since cycling has countless physical, mental, and health benefits [1]. However, the number of bicycle-related accidents is still adding up every day [2]. Various factors have led to accidents, and one common factor is riding the bike improperly [3]. Casual riders usually commit mistakes that they do not even know might cause accidents [4]. Although riding a bike is easy, riding a bike without potentially attracting accidents is surprisingly complicated. People need to learn the proper way of setting their bike up, the appropriate course of riding it, and the mistakes they need to avoid [5]. To prevent

L. G. M. Atencio · J. S. E. Cena · I. U. Opalla · R. E. Angelia (✉)
Electrical Engineering Program College of Engineering Education, University of Mindanao, Davao, Philippines
e-mail: randy_angelia@umindanao.edu.ph

© The Author(s), under exclusive license to Springer Nature Singapore Pte Ltd. 2023

M. H. A. Hassan et al. (eds.), *Technological Advancement in Instrumentation & Human Engineering*, Lecture Notes in Electrical Engineering 882,
https://doi.org/10.1007/978-981-19-1577-2_21

bicycle-related accidents caused by improper riding, bicycle manufacturing companies have been trying to build their version of the perfect drivetrain [6]. However, mechanical groupsets have a signicant aw: after a few months, or even weeks, of riding a bike, the shift cables will seem to have corroded or stretched and need to be adjusted to change gears correctly [7]. An electronic shifting system has been created to change gears more precisely using electronic switches instead of pulling cables [8]. Although the top-performing drivetrains available at the market were consistently doing great, electronic shifting was better than its mechanical counterpart. The only downside for the electronic shifters is that they are so expensive [9]. Many studies conducted directly or indirectly contributed to developing safe cycling activity, high-tech, and more fun social or sports activities in the past years. Studies like aerodynamics of cyclist posture, bicycle types and, helmet characteristics, among others like cycling positioning in reducing cycling drag force, help improve cycling activities [10–12]. Furthermore, aside from research in improving riding comfort, many studies also focus on integrating modern cycling technologies—integration of digital image processing in increasing the performance of the electronic shifting system. Installed with a high-speed camera, the motor's shifting position and shifting time is automatically measured [13, 14].

On another note, more and more automobiles nowadays have been congured with an automatic transmission to help ensure the driver's complete focus on the road. Some people, especially beginners, who use automated vehicles claim that they are generally safer, more comfortable, and fun to ride than the classic manual transmission. The driver did not need to think every few seconds about what they would do concerning changing gears. Nonetheless, some people accustomed to the driving manual still thought it was better than automatic [15]. Accidents have happened to countless people.

Nevertheless, bike accidents happened because of mistakes that might have been caused by the bike or the rider themselves [16]. Mechanical shifting uses much force to push the shift levers in changing gears; this diverts the rider's attention from the road into shifting the bike gears [17]. A rider is expected to deal with all the jumps and turns on the terrain in a cycling race, especially in mountain bike racing, where the trail can be too narrow only to accommodate one bike. In this case, the rider's focus could be divided between following the path and deciding when to shift gears [18]. Another issue is when the bike gears that are being used are not compatible with each other, like the most extensive front and rear cogs or the most undersized front and rear cogs, which causes the chain to receive more tension than it can handle, causing it to break off; this is called cross-chaining [19]. In some cases, the rider might have been over-speeding while making a hard turn, standing while biking on a relatively fast cadence, or riding at a slow pace while sitting down; these and all the different scenarios mentioned above have probably taken countless lives. Fortunately, all the potential disasters stated above can be avoided by creating a device that will prevent terrible things and hopefully provide a more enjoyable time while riding the bicycle.

In general, this study's objectives are to create an electronic bicycle transmission switching system based on an Arduino. To accomplish this, the proponents will need to design a manual electronic shifting system triggered by buttons, an automatic shifting system activated by speed and cadence sensors, and a control panel that will display basic information to the rider. Subsequently, the researchers will nd a way of switching between the automatic and manual mode, place everything together, and construct the device in a way that will not be a burden on the rider while driving, invulnerable to as many kinds of riding conditions as possible. The device will most likely benet the people who love riding bikes, especially casual riders. It will make riding a bike a lot easier, enticing people to ride bikes more often. It might even attract newcomers to the world of cycling, and the device will make their riding experience a memorable one. Bike enthusiasts searching for new bicycles will never want to miss how it feels to ride a bike with the device installed.

Furthermore, Arduino nerds or technology geeks might be thrilled to learn how a microprocessor controls a bicycle's drivetrain. The proponents will focus solely on creating a transmission system that includes a rear derailleur, shifter buttons, a control panel, and the microprocessor that controls shifting patterns. However, it has nothing to do with the design of the frame, wheels, or the bike itself. Furthermore, despite having electronics, this project is not related to e-bikes, which means that the rider, not a motor, power the bike.

2 Materials and Methods

The device's structure can be divided into three main parts: the control panel, the control units, and the shifting mechanism. The control panel comprises the Arduino Nano, an LCD, and the mode switch. The control units consist of up and downshift buttons and sensors. The shifting mechanism is composed of the Motor and the rear derailleur to be connected in a systematically stable way. The control units will transmit signals to the control panel, displaying current information and commanding the shifting mechanism to change gears accordingly.

2.1 Conceptual Framework

The conceptual framework diagram of Fig. 1 consists of the input, process, and output. The inputs are the things that the system will gather information about. These inputs can either be coming from the user or the sensors. The process is what the system will do when it has received information coming from the inputs. Finally, the device's output will happen after the system has processed the input data. This device's desired result is, just as the title suggests, an Electronic Bicycle Transmission Switching System.

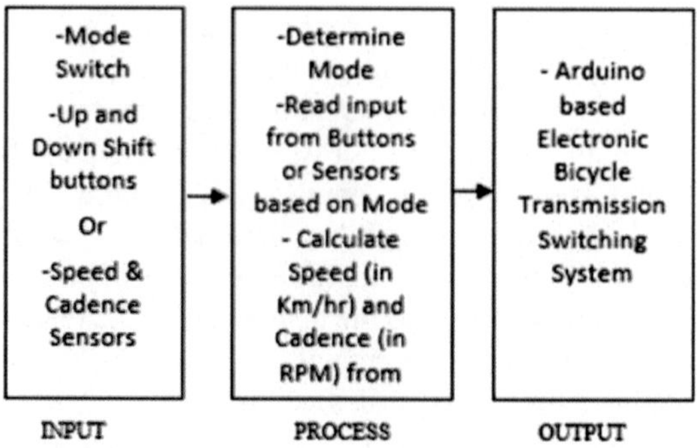

Fig. 1 Conceptual framework

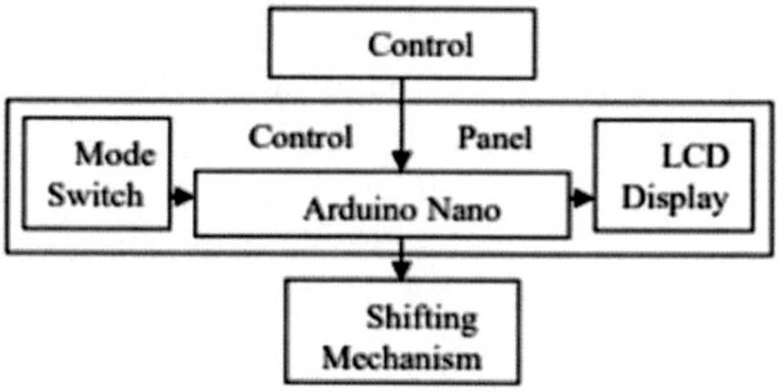

Fig. 2 Control panel block diagram

2.2 Control Panel

Figure 2 shows a block diagram of the control panel, which, as stated above, comprises the Arduino, an LCD, and the mode switch. The proponents chose the Arduino Nano as its main microcontroller due to its small size compared to the Uno while still having the same functionality. It will serve as the brain of the system, thus connecting everything to it. An LCD will display the mode selected to eliminate confusion, the current gear that the biker is using, and the riding speed. The mode switch will be a simple toggle switch and will be the Arduino's priority since it will determine whether the rider wishes to utilize the automatic or the manual mode.

2.3 Control Units

Figure 3 depicts the control units' connections, consisting of the up and downshift buttons and the sensors. The control units' functionality, however, will depend on the control panel's mode switch. Regardless of the Mode chosen, the sensors will gather data from the rider's speed and cadence. If the system is set to Manual Mode,

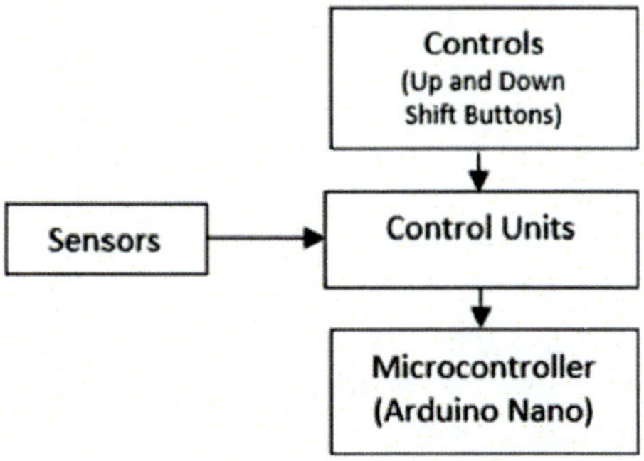

Fig. 3 Block diagram for control units

the shift buttons will function as inputs for the shifting mechanism while the sensors will not change gears. If the Automatic Mode is selected, the shift buttons will be on standby. In contrast, the data being collected by the sensors and calculated by the Arduino will be the basis of shifting gears.

2.4 Shifting Mechanism

The Shifting Mechanism comprises a motor attached to the rear derailleur and will receive signals from the control panel. Suppose the Control Panel commands the Motor to shift gears. In that case, it will turn clockwise or counterclockwise into a specific angle, pushing or pulling the rear derailleur and causing the chain to move into a different sprocket. However, it will only happen when the rear wheel, connected to the cassette, is spinning forward. Shown in Fig. 4 below is the design of the shifting mechanism, which consists of the servo motor with the metal brace (in red), which

Fig. 4 Design of the shifting mechanism

helps to mount it to the rear derailleur. The bolts on the rear derailleur (also in red) are where the two holes on the Motor's braces are supposed to be locked.

2.5 Flow of the Transmission Switching System

At first, the system will determine the Transmission Mode based on the Mode switch to decide whether the rider wishes to utilize the Automatic Mode or Manual Mode. If Manual Mode is selected, the system will receive Signals from the Speed and Cadence Sensors, convert the speed into km/hr, cadence into RPM, and wait until the up-shift button or downshift button is pressed. Pressing the up-shift button will tell the system to shift up while pressing the downshift button will tell it to shift down. If Automatic Mode is selected, the system will still receive Signals from the Speed and Cadence Sensors, convert the speed into km/hr, cadence into RPM. After that, the system will then look at the bicycle speed and compare it to the bike cadence. If the bike cadence is too fast for the bicycle speed, the device will shift up; however, if the bike cadence is too slow, the device will shift down for the rider's convenience. If the rider is freewheeling (which means that if the cadence becomes zero, but the bike is still running), the device will not shift and wait for the rider to pedal before it starts to change gear again. Shifting up will move the chain into a smaller sprocket, while gear going down will move the chain into a larger sprocket.

Using a larger sprocket will deliver more power from the pedals to the wheels, making this suitable for riding up a slope; however, this will cause the speed to decrease. On the other hand, using a smaller sprocket will make the bike faster but require more power from the rider, making downshifting suitable for riding fast or downhill. After every few milliseconds, the system will display the rider's transmission mode, cadence, speed, and current gear.

2.6 Function Testing Method

For a successful function test, a few standards should be observed. ISO 4210- 2:2015 Cycles: for the safety of the function tests, ISO 8588:2017 Sensory analysis: for the calibration of the sensors, and ASTM STP491 Servo Systems: for the calibration of the servo motor.

The first thing to test is that the control panel's functionality displays the mode, speed, cadence, and gear. During this test, the mode switch and sensors can be tried by toggling the mode switch and seeing if the mode displayed changes, turning the front wheel for the speed sensor, and spinning the pedals for the cadence sensor.

The dual functionality to test is manual mode. In this test, the device should be set in manual mode. While a proponent is spinning the pedals forward and elevating the rear wheel not to touch the ground, the shift buttons can be pushed to see whether the control panel displays the corresponding gear and whether the rear derailleur shifts into a different gear.

For the testing of the automatic mode, the bike should be ridden, and the rider should determine whether the device shifts gears as the bike travels at different speeds. For the test to be non-biased in the rider's opinion, the rider should take a video of the control panel as the bike is running since the control panel displays the mode, speed, cadence, and gear. The formula that the proponents used to measure the rear wheel's angular velocity based on the front chainset's angular velocity is the basic formula of gears and wheels.

$$N_1 = N_2 t_2 \tag{1}$$

where: N corresponds to the angular velocity (rpm), and t corresponds to the number of teeth of the selected gear. In this case, N_1 will represent the rpm of the front chainset; t1, the number of teeth of the front; N_2, the rpm of the rear cassette; and t_2, the number of teeth of the rear. Since the angular velocity of the rear cassette will be equal to that of the rear wheel, and the angular velocity of the front wheel will approximately be equal to that of the front wheel, taking the diameter of the rim (29 in) and the height of the tire (2 in) into account, we gather that the total diameter of the wheel to be 29 + 2 × 2 = 33 in, which makes the conversion of the angular velocity of the rear cassette into distance traveled by the front wheel will be as follows.

$$\begin{gathered} N_2\left(\frac{\text{rev}}{\text{min}}\right)\left(\frac{\pi(33\text{in})}{1\text{rev}}\right)\left(\frac{2.54\text{cm}}{1\text{in}}\right)\left(\frac{1\text{km}}{100,000\text{cm}}\right)\left(\frac{60\,\text{min}}{1\text{hr}}\right) \\ N_2\text{rpm} * 0.158 = \text{S kph} \end{gathered} \tag{2}$$

3 Materials and Methods

3.1 Control Panel

The control panel successfully displayed the mode, speed, cadence, and gear without errors. Shown in Fig. 5 below is a photo of the actual control panel mounted on the bike's handlebar. Inside the control panel are the electronic circuitries and the microcontroller responsible for the device's decision-making process. The control panel is securely attached to the handlebar of the bicycle together with the shifter switch. Constraints in designing the control panel to be as light, secured, and clean as possible were accomplished by proper circuit planning and accurate positioning of electronic materials.

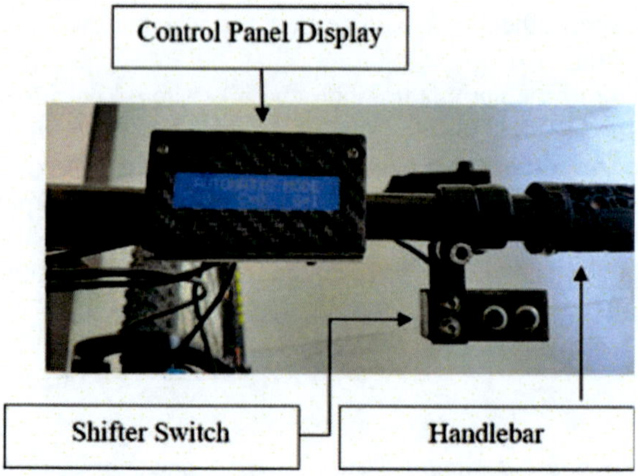

Fig. 5 Control panel

3.2 *Shifting Mechanism*

Exhibited in Fig. 6, the actual shifting mechanism developed. This consists of the servo mounted on the rear derailleur using a custom-made iron mount. The spring inside the rear derailleur was removed so that the servo would have complete control over the gear used. In this setup, there are a total of 9 gears that the shifting mechanism has access to. The wires coming out of the servo were spliced with a longer wire to be connected to the control panel placed on the front of the bicycle.

The shifting mechanism is attached to the 29 in diameter wheel of the mountain bike.

Fig. 6 Shifting mechanism

3.3 Testing

3.3.1 Manual Mode

The bicycle will function as a standard mountain bike for manual mode setup even if it is already installed with components for automatic shifting technology. The innovation in the manual operation was that the control panel was functional and set as "Manual Mode." The control panel can fetch the current gear position of the bike every time the shift button is pressed. Since the bicycle was initially in manual operation mode, the only innovation was the control panel display. Conducting the functionality test was ignored since the essential part of the study was the automatic mode function and shifting function, either manual to automatic or vice versa.

3.3.2 Automatic Mode

During the automatic mode of the bicycle, the control panel displayed the mode, speed, cadence, and gear while running as these data were fetched from sensor nodes the pushed to the microcontroller. The rear derailleur shifted into different gears for every increase and decrease in speed and bicycle cadence. Table 1 shows a table where the sensors' accuracy is measured by comparing ve actual and theoretical cadences' cadences at varying speeds achievable in each of the nine gears part of the bicycle's transmission. It also shows that the average overall accuracy of the sensors is 98%.

The theoretical cadence values are from the standard cadence for a particular speed and gear number. The theoretical cadence data were directly compared to the actual value gathered during testing. The accuracy rate in every trial is computed directly by taking its percentage, while the total accuracy was computed by taking the whole mean of all the accuracy rates.

Table 1 Accuracy of the device

Speed category	Gear setting	Trial no	Speed (km/hr)	Cadence		Accuracy rate (%)
				Theoretical	Actual	
Low	1st	1	3	20	20	100
		2	6	40	41	98
		3	9	60	61	98
		4	12	80	82	98
		5	15	101	103	98
Low	2nd	1	6	36	37	97
		2	9	54	55	98
		3	12	71	73	97
		4	15	89	91	98
		5	18	107	109	98
Low	3rd	1	9	47	48	98
		2	12	63	64	98
		3	15	78	80	98
		4	18	94	96	98
		5	21	109	111	98
Medium	4th	1	12	54	55	98
		2	15	67	68	99
		3	18	80	82	98
		4	21	94	96	98
		5	24	107	109	98
Medium	5th	1	15	59	60	98
		2	18	70	72	97
		3	21	82	84	98
		4	24	94	96	98
		5	27	106	108	98
Medium	6th	1	18	60	61	98
		2	21	70	72	97
		3	24	80	82	98
		4	27	90	92	98
		5	30	101	103	98
High	7th	1	21	59	60	98
		2	24	67	68	99
		3	27	75	77	97
		4	30	84	86	98
		5	33	92	94	98

(continued)

Table 1 (continued)

Speed category	Gear setting	Trial no	Speed (km/hr)	Cadence		Accuracy rate (%)
				Theoretical	Actual	
High	8th	1	24	58	59	98
		2	27	65	66	98
		3	30	73	75	97
		4	33	80	82	98
		5	36	87	89	98
High	9th	1	27	55	56	98
		2	30	61	62	98
		3	33	68	69	99
		4	36	74	76	97
		5	39	80	82	98
Overall accuracy						**98**

4 Conclusion and Future Works

The research concludes that it is possible to create an Electronic Bicycle Transmission System. This device made changing bicycle gears more reliable by eliminating cables and using a servo motor. It simplices changing gears in a bicycle because pushing buttons is more comfortable than learning to shift gears with mechanical drivetrain systems, especially for beginners. It can automate how the bicycle gears are changed based on the data provided by only two sensors. It also allows the riders to use two different modes with a single ick of a switch. It makes riding the bicycle safer than with an ordinary drivetrain. It gives the rider more control over the shifting of gears, and most importantly, it makes the experience of riding a bike a lot more enjoyable.

With the success of the conduct of the study gathering excellent accuracy rate during functionality test, the study is highly recommended for more testing times to test the quality of the product even in different weather conditions, which was not considered during testing. The proponents recommend the future innovation of the study by deploying wireless connectivity to eliminate the hazards brought by the cables. Lastly, additional modes to set the timing of the automatic mode shifting as well as a mode to calibrate the angle of the servos could be more interesting to include in the upcoming studies.

References

1. Oja P, Titze S, Bauman A, de Geus B, Krenn P, Reger-Nash B, Kohlberger T (2011) Health benefits of cycling: a systematic review. Scand J Med & Sci Sport 21(4)

2. Madsen TK, Christensen MB, Andersen CS, Várhelyi A, Laureshyn A, Moeslund TB, Lahrmann H (2017) Comparison of two simulation methods for testing of algorithms to detect cyclist and pedestrian accidents in naturalistic data. In: 6th International naturalistic driving research symposium, the Hague, the Netherlands, 8–9 June 2017
3. Hsiao SW, Chen RQ, Leng WL (2015) Applying riding-posture optimization on bicycle frame design. Appl Ergon 51
4. Arthurs-Brennan M (2016) 5 Most common mistakes beginners make with road bike gears. Retrieved January 16, 2019, from https://totalwomenscycling.com/roadcycling/technique/5-common-mistakes-beginners-make-road-bike-gears
5. Rashad A, Kumar V, Luff N, Habib S, Munaaf MHM, Kumar S, Noreen A (2018) Evaluate the effect of bike riding on posture. Int J Contemp Res Rev 9(01)
6. Longhurst J (2015) Bike battles: a history of sharing the American road. University of Washington Press
7. Chen Z, Wang X, Liu H, Ma Q (2016) The research progress and prospect of cable-supported structures in China. In: Proceedings of IASS annual symposia, vol 2016, no 21, pp 1–10. International Association for Shell and Spatial Structures (IASS)
8. Mechanical vs electronic shifting, Serotta design studio. Retrieved December 21, 2018, from https://www.serottadesignstudio.com/news-post/electric-vsmechanical-shifting
9. Electronic Versus Mechanical Groupsets, Evans Cycles (2018). Retrieved December 27, 2018, from https://www.evanscycles.com/coffeestop/advice/electronic-vsmechanical-groupsets
10. Chabroux V, Barelle C, Favier D (2012) Aerodynamics of cyclist posture, bicycle and helmet characteristics in time trial stage. J Appl Biomech 28:317–323
11. Chabroux V, Barelle C, Favier D (2008) Aerodynamics of time trial bicycle helmets. In: Proceedings of the international sport engineering association, Biarritz
12. Blair KB, Sidelko S (2008) Aerodynamic performance of cycling time trial helmet. The engineering of sport 7, vol 1, pp 371–377. Springer
13. Su H, Hu J, Zheng Z (2018) Design and implementation of bicycle testing platform with automatic image processing. In: IEEE international conference on consumer electronics-Taiwan (ICCE-TW)
14. Jiong L, et al (2011) Diagnosis and feedback system for bicycle pedaling technique and it application. In: International conference on future computer science and education, Xi'an, pp 205–209
15. Piersma D, de Waard D (2014) Shifting from manual to automatic gear when growing old: good advice? Results from a driving simulator study. In: Proceedings of the human factors and ergonomics society Europe, pp 43–51
16. Sadiqi S, Leenen LPH, Oner CF (2018) Injuries related to bicycle accidents: an epidemiological study in The Netherlands. Eur J Trauma Emerg Surg: Off Publ Eur Trauma Soc
17. WeLoveCycling (2016) Which shifting is better—electrical or mechanical? WeLoveCycling magazine. https://www.welovecycling.com/wide/2016/06/16/shifting-better-electrical-mechanical/. Last Accessed 23 Jan 2019. (Online)
18. Domínguez R, Correa F (2018) Risk analysis in mountain bike, using questionnaire and event tree analysis (ETA). In: International conference on applied human factors and ergonomics, pp 303–308. Springer, Cham
19. Caçador FMDF (2016) Bicycle drivetrain noise and vibration test development (Master's thesis)

A Review on the Bolted Flange Looseness Detection Method

Mohd Padzly Radzi and Mohd Hafizi Zohari

Abstract Early detection of bolted flange looseness in oil and gas industries is vital, as it may lead to severe problems such as leakage and explosion. The implications of this problem are not only focused on the system itself but the surrounding circumstances and coupled with the enormous economic losses associated with such failures. The use of bolts in flange structure itself can cause significant deficiencies or damage, e.g. self-looseness. In comparison to traditional methods, online monitoring has become more popular and promising in early bolt loosening detection in recent years. As one of the major flaws in bolted connection is looseness, this article aims to briefly review the bolted flange looseness detection in the oil and gas industry.

Keywords Fiber Bragg Grating · Structural health monitoring · Bolt looseness

1 Introduction

Structural Health Monitoring (SHM) has been in use for several decades to improve the safety and durability of flange connection. Furthermore, the implementation of SHM can reduce repair costs by providing early warning of possible damage to piping connections. In oil and gas industry, piping system is the dominant means of fluid transportation, hence, it is essential to reduce the maintenance costs and improve its safety. The usage of pipelines is indisputable due to the advantage of speed. The demand for pipeline expansion and maintenance is increasing from year to year as energy production increases. In the meantime, there were 1309 reported pipeline accidents as more than 143,000 km of pipelines occurred from 1970 to 2013 [1], which indicate the significant need of SHM for pipelines.

One of the common methods to connect between pipelines is by using bolted flange connections. Bolt has been used for clamping method for many years, and it is most energy-efficient and economically over long-distance [2]. Over the majority of joints, the bolt is convenient to be used and easy to dissemble [3]. Bolt specifications

M. P. Radzi (✉) · M. H. Zohari
Advanced Structural Integrity and Vibration Research, Faculty of Mechanical and Automotive Engineering Technology, University Malaysia Pahang, Gambang, Malaysia
e-mail: padzly@ump.edu.my

© The Author(s), under exclusive license to Springer Nature Singapore Pte Ltd. 2023
M. H. A. Hassan et al. (eds.), *Technological Advancement in Instrumentation & Human Engineering*, Lecture Notes in Electrical Engineering 882,
https://doi.org/10.1007/978-981-19-1577-2_22

and standards are based on tensile strength; however, studies have revealed that shear also plays a role in bolt failure as well. Hence, it is necessary to monitor the bolt condition, particularly in flange connection, to avoid unexpected failure to the joint.

Pipelines in the oil industry usually operate under extremely high internal pressure. If one of the bolts is loose, the worse effect will be borne by the other bolt. Catastrophe may occur such explosion, resulting in a loss of life and a detrimental impact on society [4, 5]. There are many reported incidents worldwide on the looseness of the bolt during service, not just in the oil industry but others also experiencing similar issues, such as aerospace, automotive and train industries. Because the transported material is explosive, high-pressure, and poisonous, a leak or explosion will result in a catastrophic tragedy and the situation is worse because most pipelines are buried and pass through densely populated regions [6, 7].

According to many studies, bolt may experience looseness during its applications [8–12] due to the long period of service and environment. Therefore, it is crucial to monitor bolt tension and preload joint [13–19] to prevent environmental pollution and ecosystem disruption. The technology in monitoring bolt conditions has revolutionized over several decades; and there is always a new invention to fulfil the SHM rules. For many years, bolt loosening during service on a structure or facility has been the focus of countless studies and examinations, and it has been a major research topic [20–22]. This article will briefly discuss available methods for bolted flange health monitoring, namely bolt length measurement, torque measurement, strain gauge-based sensor, piezoelectric based sensor and optical-based sensor.

2 Manual Bolt Length Measurement

Traditionally, a Vernier calliper was used to measure changes in the length of the bolt. However, this approach has been reported to give inaccurate measurement [23]. Measuring using a calliper also demands a certain level of expertise and sense. Usually, a large number of bolts are slightly bent due to tightening; and sometimes, the ends of the bolts are not flat and parallel to each other. Even if the ends of the fastener are parallel, an operator has a difficult time to determine if the anvils of the calliper are flat against the ends of the fastener on lengthy bolts. As a result, different operators will provide different conclusions.

One of the challenges of getting an accurate reading is the calliper's size. A Vernier calliper's scale usually very small. With the impaired vision or an inappropriate angle of view, it's difficult to read or discern measures. As results, more attention is required throughout the measuring process, and for a high number of bolts, a lot of time will be spent and more errors will occur [24].

Another issue to consider while inspecting bolts is the position and geometry of the bolt. If the bolt is located in the area with narrow spaces (Fig. 1), it is impossible for the calliper to access the area, resulting in failure to obtain accurate measurements. Similarly, for the bolt size extremely big, e.g. M64, the use of calliper seems technically difficult.

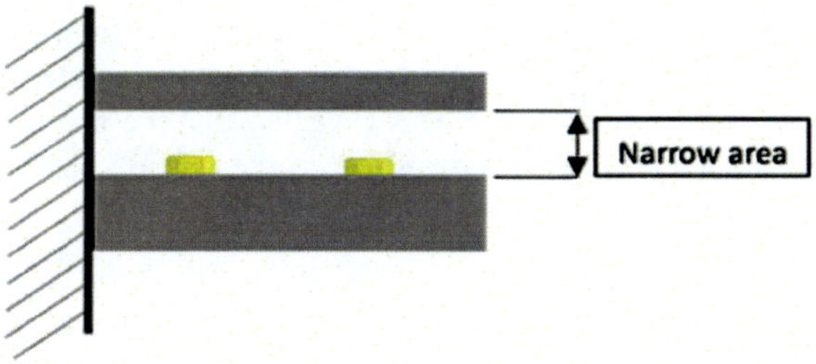

Fig. 1 Narrow area access

3 Manual Torque Measurement

A torque wrench is widely used to monitor bolt looseness in many industries. Tension in the bolt can be divided into three types which are initial preload, residual preload and the tension in a bolt service. These type of loading on bolts must be monitored frequently to guarantee the bolts integrity during services. A skilled inspector will use a torque meter to examine the torque of the bolt at specified intervals and ensures that it does not decrease over time. The torque measurement is done manually, and its effectiveness is strongly based on the inspectors' experience, therefore it is not suitable for real-time monitoring. It also necessitates periodic investigation, and human mistake is unavoidable [25].

The torque value is determined by the amount of friction in the bolt during bolted joint assembly. Variation in friction value can lead to a change in torque value. The torque wrench method relies on the linear tension–torque relationship of bolt fastening [26]. However, the tension–torque ratio is affected by a number of factors such as the coefficient of friction, thread tolerance errors and lubrication coat types. As a result, there are several inconsistencies occur [27–29], resulting in a tension force estimation error of up to 25% [30]. The influence of friction on torque can be seen in Fig. 2. Figure 2a shows that torque value is higher at the higher friction

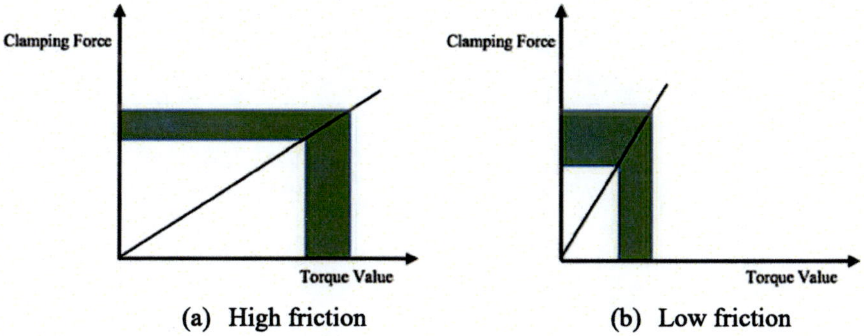

Fig. 2 Effect of friction on clamping force [31]

compare to low friction coefficient as shown by Fig. 2b. Due to high friction, torque may exceed too soon while clamping clamp still not achieving tolerable value.

Results from previous studies also have revealed that the manual method using torque wrench is not accurate due to the effect of the friction between the bolt and part [32]. As a consequence, the bolts are rotated using only 10–15% of the provided torque [27, 33].

4 Strain Gauge Sensors

Another popular method for bolt looseness monitoring is via strain gauge sensor. A strain gauge has a conductive grid that when distorted varies its electrical resistance. Forces from a loaded object create grid deformation, which is caused by connecting the strain gauge to it. The output of a strain gauge is a change in resistance, which corresponds to a physical change in the object being probed [34]. This electrical based sensor is suitable for real-time monitoring.

A strain gauge sensor is frequently utilized by embedding or directly patch to the body. This approach has proven to be accurate in monitoring the bolt strain, however, it is limited by the service environment of the bolt and how the sensor was installed.

Strain gauges can be utilized in one of two ways, as washers or as integrated bolts. The washer is also known as a compression fastener as it distributes the clamp load over a larger bearing area. A washer with an attached strain gauge can measure the amount of load or pressure when placed between plate and bolt. When torque is applied to the bolt, the axial force on the join increases and the washer continuously measures the applied force [25].

With the advancement of this sensing technology, many researchers have started to place strain gauges at the different study of points. It's possible to measure not only the load but also the length change. A strain gauge should also be installed on both sides of the bolt surface to measure the change in bolt length. Strain gauges, on the other hand, have the drawback of being difficult to install [35] and requiring precise sensor mounting positions [34]. For the case of the bolted flange, it is impossible to install the strain gauge on a bolt if the gap between bolt and flange are small as shown in Fig. 3.

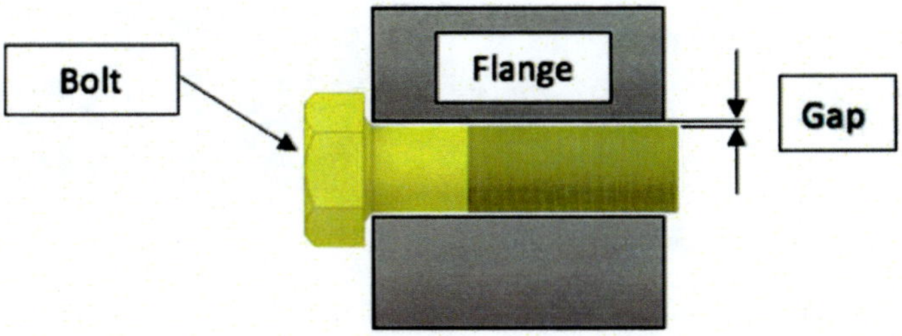

Fig. 3 Gap between bolt and flange

5 Piezo Active Sensing Method

A piezoceramic materials have been utilised as transducers to monitor bolted connections because they can provide a variety of functions, including actuation and sensing, wide bandwidth, energy harvesting [36–40], communication [41, 42], and the generation of guided waves [43]. Furthermore, since the deterioration of a structure often attenuates wave propagation, the piezoceramic transducer enabled active sensing approach [44, 45] is gaining popularity in structural health monitoring (SHM). An attenuation-based approach based on contact mechanics was utilised to monitor bolted looseness and energy dissipation as waves propagated through the bolted joint surface [46].

There are few studies utilized the piezo active sensing method for bolt looseness detection. In this method, two piezoceramics bonded at two different of a bolted part. One sensor serves as an actuator, generating ultrasonic waves, while the other serves as a receiver, receiving waves transmitted through the bolt connection as shown in Fig. 4. During torque application, the waves are caused by a rough surface between

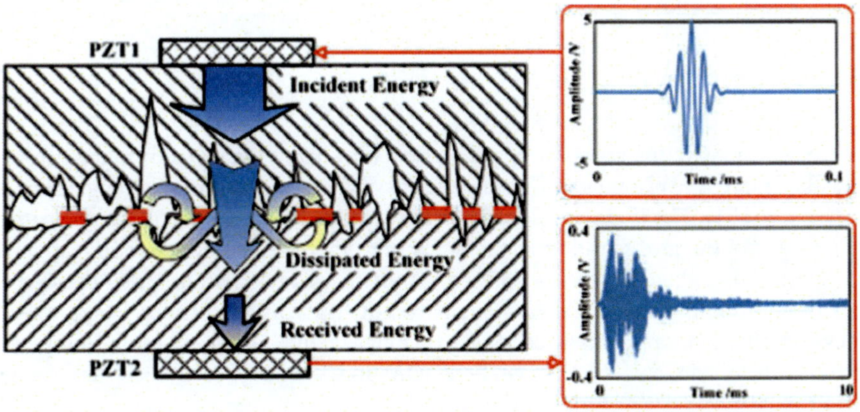

Fig. 4 Energy transmission between connection [47]

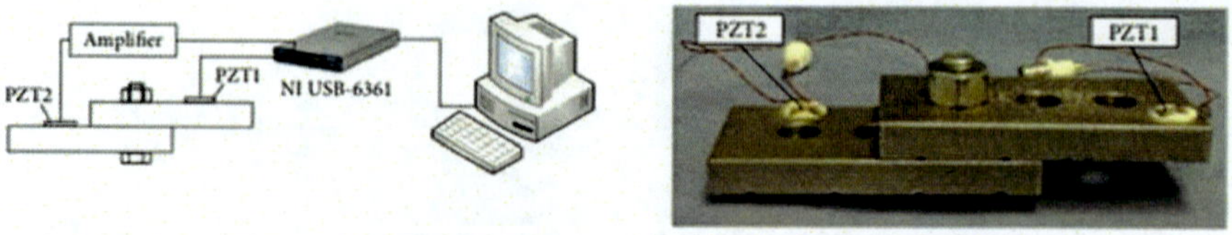

(a) Bolt connection monitoring system (b) Test specimen

Fig. 5 Device for checking the integrity of bolt connections [50]

the part and the bolt surface. By analyzing the received energy, the status of bolt connection and bolt loosening can be analyzed [47, 48].

Due to its high resolution, piezo active sensing method has the advantage of allowing instantaneous monitoring of the situation. Moreover, because the frequency range used for stimulation in this method is very low, it will not require high-frequency sampling equipment, resulting in lower costs. However, this method has drawbacks when dealing with a group of bolts to be investigated. Prices will increase as more sensors and equipment are used [25, 49].

Yang and Chang [46] give a diagnostic method for identifying the location of a loosened bolt and predicting the bolt's torque level. The method, in fact, is based on the decrease of ultrasonic wave energy levels, which is subsequently employed as the theoretical basis for active piezoelectric detection systems. The experiment has been proposed by Wang et al. [47] as shown in Fig. 5, and the result from the experiment revealed that second piezoelectric elements were affected by the changing of torque value. The exciting thing is that the increase in torque will not always increase the energy level. In fact, if the total torque rises above a certain level, the joint is saturated somehow because all touching surfaces have a tendency for accurate and perfect surface contact. The increase in torque level will not much affect the energy signal obtained.

6 Fibre Bragg Grating Sensor (FBG)

A Fiber Bragg Grating sensor is based on periodic changes in refractive index along the fiber's length. When a specific wavelength of light travels through the grating, it is reflected [51]. The fiber wavelength sensor detection mechanism can be attributed to changes in the grid duration or the index scar fiber that are effective in response to external disturbances, such as pressure, tension and temperature, that cause the Bragg wavelength of the fiber wavelength to shift. Thus, by performing Bragg wavelength shift monitoring, measurements of range values, such as strain, temperature, displacement, current, voltage, flow, vibration, intensity, or acceleration can be determined.

For monitoring of clamping connections, the FBG sensor can be embedded in a bolt or affixed to a washer. An FBG smart washer capable to monitor force or

Table 1 Experiment results

Applied shear force (kN)	Measured shear force (kN)	Measurement error (%)
5	4.78	4.4
10	9.59	4.1
15	14.54	3.2
20	19.42	2.9
25	24.35	2.6
30	29.25	2.5
35	34.16	2.4

load imposed on the bolt. Meanwhile, the embedded FBG in bolt, or also called as FBG smart bolt, can simultaneously monitor strain, temperature, load, and other data [52]. When a bolt is sheared, it produces shear stress as well as bending stress, which is linearly associated with shear stress. During deployment, axial tensions will accompany shear and bending stresses, due to initial pretension and additional applied axial loads [53].

According to Liang Ren's [53]experiment, the bolts provided with the FBG sensor exhibit excellent agreement between the measured and actual load, as shown in Table 1. To assess the link between wavelength variation of the FBG sensors and stresses, a calibration test is required. A universal testing machine applies an axial load to the smart bolt at a loading speed of 100 N/s up to a maximum load of 15 kN. All measurements were done numerous times to ensure repeatability and better averaging of the data.

To date, there are three categories of FBG sensors, distributed sensors, single point sensors and quasi distributed (multiplexed) [39]. A single-point sensor is a small, durable, and highly precise sensor unit typically mounted on a high-bandwidth fiber optic cable. These single-point sensors can multiplex and be placed at specific locations along the fiber to make partially distributed measurements. Distributed sensing uses fiber optic cables to monitor the tension along with the structure under investigation. To overcome the limitations of conventional sensors, optical fiber sensor such as the FBG sensor can be utilized because it has dominant capabilities compared to traditional sensor to operate in a harsh environment.

FBG sensors are not only a highly established field of research, but they are also gaining a larger market share due to their height sensitivity [54]. Another advantage of the FBG sensor is the small size of the sensor itself; a single installation of FBG sensor capable of measuring strain, temperature; and no need for electricity for the connection. This sensor technology can provide high performance and long-term durability.

7 Summary of Bolt Looseness Method

Methods	Advantages	Disadvantages
Manual caliper methods	Variety of measuring jaw shapes, allowing for the study of a wide range of surfaces and making it a universal tool [55]. Low cost at any number of bolt monitoring.	Measurement is inaccurate as precision is about 0.02 mm and requires experienced workers [56]
Manual torque methods	The maximum torque value can be set to prevent over-torque [57]. Low cost at any number of bolt monitoring	Measurement only focuses on the tightening process not capable to measure other directions [42]
Strain gauge sensors	Capability to provide instantaneous monitoring of the situation. Bolt monitoring at a low cost for a small number of bolts	The strain gauge placement on the bolt can be tricky in some circumstances [35]
Piezo active sensing method	This approach requires no high-frequency sampling equipment because the frequency range used for stimulation is quite low [25]	Costly bolt monitoring for a large number of bolts. Besides, the difficulty of practical integration with bolts is a frequent drawback of the sensors, and so there is still a lack of a simple approach for bolt shear force monitoring [53]
FBG sensor	FBG strain sensors have the benefits of being small, accurate, and stable. FBG sensors can also be multiplexed [49, 59]. Bolt monitoring at a low cost for huge quantities	Bare optical fibers with FBG are very brittle and easily damaged [60]

8 Conclusions

This article has reviewed some of the known approaches that frequently use in oil and gas industries for bolted flange looseness detection. In can be concluded that the FBG sensor has advantages over the other approaches covered in this article. Studies using the FBG method to detect loose bolts are still new and worthwhile still at the laboratory stage. Embedded methods are expected to be suitable for bolted flange monitoring purposes. Based on the data obtained, a smart bolt integrated with the distributed FBG is an effective and reliable tool for monitoring both bolts axial and shear load and can be used for structural health monitoring with bolts connection.

Acknowledgements The authors would like to acknowledge the UMP Research and Innovation Department for providing the internal research grant under grant no. PDU213222.The authors also would like to thank the Faculty of Mechanical & Automotive Engineering Technology, Universiti Malaysia Pahang (http://www.ump.edu.my/) for providing the laboratory facilities.

References

1. EGIG (2015) Gas pipeline incidents. In: 9th Report of the Europena gas pipeline incident data group 1970–2013, February 2015, p 61. https://www.egig.eu/uploads/bestanden/ba6dfd62-4044-4a4d-933c-07bf56b82383. (Online)
2. Dey PK (2001) Oil pipelines. Oil Gas J 99(36):81–83. https://doi.org/10.1016/b0-12-176480-x/00260-6
3. Li Q, Jing X (2020) Fault diagnosis of bolt loosening in structures with a novel second-order output spectrum–based method. Struct Health Monit 19(1):123–141. https://doi.org/10.1177/1475921719836379
4. Guo X, Zhang L, Liang W, Haugen S (2018) Risk identification of third-party damage on oil and gas pipelines through the Bayesian network. J Loss Prev Process Ind 54(18):163–178. https://doi.org/10.1016/j.jlp.2018.03.012
5. Peng XY, Yao DC, Liang GC, Yu JS, He S (2016) Overall reliability analysis on oil/gas pipeline under typical third-party actions based on fragility theory. J Nat Gas Sci Eng 34:993–1003. https://doi.org/10.1016/j.jngse.2016.07.060
6. Huang Z, Li J (2012) Assessment of fire risk of gas pipeline leakage in cities and towns. Procedia Eng 45:77–82. https://doi.org/10.1016/j.proeng.2012.08.124
7. Pipelines and the population explosion. https://onepetro.org/APIDPP/proceedings-abstract/API69/All-API69/API-69-223/51414. Last Accessed 23 Aug 2021
8. Park JH, Kim TH, Kim JT (2015) Image-based bolt-loosening detection technique of bolt joint in steel bridges. In: International conference on advances in experimental structural engineering, vol 2015, August, no 1
9. Nazarko P, Ziemianski L, Brandão P, Infante V, Deus AM (2017) ScienceDirect flange connections structural. Force identification in bolts of flange connections for structural health monitoring and failure prevention health monitoring and failure prevention. In: Thermo-mechanical modeling of a hi
10. Sah SM, Thomsen JJ, Brøns M, Fidlin A, Tcherniak D (2018) Estimating bolt tightness using transverse natural frequencies. J Sound Vib 431:137–149. https://doi.org/10.1016/j.jsv.2018.05.040
11. Qian G, González-Albuixech VF, Niffenegger M (2018) Effects of embedded cracks and residual stresses on the integrity of a reactor pressure vessel. Eng Fail Anal 90(January):451–462. https://doi.org/10.1016/j.engfailanal.2018.04.009
12. You JM, Hong YC, Jeon SH, Huh J, Ahn JH (2019) Behavior of bolt-connected steel plate girder attributable to bolt loosening failure in the lower flange. Eng Fail Anal 107(June):104208. https://doi.org/10.1016/j.engfailanal.2019.104208
13. Horváth P, Tóth P (2018) Nondestructive bolt preload measurement. Athens J Technology & Eng 5(2):91–110. https://doi.org/10.30958/ajte.5-2-1
14. Matos R, Shah Mohammadi MR, Rebelo C (2018) A year-long monitoring of preloaded free-maintenance bolts—estimation of preload loss on BobTail bolts. Renew Energy 116(123–135). https://doi.org/10.1016/j.renene.2017.05.092
15. Wang YQ, Wu JK, Liu HB, Kuang K, Cui XW, Han LS (2017) Analysis of elastic interaction stiffness and its effect on bolt preloading. Int J Mech Sci 130(June):307–314. https://doi.org/10.1016/j.ijmecsci.2017.05.032

16. Croccolo D, de Agostinis M, Vincenzi N (2011) Failure analysis of bolted joints: effect of friction coefficients in torque-preloading relationship. Eng Fail Anal 18(1):364–373. https://doi.org/10.1016/j.engfailanal.2010.09.015
17. Braithwaite J, Mehmanparast A (2019) Analysis of tightening sequence effects on preload behaviour of offshore wind turbine M72 bolted connections. Energies 12(23). https://doi.org/10.3390/en12234406
18. Li Y, Liu Z, Wang Y, Cai L, Xu W (2019) Research on preload of bolted joints tightening sequence-related relaxation considering elastic interaction between bolts. J Constr Steel Res 160:45–53. https://doi.org/10.1016/j.jcsr.2019.01.016
19. Abid M, Hussain S (2008) Bolt preload scatter and relaxation behaviour during tightening a 4 in-900# flange joint with spiral wound gasket. Proc Inst Mech Eng Part E: J Process Mech Eng 222(2):123–134. https://doi.org/10.1243/09544089JPME160
20. Kong X, Li J (2018) Vision-based fatigue crack detection of steel structures using video feature tracking. Comput-Aided Civ Infrastruct Eng 33(9):783–799. https://doi.org/10.1111/mice.12353
21. Chen Y, Gao Q, Guan Z (2017) Self-Loosening failure analysis of bolt joints under vibration considering the tightening process. Shock Vib 2017. https://doi.org/10.1155/2017/2038421
22. Kong X, Li J (2018) Image registration-based bolt loosening detection of steel joints. Sens (Switz) 18(4). https://doi.org/10.3390/s18041000
23. Motra HB, Hildebrand J, Dimmig-Osburg A (2014) Assessment of strain measurement techniques to characterise mechanical properties of structural steel. Eng Sci Technol, Int J 17(4):260–269. https://doi.org/10.1016/j.jestch.2014.07.006
24. Etingof MI (2012) Modern vernier caliper
25. Nikravesh SMY, Goudarzi M (2017) A review paper on looseness detection methods in bolted structures. Lat Am J Solids Struct 14(12):2153–2176. https://doi.org/10.1590/1679-78254231
26. Miao R, Shen R, Zhang S, Xue S (2020) A review of bolt tightening force measurement and loosening detection. Sens (Switz) 20(11). https://doi.org/10.3390/s20113165. (MDPI AG, June 01)
27. Zou Q, Sun TS, Nassar SA, Barber GC, Gumul AK (2007) Effect of lubrication on friction and torque-tension relationship in threaded fasteners. Tribol Trans 50(1):127–136. https://doi.org/10.1080/10402000601105490
28. Nassar SA, Sun T, Zou Q, The effect of coating and tightening speed on the torque-tension relationship in threaded fasteners
29. Wang Z, Zhang Z, Cui C (2018) Measurement and control method for miniature fastener tension in precision instrument. Procedia CIRP 76:110–114. https://doi.org/10.1016/j.procir.2018.01.018
30. Nazarko P, Ziemianski L (2017) Force identification in bolts of flange connections for structural health monitoring and failure prevention. Procedia Struct Integr 5:460–467. https://doi.org/10.1016/j.prostr.2017.07.142
31. Mortensen J, Friction analysis of bolts
32. Nassar SA, Sun TS (2007) Surface roughness effect on the torque-tension relationship in threaded fasteners. Proc Inst Mech Eng Part J: J Eng Tribol 221(2):95–103. https://doi.org/10.1243/13506501JET192
33. Mehmanparast A, Lotfian S, Vipin SP (2020) A review of challenges and opportunities associated with bolted flange connections in the offshore wind industry. Metals 10(6). https://doi.org/10.3390/met10060732
34. Tutak P (2015) Application of strain gauges in measurements of strain distribution in complex objects. J Appl Comput Sci Methods 6(2):135–145. https://doi.org/10.1515/jacsm-2015-0004
35. Wang T, Tan B, Lu G, Liu B, Yang D (2020) Bolt Pretightening force measurement based on strain distribution of bolt head surface. J Aerosp Eng 33(4):04020034. https://doi.org/10.1061/(asce)as.1943-5525.0001147
36. Kim HS, Kim JH, Kim J (2011) A review of piezoelectric energy harvesting based on vibration. Int J Precis Eng Manuf 12(6):1129–1141. https://doi.org/10.1007/s12541-011-0151-3

37. Chen YY, Vasic D, Costa F, Wu WJ, Lee CK (2010) Self-powered piezoelectric energy harvesting device using velocity control synchronized switching technique. In: IECON proceedings (industrial electronics conference), pp 1785–1790. https://doi.org/10.1109/IECON.2010.5675406
38. Liang J, Liao WH (2012) Improved design and analysis of self-powered synchronized switch interface circuit for piezoelectric energy harvesting systems. IEEE Trans Industr Electron 59(4):1950–1960. https://doi.org/10.1109/TIE.2011.2167116
39. Malakooti MH, Sodano HA (2015) Piezoelectric energy harvesting through shear mode operation. Smart Mater Struct 24(5). https://doi.org/10.1088/0964-1726/24/5/055005
40. Lin SC, Wu WJ (2013) Piezoelectric micro energy harvesters based on stainless-steel substrates. Smart Mater Struct 22(4). https://doi.org/10.1088/0964-1726/22/4/045016
41. Siu S, Qing J, Wang K, Song G, Ding Z (2014) "Stress wave communication in concrete: II. Evaluation of low voltage concrete stress wave communications utilizing spectrally efficient modulation schemes with PZT transducers. Smart Mater Struct 23(12):125031. https://doi.org/10.1088/0964-1726/23/12/125031
42. Siu S, Ji Q, Wu W, Song G, Ding Z (2014) Stress wave communication in concrete: I. Characterization of a smart aggregate based concrete channel. Smart Mater Struct 23(12). https://doi.org/10.1088/0964-1726/23/12/125030
43. Pierce SG, Culshaw B, Manson G, Worden K, Staszewski WJ (2000) Application of ultrasonic Lamb wave techniques to the evaluation of advanced composite structures. Smart Struct Mater 2000: Sens Phenom Meas Instrum Smart Struct Mater 3986(0):93. https://doi.org/10.1117/12.388096
44. Boller C, Biemans C, Staszewski WJ, Worden K, Tomlinson GR (1999) Structural damage monitoring based on an actuator-sensor system. Smart Struct Mater 1999: Smart Struct Integr Syst 3668(March):285–294. https://doi.org/10.1117/12.350707
45. S. P. Z. T. Actuator-sensors et al (2016) Identification of the location and level of damage in multiple-bolted-joint categories of SHM deployments: technologies and capabilities, 9445(March). https://doi.org/10.1061/(ASCE)0733-9445(2006)132
46. Yang J, Chang FK (2006) Detection of bolt loosening in C–C composite thermal protection panels: I. Diagnostic principle. Smart Mater Struct 15(2):581–590. https://doi.org/10.1088/0964-1726/15/2/041
47. Wang T, Song G, Wang Z, Li Y (2013) Proof-of-concept study of monitoring bolt connection status using a piezoelectric based active sensing method. Smart Mater Struct 22(8). https://doi.org/10.1088/0964-1726/22/8/087001
48. Huo L, Wang F, Li H, Song G (2017) A fractal contact theory-based model for bolted connection looseness monitoring using piezoceramic transducers. Smart Mater Struct 26(10). https://doi.org/10.1088/1361-665X/aa6e93
49. Wang Z, et al (2019) Clamp looseness detection using modal strain estimated from FBG based operational modal analysis. Meas: J Int Meas Confed 137:82–97. https://doi.org/10.1016/j.measurement.2019.01.051
50. Wang T, Song G, Liu S, Li Y, Xiao H (2013) Review of bolted connection monitoring. Int J Distrib Sens Netw 2013. https://doi.org/10.1155/2013/871213
51. Majumder M, Gangopadhyay TK, Chakraborty AK, Dasgupta K, Bhattacharya DK (2008) Fibre Bragg Gratings in structural health monitoring-present status and applications. Sens Actuators, A 147(1):150–164. https://doi.org/10.1016/j.sna.2008.04.008
52. Zhang H, et al (2019) FBG smart bolts and its application in power grid. I2MTC 2019–2019 IEEE Int Instrum Meas Technol Conf, Proc 2019(May):1–5. https://doi.org/10.1109/I2MTC.2019.8826850
53. Ren L, Feng T, Ho M, Jiang T, Song G (2018) A smart 'shear sensing' bolt based on FBG sensors. Meas: J Int Meas Confed 122(March):240–246. https://doi.org/10.1016/j.measurement.2018.03.026
54. Campanella CE, Cuccovillo A, Campanella C, Yurt A, Passaro VMN (2018) Fibre Bragg Grating based strain sensors: review of technology and applications. Sens (Switz) 18(9). https://doi.org/10.3390/s18093115

55. Etingof MI (2012) Modern vernier caliper. Meas Tech 55(8):890–893. https://doi.org/10.1007/s11018-012-0055-3
56. Yuan S, Zhang Y, Fan Y, Zhang Y (2016) A method to achieve uniform clamp force in a bolted rotor with curvic couplings. Proc Inst Mech Eng Part E: J Process Mech Eng 230(5):335–344. https://doi.org/10.1177/0954408914550017
57. Bickford JH (1998) Handbook of bolts and bolted joints. CRC Press, New York
58. Chen D, Huo L, Li H, Song G (2018) A fiber bragg grating (FBG)-enabled smart washer for bolt pre-load measurement: design, analysis, calibration, and experimental validation. Sens (Switz) 18(8). https://doi.org/10.3390/s18082586
59. Niewczas P, Mcmillan D, Rubert T, Niewczas P, McMillan D (2016) Life extension for wind turbine structures and foundations. In : International conference on offshore renewable energy 2016, 2016-09-12–2016-09-14. Thistle Glasgow City Hotel, no. September 2016, 2019. (In Press)
60. Cápová K, Velebil L, Včelák J (2020) Sensors for SHM for concrete and timber structures. MDPI

Pedestrian Lane Control System with Alert System

Emmanuel V. Galang, Noel R. Portillo, and Jetron J. Adtoon

Abstract Pedestrian accidents are inevitable despite the inclusion of pedestrian lanes. Vehicle users tend to neglect pedestrian lanes unless law enforcers are watching. On the other hand, pedestrian users also have lapses in the use of the lanes. This innovative study aims to secure the people and vehicles who are crossing the road. With image processing, a new pedestrian lane system can distinguish pedestrians and cars as inputs for the system to work. The algorithm of the system could process the matching scenarios provided in the fuzzy logic decision table. A prototype design of the project was fabricated along with traffic lights and cameras. The accuracy of the whole system yielded a result of 96.67% for the entire system, with 30 samples covering six scenarios simulated randomly.

Keywords Image detection · Barrier gate · Confidence level · Fuzzy logic · OpenCV

1 Introduction

Technological and industrial advancements have made significant impacts on our society today. An excellent example of this is vehicular and pedestrian traffic, which is increasing rapidly worldwide. Because of this increase, accidents are likely to happen more often than before. Traffic control better ensures the safety of pedestrian crossers and vehicle users. The idea of using barrier gates is not new due to its common usage in controlling the flow of vehicles on roads and secured areas; however, it might prove to be a valuable tool in controlling pedestrian flow and reduce accidents on the road.

Due to the changing circumstances of our economy, the purchasing capacity of citizens also increases, leading to a massive increase in vehicles used on the road [1]. This increase is directly proportional to the rise in road accidents. Road accidents have become a vital problem because they directly affect human life. On several

E. V. Galang · N. R. Portillo · J. J. Adtoon (✉)
Computer Engineering Program, College of Engineering Education, University of Mindanao, 8000 Davao City, Philippines
e-mail: jetron.adtoon22@gmail.com; jadtoon@umindanao.edu.ph

© The Author(s), under exclusive license to Springer Nature Singapore Pte Ltd. 2023

M. H. A. Hassan et al. (eds.), *Technological Advancement in Instrumentation & Human Engineering*, Lecture Notes in Electrical Engineering 882,
https://doi.org/10.1007/978-981-19-1577-2_23

occasions, people experience life-threatening injuries due to accidents. That leads to the need to develop a system, which includes traffic control signals to avoid road accidents [2]. Traffic signals worldwide exist to organized traffic flow, and they do not consider the varying number of pedestrians crossing at a given point in time. There is a possibility that more pedestrians might cross the street with the same amount of time given. That scenario often leads to accidents, resulting in loss of life and infrastructure [3]. Another critical factor to consider in the problem regarding pedestrian accidents is the behavior of pedestrians about violating traffic laws to cross the road, which could also result in pedestrian traffic accidents [4]. Another factor is the reaction time of humans. Several accounts of automobile collision cases set up conditions between the mishap and the response time of the pedestrian and the driver. It is predominant because of moment responses brought about by panic while confronting a possibly unsafe situation [5]. Due to the lack of traffic control devices and the apparent absence of knowledge by the citizens about pedestrian rights when crossing the road, the percentage of pedestrian victims of road accidents is high. Such means that there are exceptional safety needs requiring solutions for the benefit of pedestrians [6]. Also, it led to the development of different pedestrian detection methods. Pedestrian detection is needed for many vision applications, including surveillance, Advanced Driver Assistance Systems (ADAS), and Intelligent Transport System (ITS) [7]. An exciting approach to pedestrian detection systems is the deep convolutional network because good results were obtained from this method when used in vision tasks, thus making it an attractive tool to improve the capacities of pedestrian detection systems [8].

Few local governments tried mounting visible self-illuminating bollards to provide safety and control and detect pedestrians, especially for the aging and those with poor eyesight [9]. Also, to minimize the chances of people getting hit by vehicles, boom barriers are used at a crossing in India. The said barriers control pedestrian flow and serve as a guide and reminder about road safety policies [10]. Another possible approach to a solution, such as a fuzzy logic approach, may prove to be useful [11]. Another problem that requires an answer is the factor of distraction and attentiveness. Several safety experts worldwide have considered that people's dependency on mobile phones may directly threaten pedestrians [12]. The safety of pedestrians is still a significant concern in many urban societies, with distraction being one of the main factors behind grave incidents involving pedestrians [13]. It is prevalent to see pedestrians today that are running and walking while using their phones. Because of this, they are more deprived of their auditory senses. These senses would have provided essential signals to dangers, reducing the risk of being hit by vehicles [14]. Attentiveness is necessary because road traffic incidents result from different factors. It can be related to the design of the road, the cars, the pedestrians, and the way they interrelate [15].

Proper visual alarm systems are crucial in road safety. Implementing an organized pedestrian control and traffic caution system may solve the lack of road attentiveness by both the drivers and the pedestrians. Physical safety systems are also helpful because they serve as an extra line of defense between life and certain death for road users. Given those instances, this study aimed to design, develop, and implement an automated system that could control and monitor the flow of pedestrians who are crossing the pedestrian lane without compromising their safety and rights.

It is important to enhance the pedestrian lane to avoid accidents. It makes people more comfortable to cross and drive near the pedestrian lane. It makes people obey traffic rules to cross the pedestrian lane accordingly. Even vehicle drivers can be more disciplined in driving as traffic laws must be followed. This study is important as it adds more protection to the people. Also, it can help people who are pregnant and elderly as it avoids situations such as blind spots or over-speed vehicles.

The system detects multiple vehicles and pedestrians simultaneously using the same camera. This study uses visible-light communication (VLC) with light-emitting diodes (LEDs) for pedestrians and motorists' continuous and appropriate vision. The system also has a traffic light system with an override function that would extend the time for pedestrians to cross the street when activated. The study would not include an auditory alarm system or alarms that would produce sounds when triggered. It would also not involve counting pedestrians that would cross the pedestrian lane or the vehicles that would pass the street and will not detect minor modes of transportation like bicycles, motorcycles, scooters, and skateboards.

2 Materials and Methods

2.1 Conceptual Framework

Figure 1 explains that people close to the boom gate and vehicles near the pedestrian will be considered inputs of the study. Image detection will detect people and car entities; the camera will capture images then processes them. Same as image prediction, the camera will wait for the vehicles, then it captures their image. The timer of traffic also works simultaneously. After image detection and prediction, the camera will capture vehicles and persons as it automates the boom gate to open or close.

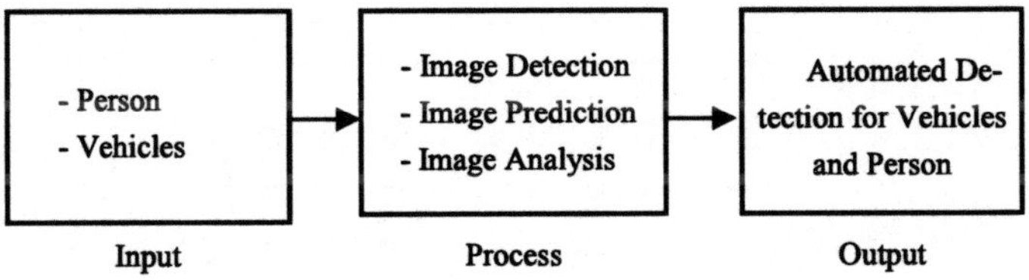

Fig. 1 Conceptual framework

2.2 Block Diagram

Figure 2 shows a block diagram of the whole system. After the Raspberry Pi 3 has been on, Arduino Mega 2560 then starts. The camera will automatically start and will automatically capture single or multiple detections for car and person. After the camera is on, Raspberry Pi and Arduino Mega 2560 will automatically connect. The traffic light for vehicles and humans will start, and the barrier gate is activated. After the countdown timer of the traffic light, the barrier gate will close the vehicle or person lane, and the system will repeat the countdown timer of the traffic light for vehicles and persons. If the camera detects single or multiple detections for car and person, the traffic light will not switch, and the barrier gate is closed. The simulation of the traffic light and the barrier gate will depend on the capture images when the countdown timer of the traffic light reaches zero. The system would rely on the

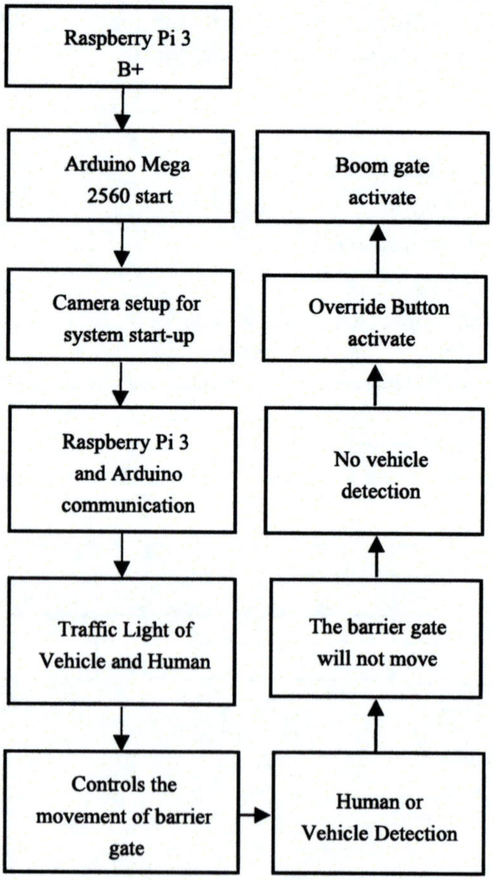

Fig. 2 System block diagram

raspberry camera for an advanced scenario to detect inputs like vehicles and persons near the pedestrian lane. An example of this scenario is that if the camera does not have image detection for vehicles within 10 s, the boom gate is activated using an override button to start the pedestrian lane and make people cross.

2.3 Materials and Resources

Raspberry Pi 3 B+ is a device that will simulate the system for traffic lights and cameras. It is used to control the motors of the barrier gates and acts as the server of the system. A Logitech c920 web camera was used for image detection of vehicles and humans. The camera is 3-megapixel with a 78° diagonal field of view and has a 1080 p/30–720p/30 fps maximum resolution. LED strips are used for traffic lights countdown timers and pedestrian and vehicle indicators. The 8 Channel relay module is an electromagnetic switch operated by a relatively small electric current that can turn on or off a much larger electric current. It will serve as a switch of the LED strips. The 7 Segment or seven-segment displays are used for the traffic light countdown timers. CD4511 is used to trigger the 7-segment for a specific number for the countdown timer. In controlling the movement of the arms of the barrier gate, a 30 kg-cm stepper motor is utilized. Also, an AC 220 V cooling fan is used to manage the temperature of the motors. Marine wood for the frame serves as a platform for the control panel and the stand for the traffic light timer and other indicators.

2.4 Methods and Procedures

Figure 3 shows the setup on how the camera will detect inputs like vehicles and people. The output or result of the camera's analysis will be the triggering input for the boom gate to activate or deactivate. The camera boosts pedestrian safety to

Fig. 3 Pedestrian image analysis

avoid any danger, accidents, collisions, etc. The camera was installed on top of the traffic light and the boom gate. The boom gate will automatically open if there are no detected vehicles within ten seconds. If there is zero detection for both cars and persons, the boom gate remains closed. The traffic will remain the same with or without vehicles and passersby.

2.5 *Software Development*

Image Processing. The triggering factor of the changes in the flow of traffic lights and the barrier gate is image detection. Digital image processing is an area characterized by the need for extensive experimental work to establish the viability of the proposed solution to a given problem [16]. The combination of single-shot detectors (SSDs) framework, which compresses all calculations in a single network and removes generation of the proposal and feature resampling, makes a faster object detector [17]. Mobilenets architecture, used for mobile and embedded vision systems [18], has been utilized for quick, efficient deep-learning-based object detection. MobileNet SSD was first trained on the datasets provided by Microsoft Common Objects in Context (COCO), large-scale object detection, segmentation, and captioning dataset [19]. The model used is a Caffe model.

The first step is image collection, and after that, it goes to category labeling. This step is deciding which object classifications are available in each picture based on the datasets present. Next, it arrives at Instance spotting. It is where all the samples of the objects were labeled based on their categories. Then it proceeds to Instance Segmentation. After classifying the thing, if it satisfies all conditions, it may be classified as a person or a vehicle.

Image Classification. For image classification, the captured image is classified and processed by the system. With the help of Raspbian Stretch OS and OpenCV 2.0, the Logitech C920 camera is connected to the Raspberry Pi using a Raspbian Stretch OS. Caffe model was used to train the datasets, and MobileNets combined with SSDs are used for object detection. Figure 5 will show the process for classifying an image to MobileNets. The captured image is sent to the Single-shot detectors (SSDs), and MobileNets classify the image.

Figure 4 shows the methods of the captured image using Geany Python 2.7. Using Common Objects in Context (COCO), all captured images are processed by MobileNets. Thus, the images that will be considered as inputs are the images captured by the camera. MobileNets is a file that stores the datasets for the system to recognize the type of image. With the help of depthwise separable convolution, it produces pointwise convolution. Combining pointwise and depth-wise convolution creates standard convolution. Standard convolution combines the input/s of an image to make a new output image. Fine-grain classification classifies an image such as dogs. Landmark Recognition is a recognition that helps divides the images and trains them

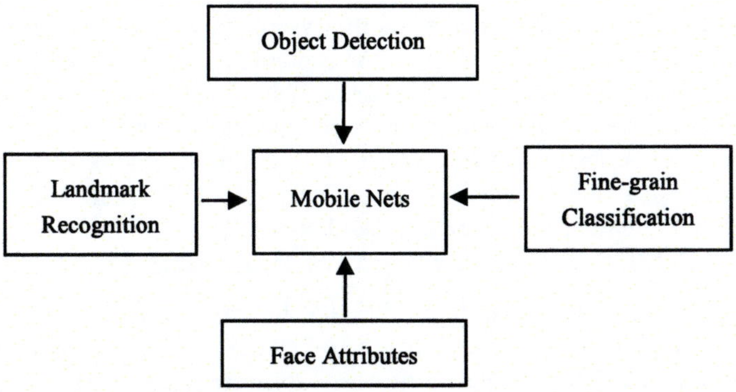

Fig. 4 Process flow for image classification

to the convolutional neural networks. Lastly, Face Attribute is a part of the datasets that helps classify the specific object together with the accuracy of the image [18].

Confidence Interval Classification. Statistical confidence interval is classified through image segmentation. The system is processing an image captured by the camera. Image segmentation is the extraction of important features of an image

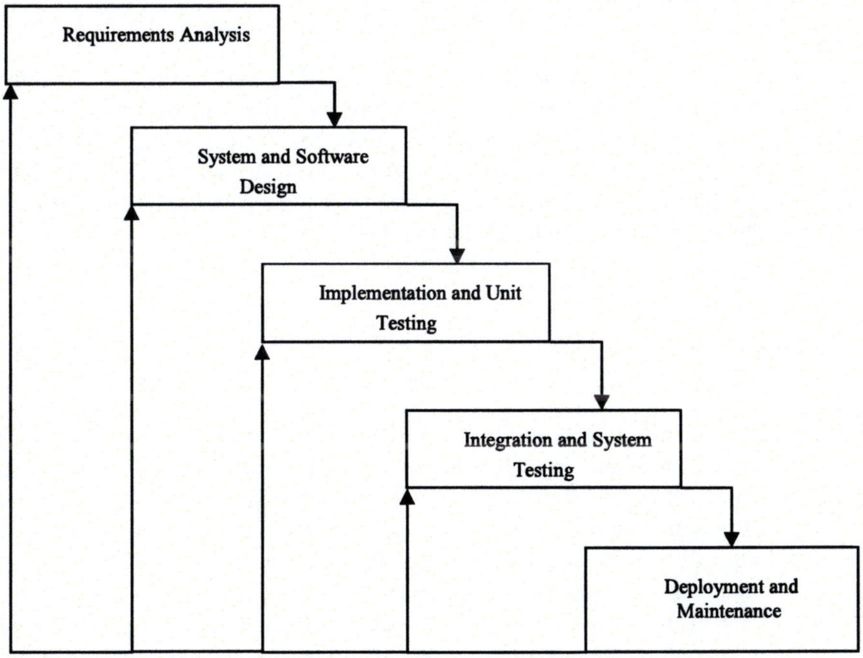

Fig. 5 Software development cycle

from the originally captured image. As the system receive the captured image of the camera, the system will pre-process the captured image with the help of convolution and Mobile Nets. Upon pre-processing, the processed image will get two or more variables. The variable is the identity and the confidence level of the captured image. Thus, the system holds the threshold values of the processed image. After the pre-processing stage, together with the multivariate threshold, the system will produce and show the result of the processed image. The confidence level will have a threshold of 95% because it is the typical confidence interval level [20].

Software Development Cycle. Figure 5 shows the software development cycle that the study utilized. Requirements Analysis is the phase where all possible methods used are captured and documented. System and software design is where the methods listed in the first part are analyzed and studied. Implementation and Unit Testing is where the developed drafts are being simulated and tested. Identification of what is lacking in the software takes place in this phase. Integration and system testing are where the software produced from the last phase is integrated with the whole system. This part lets the researchers identify the faults and holes in the study. The development and maintenance phase is the last part where the entire system is deployed and maintained.

2.6 Hardware Development

The project uses one camera for the crosswalk and the road rotating 90° from the road to the pedestrian lane through an MG995 servo motor. The prototype uses 12 volts LEDs to act as an alert for both the motorist and the pedestrian. The barrier gate part moves with the help of a stepper motor controlled by the Arduino Mega 2560 and the Raspberry Pi 3. The circuitry is encased in marine wood supported by hollow metal bars and painted in yellow for visibility.

2.7 Gathering of Data

Fuzzy Logic. Fuzzy Logic is used to gather data of the system. Fuzzy Logic is a form of many-valued logic in which the truth values of variables may be any real number between 0 and 1. The fuzzy logic would be the most appropriate method since the system follows six scenarios. Scenario 1 is where the traffic light stops, and the countdown timer for the pedestrian starts. For Scenario 2, the traffic light starts, and the countdown timer stops with a countdown timer set to 00. For Scenario 3, the override button cannot be activated when there is vehicle detection, and the boom gate closes the path of the pedestrian lane. For Scenario 4, if a person is detected within a countdown timer of 00, the boom gate is deactivated, and the traffic light is red. For Scenario 5, if there is no vehicle detection, the button can be activated for the boom gate to close the vehicle lane. For Scenario 6, if there is no vehicle detection and the button is pressed, the button will blink for 10 s. The traffic light turns red, and the boom gate closes the vehicle lane.

2.8 Interpretation of Data

Confusion Matrix. The data were interpreted using a confusion matrix. Using a confusion matrix for the project will be the most appropriate choice because the research utilizes fuzzy logic. A table was used to describe the performance of a model on a set of test data with known true values. With this, getting the overall accuracy of the system will be on point.

3 Results and Discussions

3.1 Hardware Result

Figure 6 shows the design of the project. The traffic light post is designed to carry the timer for the pedestrian lane, the camera, and the barrier gate. The barrier moves horizontally to block either the pedestrian lane or the vehicles based on the result of the camera detection. The sample in Fig. 6 is a functional prototype of the study.

The barrier gate moves using a motor situated in the internal part of the traffic light post. Its movement is based on the six scenarios as discussed in the data gathering procedure.

Fig. 6 Actual prototype of the project

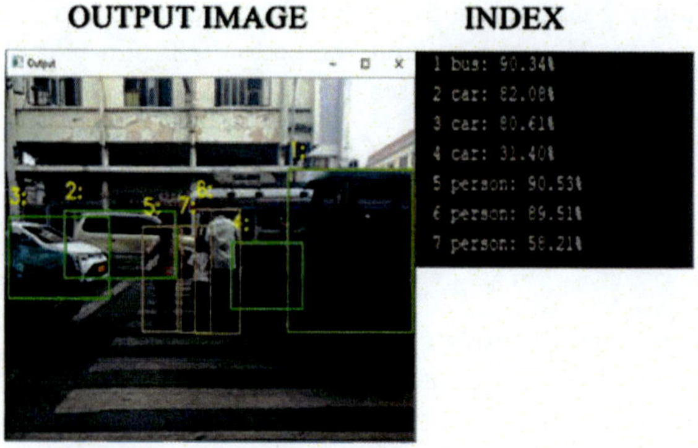

Fig. 7 Multiple human and vehicle recognition in one shot

Software. Geany Python 2.7 is an application utilized to execute the file located at home/pi. With the help of OpenCV 2, the camera will show a serial that produces an output. The program built on the Raspberry Pi 3 will auto-start when the system is on. As the program starts, it will wait for the camera to turn on. When the camera is on, it will start capturing. After capturing an image, the system will show the output. In the figure below, the images will show the output that was processed by the system. The capturing time would take about 2 s, and the processing of the image would take 3 s, then it will produce the output.

Figure 7 shows multiple human and vehicle recognition in one shot. The confidence level shows how specific the image detection is when it identifies the object in a captured image. The object detected is identified with a colored box with a confidence level. Confidence levels are established to ensure that the program is recognizing the corresponding objects in an image to classify what type of object. The objects of an image enclosed in a box with colored perimeters are the detection and the emphasize percentage. The image detected four vehicles and three people. The car with 31.40% detection detected the wheel of the vehicle. While the person detected is above 80%.

3.2 Functionality Test of the System

The system was tested in a controlled environment at the Research Hangar of the university. There were 30 trials for each threshold confidence level to gather data for the whole system. The data analysis result will explain all the data collected during the functionality test of the system.

The project is implemented in a pedestrian lane set up to inspect the system's flow in the pedestrian lane. The camera must be able to detect single and multiple detections for both vehicles and humans. The boom gate must trigger accordingly to the output of the traffic light. Also, there is an override button on both sides of the

Table 1 Confusion matrix with a threshold of 95% confidence level

Predictes	95% confidential level n = 30						
	Actual						
	Scenarios	1	2	3	4	5	6
	1	8					
	2		8				1
	3			3			1
	4				4		
	5					3	
	6						2

boom gate. The override button must only be trigger if there is no vehicle detected within 10 s.

3.3 Data Analysis Result

For the gathering of data, image processing was used to get the accuracy for humans and vehicles. Formulas were used to compute the overall accuracy of the system. The image's capturing and processing takes about 3–5 s, depending on how many are caught in the frame. Also, fuzzy logic was used to gather data. The collected data was plotted in a confusion matrix for further analysis.

Table 1 explains the system's response with a confidence level percentage threshold of 95% and a sample space of 30. The number of the sample spaces retrieved from how many times the scenarios were simulated randomly. It was based on the fuzzy logic of the situations that may occur. This yielded an accuracy of 96.67%. It produced this accuracy because the person did not have a confidence level of 95% upon the reading of the system. This occurred because the image captured was an unclear image of a person, and it also happened on the vehicle producing a misread of the software. The said errors were observed because of the lack of light intensity. Previous simulations outside the Research Hangar showed acceptable results. The testing was done in the daytime but with minimal lighting.

4 Conclusion and Future Works

The project has successfully demonstrated its ability to classify a person or vehicle using image classification. Also, the confidence level is automatically calculated based on how identical the image is compared to the datasets. The system keeps on capturing and can send data as quickly as possible to the Arduino Mega 2560. The system with the integrated image processing and algorithms that come with it helps keep the roads safer for both pedestrians and motorists. It enables the pedestrians in a way that ensures their safety. The accuracy of the image processing has a significant role in ensuring the safety of the persons using it because it is the deciding factor of the control system.

In the future, this project can be expounded by adding a density-based image processing that changes the timings, the lights would change, and the gate opens. In this way, no traffic congestion would bother the vehicles and still not compromise pedestrians' safety. Detection specifically for emergency reasons can also be added to the system. The hardware mechanism can also be improved as well as the components used especially the embedded systems.

References

1. Mokashi N (2015) Intelligent traffic signal control using image processing. Int J Adv Res Comput Sci Manag Stud 3(10):137–143
2. Kharat SB, Sawant SD (2017) Travolution: system for road safety. In: 2017 international conference on intelligent computing and control systems (ICICCS), Madurai, pp 1123–1125
3. Kumari A, Satish BA (2017) Automated traffic control for pedestrian safety. In: 2017 international conference on innovative mechanisms for industry applications (ICIMIA), Bangalore, pp 145–149
4. Guiying J (2015) Traffic control at signalized intersections based on pedestrians' group psychology crossing behaviour. In: 2015 IEEE international conference on computational intelligence & communication technology, Ghaziabad, pp 780–784
5. Sam D, Evangelin E, Raj VC (2015) A novel idea to improve pedestrian safety in Black Spots using a hybrid VANET of vehicular and body sensors. In: International conference on innovation information in computing technologies, Chennai, pp 1–6
6. Wehbe R, Massaad Z, Otayek E (2017) Using intelligent transportation systems to enhance pedestrian safety at Beirut signalized intersection. In: 2017 5th IEEE international conference on models and technologies for intelligent transportation systems (MT-ITS), Naples, pp 574–579
7. Wang M, Zhang Z (2018) FPGA implementation of HOG based multi-scale pedestrian detection. In: 2018 IEEE international conference on applied system invention (ICASI), Chiba, pp 1099–1102
8. Orozco CI, Buemi ME, Belles JJ (2017) New deep convolutional neural network architecture for pedestrian detection. In: 8th international conference of pattern recognition systems (ICPRS 2017), Madrid, pp 1–6
9. Oshiba S et al (2016) Visibility evaluation experiments of optical wireless pedestrian-support system using self-illuminating bollard. In: 2016 IEEE/ACIS 15th international conference on computer and information science (ICIS), Okayama, pp 1–6
10. (2020) Boom Barrier to pedestrian discipline. The telegraph, West Bengal
11. Olanrewaju OM, Obiniyi AA, JunaiduSB (2017) Fuzzy logic concept for safety driven vehicle-pedestrian traffic interaction. Int J Comput Appl 167(1):975–8887
12. Kumar A, Singh A, Sidhu E (2015) Smart-cap for pedestrian safety. In: 2015 international conference on control, instrumentation, communication and computational technologies (ICCICCT), Kumaracoil, pp 583–587
13. Vinayaga-Sureshkanth N, Maiti A, Jadliwala M, Crager K, He J, Rathore H (2018) Towards a practical pedestrian distraction detection framework using wearables. In: 2018 IEEE international conference on pervasive computing and communications workshops (PerCom Workshops), Athens, pp 239–245

14. Xia S et al (2018) A smartphone-based system for improving pedestrian safety. In: 2018 IEEE vehicular networking conference (VNC), Taipei, Taiwan, pp 1–2
15. Makó E (2015) Evaluation of human behaviour at pedestrian crossings. In: 2015 6th IEEE international conference on cognitive infocommunications (CogInfoCom), Gyor, pp 443–447
16. Alagao S, Alolino J, Ybanez M, Rubio E, Caya M (2018) Wireless electric consumption acquisition using image processing. In: 2018 IEEE 10th international conference on humanoid, nanotechnology, information technology, communication, and control, environment, and management (HNICEM)
17. Liu W, Anguelov D, Erhan D, Szegedy C, Reed S, Fu C-Y, Berg AC (2016) SSD: single shot multibox detector. In: Computer vision–ECCV 2016. Lecture notes in computer science, pp 21–37
18. Howard AG et al (2017) MobileNets: efficient convolutional neural networks for vision applications. Google Inc.
19. Lin T-Y, Maire M, Belongie S, Hays J, Perona P, Ramanan D, Dollár P, Zitnick CL (2015) Microsoft COCO: common objects in context. In: Computer vision–ECCV 2014. Lecture notes in computer science, pp 740–755
20. Buenestado P, Acho L (11 Jan 2018) Image segmentation based on statistical confidence intervals. MDPI. https://www.mdpi.com/1099-4300/20/1/46/htm. Accessed 27 Jun 2020

Design and Analysis of 6 Degree of Freedom Robotic Whiteboard Cleaner

Mark Ryan M. Estrera and Cresencio P. Genobiagon

Abstract The design and analysis of a six DOF robotic white board cleaner is the focus of this project. A kinematic model for a six-degree-of-freedom (DOF) white board cleaner is presented in this study. Moreover, forward kinematics and inversed kinematics model is developed, and Denavit-Hartenberg (D-H) parameters are used to find solutions. The forward model is used to determine the position and orientation of the end-effector whereas inversed kinematics is used analyze its results associated with solidworks. The experiment is then repeated to confirm the simulation results. Results exhibit that the analytical data are consistent with the measured points and orientation.

Keywords Robotic arm · D-H parameters · Eraser · White board · Kinematics

1 Introduction

In the educational system of the society, having mechanisms that provides convenience is one necessary tool so that the students and the educators can focus more on discussions or lessons [1]. In this study, a robotic whiteboard eraser is proposed, and this mechanism allows the user to erase the whiteboard without difficulties. The difficulties that are mostly considered are: there are times that it could be difficult to the user to erase some parts of writings of the board due to lack of reach or height and it would take manpower or effort to erase the whole writings of the entire board system.

A whiteboard is a wipe- able board with a white surface used for teaching and presentations [3]. According to [4], whiteboard was mostly used in offices, meeting rooms, school classrooms and other workplaces. In this study a board cleaning system will be used composed of different mechanisms but with the same aims compared to other existing research designs that is meant to reduce the stress of cleaning, reduce the effort of the board user, and a great replacement of manual dusting [5, 6]. The

M. R. M. Estrera · C. P. Genobiagon (✉)
Mechanical Engineering Program, College of Engineering Education, University of Mindanao, Matina, Davao City, Philippines
e-mail: cresencio_genobiagon@umindanao.edu.ph

© The Author(s), under exclusive license to Springer Nature Singapore Pte Ltd. 2023

M. H. A. Hassan et al. (eds.), *Technological Advancement in Instrumentation & Human Engineering*, Lecture Notes in Electrical Engineering 882,
https://doi.org/10.1007/978-981-19-1577-2_24

robotic white board cleaner will have joints with servo motor attached which allow the robotic arm to operate, move and rotate freely- performing the erasing process as it moves along the whole board system and wiping out writings. Typically, the servo motor only needs to drive the robotic arm so that the mechanism will function. And with this, the study will be able to design a mechanism by using robotic whiteboard eraser [7, 8]. The difference between this design compared to existing ones [2, 5] is that this type of design allows the user to choose freely parts of the whiteboard to erase writings.

Advance technologies are very beneficial nowadays to help lessen exhaustion of workers and makes task simpler and much easier. [9]. In this study it aims to pursue and achieve the former and the latter which lessens exhaustion, gives convenience to the user, and at the same time it provides portability to the user that will help lessen going back and forth in cleaning the whole board system [10]. The mechanism that is being developed and designed would be a great benefit to the users because of the convenience it gives [11, 12] and thus would lead to a more effective learning engagement [13].

The objective of this study is to design and analyze the kinematics of 6° of Freedom robotic whiteboard cleaner. Here the forward kinematics on a general Denavit Hartenberg parametric scheme and inversed kinematics with the aid of solidworks were utilized to model and give a summary of calculation as based.

2 Materials and Methods

The manipulator was designed and modeled using solid works software shown in Figs. 1 and 2 with mechanism in a 3-dimensional space and comes with more advanced movements. This design is composed of a base, 4 links and a wrist that holds the eraser with 8 servomotors in total. The base is mounted at the upper left of the whiteboard making a downright resting position.

Figure 3 then shows the 2D representation of the link lengths in comparison to resting position of the whiteboard cleaner. These link lengths are necessary parameters to mathematically model the position and orientation of the arm with respect to its application as presented in Table 1.

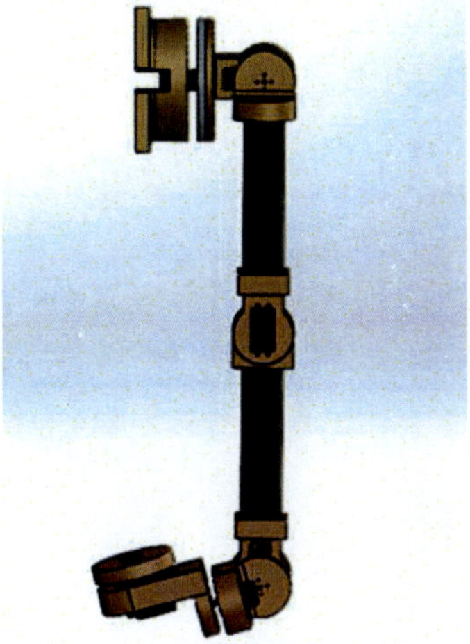

Fig. 1 Solid works model of 6DOF robotic whiteboard cleaner-resting position

Fig. 2 Solid works model of 6DOF robotic whiteboard cleaner-perspective

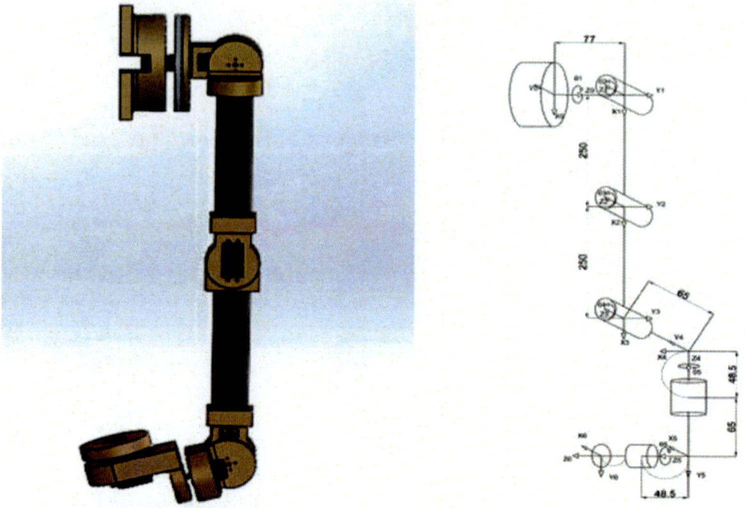

Fig. 3 2D Representation of the link

Table 1 Denavit-Hartenberg parameters of the 6DOF white board cleaner

i	A	a	d	q
1	0	90	77	q_1
2	250	0	0	q_2
3	250	0	0	q_3
4	0	90	−65	q_4
5	0	−90	113.5	q_5
6	0	0	48.5	q_6

2.1 *Forward Kinematics*

The forward kinematics problem can be stated as follows: Given the joint variables of the robot, determine the position and orientation of the end-effector. Since each joint has a single degree of freedom, the action of each joint can be described by a single number, i.e., the angle of rotation in the case of a revolute joint. The objective of forward kinematic analysis is to determine the cumulative effect of the joint variables.

$$T_{i-1}^{i} = \begin{bmatrix} Cos_{\theta i} & -Cos_{\alpha i}Sine_{\theta i} & Sine_{\alpha i}Sine_{\theta i} & A_i Cose_{\theta i} \\ Sine_{\theta i} & Cos_{\alpha i}Cos_{\theta i} & -Sine_{\alpha i}Cos_{\theta i} & A_i S_{ine\theta i} \\ 0 & Sine_{\alpha i} & Cos_{\alpha i} & d_i \\ 0 & 0 & 0 & 1 \end{bmatrix} \quad (1)$$

$$T_0^1 = \begin{bmatrix} Cos_{\theta 1} & 0 & Sine_{\theta 1} & 0 \\ Sine_{\theta 1} & 0 & -Cos_{\theta 1} & 0 \\ 0 & 1 & 0 & 77 \\ 0 & 0 & 0 & 1 \end{bmatrix} \tag{2}$$

$$T_1^2 = \begin{bmatrix} Cos_{\theta 2} & -Sine_{\theta 2} & 0 & 250Cos_{\theta 2} \\ Sine_{\theta 2} & Cos_{\theta 2} & 0 & 250Sine_{\theta 2} \\ 0 & 0 & 1 & 0 \\ 0 & 0 & 0 & 1 \end{bmatrix} \tag{3}$$

$$T_2^3 = \begin{bmatrix} Cos_{\theta 3} & -Sine_{\theta 3} & 0 & 250Cos_{\theta 3} \\ Sine_{\theta 3} & Cos_{\theta 3} & 0 & 250Sine_{\theta 3} \\ 0 & 0 & 1 & 0 \\ 0 & 0 & 0 & 1 \end{bmatrix} \tag{4}$$

$$T_3^4 = \begin{bmatrix} Cos_{\theta 4} & 0 & Sine_{\theta 4} & 0 \\ Sine_{\theta 4} & 0 & -Cos_{\theta 4} & 0 \\ 0 & 1 & 0 & -65 \\ 0 & 0 & 0 & 1 \end{bmatrix} \tag{5}$$

$$T_4^5 = \begin{bmatrix} Cos_{\theta 5} & 0 & -Sine_{\theta 5} & 0 \\ Sine_{\theta 5} & 0 & Cos_{\theta 5} & 0 \\ 0 & -1 & 0 & 113.5 \\ 0 & 0 & 0 & 1 \end{bmatrix} \tag{6}$$

$$T_5^6 = \begin{bmatrix} Cos_{\theta 6} & -Sine_{\theta 6} & 0 & 0 \\ Sine_{\theta 6} & Cos_{\theta 6} & 0 & 0 \\ 0 & 0 & 1 & 48.5 \\ 0 & 0 & 0 & 1 \end{bmatrix} \tag{7}$$

$$T_0^6 = T_0^1 T_1^2 T_2^3 T_3^4 T_4^5 T_5^6 \tag{8}$$

where the product of the above equation is equal to $\begin{bmatrix} n_x & O_x & A_x & P_x \\ n_y & O_y & A_y & P_y \\ n_z & O_z & A_z & P_z \\ 0 & 0 & 0 & 1 \end{bmatrix}$

Where: $\begin{bmatrix} P_x \\ P_y \\ P_z \end{bmatrix}$ is the position of the robot tool frame relative to frame 0 or the base.
The orientation of the end effectors is represented by the first three columns in the matrix.

2.2 *Inverse Kinematics*

Inverse Kinematics analysis determines the joint angles for the intended position and orientation of the robot end-effect in Cartesian space; consequently, each joint position must be understood to acquire the requisite robot motion that accomplishes the desired end-effector location [15]. In relation from the equations for the end-effectors position generated from the forward kinematics, we can then derive another equation to solve for the inverse kinematics and specify the variables with respect to a given end-effector position. Hence, upon thorough analysis prior to the geometric schema presented in Fig. 7 through the algebraic approach, the following equations is then formulated by which analysis of end effector is projected thru 4 quadrant-orientation in cartesian plane and with the help of Solidworks' point analysis of specific angles.

Eight (8) points on the first quadrant were projected for the end-effectors-top view position and geometrically calculated. These geometric calculations are used in all points projected. Solidworks was then used to analyze additional projection in terms of θ2, θ3, θ4 (as described in Fig. 4) to complete the 3D frame position of end effector (see Fig. 5).

$$R = \surd(X2 + Y) \tag{9}$$

$$\theta 1 = \tan^{-1}(Y/X) \tag{10}$$

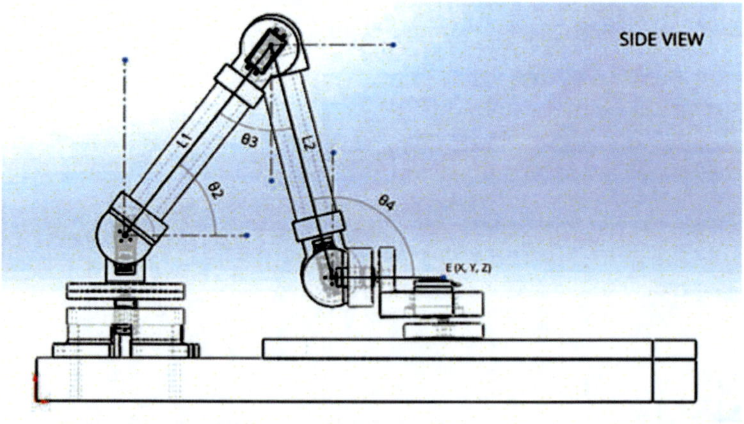

Fig. 4 2D sideview of robotic whiteboard cleaner arm

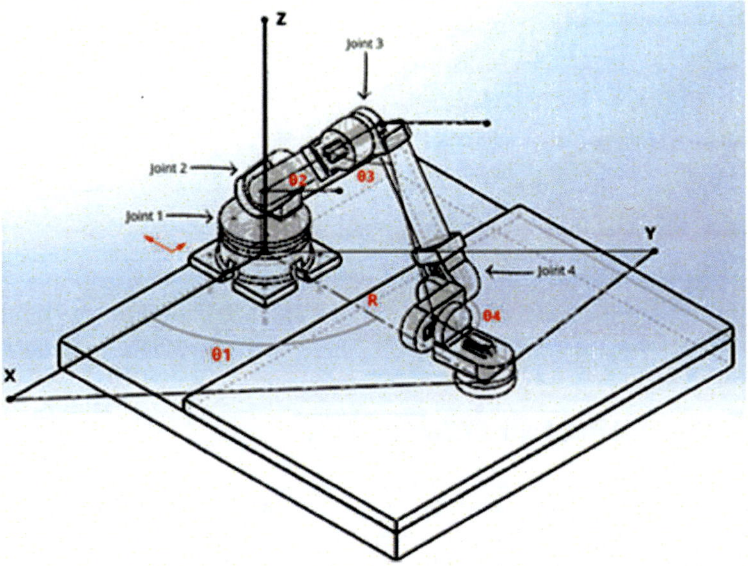

Fig. 5 3D wireframe of robotic whiteboard cleaner arm

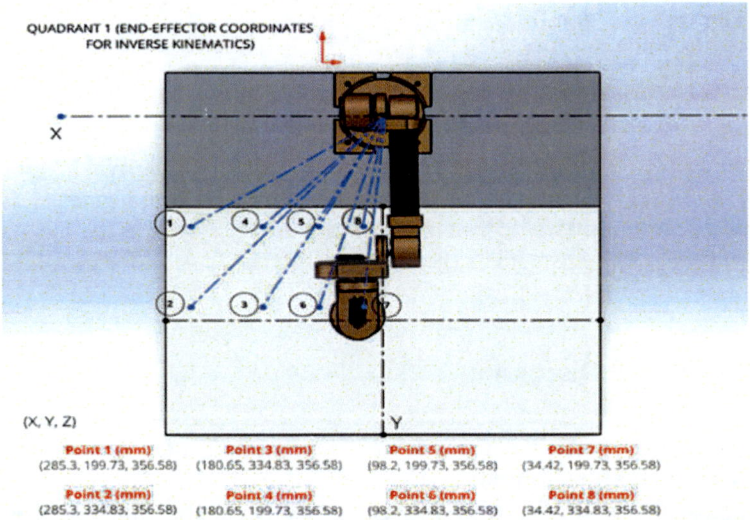

Fig. 6 Quadrant 1 end effector coordinates for inverse kinematics

Table 2 Arduino ATMega 2560

Microcontroller	CH340
Digital I/O pins	54
Analog input pins	16
Operating voltage	5 V
Input voltage	7–12 V
Serial ports	4
Flash memory	256 KB
SRAM	8 KB
EEPROM	4 KB
Clock speed	16 Hz

2.3 *Robotic Whiteboard Cleaner Arm*

The 6° of Freedom (DOF) robotic arm is to be built out of 3D printed parts with a significant emphasis on PLA or polylactic acid as its material with an average tensile stress of 54 MPa that is tested on different raster directions [14]. In the joints, it would utilize a 626zz ball bearing to smoothen its rotation by dealing with the axial and radial loadings. With the same purpose, the base is also set to have a slew bearing.

2.4 *Electronic Hardware*

An Arduino microcontroller will be utilized as it contains the Pulse Width Modulation (PWM) pins to drive the servo motors used in this project. With regards to that, the Arduino ATMega 2560 would then be a perfect fit for the system as it is incorporated with multiple PWM pins with its specification listed in Table 2.

3 Results and Discussion

This chapter will present the data gathered in the analysis of 6-DOF White Board Cleaner as described in the previous chapter.

Using the conditions described above, the solution of the inverse kinematics is obtained and summarized using the Tables 3, 4, 5, 6 and Figs. 6, 7, 8 and 9.

Table 3 Quadrant 1 end effector coordinates solution

	(X, Y)	θ1	θ2	θ3	θ4	R (mm)
Point 1	(285.3,199.73)	34.99°	50.61°	56.13°	106.73°	348.26
Point 2	(285.3,334.83)	49.57°	39.67°	82.4°	122.07°	439.89
Point 3	(180.65,334.83)	61.65°	44.02°	65.15°	109.17°	380.45
Point 4	(180.65,199.73)	47.87°	54.97°	36.69°	91.66°	269.31
Point 5	(98.2,199.73)	63.81°	53.4°	25.87°	79.26°	222.57
Point 6	(98.2,334.83)	73.65°	50.43°	57.51°	107.95°	348.93
Point 7	(34.42,199.73)	80.22°	50.96°	55.8°	106.76°	202.67
Point 8	(34.42,334.83)	84.13°	50.91°	22.21°	73.12°	336.59

Table 4 Quadrant 2 end effector coordinates solution

	(X, Y)	θ1	θ2	θ3	θ4	R (mm)
Point 1	(285.4,409.99	55.16°	32.69°	101.86°	134.54°	499.54
Point 2	(285.4,528.57)	61.63°	7.82°	154.2°	162.01°	600.7
Point 3	(199.85,528,57)	69.29°	19.55°	129.3°	148.84°	565.09
Point 4	(199.85,409.99)	64.01°	39.22°	86.84°	126.06°	456.11
Point 5	(113,409.99)	74.59°	43.54°	77.11°	120.65°	425.28
Point 6	(113,528.57)	77.93°	25.08°	118.29°	143.37°	540.51
Point 7	(34.42,528.57)	86.27°	27.79°	113.09°	140.88°	529.69
Point 8	(34.42,409.99)	85.2°	45.44°	73.16°	118.6°	411.43

Table 5 Quadrant 3 end effector coordinates solution

	(X, Y)	θ1	θ2	θ3	θ4	R (mm)
Point 1	(−275.05,409.99)	56.14°	33.64°	99.76°	133.39°	493.7
Point 2	(−275.05,528.57)	62.51°	8.42°	151.91°	160.33°	595.85
Point 3	(−197.75,528.57)	69.49°	19.66°	129.74°	149.4°	564.35
Point 4	(−197.75,409.99)	64.25	39.39°	86.27°	125.66°	455.19
Point 5	(–107.53,409.99)	75.3°	43.37°	76.74°	120.11°	423.86
Point 6	(–107.53,528.58)	78.5°	25.24°	117.37°	142.61°	539.41
Point 7	(–33.43,528.57)	86.38°	27.23°	113.1°	140.33°	529.63
Point 8	(–33.43,409.99)	85.34°	44.86°	74.02°	118.88°	411.35

Table 6 Quadrant 4 end effector coordinates solution

	(X, Y)	θ1	θ2	θ3	θ4	R (mm)
Point 1	(−275.05,196.36)	35.52°	**51.48°**	55.62°	107.1°	337.966
Point 2	(−275.05,322.53)	49.54°	**42.63°**	79.61°	122.25°	423.88
Point 3	(−197.75,322.53)	58.49°	**47.31°**	66.38°	113.69°	378.33
Point 4	(−197.75,196.39)	44.8°	**55.22°**	41.02°	96.25°	278.7
Point 5	(−107.53,196.39)	61.3°	**54.42°**	28.41°	82.83°	223.9
Point 6	(−107.53,322.53)	71.56°	**50.78°**	56.24°	107.02°	339.98
Point 7	(−33.43,322.53)	84.08°	**52.95°**	51.45°	104.4°	324.26
Point 8	(−33.43,196.39)	80.34°	**50.99°**	23.63°	74.62°	199.21

***Note: θ2, θ3, and θ4 are obtained in solidworks.**

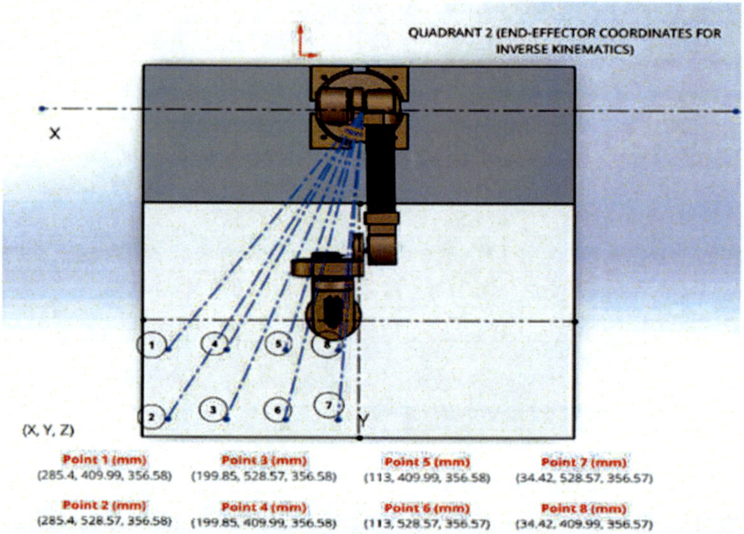

Fig. 7 Quadrant 2 end effector coordinates for inverse kinematics

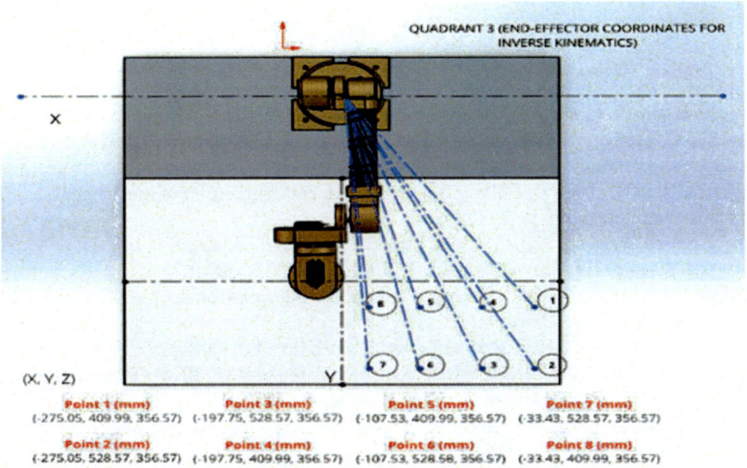

Fig. 8 Quadrant 3 end effector coordinates for inverse kinematics

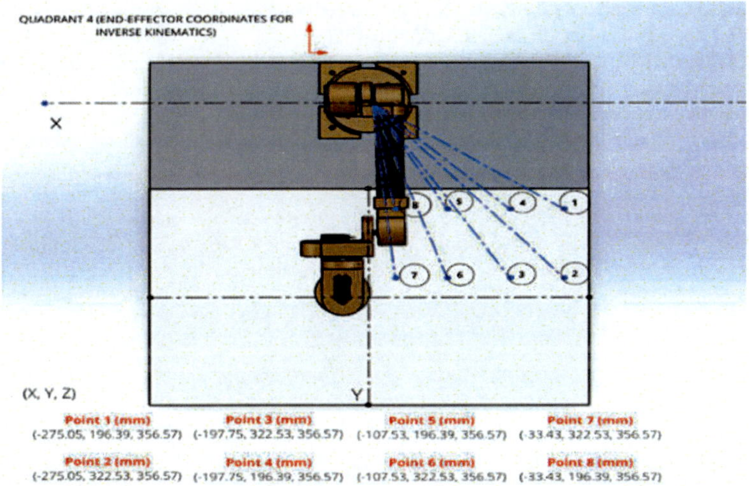

Fig. 9 Quadrant 4 end effector coordinates for inverse kinematics

4 Conclusion

Both the forward and inverse kinematic problems with the aid of solidworks are presented for the 6DOF whiteboard cleaner could analyze further forward and inverse kinematics thru simulation for specific positioning. The position and orientation of end-effector are determined by the analysis.

References

1. Arshad A, Badshah S, Soori PK (2017) Design and fabrication of smart robots. In: International conference on electronic devices, system and application, vol 6, no 2, pp 597–604
2. Sharmila C, Ramya M, Sarnya S (2016) Automatic chalk board cleaner and energy harvesting for smart classrooms, vol 5, no 1, pp 399–401
3. Ragha D, Raj PS, Punia R, Pillai P (2018) Design and fabrication of mechanical teaching aid for white board cleaning, pp 7317–7325
4. Ali F, Sintar J, Aras M, Zakishukor A (2019) Design and construction of 4-DOF EMG-based robot arm system. Int J Innov Technol Explor Eng 8(12S2):669–674
5. Vaishali MJ et al (2016) Mukteshvari, "Automatic white board eraser," vol 2, no 4, pp 812–816
6. Hanon MM, Marczis R, Zsidai L (2021) Influence of the 3D printing process settings on tensile strength of PLA and HT-PLA. Period Polytech Mech Eng. https://doi.org/10.3311/PPme.13683
7. Hasan MT, Parvez S, Roy S, Sakib MA, Khalidur Rahman KM (2018) Design of an automated white board profiler. In: MATEC web conference, vol 221, pp 1–4
8. Anish N, Roshan N, Kushal N, Sneha S (2019) Basic of automation and robotics using bluetooth controlled robotic arm, pp 3916–3922
9. Simolowo O, Ngana O (2014) Preliminary design of an automated white board cleaner. African Res Rev 8(2):68
10. Vignesh S, Vinith K, Mouleswaran K, Sanjay HM, Logeshwaran M (2018) Automated board duster. Int J Res Appl Sci Eng Technol 6(3):1743–1750
11. Rubhini B, Mrunalini T (2014) Real-time automated blackboard eraser using embedded system. Int J Innov Res Sci Eng Technol 3(8):15501–15509
12. Megalingam RK, Vivek GV, Bandyopadhyay S, Rahi MJ (2017) Robotic arm design, development and control for agriculture applications. In: 2017 4th international conference on advanced computing and communication systems ICACCS 2017
13. Chavan S, Shinde V, Murade N, Jagtap A, Degaonkar V (2019) Automatic white board cleaner (5):427–430
14. Dhanjal SS, Pratihar DK (2016) Design and development of board cleaning serial manipulator: Board cleaning manipulator. In: 2015 IEEE UP section conference on electrical. UPCON 2015
15. Wang Z, Chen D, Xiao P (2019) Design of a voice control 6DOF grasping robotic arm based on ultrasonic sensor, computer vision and Alexa voice assistance. https://doi.org/10.1109/ITME.2019.00150

Materials

Study on Mangrove Barks Activated Carbon (MBAC) for Fibre Reinforce Plastic (FRP) Rehabilitation Pile Structure

Z. Salleh and T. M. I. A. T. Mazlan

Abstract This study deals with the fabrication of composite matrix from Mangrove Bark Activated Carbon (MBAC) and epoxy resin with hardener. The MBAC were made from mangrove bark cut into pieces burn into oven to become charcoal. Then it is drained into strainer to get the small grain size of MBAC. The MBAC with different weight percentage (10–50wt%) are mixing with resin and hardener and laminated on the concrete prototype pile using PVC mold. The investigation on the barnacle growth on MBAC piles were monitored and water absorption was assessed as a function of mixing ratios according to the ASTM D570-98 standard. The barnacles were monitored frequently until 3 months and has been analysed using SCILAB image processing application. Water absorption has been measured after 24 h of immersion in water at 25 °C temperature by using two types of water such as seawater and freshwater. From observation, the results showed that barnacle's growth is slowed with reduction 64% when MBAC content has been increased from 10 to 50 wt.% into the matrix. Types of seawater shows higher water absorption with increased 18% compared to freshwater as the nature mangrove that absorption when compared with freshwater type. Thus, these results could be indicated that the fabrication of MBAC and epoxy resin would technically feasible use for pilejax for rehabilitation jetty and others marine application.

Keywords Mangrove · Activated carbon · Image processing · SCILAB

1 Introduction

From the last 2 decades, fibre reinforced polymer (FRP) composite materials are popular not only due to excellent in mechanical properties but also significant improvised in rehabilitate existing structures, especially in harsh marine environment [1]. These preferable materials have been chosen as first option because of tremendous properties such as higher strength, deterioration resistance, lightweight, high fatigue

Z. Salleh (✉) · T. M. I. A. T. Mazlan
Universiti Kuala Lumpur, Malaysian Institute of Marine Engineering Technology (MIMET), Bandar Teknologi Maritim, Jalan Pantai Remis, 32200 Lumut, Perak, Malaysia
e-mail: zulzamri@unikl.edu.my

© The Author(s), under exclusive license to Springer Nature Singapore Pte Ltd. 2023
M. H. A. Hassan et al. (eds.), *Technological Advancement in Instrumentation & Human Engineering*, Lecture Notes in Electrical Engineering 882,
https://doi.org/10.1007/978-981-19-1577-2_25

resistance, nonmagnetic, high impact resistance, and higher reliability. Moreover, reinforcement and establishment of civil infrastructures using fibre composites has been a good topic of research and development for years [2–4]. Practically the implementations on using FRP composites as laminated cover sheet has been strengthening in many engineering application such as bridges, jetty pile and can be used for rehabilitating damaged concrete, steel and timber structures to certain extend their functionality [5–7]. Several work has been done on activated carbon particularly in preparation and characterisation which has many significant level for achievement and usage. Nevertheless, an external and internal properties such as chemical properties of insulators will be differed from one to the other [8] due to the difference in the base materials and it preparations. One of based materials such as activated carbon could be prepared from coconut shell for example, it also can be derived from the palm fruit shell or rice hull by using the same method [9]. As a results, not many activated carbons are suitable to be used for a specific application. The activated carbon can be developed from a particular source may only be useful for a certain application [10]. Potential application of activated carbon from a particular source can be realized by detailed studies of its properties. The activated carbon should be characterized to identify its properties and abilities to match to a particular application. Hence, mangrove bark or stem has been selected to identify the characteristic features the useful of activated carbon. The resources from this 'particular species will gain a new source of activated carbon for future applications. Furthermore, beside useful of mangrove tree as wave resistance during tsunami disaster but with this an abundant natural resource which has also been used commercially as fire starter in charcoal-based manufacturing industries including as chromatographic stationary phase [11].

In this wider range of potentiality also can be realized by more detailed studies for future applications. This potential waste can be utilized as activated carbon as coating surface for pile structure. Despite the fact that concrete is a reliable structural material with good durability performance, exposure to severe environments makes it vulnerable [12]. Mainly due to degradation of concrete has been subjected to chemical ruin by environmental factors are lack of specifications and poor workmanship. The lack of knowledge of the corrosion mechanisms results in insufficient planning and accounting for the environmental effects. Corrosion of reinforcing bars persuaded by chloride ion ingress is a major cause of damage in marine environments [13]. Figure 1 shows some example corrosion at pile jack. Previously, many glass-FRP (GFRP) repair systems have already been used worldwide for rehabilitating damaged concrete, steel and timber structures and extending their service lives [6, 14, 15]. Alternatively, another materials such as Carbon-FRP (CFRP) is also good alternative to be used in seismic repairs to achieve enhanced structural capacities due to their higher mechanical properties compared to GFRP jackets [16–20].

Among the possibility upgrading strategies for deteriorated reinforced concrete (RC) piles in marine environment, the use of FRP wraps or jackets to prevent the lateral (radial) expansion of concrete, is gaining widespread acceptance. Many researchers have been reported in the last decade on the mechanical behavior and failure of FRP wrapped with RC columns. These are studies particularly on FRP

(a) Degradation of concrete pile showing corroded steel reinforcement (Source: http://www.substructure.com)

(b) Corrosion of steel piles (Source: http://www.watimas.com)

Fig. 1 Several types of corrosion at pile jack [21]

confinement of column structures but not in marine environment. While the pile repair and retrofit of damaged columns are using FRP system and strengthening of columns structures by using FRP composites. During this rehabilitation, confinement is the important aspect that strengthens the quality of columns or piles itself and has been observed in numerous investigations to date, e.g. [22–24]. It will be exhibited an enhanced ductility, load carrying capacity, and lateral deflection capability certain mechanical properties materials including hoop strength, longitudinal strength, and shear resistance of the wrapped piles are increased. The advantages of all these requirements beside the light weight characteristic of FRP it is also makes it easy to install in any applications and it serves as a shield that prevents deterioration due to environmental attacks. From this study, there are several significant target will be achieved at the end of the project such as MBAC can be used to prevent marine structures from corrosion and others affected when apply in salt water condition. This application also will be monitored the barnacle growth by using image processing method to check the barnacle affected with MBAC insulate in FRP.

2 Experimental Setup

2.1 Specimen for Barnacles Monitoring

Initially, the mangrove barks are cleaned repeatedly to remove dust and soluble impurities after has been purchased from the supplier. Secondly, it has been dried under the sunlight within 4 h until it totally dried. Then, it has been chopped into small pieces and burning in the oven to produce the activated carbon powder. The quality and properties charcoal activated carbon also been monitored their humidity within 55–60% RH with temperature 25 °C. The activated carbon can be seen when the mangrove bark has been turned into charcoal type. The grain type also been changed into small pieces and pour into the strainer to filter the pure activated carbon fine powder. Next staged is to prepare required materials for fibre glass composite materials. Fibre glass layers size are cut by using scissor in the warp direction (0 and 90°) according to height of PVC which is acting as structure of pile. In a total, there are 20 fiber plies has been acquired in this research. There are two types different diameter of PVC which is 15 mm (inner PVC) and 40 mm (outer PVC). The outer PVC has been inserted with the fiber, activated carbon, cement and hanging it with string from jetty shows in Fig. 2.

Once it has already been mixed and stick together, the outer PVC pipe has been cut and removed to let the surface of fiber glass composite has been exposed with seawater. The test specimen has been immersed in seawater for all specimens with the same depth and hanging with the string to ensure not loose during high tide sea condition. There are five test specimen has been immersed in seawater with different percentage of mangrove barks activated carbon, 10, 20, 30, 40 and 50%. Numerical computation providing a powerful computing environment for engineering and scientific applications is called as SCILAB. This free and open source software is including with hundreds of mathematical functions. Another functionality in this software embedded with a higher level programming language such as advanced data structures, 2-D and 3-D graphical functions. The specimen has been monitored by taking their photos for 3 months and upload into the SCILAB software to process image processing tools for investigating barnacle growth. The photos were taken using Canon DSLR EOS 80D type for better imaging results.

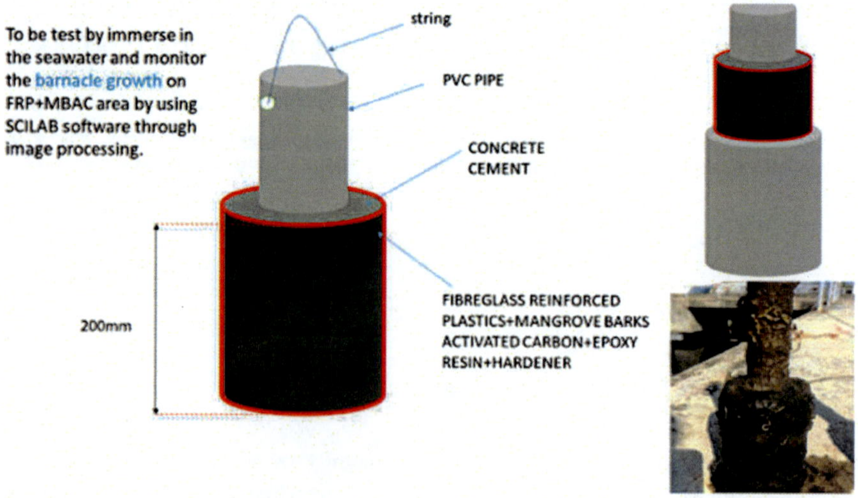

Fig. 2 Specimen for monitoring barnacles growth

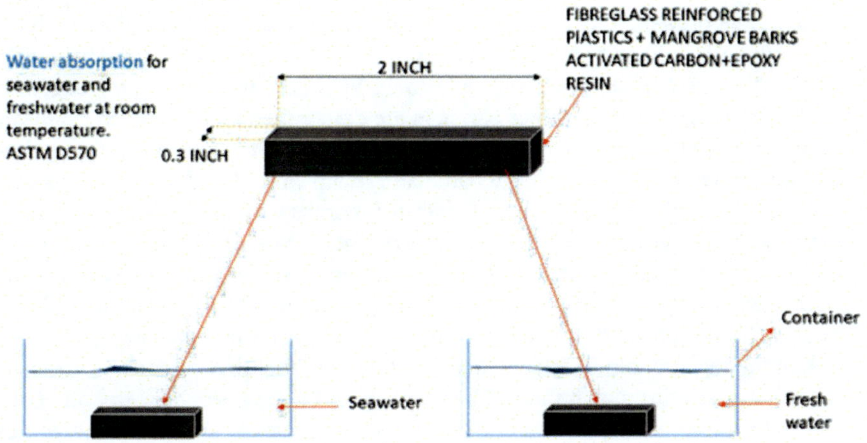

Fig. 3 Specimen for water absorption

2.2 *Specimen for Water Absorption*

The materials are mix together (MBAC + Resin + Hardener) and poured into mould to get the test specimen according to ASTM D570 standard. The test specimen size (length, width, thickness) has been measured by using Vernier caliper to get the accuracy results of specimen. The test specimens have been divided into five different percentage of mangrove barks activated carbon which is same with barnacles monitoring specimens. Each percentage have been consisting with three test specimens sample size. There are fifteen test specimen that will immerse in freshwater and seawater at room temperature. The test specimens are being weight and record before immerse in container. The container is fill with same volume of freshwater and seawater as well. Test specimens are immersing in both types of water for one day. Test specimens are wipe with rags and being weight by using electronic weighting machine. Record initial and final weight test specimen is being compare to determine water absorption. The overview of the water absorption specimen shows in Fig. 3.

3 Results and Discussion

3.1 *Barnacles Growth Monitoring*

A sample photo was taking from six test specimen at the different percentage for 3 months. A sample photo then was analyse on the SCILAB application on section image processing tools which is blob analysis. The purpose of blob analysis is to find the difference between the barnacle growth and surface of specimen. This analysis was process by command to form the grey image. The result was collected by taking photo on the surface specimen and process in SCILAB, then compare from each months. From the image, it shows barnacle are darker than surface which can been seen on dark sport or worm marks on the image. The main purpose edge detection analysis is to define the edge of the barnacle growth from the sample photo taken, the difference of edge barnacle can see within 3 months with five different specimen of percentage mangrove bark activated carbon. There are exactly two regions corresponding to the objects to be detected which is gradient and edge. Grey in colour is barnacles and black in colour are surface mangrove bark activated carbon of test specimen. Table shows the observation for specimens for 3 months monitoring.

From the result photo of edge detection, it can see that that barnacle is growing from month one until month three. The grey colour on the black background photo increase from month one to month three. Grey in colour is barnacles and black in colour are surface mangrove bark activated carbon of test specimen. From the table comparison from other researcher, it shows that the grey colour marks on the photo of mangrove barks activated carbon surface are less and smaller than the photo on Fibreglass hull. It shows that the growth of barnacle on MBAC specimen are less compare to the use of Fibreglass only (Table 1).

From the result, it be can be seen that the different of three months on sample photo taken that have been process on SCILAB, blob analysis, photos sample 10 wt.% shows the barnacles are start to grow on the first month, the dark spots and worm marks show the growth of barnacle on the surface tested. On the second month the sample with 10–40 wt.% shows the barnacles growth started to change into worm marks, however on the 50% mangrove barks activated carbon it only shows dark spot. This result also has similar explanation finding by Ismail et al. [25].

From the graph result of three months, the average peak is increase from month one to month three. However, on the 50 wt.% mangrove barks activated carbon specimen, it shows less growth of barnacle approximately 24% with low peak range compare to the others percentage of mangrove barks activated carbon. There are high and low value of peak that means the existing barnacles are growth in piece. This may cause by several factors which is due to water condition, the water contain oil and chemical that effect the life cycle and growth of the barnacles. Weather condition, rains and summer. The specimen should be on hang while immerse in the seawater, there are some specimen that immerse to the bottom of seawater. This will cause an impurities or mud from bottom sea stick to the specimen. The pixel and size of three months' image should be the same for a better processing in SCILAB application. Lastly, the camera use for taking photo should be the same for correct brightness and alignment size of photo take. From the Fig. 4 shows the comparison of graph analysis other researcher, the mangrove bark activated carbon shows low peak range compare to fibreglass hull. From the three months of test, the range of peak of MBAC is lowest with 1500 pixel which is lower than first and second month's, 4200-pixel peak range. Barnacle are slowly growth on the MBAC specimen. Similar to the previous research by Ismail et al. [25] where highest maximum pixels were found at 3000 pixel when observed on the vessel.

Table 1 Comparison barnacles growth by month

Sample	Month			Blob Analysis
(wt.%)	1st month	2nd month	3rd month	(3rd month)
Pure				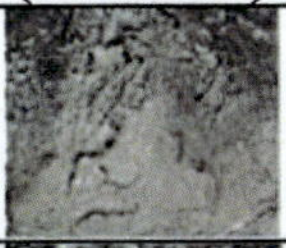
10				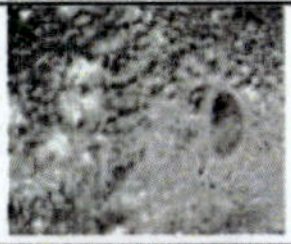
20				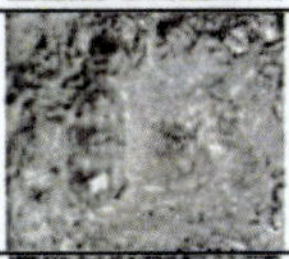
30				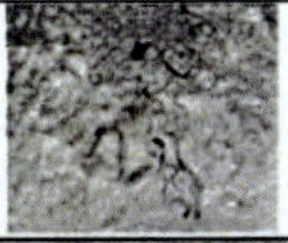
40				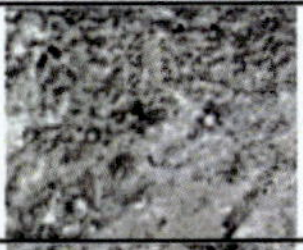
50				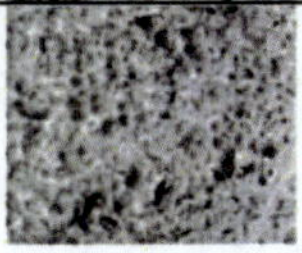

3.2 Water Absorption Measurement

The average weight of specimen immersed in seawater increase as the increasing amount of mangrove barks activate carbon in specimen. The zero percent of mangrove barks activate carbon specimen shows low percentage of water absorption and weight gain, this is because the use of epoxy and hardener that mix together with fiberglass are water resistant. The increasing percentage of water absorption MBAC specimen in seawater happen because the nature of mangrove plant that can adapt to the salty water which its extract freshwater from the seawater. Mangrove survive and can adapt by filtering out the salt found in seawater as said in article written by [26]. The more the percentage mangrove barks activated carbon in specimen, the more percentage water absorption on specimen. On 50% of MBAC specimen, its shows the highest water absorption between the others percentage specimen. The average weight before and after immerse specimen in seawater was collected and recorded

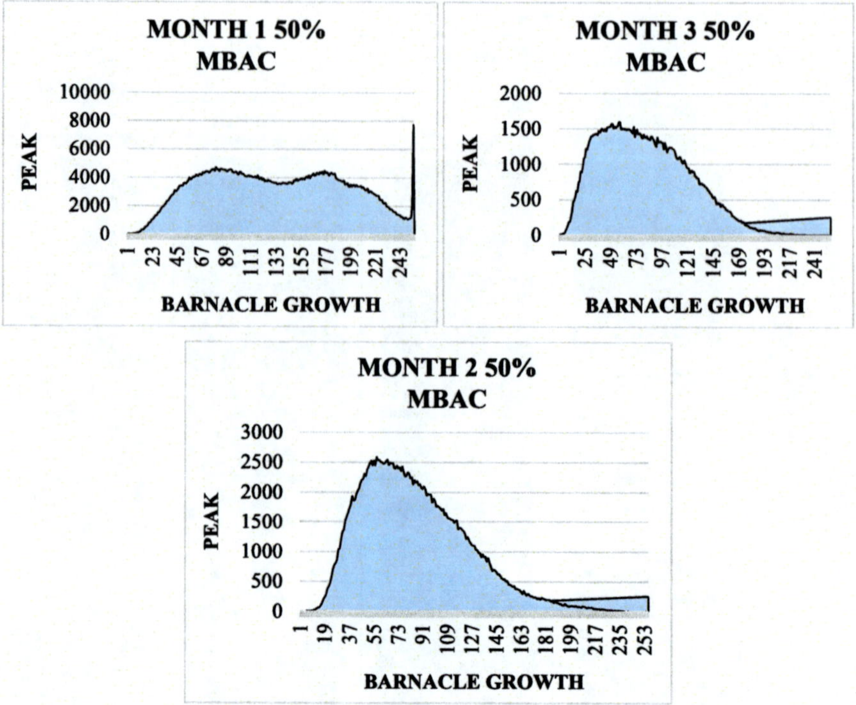

Fig. 4 Graph analysis barnacles growth with different pixel

in Table 2. The average weight before and final weight were minus together and the percentage weight and water absorption result gained.

The increasing of water absorption in in all percentage specimen is because the mangrove barks also absorb low salinity water which is write in the article [27] said that when the less saline water sources are available, it become the main source of mangrove which can increase survivorship, growth and productivity under condition which salinity is limit. The more the percentage of mangrove barks activated the

Table 2 Average weight before immerse seawater and freshwater

MBAC%	Water absorption	
	Seawater (%)	Freshwater (%)
0	4	5
10	34	18
20	52	24
30	62	34
40	80	39
50	99	46

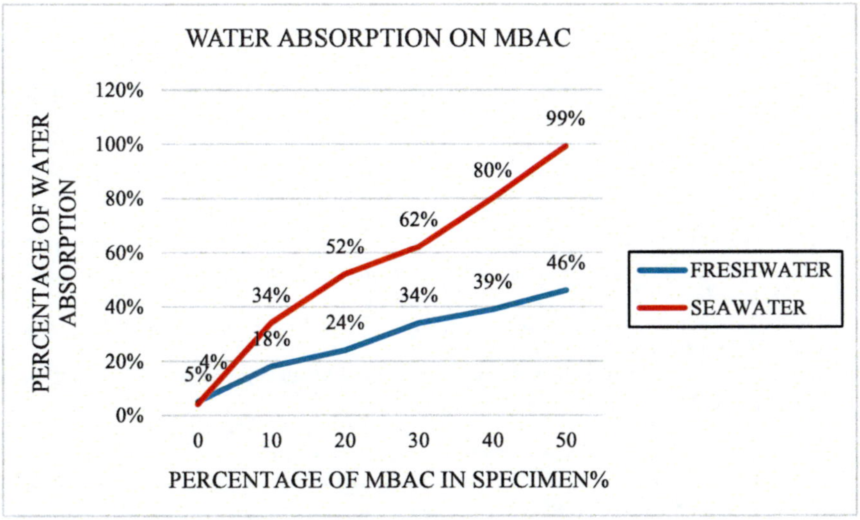

Fig. 5 Comparison water absorption between sea and fresh waters

more of water absorption. Water absorption is expressed as increase in final weight percentage. The data from two type of water shows an increase in water absorption in Fig. 5. Seawater shows higher water absorption than freshwater as the nature mangrove that absorb the seawater and extract freshwater from it. Percentage water absorption of MBAC specimen in fresh water is lower than seawater as mangrove barks absorb less fresh water in low salinity water.

4 Conclusions

In this paper, barnacle growing slowly on the surface of mangrove bark activated carbon specimen compare to the other research. The combination fabricated of FRP and MBAC shows an advantage of decreasing the growth of barnacles. Water absorption of mangrove bark activated carbon shows an increase in seawater than freshwater. Based on these finding it is showed that the pile rehabilitation can be treated damaging from barnacles with this latest anti-fouling agent which could be applied for marine industries. By introducing these materials in marine industry should need further research as we know that are many factor that effect the growth of barnacles in marine environment.

Acknowledgements The authors would like to be obliged to UniKL-MIMET materials laboratory for providing laboratory facilities for final year project.

References

1. Manalo AC, et al (2013) Fibre composite pile rehabilitation and concrete form work jacket-concept development and finite elements analysis. In: Fourth Asia-pacific conference on FRP in structures (APFIS 2013), Melbourne, Australia
2. Mangi SA et al (2019) Performances of concrete containing coal bottom ash with different fineness as a supplementary cementitious material exposed to seawater. Eng Sci Technol Int J 22(3):8
3. Kim YJ (2019) State of the practice of FRP composites in highway bridges. Eng Struct 179:8
4. Demir I et al (2018) Effects of sulfate on cement mortar with hybrid pozzolan substitution. Eng Sci Technol Int J 21(3):8
5. Monti G, Nistico N, Santini S (2001) Design of FRP jackets for upgrade of circular bridge piers. ASCE J Compos Constr 5(2):101
6. Faleschini F et al (2019) Repair of severely-damaged RC exterior beam-column joints with FRP and FRCM composites. Compos Struct 207:11
7. Yurdakul Ö et al (2019) Stochastic-based nonlinear numerical modeling of shear critical RC beam repaired with bonded CFRP sheets. J Compos Constr 23(5):18
8. Mattson JS, Jr Mark HBM (1971) Surface Chemistry and Absorption from Solution, in Activated Carbon. Marcel Decker, New York
9. Salleh MMBH, Salleh Z, Rosdi MS, Sapuan SM (2013) Mechanical properties of coconut carbon fibre/epoxy composite material. Int J Mech Eng (IJME) 2(3):55–62
10. Salleh Z, Yusop MYM, Rosdi MS (2013) Mechanical properties of activated carbon (AC) coir fibers reinforced with epoxy resin. J Mech Eng Sci 5:631–638
11. Daud JM (1989) Penggunaan arang kayu bakau (Rizophom mucronata) teraktif sebagai fasa pegun dalam kromatografi gas. In: Prosiding Kimia Analisis Kebangsaan ke III, UTM, Johor Baharu
12. Guettala A, Abibsi A (2006) Corrosion degradation and repair of a concrete bridge. Mater Struct 394:478
13. Shekarchi M, Moradi-Marani F, Pargar F (2009) Corrosion damage of a reinforced concrete jetty structure in the Persian Gulf: a case study. Struct Infrastruct Eng 7:701–713
14. Vijay P et al (2016) Repair and strengthening of submerged steel piles using GFRP composites. J Bridge Eng 21(7):04016038
15. Mohammed AA et al (2019) Effectiveness of a novel composite jacket in repairing damaged reinforced concrete structures subject to flexural loads. Compos Struct 233:111634
16. Wu R-Y, Pantelides CP (2017) Rapid seismic repair of reinforced concrete bridge columns. ACI Struct J 114(5):1–13
17. Wu R-Y, Pantelides CP (2017) Rapid repair and replacement of earthquakedamaged concrete columns using plastic hinge relocation. Compos Struct 180:16
18. Wu R-Y, Pantelides CP (2019) Seismic evaluation of repaired multi-column bridge bent using static and dynamic analysis. Constr Build Mater 208:15
19. Ferdous W et al (2019) New advancements, challenges and opportunities of multi-storey modular buildings–a state-of the-art review. Eng Struct 183:10
20. Ribeiro F et al (2018) Hybrid FRP jacketing for enhanced confinement of circular concrete columns in compression. Constr Build Mater 184:23
21. Guades E, Aravinthan MIT, Manalo A (2012) A review on the driving performance of FRP composite piles. Compos Struct 94:1942
22. Sen R, Mullins G (2007) Application of FRP composites for underwater piles repair. Compos B 38:751
23. Karagah H, Dawood M, Belarbi A (2018) Experimental study of full-scale corroded steel bridge piles repaired underwater with grout-filled fiber-reinforced polymer jackets. J Compos Constr 22(3):18
24. Moran DA, Pantelides CP, Reaveley LD (2019) Mohr-coulomb model for rectangular and square FRP-confined concrete. Compos Struct 209:15

25. Ismail S et al (2013) Monitoring of barnacle growth on the underwater hull of an FRP boat using image processing. Procedia Comput Sci 23:6
26. Arfi Y et al (2013) Characterization of salt-adapted secreted lignocellulolytic enzymes from the mangrove fungus Pestalotiopsis sp. Nat Commun 4:9
27. Reef R, Lovelock CE (2015) Review: part of a special issue on halophytes and saline adaptations regulation of water balance in mangroves. Ann Bot 115:395

Preliminary Tensile Investigation of FDM Printed PLA/Coconut Wood Composite

J. Kananathan, K. Rajan, M. Samykano, K. Kadirgama, K. Moorthy, and M. M. Rahman

Abstract Fused deposition modeling (FDM) is a technique in additive manufacturing that has been used to produce various components using a variety of materials for a broad range of applications adapting layer-by-layer deposition process. Thermoplastic polymers are commonly used raw material that comes in the form of a filament. Coconut wood is highly recognized for its naturally affable ecological components, thermal resilience, and corrosion resistance. However, PLA's qualities embedded in coconut wood remain inadequate. The objective of this analysis is to investigate and analyze the tensile properties of the 3D printed specimens with varying infill percentages (25, 50, and 75%) and the infill patterns (grid, rectilinear, concentric, honeycomb, and triangle) on coconut wood reinforced PLA using the FDM technique. The specimen is printed in accordance with the ASTM standard for tensile testing, which is ASTM D638 type 1. Following that, the tensile properties of the PLA and PLA/Coconut wood were analyzed. The results demonstrate that the concentric infill pattern with a 75% infill percentage provides the strongest structure in PLA and PLA/Coconut wood composites. The PLA has a maximum tensile strength of 37.55 MPa, whereas the PLA/Coconut wood has 19.35 MPa. PLA has a maximum elastic modulus of 1.148 GPa, while the PLA/Coconut wood composite has a maximum elastic modulus of 1.121 GPa. PLA has a yield strength of 23.33 MPa,

Supplementary Information The online version contains supplementary material available at https://doi.org/10.1007/978-981-19-1577-2_26.

J. Kananathan
Green Kingdom Solutions Sdn. Bhd., Taman Salak Selatan, 57100 Kuala Lumpur, Malaysia

J. Kananathan · M. Samykano (✉) · M. M. Rahman
College of Engineering, Universiti Malaysia Pahang, 26300 Gambang, Pahang, Malaysia
e-mail: mahendran@ump.edu.my

K. Rajan · K. Kadirgama
Faculty of Mechanical and Automotive Engineering Technology, Universiti Malaysia Pahang, 26600 Pekan, Pahang, Malaysia

K. Moorthy (✉)
College of Computing and Applied Sciences, Faculty of Computing, Universiti Malaysia Pahang, 26600 Pekan, Pahang, Malaysia
e-mail: kohbalan@ump.edu.my

© The Author(s), under exclusive license to Springer Nature Singapore Pte Ltd. 2023
M. H. A. Hassan et al. (eds.), *Technological Advancement in Instrumentation & Human Engineering*, Lecture Notes in Electrical Engineering 882,
https://doi.org/10.1007/978-981-19-1577-2_26

while the PLA/Coconut wood composite has a yield strength of 15.33 MPa in tensile testing. Meanwhile, the grid pattern has the weakest properties for both materials.

Keywords Coconut wood · PLA-based composite · 3D printing · FDM · Mechanical properties

1 Introduction

According to the American Society for Testing and Materials (ASTM), additive manufacturing (AM) is a process that builds three-dimensional (3D) solid with tailored designs in a single step layer by layer from digital models [1]. Rapid prototyping (RP) has recently emerged as a manufacturing focus point to tolerate component complexity while producing parts under tight time and cost constraints. Rapid prototyping (RP) is a technique for creating prototypes quickly and efficiently [2–4]. This technology is widely used in automobile, aerospace, medical, manufacturing and consumer products [5–8]. The AM process is classified into seven types: Material Extrusion, Sheet lamination, Material Jetting, Binder Jetting, Direct energy deposition, VAT polymerization, and Powder Bed Fusion. The mandate of using 3D printing technology has increased these days [9–11].

Fused Deposition Modeling (FDM) is one of the most frequently used RP techniques for producing components with three-dimensional (3D) complexity. It involves depositing the material layer by layer via a tiny liquefying nozzle that travels in the X and Y directions as the material is deposited (in the plane of the build platform). The build platform (or worktable) is lowered in the Z direction after depositing a layer and adding the following layer [7, 12]. The model and support materials are deposited using FDM through separate liquefier nozzles placed on the extrusion head. Simultaneous deposition of support materials at specified places enables the production of geometrically accurate components. Once fabrication is complete and the component is removed from the mold, the support material used during fabrication may be readily removed. FDM processes typically produce items with high efficiency, accuracy, and, most critically, mechanical strength [13, 14]. The resulting goods' mechanical qualities are determined by the printing parameters selected by the user. Nowadays, the FDM method may be utilized with a variety of materials, including wood particle-filled filament, carbon fibers, and ceramics [12, 15–17]. The use of FDM technology has increased dramatically in a variety of sectors throughout the world. Compared to the older, more traditional process, FDM technology offers significant benefits in terms of time savings, cost-effectiveness, and raw material use [18]. Injection molding and blow molding are real examples of conventional methods utilized in the past, and the procedure is extremely time-consuming [19, 20]. Wood grains are a sort of organic particle that has been widely used in a variety of applications, most notably in the biomedical field. However, information on the mechanical properties of natural wood products printed using a variety of printing methods is still limited.

Environmentally friendly, biodegradable, corrosion resistance and temperature resistance are well-known characteristics of coconut wood particle technology. When subjected to a range of printing circumstances, the mechanical behavior of coconut wood particles is still limited. It is critical to remember that printing factors such as layer thickness, raster angle, operating temperature, printing speed, and infill density significantly affect the mechanical properties. The infill density refers to the quantity of material filled in each layer and directly affects the product's strength. Some studies focus on the consequence of infill patterns on FDM process mechanical behavior. The purpose of the research is to examine the tensile properties of the PLA and PLA/Coconut wood filament by various infill patterns (grid, rectilinear, concentric, honeycomb, and triangle) and the infill percentage (25, 50, 75%).

2 Materials and Methods

Raise 3D N2 Plus FDM 3d printer with 0.4 mm nozzle diameter utilized to produce the tensile sample using 1.75 mm filament. The PLA filament and the coconut wood reinforced PLA filament was purchased in (Easy wood) Magma filaments, Malaysia. The composite filament contains 60% of PLA and 40% of coconut wood. The operating temperature of the filaments is 200 to 240 °C. Few factors have been persistent during the printing process, as indicated in Table 1. The temperature of the room is to be in 20 to 25 °C and the humidity of the surrounding must be between 70–80%.

2.1 Fabrication of Specimen

The design of the tensile test sample is performed using Solid works 2019. The design and dimension of the specimen were created according to the ASTM D638 Type 1 standard. The design model is initially stored in STL format, and the file is imported into slicing software. Ideamaker is the standard software for the Raise 3D N2 plus

Table 1 Process parameters for the FDM printing

No	Parameter	Values
1	Layer thickness	0.1 mm
2	First layer thickness	0.3 mm
3	Solid layers	Three layers (top and bottom)
4	Diameter of the nozzle	0.4 mm
5	Diameter of filament	1.75 mm
6	Temperature of nozzle	210 °C
7	Temperature of bed	60 °C
8	Printing speed	40 mm/sec

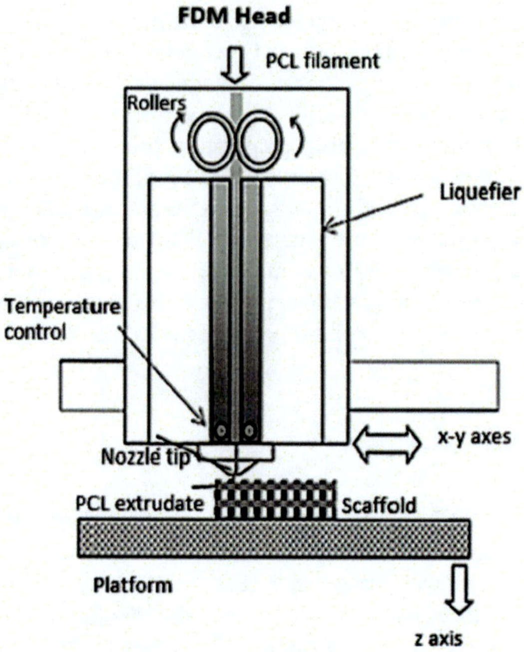

Fig. 1 FDM 3D printing process

printer. The slicing software generates the G- codes and the file is transferred to the FDM printer to print the specimens using flash drives. However, before the printing process, the heated bed is heated to a specified temperature. The nozzle temperature and the temperature of the bed remain constant throughout the printing. At extrude the filament, the extruder's temperature is set to 210 °C. Figure 1 illustrates the typical operating process for an FDM 3D printer.

2.2 *Tensile Testing*

The tensile test is performed using an INSTRON 3367 computer controller. The maximum force that the computer can exert is 30kN. For the Type 1 specimen geometry, the recommended speed for ASTM D368 is 10 mm/min with a tolerance of 25%. Figure 2 and Table 2 display the model and the Type 1 specimen in greater detail.

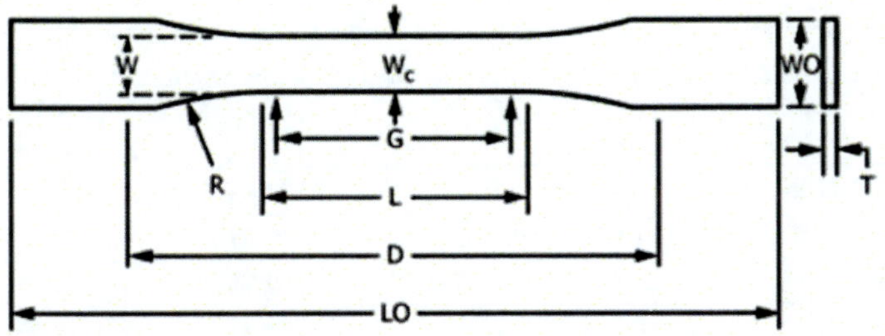

Fig. 2 ASTM D638 type 1 specimens

Table 2 ASTM D638 specimen 1 geometry

Parameters	Dimensions (mm)
W_c	13
L	57
W_o	19
L_o	165
G	50
D	115
R	76
T	3

3 Result and Discussion

The Ultimate Tensile Strength (UTS), Elastic Modulus (E), and Yield Strength (0.2% offset) are the main characteristics to be examined in the tensile test results. The Ultimate Tensile Stress (UTS) of a material is the highest stress it can take without breaking. Elastic modulus, most often referred to as Young's modulus, is used to characterize a material's rigidity. Numerous materials will follow Hooke's law in the first stage of the stress–strain curve or at low strain. During this stage, the stress is proportional to the strain using the proportionality constant. In layman's terms, the elastic modulus is a material property that undergoes stress, deforms, and tends to come back to its initial structure after the force is taken out. On the other hand, yield strength represents the stress needed to cause a modest amount of plastic distortion. The raw data created after testing with the Blue hill program can be used to calculate all of the tensile properties collected. Table 3 below shows the tensile properties of each pattern for PLA for the various infill percentage and Table 4 shows the average tensile properties of each pattern for PLA/Coconut wood for the various infill percentage. Figure 3 shows the PLA/Coconut wood composite specimens samples before printing and Fig. 4 shows the samples of PLA/Coconut wood composite specimens after tensile testing.

Table 3 Tensile properties of each pattern for PLA for the various infill percentage

Infill pattern	Infill percentage (%)	Young's modulus (GPa)	Tensile strength (MPa)	Tensile stress at break (MPa)	Tensile stress at yield (0.2%) (MPa)
Concentric	25	0.86151	25.38	24.45	11.36
Concentric	50	1.03917	32.66	29.37	20.56
Concentric	75	1.14776	37.55	32.78	23.33
Grid	25	0.835.43	25.62	24.81	16.59
Grid	50	0.83997	25.98	25.15	16.28
Grid	75	0.87069	27.73	27.19	17.12
Honeycomb	25	0.85874	25.33	23.53	16.74
Honeycomb	50	0.88192	26.11	25.05	17.27
Honeycomb	75	0.92105	27.64	26.62	17.74
Rectilinear	25	0.82934	24.8	23.69	16.41
Rectilinear	50	0.86439	26.24	25.21	16.69
Rectilinear	75	0.90865	27.78	27.11	17.38
Triangle	25	0.84259	25.42	24.51	16.66
Triangle	50	0.91949	27.29	26.48	18.14
Triangle	75	1.02473	30.66	29.3	19.99

Table 4 Tensile properties of each pattern for PLA/Coconut wood for the various infill percentage

Infill pattern	Infill percentage (%)	Young's modulus (GPa)	Tensile strength (MPa)	Tensile stress at break (MPa)	Tensile stress at yield (0.2%) (MPa)
Concentric	25	0.88841	28.57	26.39	18.03
Concentric	50	1.02498	32.87	28.29	20.24
Concentric	75	1.12134	37.21	33.55	22.58
Grid	25	0.84422	25.88	24.59	16.96
Grid	50	0.82298	25.51	24.85	16.67
Grid	75	0.88639	27.89	27.3	17.92
Honeycomb	25	0.85105	26.48	24.57	17.31
Honeycomb	50	0.86795	26.17	25.03	17.25
Honeycomb	75	0.90744	27.5	26.58	17.61
Rectilinear	25	0.81867	25.74	25.03	16.09
Rectilinear	50	0.83861	26	25.31	15.89
Rectilinear	75	0.89268	28.08	26.96	17.63
Triangle	25	0.83133	25.98	25.29	16.95
Triangle	50	0.91847	28.18	27.21	18.33
Triangle	75	1.02756	30.76	30.2	20.14

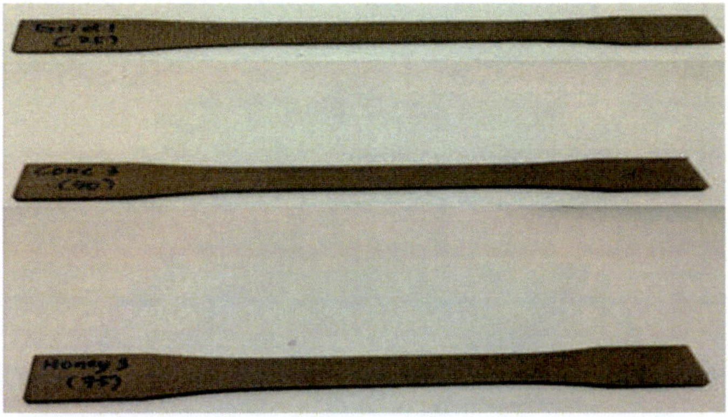

Fig. 3 FDM printed PLA/Coconut wood composite before testing

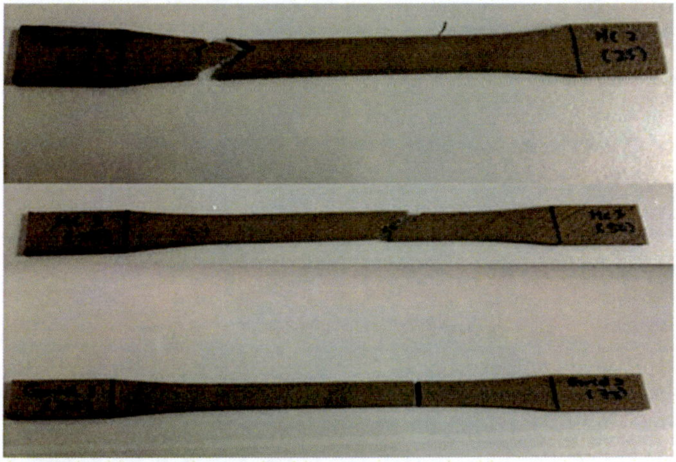

Fig. 4 FDM printed PLA/Coconut wood composite after tensile testing

3.1 Ultimate Tensile Strength

According to Tables 3, 4 and Fig. 5, for the PLA/Coconut wood samples with 75% infill percentage and concentric pattern has the maximum ultimate tensile strength of 19.35 MPa meanwhile for the PLA sample with 75% infill percentage and concentric pattern has the maximum ultimate tensile strength of 37.55 MPa. The obtained results demonstrate unequivocally that the ultimate tensile strength improves with increasing infill percentage for each infill pattern. Meanwhile, the results indicate that the greatest ultimate tensile strength was obtained at a 75% infill percentage, followed by 50% and 25% infill percentages. In terms of ultimate tensile strength, the specimens with the concentric pattern have the greatest strength, followed by

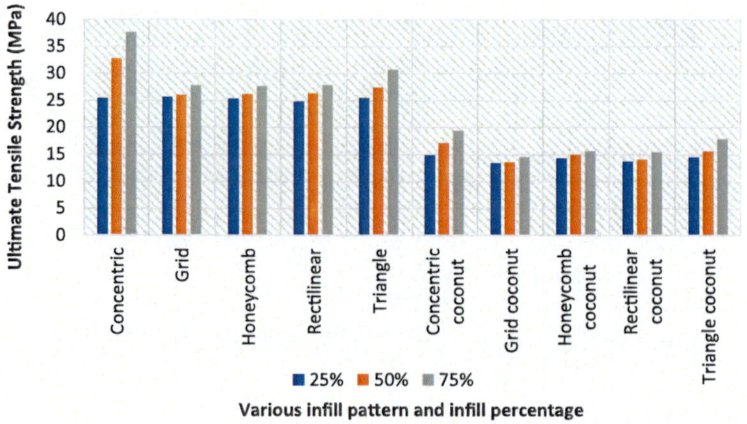

Fig. 5 Ultimate tensile strength of PLA and PLA/Coconut wood composite with various infill patterns and infill density

those with the triangular, rectilinear, grid, and honeycomb patterns for both PLA and PLA/Coconut wood. The ultimate tensile strength of PLA and PLA/Coconut wood is shown in Fig. 5.

3.2 Elastic Modulus

As reported by Tables 3, 4 and Fig. 6, the PLA/Coconut wood specimen with the highest elastic modulus was from 75% infill percentage and concentric infill pattern of 1.12134 GPa among all PLA/Coconut wood samples, whereas the PLA specimen with 75% infill percentage and concentric infill pattern has a maximum ultimate tensile strength of 1.14776 GPa. According to Fig. 6, as the infill percentage rises for each infill design, the sample's elastic modulus increases as well. The greatest elastic modulus was obtained at a 75% infill percentage, followed by 50 and 25% infill percentages. Though, when comparing infill patterns, the samples with a concentric pattern have the maximum elastic modulus across all infill percentages, followed by triangle, rectilinear, grid and finally the honeycomb pattern for both PLA and PLA/Coconut wood. Figure 6 below shows the elastic modulus of PLA and PLA/Coconut wood.

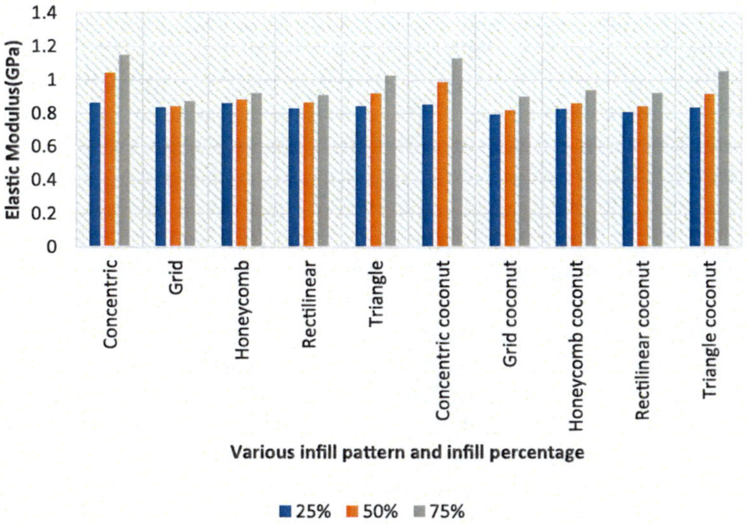

Fig. 6 Elastic modulus of PLA and PLA/Coconut wood composite with various infill patterns and infill density

3.3 *Yield Strength (0.2% Offset)*

According to Tables 3, 4 and Fig. 7, the PLA/Coconut wood specimen with the highest yield strength (0.2% offset) and concentric infill pattern provides 15.33 MPa

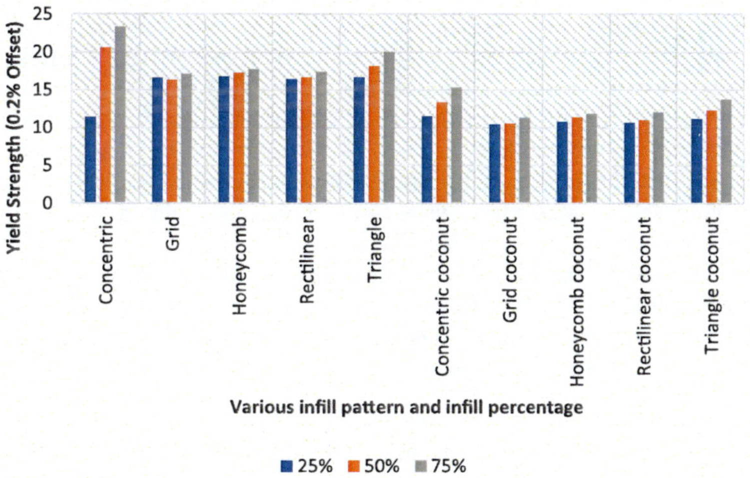

Fig. 7 Yield strength (0.2% offset) of PLA and PLA/Coconut wood composite with various infill patterns and infill density

out of all the PLA/Coconut wood specimens, meanwhile for the PLA sample with a concentric pattern at 75% infill percentage has the maximum yield strength of 23.33 MPa. As shown in Fig. 7, as the amount of infill rises for each infill design, the yield strength of the specimen (0.2% offset) improves as well. The greatest yield strength (0.2% offset) occurs at 75% infill, followed by 50% and, lastly, 25% infill. When evaluating infill designs, the sample with a concentric pattern had the highest yield strength (0.2% offset) across all infill percentages, followed by the triangular, rectilinear, grid, and honeycomb patterns for both PLA and PLA/Coconut wood. Figure 7 below show the yield strength (0.2% offset) of PLA and PLA/Coconut wood.

In general, the pattern of infill and the proportion of infill have a substantial impact on the specimen's tensile strength, elastic modulus, and yield strength. Concentric patterns have the greatest rates in PLA and PLA/Coconut wood composites. Additionally, when the specimen's infill % increases, the strength rate increases. Additionally, prior research indicates that increasing the infill density improves the product's strength. Finally, the concentric design with a percentage of infill of 75% is preferable for increasing the specimen's tensile strength, followed by the triangular infill pattern, rectilinear, grid, and honeycomb patterns.

4 Conclusion

The experimental study performed in the present work elucidates the influence of infill pattern and infill percentage on the tensile characteristics of FDM printed PLA and PLA/Coconut wood composites. All of the findings and analysis indicated that all of the study's goals were fulfilled. Finally, tensile analysis findings show that the infill percentage and infill pattern and their interaction effect have a significant influence on ultimate tensile strength, elastic modulus, and yield strength (0.2% offset). In all of the tests, the greatest tensile characteristics are achieved with the concentric infill pattern and the 75% infill percentage. The PLA/Coconut wood composite has the lowest tensile characteristics than the PLA. The outcomes indicate that the concentric infill pattern with 75% infill has the maximum strength in both PLA and PLA/Coconut wood composite. The maximum tensile strength obtained in the PLA is 37.55 MPa and in the PLA/Coconut wood is 19.35 MPa. The maximum elastic modulus of PLA is 1.148 GPa and for the PLA/Coconut wood is 1.121 GPa. The yield strength of PLA was 23.33 MPa. The PLA/Coconut wood composite has 15.33 MPa in tensile testing following the triangle, rectilinear and honeycomb infill patterns, while the Grid pattern has the weakest properties among all the patterns.

Acknowledgements The authors are grateful to Universiti Malaysia Pahang (www.ump.edu.my) for the financial support provided under the grants RDU192218, RDU190350, and RDU19402.

References

1. Standard A (2012) Standard terminology for additive manufacturing technologies. ASTM Int. F2792–12a
2. Li N, Li Y, Liu S (2016) Rapid prototyping of continuous carbon fiber reinforced polylactic acid composites by 3D printing. J Mater Process Technol 238:218–225. https://doi.org/10.1016/j.jmatprotec.2016.07.025
3. Es-Said OS, Foyos J, Noorani R, Mendelson M, Marloth R, Pregger BA (2000) Effect of layer orientation on mechanical properties of rapid prototyped samples. Mater Manuf Process 15:107–122. https://doi.org/10.1080/10426910008912976
4. Rajpurohit SR, Dave HK (2018) Effect of process parameters on tensile strength of FDM printed PLA part. Rapid Prototyp J 24:1317–1324. https://doi.org/10.1108/RPJ-06-2017-0134
5. Javaid M, Haleem A (2019) Using additive manufacturing applications for design and development of food and agricultural equipments. Int J Mater Prod Technol 58:225–238. https://doi.org/10.1504/IJMPT.2019.097662
6. Vaidya AA, Collet C, Gaugler M, Lloyd-Jones G (2019) Integrating softwood biorefinery lignin into polyhydroxybutyrate composites and application in 3D printing. Mater Today Commun 19:286–296. https://doi.org/10.1016/j.mtcomm.2019.02.008
7. Sathies T, Senthil P, Anoop MS (2020) A review on advancements in applications of fused deposition modelling process. Rapid Prototyp J 26:669–687. https://doi.org/10.1108/RPJ-08-2018-0199
8. Le Duigou A, Barbé A, Guillou E, Castro M (2019) 3D printing of continuous flax fibre reinforced biocomposites for structural applications. Mater Des 180:107884. https://doi.org/10.1016/j.matdes.2019.107884
9. Calignano F, Manfredi D, Ambrosio EP, Biamino S, Lombardi M, Atzeni E, Salmi A, Minetola P, Iuliano L, Fino P (2017) Overview on additive manufacturing technologies. Proc IEEE 105:593–612. https://doi.org/10.1109/JPROC.2016.2625098
10. Razavykia A, Brusa E, Delprete C, Yavari R (2020) An overview of additive manufacturing technologies-a review to technical synthesis in numerical study of selective laser melting. Materials (Basel). 13:1–21. https://doi.org/10.3390/ma13173895
11. Subramaniam SR, Samykano M, Selvamani SK, Ngui WK, Kadirgama K, Sudhakar K, Idris MS (2019) 3D printing: overview of PLA progress. In: AIP conference proceedings, vol 2059. https://doi.org/10.1063/1.5085958
12. Mohan N, Senthil P, Vinodh S, Jayanth N (2017) A review on composite materials and process parameters optimisation for the fused deposition modelling process. Virtual Phys Prototyp 12:47–59. https://doi.org/10.1080/17452759.2016.1274490
13. Jaisingh Sheoran A, Kumar H (2020) Fused Deposition modeling process parameters optimization and effect on mechanical properties and part quality: review and reflection on present research. Mater Today Proc 21:1659–1672. https://doi.org/10.1016/j.matpr.2019.11.296
14. Popescu D, Zapciu A, Amza C, Baciu F, Marinescu R (2018) FDM process parameters influence over the mechanical properties of polymer specimens: a review. Polym Testing 69:157–166. https://doi.org/10.1016/j.polymertesting.2018.05.020
15. Liu Z, Lei Q, Xing S (2019) Mechanical characteristics of wood, ceramic, metal and carbon fiber-based PLA composites fabricated by FDM. J Mater Res Technol 8(5):3741–3751. https://doi.org/10.1016/j.jmrt.2019.06.034
16. Andrzejewski J, Marciniak-Podsadna L (2020) Development of thermal resistant FDM printed blends. the preparation of GPET/PC blends and evaluation of material performance. Materials 13(9):2057. https://doi.org/10.3390/ma13092057
17. Wankhede V, Jagetiya D, Joshi A, Chaudhari R (2019) Experimental investigation of FDM process parameters using Taguchi analysis. Mater Today Proc 27:2117–2120. https://doi.org/10.1016/j.matpr.2019.09.078
18. Bryll K, Piesowicz E, Szymański P, Slaczka W, Pijanowski M (2018) Polymer composite manufacturing by FDM 3D printing technology. In: MATEC web of conference, vol 237. https://doi.org/10.1051/matecconf/201823702006

19. Rahim TNAT, Abdullah AM, Akil HM, Mohamad D, Rajion ZA (2017) The improvement of mechanical and thermal properties of polyamide 12 3D printed parts by fused deposition modelling. Express Polym Lett 11:963–982. https://doi.org/10.3144/expresspolymlett.2017.92
20. Dawoud M, Taha I, Ebeid SJ (2016) Mechanical behaviour of ABS: an experimental study using FDM and injection moulding techniques. J Manuf Process 21:39–45. https://doi.org/10.1016/j.jmapro.2015.11.002

Preliminary Tensile Investigation of FDM Printed PLA/Copper Composite

A. Kottasamy, K. Rajan, M. Samykano, K. Kadirgama, K. Moorthy, and M. M. Rahman

Abstract Fused Deposition Modeling (FDM) is a rapid prototyping technique that has been used to create components from a range of materials for a wide range of products. Thermoplastic polymers were used as a material that comes in the form of a filament. Copper is highly known for its electrical conductivity, high thermal resistance, and corrosion resistance. However, PLA's characteristics embedded with Copper particles still remain limited. The aim of this study is to investigate and analyze the tensile properties of the specimens with varying infill percentages (25, 50, and 75%) and infill patterns (Honeycomb, rectilinear, concentric, grid, and triangle) on Copper reinforced PLA using the FDM technique. The specimen is printed in accordance with the ASTM standard for tensile testing. Following that, the tensile properties of the PLA and the PLA/Copper composite were analyzed. The triangle pattern for the PLA/Copper composite has the same strength as the Concentric pattern of pure PLA. The results demonstrate that the concentric infill pattern with a 75% percentage of infill of pure PLA and the PLA/Copper composite has the same value in the overall tensile strength. The ultimate tensile strength of the PLA and PLA/Copper of 37.55 Mpa, the elastic modulus of 1.148 GPa, and yield strength of 23.33 MPa. Meanwhile, the grid infill pattern of PLA/Copper has the lowest tensile strength.

Supplementary Information The online version contains supplementary material available at https://doi.org/10.1007/978-981-19-1577-2_27.

A. Kottasamy
Green Kingdom Solutions Sdn. Bhd., Taman Salak Selatan, 57100 Kuala Lumpur, Malaysia

A. Kottasamy · M. Samykano (✉) · M. M. Rahman
College of Engineering, Universiti Malaysia Pahang, 26300 Gambang, Pahang, Malaysia
e-mail: mahendran@ump.edu.my

K. Rajan · K. Kadirgama
Faculty of Mechanical and Automotive Engineering Technology, Universiti Malaysia Pahang, 26600 Pekan, Pahang, Malaysia

K. Moorthy (✉)
College of Computing and Applied Sciences, Faculty of Computing, Universiti Malaysia Pahang, Gambang, Malaysia
e-mail: kohbalan@ump.edu.my

© The Author(s), under exclusive license to Springer Nature Singapore Pte Ltd. 2023
M. H. A. Hassan et al. (eds.), *Technological Advancement in Instrumentation & Human Engineering*, Lecture Notes in Electrical Engineering 882,
https://doi.org/10.1007/978-981-19-1577-2_27

The specimens' strength varies with the infill percentage, the 25% infill has the low strength in all infill patterns and the 75% has the maximum strength.

Keywords Cu-PLA composite · Tensile strength · 3D printing · FDM · Mechanical properties · Additive manufacturing

1 Introduction

The process of additive manufacturing (AM) is defined by the American Society for Testing and Materials (ASTM) as a method of creating three-dimensional (3D) solid products from digital models layer by layer [1]. Recently, rapid prototyping (RP) has emerged as a manufacturing focal point as a technique of tolerating part complexity while creating parts within strict time and cost constraints [2, 3]. This technology is widely employed in a variety of industries, including automobiles, aerospace, medicine, manufacturing, and consumer products [4–7]. AM processes are divided into seven types: material extrusion, binder jetting, material jetting, sheet lamination, direct energy deposition, VAT Photopolymerization, and powder bed fusion. Material extrusion is the most common form of AM process. The requirement for the use of 3D printing technology is becoming more prevalent [8, 9].

Fused Deposition Modeling (FDM) is one of the most commonly used RP methods for manufacturing components with three-dimensional (3D) complexity. It involves depositing the material layer by layer through a small liquefying pin that moves in the X and Y directions as the component is being manufactured (in the plane of the build platform). Immediately following the deposit of a layer, the addition of the subsequent layer initiates and the build platform (or worktable) is lowered in the Z direction [10–12]. A separate liquefier nozzle is placed on the extrusion head to deposit the support materials during the printing process. Components are manufactured by depositing support materials at strategic locations throughout the component as it is being assembled to minimize substantial geometric deformation. Once the fabrication is complete, the item is withdrawn from the platform. The deposited support material is also removed. FDM methods are often used to create products with a high degree of efficiency, accuracy, and, most importantly, mechanical strength, among other characteristics [13–15]. The mechanical properties of the produced specimen are determined by the printing parameters that the user has specified during printing. FDM may now be used with a wide range of materials, including wood particle-filled filament, carbon fibers, and ceramics, to create complex shapes [16, 17]. The application of FDM technology has risen rapidly in a wide range of industries worldwide in recent years. Compared to the older, more traditional technique, FDM technology provides substantial advantages in time savings, cost reductions, and raw material utilization. Traditional methods such as injection molding, sand casting, and blow molding are examples of old-fashioned methods that were once popular, but the process is extremely time-consuming and labor-intensive. Wood particles are a type of natural particle that has been widely used in a range of applications, the most notable of which has been in the biomedical industry [18, 19]. On the other hand, a scarcity of knowledge exists about the mechanical characteristics of pure wood items that have been printed using a variety of printing techniques.

Copper particles are well known for their electrical conductivity, corrosion resistance and thermal resistance. The mechanical behavior of copper particles, when subjected to a range of printing circumstances, is still limited. It is important to note that the mechanical characteristics of the product are significantly influenced by the printing parameters such as layer thickness, raster angle, operating temperature, printing speed and infill density. The infill density is the amount of material filled in each layer, and it directly affects the product's strength. Few studies have focused on the investigation of the consequence of infill patterns on FDM process mechanical behavior. As such, the present study aims to investigate the tensile properties of the PLA and PLA/Copper filament by various infill patterns (Grid, rectilinear, honeycomb, concentric, triangular) and infill percentages (25, 50, 75%).

2 Materials and Methods

The printing process is done using the Raise 3D N2 Plus FDM 3D printer with 0.4 mm nozzle diameter and 1.75 mm filament for the tensile test specimen. The PLA filament and the copper reinforced PLA filament were purchased from Magma filaments. The composite filament contains 60% of PLA and 40% of copper. The working temperature of the filaments is 200 to 240 °C. Few parameters were maintained fixed through the printing process is shown in Table 1. The humidity in the surrounding area should be between 70 and 80%, and the room temperature maintained to be between 20 to 25 °C.

Table 1 Process parameters for the FDM printing

No.	Parameter	Values
1	Layer thickness	0.1 mm
2	First layer thickness	0.3 mm
3	Solid layers	Three layers (top and bottom)
4	Diameter of the nozzle	0.4 mm
5	Diameter of filament	1.75 mm
6	Nozzle temperature	210 °C
7	Bed temperature	60 °C
8	Printing speed	40 mm/s

2.1 Fabrication of Specimen

The design of the tensile test sample is performed using Solid works 2019. The design and dimension of the specimen were performed according to the ASTM D638 Type 1 standard. The design model is initially stored in STL format, and the file is imported into slicing software. Ideamaker is the standard software for the Raise 3D N2 plus printer. The slicing software generates the G- codes and the file is transferred to the FDM printer to print the specimens using flash drives. However, before the printing process, the heated bed is heated to a specified temperature. The nozzle temperature and the temperature of the bed remain constant throughout the printing. At extrude the filament, the extruder's temperature is set to 210 °C. For each composition, the number of samples n = 5 and a total of 75 samples were printed. The Raise 3D N2 Plus, FDM 3D printers image, is shown in Fig. 1.

Fig. 1 Raise 3D N2 Plus FDM 3D printer

2.2 *Tensile Testing*

The tensile test is performed using an INSTRON 3367 computer controller. The maximum force that the computer can exert is 30 kN. For the Type 1 specimen geometry, the recommended speed for ASTM D368 is 10 mm/min with a tolerance of 25%. Figure 2 and Table 2 display the model and the Type 1 specimen in greater detail.

3 Result and Discussion

The major properties to be evaluated in the tensile test findings are the Ultimate Tensile Strength (UTS), Elastic modulus (E), and yield strength (0.2% offset). The Ultimate Tensile Strength (UTS) of a material is the highest stress that it can take without breaking. Elastic modulus, commonly known as Young's modulus, is a word used to describe the stiffness of a material. Many materials will obey Hooke's law in the early stage or low strain of the stress–strain curve. In layman's terms, the elastic modulus is a material property that undergoes stress, deforms, and tends to return to its original shape after the force is taken out. On the other hand, yield strength

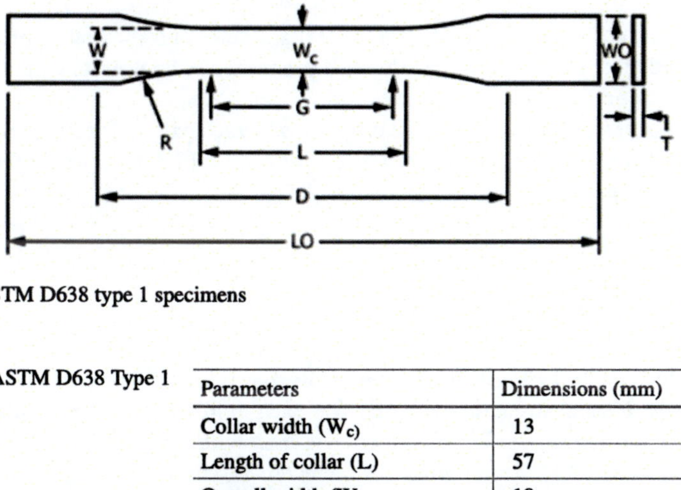

Fig. 2 ASTM D638 type 1 specimens

Table 2 ASTM D638 Type 1 geometry

Parameters	Dimensions (mm)
Collar width ($W_{c)}$)	13
Length of collar (L)	57
Overall width ($W_{o)}$)	19
Overall length ($L_{o)}$)	165
Gage length (G)	50
Grip distance (D)	115
Fillet radius (R)	76
Thickness (T)	3

Table 3 Tensile properties of each pattern for PLA for the various infill percentage

Infill pattern	Infill percentage (%)	Young's modulus (GPa)	Tensile strength (MPa)	Tensile stress at break (MPa)	Tensile stress at yield (0.2%) (MPa)
Concentric	25	0.861	25.38	24.45	11.36
Concentric	50	1.039	32.66	29.37	20.56
Concentric	75	1.147	37.55	32.78	23.33
Grid	25	0.835	25.62	24.81	16.59
Grid	50	0.839	25.98	25.15	16.28
Grid	75	0.870	27.73	27.19	17.12
Honeycomb	25	0.858	25.33	23.53	16.74
Honeycomb	50	0.881	26.11	25.05	17.27
Honeycomb	75	0.921	27.64	26.62	17.74
Rectilinear	25	0.829	24.80	23.69	16.41
Rectilinear	50	0.864	26.24	25.21	16.69
Rectilinear	75	0.908	27.78	27.11	17.38
Triangle	25	0.842	25.42	24.51	16.66
Triangle	50	0.919	27.29	26.48	18.14
Triangle	75	1.024	30.66	29.30	19.99

represents the stress needed to cause a modest amount of plastic distortion. The raw data created after testing with the Blue hill program can be used to calculate all of the tensile properties collected.

Table 3 below indicates the tensile properties of each pattern for PLA and Table 4 shows the average tensile properties of each pattern for PLA/Copper for the various infill percentage.

3.1 *Ultimate Tensile Strength*

The obtained results demonstrate unequivocally that as the infill percentage increases, the ultimate tensile strength increases for each infill design. Corresponding to Tables 3, 4 and Fig. 3, the PLA/Copper specimen with a triangular pattern at 75% of infill percentage has the maximum ultimate tensile strength of 37.55 MPa. Meanwhile, the PLA specimen with the concentric infill pattern at 75% infill percentage has the maximum ultimate tensile strength of 37.55 MPa. Also, according to these findings, the 75% infill percentage gets the highest ultimate tensile strength, followed by 50% infill percentage and 25% infill percentage. In terms of ultimate tensile strength, the specimen with the concentric pattern for PLA and triangular pattern for the PLA/Copper composite has the maximum, followed by the rectilinear grid

Table 4 Tensile properties of each pattern for PLA/Copper for the various infill percentage

Infill pattern	Infill percentage (%)	Young's moduls (GPa)	Tensile strength (MPa)	Tensile stress at break (MPa)	Tensile stress at yield (0.2%) (MPa)
Concentric	25	0.780	21.12	18.79	14.41
Concentric	50	0.882	24.38	21.94	16.66
Concentric	75	1.022	28.35	23.94	19.00
Grid	25	0.700	18.43	17.44	12.70
Grid	50	0.701	17.49	16.20	12.33
Grid	75	0.744	18.98	18.06	13.07
Honeycomb	25	0.715	19.68	18.63	13.49
Honeycomb	50	0.752	19.48	18.13	13.54
Honeycomb	75	0.777	19.98	18.94	13.87
Rectilinear	25	0.716	18.39	17.66	12.92
Rectilinear	50	0.700	17.63	16.59	12.29
Rectilinear	75	0.748	19.11	17.97	13.16
Triangle	25	0.861	25.38	24.45	11.36
Triangle	50	1.039	32.66	29.37	20.56
Triangle	75	1.147	37.55	32.78	23.33

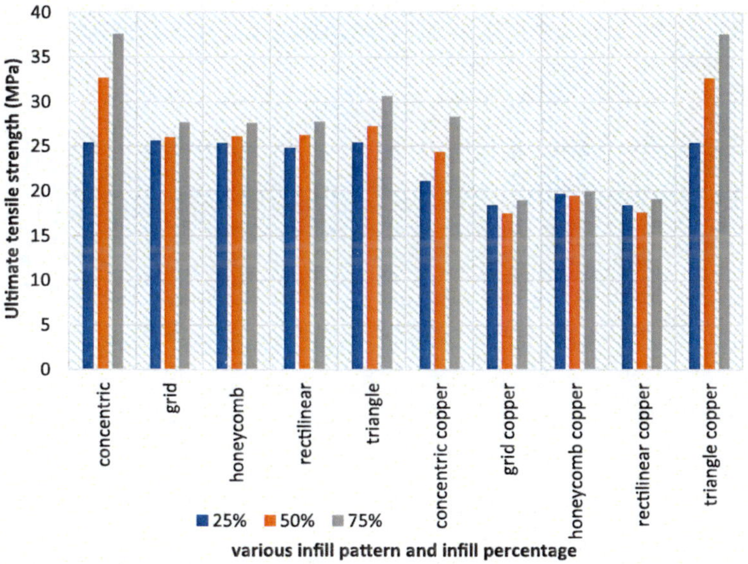

Fig. 3 Ultimate tensile strength of PLA and PLA/Copper composite with various infill patterns and infill density

and the honeycomb pattern for both PLA and PLA/Copper. Figure 3 below show the ultimate tensile strength of PLA and PLA/Copper.

3.2 Elastic Modulus

From Tabless 3, 4 and Fig. 4, the PLA/Copper specimen with the highest elastic modulus at concentric infill pattern and 75% infill percentage has 1.147 GPa among all PLA/Copper specimens. Meanwhile, the PLA specimen with the concentric pattern and 75% infill percentage has the maximum elastic modulus of 1.147 GPa. Figure 4 clearly shows that as the infill % increases at each infill pattern, the elastic modulus of the specimen increases as well. This was demonstrated in each pattern, where the 75% infill percentage achieves the maximum elastic modulus, followed by the 50% and 25% infill percentage. However, when comparing infill patterns, a specimen with a concentric pattern at PLA and for the PLA/Copper composite materials triangular pattern have the maximum elastic modulus across all infill percentages, continued by a rectilinear, grid, and the honeycomb pattern for both PLA and PLA/Copper. Figure 4 below shows the elastic modulus of PLA and PLA/Copper.

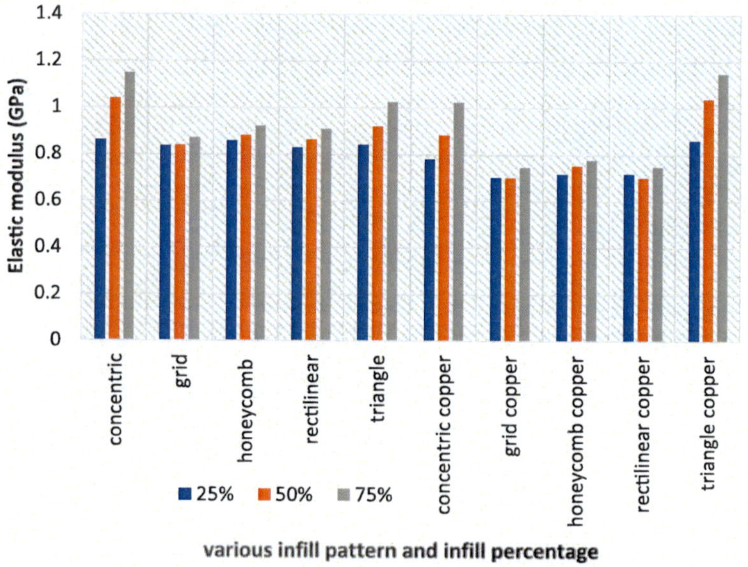

Fig. 4 Elastic modulus of PLA and PLA/Copper composite with various infill patterns and infill density

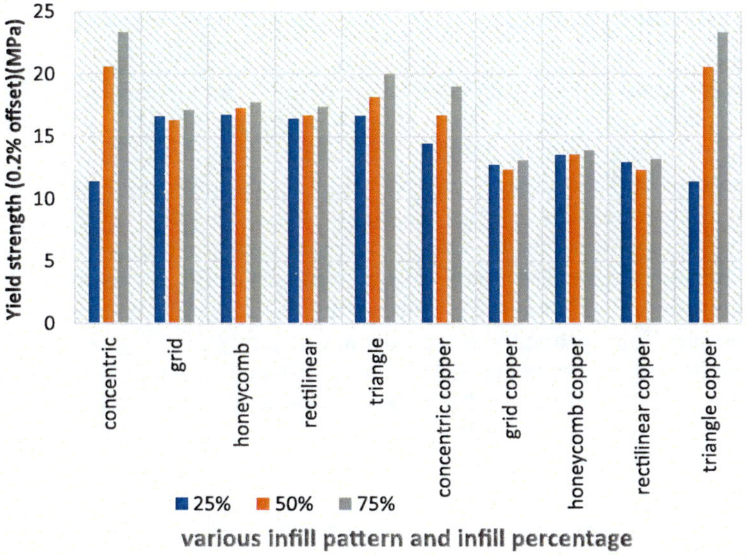

Fig. 5 Yield strength (0.2% offset) of PLA and PLA/Copper composite with various infill patterns and infill density

3.3 *Yield Strength (0.2% Offset)*

According to Tables 3, 4 and Fig. 5, the PLA/Copper specimen with the highest yield strength (0.2% offset) was from a concentric infill pattern revealing 23.33 MPa. Meanwhile, the concentric pattern and 75% infill percentage of the PLA specimen have the maximum ultimate tensile strength of 23.33 MPa. Figure 5 clearly shows that as the infill percentage increases at each infill pattern, the value of the specimen's yield strength (0.2% offset) increases as well. This is demonstrated in each infill pattern, where the 75% infill percentage has the maximum yield strength (0.2% offset), continued by the 50% infill percentage and 25% infill percentage. In terms of infill patterns, the specimen with the concentric pattern for PLA and the triangular pattern for PLA/Copper composite have the maximum yield strength (0.2% offset), followed by the rectilinear, grid and the honeycomb pattern for both PLA and PLA/Copper. Figure 5 below shows the yield strength (0.2% offset) of PLA and PLA/Copper.

Overall, the concentric infill pattern in pure PLA and triangle infill pattern in the PLA/Copper composite has an enormous impact on the tensile strength of the specimen. Also, with respect to the infill density, the strength of the specimen was changed. Compared to the results with Z. Liu et al. [20], 75% of infill density and triangular infill pattern have better tensile strength. The 25% of infill density has a low tensile strength ascending 50 and 75%.

4 Conclusion

The experimental analysis performed in the present study elucidates the effect of the infill pattern and infill percentage on the tensile properties of FDM printed PLA and PLA/Copper composite. All the results and analyses demonstrated that all of the objectives of this study project were met. The concentric infill pattern and the 75% infill percentage attain all of the highest tensile qualities for the pure PLA. The tensile properties of the PLA/Copper are lower compared with the PLA except for the triangle pattern. The triangle pattern for the PLA/Copper composite has the same strength as the Concentric pattern of pure PLA. The results demonstrate that the concentric infill pattern with a percentage of infill 75% of pure PLA and the PLA/Copper composite produced 37.55 MPa of Ultimate tensile strength and the maximum elastic modulus of 1.148 GPa and yield strength of 23.33 Mpa in tensile testing. Followed by triangle, rectilinear and honeycomb. Finally, tensile test findings show the interaction impact of infill pattern and infill percentage have a considerable effect on ultimate tensile strength, elastic modulus, and yield strength (0.2% offset). Meanwhile, the grid pattern has the weakest properties among all the patterns. The strength of the specimens also observed varies with infill percentage.

Acknowledgements The authors are grateful to Universiti Malaysia Pahang (www.ump.edu.my) for the financial support provided under the grants RDU192218, RDU190351, and RDU19401.

References

1. ASTM: ASTM International Committee F42 on Additive Manufacturing Technologies, ASTM F2792–10 Standard Terminology for Additive Manufacturing Technologies. Presented at the (2009)
2. Li N, Li Y, Liu S (2016) Rapid prototyping of continuous carbon fiber reinforced polylactic acid composites by 3D printing. J Mater Process Technol 238:218–225. https://doi.org/10.1016/j.jmatprotec.2016.07.025
3. Bertsch A, Bernhard P, Vogt C, Renaud P (2000) Rapid prototyping of small size objects. Rapid Prototyp J 6:259–266
4. Mohamed OA, Masood SH, Bhowmik JL (2016) Mathematical modeling and FDM process parameters optimization using response surface methodology based on Q-optimal design. Appl Math Model 40:10052–10073. https://doi.org/10.1016/j.apm.2016.06.055
5. Zhang X, Fan W, Liu T (2020) Fused deposition modeling 3D printing of polyamide-based composites and its applications. Compos Commun 21:100413. https://doi.org/10.1016/j.coco.2020.100413
6. Sathies T, Senthil P, Anoop MS (2020) A review on advancements in applications of fused deposition modelling process. Rapid Prototyp J 26:669–687. https://doi.org/10.1108/RPJ-08-2018-0199
7. Mohd NAS, Abdul RH, Mohd H, Abdullah HZ, Idris MI, Lee TC (2020) Review on the fabrication of fused deposition modelling (FDM) composite filament for biomedical applications. Mater Today Proc 29:228–232. https://doi.org/10.1016/j.matpr.2020.05.535
8. Chen Z, Li Z, Li J, Liu C, Lao L, Fu Y, Liu C, Li Y, Wang P, Yi H (2019) 3D printing of ceramics: a review. J Eur Ceram Soc 39:115–143

9. Rasoulianboroujeni M, Fahimipour F, Shah P, Khoshroo K, Tahriri M, Eslami H, Yadegari A, Dashtimoghadam E, Tayebi L (2019) Development of 3D-printed PLGA/TiO 2 nanocomposite scaffolds for bone tissue engineering applications. Mater Sci Eng C 96:105–113. https://doi.org/10.1016/j.msec.2018.10.077
10. Liu Z, Wang Y, Wu B, Cui C, Guo Y, Yan C (2019) A critical review of fused deposition modeling 3D printing technology in manufacturing polylactic acid parts. Int J Adv Manuf Technol 102:2877–2889. https://doi.org/10.1007/s00170-019-03332-x
11. Kumar P, Ahuja IPS, Singh R (2012) Application of fusion deposition modelling for rapid investment casting - a review. Int J Mater Eng Innov 3:204–227. https://doi.org/10.1504/IJMATEI.2012.049254
12. Subramaniam SR, Samykano M, Selvamani SK, Ngui WK, Kadirgama K, Sudhakar K, Idris MS (2019) 3D printing: overview of PLA progress. In: AIP conference proceedings, vol 2059. https://doi.org/10.1063/1.5085958
13. Popescu D, Zapciu A, Amza C, Baciu F, Marinescu R (2018) FDM process parameters influence over the mechanical properties of polymer specimens: a review. Polym Test 69:157–166. https://doi.org/10.1016/j.polymertesting.2018.05.020
14. Mazzanti V, Malagutti L, Mollica F (2019) FDM 3D printing of polymers containing natural fillers: a review of their mechanical properties. Polymers (Basel) 11:1094. https://doi.org/10.3390/polym11071094
15. Vyavahare S, Teraiya S, Panghal D, Kumar S (2020) Fused deposition modelling: a review. Rapid Prototyp J 26:176–201. https://doi.org/10.1108/RPJ-04-2019-0106
16. Li Q, Zhao W, Li Y, Yang W, Wang G (2019) Flexural properties and fracture behavior of CF/PEEK in orthogonal building orientation by FDM: microstructure and mechanism. Polymers (Basel) 11:656. https://doi.org/10.3390/polym11040656
17. Heidari-Rarani M, Rafiee-Afarani M, Zahedi AM (2019) Mechanical characterization of FDM 3D printing of continuous carbon fiber reinforced PLA composites. Compos Part B Eng 175:10714. https://doi.org/10.1016/j.compositesb.2019.107147
18. Singh S, Singh R (2020) Mechanical characterization and comparison of additive manufactured ABS, Polyflex™ and ABS/Polyflex™ blended functional prototypes. Rapid Prototyp J 26:225–237. https://doi.org/10.1108/RPJ-11-2017-0234
19. Jaisingh A, Kumar H (2020) Fused Deposition modeling process parameters optimization and effect on mechanical properties and part quality: review and reflection on present research. Mate. Today Proc 21:1659–1672. https://doi.org/10.1016/j.matpr.2019.11.296
20. Liu Z, Lei Q, Xing S (2019) Mechanical characteristics of wood, ceramic, metal and carbon fiber-based PLA composites fabricated by FDM. J Mater Res Technol 8:3743–3753. https://doi.org/10.1016/j.jmrt.2019.06.034

A Narrative Review: Bamboo Fiber as an Alternative Source for Pulp and Paper

Mohd Faizal Esa, Nor Mazlana Main, Mohd Nazrul Roslan, Noraini Marsi, Khairu Kamarudin, and Latifah Jasmani

Abstract This review was done to investigate the bamboo fiber properties suitable for replace hardwood fiber for Pulp and paper industry in Malaysia. Hardwood fiber sourcing from the forest had a challenge to an ecosystem and need to reduce. Bamboo fiber morphology and chemical characterization by 13 research papers from 1994 to 2021 were reviewed to find suitability for the paper-making process. Gigantochloa scortechinii and Bambusa vulgaris have become favorite species for a researcher. These two species of bamboo can be categories as long fiber, Runkle ratio around 1, and slenderness ratio above 33. Chemical characterization for both species also above 34% content for alpha-cellulose, more than 50% holocellulose, ash, and lignin content same as hardwood fiber. Investigation for other species also can vary the fiber source for the pulp and paper industry. Hence, these two species of bamboo fiber potential to replace hardwood fiber.

Keywords Bamboo · Pulp and paper · Morphology · Chemical composition

Supplementary Information The online version contains supplementary material available at https://doi.org/10.1007/978-981-19-1577-2_28.

M. F. Esa (✉) · N. M. Main · M. N. Roslan · N. Marsi · K. Kamarudin
Faculty of Engineering Technology, University Tun Hussein Onn Malaysia (UTHM) Pagoh Campus, Pagoh Higher Education Hub, KM1 Jalan Panchor, 84600 Panchor, Johor, Malaysia
e-mail: faizalesa@uthm.edu.my

N. M. Main · M. N. Roslan
Bamboo Research Center, Faculty of Engineering Technology, University Tun Hussein Onn Malaysia (UTHM) Pagoh Campus, Pagoh Higher Education Hub, KM1 Jalan Panchor, 84600 Panchor, Batu Pahat, Johor, Malaysia

L. Jasmani
Pulp and Paper Laboratory, Forest Products Division, Forest Research Institute Malaysia (FRIM), 52109 Kepong, Selangor, Malaysia

© The Author(s), under exclusive license to Springer Nature Singapore Pte Ltd. 2023
M. H. A. Hassan et al. (eds.), *Technological Advancement in Instrumentation & Human Engineering*, Lecture Notes in Electrical Engineering 882,
https://doi.org/10.1007/978-981-19-1577-2_28

1 Introduction

In the modern economy, pulp and paper products are still viable and relevant to the world. Paper and papermaking are among the pioneer industries in the world and are still important today [1]. Globalization in the pulp and paper industry significantly improves the global economy and produces high-quality products demand [2, 3]. Paperless in daily routine is a voice to support a sustainable environment and avoid global warming issues [4, 5]. Despite that, in 2025, global paper consumption will increase around 1.6% a year, equally 500 million tones [6].

Many companies are also in line to develop more efficient resources and strategies to support the bio-based economy [7]. Hence, an alternative source called biomass need to provide for this industry to avoid wood-based dependence [8].

In Malaysia, mainly secondary fiber and wastepaper become raw materials for the paper industry [3]. However, increasing demand for packaging and online purchasing shows an increase in the usage of paper products. These make a packaging company need to maintain the quality of product safety and value. Thus, demand for paper products and plastic packaging automatically affects. Nevertheless, plastic products for packaging can cause many challenges for the future if we did not provide biodegradable products based on a sustainable future [9, 10]. Malaysia still imports pulp of wood or other fibrous cellulose material for pulp and paper industries [11, 12].

Pulp and paper industries need an alternative source for a sustainable future and reduce the wood-based source dependence.

1.1 Bamboo

Bamboo is a member of a particular taxonomic group of large woody grasses (Poaceae) and comprises the subfamily Bambusoideae [13, 14]. Bamboo encompasses 1250 species within 75 genera. About 220 000 km^2 of bamboo resource is a vital component of plant and forest resources in the world [15]. Figure 1 show the

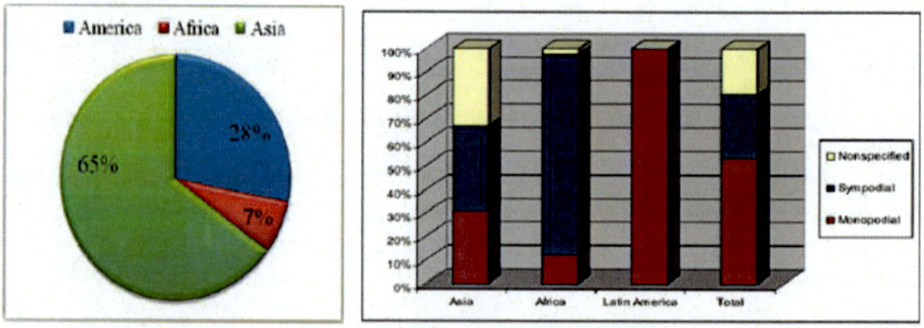

Fig. 1 Bamboo resource according to continents [16, 17]

Table 1 Species, genera and area of bamboo forest in some country[15]

Country	Species	Genera	Area (km^2)
China	500	40	61,586
India	136	23	108,630
Myanmar	100	17	8,950
Thailand	60	13	8,100
Bangladesh	30	13	863
Cambodia	10	4	2,870
Vietnam	101	15	10,000
Malaysia	44	7	5,920
The Philippines	55	12	1,560
Japan	230	13	1,413

bamboo distribution worldwide dominate by the Asia region [16, 17]. Most of which are relatively fast-growing and attain maturity within 4 to 5 years for 10 to 40 m in height. Bamboo can be harvest annually and potential to become a sustainable material [18].

In Asia, about 25 million people used bamboo as a portion of food and fiber for daily use [19]. Bamboo is distributed mainly in the tropical and occur naturally in the subtropical and temperate zone except for Europe. Asia accounts for about 1000 species and covering about 180 000 km^2. China alone has about 300 species in 44 genera and occupying about 601 000 km^2 area from their total forest area as bamboo resources [20].

Another major country in producing bamboo is India. They occupy 96 000 km^2 of the area and about 13% of the total forested area [21]. In India, about 113 bamboo species are in the forest, and the main species belong to Bambusa balcooa, Melocanna baccifera, and Dendrocalamus hamiltonii [22]. Other countries that contributed to major bamboo producers can refer to Table 1.

Bamboo species growth can be divided into two types which are sympodial or call clump and monopodial or single stem growth [23]. In Malaysia, bamboo growth belongs to the sympodial type [24]. Table 2 show the difference between monopodial and sympodial type. There are about 59 bamboo species in Peninsular Malaysia which are from 7 genera of Bambusa, Dendrocalamus, Dinochloa, Racemobamboos, Schizostachyum, Thyrsostachys, and Gigantochloa [25].

Bamboo plays an important economic role and creates an increasing demand yearly from the rural population, urban centers, and international trade. This issue affects the country stock and increasing bamboo conservation [25, 26]. In Asia, bamboo is sometimes used for housing, crafts, flooring, roofing fabrics, vegetable (bamboo shoot), pulp, paper, and composite panel [27]. However, lack of reliable, comprehensive data and utilization suitability make bamboo had a limit for their potential to contribute as an important source to the country [28].

In Malaysia, bamboo is used as a craft product, chopstick, frame, basket, and furniture only [29–31]. There are many potentials of bamboo product need to expand

Table 2 Difference between monopodial and sympodial [23, 24]

Monopodial	Sympodial

Growth in single	Growth in clump
Growth in temperate climate	Growth in tropical climate

for more versatile the bamboo source. Bamboo farming in Malaysia also abandons because lack of knowledge of bamboo strength and support from the industry.

So, this paper reviews the potential strength of bamboo as an alternative to Malaysia as a replacement of wood base fiber in the pulp and paper industry. Also, a decline of forest resources requires non-wood material in many structural and nonstructural applications [32]. A lot of research had been done to prove the potential of bamboo as an alternative source for the replacement of wood base products [33–40].

In 2019, the Minister of Main Industry planned to explore the potential of bamboo as an important commodity crop in the future, and on 2020 Bamboo Industry Development Action Plan 2.0 (2021–2030) already release to realize the bamboo industry as a new source of the national economy [31, 41, 42]. Ministry of Plantation Industries and Commodities Malaysia (MPIC) media release also enforces the government to support the bamboo industry [43]. All this shows the Malaysian government's serious action to find a new source of commodity crop in the future. Malaysian Timber Industry Board (MTIB) also done a seminar on bamboo to explore the strength of bamboo in Malaysia and release a huge amount of bamboo plant in Malaysia forest, Fig. 2 [44].

Bamboo Morphology

Bamboo is a unique plant structure compare to other plants. The plant can divide into rhizome, root, culms, branches, leaves, and flowers [35, 45]. Figure 3 displays the nomenclature of a bamboo plant. Bamboo is a faster-growing plant and used poor soil for planting. Bamboo need 3–4 year to matured and harvesting [46]. The stems of bamboo are hollow but very strong, and these provide great popularities in everyday used [18].

- Rhizomes stem extend underground from the main plant in a horizontal direction and travel for new territory. During travel through the soil, the primary nutrient will collect by rhizomes for growth and energy [47–49].

Species	Numbers of Bamboo Clumps		Total Clumps
	Permanent Forest Reserve	Government Land	
Gigantochloa scortechinii (Buluh Semantan)	4,528,684	614,059	5,142,743
Gigantochloa wrayi (Buluh Beti)	13,739,567	921,088	14,660,656
Gigantochloa levis (Buluh beting)	21,722,333	767,574	22,489,906
Dendrocalamus asper (Buluh betong)	8,136,280	-	8,136,280
Schizostachyum grande (Buluh semeliang)	19,880,156	307,029	20,187,186
Schizostachyum zollingeri (Buluh dinding)	4,375,170	-	4,375,170
Total	72,382,190	2,609,750	74,991,941

Fig. 2 Distribution of six (6) bamboo species according to forest status in Peninsular Malaysia [44]

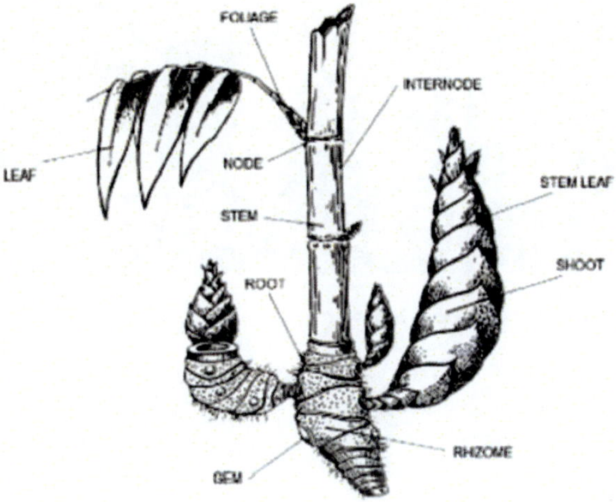

Fig. 3 Bamboo part [45]

- Culms are the most important part of a bamboo plant for the new source of fiber. Culm color majority are green but sometimes can also be brown, yellow, black, or stripe [30]. Culm shape mostly in around and have a different diameter. Culm is always straight and divides into nodes and internodes, Fig. 4 [50].

The others important part of the bamboo is the anatomy of the leaf. Leaf plays an essential function as a protective cover to rhizome as it travels underground and has a commercial value in the industry [51, 52]. The appearance of leaves plays a significant role in identifying bamboo but is sometimes disposable as waste after harvesting [53]. The major peculiarity of bamboo is the flower, which flowering about 60 to 120 intervals [54]. It makes bamboo species slow in commercialization

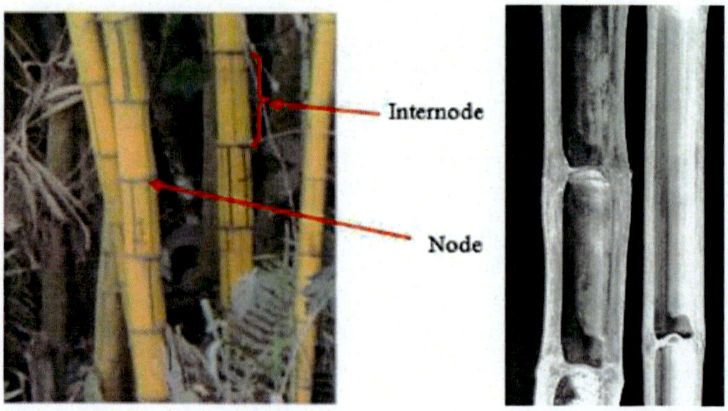

Fig. 4 Internode and node at bamboo culm [50]

Fig. 5 Example of **a** bamboo leaf and **b** flower

and sometimes can course bamboo cells to die [55]. Refer to Fig. 5 as an example of bamboo leaf (a), and flower (b).

2 Pulp and Paper

The properties of paper in the pulp and paper industry are dependent on fiber dimensions. The suitability of the papermaking process is dependent on chemical composition. This two-parameter importance for pulp and paper industry to make sure the new fiber is suitable for the process and came out with good quality paper same as wood base paper [56, 57].

According to Suhaimi et al. [60], Gigantochloa levis and Gigantochloa scortechinii have a long fiber type and are suitable for the pulp and paper industry. Razak et al. [70], investigated the chemical composition of Gigantochloa brang, levis, scortechinii and wrayi found all species have acceptable chemical content and are suitable for the papermaking process. Then Sadiku et al. [69], found Bambusa vulgaris chemical content and fiber length suitable for pulp and paper making. Also, Razak et al. [64]

Fig. 6 Pulp and paper making process [58]

found that Bambusa vulgaris fiber length in the category long fiber. These three researchers show bamboo fiber has potential for the pulp and paper industry.

So, bamboo fiber can promise to become a new source of pulp and paper. Besides that, other species of bamboo still do not fully investigate the morphology and chemical content and maybe can be a new source also. Figure 6 below shows the process of pulp and paper making. The primary step of the process is pulping and papermaking [58]. The process for investigating the fiber length and chemical needs to be done at stage chipping before starting the pulping process.

2.1 *Fiber Morphology*

Bamboo fiber chooses as an alternative to a new source of Malaysian pulp and paper because of its fast-growing, abundance and shorter time of harvesting [58, 59]. The fiber morphology and chemical composition similar to hardwood make bamboo fiber suitable to replace long fiber from the wood base.

The high quality of pulp depends on fiber dimension and can affect the properties of the end product. The important parameter in fiber morphology is Fiber length (FL), Fiber Diameter (FD), Lumen Diameter (LD) and Cell wall thickness (CWT). The length of fiber usually will be affected by paper strength and thickness. Meanwhile, the diameter of the fiber wall affected the flexibility of the paper [60]. The review data of fiber morphology by 12 researchers show in Table 3.

Finding for review on bamboo morphology for Malaysia researchers start from 1994 to 2021. The data of review divide into age to understand the fiber morphology behaviors according to year planting. There are 10 researchers investigate the fiber morphology at 4 years old and above. The 4 years old bamboo choose for research because the maturity of bamboo starts after this year of planting. There are 5 researchers done the morphology for Bambusa Vulgaris species. 4 researchers done

Table 3 Bamboo fiber morphology at difference year of planting

Species	Fiber length (μm)	Fiber diameter(μm)	Lumen diameter(μm)	Fiber wall thickness(μm)	Runkel ratio	Felting power	References
4 year and above							
Bambusa vulgaris – 5 yr	3120	14.40	5.61	4.40	1.57	216.67	[60]
Gigantochloa levis_5yr	3470	18.71	7.68	5.52	1.44	185.46	
Gigantochloa scortechinii_5yr	4000	17.05	5.27	5.89	2.24	234.60	
Gigantochloa thoii	4071	25.2	5.0	12.2	4.88	161.54	[61]
Gigantochloa scortechinii	2326	14.8	8.1	3.3	0.815	157.16	
Gigantochloa ligulata	3930	22.6	4.3	9.2	4.28	173.90	
Gigantochloa wrayi	2753	24	10.3	7.0	4.28	114.71	
Gigantochloa brang	3543	21.4	6	7.7	2.57	165.56	
Bambusa vulgaris	2494	14.1	3.5	7.1	4.05	176.88	
Bambusa blumeana	2905	18.9	7.6	5.7	1.5	153.70	
Bambusa heterostacya	3764	26.8	5.8	10.5	3.62	140.45	
Bambusa vulgaris cv vitata	3592	20.3	6.9	6.3	1.83	176.95	
Dendracalamus asper	2998	23.3	6.1	8.6	2.82	128.67	
Scizotachyum brachycladum	2840	22.2	6.4	7.9	2.47	127.93	
Scizotachyum grande	2451	15	3.1	6.1	3.94	163.4	
Scizotachyum zolingeri	2326	14.8	8.1	3.3	0.814	157.16	
Bambusa vulgaris	2106	14.43	12.41	1.01	0.84	145.95	[38]

(continued)

Table 3 (continued)

Species	Fiber length (µm)	Fiber diameter(µm)	Lumen diameter(µm)	Fiber wall thickness(µm)	Runkel ratio	Felting power	References
Scizotachyum grande	2451	15.1	3.3	6.5	3.9	162.31	[62]
Scizotachyum brachycladum	2840	23.7	6.4	9.2	2.88	119.83	
Gigantochloa atter – 4 yr	2380	13.43	8.78	2.32	0.53	177.22	[63]
Gigantochloa atter – 16 yr	2390	11.72	7.16	2.28	0.64	203.92	
Gigantochloa atter – 40 yr	2450	14.45	8.58	2.94	0.69	169.55	
Bambusa vulgaris	4200	18.0	2.4	7.6	6.33	233.33	[64]
Schizostachyum brachycladum	2840	23.73	6.43	9.18	2.86	119.68	[65]
Gigantochloa scortechinii	4400	18.3	2.5	7.9	6.32	338.46	[66]
Gigantochloa scortechnii–5.5 yr	2380	26	9	8	1.78	91.53	[67]
Bambusa vulgaris	2820	13	2.7	5.7	4.22	216.92	[68]
Schizostachyum brachycladum	2840	23.71	6.4	9.2	2.88	119.78	[68]
Schizostachyum grande	2451	15.1	3.3	6.5	3.94	162.32	
3 year and above							
Bambusa vulgaris	3070	12.87	5.80	3.53	1.21	238.54	[60]
Gigantochloa levis	3270	14.42	8.95	2.73	0.61	226.77	

(continued)

Table 3 (continued)

Species	Fiber length (μm)	Fiber diameter(μm)	Lumen diameter(μm)	Fiber wall thickness(μm)	Runkel ratio	Felting power	References
Gigantochloa scortechinii	3750	15.35	7.16	4.10	1.14	244.30	
Bambusa vulgaris	2153	17.31	12.66	2.327	0.38	124.38	[69]
Gigantochloa brang	1909.6	22.75	4.75	9.02	3.79	83.91	[70]
Gigantochloa levis	2039.9	22.67	4.00	9.34	4.67	89.99	
Gigantochloa scortechinii	1745.2	17.26	8.66	4.30	0.99	101.11	
Gigantochloa Wrayi	1798.79	17.86	3.83	7.02	3.65	100.72	
Gigantochloa scortechinii–3.5 yr	2500	26	8	9	2.25	96.2	[67]
Gigantochloa scortechinii_	4400	28.44	14.47	6.98	0.96	154.71	[71]
2 year and above							
Bambusa vulgaris	2029	13.95	12.19	0.896	0.147	145.45	[69]
Bambusa vulgaris	3600	16.9	2.5	7.1	5.68	213.02	[64]
Gigantochloa scortechinii	3800	17.3	2.6	7.4	5.69	223.53	[66]
Gigantochloa scortechinii	3780	30.82	11.79	9.51	1.61	122.65	[71]
1 year and above							
Bambusa vulgaris	2980	12.61	6.28	3.16	1.00	236.32	[60]
Gigantochloa levis	3090	14.38	8.96	2.71	0.60	214.88	

(continued)

Table 3 (continued)

Species	Fiber length (µm)	Fiber diameter(µm)	Lumen diameter(µm)	Fiber wall thickness(µm)	Runkel ratio	Felting power	References
Gigantochloa scortechinii	3350	13.90	7.22	3.34	0.925	241.00	
Gigantochloa scortechinii–1.5 yr	2630	26.0	9	9	2.0	101.15	[67]
Gigantochloa scortechinii	3480	32.41	9.61	11.39	2.37	107.37	[71]
0.5 year and above							
Gigantochloa scortechinii	2230	26	10	8	1.6	85.77	[67]

for Gigantochloa scortechcinii and Schzotachyum brachycladum. Species Schzotachyum grande done by 3 researchers and other species only 1 researcher. So, for 4 years old bamboo only Bambusa vulgaris is a favorite species done for morphology.

Data for 4 years old species shows all species have long type fiber range from 2140 mm Bambusa vulgaris by Sadiku et al. [69] to 4400 mm for Gigantochloa Scortechinii by Razak et al. [66]. According to Husna et al. [80], the fiber produces from Bamboo can be categories as long fiber because its length is more than 2000 mm compare to hardwood fiber. Long fiber is important in pulp and paper for producing high paper tearing and good tensile strength [72]. So, all bamboo species in this review are suitable for the pulp and paper industry.

There are 5 researchers done morphology checks for 3 years old bamboo species. Gigantochloa Scortechinii becomes a favorite for 4 researchers to investigate. Follow by Bambusa vulgaris and Gigantochloa brang and wrayi. Only Tamizi et al. [70] shows the fiber length below 2000 mm. Others bamboo fiber by 4 researchers categorizes into long fiber that is the longer produce by Mohmod et al. [71], about 4400 mm for Gigantochloa scortechinii.

Bambusa vulgaris and Gigantochloa scortechinii still become a favorite species for 2 years old bamboo morphology. Four researchers have done for these two species and long fiber produce from this species range from 2029 mm by Sadiku et al. [69] to 3810 mm by Mohmod et al. [71]. Gigantochloa scortechinii is still famous for research at 1-year-old planting and done by 3 researchers. All fiber is still in categories long fiber by range from 2630 to 3480 mm. Only Hisham et al. [67] done fiber morphology for 0.5 years old and the long fiber produce for 2230 mm by Gigantochloa scortechinii species.

The overall length of bamboo fiber produce by 13 researchers shows the majority of long fiber and can promise to produce strong paper in Malaysia pulp and paper industry. So, from this, the paper produces bulky and poor fiber bonding. Despite that, the bulky paper also has the potential to become a valuable product like molding pulp products. The thickness of paper functions to absorb the energy when transportation dynamic or drop.

A stronger and smoother sheet can be produced if the thin fiber wall uses in papermaking. This is due to a thin fiber wall makes its fibers bonding to each other easily and fix [73, 74]. The Runkel ratio and Felting power are used to measure the suitability of fiber in papermaking.

Runkel ratio is used to predict the suitability of a material for paper production, and it can be derived from the thickness of the cell wall and lumen diameter (2 $\times$ CWT/LD) [37, 57]. Meanwhile, measuring the strength of fiber bonding in the paper depends on Felting Power (FL/FD) [60, 70]. Figures 7 and 8 shows the Runkel ratio and Felting power (slenderness ratio) for 4 years old and above by 10 researchers.

Runkel ratio for 4 years fiber from Suhaimi et al. [60] measures that all fiber is more than 1. Siam et al. [61] measures Gigantochloa scortechiniii and Scizotachyum zolingeri have the potential to become a good material for pulp and paper. Bambusa vulgaris fiber by Sadiku et al. [38] and Gigantochloa atter by Marsoem et al. [63] also can become a good material after show the Runkel ratio value range between 0.53 to 0.84. Wahab et al. [64], Nordahlia et al. [65], Razak et al. [66], Hisham et al.

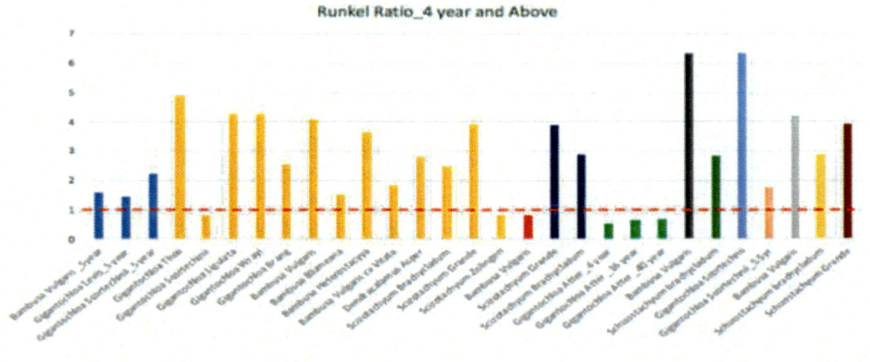

Fig. 7 Runkel ratio by 10 researchers

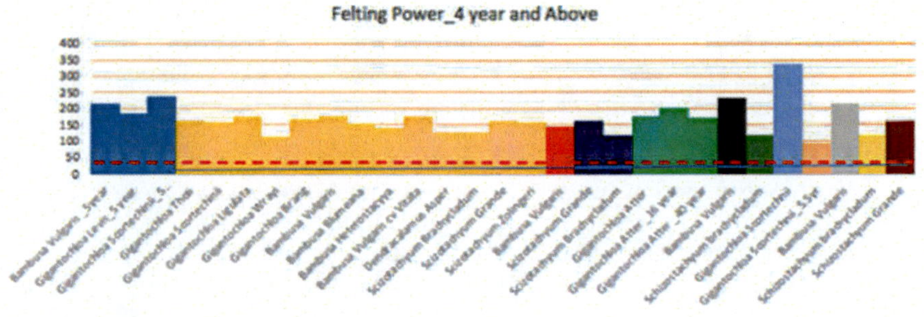

Fig. 8 Felting power by 10 researchers

[67], and Kassim et al. [68] show over than 1 and maybe cannot produce good paper strength.

Gigantochloa levis by Suhaimi et al. [60], Bambusa vulgaris by Sadiku et al. [69], and Gigantochloa scortechinii by Tamizi et al. [70] and Mohmod et al. [71] show Runkel ratio below than 1 for 3 years old bamboo at Fig. 9. These 3 species of fiber are suitable for produces strong paper. Bambusa vulgaris by Sadiku et al. [69] is suitable for making paper for 2 years old planting bamboo. Gigantochloa scortechinii and levis for 1-year-old by Suhaimi et al. [60] show Runkel ratio below 1. Gigantochloa scortechinii by Hisham et al. [67] and Mohmod et al. [71] show Runkel ratio more than 1 and not suitable for making g strong paper.

The value of the Runkel ratio must be less than 1 to be good material for pulp and paper application [57]. High Runkel ratio will produce bulkier paper because fiber stiff and less flexible, can produce poor bonding ability.

Felting power or slenderness ratio by all researchers and years show the value of more than 33 and suitable to produce high tearing resistance paper. This ratio is important for the pulp and paper industry making to make sure strong paper produce. This value match with the finding on previous paper [57, 60, 75]. So, all species are suitable to use as a new fiber for pulp and paper. The type of paper products will

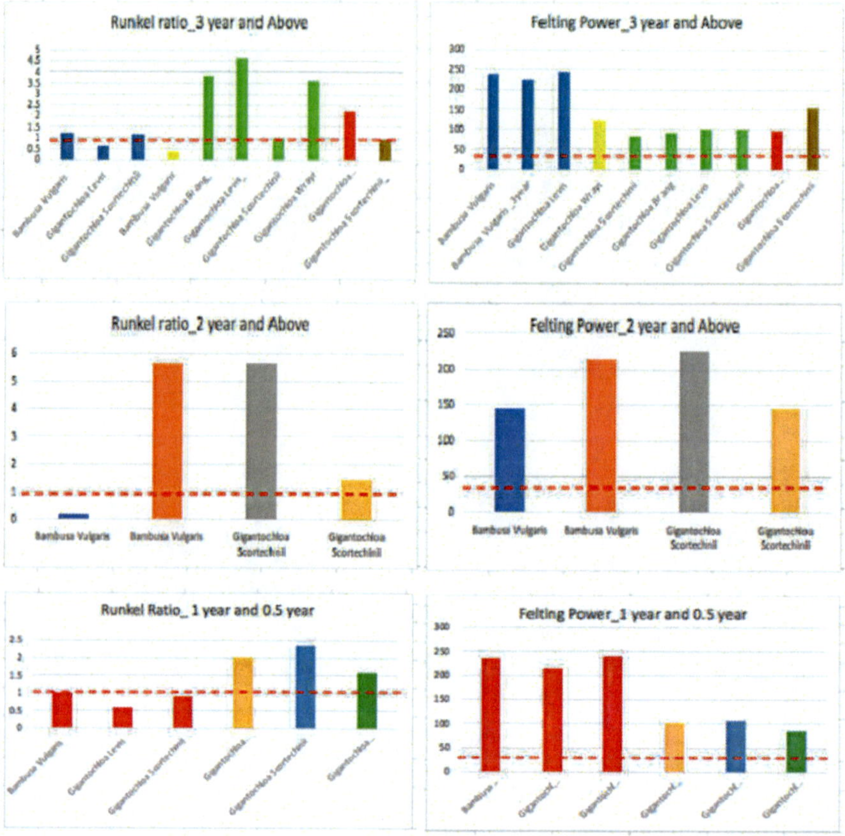

Fig. 9 Runkel ratio and Felting power for 3, 2, 1 and 0.5 years of fiber

have a high tearing resistance because of the low stiffness and high fiber flexibility. The investigation for other species should be recommended for understanding the suitability and adding a new source of raw material for pulp and paper.

3 Chemical Composition

The process for papermaking and the suitability of new fiber for paper also depends on chemical composition. The chemical constituents of a bamboo culm are very complicated. Like wood and agricultural residue, bamboo is mainly composed of cellulose, hemicelluloses, and lignin, even though the contents of these compositions are different.

There are 8 researchers investigate chemical characterization for bamboo fiber in Malaysia. The review starts from 1994 until 2021. The focus species are Gigantochloa albociliata, Bambusa Vulgaris, Gigantochloa scortechinii, wrayi and brang. From review data, majority researcher focuses on 4- and 3-years old planting bamboo. For 2- and 1-years bamboo, only 2 researchers focused on Bambusa vulgaris and Gigantochloa scortechinii only. The data for chemical composition shows in Table 4.

Gigantochloa albociliata, scortechinii and Bambusa vulgaris become favorite species for research at 4 and above years old bamboo. From the table, these three species have content for alpha-cellulose range between 57.33 to 67.62%. Gigantochloa albociliata have low alpha-cellulose by Yusoff et al. [77] and Bambusa vulgaris by Sadiku et al. [69] have high cellulose content. The ash content for 4 years old bamboo ranges from 1.59 to 3.5 still acceptable for the pulp and paper industry but needs some extra effort at material handling and washing operation. The lignin content ranges from 22.7 to 29.24% can consider high and maybe can cause challenges in the cooking process, hardness, and bleaching. The high content of Holocellulose range from 57.33 to 82.3% may be a good signal for the pulp and paper industry that bamboo fiber has the potential for replacing the hardwood fiber.

Three years old bamboo same as 4 years but Wahab et al. [78] had advanced the research to the species Gigantochloa wrayi, levis, and brang. 6 researchers investigate the fiber and mainly focus on Bambusa vulgaris and Gigantochloa scortechinii. Holocellulose content from 53.18 to 84.53% shows all the species suitable for replacing the current fiber in industry. Alpha cellulose content ranges from 41.1 to 67.07% is quite suitable for pulp and paper. However, Gigantochloa levis by Wahab et al. [78] may be the only fiber below the standard. The ash and Lignin content same as 4 years and can consider the same challenge in the process.

For 2, 1, and 0.5 years, bamboo fiber shows holocellulose content between 66.7 to 80.1%. The species focus only on Gigantochloa scortechinii and Bambusa vulgaris. The alpha-cellulose content ranges from 40.7 to 75.68% is quite suitable for pulp and paper. This range is acceptable for the pulp and paper process. The ash and lignin content are still the same as the 3- and 4-years content and may challenge the manufacturing process.

Overall, all species suitable for the pulp and paper industry cause chemical content of over 34% of alpha cellulose [57, 69, 72, 76].

Table 4 Chemical composition for bamboo species by 7 researchers

Species	Extractive (%)	Extractive content _cold eater (%)	Extractive content _hot eater (%)	Holo cellulose (%)	Alpha cellulose (%)	Hemi cellulose (%) (Pentosan)	Ash (%)	Lignin (%)	References
4 year and above									
Gigantochloa albociliata-5 yr	–	–	9.71	57.33	43.12	–	1.59	24.67	[77]
Bambusa vulgaris	–	4.27	7.89	79.01	67.62	11.41	2.86	29.24	[69]
Bambusa vulgaris		8.5	10.4	77.4	55.4	–	2.7	22.7	[68]
Gigantochloa scortohcinii–6.5 yr	5.6	–	–	82.3	64.4	–	3.5	29	[67]
Gigantochloa scortohcinii–5.5 yr	3.5	–	–	81.5	63.3	–	3.0	28.7	
3 year and above									
Gigantochloa albociliata	–	–	12.04	53.18	42.98	–	1.87	22.82	[77]
Bambusa vulgaris	–	3.83	9.28	80.62	67.07	13.55	2.52	36.40	[38]
Gigantochloa brang	8.30	–	–	79.70	51.58	–	1.26	24.83	[78]
Gigantochloa levis	9.23	–	–	84.52	33.81	–	1.30	26.50	

(continued)

Table 4 (continued)

Species	Extractive (%)	Extractive content _cold eater (%)	Extractive content _hot eater (%)	Holo cellulose (%)	Alpha cellulose (%)	Hemi cellulose (%) (Pentosan)	Ash (%)	Lignin (%)	References
Gigantochloa scortechinii	8.00	–	–	74.63	46.87	–	2.84	32.55	
Gigantochloa Wrayi	8.62	–	–	84.53	37.66	–	2.84	32.55	
Gigantochloa scortechinii	–	–	–	81.4	55.2	26.2	–	22.3	[79]
Gigantochloa scortechinii	5.3	–	–	80.6	64.6	–	2.8	27.8	[67]
Gigantochloa scortechinii	–	5.3	5.4	68.0	41.1	–	1.14	28.0	[71]
2 year and above									
Bambusa vulgaris	–	5.21	9.03	79.81	75.68	4.13	2.36	45.90	[38]
Gigantochloa scortechinii	–	4.4	5.9	67.8	41.41	–	1.11	24.9	[71]
1 year and above									
Gigantochloa scortechinii–1.5 yr	3.4	–	–	80.1	64.1	–	2.5	26.8	[67]
Gigantochloa scortechinii	–	4.3	6.3	66.7	40.7	–	1.10	25.73	[71]
0.5 year and above									

(continued)

Table 4 (continued)

Species	Extractive (%)	Extractive content _cold eater (%)	Extractive content _hot eater (%)	Holo cellulose (%)	Alpha cellulose (%)	Hemi cellulose (%) (Pentosan)	Ash (%)	Lignin (%)	References
Gigantochloa scortechinii	5.8	–	–	78.6	64.6	–	1.9	23.4	[67]

4 Conclusion

According to previous research data, all bamboo species are suitable for the pulp and paper industry at different years. The favorite species are Bambusa vulgaris and Gigantochloa scortechinii and show acceptable morphology value and good chemical content. The challenge may be coming from the value of the Runkel ratio and will affect paper making process. Nevertheless, besides paper, bamboo is an alternative for other products like paperboard, composite, and molded pulp product, as can be seen from the morphology data. The data show, bulky paper can be produced with high bonding inter fiber. These two features suitable for others paper product that important in high thickness, high tension strength, and can react to energy absorption parameter. Bambusa vulgaris and Gigantochloa scortechinii maybe can be a promising material to fulfill the requirement. So, bamboo fiber can be a replacement for current hardwood fiber. Besides that, 5 other genera or type of bamboo in Malaysia still do not investigate the potential and can be a focus for the researcher to multiple the source of long fiber choice. Overall, bamboo species from Gigantocholoa and Bambusa are suitable for the pulp and paper industry according to morphology, chemical composition, and source of bamboo plants in Malaysia.

Acknowledgements The authors express their appreciation to University Tun Hussein Onn Malaysia through Internal University Grant (TIER 1-Vot H795) for providing the financial and Bamboo Research Center for reference information.

References

1. Setiawan PA (2020) Augmented reality as potential alternative media for paperless advertising. Bus Econ Commun Soc Sci J 2(1):89–96. https://doi.org/10.21512/becossjournal.v2i1.6077
2. Lantto R (2001) 8 international conference on biotechnology in the pulp and paper, in prevention, pp 15–16
3. Main NM, Talib RA, Ibrahim R, Rahman RA, Mohamed AZ (2014) Suitability of coir fibers as pulp and paper. Agric Agric Sci Procedia 2:304–311. https://doi.org/10.1016/j.aaspro.2014.11.043
4. Al Sadoon GM, Makki HA, Saleh AR (2017) Green computing system, health and secure environment management system. In: 4th IEEE international conference on engineering technology and applied science, ICETAS 2017, vol 2018-Janua, pp 1–6. https://doi.org/10.1109/ICETAS.2017.8277847
5. Nayyar N, Arora S (2019) Paperless technology - a solution to global warming. In: 2019 2nd international conference on power energy environment and intelligent control, PEEIC 2019, pp 486–488. https://doi.org/10.1109/PEEIC47157.2019.8976599
6. Bajpai P (2015) Paper and paperboard industry. Pulp and Paper Industry. Elsevier, Amsterdam, pp 13–24. https://doi.org/10.1016/B978-0-12-803408-8.00002-0
7. Johansson MT, Broberg S, Ottosson M (2021) Energy strategies in the pulp and paper industry in Sweden: interactions between efficient resource utilisation and increased product diversification. J Clean Prod 311:127681. https://doi.org/10.1016/j.jclepro.2021.127681
8. Del M, Palenzuela V, Eugenia M, Martín E, Pascual AR, Raya AR (2020) Evaluation of the potential of alternative vegetable materials for production of paper through kraft processes. Cellul Chem Technol 54:73–81

9. Liu C et al (2020) Biodegradable, hygienic, and compostable tableware from hybrid sugarcane and bamboo fibers as plastic alternative. Matter 3(6):2066–2079. https://doi.org/10.1016/j.matt.2020.10.004
10. Su Y et al (2018) Prospects for replacement of some plastics in packaging with lignocellulose materials: a brief review. BioResources 13(2):4550–4576. https://doi.org/10.15376/biores.13.2.su
11. TrendEconomy (2019). Annual International Trade Statistics by Country. https://trendeconomy.com/data/h2/Malaysia/47, Accessed 10 June 2021
12. Omar NN (2008) Kenaf as a new raw material in pulp and paper industries in Malaysia. In: Intropica, UPM. https://introp.upm.edu.my/upload/dokumen/201809131055453._yat.pdf.
13. Clark LG, Londoño X, Ruiz-Sanchez E (2015) Bamboo taxonomy and habitat, pp 1–30. https://doi.org/10.1007/978-3-319-14133-6_1.
14. Ohrnberger D (1999) The bamboos of the world: annotated nomenclature and literature of the species. Choice Rev Online 37(02):37-0926-37–0926. https://doi.org/10.5860/choice.37-0926
15. Fei B, Gao Z, Wang J, Liu Z (2016) Biological, anatomical, and chemical characteristics of bamboo. Secondary Xylem Biology. Elsevier, pp 283–306. https://doi.org/10.1016/B978-0-12-802185-9.00014-0
16. Jawaid M (2012) Bamboo: versatile material for polymer composite. Bamboo World. 7:2–3
17. Lobovikov M, Paudel S, Piazza M, Ren H, Wu J (2007) World bamboo resources. In: Inbar, p 73
18. Nguyen TBV (2018) Bamboo - the eco-friendly material - one of the material solutions of the sustainable interior design in Viet Nam. MATEC Web Conf 193:04014. https://doi.org/10.1051/matecconf/201819304014
19. Dayton DC, Scurlock BHJMO (2000) Bamboo: an overlooked biomass resource ? Biomass Bioenergy 19(2000):229–244
20. Mera FAT, Xu C (2014) Plantation management and bamboo resource economics in china. Cienc y Tecnol 7(1):1–12. https://doi.org/10.18779/cyt.v7i1.93
21. Shanmughavel P, Francis K (1996) Above ground biomass production and nutrient distribution in growing bamboo (Bambusa bambos (L.) Voss). Biomass Bioenergy 10(5–6):383–391. https://doi.org/10.1016/0961-9534(95)00124-7
22. Bhatt BP, Singha LB, Singh K, Sachan MS (2003) Some commercial edible bamboo species of North East India: production, indigenous uses, cost-benefit and management strategies. Bamboo Sci Cult 17(1):4–20
23. Makita A (1998) The significance of the mode of clonal growth in the life history of bamboos. Plant Species Biol 13(2–3):85–92. https://doi.org/10.1111/j.1442-1984.1998.tb00251.x
24. Forestry Department of peninsular Malaysia (2016) Rattan and bamboo. Official website forestry department peninsular Malaysia. https://www.forestry.gov.my/index.php/en/buluh-dan-rotan.
25. Imadi SR, Kazi AG (2015) Agricultural biomass based potential materials. In: Extraction of Lignin from Biomass for Biofuel Production, 1st edn. Springer International Publishing, Heidelberg
26. Khan KK, Hemalatha E (2015) A review on the genus bambusa and one particular species bambusa vulgaris in sabah (Malaysia). Int Res J Pharm 6(9):580–584. https://doi.org/10.7897/2230-8407.069114
27. Prem Kumar V, Vasugi V (2020) Bamboo materials in cement, geopolymer and reinforced concrete as sustainable solutions for better tomorrow. IOP Conf Ser Earth Environ Sci 573(1):012036. https://doi.org/10.1088/1755-1315/573/1/012036
28. Singh G, Richa MLS (2017) Bamboo - a miracle plant. Int J Curr Res Biosci Plant Biol 4(1):110–112. https://doi.org/10.20546/ijcrbp.2017.401.013
29. Lias H, Ismail AR, Jusoh N, Abd Hamid H (2020) Technological development of bamboo for "batik" stamp in malaysian textile industries. IOP Conf Ser Earth Environ Sci 616(1):012037. https://doi.org/10.1088/1755-1315/616/1/012037
30. Liese W, Kohl M (2015) Bamboo: The plant and its uses, 1st edn. Springer, Heidelberg

31. Azmy HM, Razak W, Hashim WS, Sulaiman O (2004) Application of organic fertilisers on natural stand bamboos for sustainable management in peninsular Malaysia. J Bamboo Ratt 3(4):309–317. https://doi.org/10.1163/1569159042464617
32. Chaowana P (2013) Bamboo: an alternative raw material for wood and wood-based composites. J Mater Sci Res 2(2):90. https://doi.org/10.5539/jmsr.v2n2p90
33. Awalluddin D et al (2017) Mechanical properties of different bamboo species. MATEC Web Conf 138:01024. https://doi.org/10.1051/matecconf/201713801024
34. Wahab R, Mustafa MT, Salam MA, Tabert TA, Sulaiman O, Sudin M (2012) Potential and structural variation of some selected cultivated bamboo species in Peninsular Malaysia. Int J Biol 4(3):102–116. https://doi.org/10.5539/ijb.v4n3p102
35. Liese HW (1987) Research on Bamboo. Wood Sci Technol 27:189–209
36. Mohamed A, Abdullah N, Halis R, Wan Abd Kadir W (2009) Early performance trial of four Malaysian commercial bamboos in Shouthern Peninsular Malaysia. Borneo Sci 25:81–86
37. Cao S, Ma X, Lin L, Huang F, Huang L, Chen L (2014) Morphological and chemical characterization of green bamboo (dendrocalamopsis oldhami (munro) keng f.) for dissolving pulp production. BioResources 93:4528–4539. https://doi.org/10.15376/biores.9.3.4528-4539
38. Sadiku AOO (2016) Nusirat, "Fibre dimension and chemical characterisation of naturally grown Bambusa vulgaris for pulp and paper production." J. Bamboo Ratt 15:33–43
39. Sugesty S, Kardiansyah T, Hardiani H (2015) Bamboo as raw materials for dissolving pulp with environmental friendly technology for rayon fiber. Procedia Chem. https://doi.org/10.1016/j.proche.2015.12.122
40. Gupta A, Kumar A (2008) Potential of bamboo in sustainable development. Asia Pacific Bus Rev 4(3):100–107. https://doi.org/10.1177/097324700800400312
41. Harian BB (2019) Kerajaan sedia pinjaman mudah galak penanaman buluh - Kok Watch Takes Malaysia by Storm
42. Harian BB (2020) Industri buluh antara tumpuan kerajaan dalam RMK-12, pp 10–13
43. MPIC (2021) Kenyataan Media: Penghasilan Buluh Pemangkin Program Pertanian Komuniti Buluh 2021–2025. https://www.mpic.gov.my/mpi/info-siaran-media/siaran-media-2021/siaran-media-30-mac-2021, Accessed 29 June 2021
44. Forest Department of Peninsular Malaysia (2021) Policy and Legislation Industry Related Bamboo in Peninsula Malaysia. Official website of the forestry department of peninsular Malaysia. https://www.forestry.gov.my/my/perkhidmatan/info-perhutanan/kesalahan-hutan, Accessed 22 Apr 2021
45. Barbosa JC, Morales E, Antonio F, Lahr R (2015) Bamboo Particulate waste-Production of High Performance Structurals Panel, 1st edn. SP, Bauru
46. Ibrahim FH, Tombuloglu H, Hakeem KR, Ibrahim S (2015) Bamboo biomass: various studies and potential applications for value-added products. In: Ibrahim FH, Tombuloglu H, Hakeem KR, Ibrahim S (eds) Agricultural Biomass Based Potential Materials. Springer, Switzerland, pp 231–244
47. Anokye R, Bakar ES, Ratnasingam J, Awang BK (2016) Bamboo properties and suitability as a replacement for wood. PJSRR Pertanika J Sch Res Rev 2(1):63–79
48. Zhang M, Chen S, Jiang H, Cao Q (2020) The water transport profile of Phyllostachys edulis during the explosive growth phase of bamboo shoots. Global Ecol Cons 24:e01251. https://doi.org/10.1016/j.gecco.2020.e01251
49. Long C, Liu Y, Xue J, Li D (2003) A bamboo germplasm collection for community development in Central Yunnan, China. J Bamboo Ratt 2(1):3–11. https://doi.org/10.1163/156915903321908512
50. Liese W, Tang TKH (2015) Properties of the bamboo culm. In: Bamboo, pp 227–256
51. Wang CZ, Zhang HY, Li WJ, Ye JZ (2015) Chemical constituents and structural characterization of polysaccharides from four typical bamboo species leaves. Molecules 20(3):4162–4179. https://doi.org/10.3390/molecules20034162
52. Sundriyal RC, Upreti TC, Varuni R (2002) Bamboo and cane resource utilisation and conservation in the Apatani plateau, Arunachal Pradesh, India: implications for management. J Bamboo Ratt 1(3):205–246. https://doi.org/10.1163/156915902760184277

53. Frías M, Savastano H, Villar E, Sánchez De Rojas MI, Santos S (2012) Characterization and properties of blended cement matrices containing activated bamboo leaf wastes. Cem Concr Compos 34(9):1019–1023. https://doi.org/10.1016/j.cemconcomp.2012.05.005
54. Biswas P, Chakraborty S, Dutta S, Pal A, Das M (2016) Bamboo flowering from the perspective of comparative genomics and transcriptomics. Front Plant Sci 7:1900. https://doi.org/10.3389/fpls.2016.01900
55. Rai V, Dey N (2012) Identification of programmed cell death related genes in bamboo. Gene 497(2):243–248. https://doi.org/10.1016/j.gene.2012.01.018
56. Takeuchi R et al (2016) Wood properties related to pulp and paper quality in two *Macaranga* species naturally regenerated in secondary forests, Central Kalimantan, Indonesia. Tropics 25(3):107–115. https://doi.org/10.3759/tropics.ms15-23
57. Bidin N, Zakaria MH, Bujang JS, Aziz NAA (2015) Suitability of aquatic plant fibers for handmade papermaking. Int J Polymer Sci 2015:1–9. https://doi.org/10.1155/2015/165868
58. Ali MM, Muhadi NA, Hashim N, Abdullah AF, Mahadi MR (2020) Pulp and paper production from oil palm empty fruit bunches: a current direction in Malaysia. J Agric Food Eng 1(2):1–9. https://doi.org/10.37865/jafe.2020.0017
59. Chen Z, Zhang H, He Z, Zhang L, Yue X (2019) Bamboo as an emerging resource for worldwide pulping and papermaking. BioResources 14(1):3–5. https://doi.org/10.15376/biores.14.1.3-5
60. Suhaimi NM, Hassan NHM, Ibrahim R, Jasmani L (2021) Fiber morphology of different bamboo species and age. Mater Sci Forum 1025:312–318. https://doi.org/10.4028/www.scientific.net/MSF.1025.312
61. Siam NA, Uyup MKA, Husain H, Mohmod AL, Awalludin MF (2019) Anatomical, physical, and mechanical properties of thirteen Malaysian bamboo species. BioResources 14(2):3925–3943. https://doi.org/10.15376/biores.14.2.3925-3943
62. Siam NA, Uyup MKA, Husain H, Mustafa MT (2015) Selected properties of Malaysian Bamboo Schizostachyum grande (Buluh Semeliang) and Schizostachyum brachycladum (Buluh Lemang). In: 7th international symposium Indonesia, wood research society, pp 8–15
63. Marsoem SN et al (2015) Fiber morphology and physical characteristics of Gigantochloa atter at three different ages and heights of culms for better utilization. J Korean Wood Sci Technol 43(2):145–155. https://doi.org/10.5658/WOOD.2015.43.2.145
64. Wahab R, Mustapa MT, Sulaiman O, Mohamed A, Hassan A, Khalid I (2010) Anatomical and physical properties of cultivated two- and four-year-old bambusa vulgaris. Sains Malaysiana 39(4):571–579
65. Nordahlia AS, Anwar UMK, Hamdan H, Latif MA, Mahanim SMA (2011) Anatomical, physical and strength properties of Shizostachyum brachycladum Buluh lemang. J Bamboo Ratt 10(3–4):111–122
66. Razak W, Janshah M, Hashim WS, Shirley B (2007) Morphological and anatomical characteristics of managed natural bamboo stands - Gigantochloa scortechinii. J Bamboo Ratt 6(1–2):115–121
67. Hisham HN, Othman S, Rokiah H, Latif MA, Ani S, Tamizi MM (2006) Characterization of bamboo gigantochloa scortechinii at different ages. J Trop For Sci 18(4):236–242
68. Kasim J, Ahmad AJ (1994) A note on the proximate chemical composition and fiber morphology of Bambusa vulgaris. J Trop For Sci 6(3):356–358
69. Sadiku NA, Oluyege AO, Ajayi B (2016) Fibre dimension and chemical charcterisation of naturally grown bambusavulgaris for pulp and paper production. J Bamboo Ratt 15:33–43
70. Tamizi MM, Razak W, Sudin M, Khalid I, Kamal NAM (2011) Anatomical properties and microstructures features of four cultivated bamboo gigantochloa species abstract. J Asian Sci Res 1(7):328–339
71. Mohmod AL, Khoo K, Kasim J, Ahmad AJ (1994) Fiber morphology and chemical properties of Gigantochloa scortechinii. J Trop For/Sci 6(4):397–407
72. Ververis C, Georghiou K, Christodoulakis N, Santas P, Santas R (2004) Fiber dimensions, lignin and cellulose content of various plant materials and their suitability for paper production. Ind Crops Prod 19(3):245–254. https://doi.org/10.1016/j.indcrop.2003.10.006

73. Masrol SR et al (2018) Durian rind soda-anthraquinone pulp and paper: effects of elemental chlorine-free bleaching and beating. J Trop For Sci 30(1):106–116. https://doi.org/10.26525/jtfs2018.30.1.106116
74. Main NM, Talib RA, Ibrahim R, Rahman RA, Mohamed AZ (2015) Linerboard made from soda-anthraquinone (Soda-AQ) treated coconut coir fiber and effect of pulp beating. BioResources 10(4):6975–6992. https://doi.org/10.15376/biores.10.4.6975-6992
75. Khan MZH, Sarkar MAR, Imam FIA, Zahid M, Khan H, Malinen RO (2014) Paper making from banana pseudo-stem: characterization and comparison. J Nat Fib 11(3):199–211. https://doi.org/10.1080/15440478.2013.874962
76. Daud Z, Hatta MZM, Kassi ASM, Aripi AM (2014) Analysis of the chemical compositions and fiber morphology of pineapple (ananas comosus) leaves in Malaysia. J Appl Sci 14(12):1355–1358. https://doi.org/10.3923/jas.2014.1355.1358
77. Yusoff MSR, Bahari SA, Haliffuddin RMAA, Zakaria MN, Jamaluddin MA, Rashid NHMN (2021) Chemical contents and thermal stability of Madu bamboo (Gigantochloa albociliata) for natural-bonded fiber composites. IOP Conf Ser Earth Environ Sci 644(1):012009. https://doi.org/10.1088/1755-1315/644/1/012009
78. Wahab R, Mustafa MT, Salam MA, Sudin M, Samsi HW, Rasat MSM (2013) Chemical composition of four cultivated tropical bamboo in genus gigantochloa. J Agric Sci 5(8):66–75. https://doi.org/10.5539/jas.v5n8p66
79. Salim HWSR, Wahab R, Ashaari Z (2009) Chemical constituents of oil cured tropical bamboo gigantochloa scortechinii. J Appl Sci 9(1):149–154
80. Husna N, Mohd B, Ibrahim R (2015) Properties of corrugated paper from recycled paper blended with semantan bamboo pulp. Aust J Basic Appl Sci 9:113–117

Investigation of Curing Process of Silver Conductive Ink on Polymer Substrates Using Halogen Lamp and Oven

Muhammad Najmi Zainal, Rd Khairilhijra Khirotdin, Siti Nur Elida Eraman, Muhamad Fahrul Nizam Suhaimi, and Nurhafizzah Hassan

Abstract Conductive silver ink has lately become popular in various electrical applications. It often comes in liquid form and requires curing normally by heating to reveal its metallic components. The solvent residues in conductive ink are vaporized thoroughly after curing for a few hours thus prolonging the printing process. Conventional curing methods such as oven and laser curing are frequently used but these methods require longer curing time and sometimes costly hence a better alternative method is eminent. Halogen lamp has been seen generated faster higher temperature which resulting quicker curing process and potentially overcome the problems. Thus, this study is to evaluate the performance of halogen lamp as a comparison to oven in curing silver conductive ink on polymer substrate. The significant curing parameters were determined, and the impact of on resistance were analyzed. The relationship between the significant parameters including curing time and temperature has also been established which demonstrate the resistance of the ink track is proportionate to the length of curing time and how high the temperature is applied. When the temperature is raised, the time to cure the ink is lowered significantly thus producing better conductive properties and lower resistance value in the range of 1 and 2 Ω for both curing methods. This concludes that halogen lamp is capable of generating a comparable curing process and could be used as alternate method to properly cure conductive ink.

Keywords Curing · Halogen lamp · Oven · Conductive ink · Polymer substrate · Resistance · Time · Temperature

M. Najmi Zainal · R. Khairilhijra Khirotdin (✉) · S. Nur Elida Eraman · M. Fahrul Nizam Suhaimi
Additive Manufacturing Research Group, Faculty of Mechanical and Manufacturing Engineering, University of Tun Hussein Onn Malaysia, 86400 Parit Raja, Batu Pahat, Johor, Malaysia
e-mail: khairil@uthm.edu.my

N. Hassan
Department of Mechanical Engineering Technology, Faculty of Engineering Technology, 84200 Pagoh, Muar, Johor, Malaysia

© The Author(s), under exclusive license to Springer Nature Singapore Pte Ltd. 2023
M. H. A. Hassan et al. (eds.), *Technological Advancement in Instrumentation & Human Engineering*, Lecture Notes in Electrical Engineering 882,
https://doi.org/10.1007/978-981-19-1577-2_29

1 Introduction

Printed electronics is a technology that has been improving and progressing throughout this decade. This technology is used to create electrical devices on different types of substrates. The ink is the core of the technology because all final material properties as well as drawbacks are dictated by its chemistry. Different printing ink drying mechanisms are used useable, depending on the following factors: printing process, type of printing ink (liquid, paste, water or organic solvent-based, radiation curing ink, etc.), speed of printing machines, substrates of printing materials, type of printing product and characteristics of drying systems (hot air, curing systems, infrared drying) [1]. Silver is one of the widely known metals to be used in conductive ink applications due to the advantage of having the most conductive element (6.30×10^7 S/m at 20 °C), well-known for its thermal conductivity that can withstand in extreme temperature conditions, metal with very good reflectance, anti-bacterial nature and corrosion free ability [2]. The development of conductive inks is a complex task in which it is often necessary to take into account many factors. It is desirable that the inks be stable during storage and have low viscosity with high metal content. When printing on thermal sensitive polymeric materials, it is recommended to perform curing at low temperatures or to use other post-treatment techniques in order to remove organic components and to form a proper conductive layers without any damage [3].

Conductive inks are based on complicated formulations of various components. Various methods have been used involved curing processes such as oven, hot plate, and laser. Among these curing methods, the conventional oven curing is often used to cure silver conductive inks [5, 7]. Unfortunately, this method takes a very long time to pre-heat and cure at designated high temperature thus extending the time to cure the sample [1, 5, 7]. Besides, the limitation of this technique is that it might physically deform cheap and thin substrate as the close system of the oven cause overheating towards the substrates. On the other hand, using a laser to cure the conductive ink gives faster result [4] but the downside is it has cost more than other technologies where sophisticated equipment are required and needed proper skills to operate it [8]. Furthermore, the curing time by localized laser curing process is proportional to the total length and width of the traces which means that the longer the length of the sample, the longer the period of curing [5]. Microwave heating is not suitable for heating metals as metals are known to reflect microwaves and have a short penetration depth of a few microns [6]. Special precautions also needed to be taken when putting metal in a microwave to avoid the formation of electrical discharge that can damage the magnetron. Hot presses involving high internal stream pressures may damage the sample if not properly control [9] while hot air drying takes longer curing time despite producing high electrical conductivity [10].

The Halogen lamp is chosen to be a better replacement to cure silver conductive ink due to it is low in cost, shorter period of pre-heating time and has the ability to produce designated high temperature by controlling the input power without damaging the substrates [7]. The objective of the study is to evaluate the performance of cured ink track between halogen lamp and oven curing process. At the end of the study, a comparison between the halogen lamp and oven curing process was performed and analysed and likewise a correlation between the variation of curing time and temperature to resistivity was established and the capability of halogen lamp in reducing curing time without damaging the substrate was determined.

2 Methodology

A straight line of silver ink track having a length of 30 mm and a width of 0.8 mm is designed due to its simplicity as shown in Fig. 1a and the sample is printed via automated syringe-based deposition technique (Model: F4200N.1) shown in Fig. 2a. The technique is precise, fast, and capable of working with a broad range of materials, including conductive ink. The optimum printing parameters including printing speed and pressure as listed in Table 1 were utilized while other printing parameters were kept constant throughout the experiment.

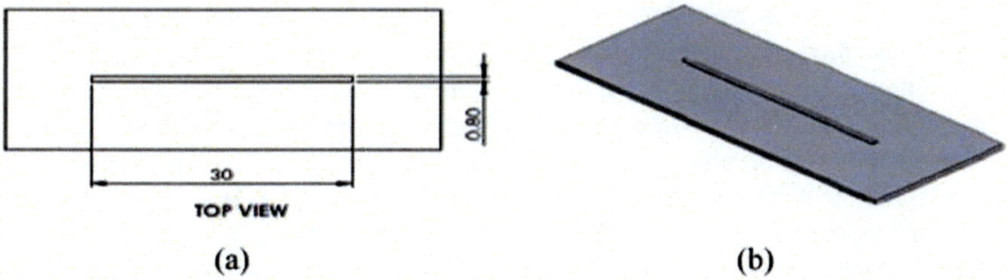

Fig. 1 A straight line ink track on a polymer substrate **a** Top view **b** Isometric view (unit in mm)

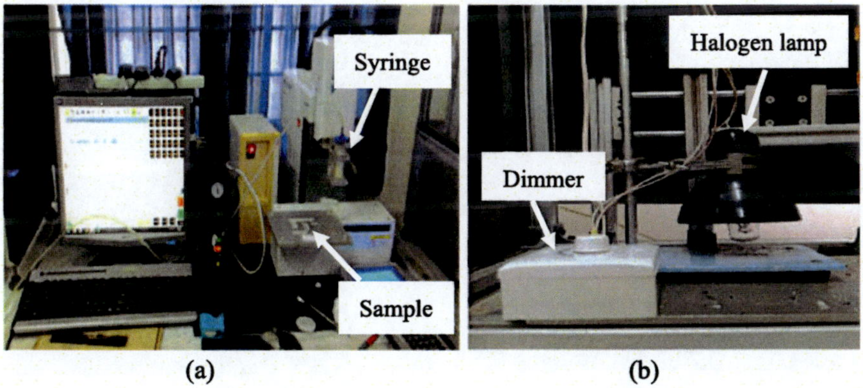

Fig. 2 **a** Automated syringe-based deposition **b** Halogen lamp curing system

Table 1 Printing and curing parameters

Printing parameter				
Pressure (kPa)	Speed (mm/s)	Deposition height (mm)	Nozzle diameter (mm)	Viscosity (Pa.s)
100.00	17.00	1.00	0.51	2.40
Curing parameter				
Temperature (°C)		Time (minute)		
110–190		10–90		

A curing process is necessary in removing solvents from the ink, allowing the metal content to be visible while also promoting adhesion to the polymer materials. The curing process was performed using the halogen lamp as shown in Fig. 2b and oven using the selected optimum curing parameters tabulated in Table 1. The curing temperatures were ranged between 110 and 190 °C and the curing times were varied in the range of 10 to 90 min. The process of curing with a halogen lamp was started by adjusting the distance between the lamp and the work area to get an optimal space for effective curing process. The temperature was measured using a Fluke thermal imager and it was discovered that the temperature acquired was directly related to the distance specified. A sufficient length of 1 cm between the halogen lamp and sample was maintained throughout the experiment. The analysis of the cured sample is done using a digital multi-meter to measure its electrical properties. The performance of the curing method is evaluated by analyzing the curing time and temperature applied on the conductive ink and the resultant electrical resistance of the printed ink track.

A total of 30 experimental runs were conducted which 15 runs for each oven curing and halogen lamp respectively. A polymer sheet via a Polyethylene Terephthalate (PET) transparency sheet was used as a substrate as depicted in Fig. 3a and the material used throughout this research is silver epoxy-based ink substance (Model:

(a) (b)

Fig. 3 **a** PET transparency sheet **b** Silver conductive ink (Model: AG 806)

AG806) as shown in Fig. 3b. Due to the high solid fraction of the ink, the viscosity was adjusted to the recommended viscosity (refer to Table 1) through dilution with toluene's solvent for ease of deposition process.

3 Result and Discussion

Figure 4a shows one of the successful printed sample after curing. The straight line track printed was observed to be continuous and smooth as expected. Morphological analysis was conducted to measure the physical characteristic of the silver conductive ink track using an optical microscope. The average width obtained for all samples was approximately 0.8 mm as predicted (refer Fig. 4b). A relationship was formed in which higher printing speed was resulted in thinner ink track when the printing pressure is kept constant. It was also observed that most of the ink track has a constant thickness proved that a smooth printing process was successfully conducted. The viscosity adjusted earlier played a vital role in giving a smooth printing process. It was also found that the printing parameters used were suitable to produce the required physical characteristic of the silver conductive ink track.

In the meantime, Table 2 shows that oven had an overall lower resistance values compared to halogen lamp where the lowest resistance value obtained was 1.03 Ω

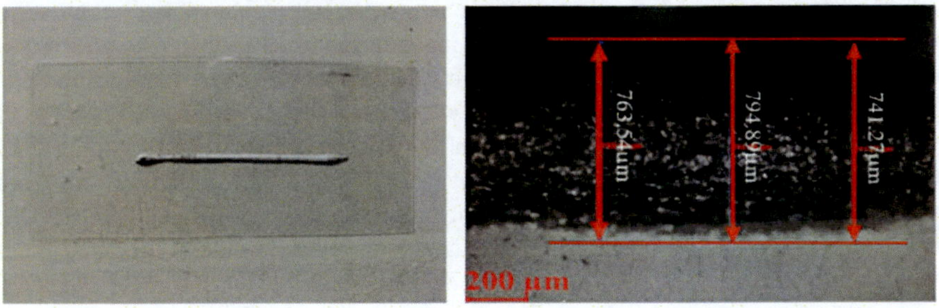

(a) (b)

Fig. 4 **a** One of the successful printed sample after curing **b** The width of the silver conductive ink track

Table 2 Average resistance value between halogen lamp and oven

Experiment	Temperature (°C)	Time (minute)	Average width (mm)		Average resistance (Ω)	
			Halogen lamp	Oven	Halogen lamp	Oven
1	110	90	0.801	0.796	2.13	1.73
2	130	70	0.775	0.777	2.27	1.47
3	150	50	0.811	0.732	2.00	1.40
4	170	30	0.753	0.778	1.70	1.07
5	190	10	0.792	0.760	1.60	1.03

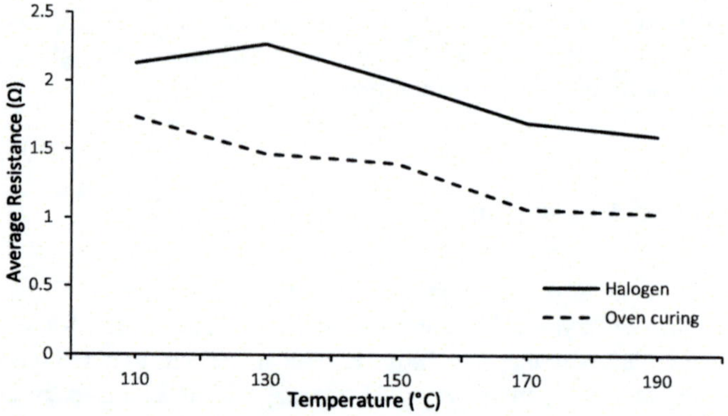

Fig. 5 Graph comparison between the changes of average resistance value against temperature between halogen lamp and oven curing process

while for halogen, it was 1.60 Ω. When the temperature was set at 110 °C, the highest resistance value obtained for oven was at 1.73 Ω while halogen lamp recorded at 2.13 Ω at 110 °C.

Figure 5 represents the variation of resistance transition between inks cured with halogen lamp and oven as a function of curing time and temperature. Both results show that the resistance value decreased as the temperature was increased from 110 to 190 °C. The time for curing process was also becoming shorter as higher temperature was set with decreasing resistance value. Longer curing time tends to lower the initial resistance value of silver conductive ink track [12]. It was observed that the trend line for halogen lamp shared the same pattern with oven curing, albeit slightly higher. It was evident that for halogen lamp, 90 min of curing time was required to get 2.13 Ω when the curing temperature set was at 190 °C while 50, 30 and 10 min were needed to obtain 2.0, 1.7 and 1.6 Ω with the temperature set at 150, 170 and 190 °C respectively. This shows the resistance value decreased as higher temperature is set with shorter curing time. The resistance value for the halogen lamp was observed to be in the range of 1 to 2 Ω similar with the oven curing range. This concludes that halogen lamp was capable of becoming a proper alternate curing process and comparable to oven curing process. Previous study has shown that post thermal treatment after the standard printing process can improve the electrical conductivity by reducing the resistance of printed devices. Higher curing temperature led to a more dramatic resistance drop, especially within the first hour of heat treatment [13].

Besides, the pre-heat process is needed for most of curing methods and this is also applied when utilizing halogen lamp and oven. It was observed previously that halogen lamp is capable of reaching higher temperature faster than oven during the pre-heating process [7] as shown in Fig. 6. Additionally, the temperature of the halogen lamp is also proportional to the distance between the halogen lamp and the sample therefore a proper temperature to cure the ink track can be achieved by

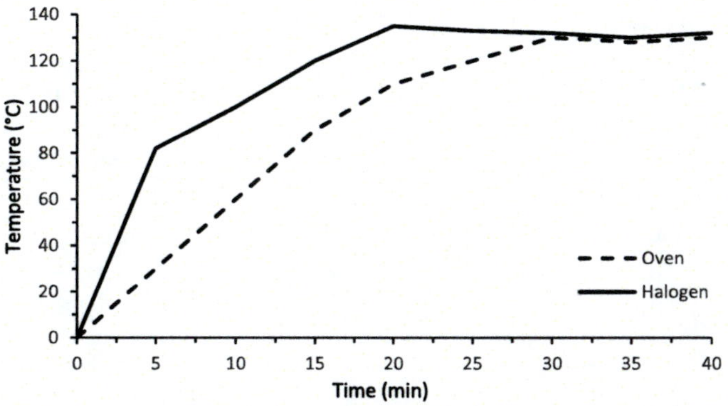

Fig. 6 Comparison between of pre-heat time between halogen (at 1 cm) and oven [7]

properly set the sample as close as possible to the lamp. Furthermore, the substrates used in experiment were not badly damaged at a high temperature of 190 °C.

The hardness test was conducted by utilizing the micro-hardness tester. A minimum and maximum load were identified first in which the minimum load of 980.7 mN (HV0.1) and 196.1 mN (HV2) were selected and applied. It was done to ensure a required diagonal shape mark is presented on the surface of the ink track through the use of indenter. It was found out that the suitable test load was 980.7 mN (HV0.1) as diagonal shape was presence on the ink track during testing as shown in Fig. 7a. Table 3 shows the overall result of the average hardness level while Fig. 8 shows the comparison between the hardness level between halogen lamp and oven. It was observed that for oven curing, the hardness value obtained was 48.24 HV at 110 °C and increased to 48.67 HV at 130 °C. The hardness value decreased to 46.86 HV at 150 °C but increased onwards to 49.70 and 51.20 HV for sample 170 °C

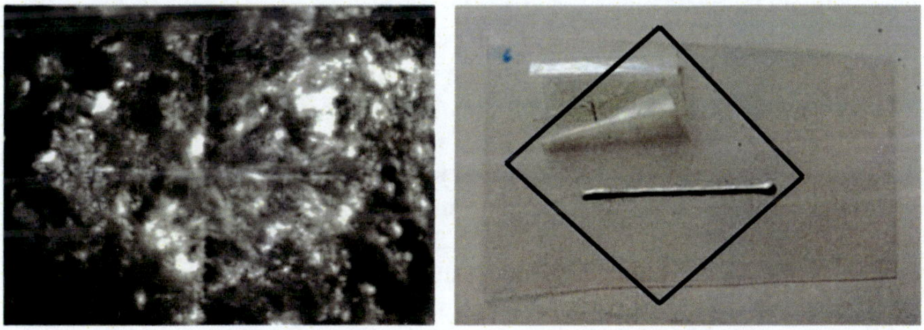

(a) (b)

Fig. 7 **a** Diagonal shape left by the indenter **b** Ink residues removed from a sample and stick on the tape during adhesion test

Table 3 Result of adhesion and hardness level

Experiment	Temperature (°C)	Time (minute)	Average Hardness level (HV)		Average Adhesion level (B)	
			Halogen lamp	Oven	Halogen lamp	Oven
1	110	90	32.35	48.24	2.00	4.00
2	130	70	37.21	48.67	2.00	4.00
3	150	50	38.50	46.86	3.00	4.00
4	170	30	38.39	49.70	3.00	4.00
5	190	10	44.50	51.20	3.00	4.00

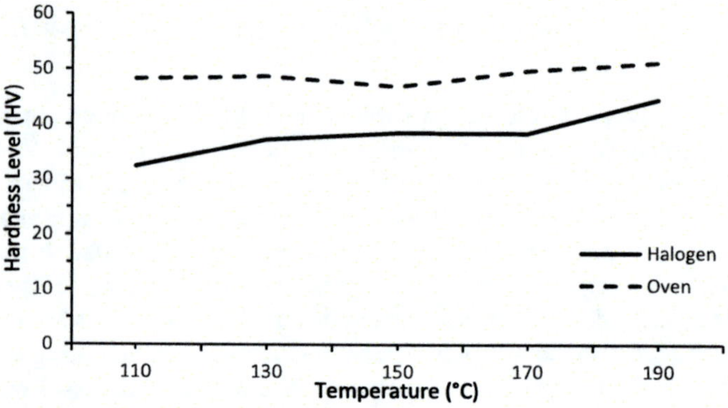

Fig. 8 Graph comparison of changes of average hardness level between halogen lamp and oven curing

(30 min) and 190 °C (10 min) respectively. It can be said that oven curing maintains relatively the same high hardness value of ink track thus proving it that the ink tracks were properly cured. Furthermore, it was reported previously that oven curing increased the level of hardness when more of unwanted solvent in conductive ink track are evaporated during curing process [14]. In addition, previous study stated that the hardness level of sample at 190 °C is higher than 170 °C which means that increasing temperature will increase the hardness level [15]. For halogen lamp, it was observed that the hardness value varied from the sample 110 °C (90 min) to 190 °C (10 min). The lowest hardness value was 32.35 HV at 110 °C in 90 min but increased to 37.21 and 38.50 HV as the temperature increased to 130 °C (70 min) and 150 °C (50 min) respectively. But at temperature of 170 °C (30 min), the hardness value slightly decreased to 38.39 HV but increased to 44.50 HV at 190 °C. The sample of 190 °C (10 min) has the highest value of 44.50 HV from curing using halogen. In summary, increasing the temperature of the halogen lamp cause the silver ink track to become more harden as more solvent were removed, leaving only a strong bond of metallic particle.

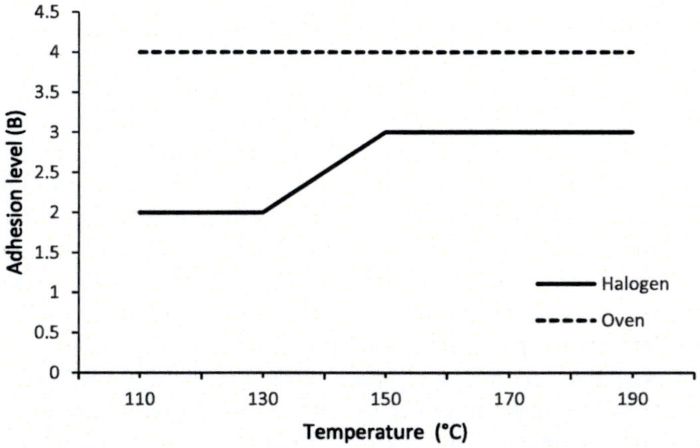

Fig. 9 Graph comparison of changes of adhesion level between halogen lamp and oven curing

Furthermore, adhesion test was executed on samples to measure the adhesion level of the cured silver ink track. The adhesion level was analyzed based on the ASTM D3359. A crosshatch pattern was applied to the samples and then compared to its classification based on the percentage of removal of ink residues as shown in Fig. 7b. Table 3 shows the overall results of the adhesion test while Fig. 9 shows the comparison result of adhesion level between halogen lamp and oven. It was observed that oven curing maintained constant adhesion level of 4B in which a less than 5% of the ink residues were removed. It means that the ink tracks were properly cured due to the metallic particle adhering more onto the substrates surface. Previous study has also stated that samples are completely cured when the abscission area is less than 5% [16]. For halogen lamp, the adhesion level obtained were varied at 2B for sample from 110 °C (90 min) to 130 °C (70 min) but increases to 3B and maintained onwards for sample of 150 °C (50 min), 170 °C (30 min) and 190 °C (10 min) respectively. 5% to 15% of the silver ink particulates were removed. The curing parameter has also been observed to be effecting the adhesion as higher curing temperatures produce a better adhesion level [11]. In summary, the samples have a low adherence at 110 and 130 °C but increased as the temperature was increased. Thus it can be said that increasing the temperature of the halogen albeit by increasing the power input, will increase the adherence intensity of the silver conductive ink track as the curing time is reduced.

4 Conclusion

As a conclusion, it was proven that halogen lamp could be used as a replacement to oven curing due to its capability to achieve the expected resistance level which is comparable to the oven curing process. This means that halogen lamp could adequately remove unwanted material from the silver conductive solvent resulting in more conductive properties. Furthermore, the more heat is supplied from the halogen lamp to the conductive ink, the better the conductivity of the ink track obtained without damaging the substrates. Halogen lamp also shows a better performance in pre-heating compared to oven where halogen lamp could reach a higher temperature quicker than oven thus shorten the manufacturing time while helps increasing the productivity respectively. Morphological analysis through the optical microscope also shows that the printing parameters used are suitable to acquire the required physical characteristic of the ink track. Oven curing had produced a good adhesion result while for halogen lamp, increasing the temperature will evaporate more the unwanted solvent thus resulted in additional force of metallic particles to adhere on substrate under a short amount of curing time. Moreover, hardness test shows that curing using high temperature for halogen lamp had resulted in increased hardness value due to the unwanted solvent being evaporated thus leaving metallic particle with strong bond in them. This prove that the performance of halogen lamp is comparable to oven curing and could be used as an alternative for oven curing.

Acknowledgements The authors gratefully acknowledged the support by Ministry of Higher Education (MOHE) through Fundamental Research Grant Scheme (FRGS/1/2019/ICT02/UTHM/02/4). We also want to thank the Universiti Tun Hussein Onn Malaysia for giving us the opportunity to attend and present at this conference.

References

1. Saad A (2007) Environmental pollution reduction by using VOC-free water-based gravure inks and drying them with a new drying system based on dielectric heating. Dr. Thesis
2. Abhinav KV, Rao RV, Karthik P, Singh S (2015) Copper conductive inks: synthesis and utilization in flexible electronics. RSC Adv 5(79):63985–64030
3. Titkov A, Bukhanets O, Gadirov R, Yukhin Y, Lyakhov N (2015) Conductive inks for inkjet printing based on composition of nanoparticles and organic silver salt. Inorg Mater Appl Res 6(4):375–381
4. Sato T, Fearon E, Curran C, Watkins K, Dearden G, Eckford D (2010) Laser-assisted direct write for aerospace applications. Proc Inst Mech Eng Part G J Aeros Eng 224(4)
5. Roberson D, Wicker R, MacDonald E (2012) Ohmic curing of printed silver conductive traces. J Electron Mater 41(9):2553–2566
6. Eugene W, Gupta M (2010) Characteristics of aluminum and magnesium based nanocomposites processed using hybrid microwave sintering. J Microw Power Electromagn Energy 44(1):14–27
7. Khirotdin R, Ahmad M, Nizam M, Hassan N (2021) Investigation of reducing curing time of silver conductive ink using halogen lamp on polymer substrate. In:Proceeding of international conference of mechanical engineering ICME21, pp. 5–9
8. Mustapha S, Ye L (2015) Bonding piezoelectric wafers for application in structural health monitoring-adhesive selection. Res Nondestr Eval 26(1):23–42
9. Taghiyari H, Rangavar H (2007) Effect of nano-silver on reduction of hot-pressing time and improvement in physical and mechanical properties of particleboard
10. Park J, Rhee S (2010) A study of hybrid drying process to improve drying speed and electrical conductivity for roll-to-roll printing. Japan J Appl Phys 49(5 PART 2)

11. Khinda G, Kokash M, Alhendi M, Yadav M, Lombardi J, Weerawarne D, Poliks M, Borgesen P, Stoffel N (2019) Effects of oven and laser sintering parameters on the electrical resistance of IJP nano-silver traces on mesoporous PET before and during fatigue cycling. In: Electronic components and technology conference
12. Sun H, Xiao GG, Lang S, Zhang Z, Tao Y (2018) Optimizing the electrical conductivity of screen printed silver conductive tracks by post treatment. In: 2018 international flexible electronics technology conference (IFETC), pp 1–3
13. Sun H, Xiao G, Lang S, Zhang Z, Tao Y (2019) Screen printed HF RFID antennas on polyethylene terephthalate film. IEEE J Radio Freq Ident PP(c):1
14. Khirotdin R, Cheng T, Mokhtar K (2016) Printing of conductive ink tracks on textiles using silkscreen printing. ARPN J Eng Appl Sci 11(10):6619–6624g
15. Kim NW, Lee D-G, Kim K-S, Hur S (2020) Effects of curing temperature on bending durability of inkjet-printed flexible silver electrode. Nanomaterials
16. Wang XQ, Gan WP, Zhou J, Xiang F, Long M, Peng D (2018) Effect of silver powders on a low curing temperature silver conductive adhesive. J Mater Sci Mater Electron 29(21):18547–18552

Review: Two-Dimensional Layered Material Based Electrodes for Lithium Ion and Sodium Ion Batteries

Omama Javed and Radhiyah Binti Abd Aziz

Abstract Energy crisis is a worldwide problem due to the current conventional resources which are depleting and are causing great environmental concerns. Among many feasible solutions, rechargeable batteries are considered as a powerful alternative to these conventional energy resources. Lithium and sodium ion batteries are one of the best energy storage systems which provide greater cyclic stability and better charge–discharge capacity. These metal ion batteries have achieved great attention and are being used in a range of applications from small electronic devices to electric vehicles. An overall enhancement of metallic ion storage and transportation is the current concern and focus of the researchers. Many electrode materials are tried and tested in this regard which provided great deal of structural and functional improvement. Two dimensional layered materials have also gained much attraction recently in energy storage application due to their higher surface to volume ratio. There are a number of layered structures which have been developed and electrochemically tested as anode or cathode for both lithium and sodium ion batteries. Materials like graphene based structures, transition metal chalcogenides (TMDCs), MXenes, nitrides, Molybdenum Sulfide and organic frameworks showed promising results as electrode for lithium and sodium ion batteries. In this work, an effort is done to cover all these electrode materials along with their complete structural analysis and a thorough evaluation of their electrochemical activity in energy storage systems.

Keywords Two dimensional layered materials · Lithium-ion batteries · Sodium-ion batteries · Electrode · Efficiency · Charge–discharge capacity

O. Javed · R. B. Abd Aziz (✉)
Faculty of Manufacturing and Mechatronic Engineering Technology, Univesriti Malaysia Pahang, 26600 Pahang, Malaysia
e-mail: radhiyah@ump.edu.my

© The Author(s), under exclusive license to Springer Nature Singapore Pte Ltd. 2023

M. H. A. Hassan et al. (eds.), *Technological Advancement in Instrumentation & Human Engineering*, Lecture Notes in Electrical Engineering 882,
https://doi.org/10.1007/978-981-19-1577-2_30

1 Introduction

The energy storage systems are those devices that are charged when they absorb some form of energy either directly from renewable generation devices or indirectly from the electricity grid. They discharge when they deliver the stored energy back into the grid. The process of charging and discharging normally requires power conversion devices, to transform electrical energy into a different form of electrical, thermal, mechanical or chemical energy. Among all of these various types of energy storage systems, electrochemical energy storage is our main concern. In this type of energy storage, the energy is stored in chemical form and later used in electrical form. It involves the shifting of electrons around and stacking them chemically for later use.

Batteries can be either high power or high energy. Therefore, these features are considered to be the classifying categories. Other common classifications are high durability, meaning that the chemistry has been modified to provide higher battery life at the expense of power and energy. To reduce the cost, many cheap alternatives have been introduced by replacing the alkaline metals and making their hybrids with transition metals. This led to the development of many popular types of batteries. For example, Nickle cadmium battery, Mercury silver button batteries, Lithium manganese batteries, Nickle metal hydride batteries etc. Among these lithium ion batteries offered maximum capacity. The reserves of lithium are very low and the cost is also still very high. Therefore, the lithium batteries were later switch to lithium ion batteries in which the amount of lithium in one redox reaction is almost half the others. This led to reduction of cost and improvement in functionality, but the overall cost of the battery system is still so high that it cannot be commercialized on larger scales. The solution of this was to look for an alkali metal option that has similar properties but is cheap. The answer is Sodium (Na). The sodium offers same electrochemical properties as those of lithium (Li) but is very cheap as its reserves are almost everywhere around the world. The overall capacity and cyclic stability of sodium batteries is also very promising and it is being considered to be able to replace the lithium based batteries soon. Only problem associated to the sodium ion batteries is the size of sodium ion. The large size of sodium ion often disrupts the structure of the electrode material during the sodiation and desodiation process. This results in the loss of active material and the overall capacity is reduced (Table 1).

Almost in 400 C.E., people used layered clays to make dyes, people utilized the properties of layered materials. The progressively expanded scientific research leads to the clarification of the laminar structure of layered materials and comprehensive awareness of their properties. This capped in the breakthrough of graphene, causing in a new outbreak of interest in two-dimensional materials [1].

Layered or two dimensional (2D) layered materials are those materials which are arranged in stack or layers [2]. In such a structure, a material has a chain of sub material which can be placed transversely on the surface. The layering is generated due to the high conductivity, higher mobility, greater mechanical strength, and longer spin diffusion [3]. In this regard the first material which was a wonder material, especially in the field of layered materials, was Graphene [4]. After the discovery

Table 1 Comparison of properties of Sodium and Lithium

Factor	Lithium	Sodium
Ratio of reserves	1	1000
Cost (for carbonate) ($/Ton)	5000	150
Atomic weight (G Mol-1)	6.9	23
Ionic volume ($Å^3$)	1.84	4.44
Theoretical capacity (MAh/g)	3829	1165
Normal electrode potential vs she (V)	– 3.045	– 2.714
Distribution	70% in South America	Everywhere

of graphene, the research and development sector has opened many doors in the field of materials and their further studies. Such materials offered high surface area, higher young's modulus, and extraordinary thermal electrical properties [5]. They were also considered ideal materials for making composites materials [6]. After graphene, researchers started focusing on discovering alternative two-dimensional layered materials with same or better properties.

For the purpose of study, these materials are divided into following five groups [6]:

These materials have gained attention in various applications because of the fact that one can control the sites and properties, tune-able pore size and volume [7].

Layered materials are of two types, van der Waals materials and ionic materials [8]. The materials which have weak interlayer bonding lies in the category of van der Waals and those materials which have strong interlayer bonding falls in the ionic category[8]. The ionic category has been studied recently. And the single example of ionic category is extension of the topology scaling method to investigate Na containing materials for battery application [9].

On the basis of formation method of layers, the layered materials may be distributed in three main branches [7, 10]:

1. Multi-layer catalysts which are developed by layering methods such as ultra-thin films methods, atomic layer deposition/epitaxy etc. (Thin Films)
2. When layered structures are so developed that only along one main crystallographic axis van-der Waals forces are involved, this leads to anisotropy of crystal bonding energy in lattice directions. (Nano Scrolls)
3. Delaminated materials obtained by swelling of layered materials developed as first two types. (Aggregated Nanosheets) (Fig. 1).

Layered structures, based on charge in layers, can also be classified as [7, 12, 13]

1. Negatively charged layers which have compensating positive ions in the interlayer spaces, e.g., widespread lamellar compounds in nature such as cationic clays (montmorillonite, hectorite and beidellite, etc.)

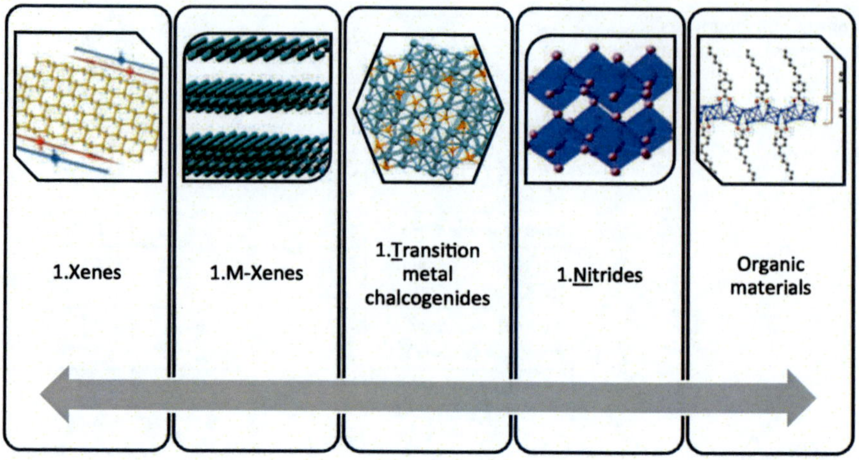

Fig. 1 List of two-dimensional layered material

2. Positively charged layers with compensating negative ions in the interlayer spaces, the most common of which are the layered double hydroxides known as LDH
3. Neutral layers, e.g., brucite ($Mg(OH)_2$) and other hydroxides, phosphates and chalcogenides, and various metal oxides such as V_2O_5 (Fig. 2).

Another important and rather new concept is multi-layer catalysts which are at a too early stage of development [7]. It is also note-worthy that these catalysts show a layered structure on which their catalytic properties depend, mainly because the surface structure reflects the order–disorder transformations in the structure [14]. Examples are the $(VO)_2P_2O_7$ for n-butane oxidation to maleic anhydride and Molybdenum (Mo), Vanadium (V), and Tungsten (W) oxides for conversion of acrolein to acrylic acid.

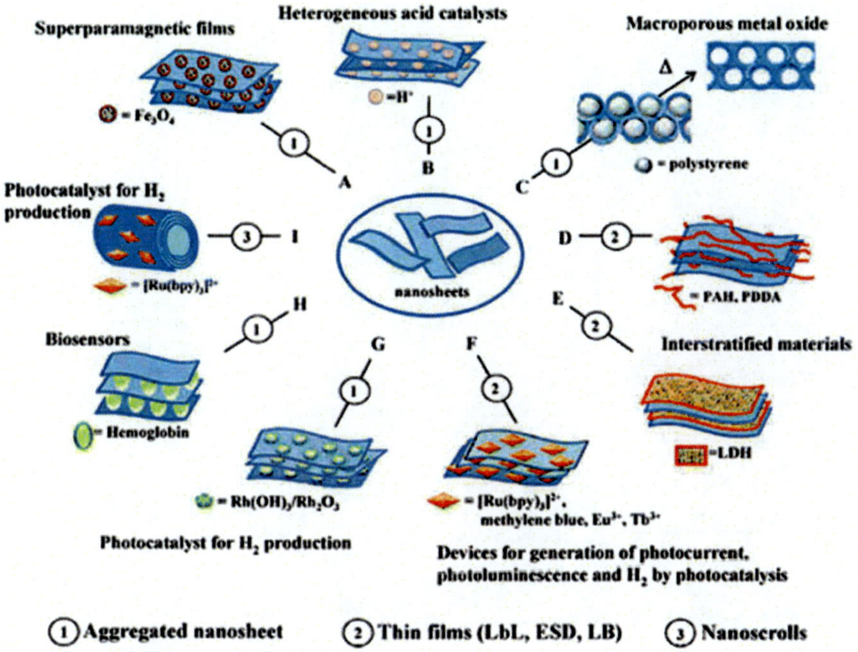

Fig. 2 Types of layered materials [11]

2 Two Dimensional Layered Electrode Materials for Lithium Ion Batteries

In 1990s, rapidly growing market of portable devices led to the concept of rechargeable cell which cause the introduction of the lithium ion batteries. The factors which make the lithium ion batteries a desirable solution is high number of discharge cycles and high energy densities. The other advantage of lithium ion batteries is that they don't have effect on the memory of the battery to remember the charge/discharge cycles which can affect a battery to remember the low densities. Owing to these advantages, lithium has replaced the Ni–Cd batteries in the market as rechargeable powerful device.

2.1 Graphene

All the materials with graphene as structural part are very important for batteries and other storage applications [15]. Reasons behind this fact are the two very important features of graphene sheets i.e., elevated electrical conductivity and active edge-plane sites [16]. The complexed structure of branched graphene nano capsule and graphene

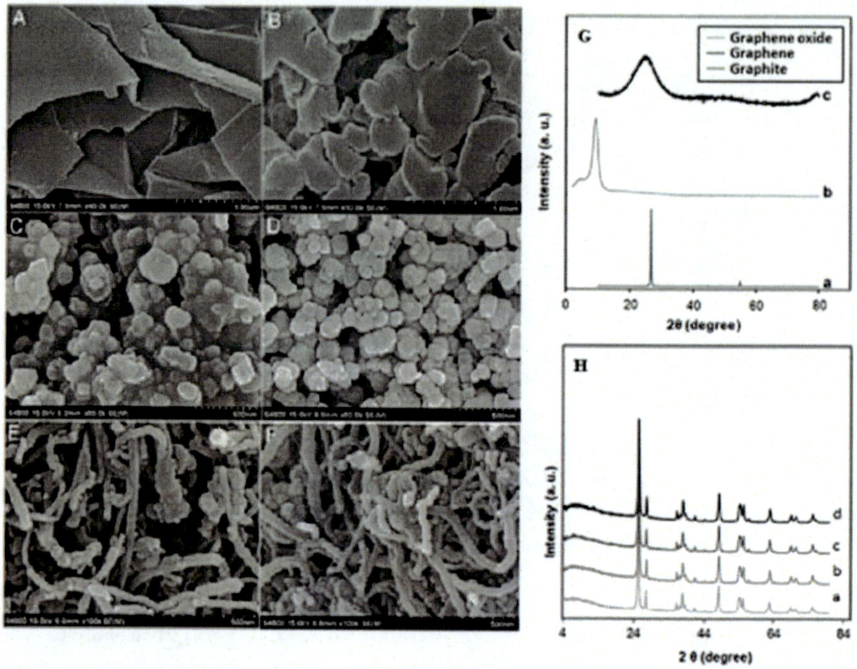

Fig. 3 SEM images of graphene [21]

sheets formed from in-situ method with edges and defects on sheets and capsule respectively helped in achieving the great level of storage in Li-ion batteries with high performance of lithiation and de-lithiation [17]. Graphene sheets in LIB applications develop some different and interesting properties when doped with heteroatom [18]. Graphene anode, used for lithium ion storage can show high energy densities and power when doped with sodium or boron [19]. The quick absorption of Li-ion on the surface, the ultrafast diffusion of Li-ion through the structure, and the transportation of electrons through the doped graphene are enabled by heteroatom defects. Apart from that heteroatom also provides good electrode/electrolyte wettability, improved electrical conductivity, and greater inter-sheet distances. Greater Li-ion storage has been shown by the porous graphene sheets as well [20]. The (Fig. 3a-b) shows the pure SEM images of graphene, nanoparticles can be seen in (Fig. 3c-d), and nanotubes in (Fig. 3e-f). In XRD the peak at (002) planes indicate the presence of graphene.

2.2 *Mesoporous Carbon Nanosheets*

Another variation in this regard is the two-dimensional ordered mesoporous carbon nanosheet which can be synthesized by the mono-micelle close packing assembly

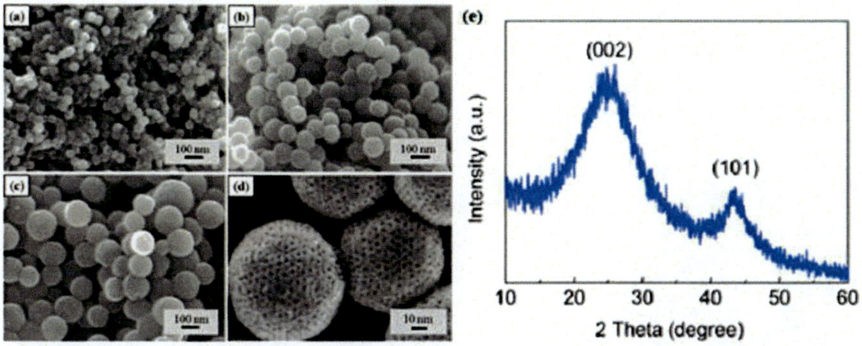

Fig. 4 SEM images of Two-Dimensional Carbon Mesoporous **a** OMCS7, **b** $OMCS_9$, **c** and **d** $OMCS_{11}$ [26] and **e** XRD pattern of Two-Dimensional Carbon Mesoporous [27]

in controlled low concentration [22]. These ordered mesopores on the surface of the carbon substrate, have only one bulk of carbon and can be changed into nanosheets of mesoporous graphene [20]. The high surface area of such a structure attributed by the flat mesoporous graphene layers enables the efficient Li-ion absorption and intercalation while a great deal of ion-transportation and volume expansion are enabled by the ordered mesopores [23]. To increase the rate capability and cycling stability, graphene can be hybridized with inorganic materials due to their high electrical conductivity [24]. This method was initially reported for $ZnMn_2O_4$-Graphene hybrid for Li-ion batteries [25]. SEM images shown in (Fig. 4a-d), proved that homogeneous spherical structure is formed. In (Fig. 4e) XRD pattern is given. The peak at (002) is intensive and shows the highly graphitic ordering.

2.3 *Two-Dimensional Transition Metal Oxides*

Two-dimensional Transition Metal Oxides based nanomaterials are also considered a great option as electrodes for Li-ion batteries due to their high redox performance [28]. The method of chemical bath deposition was used for the synthesis of porous ZnO nanosheets on copper substrate following the technique of heat treatment [29]. This nano alternative showed greater reversible capacities and cyclabilities than the trivial ZnO powders available commercially [29]. Likewise, synthesis of ultra-dispersed TiO_2 nanoparticles on rGO with fine structural control was done by sol–gel method. Higher cyclic performance, rate capability, high specific capacity during the lithiation process, zero volume change and enhanced safety are the norms of such hybrid materials [30, 31]. The SEM images shown in (Fig. 5) depicts that the nanosheets were formed successfully and the edges of the materials are rolled up because of the surface tension. In (Fig. 5e), XRD pattern shows the two crystal

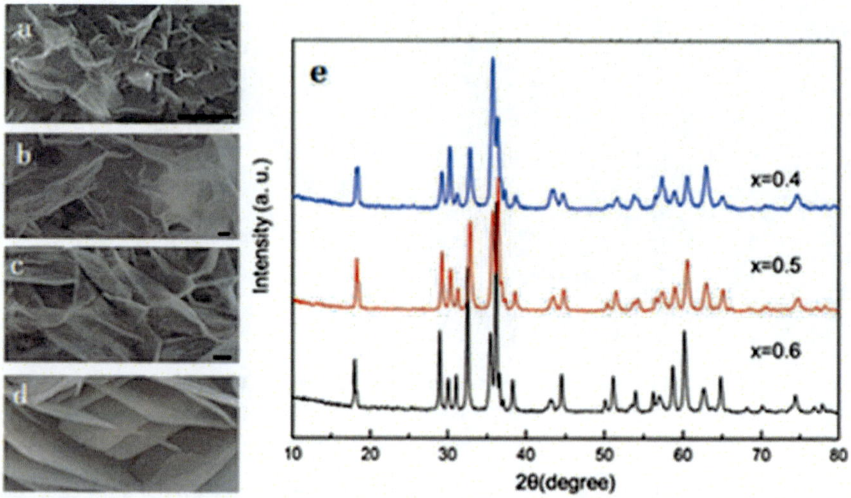

Fig. 5 SEM images of Two-dimensional Transition Metal Oxide **a** TiO_2 **b** ZnO **c** CO_3O_4 **d** WO_3 [32], **e** XRD pattern of Two-Dimensional Transition Metal Oxide $xLi_2MnO_{3\text{-}(1-x)}LiNi_{0.33}Co_{0.33}Mn_{0.33}O_2$ ($x = 0.4$, 0.5 and 0.6) [33]

phases of the $CoMn_2O_4$ and $NiMn_2O_4$ are formed, and these results showed the phase separation of metal oxide.

2.4 *Spinel Lithium Titanate Structures*

Spinel lithium titanate is also highly regarded as the anode material for li-ion batteries [34]. A hydrothermal technique, involving the titanium (Ti) foil in a Lithium hydroxide (LiOH) solution, is used to directly grow the self-supported $Li_4Ti_5O_{12}$ nanosheet arrays on the titanium substrate [35]. This self-supported lithium titanate nanosheets based anode for Li-ion batteries came out with excellent cyclic performance and outstanding rate capability. These characteristics are attributed to the excellent conductivity due to high surface area and highly stable structure [36]. The SEM images shows that the titanium oxide possess the well-defined sphere and the surface is smooth (Fig. 6a). The XRD pattern shows no diffraction lines in low angles which predicts that there are no ordered arrangements of mesopores (Fig. 6b).

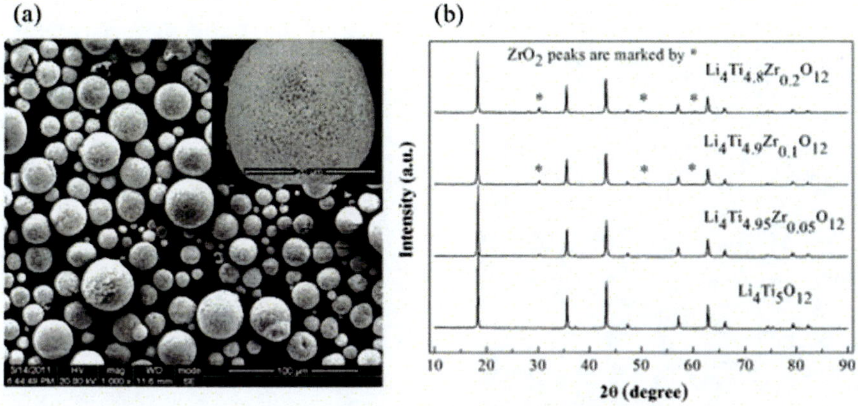

Fig. 6 **a** SEM images and **b** XRD pattern of Spinel Lithium Titanate $Li_4Ti_5O_{12}$ [37]

2.5 *Transition Metal Dichalcogenides*

Another class of compounds that proved to be a potential anode material for li-ion batteries is transition metal dichalcogenides. These compounds garnered great attention due to their low operation voltage and their ability to electrochemically store lithium ions. All this is possible due to their large surface areas and a considerably good interlayer spacing. To check their activity as anode material for Li-ion batteries, MoS_2 nanoplates were synthesized by solvothermal method [38]. MoS_2 plates have an advantage of storing the lithium entities with high capacity and it comes with the high-rate performance. For improved cycling and rate performance, graphene like MoS_2 nanosheets as composite with the ultrasmall Fe_3O_4 nanoparticles are used for li-ion batteries as anode material. They have better performance than bare MoS_2 nanosheets [39]. Two dimensional WS_2/N-doped graphene sheet nanocomposite is another example of such compounds that are used as anode for lithium ion batteries and represented great cyclic performance and rate capabilities [40]. In XRD pattern (Fig. 7b), the peak at (002) plane suggest that few layers are present in c-axis of the $MoSe_2$.

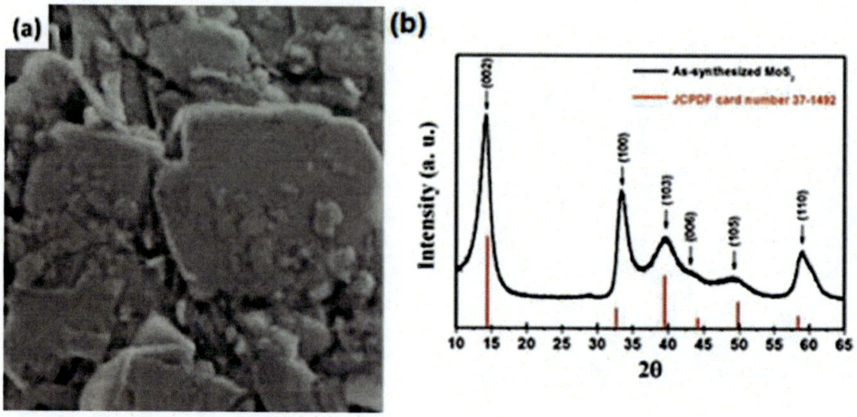

Fig. 7 **a** SEM and **b** XRD pattern of two-dimensional transition metal dichalcogenides [41]

2.6 *MXenes*

MXenes are the ceramic materials which comprise one of the largest groups of two-dimensional layered materials. Synthesized from a bulk crystal, these materials have inherently good conductivity and outstanding volumetric capacitance. These attributes make them great option for many applications including energy storage ones. MXenes are basically the transition metal (Molybdenum or Titanium) carbide and nitride materials like Ti_2CT_x, $Ti_3C_2T_x$, etc., where T is the surface terminating functional group [42, 43]. MXenes has promising properties of metallic conductivity and hydrophilicity for Li-ion batteries as anode materials with low metal diffusion on the surface barrier. Mo_2TiC_2 is another ordered MXene which is used for electrochemical energy storage [44]. The Sem images of (Fig. 8a-b) shows the layered structure and the (Fig. 8c-d) shows the compact layered structure after introducing

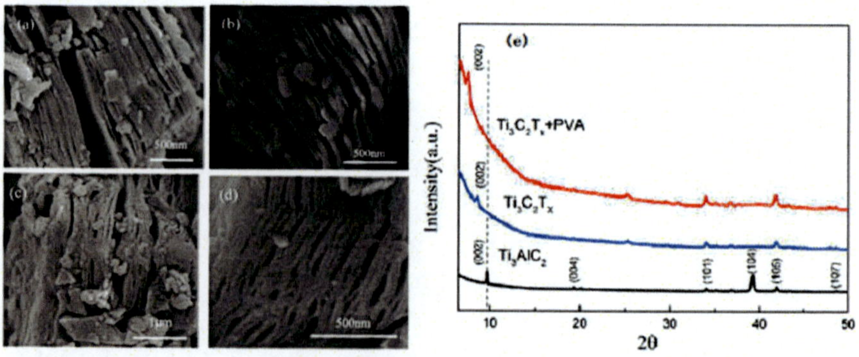

Fig. 8 **a-d** SEM images and **e** XRD pattern of MXene [45]

the PVA. In the XRD pattern of the (Fig. 8e), the peak (002) shifts towards the lower angle increasing the *c* parameter. It indicates that PVA can interact with $Ti_3C_2T_x$ MXene.

A comparison of all the discussed two dimensional layered material as anode is given in Table 2.

As far as cathode materials for lithium ion battery are concerned, high energy density, thermal stability, environmental friendliness, and low cost of the raw material are the characteristics that are most desirable. Such properties are found potentially in olivine type lithium transition metal phosphates which have attracted the researchers' attention as potential cathode material for li-ion batteries. With exposed surface facet, two-dimensional single crystal $LiFePO_4$ nanosheets are in good favor to obtain high-rate capability with fast lithium transport and storage. Solvothermal process with diethylene glycol as solvent is appropriate for preparing single crystalline $LiFePO_4$ with the thickness of 30–60 nm [48] (Table 3).

Table 2 List of anode materials for lithium ion batteries

Active material	Capacity/Current /C-rate (mAh/g)/(mA/g)	Cycles	Rate	Reference
Mesoporous carbon nanosheets	770/100/–	—	370	[20]
Graphene nano-capsules	1373/500/–	200	750	[46]
Porous ZnO nanosheets	750/50/–	100	195	[29]
$ZnMn_2O_4$/Graphene composites	730/500/–	100	568	[25]
$Li_4Ti_5\ O_{12}$ nanosheet arrays	163/–/20C	3000	163	[35]
WS_2/N-doped graphene	905/100/–	100	700	[39]
MXene/CNT paper	420/–/0.5C	300	270	[43]
TiO_2 Nanocrystals/rGO sheets	189/100/–	100	94	[31]
Fe_3O_4/MoS_2 composites	1079/100/–	500	569	[40]
Mo_2TiC_2 (MXene)	269/–/C/10	25	176	[47]
B-doped graphene nanosheets	1549/50/–	30	380	[19]
N-doped graphene nanosheets	1043/50/–	30	296	[19]
MoS2 nanoplates	912/–/1C	50	658	[38]
$Li_4Ti_5O_{12}$	114/–/30C	200	—	[37]

Table 3 List of cathode materials for lithium ion batteries

Active material	Capacity/ C-Rate (mAh/g)	Cycles	Rate	Reference
$LiFePO_4$/C Nanosheets	164/0.2C	50	70	[49]
$LiMnPO_4$/C Nanosheets	157/0.2C	50	40	[49]
$LiCoPO_4$/C Nanosheets	153/0.2C	50	53	[49]
$LiFePO_4$ Nanosheets	151/0.5C	1000	120	[48]

3 Two-Dimensional Layered Electrode Materials for Sodium Ion Batteries

When li-ion batteries were announced in the 1970s, Na-ion batteries were also there but due to their inferior performance they were neglected for practical work for long time. Sodium has advantage of abundance on lithium but due to its ionic radius and redox potential. Sodium ion has very large radius of 102 pm while lithium ion has radius of 76 pm [50]. Sodium ion has also higher redox potential of ≈0.3 V [51]. Sodium has gathered large number of interests in the development of Na-ion batteries with the development of new materials and methods apart from its disadvantages and uncompetitive performance.

Two dimensional nanomaterials also show the great performance in the Na-ion batteries. They show greater specific capacity, enhanced rate capability, and longer cycle life. High electrical conductivity, high surface area, large quantity of active sites for ion storage, and short ion diffusion distance has enhanced the gravimetric capacity and energy density of graphene and graphene-based composites as compared to earlier technologies [15, 52, 53]. High working voltage and specific capacity make layered transition metal oxides an important cathode material for Na-ion batteries [54].

3.1 *Graphene Sheets*

Graphene is a single or multiple layers two-dimensional material made up of carbon atoms. Graphene is typically made up from commercial graphite [55–57]. Graphene exhibits more active sites for both sodium and lithium ions due to involvement of carbon sheets, whose both sides participate actively. Not only both sides of carbon sheet, its edges and covalent sites are also involved in the reactions [57]. Graphene can be obtained from many methods which includes, chemical vapor deposition, mechanical exfoliation, bottom-up synthesis, reduction of graphene oxide, liquid phase exfoliation are the ways from which single layer and multiple layers graphene can be obtained [15]. Graphene has many advantages like, exceptional mechanical properties, and higher electronic conductivity made graphene extremely acclaimed material as anode for both sodium and lithium-ion batteries [15, 55–61]. Graphene is used to improve the properties of different materials, transition metal disulfide [60, 62–67], Si, Sn, and P [47, 68–73], and transition metal oxides [64, 74–82]. These materials have higher theoretical capacities but endure weak cyclic stability. After adding graphene with active material, it not only increased the electrochemical properties but also enhanced the structure stability [25, 63, 67, 71].

The insertion of lithium-ion between the atomic layers of graphite and the development of Lithium-Graphite intercalation compounds, lithium-ions can be store in graphite with the rescindable capacity of greater than 360 mAh/g which is near the theoretical value of 372 mAh/g [57, 83]. Hard carbon (bulk material), due to the structural disorder is more effective in Na-ion storage than graphite as staging transitions can be avoid during Na-ions insertion and extraction in the hard carbon [84]. Sodium insertion and extraction of hard carbon at room temperature was also reported [85, 86].

The higher electrical conductivity makes the graphene materials higher in specific capacity compared to other materials. Reduced graphene oxide was first used in sodium ion batteries as anode material in 2003 by Wang. Although graphene plays no active part in the energy storage process, but it can be key part in the EES applications. Besides high electrical conductivity, heat dissipation also helps when greater current load is needed due to its greater thermal conductivity. It is also an auspicious functional substrate due to its hybrid electrode material for sodium ion batteries [87, 88].

3.2 Layered MoS_2

MoS_2, (Transition Metal Disulfides) [39, 63, 89–98] has been rigorously experimented in past decades due to their distinctive stuffing layered structure [71]. MoS_2 transferred four electrons during charge and discharge which gives great theoretical capacity of 670 mAh/g as compared with commercial graphene 372 mAh/g [71, 73, 99–101]. MoS2 has many drawbacks which makes it less favorable candidate in the field of research. It suffers weak electronic conductivity among the layers and all this is caused by the wea Van der Waals interactions. MoS_2, used in lithium sulfur batteries used as anode suffers the degradation of electrons which is due to the condition called polysulfide shuttling effect [67, 71, 93, 99]. To overcome the issues, researchers are finding the better solution for MoS2 by optimizing the morphology [75, 89], nanostructure [73, 93], and its combination with carbon and non-carbon materials [39, 60, 77, 95, 97, 98, 100–106].

Researchers are trying to improve the electrochemical behavior of MoS_2. Researchers are redefining the structure of the material by controlling the nano-size of the material, reducing the crystal defects, and increasing the layer size of the material [73, 75, 89, 93].

Like in lithium ion batteries, MoS_2 exhibits high theoretical value with same chemistry in sodium ion batteries which is 670 mAh/g [63, 90–92, 107–109]. In half cell of Na-MoS_2, intercalation and conversion are done at different voltage in two steps. Phase transition occurs in the process of conversion in which MoS_2 changes from 2H-phase to 1 T-phase. This phase transition occurs in both sodium ion batteries and as well in lithium ion batteries [71, 90, 99]. During the intercalation if 1.5 or more than this sodium ions participate then the 1 T- phase cannot be retrieved. This is one of the main reasons that anode capacity of MoS_2 progressively decreases [90, 99].

Sodium has much larger ionic radius as compared to lithium which causes the rapid decrease in kinetic reactions [63, 91, 110]. MoS_2 exhibits obstacles in achieving the high performance of electrode materials in sodium ion battery. Degradation within the structure of electrode material, improper electronic conductivity and large volume expansions are the main reasons which are the main barriers in achieving the high rate performance of electrode material in sodium ion batteries [63, 91, 97, 107, 110–112].

3.3 Two-Dimensional Layered Oxide for SIBs

Layered oxides are also massively studied for sodium ion batteries as positive electrode and early contribution was made by Delmas in 1980s for sodium insertion and extraction [113]. During the process of lithium extraction, the layered oxides of lithium whether went for irretrievable level transition or are sedentary, while the layered oxides of sodium during insertion are extremely effective.

Na_xMO_2 layered oxide containing sodium is made of pile of edge sharing octahedra MO6 sheets. It can also be divided into two main groups O3- type and P2-type suggested by Delmas. Sodium ions placed at octahedral sites are O3-type and sodium ions placed at prismatic sites are P2-type [114]. When sodium ions sandwiched in the layers of MO_2 at octahedral sites, this type of structure is called O3-type layered structure. Sodium ions are so big in size that they are also placed at trigonal sites which is also referred as prismatic sites and called P2-type layered structure. Sodium ions are highly stable at octahedral sites (O3-type level) and at prismatic sites (P2-type level) due to the empty spaces, sodium ions also became dynamically stable [113]. O3-type can be changed into P3-type and P2-type can be changed into O2-type without destroying the M–O bonds by gliding the slabs of MO2. Study shows that O3-type level shows better performance in the first cycle and P2-type level shows better performance in the second cycle this is due to the greater ionic conductivity and lower diffusion barrier of P2-type level. Due to the slab gliding in the O3-type level, P2-type level deemed to be highly stable in the structure.

High-level specific capacity and enhanced cyclic performance are obtained by creating nanosheets of layered oxide for cathode material by ordinary precipitation and annealing of $Na_{0.4}Mn_{0.54}Co_{0.46}O_2$ [115]. The P2-type of $Na_{0.4}Mn_{0.54}Co_{0.46}O_2$ nanosheets are high in reversible capacity with the value of 151 mAh/g at a current density of 20 mAh/g and it maintains its capacity at 120 mAh/g after the 65 charge discharge cycles.

3.4 Other Materials for Sodium Ion Batteries

There are many other materials for sodium ion batteries which have attracted the researchers for the better development of sodium ion batteries. Interlayer expanded MoS_2 nanosheets, Ultrathin MPO_4 nanosheets, porous carbon nanosheets, TMD nanosheets, and MXene nanosheets are the areas of research for the development of sodium ion batteries. In interlayered expanded method, exfoliation restacking process is used during the synthesis of MoS_2 composites and used in sodium ion batteries as anode material. Highly thermal stability and higher operating potentials are the characteristics which made metal ion phosphates strong contender as cathode material for sodium ion batteries. As the number of cycles increases, the shrinkage and expansion of lattice also increases during the insertion and extraction of sodium ions which leads towards the loss in capacity. To solve this issue, mesoporous materials are used to improve the transportation of sodium ions.

4 Conclusion

Owing the high-level particular surface area, all the two dimensional materials have the advantage of large number of active sites for the storage of ions. The two dimensional materials for both lithium ion batteries and sodium ion batteries perform the part in improving the electrical conductivity and overall performance due to the functional substrate. All the two dimensional materials especially graphene urges to increase the power density and energy of all the battery systems, but the effort is required to resolve the mechanism of two dimensional materials for the future battery technologies. For example, graphene is a promising material in lithium sulfur batteries but, graphene itself and the other two dimensional materials need to be as much effective in other metal sulfur batteries. for the future perspective all the two dimensional materials can be good expansion after the understanding of interface and surface reactions.

Acknowledgements The author would like to thank the Ministry of Higher Education Malaysia for the financial aids and Universiti Malaysia Pahang and its staffs for the laboratory facilities and financial supports from the Fundamental Research Grants Scheme FRGS/1/2019/STG07/UMP/02/7 (University reference RDU1901205) and PGRS2003152.

References

1. Nicolosi V et al (2013) Liquid exfoliation of layered materials. Science 340(6139):1226419
2. Wu C (2017) Inorganic two-dimensional nanomaterials: fundamental understanding, characterizations and energy applications. Royal Society of Chemistry
3. Jacobs JA, Kilduff TF (2005) Engineering Materials Technology: Structures, Processing, Properties, and Selection. Pearson/Prentice Hall, Hoboken
4. Geim AK (2009) Graphene: status and prospects. Science 324(5934):1530–1534
5. Kumar P et al (2015) Large-area reduced graphene oxide thin film with excellent thermal conductivity and electromagnetic interference shielding effectiveness. Carbon 94:494–500
6. Geng D, Cheng Y, Zhang G (2019) Layered Materials for Energy Storage and Conversion. Royal Society of Chemistry
7. Centi G, Perathoner S (2008) Catalysis by layered materials: a review. Microporous Mesoporous Mater 107(1–2):3–15
8. McKinney RW et al (2018) Ionic vs. van der Waals layered materials: identification and comparison of elastic anisotropy. J Mater Chem A 6(32):15828–15838
9. Meng FL et al (2018) Recent advances toward the rational design of efficient bifunctional air electrodes for rechargeable Zn–air batteries. Small 14(32):1703843

10. Corma A et al (1998) Delaminated zeolite precursors as selective acidic catalysts. Nature 396(6709):353–356
11. Themed issue (2009) Layered materials: structure and properties. J Mater Chem 19(17):2453–2456
12. Tichit D, Coq B (2003) Catalysis by hydrotalcites and related materials. CATTECH 7(6):206–217
13. Wang Q, O'Hare D (2012) Recent advances in the synthesis and application of layered double hydroxide (LDH) nanosheets. Chem Rev 112(7):4124–4155
14. Somorjai GA, Li Y (2010) Introduction to Surface Chemistry and Catalysis. John Wiley & Sons, Hoboken
15. Raccichini R et al (2015) The role of graphene for electrochemical energy storage. Nat Mater 14(3):271–279
16. Pumera M (2009) Electrochemistry of graphene: new horizons for sensing and energy storage. Chem Rec 9(4):211–223
17. Ying W, Limin Z, Tianjun H (2015) Progress in oxygen reduction reaction electrocatalysts for metal-air batteries. Acta Chim Sin 73(4):316–325
18. Ai W et al (2014) Nitrogen and sulfur codoped graphene: multifunctional electrode materials for high-performance Li-ion batteries and oxygen reduction reaction. Adv Mater 26(35):6186–6192
19. Wu Z-S et al (2011) Doped graphene sheets as anode materials with superhigh rate and large capacity for lithium ion batteries. ACS Nano 5(7):5463–5471
20. Fang Y et al (2013) Two-dimensional mesoporous carbon nanosheets and their derived graphene nanosheets: synthesis and efficient lithium ion storage. J Am Chem Soc 135(4):1524–1530
21. Gong A et al (2017) Moving and unsinkable graphene sheets immobilized enzyme for microfluidic biocatalysis. Sci Rep 7(1):1–15
22. Lan K et al (2019) Two-dimensional mesoporous heterostructure delivering superior pseudocapacitive sodium storage via bottom-up monomicelle assembly. J Am Chem Soc 141(42):16755–16762
23. Lotfabad EM et al (2014) High-density sodium and lithium ion battery anodes from banana peels. ACS Nano 8(7):7115–7129
24. Wang H et al (2010) Mn3O4–graphene hybrid as a high-capacity anode material for lithium ion batteries. J Am Chem Soc 132(40):13978–13980
25. Xiong P et al (2014) Chemically integrated two-dimensional hybrid zinc manganate/graphene nanosheets with enhanced lithium storage capability. ACS Nano 8(8):8610–8616
26. Chang P-Y, Bindumadhavan K, Doong R-A (2015) Size effect of ordered mesoporous carbon nanospheres for anodes in Li-ion battery. Nanomaterials 5(4):2348–2358
27. Wang J et al (2018) Confined self-assembly in two-dimensional interlayer space: monolayered mesoporous carbon nanosheets with in-plane orderly arranged mesopores and a highly graphitized framework. Angew Chem Int Ed 57(11):2894–2898
28. Zhang J, Yu A (2015) Nanostructured transition metal oxides as advanced anodes for lithium-ion batteries. Sci Bull 60(9):823–838
29. Huang X et al (2011) Porous ZnO nanosheets grown on copper substrates as anodes for lithium ion batteries. Electrochim Acta 56(14):4960–4965
30. Han M, Chen G (2016) Optimized dispersion of conductive agents for enhanced Li-storage performance of TiO2. Appl Surf Sci 388:401–405
31. Li W et al (2013) Sol–gel design strategy for ultradispersed TiO2 nanoparticles on graphene for high-performance lithium ion batteries. J Am Chem Soc 135(49):18300–18303
32. Sun Z et al (2014) Generalized self-assembly of scalable two-dimensional transition metal oxide nanosheets. Nat Commun 5(1):1–9
33. Hou C et al (2016) Bio-inspired synthesis of xLi2MnO3-(1−x) LiNi0. 33Co0. 33Mn0. 33O2 lithium-rich layered cathode materials. Mater Des 109:718–725
34. Sandhya C, John B, Gouri C (2014) Lithium titanate as anode material for lithium-ion cells: a review. Ionics 20(5):601–620

35. Chen S et al (2014) Self-supported Li 4 Ti 5 O 12 nanosheet arrays for lithium ion batteries with excellent rate capability and ultralong cycle life. Energy Environ Sci 7(6):1924–1930
36. Zhang F, Qi L (2016) Recent progress in self-supported metal oxide nanoarray electrodes for advanced lithium-ion batteries. Adv Sci 3(9):1600049
37. Li X, Tang S, Qu M, Huang P, Li W, Yu Z (2014) A novel spherically porous Zr-doped spinel lithium titanate (Li4Ti5− xZrxO12) for high rate lithium ion batteries. J Alloy Compd 5(588):17–24
38. Hwang H, Kim H, Cho J (2011) MoS2 nanoplates consisting of disordered graphene-like layers for high rate lithium battery anode materials. Nano Lett 11(11):4826–4830
39. Chen Y et al (2014) Ultrasmall Fe3O4 nanoparticle/MoS2 nanosheet composites with superior performances for lithium ion batteries. Small 10(8):1536–1543
40. Chen D et al (2013) In situ nitrogenated graphene–few-layer WS 2 composites for fast and reversible Li+ storage. Nanoscale 5(17):7890–7896
41. Rojas D, Della Pelle F, Del Carlo M, Compagnone D, Escarpa A (2020) Group VI transition metal dichalcogenides as antifouling transducers for electrochemical oxidation of catechol-containing structures. Electrochem Commun 115:106718
42. Naguib M et al (2011) ACS nano. Adv Mater 23:4248
43. Naguib M et al (2013) New two-dimensional niobium and vanadium carbides as promising materials for Li-ion batteries. J Am Chem Soc 135(43):15966–15969
44. Anasori B et al (2015) Two-dimensional, ordered, double transition metals carbides (MXenes). ACS Nano 9(10):9507–9516
45. Liu R, Li W (2018) High-thermal-stability and high thermal-conductivity Ti3C2T x MXene/poly (vinyl alcohol)(PVA) composites. ACS Omega 3(3):2609–2617
46. Hu C et al (2015) Branched graphene nanocapsules for anode material of lithium-ion batteries. Chem Mater 27(15):5253–5260
47. Naguib M et al (2011) Two-dimensional nanocrystals produced by exfoliation of Ti3AlC2. Adv Mater 23(37):4248–4253
48. Zhao Y et al (2014) Single-crystalline LiFePO4 nanosheets for high-rate Li-ion batteries. Nano Lett 14(5):2849–2853
49. Rui X et al (2013) Olivine-type nanosheets for lithium ion battery cathodes. ACS Nano 7(6):5637–5646
50. Li J-Y et al (2015) Romanechite-structured Na 0.31 MnO 1.9 nanofibers as high-performance cathode material for a sodium-ion battery. Chem Commun 51(80):14848–14851
51. Wang L et al (2017) Copper-substituted Na 0.67 Ni 0.3− x Cu x Mn 0.7 O 2 cathode materials for sodium-ion batteries with suppressed P2–O2 phase transition. J Mater Chem A 5(18):8752–8761
52. Liu J, Liu XW (2012) Two-dimensional nanoarchitectures for lithium storage. Adv Mater 24(30):4097–4111
53. Bonaccorso F et al (2015) Graphene, related two-dimensional crystals, and hybrid systems for energy conversion and storage. Science 347(6217):1246501
54. Han M et al (2010) Energy environ. Angew Sci Chem Int Ed 49:8738. 2015, 8, 81; b) R. Tripathi, TN Ramesh, BL Ellis, LF Nazar
55. Chen S et al (2013) Large-scale and low cost synthesis of graphene as high capacity anode materials for lithium-ion batteries. Carbon 64:158–169
56. Hassoun J et al (2014) An advanced lithium-ion battery based on a graphene anode and a lithium iron phosphate cathode. Nano Lett 14(8):4901–4906
57. Wang G et al (2009) Graphene nanosheets for enhanced lithium storage in lithium ion batteries. Carbon 47(8):2049–2053
58. Huang S et al (2015) Nitrogen and fluorine co-doped graphene as a high-performance anode material for lithium-ion batteries. J Mater Chem A 3(46):23095–23105
59. Jeon IY et al (2015) Edge-fluorinated graphene nanoplatelets as high performance electrodes for dye-sensitized solar cells and lithium ion batteries. Adv Func Mater 25(8):1170–1179
60. Qu B et al (2014) Layered SnS2-reduced graphene oxide composite–A high-capacity, high-rate, and long-cycle life sodium-ion battery anode material. Adv Mater 26(23):3854–3859

61. Sahoo M et al (2015) Green synthesis of boron doped graphene and its application as high performance anode material in Li ion battery. Mater Res Bull 61:383–390
62. Chen R et al (2014) Free-standing hierarchically sandwich-type tungsten disulfide nanotubes/graphene anode for lithium-ion batteries. Nano Lett 14(10):5899–5904
63. Xie X et al (2015) MoS2/Graphene composite anodes with enhanced performance for sodium-ion batteries: the role of the two-dimensional heterointerface. Adv Func Mater 25(9):1393–1403
64. Chen Y et al (2014) Reduced graphene oxide networks as an effective buffer matrix to improve the electrode performance of porous NiCo 2 O 4 nanoplates for lithium-ion batteries. J Mater Chem A 2(12):4449–4456
65. Masa J et al (2014) MnxOy/NC and CoxOy/NC nanoparticles embedded in a nitrogen-doped carbon matrix for high-performance bifunctional oxygen electrodes. Angew Chem Int Ed 53(32):8508–8512
66. Wang J et al (2014) Self-assembly of honeycomb-like MoS2 nanoarchitectures anchored into graphene foam for enhanced lithium-ion storage. Adv Mater 26(42):7162–7169
67. Xie X et al (2014) SnS2 Nanoplatelet@ graphene nanocomposites as high-capacity anode materials for sodium-ion batteries. Chem Asian J 9(6):1611–1617
68. Chang J et al (2014) Multilayered Si nanoparticle/reduced graphene oxide hybrid as a high-performance lithium-ion battery anode. Adv Mater 26(5):758–764
69. Chen S et al (2014) Hierarchical 3D mesoporous silicon@ graphene nanoarchitectures for lithium ion batteries with superior performance. Nano Res 7(1):85–94
70. Sun J et al (2015) A phosphorene–graphene hybrid material as a high-capacity anode for sodium-ion batteries. Nat Nanotechnol 10(11):980–985
71. Xie X et al (2017) Two-dimensional layered compound based anode materials for lithium-ion batteries and sodium-ion batteries. J Colloid Interface Sci 499:17–32
72. Urbankowski P et al (2016) Synthesis of two-dimensional titanium nitride Ti 4 N 3 (MXene). Nanoscale 8(22):11385–11391
73. Wang G et al (2009) Sn/graphene nanocomposite with 3D architecture for enhanced reversible lithium storage in lithium ion batteries. J Mater Chem 19(44):8378–8384
74. Su D, Ahn H-J, Wang G (2013) SnO 2@ graphene nanocomposites as anode materials for Na-ion batteries with superior electrochemical performance. Chem Commun 49(30):3131–3133
75. Xie X et al (2015) 3D Networked tin oxide/graphene aerogel with a hierarchically porous architecture for high-rate performance sodium-ion batteries. Chemsuschem 8(17):2948–2955
76. Zhu J et al (2014) Graphene double protection strategy to improve the SnO2 electrode performance anodes for lithium-ion batteries. Nano Energy 3:80–87
77. Jian Z et al (2014) Fe2 O3 nanocrystals anchored onto graphene nanosheets as the anode material for low-cost sodium-ion batteries. Chem Commun 50(10):1215–1217
78. Lin J et al (2014) Iron oxide nanoparticle and graphene nanoribbon composite as an anode material for high-performance Li-ion batteries. Adv Func Mater 24(14):2044–2048
79. Gao G, Wu HB, Lou XW (2014) Citrate-assisted growth of NiCo2O4 nanosheets on reduced graphene oxide for highly reversible lithium storage. Adv Energy Mater 4(14):1400422
80. Chen Y et al (2014) Graphene improving lithium-ion battery performance by construction of NiCo2O4/graphene hybrid nanosheet arrays. Nano Energy 3:88–94
81. Huang X et al (2014) Self-assembling synthesis of free-standing nanoporous graphene–transition-metal oxide flexible electrodes for high-performance lithium-ion batteries and supercapacitors. Chem Asian J 9(1):206–211
82. Wang H-W et al (2016) Resolving the structure of Ti3C2T x MXenes through multilevel structural modeling of the atomic pair distribution function. Chem Mater 28(1):349–359
83. Ohzuku T, Iwakoshi Y, Sawai K (1993) Formation of lithium-graphite intercalation compounds in nonaqueous electrolytes and their application as a negative electrode for a lithium ion (shuttlecock) cell. J Electrochem Soc 140(9):2490
84. Yata S et al (1994) Structure and properties of deeply Li-doped polyacenic semiconductor materials beyond C6Li stage. Synth Met 62(2):153–158

85. Stevens D, Dahn J (2000) High capacity anode materials for rechargeable sodium-ion batteries. J Electrochem Soc 147(4):1271
86. Dahn J, Seel J (2000) Energy and capacity projections for practical dual-graphite cells. J Electrochem Soc 147(3):899
87. Huang X et al (2012) Graphene-based composites. Chem Soc Rev 41(2):666–686
88. Huang C, Li C, Shi G (2012) Graphene based catalysts. Energy Environ Sci 5(10):8848–8868
89. Zhang L et al (2014) Hierarchical MoS2 microboxes constructed by nanosheets with enhanced electrochemical properties for lithium storage and water splitting
90. Wang X et al (2014) Atomic-scale clarification of structural transition of MoS2 upon sodium intercalation. ACS Nano 8(11):11394–11400
91. Su D, Dou S, Wang G (2015) Ultrathin MoS2 nanosheets as anode materials for sodium-ion batteries with superior performance. Adv Energy Mater 5(6):1401205
92. Xie X et al (2016) MoS2 nanosheets vertically aligned on carbon paper: a freestanding electrode for highly reversible sodium-ion batteries. Adv Energy Mater 6(5):1502161
93. Liu H et al (2012) Highly ordered mesoporous MoS2 with expanded spacing of the (002) crystal plane for ultrafast lithium ion storage. Adv Energy Mater 2(8):970–975
94. Shaju K, Rao GS, Chowdari B (2002) Performance of layered Li (Ni1/3Co1/3Mn1/3) O2 as cathode for Li-ion batteries. Electrochim Acta 48(2):145–151
95. Mao M et al (2014) High electrochemical performance based on the TiO 2 nanobelt@ few-layered MoS 2 structure for lithium-ion batteries. Nanoscale 6(21):12350–12353
96. Bruce PG, Scrosati B, Tarascon JM (2008) Nanomaterials for rechargeable lithium batteries. Angew Chem Int Ed 47(16):2930–2946
97. Zhou J et al (2015) 2D space-confined synthesis of few-layer MoS2 anchored on carbon nanosheet for lithium-ion battery anode. ACS Nano 9(4):3837–3848
98. Yoo H et al (2015) Cylindrical nanostructured MoS 2 directly grown on CNT composites for lithium-ion batteries. Nanoscale 7(8):3404–3409
99. Oakes L et al (2016) Interface strain in vertically stacked two-dimensional heterostructured carbon-MoS 2 nanosheets controls electrochemical reactivity. Nat Commun 7(1):1–7
100. Jiang H et al (2015) 2D monolayer MoS2–carbon interoverlapped superstructure: engineering ideal atomic interface for lithium ion storage. Adv Mater 27(24):3687–3695
101. Li J et al (2015) A three-dimensionally interconnected carbon nanotube/layered MoS2 nanohybrid network for lithium ion battery anode with superior rate capacity and long-cycle-life. Nano Energy 16:10–18
102. Taberna P-L et al (2006) High rate capabilities Fe 3 O 4-based Cu nano-architectured electrodes for lithium-ion battery applications. Nat Mater 5(7):567–573
103. Wang X et al (2015) Sulfur atoms bridging few-layered MoS2 with S-doped graphene enable highly robust anode for lithium-ion batteries. Adv Energy Mater 5(23):1501106
104. Miao Y-E et al (2015) Electrospun porous carbon nanofiber@ MoS 2 core/sheath fiber membranes as highly flexible and binder-free anodes for lithium-ion batteries. Nanoscale 7(25):11093–11101
105. Kong D et al (2014) Rational design of MoS 2@ graphene nanocables: towards high performance electrode materials for lithium ion batteries. Energy Environ Sci 7(10):3320–3325
106. Yang L et al (2013) Hierarchical MoS2/polyaniline nanowires with excellent electrochemical performance for lithium-ion batteries. Adv Mater 25(8):1180–1184
107. Hu Z et al (2014) MoS2 nanoflowers with expanded interlayers as high-performance anodes for sodium-ion batteries. Angew Chem Int Ed 53(47):12794–12798
108. Wang X et al (2015) Guest–host interactions and their impacts on structure and performance of nano-MoS 2. Nanoscale 7(2):637–641
109. Choi SH et al (2015) 3D MoS2–graphene microspheres consisting of multiple nanospheres with superior sodium ion storage properties. Adv Func Mater 25(12):1780–1788
110. Hu L et al (2015) MoS 2 ultrathin nanosheets obtained under a high magnetic field for lithium storage with stable and high capacity. Nanoscale 7(25):10925–10930
111. Zhou R et al (2016) Determination of the electron transfer number for the oxygen reduction reaction: from theory to experiment. ACS Catal 6(7):4720–4728

112. Wang J et al (2015) An advanced MoS2/carbon anode for high-performance sodium-ion batteries. Small 11(4):473–481
113. Delmas C, Fouassier C, Hagenmuller P (1980) Structural classification and properties of the layered oxides. Physica B+c 99(1–4):81–85
114. Yasuda K et al (2012) Acta crystallogr. Bull Chem Soc Japan 85(4):522–526. Sect. A Acta Crystallogr., Sect. A 32, 751, 1976
115. Xu X et al (2015) Facile synthesis of P2-type Na 0.4 Mn 0.54 Co 0.46 O 2 as a high capacity cathode material for sodium-ion batteries. RSC Adv 5(63):51454–51460

Design of an Experimental Test Rig to Determine the Battery Internal Resistance Beyond Normal Operating Condition

Abdullah Jubair, Zul Hilmi Che Daud, Izhari Izmi Mazali, Zainab Asus, and Mohd Kameil Abdul Hamid

Abstract The necessity of a precise battery thermal management system is to maximize the efficiency of energy storage capacity, driving range, cell longevity and system safety which requires a better understanding on battery thermal distribution and behavior. Since most of the research work is conducted based on the normal operating temperature, there is lack of data of the batter internal resistance at high temperature mostly above 60 °C. Hence, it is necessary to design a test rig where the battery can be tested for a wide range of temperature. During the design process, several paramcters were considered such as battery box shape, material and strength, air tunnel material and efficiency, chassis strength and stability, alignment and battery positioning. This test rig was designed to test the battery up to 80 °C of temperature. Finalized design was validated through finite element analysis and flow simulation in Solidworks. Justification was made through simulation results that the battery box fulfills the design requirements of safety and efficiency with 0.05 mm of displacement at maximum load and 28 m/s of maximum air velocity during operation.

Keywords Electric vehicle · Hybrid electric vehicle · Battery thermal management system · Lithium- ion battery

1 Introduction

The increasing demand of energy supply without the expansion of oil production leads to shortage of crude oil as a source on energy [1]. Also, the destructive environmental impact caused by oil consumption means that faster development and utilization of alternative powertrain system are urgently required [2]. Notably, transportation industry using combustion technology is responsible for more than 20% of

Supplementary Information The online version contains supplementary material available at https://doi.org/10.1007/978-981-19-1577-2_31.

A. Jubair · Z. H. C. Daud (✉) · I. I. Mazali · Z. Asus · M. K. A. Hamid
School of Mechanical Engineering, Faculty of Engineering, Universiti Teknologi Malaysia, 81310 Johor Bahru, Johor, Malaysia
e-mail: hilmi@mail.fkm.utm.my

© The Author(s), under exclusive license to Springer Nature Singapore Pte Ltd. 2023
M. H. A. Hassan et al. (eds.), *Technological Advancement in Instrumentation & Human Engineering*, Lecture Notes in Electrical Engineering 882,
https://doi.org/10.1007/978-981-19-1577-2_31

total carbon emissions causing significant damage to the environment [2, 3]. Hybrid electric vehicle (HEV) and electric vehicle (EV) are considered as alternatives of conventional vehicles due to their better fuel efficiency and minimum damage to the environment [4]. Lithium ion battery is one of the most preferred power sources in hybrid electric vehicles (HEVs) and electric vehicles (EVs) compared to the other types of battery due to its higher energy density, power density and no memory effect [5]. Fast acceleration capability and long driving range of lithium ion batteries makes them suitable for HEVs and EVs application [4, 5]. Battery temperature usually dependent of several parameters such as state of charge (SOC), charge and discharge rate, battery internal resistance and cooling system [6, 9]. Subsequently, the battery temperature has a great influence on efficiency, cell degradation and the battery life time, it needs to be carefully controlled [5, 6]. Uncontrolled variation of temperature may lead the battery to over- charge or over- discharge during cycling resulting premature failure or thermal runaway [7, 8].

Since the battery internal resistance plays a significant role in heat generation, it is necessary to understand the internal resistance behavior so that an efficient battery management system can be implemented. In the work by Sen and Kar [4], a small linear decrease of internal resistance with temperature from 25 °C up to 45 °Cwas observed together with a drastic increase at 45 °C and linear increase was found afterwards up to 150 °C. Whereas a constant decrease of battery internal resistance with temperature was found varying with the built material of the battery [10, 11]. Therefore, this paper explained a process to design a proper test rig so that the relationship of battery internal resistance to temperature for a wide temperature range, especially beyond normal operating condition, can be studied effectively. Thermal runway temperature is found to be 130, 170 °C for polyethylene and 170 °C for polypropylene based separator [12], 231 °C for polyethylene terephthalate ceramic separator [13], 244.1 °C for a 50 Ah NMC battery [14]. For a precise understanding of battery thermal behavior, developing an efficient test rig is necessary. In this paper, several designs of the experimental test rig have been proposed considering the thermal behavior during operation and thermal runaway under different condition such as environmental or adiabatic [14–19].

2 Methodology

2.1 Design Process

The main objective of the test rig designed in this paper is to analyze the thermal behavior of a large scale lithium ion battery beyond normal operating condition. Beyond normal operating condition is defined here as the battery temperature of over 60 °C and up to 80 °C. The design of the test rig is divided into three important components, namely, the battery box, air tunnel chassis, and, battery positioning. The design criteria for the battery box are high strength (to ensure that it can hold the battery cell properly), high thermal resistivity (to ensure that it can sustain battery's temperature up to 80 °C) and high transparency (to ensure that the battery cell can be observed during testing). For the air tunnel chassis, it is used to provide externally controlled air flow to cool down the battery. Its criteria are stability, ability to achieve high velocity air flow and cost efficient. The dimension and parameters of the air tunnel was decided considering the strength, minimum air flow resistance and cost efficiency. Finally, for the battery positioning, the main criteria are to make sure that uniform air flow can be distributed to the battery's terminals and to ensure that external connection facility can be provided so that proper sensors and data acquisition system can be connected to the tested battery.

In terms of analysis, strength analysis and flow analysis are carried out to check the final design's workability. In the strength analysis, conducted using finite element analysis in Solidworks maximum battery load has been applied on the battery terminal location and in the flow analysis, maximum air flow rate of 1095 m^3/h has been applied on the rig and measured air flow velocity and pressure.

2.2 *Material Selection*

Several materials are considered for the components of the test rig. The first component, which is the battery box, is required to provide safety and transparency during testing of the battery beyond the normal temperature condition. Thus, acrylic material is selected because it fulfills the strength and melting point requirements with its properties shown in Table 1 [20]. In terms of transparency, acrylic provides 92% light transparency per ASTM D1003 as shown in Fig. 1 [21]. Besides, material for adhesion must also be evaluated properly since it is responsible to fix the parts of the battery box together. The adhesion must also ensure that the box does not fail within the specified testing conditions. Thus, epoxy adhesive is chosen since it provides the maximum strength with 3.3 GPa of elastic modulus and Poisson's ratio of 0.3 for battery box material shown in Table 2 [23].

Next, the second component of the test rig, namely the air tunnel, is responsible to provide guided air flow through the battery box. Hence, its material must be cost

Table 1 Material properties of Acrylic battery box [20]

Property	Value	Units
Mass density	1200	kg/m^3
Elastic modulus	3000	N/mm^2
Poisson's ratio	0.35	–
Shear modulus	890	N/mm^2
Tensile strength	73	N/mm^2
Yield strength	45	N/mm^2
Melting point	160	°C

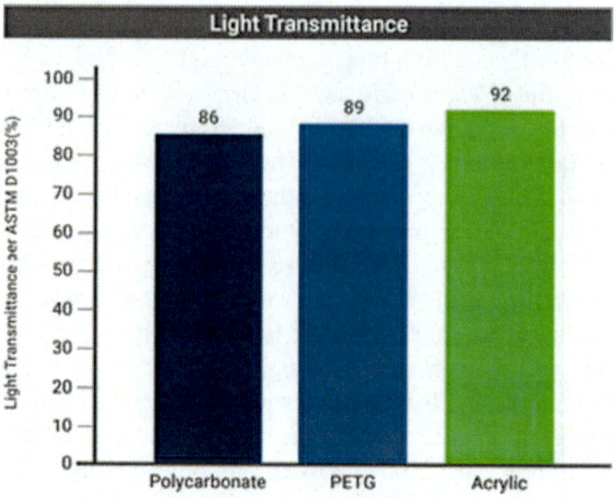

Fig. 1 Transparency of acrylic material [21]

Table 2 Material properties of Epoxy Adhesive [23]

Property	Value	Units
Density	1264	kg/m^3
Specific heat	1046	J/kgK
Thermal conductivity	0.179	W/mK
Elasticity modulus	3.3	GPa
Poisson's ratio	0.30	–
Thermal expansion coefficient	43.3	$\mu m/m°C$

effective and efficiency. As a result, PVC plastic is selected because it has the density of 1.4 g/cm^3 and operating temperature of up to 70 °C with its complete data shown in Table 3 [22]. The chassis of the air tunnel, on the other hand, is responsible for providing stability of the rig and to mount battery box and air tunnel together. Because of that, steel alloy with its properties provided in Table 4 is chosen.

Table 3 Material properties PVC for air tunnel [22]

Property	Value	Units
Density	1.5	g/cm^3
Insulation thickness	1.0	mm
Maximum temperature	60–70	°C
Softening temperature	120	°C
Insulation short-circuit temperature	135	°C
Insulation resistance	20	$m\Omega.km^{-1}$
Volume resistivity	1012–1015	Ω.cm
Dielectric strength	20–35	$Kv.cm^{-1}$

Table 4 Property table for steel alloys [24]

Property	Carbon steel	Aluminium alloy	Stainless steel	Units
Material yield strength	275	110	220	N/mm^2
Young's modulus	210,000	70,000	200,000	N/mm^2
Strain at fracture	24	12	45	%
Density	7850	2700	8000	kg/m^3
Thermal expansion coefficient	12×10^{-6}	23.2×10^{-6}	16×10^{-6}	K^{-1}
Thermal conductivity	54	250	16	W/mK
Total amount of material recycled	60	70	70	%

3 Results and Discussions

3.1 Final Design of the Test Rig

Table 5 shows the final design selected for the aforementioned test rig components of battery box, air tunnel chassis and battery positioning. For the battery box, acrylic is chosen as its material due to its high thermal resistance (melting point at 160 °C) and its transparency. The material most suitable for the air tunnel is PVC because of its cost and performance as an air tunnel whereas chassis material is steel alloy due to its strength and durability. In terms of battery positioning, it is decided that the vertical positioning is the most suitable since the air flow can be distributed uniformly

Table 5 Design criteria and selection process

	Battery Box	Air Tunnel & chassis	Battery positioning
Final Design		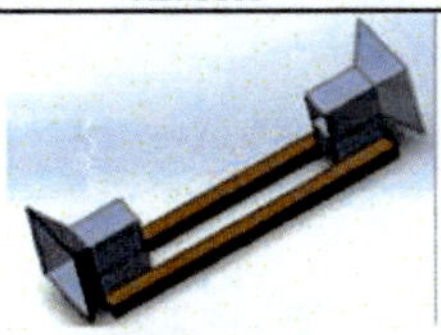	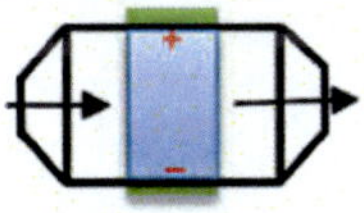
Selection criteria	• Strength • Transparency • Air flow rate • Axial Fan and tunnel mounting	• Stability and strength • Mounting and maximum air flow • Cost efficient	• Battery terminal location • Vertically aligned • Uniform air flow on the terminals • Replicability • External connection facility
Material	• Acrylic	• PVC (tunnel) and steel alloy (chassis)	-

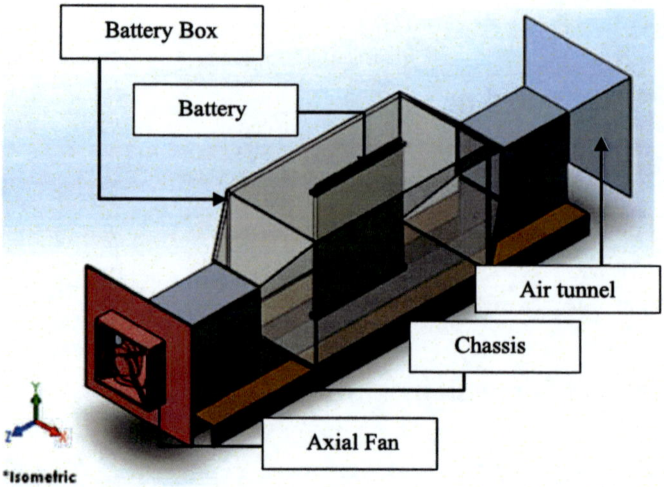

Fig. 2 Test rig isometric view

at the same velocity to both the positive and negative terminals. Upon considering three different position of the battery which are horizontal, vertical and 90 with surface, the vertical position of the battery can achieve maximum and uniform air flow, uniform terminal locations, zero engagement of connecting cable inside the box. This positioning also helps in terms of the electrical and censoring connections for the battery. Figure 2 shows CAD model of the test rig's final design which is particularly designed for 3.6 V 40 Ah NMC battery cell. The axial fan is mounted on one side of the air tunnel which is powered by 24 V DC to produce maximum air flow rate of 1095 m^3/h. Theoretically the air velocity and pressure can be measured using the Bernoulli's equations:

$$\frac{v^2}{2} + \frac{p}{\rho} + gz = constant \tag{1}$$

where, "v" is fluid velocity, "p" is fluid pressure, "ρ" is fluid density, "g" is gravitational constant and "z" is elevation.

In the final design, one battery cell is to be placed inside the battery box. Battery is then connected to the dynaload which is controlled by the system control cRIO 9074 to control the battery charge and discharge process. Then, two RTD has been used in this experiment which are located on the positive and negative terminals of the battery as the terminals are expected to be the most heated location. In addition, the anemometer is to be placed on the proper location where the air velocity is to be measured. In this experiment the anemometer has been placed at the nearest location on the battery air inlet end. Once the anemometer is placed accordingly is it connected to the system control. The anemometer and the RTDs are connected to the LabVIEW

Table 6 System control and data acquisition

Card name	Device name	System control/data acquisition	Input/ Output
NI 9217	RTD thermocouple	Data acquisition: cell surface temperature	RTD analog input
NI 9210	Anemometer	Data acquisition: Cooling air velocity	±10 V analog input
	Battery cell	Data acquisition: cell velocity	±10 V analog input
	Dynaload	Data acquisition: Discharge current	±10 V analog input
NI 9263	Electric fan	Control: Fan speed	±10 V analog ouput
	Dynaload	Control: Discharge Current	±10 V analog output

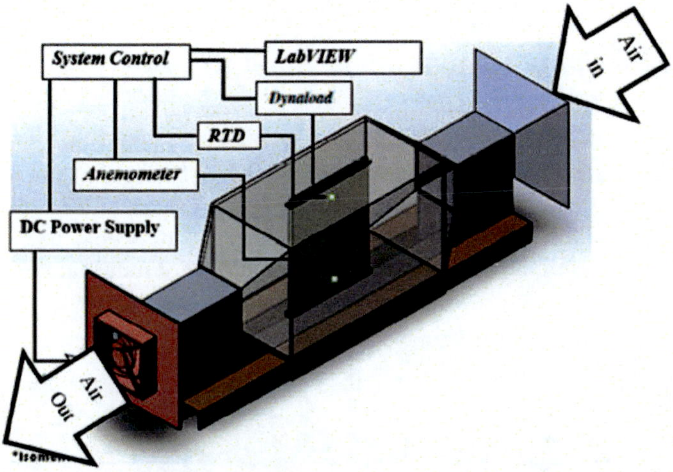

Fig. 3 Test rig set up

software through USB has been presented below in Table 6 [1]. Connections of the sensoring system are shown in Fig. 3.

3.2 Proposed Final Testing Procedure

This testing procedure is proposed to study the relation of battery internal resistance with temperature by using the final test rig's design:

i. The battery is fully charged, it is considered fully charged when voltage is 4 V.
ii. Measure and record the battery initial temperature at critical locations. Initially the temperature should be at environment temperature.
iii. Apply discharge current for 30 s

iv. Based on the voltage response in this step, the battery resistance in function of temperature calculate $R_{s,T}$ using the equation below:

$$R_{s,T} = \frac{\Delta V_s}{I} \tag{2}$$

v. Turn on the axial fan and provide maximum air flow. Allow a relaxation time for the battery to stabilize.
vi. Continue the experiment at higher temperature. In order to achieve this temperature, discharge the battery at higher rate state like 5 °C without cooling system. Using this process the cell temperature will increase rapidly.
vii. Then repeat experimental procedure 2 to 4 and calculate internal resistance up to 80 °C.

3.3 Results of the Analysis on the Final Design

For the battery box, it is necessary to analyze the box strength to make sure the battery box is safely capable of maintaining the overall weight of the battery and provide sufficient air flow to the battery during experiment. Since the battery pack is the heaviest component which will be placed inside the battery box, the most critical location is on the terminal gapes on the battery surface. From Fig. 4, Solidworks

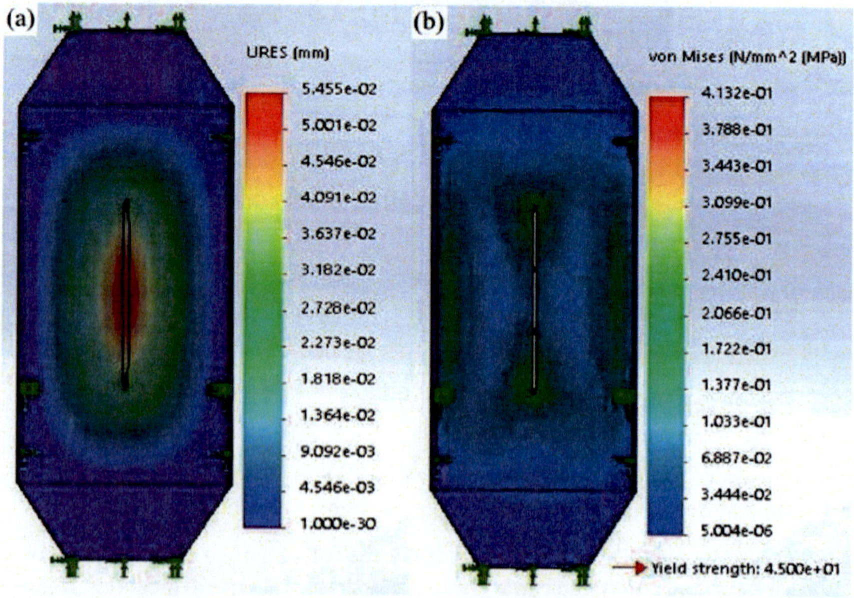

Fig. 4 Strength analysis **a** max. Displacement **b** Yield strength

stress analysis considering weight of large scale battery pack shows that the battery box has a safety factor of 143 with 0.05 mm displacement and 0.27 MPa of yield stress which concludes the battery box satisfies the strength requirements.

The battery box is designed in such a way which will provide minimum air flow resistance. Also, the air tunnel was designed to provide maximum efficiency. With the selected axial fan which can provide maximum air flow rate of 1095 m^3/h a simulation was conducted using Solidworks flow simulation. It was found that, a maximum velocity of 37 m/s can be reached in the tunnel but the battery box will reach up to 28 m/s which is because of its dimensional difference. A minor air resistance is also noticed which can be neglected as a high velocity of air is being provided to the battery. This will help the battery to a fast cool down and stabilization. Figure 5 below provides the simulation results of air velocity profile inside the rig.

In addition, from the air pressure profile throughout the test rig, it is noticed that the maximum air pressure is felt at the inlet and the minimum at outlet. Maximum air pressure is found to be approximately 101325 Pa and minimum of 99950 Pa.

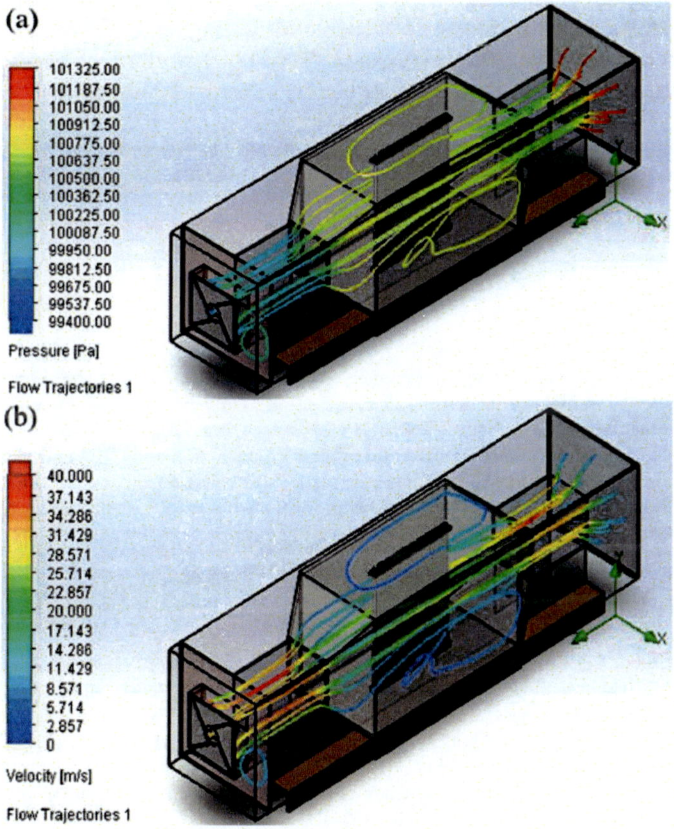

Fig. 5 Air flow simulation profile **a** Velocity **b** Pressure

However, inside the battery box air pressure is approximately 100637 Pa. So it can be concluded that the maximum air pressure will be at inlet, minimum air pressure is in outlet and battery box air pressure is in between the maximum and minimum values. Figure 5 shows the air pressure simulation profile of the rig. Finally, it can be said that the test rig especially the battery box is strong enough to maintain the battery and additional components with proper safety. The rig consisting battery box, air tunnel and axial fan can provide maximum efficient air flow to assist the battery a fast cool down and stabilization process. It can be conclude that, justification of design selection criteria has been made through simulation which fulfills the strength and air flow requirements.

4 Conclusion

As a conclusion, the final design of the test rig has been successfully generated with the ambition of studying the relation between battery internal resistances with temperature. The design proposed is suitable for testing a HEV/EV battery beyond the normal temperature condition (up to 80 °C). Future work of this research would be fabrication of the rig with the aim of battery experimental investigation.

Acknowledgements This work was funded by the Ministry of Higher Education under Fundamental Research Grand Scheme (FRGS/1/2018/TK03/UTM/02/18).

References

1. Zul HCD (2014) Contribution to thermal behavior study of lithium-ion battery for electric and hybrid electric vehicles, Mechanics [physics.med-ph], Université de Bourgogne
2. The International Energy Agency (2016) World Energy Outlook 2016, IEA
3. Zheng Y, Li S, Xu S (2019) Transport oil product consumption and GHG emission reduction potential in China: an electric vehicle-based scenario analysis. PLoS ONE 14(9):e0222448
4. Sen C, Kar NC (2009) Battery pack modeling for the analysis of battery management system of a hybrid electric vehicle. In: Vehicle power and propulsion conferences (VPPC), Dearborn, MI, pp 207–212
5. Watrin N, Roche R, Ostermann H, Blunier B, Miraoui A (2012) Multiphysical lithium-based battery model for use in state-of-charge determination. IEEE Trans Veh Technol 61(8):3420–3429
6. Xiao M, Choe S-Y (2013) Theoretical and experimental analysis of heat generations of a pouch type LiMn2O4/carbon high power Li-polymer battery. J Power Sources 241:46–55
7. Feng X, He X, Ouyang M, Lu L, Wu P, Kulp C et al (2015) Thermal runaway propagation model for designing a safer battery pack with 25 Ah LiNixCoyMnzO2 large format lithium ion battery. Apple Energy 154:7491
8. Zeng Y, Wu Z, Wang D, Wang Z, Chen L (2006) Overcharge investigation of lithium-ion polymer batteries. J Power Source 160:1302–1307
9. Lin C, Xu S, Liu J (2018) Measurement of heat generation in a 40Ah LiFePO4 Prismatic battery using accelerating rate calorimetry. Int J Hydrogen Energy 43(17):8375–8384

10. Lin C, Xu S, Liu J (2018) Measurement of heat generation in a 40 Ah Prismatic battery using accelerating rate calorimetry
11. Amirul HM, Zul HCD, Zainab A (2017) The impact of battery operating temperature and state of charge on the lithium-ion battery internal resistance. J Mek 40:01–08
12. Feng X, Zheng S, Ren D et al (2019) Investigating the thermal runaway mechanisms of lithium-ion batteries based on thermal analysis database. Appl Energy 246:53–64
13. Liu X, Ren D, Hsu H et al (2018) Thermal runaway of Lithium-Ion batteries without internal short circuit. Joule 2(10):1923–1924
14. Jin C, Sun Y et al (2021) Model and experiments to investigate thermal runaway characteristics of lithium ion batteries induced by external heating. J Power Sources 504:230065
15. Fan Y, Bao Y et al (2019) Experimental study on the thermal management performance of air cooling for high denslity cylindrical lithium-ion batteries. Appl Therm Eng 155:96–109
16. Huang Z, Liu J et al (2021) Experimental investigation on the characteristics of thermal runaway and its propagation of large-format lithium-ion batteries under overcharging and overheating conditions. Energy 233:121103
17. Sheng L, Zhang Z et al (2021) Experimental and numerical approach for analyzing thermal behaviors of a prismatic hard-cased lithium-ion battery. J Energy Storage 35:102313
18. Zhao C, Wang T et al (2021) Experimental study on thermal runaway of fully charged and overcharged lithium-ion batteries under adiabatic and side-heating test. J Energy Storage 38:102519
19. Troxler Y, Wu B et al (2013) The effect of thermal gradients on the performance of lithium-ion batteries. J Energy Storage 247:1018–1025
20. Taib M (2018) Study the effect of the shape of the miniplate on the stability of mandibular symphyseal fracture using finite element analysis. Egypt J Oral Maxillofac Surg 9(3):135–149
21. Plastics C. Building a better machine guard with plastics. https://www.curbellplastics.com/Research-Solutions/Resources/Articles/Building-a-Better-Machine-Guard-With-Plastic, Accessed 23 Sept 2021
22. Shan-jun M, zhang J, et al (2013) Study on pyrolysis characteristics of cross-linked polyethylene material cable. Proc. Eng 52:588–892
23. Kamel A, Faruk S (2013) Stress analysis of hybrid joints of metal and composite plates via 3D-FEM. Indian J Eng Mater Sci 20:92–100
24. Leroy G et al (2007) Life-cycle costing of metallic structure. Eng Sustain 160(4):167–177

Mechanical Performance of Cornstalk Fibre/Fibreglass/Polyphenylene Sulfide (PPS) Composites Bicycle Frame Using Finite Element Analysis

Humeshvaren Maganathan, Cik Suhana Hassan, Muhamad Faliq Mohamad Nazer, and Nor Fazilah Abdullah

Abstract Bicycle is a widely used human powered machine that is not just use as a mode of transportation but also serves purpose for recreational and sports activity. The main aim of this paper is to evaluate the adaptability of natural fiber-based composites as bicycle frame material. The mechanical performance of a standard road bicycle frames was evaluated under riding and climbing conditions where the load mainly acting at the hub, bottom, handlebars, seat post and saddle. The analysis was conducted using ANSYS finite element software employing hybrid corn stalk fibre/fiberglass reinforced Polyphenylene Sulfide composites as the frame material. The material performance was compared with conventional bicycle frame material, that is aluminium and steel. It was found that the deformation of composite material frames was generally higher than the conventional material while stresses induced were comparable. Vibration dampening properties and high strength to weight ratio of the natural fibre promises more comfortable ride as opposed to the stiff and heavy steel, thus with proper composite material design; a balance in cost-performance-sustainability could be achieved.

Keywords Biocomposites · Bicycle frame · Cornstalk fibre · Bicycle material

1 Introduction

The production of bicycles globally outranges the production of automobiles with a ratio of three to one approximately [1]. The purpose of bicycles in this twenty-first century falls under these six categories such as utility, BMX, touring, mountain, racing and hybrid. Bicycle frame is the major component of a bicycle where the wheels and other component will be installed onto the frame, hence indicating that the frame plays a major role in sustaining loads. The safety factor of a bicycle frame relies on the structural design of the frame and the material used for the frame [1]. The material that has greater stiffness will prevent flexing on the frame and prevents

H. Maganathan · C. S. Hassan (✉) · M. F. M. Nazer · N. F. Abdullah
Mechanical Engineering Department, Faculty of Engineering, Technology and Built Environment, UCSI University, 56000 Kuala Lumpur, Malaysia
e-mail: suhana@ucsiuniversity.edu.my

© The Author(s), under exclusive license to Springer Nature Singapore Pte Ltd. 2023
M. H. A. Hassan et al. (eds.), *Technological Advancement in Instrumentation & Human Engineering*, Lecture Notes in Electrical Engineering 882,
https://doi.org/10.1007/978-981-19-1577-2_32

parts collision when heavy load is applied. The conventional bicycle frame materials such as aluminium and steel are relatively heavy, costly, requires high precision metal working tools and it is a non-renewable material source. Steel is a strong material but it is heavy and prone to rust. Due to considerable demand for lightweight properties which will enhance the riding performance, manufacturers began to replace the steel with Aluminum. However, despite being light weight and corrosion resistant, Aluminum is relatively expensive and often resulted in harsher quality ride due to its high stiffness. The drawback of the conventional materials thereby leads. Researchers into exploring utilization of composites material for bicycle frame as effort to satisfy both the weight and cost reduction while also promising comfort ride. Carbon fiber and fibreglass composite are the modern light weight material that has begun to take its place in the industry with carbon fibre has also been utilized as bicycle frame material. In addition to high strength, carbon fiber offers considerable damping coefficient to absorb to dissipate vibration [2]. Carbon fiber is a synthetic material that requires petroleum based raw material to construct thus, it is a non-biodegradable material [3]. Manufacturing of carbon fiber material exhibited emission of carbon dioxide gases to the environment. On the other hand, fiberglass productions also found to emit pollutants and volatile organic compounds (VOCs) [4]. The grown of environmental concern has fueled the search for alternative materials that can reduce the usage of synthetic fibre but still retain the beneficial properties. The awareness towards a better environment and depletion of non-renewable material sources opened the opportunity for natural fibers to be used concurrently with the synthetic fibers. Natural fibers are lightweight and possess great mechanical properties, thus, offering weight reduction opportunities without compromising the expected strength and stiffness. Being both biodegradable and recyclable, the natural fibre materials such as flax, cornstalk, kenaf and many others could essentially be grown for the purpose of manufacturing composites. George and James [5] found that addition of an external flax layer into an epoxy/carbon fibre-reinforced composite considerably enhanced its damping ratio by 53.6% and by adding two layers increased it by 94%. Hybrid flax/carbon fiber composite is a mixture of flax material with carbon fiber material to form a hybrid composite material. This hybrid composite material balances the strength, stiffness and cost of a composite material. Flax fiber alone expresses poor mechanical properties as compared to synthetic fibers, thus the usage can be hybridized with carbon fiber to overcome the shortcoming [5]. In addition to enhancing degradability and recyclability, the use of natural fiber offers also capable of offering significant cost and weight reduction. Furthermore, as compared with glass fibre, natural fibres have better sound absorbing efficiency, are more shatter resistant, and have better energy management characteristics [6].

According to Azman et.al [7], the lower weight and relatively lower cost of natural fibers are the main aspects referred in consideration for the use of natural fiber composites in sports equipment. As most sports equipment relies on humans to move light equipment is desirable [8]. According to Ali Amiri et.al [9], hybridizing carbon fiber with flax fiber withstands greater vibration than synthetic materials. The study on damping ratio and deflection shows that hybrid flax/carbon fiber is slightly stiffer than carbon fiber and close to aluminium stiffness, and the material has greater damping ratio compared to all the conventional frame material [9]. The research on natural fiber composites application for bicycle frame material is still very low. This has opened opportunity for many other natural fibre to be explored such as cornstalk fiber. Corn is categorized as the world's third largest agricultural crop that produces large amount of agricultural waste. The United States of America, (USA) produces 366.6 million metric tons of corn approximately. It is reported that they are the world's largest corn producer and exporter. China is the second largest corn producer in the world with a production of 257.7 million metric tons. Besides, there are several other countries that also contributes to this global corn demand such as Brazil, Argentina, Ukraine and India with a production number of 94.5 million metric tons, 46 million metric tons, 35.5 million metric tons and 26 million metric tons approximately [10]. Malaysia produced 58 thousand tons of sweet corn in the year 2020 and it is growing at an average annual rate of 13.96% [11]. This agricultural crop produces a large amount of agricultural waste globally that consists of leaves and stalks. It is reported that, the United States produces 250 million tons of corn crop waste per year approximately [12]. This corn crop waste is just being abandoned or burned in the field thereby considered harmful to the environment due to the poor waste management. The abundantly available corn stalks can be used to design light weighted materials for automation, aerospace, bicycle frames and many more. As such, this would not only be able to create a value-added product but would also be able to solve the negative influence of the abundant agricultural waste on the environment.

In this study, an analysis on the mechanical performance corn stalk/fiberglass hybrid reinforced PPS composites have been carried out to determine the suitability of these materials as a bicycle frame material. The main idea is to achieve weight reduction of the frame by the application of those materials. The performance was evaluated and compared against the conventional bicycle frame material.

2 Methodology

Figure 1 shows the standard road bicycle frame utilized in this analysis. The modelling was done utilizing SOLIDWORKS software. The main parts of the road bicycle frame can be divided into head tube, top tube, seat tube, down tube, seat stay and chain stay. The design parameters for the bicycle frame is shown in Fig. 1 with Table 1. For this analysis, the meshing element used is 5 mm as shown in Fig. 1 below, the number of nodes and elements employed are 175,417 and 99,203 respectively.

The materials utilized for this analysis are S-glass fibre (FG), cornstalk fibre (CS) and Polyphenylene Sulfide (PPS). The properties of the materials are listed in Table 2. PPS is a high-performance, building thermoplastic characterized by an unordinary combination of properties. These properties run from high temperature execution to dimensional solidness and great electrical separator properties. It expresses great fatigue endurance and creep resistance and has low elongation to break [16].

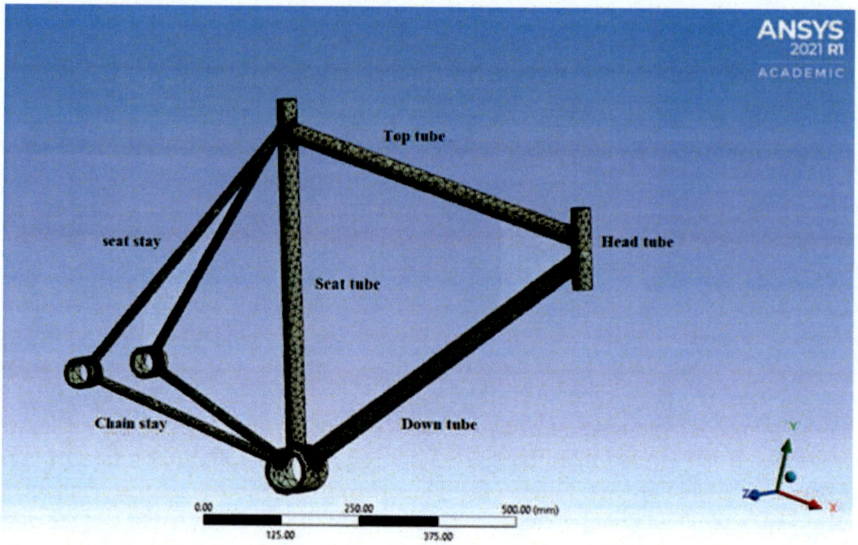

Fig. 1 Bicycle frame with 5 mm meshing

Table 1 Design parameters of bicycle frame

Parameter	Dimension
Head tube angle	73.5°
Seat tube angle	73.5°
Top tube length	570 mm
Seat tube length	543.75 mm
Chain stay length	360 mm
Head tube length	130 mm

Table 2 Mechanical properties of raw materials

Material Properties	S-glass fibre (FG) [13]	Cornstalk fibre (CS) [14, 15]	Polyphenylene sulfide (PPS) [16]
Diameter, μm	3.8–12.7	20	–
Length, mm	3.17	1.0–1.5	–
Density, g/cm^3	2.495	1.04	1.35–1.80
Tensile strength, MPa	4800	470–670	83–195
Modulus (E), GPa	55	6.3	0.0031–0.0373
Young's modulus, GPa	93	8.6	3.3–4
Poisson's ratio	0.23	0.32	0.37–0.43
Shear modulus, GPa	39	0.92	0.0558–0.097

Table 3 Mechanical properties of materials selected

Composite	Composition	Density (g/cc)	Modulus of elasticity (GPa)	Poisson's ratio	Tensile yield strength (MPa)	Shear modulus (GPa)
Hybrid 1	40% CS + 10% FG + 50% PPS	1.453	8.03	0.35	817.5	0.174
Hybrid 2	30% CS + 10% FG + 60% PPS	1.5065	7.4	0.36	764.4	0.145
Hybrid 3	20% CS + 20% FG + 60% PPS	1.652	12.272	0.35	1177.4	0.145
Hybrid 4	30% CS + 20% FG + 50% PPS	1.5985	12.9	0.34	1230.5	0.174

The mechanical properties of composite were calculated using rules of mixture method [17]. Table 3 shows the calculated mechanical properties of composites. The fibre-to-weight ratio has been analysed using the calculation as shown in the Table 2 in terms of hybrid 1, hybrid 2, hybrid 3 and hybrid 4. From the calculation, it has proved that the fibre-to-weight ratio has significant affect in the mechanical performance of the hybrid materials.

The statics structural analysis for the frame were conducted under five loading conditions based on past work [18, 19] and are applied on all the six frames individually. The load conditions applied were:

(i) *Static start up*

Static start-up simulated the condition where rider is about to start pedaling but the bicycle is in rest and vertical equilibrium position. A load of 700 N was applied on the seat tube to imply the rider's weight and the pedaling force of 200 N was applied on the bottom bracket. The front and rear forks were fixed in this condition as shown in Fig. 2.

(ii) *Steady state pedaling*

Steady state pedaling is a condition where the rider sits on the saddle and applies force constantly. The head tube and the rear end fork was fixed in this condition. A force of 200 N was applied on the bottom bracket and a force of 1000 N was applied on the head tube as shown in Fig. 3.

(iii) *Vertical impact*

The load in this condition is applied on the seat tube of the bicycle frame in vertical direction. The load applied on the frame is multiplied with predetermined amount of G factor which is 2250 N. The front tube and the rear fork end were fixed in this loading condition as shown in Fig. 4.

(iv) *Horizontal impact*

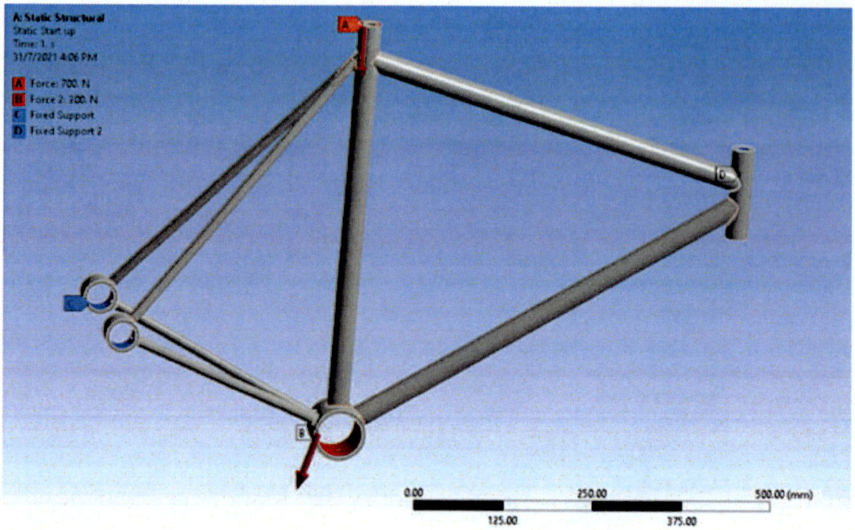

Fig. 2 Static start-up condition

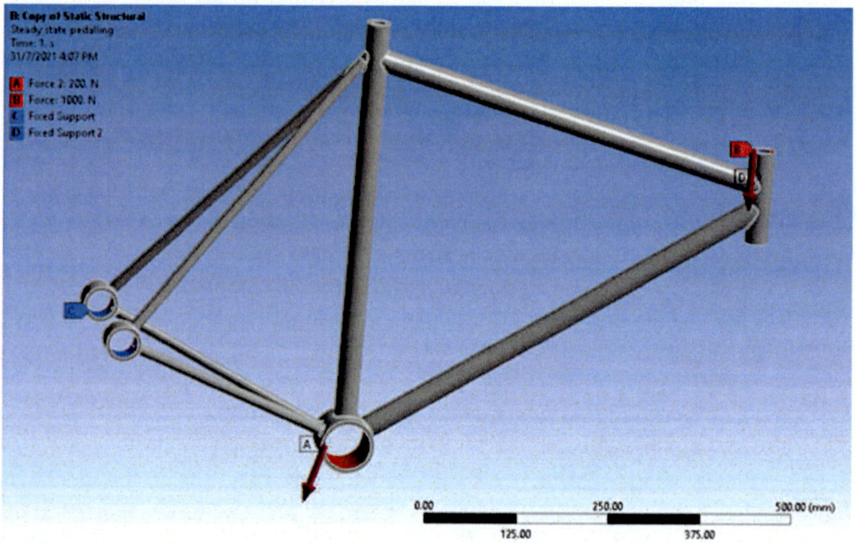

Fig. 3 Steady-state pedaling condition

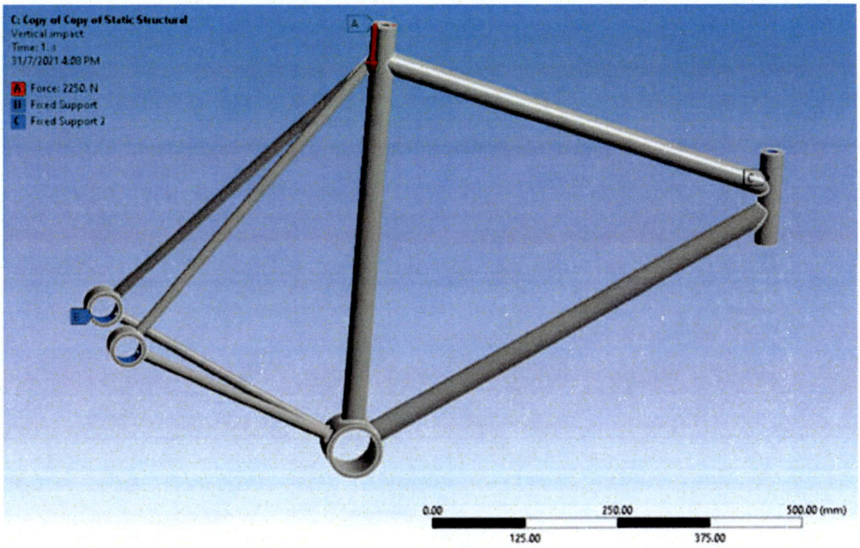

Fig. 4 Vertical impact condition

Horizontal impact loading is to determine the deformation on the head tube of the bicycle frame. The rear fork end and the head tube was fixed and a load of 2250 N was applied on the top tube in horizontal direction to determine the stress analysis for this loading condition as shown in Fig. 5.

(v) *Rear wheel braking*

To simulate rear wheel braking condition, the forces are applied on the rear ends of the fork and the bottom bracket is subjected to fixed support condition. A force of 750 N is applied on the rear end forks of the bicycle frame in horizontal direction as shown in Fig. 6.

3 Results and Discussion

The mechanical performance of composites material was evaluated against the performance of conventionally available bicycle frame materials, that is Aluminum and Steel. The maximum deformation under each loading conditions are presented in Table 4. It is observed that the deformation of the bicycle frame made of composites material generally higher than the stiffer Aluminum and Steel. It is also worth noting that the increases of the cornstalk fibre; supplemented with comparable amount of fibre glass would be able to reduce the deformation of the composites frame, as shown by Hybrid 4 in comparison to Hybrid 1, 2 and 3 frames.

Despite the large differences in the deformation, the maximum stresses induced in the composites bicycle frame materials are generally similar with the Aluminum and

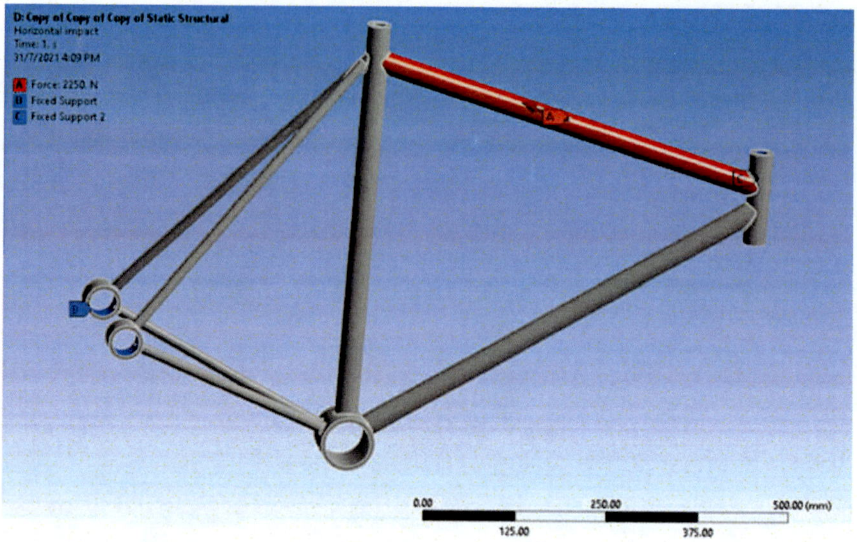

Fig. 5 Horizontal impact condition

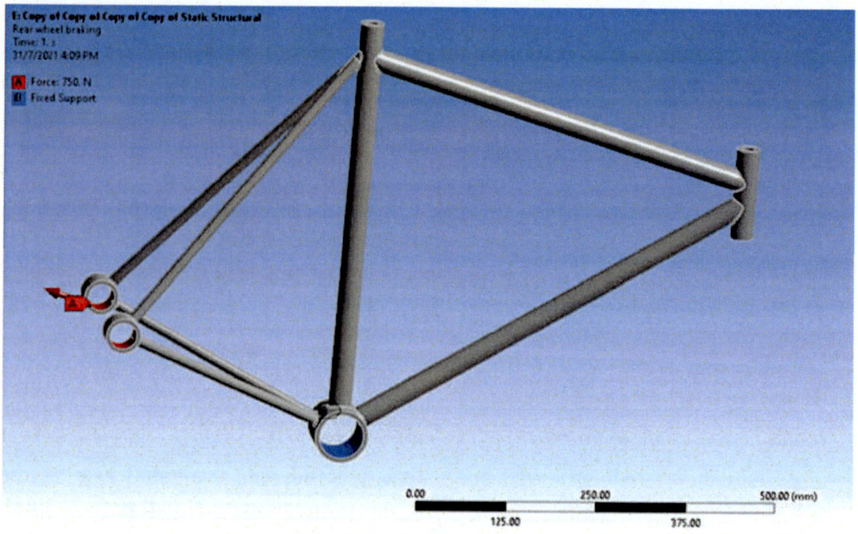

Fig. 6 Rear wheel braking condition

Table 4 Comparison of maximum deformation on composites bicycle frame under various loading condition

Condition	Maximum deformation (mm)					
	Aluminum	Steel	Hybrid 1	Hybrid 2	Hybrid 3	Hybrid 4
Static start-up	0.027181	0.00967	0.26016	0.23995	0.15701	0.14948
Steady state pedaling	0.005059	0.001801	0.048393	0.044642	0.029211	0.027817
Vertical impact	0.073468	0.026141	0.70308	0.6485	0.42433	0.40402
Horizontal impact	0.018223	0.006474	0.1747	0.16104	0.10537	0.10027
Rear wheel braking	4.6251	1.6358	44.529	40.99	26.821	25.486

Steel frames under all loading conditions, as shown in Table 5. Further research is required to analyse the effect of fibre orientation to control the deformation and understand the implications of the composite's material on safety. With proper composite material design, a balance in cost-performance-sustainability could be achieved. Furthermore, the vibration dampening properties and high strength to weight ratio of the natural fibre promises more comfortable ride as opposed to the stiff and heavy Steel.

Figures 7 and 8 shows the deformation of Hybrid 4 under static start-up and vertical impact loading condition, respectively. For both static start-up and vertical impact loading condition; the deformation found to be concentrated at the top tube of the bicycle frame. The results indicates that the top tube is the most prone to break in the event of high load imparted on the seat including rider's weight. Careful consideration on laying up of composites at the top tube hence must be designed following the load at the seat for safety and as lesser vertical bending also plays vital role in achieving smoother ride.

Figures 9 and 10 shows the deformation contour of the Hybrid 4 bicycle frame material under horizontal impact and steady-state pedaling condition, respectively.

Table 5 Comparison of maximum stress on composites bicycle frame under various loading condition

Condition	Maximum stress (MPa)					
	Aluminum	Steel	Hybrid 1	Hybrid 2	Hybrid 3	Hybrid 4
Static start-up	7.9509	7.9836	7.9155	7.9155	7.9277	7.9395
Steady state pedaling	3.527	3.6098	3.4373	3.4373	3.4681	3.498
Vertical impact	22.237	22.354	22.173	22.173	22.155	22.196
Horizontal impact	7.9069	7.8907	7.9261	7.9261	7.9195	7.9131
Rear wheel braking	106.42	105.47	107.43	107.43	107.09	106.75

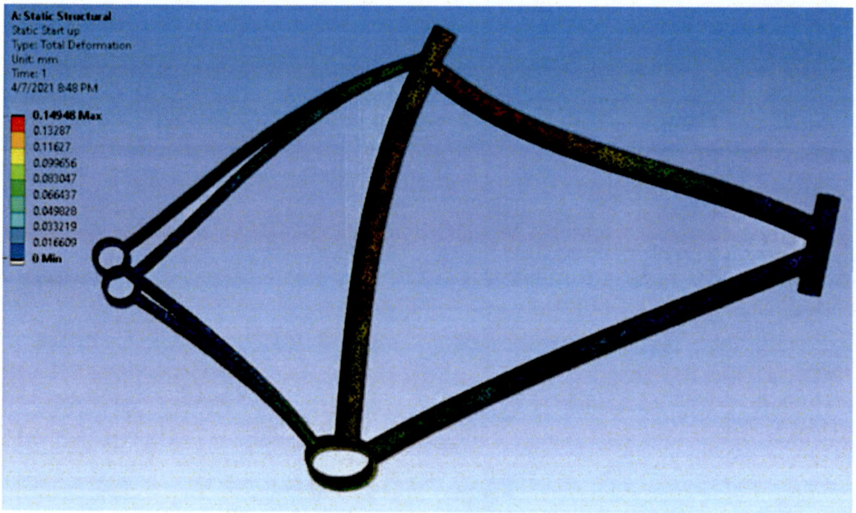

Fig. 7 Deformation contour under static start-up condition

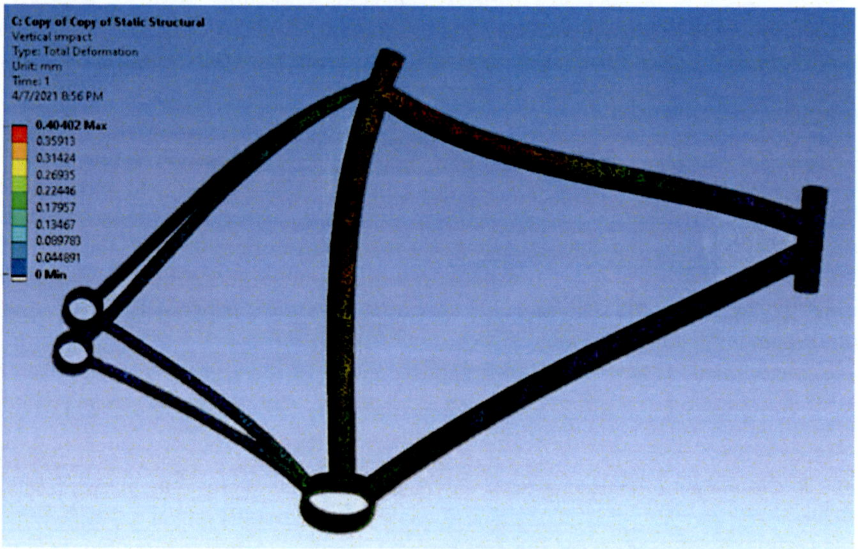

Fig. 8 Deformation contour under vertical impact condition

The deformation was observed to be induced at the upper region of top tube, seat stay and seat tube for horizontal impact condition and observed to be opposite under the steady-state pedaling condition. The deformation under steady-state pedaling however not as high as the horizontal impact condition; firstly due to the lower

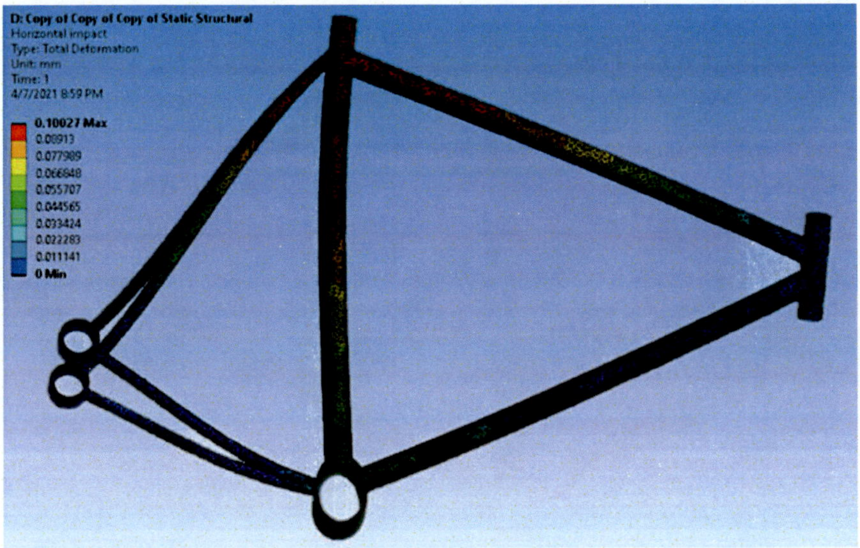

Fig. 9 Deformation contour under horizontal impact condition

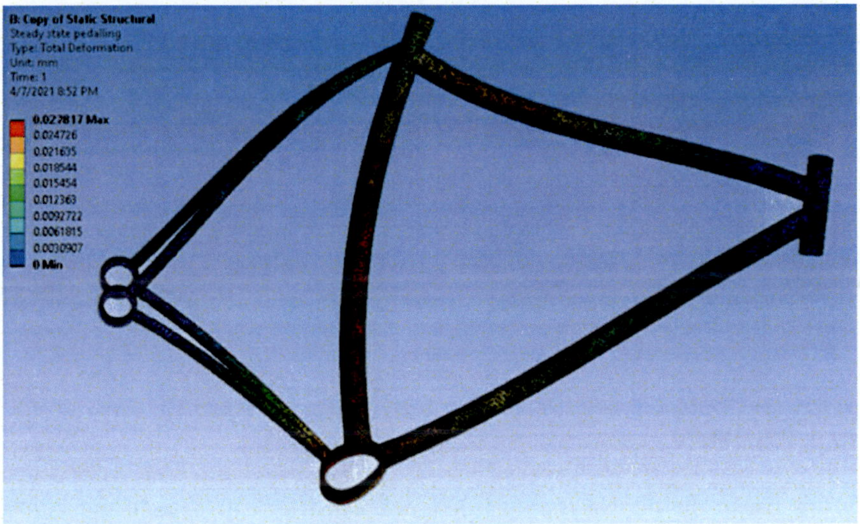

Fig. 10 Deformation contour under steady-state pedaling condition

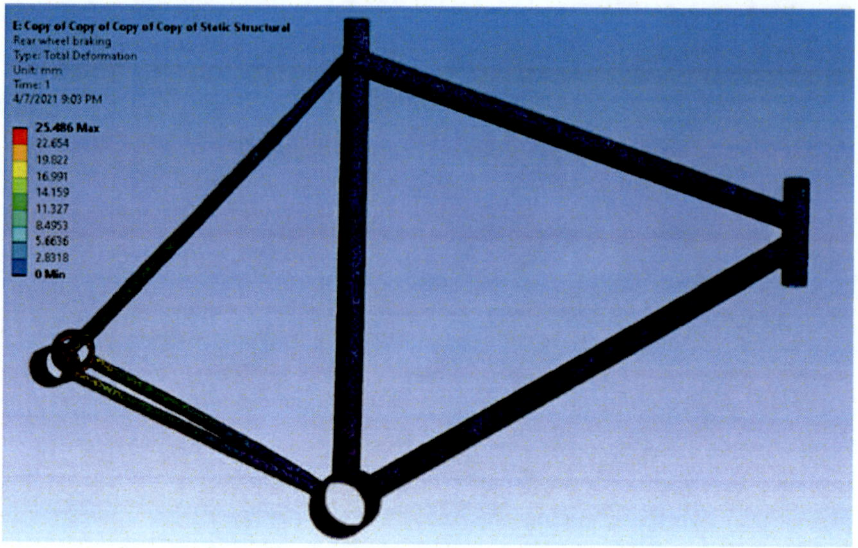

Fig. 11 Deformation contour under rear wheel braking condition

amount of load and secondly due the tapered dimension of the bottom part of the frame which may provide additional stiffness which may also applies under the horizontal impact condition.

Deformation contour of Hybrid 4 bicycle frame under rear wheel braking condition is shown in Fig. 11. The large deformation and high amount of stress as spelt in Table 5 might indicates that the bicycle frame underwent plastic deformation at the seat and chain stay and the material may permanently weaken. Further analysis on composites fibre arrangement and alterations are required order to reduce the stresses induced to ensure stiffness of each bicycle parts are catered in accordance with the load of each riding condition.

4 Conclusion

In this paper, the bicycle frame is designed with cornstalk fibre/fibreglass/PPS composites to analyze the mechanical performance of a road bicycle standard dimension for a rider with weight of 700 N. The performance was analyzed under five different loading conditions that characterized the stress and deformation developed upon load application. It was found that the cornstalk/fiberglass/PPS composites bicycle frame would deform substantially under all loading conditions as compared to the Aluminum and Steel bicycle frame. It is however worth noting that the increases of the cornstalk fibre; supplemented with comparable amount of fibre glass would be able to reduce the deformation of the composites frame, as depicted by performance of Hybrid 4 which made up of 30% CS + 20% FG + 50% PPS. With proper analysis of fibre alignment, the composites may be able to generate comparable performance as the conventional material. The values of the parameters obtained in this paper however ought to be taken with save as it ought to be deciphered as a great estimation. The concrete values ought to be confirmed by point-by-point exploratory testing of the fabricated model beneath genuine conditions.

References

1. Berto, FJ (2021) bicycle|Definition, history, types, & facts. Encyclopedia Britannica. https://www.britannica.com/technology/bicycle, Accessed 12 Jan 2021
2. Kang H, Kim C, Lee J (2020) Modal damping coefficient estimation of carbon-fiber-reinforced plastic material considering temperature condition. Materials 13(12):2872
3. Relja. Bicycle frame materials—explained|BikeGremlin. BikeGremlin. https://bike.bikegremlin.com/11144/bicycle-frame-materials-explained/, Accessed 8 Mar 2021
4. Joshi S, Drzal L, Mohanty A, Arora S (2004) Are natural fiber composites environmentally superior to glass fiber reinforced composites? Compos Part A Appl Sci Manuf 35(3):371–376
5. Fairlie G, Njuguna J (2020) Damping properties of flax/carbon hybrid epoxy/fibre-reinforced composites for automotive semi-structural applications. Fibers 8(10):64
6. Hassan CS, Durai V, Salit MS, Aziz NA, Yusoff MZM (2018) Mechanical and crash performance of unidirectional oil palm empty fruit bunch fibre-reinforced polypropylene composite. Bioresources 13:8310–8328
7. Azman M, Asyraf M, Khalina A, Petrů M, Ruzaidi C, Sapuan S, Nik W, Ishak M, Ilyas R, Suriani M (2021) Natural fiber reinforced composite material for product design: a short review. Polymers 13(12):1917
8. Zhang L (2015) The application of composite fiber materials in sports equipment. Atlantis Press, p 4
9. Amiri A, Krosbakken T, Schoen W, Theisen D, Ulven C (2017) Design and manufacturing of a hybrid flax/carbon fiber composite bicycle frame. Proc Inst Mech Eng Part P J Sport Eng Technol 232(1):28–38
10. Carol M. (2021) The World's 6 Biggest Corn Producers. Investopedia. https://www.investopedia.com/articles/markets-economy/090316/6-countries-produce-most-corn.asp, Accessed 13 July 2021
11. Adnan M (2021) An overview of the grain corn industry in Malaysia. FFTC Agricultural Policy Platform (FFTC-AP). https://ap.fftc.org.tw/article/1377#:~:text=In%202003%2C%20the%20sweet%20corn,tons%20to%2084%2C170%20tons%2C%20respectively, Accessed 1 Feb 2021
12. Penn state. Corn waste potentially more than ethanol. https://www.sciencedaily.com/. https://www.sciencedaily.com/releases/2006/07/060719091421.htm, Accessed 14 July 2021
13. AZoM.com. Properties: s-glass fibre. https://www.azom.com/properties.aspx?ArticleID=769, Accessed 10 July 2021
14. Daud Z, Mohd M, Abdul A, Awang H (2016) Corn stalk fiber material by chemical pulping process for pulp and paper industry. Adv Mater Res 1133:608–611
15. Daud Z, Mohd M, Mohd A, Awang H, Mohd A (2013) Exploring of agro waste (pineapple leaf, corn stalk, and napier grass) by chemical composition and morphological study. BioResources 9(1):872–880
16. Anil Kumar A, Dinesh BL, Gaddikeri KM, Sundaram R (2017) Challenges in processing of PPS-glass Thermoplastic composites. Research Gate, (ICEMP-2014), p 7. https://www.researchgate.net/publication/283495508_Challenges_in_Processing_of_PPS-Glass_Thermoplastic_Composites, Accessed 6 July 2021

17. Kopeliovich D (202) Estimations of composite materials properties [SubsTech]. Substech.com. https://www.substech.com/dokuwiki/doku.php?id=estimations_of_composite_materials_properties#:~:text=Rule%20of%20Mixtures,matrix%20and%20dispersed%20phase)%20properties, Accessed 31 July 2021
18. Mishra KK, Chaturvedi V(2021) Stress analysis of bicycle frame using ANSYS. Int Res J Eng Technol 8(5)
19. Sarath P (2021) Stress analysis of bicycle frame using different materials by FEA. GRD JS Glob Res Dev J Eng 6(7):14–20

Study on Low Frequency Vibration Isolation Characteristics of Transformer by Phononic Crystal

Zhaokai Sun, Bo Zhang, Yudong Pan, Shilong Lu, and Yuyang Mao

Abstract In this paper, two new types of locally resonant phononic crystal structures are proposed. The mechanism of band gap and the influence factors of start-stop frequency of phononic crystals in low frequency range are analyzed by COMSOL Multiphysics calculation. The results show that the maximum band gap of the original structure is 179–1218 Hz. The phononic crystals improved by piercing have obvious advantages in the low frequency vibration range of 50–200 Hz. In order to achieve the goal of improving the damping performance, the effects of different arrangement forms of phononic crystals on the transmission loss are compared. The results show that the mixed arrangement and combination method increases the transmission loss peak to 38 dB, has the band gap characteristics of two kinds of phononic crystal structures, and broadens the sound insulation range. This study provides a new method for the application of vibration and noise reduction of power transformer.

Keywords Local resonance · Low frequency band gap · Phononic crystals · Vibration isolation characteristics · Band broadening

1 Introduction

Low-frequency vibration and noise have become a new feature of environmental pollution because they are not easy to attenuate, have a long propagation distance and have strong penetration ability. The frequency band of vibration and noise of transformer is concentrated in 0–1000 Hz, which is the most prominent in the low frequency range of 50–200 Hz [1, 2, 21]. It is mainly the vibration and secondary noise transmitted to the base by fan, fuel tank, winding and so on. At present, the actual passive measures for shock reduction and noise reduction of power transformers are mainly through welding damping steel plate [3], installing frequency interferer [4],

Z. Sun · B. Zhang (✉) · Y. Pan · S. Lu · Y. Mao
School of Mechanical Engineering, Ningxia University, Ningxia, China
e-mail: zhangb@nxu.edu.cn

B. Zhang
CAE Key Laboratory for Intelligent Equipment of Ningxia, Ningxia, China

© The Author(s), under exclusive license to Springer Nature Singapore Pte Ltd. 2023
M. H. A. Hassan et al. (eds.), *Technological Advancement in Instrumentation & Human Engineering*, Lecture Notes in Electrical Engineering 882,
https://doi.org/10.1007/978-981-19-1577-2_33

setting up sound insulation damping wall [5], applying multi-layer composite sound absorption material [6–8] and so on.

With the development of science and technology, artificial acoustic materials such as negative equivalent mass density and negative equivalent bulk modulus have been proposed. Phononic crystals can be used in vibration isolation [13–17], waveguide [19] and positioning [9]. The band gap characteristics of phononic crystals depend on the distribution of scatterers and matrix materials in the unit structure. The factors that affect the band gap range mainly include structural parameters, material parameters, multiphase material distribution and other factors [10, 11]. In 2000, Liu [12] put forward the concept of local resonance mechanism for the first time. Local resonant phononic crystals with low frequency characteristics are favored by many scholars. In 2016, Qi [14] proposed a four-corner connected local resonant phononic crystal. By introducing the gas pump damping technology, changing the material composition of the multi-oscillator broke the symmetry of the model, and studied the sound insulation characteristics of the model. In 2020, Peng [15] proposed that two-dimensional open-hole local resonance phononic crystals achieve a complete band gap as low as 29.72–91.93 Hz by changing the opening radius of the cladding layer and the radius of the scatterer. Ruan [16] prepared spiral resonant phononic crystals [13, 17] by using Archimedes helix to reduce structural stiffness, which solved the 15–45 Hz low frequency mechanical vibration produced by warship power system and improved the subwavelength band gap range. Guo [20] designed the phonon crystal plate of the composite column with reference to the floating raft vibration isolation system. Guo [22, 23]designed a hybrid acoustoelastic metamaterial structure [22], a thin film structure [22] and a variety of supercellular plates of tungsten vibrators [23], but they are not practical. Wang [18] proposed a method for active noise control of power transformers, using HFAMM structure to prepare sound barriers. Based on the above research, in order to meet the requirements of low frequency vibration and noise reduction, the complete band gap width can be increased and the initial frequency of the first band gap can be reduced by changing the microunit structure and adjusting the material parameters.

This paper aims to cover the low frequency vibration and noise below 1000 Hz, focusing on the range of 50–200 Hz. Considering the anti-tear strength of the rubber shock absorber and the support of the oscillator, a necked phononic crystal is proposed, and the relationship between the oscillator density of the structural unit and the geometric parameters of the structure and the band gap frequency is discussed in detail. Based on the structural model, the stiffness k of the shock absorber is further reduced by opening holes in the shock absorber, and the coverage rate of 86.66% is achieved in the 50–200 Hz range. Combined with the finite element method, the transmission losses of various phononic crystal arrangements are calculated, and the effect of vibration and noise reduction is improved.

2 Microstructure Unit Model of Phononic Crystal

2.1 *Introduction of unit structure model*

Because the phononic crystal is a periodic composite material with elastic wave band gap characteristics composed of more than two elements, the phononic crystal structure (as shown in Fig. 1a is composed of a metal mass oscillator, an elastic shock absorber and a matrix. A is a metal oscillator, B is a silicone rubber necking damper, and C is an epoxy resin matrix. Oscillator is located in the center of the structure, necking B is distributed in the middle of the four edges of the matrix, and the structure is highly symmetrical.

COMSOL Multiphysic finite element commercial software has a powerful function of physical field coupling, which is used for numerical calculation in this study. The first Brillouin zone of the element structure is given, and the wave vector k is introduced to scan along the boundary "M- Γ-X –M" of the irreducible Brillouin region (the shaded part of Fig. 1b). The natural frequency and mode of the structure in this direction can be obtained. The energy band structure of the structure can be obtained by arranging the natural frequencies in different directions according to the direction. In the calculation, the periodicity of the element structure in the x and y directions is also considered, and the Bloch periodic boundary conditions are also set. The geometric and material parameters are shown in Tables 1 and 2.

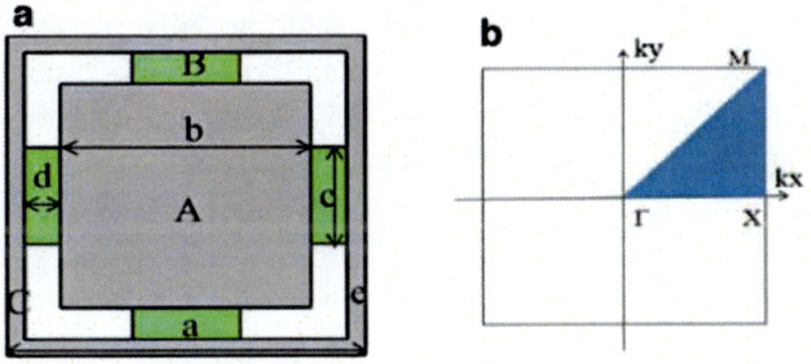

Fig. 1 **a** Microunit structure of locally resonant phononic crystals in moment form **b** First Brillouin zone

Table 1 Geometric structure parameters

a/mm	b/mm	c/mm	d/mm	e/mm
20	14	6	2	1

Table 2 Material parameters

Material	Density(kg/m^3)	Elastic modulus(E/10^{10} Pa)	Poisson's ratio(μ)
Gold	19,500	8.5	0.4214
Tungsten	19,100	35.41	0.35
Plumbum	11,600	4.08	0.369
Copper	8950	16.46	0.093
Aluminum	2730	7.76	0.352
Silicon rubber	1300	1.175e-5	0.469
Epoxy resin	1180	0.435	0.368

3 Band Gap Characteristics and Resonance Mechanism

3.1 Band gap characteristic

The energy band relationship of the localized resonant phonon crystal is shown in Fig. 2:

Low-frequency vibration and noise are the difficulties of current control, so we only pay attention to the low-frequency bandgap below 1000 Hz. There is a wide low-frequency complete band gap between 154–685 Hz, bandwidth 531 Hz, followed by three narrow sub-bands in 1365–1410 Hz and one sub-band gap in 2156–2178 Hz. The first band gap frequency of the fully coated oscillator model is 327–1090 Hz bandwidth 763 Hz. Compared with the fully coated phononic crystal, the proposed necked phononic crystal reduces the lower boundary frequency of the first band gap by 173 Hz, which not only ensures a certain volume, but also reduces the quality. Although the bandgap width of the fully coated model has 763 Hz, the

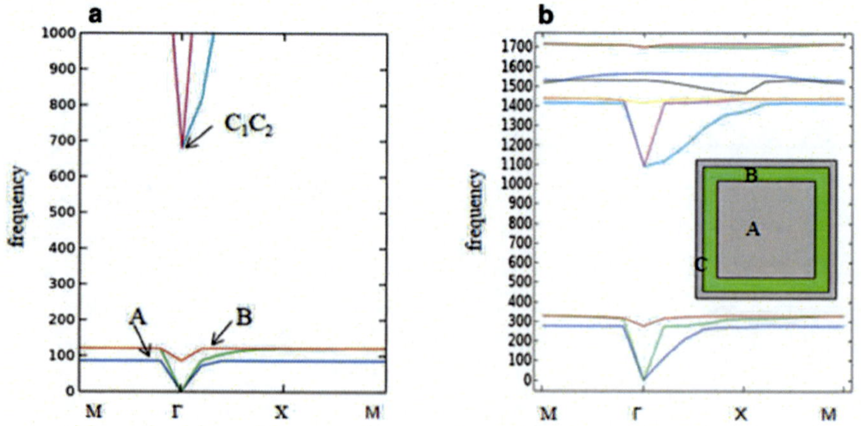

Fig. 2 Bandgap structure diagram **a** Necking structure **b** Fully clad structure

low frequency vibration of 50–200 Hz is a difficult problem in practical engineering applications. The new rectangular hole structure model based on this type of structure can effectively reduce the effective stiffness of the spring and obtain lower resonance frequency, so the model has the characteristics of low frequency, low cost and high practical application value.

In order to explore the formation mechanism of local resonance band gap, the vibration modes of the upper and lower boundary of the band gap are analyzed as shown in Fig. 3.

The vibration mode at point A of the band gap boundary can be seen in Fig. 3a. It can be seen that the vibration is mainly concentrated on the vibrator, and the oscillator undergoes counterclockwise torsion. at the same time, from the point of view of force and displacement, the necked damping block is only subjected to torsional shear deformation, and there is almost no vibration in the matrix. The vibration mode has no torque effect on the matrix, and the resultant force of the oscillator acting on the matrix in the x direction and y direction tends to 0. The long-wave traveling wave propagating in the matrix is difficult to couple with the vibration of the oscillator in this mode, and there is no band gap. At the initial band gap point B of the band gap, the vibration is mainly concentrated on the metal oscillator, and the oscillator moves to the right. In this translational vibration mode, as shown in Fig. 3b, the force in the y direction produced by the oscillator on the matrix is very small, and the vibration of the oscillator is coupled with the long wave traveling wave in the matrix, and the energy is localized at this frequency, resulting in the initial band gap. The points C_1 and C_2 on the upper boundary of the band gap are represented as the overlapping part of the bandgap, and the vibration is shown as the upward translation movement to the right and the translation movement to the left of the matrix frame, respectively, as shown in Fig. 3 (C_1, C_2). At this time, the vibration mainly occurs on the matrix frame, and the oscillator almost remains at rest. Therefore, the coupling between the long-wave traveling wave in the matrix and the vibration of the oscillator is decoupled, and the band gap is cut off.

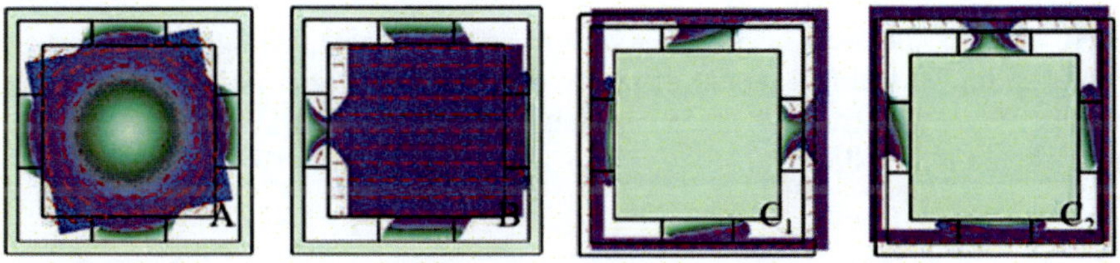

Fig. 3 Eigen mode diagram of the characteristic points at the edge of the bandgap

3.2 Mathematical Calculation Method and Equivalent Model of Band Start and Cut-off Frequency

Here we regard the intermediate scatterer as the concentrated mass and the rubber as the spring connecting the concentrated mass and the matrix. The mass M of the metal vibrator and the spring stiffness coefficient k of the rubber shock absorber can be approximately given by the following formula: M_1 is the sum of the intermediate scatterer and 1/2 rubber mass, and M_2 is the sum of the epoxy resin matrix and the remaining 1/2 rubber mass. according to the unit cell vibration mode Fig. 3, the rubber mass is evenly distributed to the scatterer and epoxy resin, the specific calculation formula is as follows:

$$M_1 = M_{pb} + 2M_{rb}\frac{\alpha}{1+\alpha} \tag{1}$$

$$M_2 = M_{ep} + 2M_{rb} + 2M_{rb}\frac{1}{1+\alpha} \tag{2}$$

In the formula, $M_{pb} = \rho_{pb}b^2$ is the scatterer mass, ρ_{pb} is density of lead, the matrix mass,$M_{ep} = \rho_{ep}4e(a - e)$ is density of epoxy resin,$M_{rb} = \rho_{rb}(a - 2e - b)c/2$ is single damping block,ρ_{rb} is density of rubber shock absorber, and α is proportional coefficient.

$$\alpha = \frac{M_2}{M_1} = \frac{M_{ep} + 4M_{rb}}{M_{pb} + 2M_{rb}} \tag{3}$$

The effective spring stiffness coefficient a is only related to the size and elastic constant λ and μ of rubber.

$$k = \frac{2(\lambda_{rb} + 2\mu_{rb})c}{a - 2e - b} \tag{4}$$

To sum up, the local resonance microstructural element can be equivalent to a "mass-spring" simplified model of single-degree-of-freedom system, as shown in Fig. 4a. When the particle M_1 resonates under the action of a spring, the initial band gap formula is as follows.

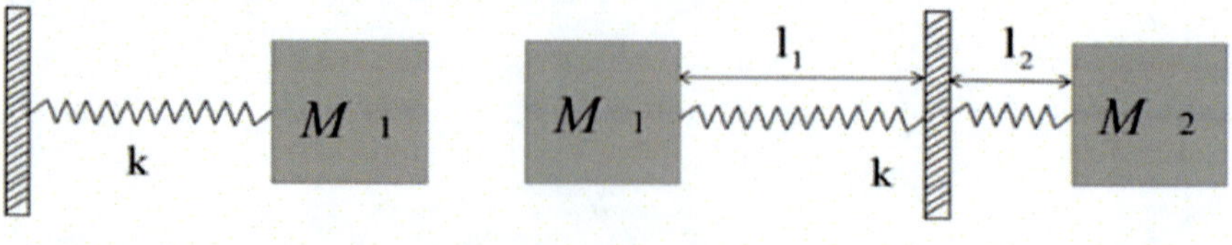

Fig. 4 Equivalent mass model of the upper and lower boundary of the first band gap **a** Initial band gap equivalent model **b** Cut-off band gap equivalent model

$$f_{little} = \frac{1}{2\pi}\sqrt{\frac{2k}{M_1}} \tag{5}$$

The simplified model corresponding to the cutoff frequency of the element structure is shown in Fig. 4b, in which the local resonance element can be equivalent to a "mass-spring-mass" two-degree-of-freedom system. M_2 is the equivalent mass of a single matrix. At the band gap cut-off frequency, the particles M_1 and M_2 resonate in the way of relative vibration under the connection of the spring k, and the spring is stationary in one position. The formula of the cut-off band gap is as follows.

$$f_{high} = \frac{1}{2\pi}\sqrt{\frac{k(M_1 + M_2)}{M_1 + M_2}} \tag{6}$$

4 The Influence of Unit Cell Related Parameters on the Band Gap

4.1 *Effect of Density of Metal Oscillator on Band Gap*

The density difference of metal oscillator is mainly reflected in the effect of material change on band gap as shown in Fig. 5a. The frequency of the upper boundary of the band gap decreases slowly, but the frequency of the lower boundary decreases rapidly and the bandwidth increases. According to formula (5) and formula (6), it can be explained that the stiffness k decreases relatively because of the increase of M_1, and the increasing speed of M_1 is faster than that of k, so that the initial frequency of the band gap moves rapidly to low frequency, and the equivalent mass M_1 dominates

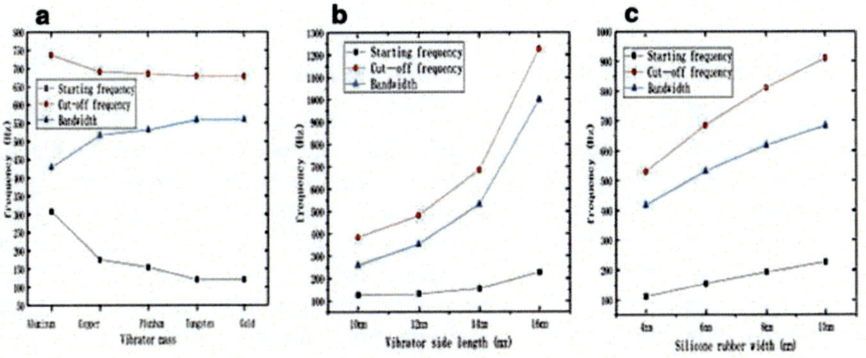

Fig. 5 **a** Effect of different material density on band gap **b** Effect of oscillator geometry on band gap **c** Effect of necking damping block width on band gap

the starting position of the band gap. Because there is little difference in density between tungsten and gold, the band gap is almost unchanged. Lead is a metal with high density and low cost, which can be used as an engineering material, so it is necessary to further discuss the unit structure of lead oscillator material.

4.2 The influence of the geometric size of the oscillator on the band gap

Other structural parameters remain unchanged, and as the edge length of the lead oscillator increases from 10 to 16 mm, the height of the damping block decreases with the increase of the oscillator. The band gap starting frequency, cut-off frequency and band gap width are shown in Fig. 5b. The numerical value shows that the band gap frequency increases, but the cutoff frequency increases faster than the starting frequency. This is because the size of the oscillator of the local resonance element directly affects the equivalent mass of the oscillator, thus affecting the natural frequency of the "lead-necked shock absorber-matrix" element, and the mass of the lead oscillator increases with the increase of the side length. The height of the rubber shock absorber also decreases with the increase of the oscillator size, and the decrease of the height of the damping block leads to the relative increase of the equivalent stiffness of the rubber shock absorber. According to the superposition of the two variables, the frequency of the cutoff band gap increases rapidly, while the mass M_1 increases more slowly than the equivalent stiffness k, so the frequency of the initial band gap increases slowly. Therefore, increasing the side length of the lead oscillator is beneficial to the expansion of the complete band gap. Because the height of the shock absorber changes with the change of the size of the oscillator, and the increase of the height is equivalent to the decrease of the stiffness, the influence relationship is consistent with the influence of the side length of the plumbum oscillator on the band gap, so I will not repeat it.

4.3 Effect of Shock Absorber Width on Band Gap

Keep other parameters unchanged and only change the width of the damping block. The width of the damping block increases from 4 to 10 mm, and the band gap changes as shown in Fig. 5c. The initial frequency of the band gap increases from 112 to 227 Hz, the cut-off frequency increases from 530 to 910 Hz, and the band gap moves to high frequency. The principle is that the stiffness k in the equivalent mass model increases due to the increase of the width of the damping block, which affects the natural frequency of the structure. When the equivalent masses M_1 and M_2 are determined, the stiffness k has a greater influence on the cutoff frequency, so the rising speed of the cutoff band gap is faster than that of the initial band gap, and the width of the band gap increases. Considering the actual structure of the phononic crystal, the aspect ratio of the damper can provide stable support for the oscillator.

5 Structural Optimization and the Influence of Parameters on Band Gap

After parameterized adjustment, the band gap of the above necked phononic crystal structure can satisfy the suppression of low frequency vibration and noise below 1000 Hz, but the low frequency vibration of 50–200 Hz is difficult to control in practical engineering applications. Therefore, it is necessary to optimize and improve the structure based on this type of structure. The improved new rectangular hole structure model can effectively deal with the low frequency range of 50–200 Hz.

The improvement method of the structure is to make a rectangular hole on the shock absorber with the size of 8 × 2 mm as shown in Fig. 6. Considering the anti-tear strength of the shock absorber, the rectangular hole is suitable to be in the middle position. In order to ensure the mechanical properties of the shock absorber, the size requirements of the four holes are the same. The low-frequency band gap with a width of 252 Hz is opened in Fig. 6, and this result has excellent damping performance for the low-frequency vibration of the following 50–200 Hz. In this paper, the bandgap results of the element structures of rectangular holes with different sizes are discussed.

Here, the energy bands of three kinds of phononic crystals with different sizes of rectangular holes are calculated as shown in Table 3. The results show that the improved structure and the structure without rectangular hole are easier to obtain low frequency band gap. This is mainly due to the fact that piercing is an effective method to reduce the stiffness of the hollow structure of the damping body. When the equivalent mass is determined, the equivalent stiffness k provided by the shock absorber decreases with the expansion of the rectangular hole parameters. It has a great influence on the starting frequency and cut-off frequency, so the band gap frequency decreases. When the size of the rectangular hole is 6 × 1 mm, the low

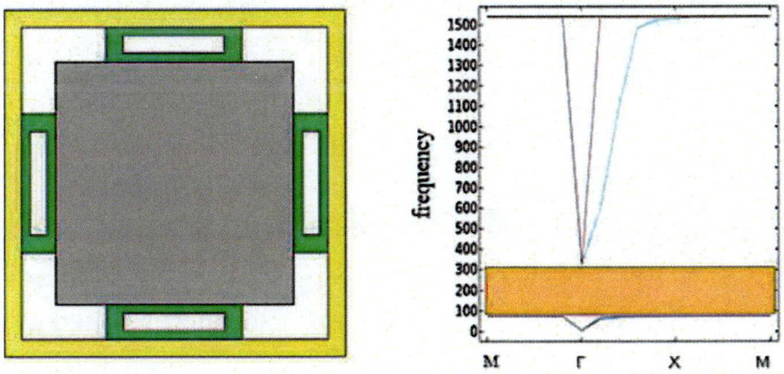

Fig. 6 Unit cell model with rectangular holes and its energy band structure

Table 3 Influence of different rectangular holes on band gap

Rectangular hole length × wide(mm)	Staring frequency(Hz)	Cut—off frequency(Hz)	Bandwidth(Hz)
6 × 1	70	322	252
4 × 1	107	486	379
2 × 1	139	621	482
Solid	193	811	618

frequency range of 70–252 Hz can be obtained to adapt to the low frequency vibration below 200 Hz. The changing trend of the above calculation results is also consistent with the law of the equivalent mass model.

6 Calculation of Vibration Transmission Loss by Finite Element Method

In practical application, the material structure can only take the finite structure as shown in Fig. 7. Ten element structures are arranged along the x direction, and symmetrical boundary conditions are imposed on the two boundaries in the y direction. The transmission loss reflects the relationship between the excitation and the corresponding time. Taking the frequency as the independent variable, the excitation and the corresponding time relationship are transformed into the amplitudes at each frequency by fourier transformation, and then the results can be obtained by comparing the amplitudes of the response and the excitation at the same frequency. The transmission loss is defined as:

$$TL = 20\log\frac{X_{out}}{X_{in}} \tag{7}$$

In the formula, the force per unit area picked up on the right side of X_{out} is excited, and X_{in} is the unit area force applied on the left side.

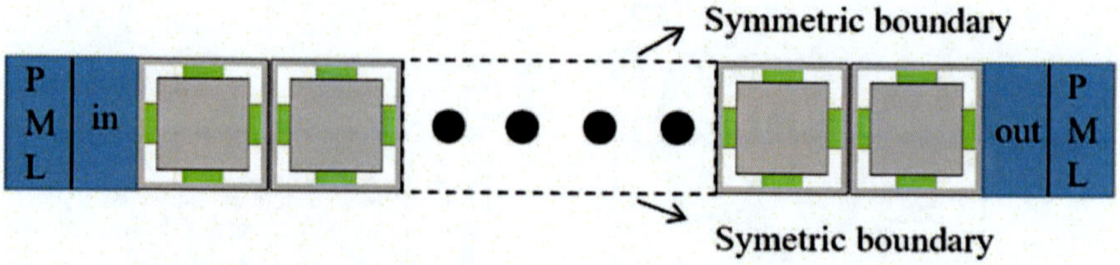

Fig. 7 Schematic diagram of transmission loss

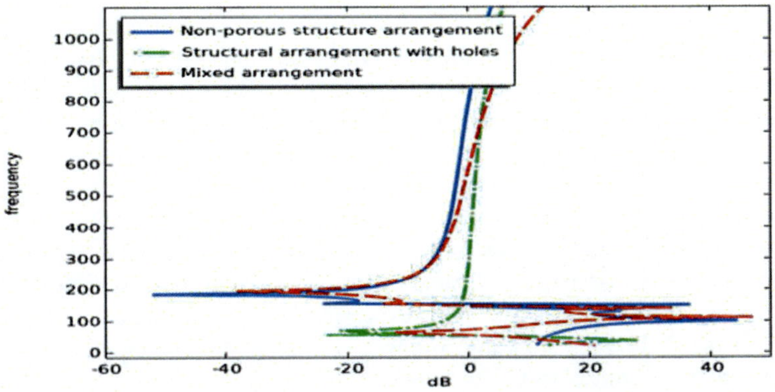

Fig. 8 Influence of different arrangement structures on transmission loss

The transmission losses of the three arrangement modes are calculated here. The band gap range of the non-rectangular hole is 193–811 Hz, and the band gap range of the rectangular hole is 70–322 Hz. According to Fig. 8, it can be found that the effective frequency and the band gap frequency range of the transmission loss of the two structures are the same. In this band gap frequency range, a better vibration isolation effect can be achieved, and the maximum vibration isolation is 52 and 23 dB. The mixed arrangement is formed by the arrangement and combination of five 6 × 1 mm rectangular hole phononic crystals and five non-rectangular hole phononic crystals, so this mixed arrangement has the transmission loss characteristics of the above two kinds of structure arrangement, and two kinds of resonance peaks appear in Fig. 8. This is due to the fact that the band gap opened by the two structures coincides with each other, and the complementary transmission loss improves the vibration isolation performance. This method improves the comprehensive performance of transmission loss below 1000 Hz, especially the maximum vibration isolation performance of the frequency band above 150 Hz reaches 38 dB. For this reason, the use of mixed collocation arrangement in engineering applications can make up for the deficiency of a single type of phononic crystals in the effective range.

7 Conclusion

In this paper, a new necking phononic crystal is proposed, which is composed of a single oscillator, four silicone rubber damping blocks and a matrix frame. The maximum bandwidth of 179–1218 Hz is obtained by parametric adjustment. At the same time, an improved structure is proposed, which greatly reduces the stiffness k of the rubber shock absorber and opens the 70–322 Hz low frequency band gap by setting a rectangular hole on the shock absorber. After a series of calculations, the following conclusions can be drawn: (1) The change of material properties affects the size of the forbidden band value by changing the equivalent mass of the model, and cannot change the position of the forbidden band appearing in the whole energy band diagram. (2) The change of the position of the forbidden band can be regulated by adjusting its structural form and geometric parameters. (3) The comprehensive vibration isolation performance can be improved by combining different types of structures to achieve complementarity at different frequencies.

The following points should be paid attention to in further application. First of all, when different phononic crystals are mixed and arranged, it is necessary to use the unit structure with a large range of band gap coincidence in order to achieve a better band gap coincidence effect. Secondly, the strength of structural matrix is the focus of application. When the structural strength is limited, the combination of passive control and active control can be considered.

Acknowledgements The authors gratefully acknowledge supports for this work from Key technology R&D project of Ningxia (Grant No. 2018BFH03001) and Project of National Natural Science Foundation of China Research Project of CAE Key Laboratory for Intelligent Equipment of Ningxia (Grant No. 51365046).

References

1. Lu X, Qian S, Wei C, Li X, Zhu X (2021) Analysis of noise and vibration control of distribution transformer. J Trans 58(01):34–38. (Chinese edition)
2. Tempest W (1976) Infrasound and low frequency vibration (London:Academic Press Inc.), p187
3. Gu X, Shen R, Xu J (2001) Study on the Vibration and noise control techniquesin large power transformer. J Noise Vibration Control (05):7–11. (Chinese edition)
4. Jiang H, Ma H, Jiang N, Wang C, Li K (2014) A new substation noise control method based on notch filter. J Electric Power 47(06):43–48.(Chinese edition)
5. Li B, Hu G (2004) Preliminary discussion on measures to reduce transformer noise. J Trans (08):40–42. (Chinese edition)
6. Shen B (2019) Application of noise reduction technology in typical 220kV indoor substation [A]. Shanghai Acoustics Society, Xi'an Acoustics Society. In: 2019 Proceedings of the sixth acoustics academic exchange conference of Shanghai-Xi'an acoustics society [C]. Shanghai society of acoustics, Xi'an society of acoustics: editorial department of acoustic technology 5. (Chinese edition)
7. Xie S, Yang S, Yang C et al (2020) Sound absorption performance of a filled honeycomb composite structure. J Appl Acoust 162:107202
8. Zhang X, Wu J, Mao Q et al (2020) Design of a honeycomb-microperforated panel with an adjustable sound absorption frequency. J Appl Acoust 164:107246
9. Kushwaha MS, Halevi P, Dobrzynski L et al (1993) Acoustic band structure of periodic elastic composites. J Phys Rev Lett 71(13):2022
10. Huang Y, Liu S, Zhao J (2016) A gradient-based optimization method for the design of layered phononic band-gap materials. J Acta Mech Solida Sin 29(4):429–443
11. Maldovan M (2013) Sound and heat revolutions in phononics. J Nature 503(7475):209–217
12. Liu Z, Zhang X, Mao Y et al (2000) Locally resonant sonic materials. J Sci 289(5485):1734–1736
13. Wu J, Zhang S, Shen L (2013) Study on low frequency vibration band gap of phonon crystal plate structure with helical local resonance element. J Mech Eng 49(10):62–69. (Chinese edition)

14. Qi P, Du J, Jiang J et al (2016) Transmission loss mechanism and characteristics of two-dimensional phononic crystals. J Chinese Ceramic Soc 44(10):1458–1464. (Chinese edition)
15. Peng Z, Li C, Gao Y (2020) Study on low-frequency band gap of two-dimensional open-hole locally resonant phononic crystals. J Mech Design 037(04):73–77. (Chinese edition)
16. Ruan Y, Liang X, Hua X et al (2021) Isolating low-frequency vibration from power systems on a ship using spiral phononic crystals. J Ocean Eng 225:108804
17. Bilal OR, Foehr A, Daraio C (2020) Enhancement of deep-subwavelength band gaps in flat spiral-based phononic metamaterials using the trampoline phenomena. J Appl Mech 87(7):071009
18. Wang T, Jin T, Lu Z et al (2018) Research on active noise control method compensating for acoustic metamaterial noise barrier in transformer noise reduction. In: 2018 IEEE 2nd international electrical and energy conference (CIEEC). IEEE, pp 648–652
19. Shao H, He H, He C et al (2020) Study on the band gap optimization and defect state of two-dimensional honeycomb phononic crystals. J Mater Res 35(21):3021–3030
20. Guo X, Cui H, Hong M (2021) Research on vibration and noise reduction of locally resonant phonon crystal plate. J Ship Mech 25(04):509–516. (Chinese edition)
21. Hou D, Duan L, Cao C et al (2020) Noise and vibration characteristics of 35 kV oil-immersed distribution transformer. J Appl Acou 39(06):964–968. (Chinese edition)
22. Guo Z (2021) Research on the mechanism of phononic crystals on low frequency noise control of substation. North China Electric Power University (Beijing). (Chinese edition)
23. Guo Z, Chen C, Qiao J, Ni Y (2021) Lightweight design and study of supercell phonon crystal plate. J Synthetic Cryst 50(01):13–19. (Chinese edition)

Study on the Band Gap Characteristics of Two-Dimensional Local Resonant Phononic Crystals

Yuan Xing, Bo Zhang, Yao Zhang, Jiaxing Song, and Meng Wang

Abstract A local resonant phononic crystal is proposed. The phononic crystal can control vibration and noise effectively. It also can control the frequency range. We can obtain elastic wave dissipation in a wide frequency range. The band structure and transmission characteristics are calculated by the finite element software COMSOL Multiphysics. Through the results, the vibration modes at the initial frequency, cut-off frequency and other typical frequencies are analyzed. So we can elaborate the mechanism of the band gap. The opening of the band gap is the result of the coupling between the long wave traveling wave in the matrix and the resonant characteristics of the oscillator. The band gap frequency range of the phononic crystal structure is 268.7 ~ 1097.9 Hz. The factors affecting the band gap of the phononic crystal were studied. The result show: The density of materials and the geometric parameters of phononic crystals will affect the band gap. The density of scatterers mainly affects the initial frequency. The density of the matrix only affects the cut-off frequency. The geometric size has an effect on the starting frequency and cut-off frequency.

Keywords Phononic crystal · Wave coupling · Band gap

1 Introduction

With the improvement of China's international competitiveness, the manufacturing industry has made great contributions. China takes "Made in China 2025"as the national action program. With the development of manufacturing industry, the problem of vibration and noise has become an unavoidable problem perplexing the development of the industry [1, 2]. Vibration poses a great challenge to the accuracy requirements of manufacturing industry. Noise affects people's life and workers' operating environment. These two problems have become important issues that designers must consider in the process of product development [3–5].

Y. Xing · B. Zhang (✉) · Y. Zhang · J. Song · M. Wang
School of Mechanical Engineering, Ningxia University, Ningxia, China
e-mail: zhangb@nxu.edu.cn

© The Author(s), under exclusive license to Springer Nature Singapore Pte Ltd. 2023

M. H. A. Hassan et al. (eds.), *Technological Advancement in Instrumentation & Human Engineering*, Lecture Notes in Electrical Engineering 882,
https://doi.org/10.1007/978-981-19-1577-2_34

Artificial periodic structure is a new structure found to effectively control this problem in recent years. Phononic crystal is a typical representative. It can be used to control the propagation of elastic waves. In national defense, national economy and daily life, acoustic metamaterials based on phononic crystals have important potential applications in the fields of vibration and noise reduction [6–9]. Phononic crystal is a new physical concept in the field of condensed matter physics. It is a composite material or structure with elastic wave band gap characteristics composed of two or more media. Phononic crystal can be regarded as the extension of the concept of crystal in solid physics in the sense of elastic wave. Its important characteristic is its attenuation domain. The research of phononic crystal can learn from the research of elastic dynamics and solid physics.

Phononic crystals are periodic composites or structures with elastic band gap characteristics. Phononic crystal is a new artificial periodic structure. Although human beings have studied the propagation characteristics of elastic waves in layered periodic media for about 80 years, the concept of phononic crystal and the related theory of phononic crystal have only been studied for more than 20 years [10–12]. In 1993, M.S. Kushwaha and others clearly put forward the concept of phononic crystal for the first time when studying nickel/aluminum two-dimensional solid periodic composite medium. At the same time, they also clearly and firstly put forward that the band gap characteristics of phononic crystals have application prospects in high-precision vibration free environment [13, 14]. In 2000, Liu studied the three-dimensional three-component phononic crystal formed by a simple cubic lattice structure composed of shot put coated with viscoelastic soft material and buried in epoxy resin [15–18]. He found that the wavelength corresponding to the band gap frequency of the phononic crystal was much larger than the lattice size. Therefore, he proposed the local resonance band gap mechanism of phononic crystal, which marked a major breakthrough in the research of phononic crystal. The band gap of local resonant phononic crystal is caused by the resonance of local oscillator [19–21]. In 2001, Vasseur et al. designed two-dimensional solid phononic crystals using steel and epoxy resin. The existence of complete band gaps of elastic wave, shear wave and longitudinal wave is found and formalized. Before that, the band gap frequency of elastic wave is generally in the same order of magnitude as the corresponding wavelength size. The formation mechanism of this band gap is called Bragg scattering mechanism. However, phononic crystals based on Bragg scattering mechanism are sensitive to the size of materials. In order to obtain low-frequency band gap, it is necessary to design large macro size, which is not conducive to practical application. The proposal of local resonance mechanism is a major theoretical breakthrough in the study of phononic crystals [22–25]. On this basis, new concepts such as acoustic black hole, acoustic cloak and acoustic metamaterial are proposed [26].

Phononic crystals are further studied in this paper. A traditional local resonant phononic crystal is proposed. The vibration modes are carefully analyzed, including whether each mode produces band gap and the mechanism of band gap. The factors affecting the band gap start frequency and cut-off frequency are deeply studied.

2 Structural Model and Calculation Method

A two-dimensional three component phononic crystal with local resonance is designed. A two-dimensional three component phononic crystal with local resonance is designed. The structure is shown in Fig. 1a.The structure is wrapped by a large mass circular vibrator a with a coating B and embedded into the matrix C. No material is filled between the cladding layers B to form a cavity. The vibrator A is made of lead with high density, the coating B is made of silicone rubber with good elasticity, and the matrix C is made of epoxy resin. The detailed geometric parameters of the phononic crystal structure are shown in Table 1. and the material parameters are shown in Table 2. The control group is also set up, as shown in Fig. 1b. The energy band structure is compared by changing the coating position without changing other parameters.

The research on the mechanism and characteristics of phononic crystal band gap depends on the corresponding calculation methods. In this paper, the band gap is calculated by finite element method, and the finite element software COMSOL Multiphysics is used for analysis and calculation.

The propagation equation of elastic wave in structure is

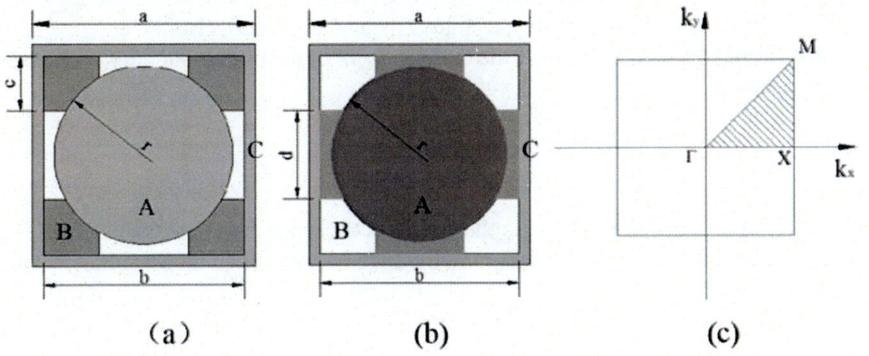

Fig. 1 Phononic crystal model **(a) (b)** and the first Brillouin zone **(c)**

Table 1 Structural parameters

a	b	c	d	r	mm
20	18	5	10	5	–

Table 2 Material parameters

Material	ρ(kg/m^3)	E(10^{10}Pa)	υ
Lead	11,600	4.08e-10	0.369
Silicon rubber	1300	1.175e-5	0.469
Epoxy resin	1180	0.435	0.368

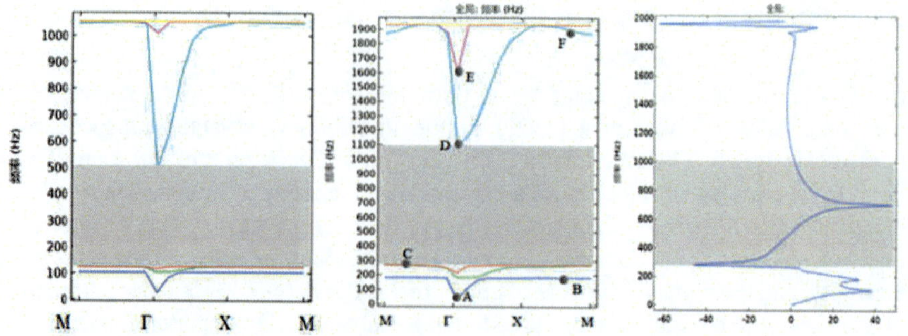

(a) (b) (c)

Fig. 2 The band gap diagram and transmission characteristic curve of Phonon Crystal

$$\sum_{i=1}^{3} \frac{\partial}{\partial x_j}\left(\sum_{l=1}^{3}\sum_{k=1}^{3} C_{ijkl}\frac{\partial u_k}{\partial x_l}\right) = \rho\frac{\partial^2 u_i}{\partial t^2} \tag{1}$$

ρ is density, u_i is the displacement, t is the time, C_{ijkl} is the elastic constant, x_j ($j = 1, 2, 3$) represents the coordinate variables x, y and z respectively.

According to Bloch's theorem, the displacement of the outer boundary of a single cell satisfies

$$\mathrm{u}(r) = e^{i(kr)}\mathrm{u}_k(r) \tag{2}$$

$\mathbf{r} = (\mathbf{x}, \mathbf{y}, \mathbf{z})$ is position vector, **k** is the wave vector of the first irreducible Brillouin region, $\mathbf{u_k(r)}$ is the node displacement field function with the same periodicity as the cell structure.

In the finite element software COMSOL Multiphysics, the boundary of the non Brillouin region M-Γ-X-M of the single cell structure of phononic crystals is scanned by using the solid mechanics module, as shown in Fig. 1c.It is solved to obtain the characteristic frequency, energy band structure and vibration mode diagram, as shown in Fig. 2.

3 Equivalent Simplified Model

The phononic crystal is a typical two-dimensional three component phononic crystal. In principle, it can be simplified into a relatively simple one-dimensional spring oscillator structure. The matrix is simplified as a thin straight beam with uniform mass. The scatterer is simplified as a mass block with fixed mass. The cladding is simplified as a spring with a certain elastic coefficient, as shown in Fig. 3.However, in

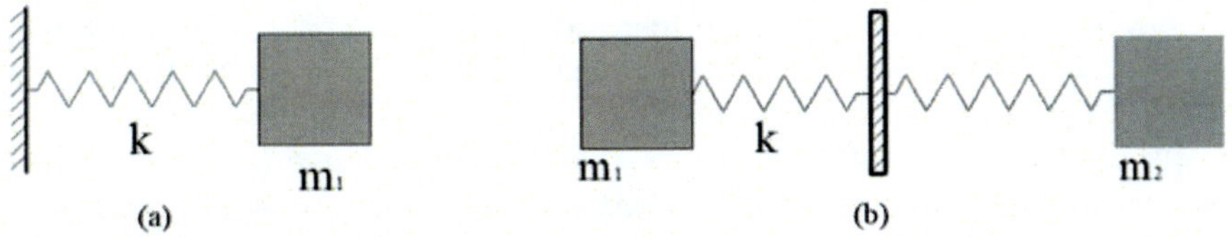

Fig. 3 Equivalent simplified model of local resonance band gap in phononic crystals

different modes, they represent different equivalent masses and elements, which also explains the reason why the band gap and modes of phononic crystals in the control group are different from those in the original group. For example, in the mode (d) of Fig. 4, the vibrator moves up and down. The upper and lower cladding layers vibrate up and down with the vibrator. At this time, it can be considered that the tension and compression of the upper and lower cladding layers play a major role. The cladding of these two parts can only be equivalent to the equivalent stiffness of the spring, and the mass of the cladding can not be ignored. According to the modal diagram, the mass of the cladding of the left and right parts can be distributed to the matrix, and the mass of the cladding translating up and down with the vibrator can be distributed to the vibrator. In Fig. 4c, the vibrator makes a translational movement to the right, and the left and right cladding layers are obviously stretched and compressed. These two parts are equivalent to the stiffness of the spring. In the vibration, according to

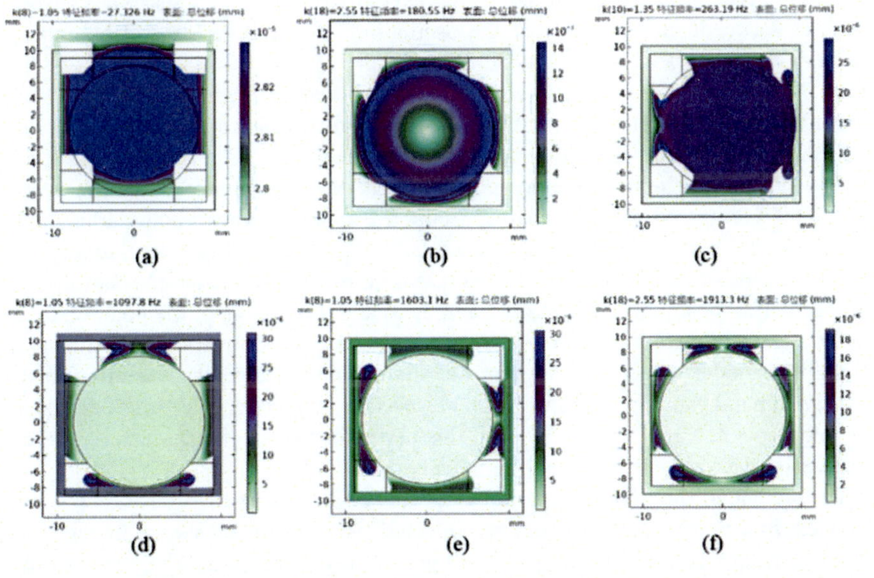

Fig. 4 Mode shapes of the key points at the upper and lower boundary of the band gap of the phononic crystal **a** A-point mode; **b** B-point mode; **c** C-point mode; **d** D-point mode; **e** E-point mode; **f** F-point mode

the values of the modal diagram, the four parts of the cladding are translating with the translation of the vibrator. Therefore, the mass of the four parts of the cladding can be distributed to the mass of the vibrator.

The initial frequency of the band gap is due to the overall translational motion of the oscillator. The vibration phases of adjacent oscillators are opposite. The cladding layer is stretched and compressed with the translation of the vibrator. The whole phononic crystal system reaches a dynamic equilibrium state. At the cut-off frequency of the band gap, under the elastic connection of the coating, the oscillator and the substrate form anti phase resonance. The vibration phase of all cells in phononic crystal is the same and reaches dynamic equilibrium. The starting frequency and cut-off frequency of the simplified model are

$$f_1 = \frac{1}{2\pi}\sqrt{\frac{k}{m_1}} \qquad f_2 = \frac{1}{2\pi}\sqrt{\frac{k(m_1 + m_2)}{m_1 m_2}} \tag{3}$$

4 Analysis of Band Gap Characteristics and Band Gap Formation Mechanism of Phononic Crystals

Figure 3 shows the energy band structure. It can be seen from the figure that the band gap of the phononic crystal shows strong asymmetry. There is a sharp attenuation peak at the resonant frequency of the oscillator. The phononic crystal belongs to local resonance type. There is a wide full band gap. The frequency range is 268.7 ~ 1097.9 Hz. The center frequency is 683.3 Hz. The relative bandgap width is 121.4% (relative bandgap width = actual bandgap width/bandgap center frequency) [27, 28].The control group also produced a complete band gap. The frequency range is 125 ~ 498.5 Hz. The center frequency is 311.75 Hz. The relative band gap width is 119.8%.Compared with the control group, the upper and lower boundaries of the band gap in the original group decreased. Although the band gap is relatively low frequency, the actual band gap width decreases more. This is because the coating area of the coating layer and the vibrator decreases, and the change of the position leads to the change of the vibration mode. The parameters of the corresponding equivalent simplified model will also change at the characteristic frequency, resulting in the change of band gap. The original group has a wider range of applications because of its low frequency and large band gap. This paper focuses on the characteristics and impression factors of primitive phononic crystals.

In this paper, the vibration modal diagrams of six points of A, B, C, D, E and F in the first non Brillouin region are selected for analysis, as shown in Fig. 4. C is the starting frequency of the full band gap. Analyze the modal diagram of point C: the outer boundary matrix basically remains unchanged. The vibrator moves in translation, and the coating produces tensile and compressive deformation. At this point, the vibration resultant force generated by the vibrator as a scatterer is coupled with the long wave traveling wave in the matrix, resulting in a band gap. Point D is the cut-off point of the band gap. The corresponding frequency is 1097.9 Hz. Analyze the vibration mode diagram of point D: the vibrator of point D basically has no displacement. The outer boundary box of the matrix moves in translation. The cladding layer then undergoes tensile and compressive elastic deformation. The long wave traveling wave of the matrix causes the vibration displacement of the matrix. The scatterer has no motion. It is not coupled with the long wave traveling wave in the matrix. So the band gap is cut off. Point F is the lower boundary of the directional band gap of the flat band part. Analyze the vibration mode diagram of point F: it can be seen that the cladding moves in translation. There is little change in the vibrator and matrix frame. There is no mutual coupling of waves. Therefore, the band gap is cut off. At point A, the matrix frame, cladding and vibrator move in translation under the action of elastic wave, without wave coupling, so there is no band gap. At point B, the vibrator has torsional motion. In this mode, the vibrator only produces torque on the base frame. There is no effect of plane internal force. The long wave traveling wave in the matrix is difficult to couple with the oscillator, so there is no band gap. The outer frame of the matrix at point E has a small translational motion. The vibrator hardly moves. The coating layer has a large translational movement and extrusion deformation with the vibrator and the matrix frame. Therefore, it is difficult to couple the long wave traveling wave in the oscillator and the matrix frame, and there is no band gap. From the above modal diagram analysis, it can be seen that the main reason for opening the band gap of the local resonant phononic crystal is the result of the coupling between the translational motion of the oscillator and the long wave traveling wave in the matrix frame. The existence of the coupling is the key factor to determine whether the band gap can be generated.

The above band gap diagram and analysis of phononic crystals are based on the ideal case. The calculation results are obtained under the assumption of infinite periodic structure. Some of the smaller attenuation will increase cumulatively due to the increase of periodic structure. But in reality, there is no infinite period. The band gap can not be directly used as the physical properties of materials in applications. In practical application, the finite periodic structure and the transmission loss of elastic wave have more practical significance. Figure 5 is a periodic arrangement of 10 phononic crystals along the X direction. The finite element simulation calculation is carried out by COMSOL Multiphysics software. The transmission loss diagram is obtained. According to the band gap characteristics of phononic crystals, the transmission loss should be positive. The magnitude of the absolute value indicates the degree of attenuation. It is basically consistent with the energy band structure. The negative peak analysis in the middle produces resonance because the natural frequency of the material is consistent with the transmission frequency. Other frequencies are not in the band gap range, so the transmission loss is very small.

Fig. 5 Calculating the finite periodic structure of transmission spectrum

5 Influencing Factors of Band Gap

The density of materials and the geometric size of structure are the key factors affecting the band gap of phononic crystals. In this paper, the effects of material density and phononic crystal geometry on phononic crystal band gap will be investigated.

Several simulations are carried out without changing the geometry and size of phononic crystals. A density variation diagram as shown in Fig. 6 is obtained. From Fig. 6a, we can see that due to the increase of scatterer density, the upper boundary of the band gap decreases slightly at the beginning, and then tends to be flat and slightly changed. The lower boundary of the band gap decreases gradually with the increase of scatterer density, and the band gap width widens gradually and moves to low frequency. This shows that the density of the scatterer mainly affects the initial frequency of the band gap and has little effect on the cut-off frequency. This is consistent with the equivalent simplified model of the phononic crystal. The particle at the initial frequency of the band gap resonates under the action of the spring. The resonant frequency is determined by the resonant frequency of the translational motion of the local oscillator. The greater the density and mass of the vibrator, the less the stiffness of the spring. And the rate at which the vibrator mass m increases is greater than the rate at which the spring stiffness K decreases, so the resonant frequency decreases. Therefore, by increasing the density of the oscillator, the starting frequency of the band gap moves to the low frequency.

The density of the substrate mainly affects the cut-off frequency of phononic crystals. Figure 6b shows that the upper boundary of the band gap moves rapidly downward as the substrate density increases. The lower boundary does not change, and the band gap width decreases gradually. This shows that the density of the matrix directly affects the band gap cut-off frequency of phononic crystals, and has little effect on the initial frequency.

When considering the influence of geometric size on the band gap of phononic crystals, the lattice size a remains unchanged, that is, the size of the matrix frame

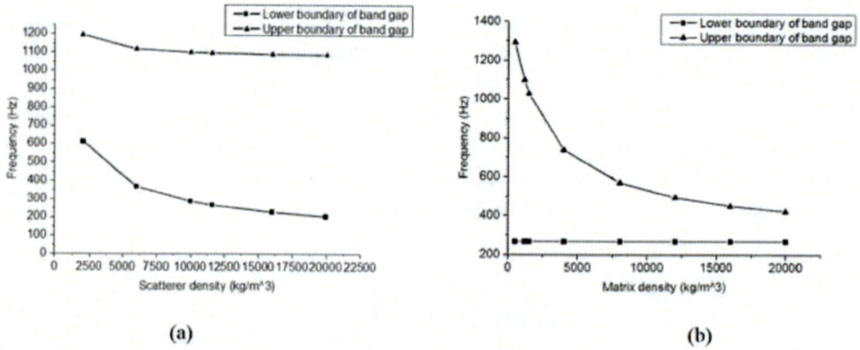

Fig. 6 Effect of material parameters on band gap

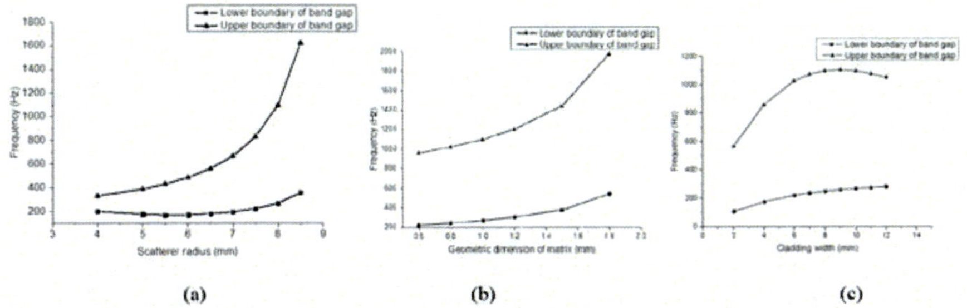

Fig. 7 The influence of the change of geometry on the band gap

remains unchanged. First, change the radius of the scatterer. The frequency variation diagram of the upper and lower boundaries of the band gap is obtained, as shown in Fig. 7a. It can be seen that in the process of increasing the radius of the scatterer, the lower boundary of the band gap first decreases slightly and then increases steadily. And the change is not great relative to the upper boundary. The upper boundary of the band gap increases with the increase of the radius of the scatterer, and the change rate is large. The band gap width increases and moves to high frequency. The mechanism is that the filling ratio of the scatterer in the phononic crystal increases with the increase of the radius of the scatterer. It has a great influence on the upper boundary of the band gap. The larger the filling rate is, the greater the frequency of the upper boundary of the band gap is, and the smaller the influence on the lower boundary is.

The effect of substrate thickness on band gap is considered. Ensure that the lattice size a remains unchanged and increase the thickness of the matrix inward to obtain Fig. 7b. It can be seen from the figure that with the increase of substrate thickness, the upper and lower boundaries of band gap move to high frequency. The upper boundary rises rapidly. The width of the band gap increases gradually. The mechanism is that with the increase of matrix thickness, the mass increases, the relative mass of scatterers decreases, and the spring stiffness pair decreases. Therefore, the band gap moves to high frequency.

Consider the influence of cladding width. Change the width of the coating and keep it in the middle of the substrate. Figure 7c is obtained. As shown in the figure, the lower boundary of the band gap rises gradually, and the lower boundary rises first and then decreases. The peak value is reached when the cladding width is 9 mm. The band gap width first increases and then decreases. The mechanism can be explained by equivalent simplified model. According to formula (3), when the width of the cladding layer increases, the equivalent stiffness K of the spring increases and the initial frequency increases. In the cut-off frequency formula, the spring stiffness plays a major role at the beginning. The principle is the same as the starting frequency. Combined with the equivalent model and modal diagram, with the increasing width of the cladding layer, the mass distribution of the cladding layer in the equivalent model at point (c) of Fig. 4 is as mentioned above. At this time, the mass plays a major role in the cut-off frequency formula of formula (3), so the cut-off frequency is reduced.

6 Conclusion

A typical local resonant phononic crystal structure is proposed in this paper. The equivalent simplified model is abstracted. The band gap characteristics and transmission loss of the phononic crystal with finite period structure are analyzed by finite element method. The band gap changes of phononic crystals under different material densities and different geometric structures are analyzed, and the following conclusions are drawn: The local resonant phononic crystal produces a wide complete band gap at 268.7 ~ 1097.9 Hz. The band gap width is 832.2 Hz. The main mechanism is the translational motion of the local resonance of the lead scatterer, which is coupled with the long wave traveling wave in the matrix. The transmission characteristics of phononic crystals with finite periodic structure also verify the ability to suppress waves in the band gap frequency range. By changing the material density and geometric parameters of phononic crystals, the band gap range of phononic crystals can be changed. The upper and lower boundaries of the band gap can be controlled purposefully.

Acknowledgements The authors gratefully acknowledge supports for this work from Key technology R&D project of Ningxia (Grant No. 2018BFH03001) and Project of National Natural Science Foundation of China Research Project of CAE Key Laboratory for Intelligent Equipment of Ningxia (Grant No. 51365046).

References

1. Xu J, Cui H, Hong M (2021) Study on sound insulation performance of phononic crystal sandwich plate structure. Vibrat Impact 40(09):285–291
2. Zhang S, Xia CG, Fang N (2011) Broadband acoustic cloak for ultrasound waves. J Phys Rev Lett 106(2):024301
3. Cheng Y, Zhou C, Yuan BG (2015) Ultra-sparse metasurface for high reflection of low-frequency sound based on artificial Mie resonances. Nature Mater 14(10):1013–1019
4. Zhang H, Xiao Y, Wen JH (2016) Ultra-thin smart acoustic metasurface for low-frequency sound insulation. Appl Phys Lett 108(14):141902
5. Lu X,Qian S, Wei C, Li X, Zhu X (2021) Analysis of noise and vibration control of distribution transformer. Transformer 58(01):34–38
6. Qi P, Du J, Jiang J, Dong Y, Zhang J (2016) Sound insulation mechanism and characteristics of two-dimensional phononic crystals. J Silicate 44(10):1458–1464
7. Peng Z, Li C, Gao Y (2020) Study on low frequency band gap of two-dimensional open-hole locally resonant phononic crystals. Mech Design 037(04):73–77
8. Sun X, Yan Q, Guo X (2021) Analysis of low frequency band gap characteristics and structural improvement of single-sided cylindrical local resonance phononic crystals. J Intraocular Lenses 50(07):1378–1385

9. Guo X, Cui H, Hong M (2021) Study on vibration and noise reduction of local resonant phononic crystal plate. Ship Mech 25(04):509–516
10. Tan P, Sun X, Song T, Wen X, Liu X, Liu Z (2021) Simulation of band gap characteristics of surface wave phononic crystals with spherical composite columns. J Phys 70(14):243–252
11. Tang R, Pan C, Zheng W, He H, Tang J (2021) Study on propagation band gap characteristics of open ring phononic crystals. J Intraocular Lenses 50(03):428–434
12. Lu Y, Cao D, Shen Y, Chen X (2021) Study on defect state band gap and energy capture characteristics of local resonant phononic crystal plate. J Mech 53(04):1114–1123
13. Huang W, Yan S, Li X, Lu M, Li Y, Wang Z, Zhang Y, Wu S, Guo Y, Fan Q, Qian S, Zhang H, Sun Y, Lu C, Chen Y (2021) Discussion on terminology of acoustic metamaterials. Progr Mater China 40(01):1–6 + 20–21
14. Chen Q, Zhang B, Bai Y, Wang L (2021) Band gap characteristics of a new composite local resonance phononic crystal. Acoustic Technol 40(02):157–166
15. Guo X, Sun X, Zhu Y (2020) Study on band gap characteristics of two-dimensional square lattice tungsten silicone rubber phononic crystals. J Intraocular Crystals 49(09):1583–1589
16. Peng Z, Li C, Gao Y (2020) Study on vibration based on band gap characteristics of two-dimensional three component phononic crystals. J Chongqing Jiaotong Univ (Natural Science Edition) 39(08):134–138
17. Kang T, Sun X, Song T, Sun W, Liao T, Tan P (2020) Low frequency band gap characteristics and formation mechanism of two-dimensional hollow scatterer phononic crystal plate . Acta Acoustic A Sinica 45(04):601–608
18. Huang H, Ran J, Chen K, Zhang Y, Liu C (2020) Study on band gap characteristics of local resonant cylindrical shell phononic crystals. J Intraocular Crystals 49(06):1078–1082 + 1106
19. Wei S(2020) Study on the effect of edge bulge of phononic crystal scatterer on band gap . Beijing Univ Technol
20. Chen L, You S, Zhao X (2020) Study on sound insulation characteristics of two-dimensional phononic crystal thin plate. Materials Guide 34(S1):90–93
21. Chen Q (2020) Study on low frequency band gap characteristics and structural design of phononic crystals [D]. Ningxia University
22. Peng Z, LI C, Gao Y (2020) Study on low frequency band gap of two-dimensional perforated local resonant phononic crystals. Mech Design 37(04):73–77
23. Sun W, Wang T, Sun X, Kang T, Tan P, Liu Z (2019) Defect states and vibration energy recovery of a new two-dimensional three component piezoelectric phononic crystal plate. J Phys 68(23):145–153
24. Jia D, Ge Y, Yuan S, Sun H (2019) Dual band acoustic topological insulator based on honeycomb lattice phononic crystal. Acta Phys 68(22):235–241
25. Zhu J, Chen J, Chen J, Xu L, Wang X (2019) Dual channel filtering of cross arranged elliptical hole phononic crystal plates. Sci Bullet 64(26):2703–2709
26. Zhong L, Chen J, Liu J, Ling C, Fu X, Cai M, Huang Z (2019). Acoustic waveguide design based on line defect phononic crystals. Phys Exper 39(08):9–13
27. Shi G, Wang Z, Hunag S, Feng F, Zhang B (2019) Study on band gap of two-dimensional internal mass structure triangular lattice phononic crystals. J Yunnan Univ (Natural Science Edition) 41(03):545–550
28. He Z, Zhao J, Yao H, Jiang J, Zhang S, Chen X (2019) Band gap characteristics and vibration isolation performance of honeycomb phononic crystals. J Silicate 47(07):983–989

Modelling and Simulation

Numerical Study of Encased RTP Behavior Under Internal Pressure

M. S. Noorazizi and N. A. H. Jasni

Abstract Reinforced Thermoplastic Pipe (RTP) rehabilitation for internally corroded pipes is unquestionably a determination to extend the pipe's lifetime and reduce maintenance costs. The host pipe effectively increases the geometric stiffness of the RTP, but the pressure capacity of the encased RTP remains unknown. A study was conducted in this paper to investigate the encased RTP behaviour under internal pressure. To carry out the study, a non-linear finite element (FE) model was created using FE software. The FE model of standalone RTP was validated against the actual burst test to ensure that the model is very close to the actual response. The validated model was then extended as an encased RTP and tested under internal pressure. The non-linear finite element analysis shows that as the internal pressure rises, the encased RTP expands. The expansion of RTP is dramatically reduced once it interacts with the host pipe wall, which governs the hoop strength and, as a result, increases the pressure capacity of the encased RTP. The established internal pressure-strain curve can be a useful engineering tool for determining the actual strain of RTP under internal pressure while avoiding the costly and time-consuming full-scale experimental setup. For future work on encased RTP against host pipe wall imperfections under internal pressure, a numerical study should be performed.

Keywords Reinforced thermoplastic pipe · Underground piping rehabilitation · Numerical method

1 Introduction

When there is corrosion or deterioration over a long service period, the buried pipe may become functionally obsolete [1, 2]. Nonetheless, it is structurally safe, and any practical rehabilitation to restore its hydraulic transit ability and protect its current structural stability [3, 4, 5] is still necessary. Trenchless technology is effective for retrofitting deteriorated underground pipes by fitting a Reinforced Thermoplastic

M. S. Noorazizi (✉) · N. A. H. Jasni
Fakulti Teknologi dan Informatik Razak, Universiti Teknologi Malaysia, Jalan Sultan Yahya Petra, 54100 Kuala Lumpur, Malaysia
e-mail: noorazizi@uitm.edu.my

© The Author(s), under exclusive license to Springer Nature Singapore Pte Ltd. 2023
M. H. A. Hassan et al. (eds.), *Technological Advancement in Instrumentation & Human Engineering*, Lecture Notes in Electrical Engineering 882,
https://doi.org/10.1007/978-981-19-1577-2_35

Pipe (RTP) closely to the pipe's inner surface [6–9]. Because of its exposure to internal and external pressures, the RTP bears the net pressure in these applications [10–14]. Although the host pipe effectively increases the geometric stiffness of the RTP, the pressure capacity of the encased RTP remains unknown.

Hu et al. [15] conducted an experimental and numerical study on the performance of carbon fibre reinforced polymer (CFRP) liners under internal pressure. A three-dimensional FE model for CFRP lined structural performance was developed, taking into account the complex interface behaviour of each layer. Because the „smeared damage modulus" approach cannot effectively capture the development and influence of local contacts between the host pipe and the CFRP liner, the analysis cannot be expected to provide reasonable estimations of diameter change or practical calculations of local bending in the liner. Furthermore, Rueda et al. [16] used a numerical and experimental approach to investigate the confined behaviour of an encased thermoplastic liner. Full three-dimensional FEM simulations were run and validated against experimental data. Polymers with non-linear properties were used to reproduce experimental behaviour. Nonetheless, the study focuses on the buckling collapse of the thermoplastic liner as a result of external pressure rather than internal pressure.

Brown et al. [17] investigated a cured-in-place pipe (CIPP) liner that spans a section of host pipe that has lost structural integrity. Finite element analysis was used to investigate a potential limit state in which pressure pipe liners transition from within an intact region of old pipe to a region that has lost its load-carrying capacity. According to the authors, uniaxial rather than multiaxial strength properties dominate the line response. Lu et al. [18] investigated the stress performance of an urban gas pipeline repaired with a liner. According to the authors, the installed liner method effectively reduced the stress of the old pipeline, which is the primary pressuring bearing component.

It has also been argued and theorised that adding such RTP with its own hoop strength capabilities could then be used to share any pressure applied to a host pipe, allowing for increased pressure capability. The behaviour of the encased RTP under internal pressure is uncertain and must be investigated. A numerical study was carried out in this paper to investigate the encased RTP behaviour under internal pressure. Its goal is to create a predictive engineering tool for industrial use based on the stress–strain curve and RTP burst pressure. The methodology developed can be used to calculate the strain experienced by RTP under internal pressure.

2 Methodology

2.1 Case Study

In May 2019, a piping rehabilitation work was carried out in Sarawak, Malaysia, by installing RTP in DN300 underground piping, see Fig. 1 for the RTP installation process at the site.

All stress–strain curves must be established in order to predict the mechanical response of the host pipe and RTP. The stress–strain curve for the ASTM A106 Grade B as shown in Fig. 2 is extracted from Nam et al. [19], meanwhile, the stress–strain curve of TPU and polyester fabric as shown in Figs. 3 and 4, respectively are obtained from the RTP manufacturer.

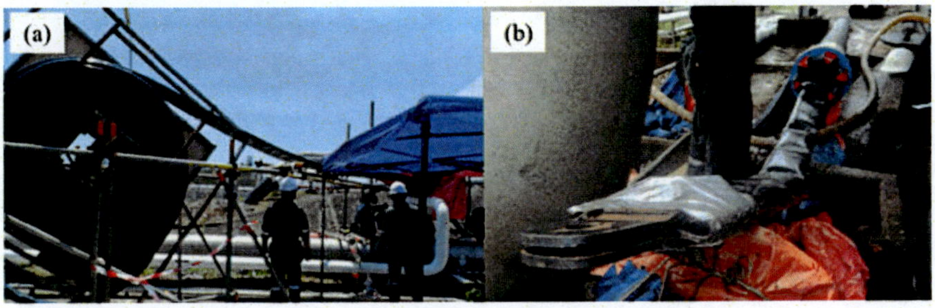

Fig. 1 **a** RTP insertion at feeding point and **b** arrived RTP at pulling point

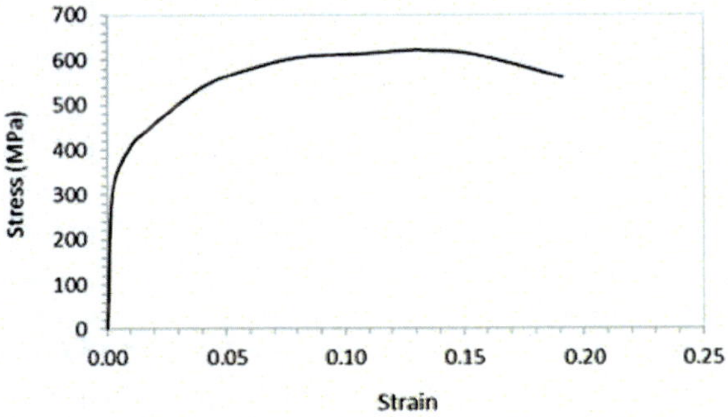

Fig. 2 Stress–strain curve of ASTM A106 Grade B

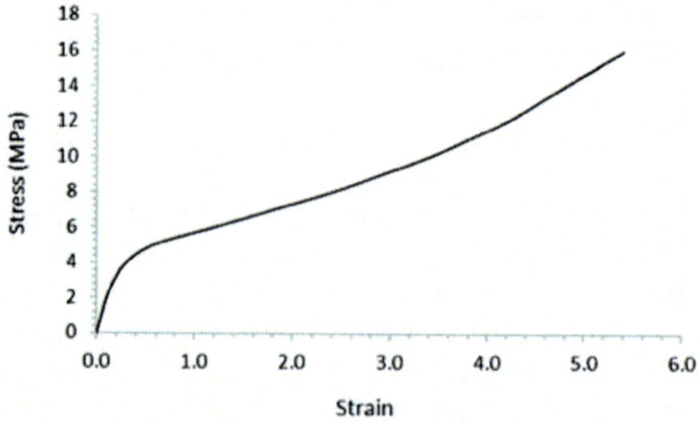

Fig. 3 Stress–strain curve of TPU

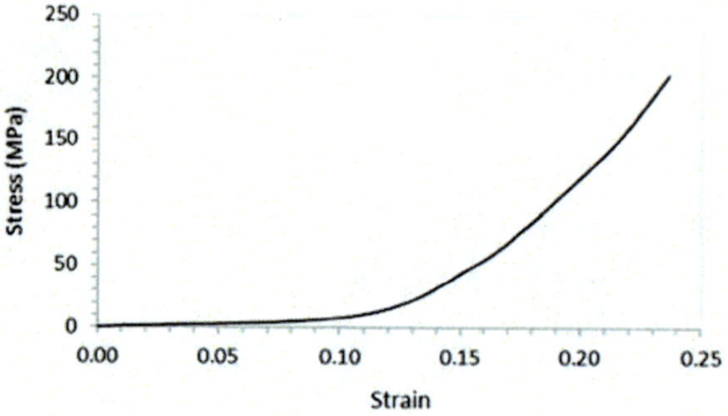

Fig. 4 Stress–strain curve of Polyester fabric

2.2 *Numerical Approach*

A three-dimensional FE model was created using FE software in accordance with DNVGL-ST-C501 to better understand the structural performance of encased RTP behavior under internal pressure. All components were included in this model and were given realistic geometrical sizes. The RTP layer boundary condition is set to bonded, while the host pipe and RTP interface is set to unbonded. When dealing with non-linear materials such as polymers, non-linear analysis should be used. The nominal layer stress as a function of nominal layer strain obtained from each material's stress–strain curve, as shown in Figs. 2, 3 and 4 is provided as input for the model. Because the polyester fabric is the governing strength of the RTP hoop, the onset of damage in the fabric material is defined by providing the fabric's failure strain above which the material response in unloading does not retract along the loading curve. The failure strain of the polyester fabric is set to $\varepsilon_f = 0.237$, as determined by the stress–strain curve in Fig. 5.

A burst pressure test in accordance with ISO 1402 is used to determine the internal pressure resistance and strength of the manufactured RTP. The 1.3 m long RTP

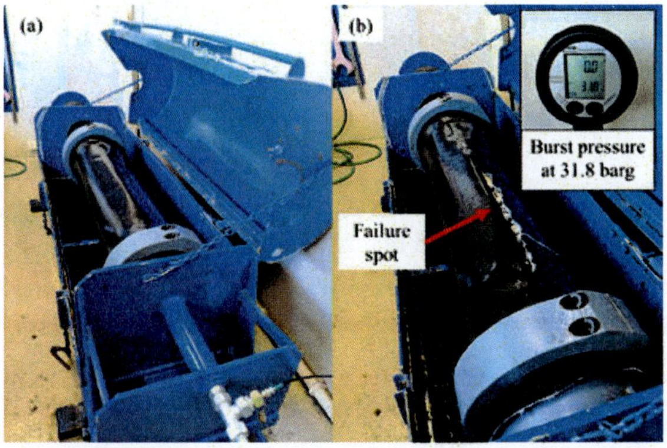

Fig. 5 **a** Before and **b** after burst test

specimen is fixed at both ends and then sealed off for pressurisation. A high-pressure pump continuously applies pressure to the specimen until it fails. Figure 5(a) shows the RTP specimen setup in the chamber before the burst test, meanwhile Fig. 5(b) shows the burst RTP specimen after the burst test, with a recorded burst pressure is 31.8 barg.

To obtain the RTP's strain value, a similar burst test was simulated in finite element software. Figure 6 shows the RTP meshed model with both ends fixed, while a finer mesh was generated in the RTP layers and an even denser mesh was generated in the polyester fabric (middle) layer, and an internal pressure load was applied on the inner surface. The loading increment studied was assumed to represent an experimental-like condition.

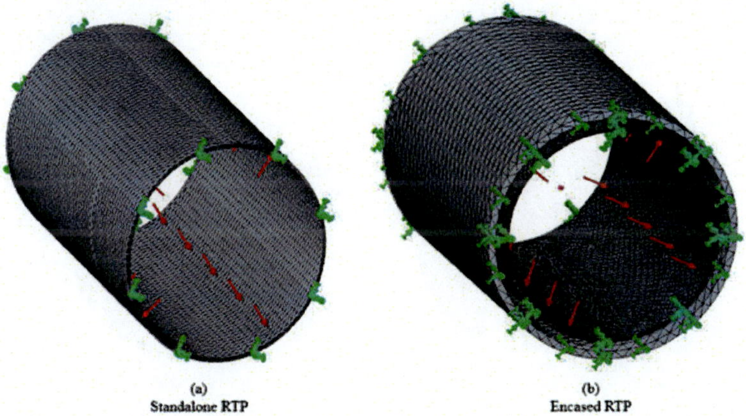

Fig. 6 **a** Standalone RTP and **b** encased RTP meshed model

3 Result and Discussion

3.1 FE Model Verification

Figure 7 shows the numerical simulation results for RTP under internal pressure, while Fig. 8 shows the strain-based results. For clarity, the inner layer of TPU is set to hidden, and the RTP model is shown in cross-sectional view to show the polyester fabric (middle) layer. According to the findings, the maximum strain experienced by RTP increases as internal pressure increases. The internal pressure vs. maximum strain of RTP, like the stress–strain curve of polyester fabric, is found to exhibit non-linear behavior. As a result, it is demonstrated that the polyester fabric governs the RTP hoop strength. At an internal pressure of P = 31.8 bar, which corresponds to the actual burst pressure, the maximum strain level of RTP reaches $\varepsilon_{max} = 0.232$, which is comparable to the actual failure strain, $\varepsilon_f = 0.237$. As a result, the FE model

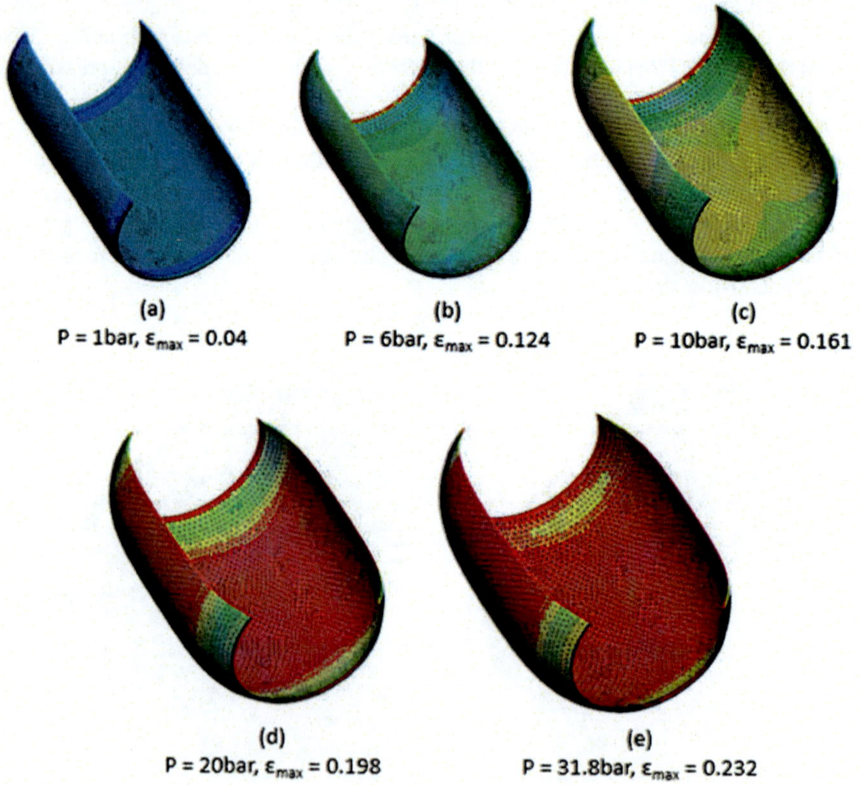

Fig. 7 Cross-sectional view of RTP model under internal pressure

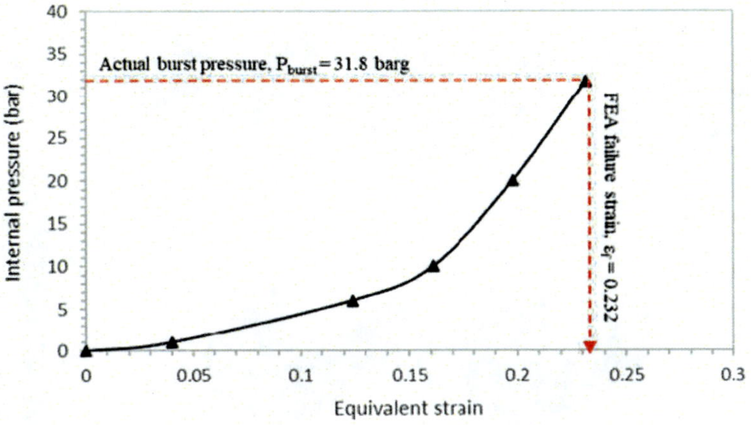

Fig. 8 Internal pressure against strain of RTP

has been validated against the actual burst test and can be used for further numerical simulation because the results are very close to the actual response.

3.2 Numerical Simulation of Encased RTP Under Internal Pressure

The numerical simulation of an encased RTP under internal pressure is shown in Fig. 9. In this case, the encased RTP moves freely outward until it comes into contact with the host pipe at an internal pressure of P = 3 bar. As the RTP expansion is restrained by the host pipe wall, the RTP strain level drops dramatically after P = 3 bar. As a result, any additional load after internal pressure of P = 3 bar must be carried by the host pipe. Figure 10 compares the internal pressure-strain curves of standalone and encased RTP. Because it is restrained by the host pipe as the internal pressure rises, encased RTP experiences less strain than standalone RTP.

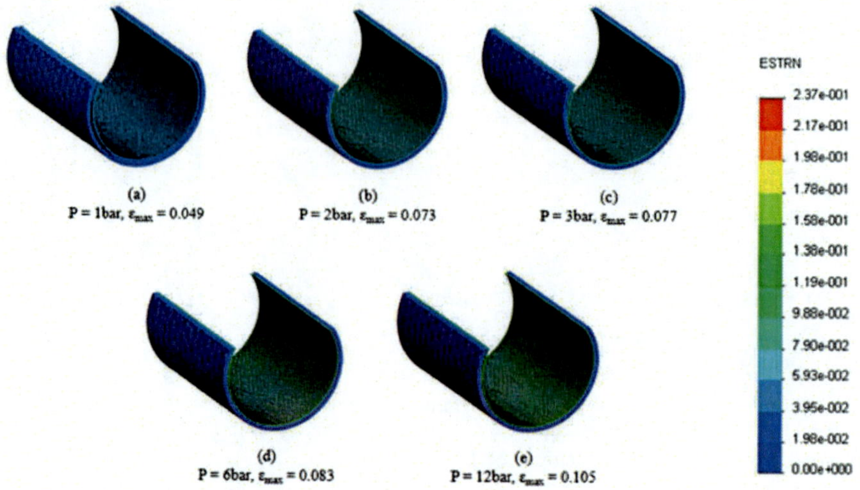

Fig. 9 Cross-sectional view of encased RTP under internal pressure

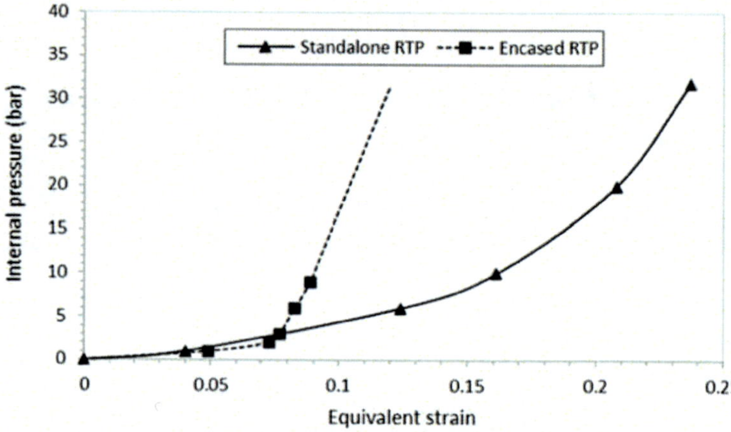

Fig. 10 Internal pressure-strain curve of standalone and encased RTP

3.3 *Hydrostatic Pressure Test post RTP Installation*

Following RTP installation, a hydrostatic pressure test is performed, as shown in Fig. 11 below. For a specified holding period, the RTP is pressurized at a test pressure of $P_{test} = 6$barg. The absence of pressure drops during the holding period indicates that the installed RTP is not leaking. The RTP strain level can be determined during the hydrostatic pressure test by using the encased RTP pressure-strain curve obtained from the numerical simulation. Regarding the internal pressure-strain curve

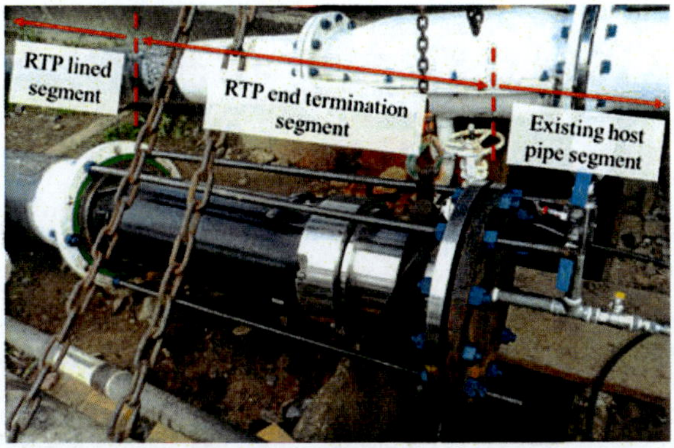

Fig. 11 Hydrostatic pressure test of RTP post installation

of encased RTP in Fig. 10, the RTP max. strain is $\varepsilon_{max} = 0.08$ at test pressure of $P_{test} = 6$barg which is significantly less than the actual failure strain, $\varepsilon_f = 0.237$.

4 Conclusion

The following conclusions can be drawn from the above consideration on the numerical study of encased RTP subjected to internal pressure:

i) The effect of internal pressure on RTP strain could be understood. The greater the internal pressure, the greater the strain on the RTP. The reinforcement layer of polyester fabric governs RTP pressure capacity because the internal pressure-strain curve has a typical trendline with the stress–strain curve of polyester fabric.
ii) Because it is restrained by the host pipe as the internal pressure rises, encased RTP experiences less strain than standalone RTP. According to the encased RTP result, the strain increments decreased dramatically after an internal pressure of 3 bar as the RTP made full contact with the host pipe wall. As a result, any additional load after an internal pressure of 3 bar is transferred to the host pipe.
iii) Based on the very good agreement between the actual RTP burst test and numerical simulation, it is possible to conclude that the established internal pressure-strain curve can be an appropriate engineering tool for determining the actual strain of RTP under internal pressure while avoiding costly and time-consuming full-scale experimental setup.
iv) As future work, a numerical study for encased RTP against host pipe wall imperfection should be performed, as all of the existing host pipe walls are severely corroded. The effect of RTP on protrusions such as well beads inside the host pipe is unknown and can be investigated in the future.

Acknowledgements This work is supported by Universiti Teknologi Malaysia under Research University Grant R.K130000.7656.4C295 and Q.K130000.2656.16J45 for the financial support provided throughout the course of this research project.

References

1. Lu H, Behbahani S, Azimi M, Matthews JC, Han S, Iseley T (2020) Trenchless construction technologies for oil and gas pipelines: state-of-the-art review. J Constr Eng Manag 146:0312001
2. Smith T, Hoult NA, Moore ID (2015) Role of grout strength and liners on the performance of slip-lined pipes. J Pipeline Syst Eng Pract 6:04015007
3. Li Z, Tang F, Chen Y, Zou X (2019) Stability of the pipe-liner system with a grouting void surrounded by the saturated soil. Eng Struct 196:109284
4. Kharazmi P (2019) Experimental assessment of the state of the lining materials used in the rehabilitation of sewers in residential buildings. Case Stud Constr Mater 11:e00291
5. Vistasp M (2015) Rehabilitation of pipelines using fiber-reinforced polymer (FRP) composites. Woodhead Publishing, Cambridge
6. Bakeer RM, Barber ME, Pechon SE, Taylor JE, Chunduru S (1999) Buckling of HDPE liners under external uniform pressure. J Mater Civ Eng 11:353–361
7. Lo K, Zhang J (1994) Collapse resistance modeling of encased pipes. Buried Plastic Pipe Technology, 2nd edn. ASTM International, West Conshohocken
8. Najafi M (2011) Service life estimation and extension of civil engineering structures. In: Vistasp M, Luke S (eds) Woodhead Publishing, Cambridge
9. Rahmaninezhad SM, Han J, Al-Naddaf M, Jawad S, Parsons RL, Liu H (2020) Field evaluation of performance of corroded corrugated steel pipe before and after sliplining rehabilitation. Tunnel Undergr Space Technol 102:103442
10. El-Sawy KM, Elshafei AL (2003) Neural network for the estimation of the inelastic buckling pressure of loosely fitted liners used for rigid pipe rehabilitation. Thin-Walled Struct 41:785–800
11. Association AWW (2001) Rehabilitation of water mains, 2nd edn, vol 28. American Water Works Association, Denver
12. Esaklul K, Mason J (2017) Trends in oil and gas corrosion research and technologies. In: El-Sherik A (ed.) Woodhead Publishing, Boston
13. Zhao W, Nassar R, Hall D (2005) Design and reliability of pipeline rehabilitation liners. Tunn Undergr Space Technol 20:203–212
14. El-Sawy KM, Sweedan AMI (2010) Effect of local wavy imperfections on the elastic stability of cylindrical liners subjected to external uniform pressure. Tunn Undergr Space Technol 25:702–713
15. Hu H, Dou T, Niu F, Zhang H, Su W (2019) Experimental and numerical study on CFRP-lined prestressed concrete cylinder pipe under internal pressure. Eng Struct 190:480–492
16. Rueda F, Marquez A, Otegui J, Frontini PM (2016) Buckling collapse of HDPE liners: Experimental set-up and FEM simulations. Thin-Walled Struct 109:103–112
17. Brown MJP, Moore ID, Fam A (2014) Performance of a cured-in-place pressure pipe liner passing through a pipe section without structural integrity. Tunn Undergr Space Technol 42:87–95
18. Lu H, Wu X, Ni H, Azimi M, Yan X, Niu Y (2020) Stress analysis of urban gas pipeline repaired by inserted hose lining method. Compos Part B: Eng 183:107657
19. Nam H-S, Je J-H, Han J-J, Kim Y-J (2014) Investigation of crack tip stress and strain fields at crack initiation of A106 Gr. B carbon steels under high strain rates. Procedia Mater Sci 3:764–771

Airflow Analysis of Contra-Rotating Fans Performance by Numerical Simulation

Teo Ting He, Muhammed Abdelfattah Sayed Abdelaal, Izzuddin Zaman, Djamal Hissein Didane, and Bukhari Manshoor

Abstract A contra-rotating fan is a mechanism that most used in aviation and marine propulsion system. It has two fan rotors sharing the same driveshaft but spinning in opposite direction by mechanical means. However, its implementation outside of the fields is scarce, possible due to its mechanical difficulty, assembly, and maintenance cost. The objective of this study is to investigate the validity of the advantages brought forth by contra-rotating fans using numerical simulation. The model of the contra-rotating fans was designed by using SolidWorks software and numerical simulation performed by using ANSYS Fluent. The inlet velocity was set at 0 m/s for the control set and 15 m/s for performance evaluation. The results showed that contra-rotating fans performed better in terms of velocity contour and flow streamlines, albeit having lower centreline velocity. From the results obtain, contra-rotating fans are shows to be able to produce a straighter and more focused airflow than a conventional single fan setup.

Keywords Airflow · Contra-rotating fan · CFD · Numerical Simulation

1 Introduction

Fans are commonly used in both commercial and industrial application, such as climate control, vehicle and machine cooling system, ventilation, etc. A fan works

---

Supplementary Information The online version contains supplementary material available at https://doi.org/10.1007/978-981-19-1577-2_36.

T. T. He
Engineering Division, Sharp Manufacturing (M) Sdn. Bhd, Batu Pahat, Johor, Malaysia

M. A. S. Abdelaal
Engineers House Academy, Manshiet El Bakry Street, Qesm Heliopolis, Egypt

I. Zaman · D. H. Didane · B. Manshoor (✉)
Faculty of Mechanical and Manufacturing Engineering, Universiti Tun Hussein Onn Malaysia, Batu Pahat, Johor, Malaysia
e-mail: bukhari@uthm.edu.my

© The Author(s), under exclusive license to Springer Nature Singapore Pte Ltd. 2023
M. H. A. Hassan et al. (eds.), *Technological Advancement in Instrumentation & Human Engineering*, Lecture Notes in Electrical Engineering 882,
https://doi.org/10.1007/978-981-19-1577-2_36

by rapidly rotating its fan blade around an axis, creating a low-pressure zone on one side of the fan blade. According to Bernoulli's principle [1], fluid will flow from a region of higher pressure to a region of lower pressure, thus creating an accelerated airflow towards the direction of the region with lower pressure. A contra-rotating fan is a type of axial flow fan with a special structure [2]. Its front and rear impellers are located on the same shaft, but they rotate in opposite direction from each other. Because of their mechanical structure, contra-rotating fans have the perks of having a high-pressure coefficient and great performance in reversing airflow. The contra-rotating fans commonly used in mine and tunnel ventilation, aircraft, and submarine propellers. Some contra-rotating fans are powered by a single motor with a planetary gear system to reverse the rotation direction of one of the fans, while some of them are powered by two motors.

Despite the hydrodynamics advantages and growing interest in the implementation of contra-rotating fans, the interest in commercialisation of contra-rotating fans is still low. There is only a very limited amount of product that utilises the contra-rotating mechanism in the market. That raises the question of why it is not popular despite its improvement in power efficiency when compared to a conventional single fan setup [3]. Although contra-rotating propellers can offer more power for a given radius compared to a conventional one-fan setup, cancelling out the heeling moment or torque effect produced by one propeller [4, 5], and generally improves the efficiency [6], it also brings forth a few disadvantages. Among the disadvantages of the counter-rotating propeller was a noisy from the fan. Even extra noise can be found in the higher frequencies. These substantial noise problems limit commercial applications [7, 8].

A study on numerical simulation of the unsteady aerodynamics in an axial counter-rotating fan stage was carried out by M. Younsi et al., [9]. The study investigated the unsteady flow behaviour in axial counter-rotating fan stage and aims to optimize the impact of the axial distance between the two rotors on the internal flow and the acoustic behaviour. The methodology involves 3D numerical simulations combined with experimental investigations. The numerical results were compared to the measurement data in terms of overall performance and local unsteady variables. The unsteady flow through an axial counter-rotating fan stage was studied and investigated using ANSYS CFX and the numerical procedure was compared locally and globally to experimental data [10, 11]. The results showed that the velocity contours are similar between the steady and the unsteady solutions and the wakes generated by the counter-rotating blades are crossing with each other's.

Another study of axial spacing effects on rotor-rotor interaction noise and vibration in a contra-rotating fans, also carried out by H. Luan et al. [12], aims to investigate the relationship between axial spacing of two-stage contra-rotating blade rows and a contra-rotating fans/compressor. The governing equations were solved with the CFX software that uses the finite volume method based on the backward Euler scheme, time marching was performed by using implicit time integration scheme and coupled with a second-order dual time stepping method [13–15]. The unsteadiness of rotor 1 is caused by the potential disturbance, and the upstream wake leads to the strong unsteadiness of rotor 2. The mean value of sound pressure level decreases by 17.2 dB in total when the axial spacing increased to 1.1 chords from 0.3 chords. With two sets of blades on one shaft and the mechanism to invert the rotating direction for one of the propellers increase the complexity of the whole system [4]. The increased part count also made the system more expensive when compared to a conventional setup. Other than that, the hydrodynamic gains are partially compensated by mechanical losses in shafting [16, 17]. Contra-rotating propellers are often used on torpedoes for their torque elimination properties. For normal ships, the increased efficiency offered by contra-rotating propellers is outweighed by the trouble of boring the outer shafts and the problems of mounting the inner shafts bearings.

The present work aims to analyse the airflow behaviour of the counter-rotating fan on the interaction process of the rotor-rotor interaction for the fans. To make up for the numerical investigation, a contra-rotating axial fan with an unsteady characteristic of two-stage rotors was investigated in the present study. Hence, the unsteady flow mechanism in the contra-rotating fan could be known profoundly. The effect of axial spacing is also investigated for the unsteady aerodynamic characteristics and finally, the performance of contra-rotating fans in terms of airflow velocity and vorticity will be compared to the other two setups which are through a conventional fan setup and two fans rotating in the same direction.

2 Methodology

2.1 Model Setup

The simulation model consists of two region which are the contra-rotation fans ducting and outer region. The ducting was made up for two cylinders of difference size. The smaller section which houses the fan blades have a diameter of 116 mm and a length of 45 mm for the single fan setup, while the contra-rotating fans setup has a length of 80 mm to contain both the front and rear fans. The outer region has a diameter of 250 mm and a length of 450 mm for both setups. Figure 1 shows the region create for the simulation work, modelled in SolidWorks.

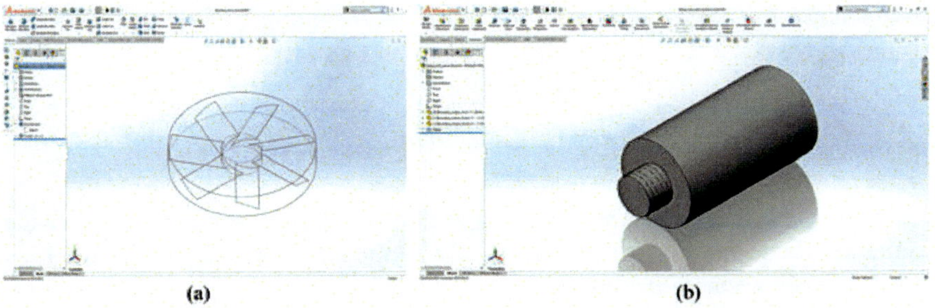

Fig. 1 Model setup by using SolidWorks **a** Rotor 1 **b** Ducting and outer region domain

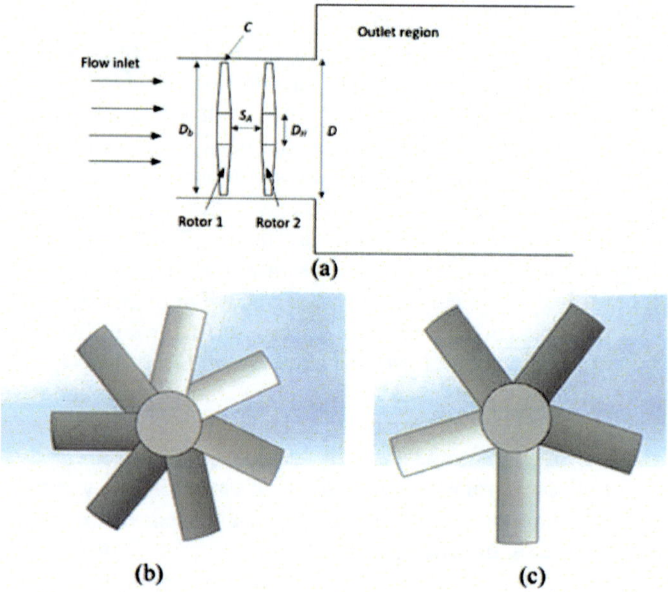

Fig. 2 Model for contra-rotating axial fan. **a** Simulation domain, **b** Rotor 1, **c** Rotor 2.

The test case considered by this research was a contra-rotating fan involved in the aerodynamic studies. The simulation model for contra-rotating axial fan is shown in Fig. 2. It consisted of a clockwise-rotating front rotor (rotor 1) and an anticlockwise rotating rear rotor (rotor 2), with the rotor 1 and rotor 2 had 7 and 5 blades respectively. This is to accommodate the fact that the front fan spins in a clockwise direction when viewed from the front, and the rear blade spins in an anti-clockwise direction, resulting in a contra-rotating system. For comparison purpose, another set of double rotating fan that rotates in the same direction was simulated. The other case simulated was a single fan which only contain of rotor 1.

The parameters of the contra-rotating fans used are summarize as in Table 1. The parameters such as blade diameter, hub diameter, duct inner diameter, tip clearance, number of blades, rotational speed as well as an axial spacing between two fans are based on previous research [2].

2.2 *Meshing and Boundary Condition*

One of the important steps in simulation process is mesh generation. Accuracy of the simulation are depending on the meshing size applied to the model. As fine meshing will give more accurate in simulation results but will cost a higher computational time. Wrong selection of the meshing size also may result in instability or even not converges of the simulation. For this study, medium mesh size was used to maintain

Table 1 Parameters involved for contra-rotating fans simulation

Parameter	Rotor 1	Rotor 2
Blade diameter, D_b (mm)	114.4	114.4
Pitch angle, θ (°)	20	20
Hub diameter, D_h (mm)	63	63
Duct inner diameter, D (mm)	116	116
Tip clearance, C (mm)	0.8	0.8
Number of blades, B	7	5
Rotational velocity, N (rpm)	3650	3180
Axial spacing, S_A (mm)	20 mm	

a balance between mesh size and calculation time. Figure 3 depicts the skewness of the mesh created for the contra-rotating fan and its meshing.

Three types of boundary operating conditions were imposed. There are two inlet velocity used, 0.1 m/s and 15 for the simulation. The inlet turbulent intensity, I_{in} = 3.5% - 4.1% was estimated for fully developed turbulent flow. At the outlet boundary, the pressure was set at atmospheric pressure (0 gage pressure). At the solid wall, the velocity was zero due to the no-slip condition. The working fluid was air at room temperature with density of 1.225 kg/m^3. The governing equations were independently solved using a double-precision pressure-based solver with a robust pressure–velocity coupling algorithm, SIMPLE been applied. Second order scheme was employed for the discretization of the pressure and momentum equations, while first order scheme for the turbulent kinetic energy and dissipation rate equations. Standard k-ε turbulence model equipped with enhanced near-wall treatments

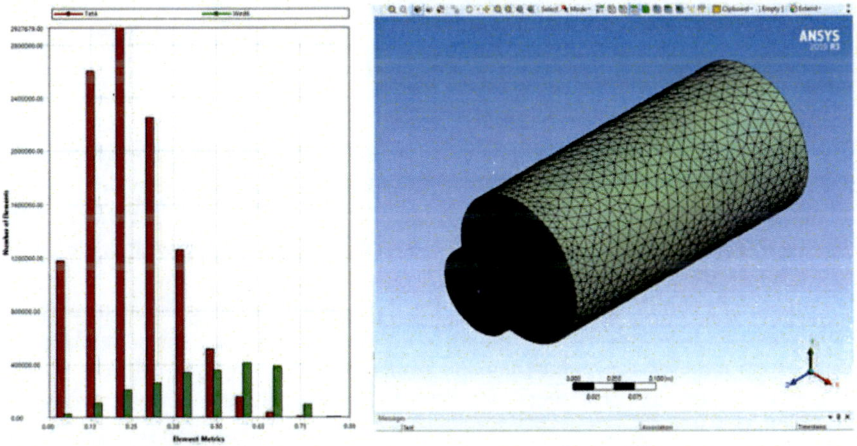

(a) Skewness (b) Meshing

Fig. 3 Skewness of the mesh created

Table 2 Solver setting

Solver scheme	SIMPLE
Gradient	Least square cell based
Pressure	Second order
Momentum	Second order upwind
Turbulent kinetic energy	First order upwind
Turbulent dissipation rate	First order upwind
Turbulence models	Standard k-ε model
Near wall treatment	Enhanced wall treatment (EWT)

was applied for the simulation. Detail for the solver setting applied summarised in Table 2.

3 Result and Discussions

The results of the simulation were obtained from the three different analysis were used to evaluate and compare the outcomes of the three setups. The three analysis included the velocity contour, streamline, and centreline velocity. Each result included the combination of two inlet velocity (0.1 and 15 m/s), and three variations of fan setups which are contra-rotating, same direction of rotating and single fan. Contra setup has two fans spinning in the opposite direction, same direction has two identical fans spinning in the same direction, while single setup has only one fan spinning in the clockwise direction.

3.1 Velocity Contour

The velocity contour used to visualise the velocity of airflow and plotted along the *yz*-plane and viewed from the *x*-axis. In Layman's terms, it is being viewed from the side of the system. All the contour maps were taken from the last frame of the 50-time steps calculation. The velocity contour of the three setups with 0.1 m/s inlet velocity is as shown below in Fig. 4.

From the figure, it was found that all the three setups failed to create a sufficient airflow that can traverse the whole length of the system. However, it is noticed that the contra setup achieved the highest velocity among the three setups, which is 41.55 m/s, despite having the same rotational velocity of 3650 rpm. This shows that the contra-rotating fans can produce better airflow than conventional single fan setup. From the figure, single setup had the second highest velocity at 28.16 m/s while the fan with same direction setup had the lowest velocity of 24.59 m/s. The low velocity achieved

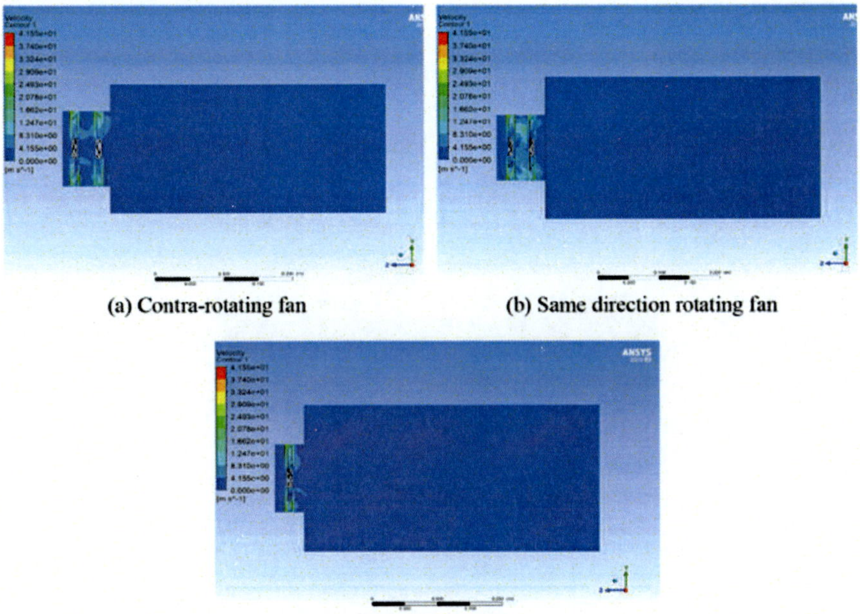

(a) Contra-rotating fan (b) Same direction rotating fan

(c) Single fan

Fig. 4 Velocity contour of the setups at 0.1 m/s

by the same direction setup is due to the fans spinning in the same direction, trapping the air in between two fans as shown in the Fig. 4(b).

For the case of 15 m/s velocity inlet, the airflow created by spinning fans, since the design are flawed and is unable to generate adequate airflow for the study. The velocity contour was taken from the last frame of the 50-time steps, and the results are as shown in Fig. 5. The result is vastly different from the 0.1 m/s setups. For 15 m/s inlet velocity, the maximum velocity value obtained by the contra-rotating fan, which is around 83.71 m/s. The result is vastly different from the 0.1 m/s setups as shown from the figure. With the case of 15 m/s inlet velocity, the maximum velocity value obtained by the contra-rotating fan, which is around 83.71 m/s. However, the maximum velocity is not suitable for performance evaluation since it occurred at the tip of the fan blades instead of travelling through the system. Thus, the velocity with the most presence will be considered.

The most noticeable difference between the three setups is the cross-sectional area of the airflow. Contra setup had an airflow that is more focused and straight, while both the same direction and single setups had their airflow spread out onto the boundary walls. So far, the results are consistent with the hypothesis that contra-rotating fans produce a straight and more focused airflow as mentioned in previous sections.

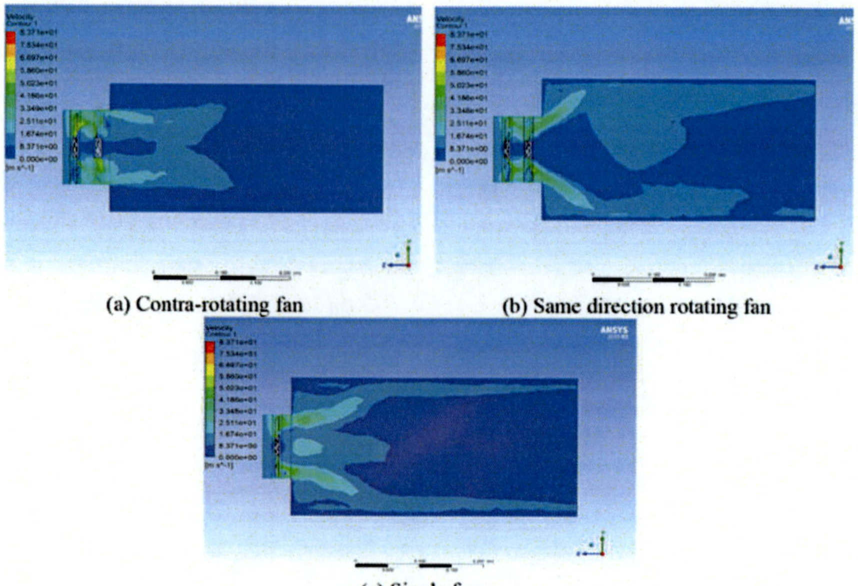

(a) Contra-rotating fan (b) Same direction rotating fan

(c) Single fan

Fig. 5 Velocity contour of the setups at 15 m/s

3.2 *Streamlines*

A streamline is a path that a particle of zero mass would take through the fluid domain. The path is calculated using a Runge–Kutta method of vector variable integration with variable time step control. Streamline is another interesting parameter to visualise the behaviour of airflow needed to be analysed for fluid flow. The method of evaluation is similar as before, taking the last frame of the calculation of the 50-time step as the analysis subject. The streamlines are created on the *yz*-plane and viewed from the *x*-axis as before. The number of points is set at 150 and the variable is velocity. Figure 6 below shows the streamline of the three setups for inlet velocity 0.1 m/s.

The airflow behaviour in all three setups is rather erratic. The maximum velocity, in this case, is not to be considered since it occurs on the spinning fan blade instead of travelling through the system. The straight lines at the rear of the system are areas that the airflow has yet to reach, possibly caused by the inability of the fans to propel the air to adequate velocity, or the analysis time is too narrow since 50-time steps of 0.01 s only equal to 0.5 s in real-time. If the time steps were to be increased, the airflow might be able to travel through the whole system.

For the 15 m/s inlet velocity, the streamline starts from the inlet instead of on the *xy*- plane. This is due to the inlet velocity was able to compensate for the inability of the fans to create a sufficient airflow that can traverse to the whole system. Figure 7 below shows the streamline of the flow at 15 m/s for the three conditions of the simulation.

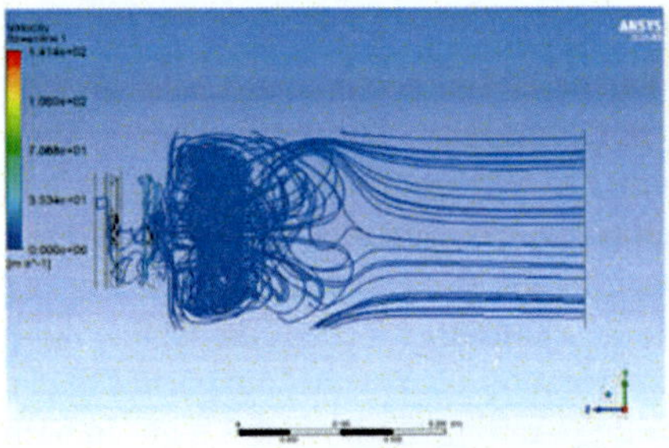

(a) Contra-rotating fan

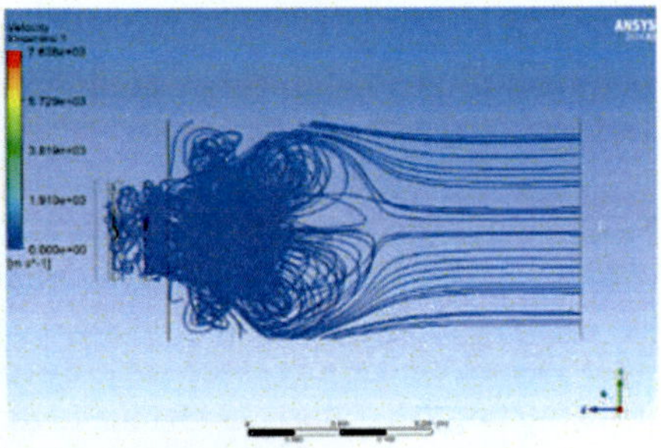

(b) Same direction rotating fan

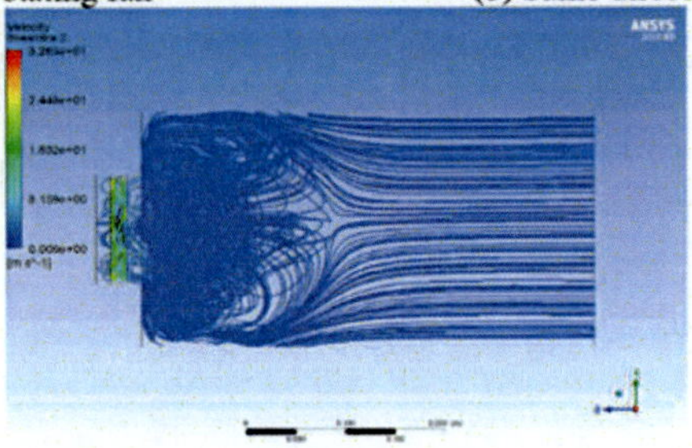

(c) Single fan

Fig. 6 Streamline of the flow with 0.1 m/s inlet velocity

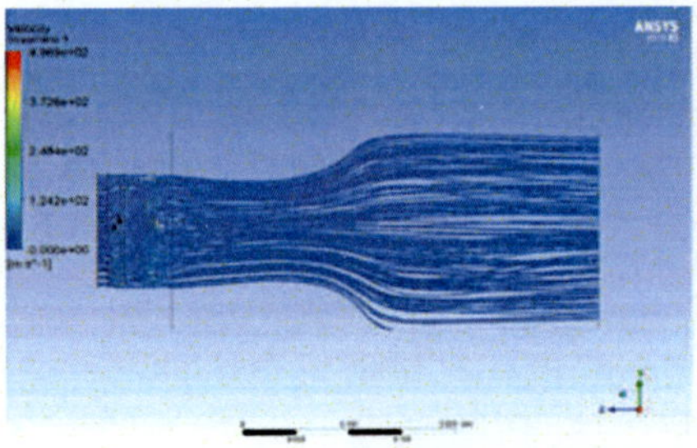

(a) Contra-rotating fan

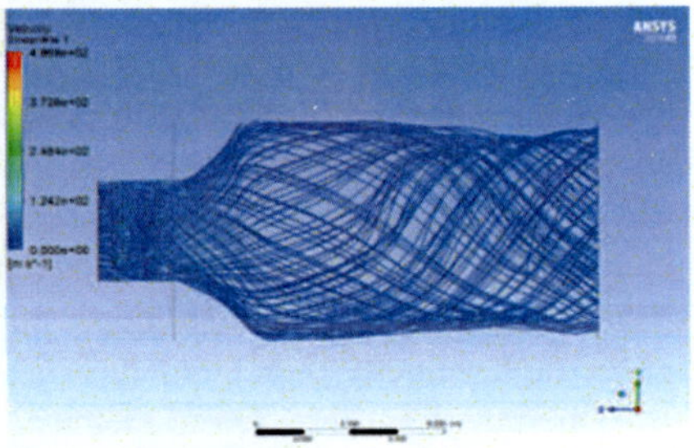

(b) Same direction rotating fan

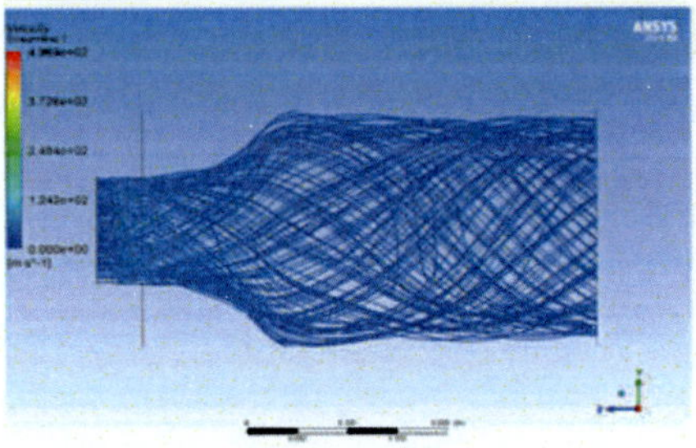

(c) Single fan

Fig. 7 Streamline of the flow with 15 m/s inlet velocity

From the figures, it was shown that the contra setup produced a narrow and vigorous airflow that fits the hypothesis from the earlier section. It is also noticed that the airflow is almost straight only in the contra-rotating setup, while the other two setups have the airflow spinning in an anti-clockwise direction due to the clockwise rotating fans. The maximum velocity is capped at 496.85 m/s achieved by contra-rotating fan evaluate the relative performance of the three simulation setups. The same direction-rotating fan and single fan create a similar pattern of the airflow with similar, albeit the airflow in the same direction rotating fan spreads out earlier than a single rotating fan. This is likely due to rotor 2 disrupting the flow of the air. In both cases, the same direction rotating, and single rotating fan had a flow of air expanded towards the boundary, and likely to spread out even further if not limited by the 250 mm diameter of the boundary.

3.3 Centerline Velocity

Centreline velocity is a commonly used parameter to measure the velocity of fluid flow in a pipe or similar setting. For this study, a line is generated at the centre of the system by using two point's method and specifying the coordinates of the two points. The x and y coordinates are 0 at the centre, and the z coordinate depends on the length of the setup since a single rotating fan setup is shorter than both contra-rotating and the same direction rotating setups. The data is presented in a velocity (m/s) against distance in z-axis, as shown in Fig. 8.

From the figure, all the simulation cases show a similar trend of centreline velocity. The velocity was increased then decrease when through the fans. The fan with the same direction rotation had given the highest value of centreline velocity while the single fan has the lowest centreline velocity, 1.176 and 0.731 m/s respectively. For the contra-rotating fan, the maximum centreline velocity is 0.772 m/s. All three cases had the high velocity then suddenly fall off quickly after flowing through the fans due to the inability of the fan to create a sufficient airflow.

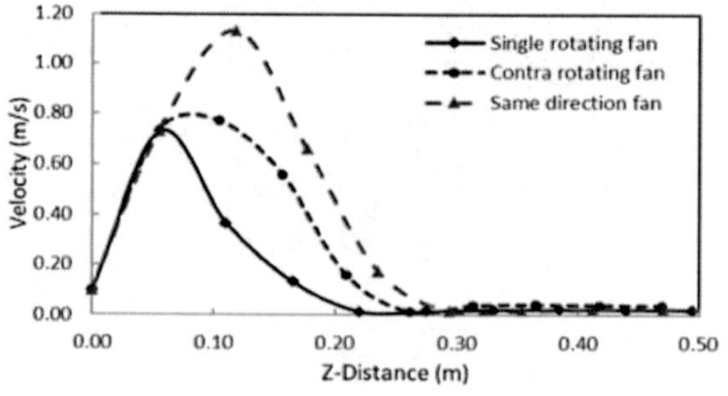

Fig. 8 Centreline velocity of the three cases with 0.1 m/s flow

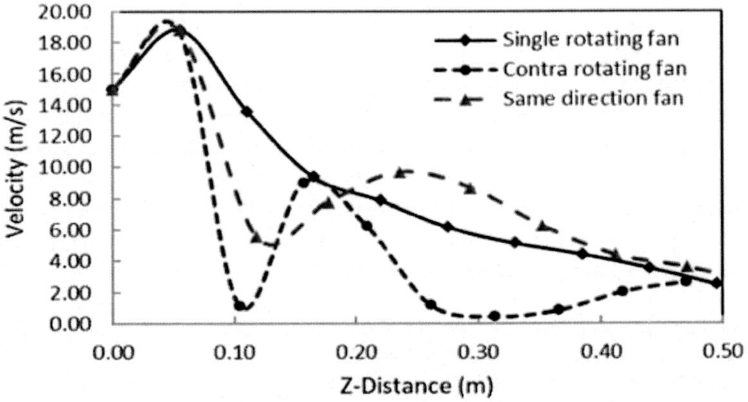

Fig. 9 Centreline velocity of the three cases with 15 m/s flow

For the cases of 15 m/s inlet velocity, the trend for centreline velocity shown in Fig. 9. Referring to this velocity inlet, the single rotating fan provides the highest velocity, which is 18.8 m/s for maximum, followed by the contra-rotating fan with 9.7 m/s. Different from the low-velocity inlet (0.1 m/s), the same direction-rotating fan has the lowest centreline velocity compare to the other two. From these results, it was showed that the contra-rotating fan and same direction-rotating fan had a downward trend right after the air goes through the inlet and rise again after travelling a certain distance in the duct. This is due to the result of air flowing around the fan hub, creating an area with reduced airflow between the two rotors. The single rotating fan, on the other hand, does not have this problem and follows a simple upward-downward trend. However, the maximum centreline velocity is not the ideal indicator of the performance of the fan setup since airflow around the hub and through the blades instead of uniformly through the system like in a ducting system, therefore can only be used as a reference.

4 Conclusion

In conclusion, contra-rotating fans can produce a straight, focused airflow and reaching higher velocity in comparison with conventional single fan design as the theory suggests. This conclusion proved by the evidence from the results of the simulation of different setups at different inlet velocity. In the contour maps section, contra setup achieved the highest maximum velocity of 0.1 and 15 m/s inlet speed condition. Contra setup created a straight and focused airflow in 15 m/s streamline simulation, while the same direction and single produced an airflow that is spread out and spinning in an anti-clockwise direction, albeit travelling uniformly. However, contra setup did not top the other setups in the centreline velocity section. This might be since centreline velocity is usually implemented for fluid flow in a pipe which has the fluid flowing uniformly across the whole cross-sectional area of the system, while in this study the fluid flow around the fan hub and across the fan blades instead of through the centre of the system. So, despite the unconvincing results in centreline velocity, contra-rotating fans are still considered to be the best performing setup within the three, judging by its performance in both contour maps and streamline. However, there are some more data that needs altering in the parameters such as axial distance, blade pitch angle so that it can be more precise in conclusion regarding the advantages of the contra-rotating fan.

Acknowledgements This research was supported by Ministry of Higher Education through Fundamental Research Grant Scheme (FRGS/1/2020/TK0/UTHM/02/17). The authors also would like to acknowledge the Faculty of Mechanical and Manufacturing Engineering, Universiti Tun Hussein Onn Malaysia (UTHM), Sharp Manufacturing (M) Sdn Bhd and Engineers House Academy for supporting data and technical advice.

References

1. Babinsky H (2003) How do wings work? Phys Educ 38:497
2. Dong B et al (2020) Theoretical characterization and modal directivity investigation of the interaction noise for a small contra-rotating fan. Mech Syst Signal Process 135:106362
3. Chen YY et al (2008) A study of speed ratio affecting the performance of a contra-rotating axial compressor. Proc Inst Mech Engineers Part G-J Aerospace Eng 222:985–991
4. Luan H, et al (2019) Axial spacing effects on rotor-rotor interaction noise and vibration in a contra-rotating fan. Int J Aerosp Eng
5. Manshoor B, Khalid A (2012) Numerical investigation of the circle grids fractal flow conditioner for orifice plate flowmeters. Appl Mech Mater 229:700–704
6. Zaman I et al (2014) The application of multiple vibration neutralizers for vibration control in aircraft. Appl Mech Mater 629:191–196
7. Dong B et al (2021) Noise attenuation and performance study of a small-sized contra-rotating fan with microperforated casing treatments. Mech Syst Signal Process 147:107086
8. Chen Q, Luan H, Weng L, Luan Y, Chen P (2017) Effect of counter-rotating fan's speed matching on stall inception and characteristics of tip clearance flow. J Vibroeng 19(6):4630–4643
9. Younsi M, et al (2016) Numerical simulation of the unsteady aerodynamics in an axial counter-rotating fan stage. In: 12th international conference on heat transfer, fluid mechanics and thermodynamics, Malaga, Spain
10. Wang Y, et al (2014) A Computational study on the aerodynamics of a 90 mm ducted contra-rotating lift fan. In: IMAV 2014: international micro air vehicle conference and competition 2014. Delft University of Technology
11. Aranake AC, Lakshminarayan VK, Duraisamy K (2015) Computational analysis of shrouded wind turbine configurations using a 3-dimensional RANS solver. Renew Energy 75:818–832
12. Luan H, Weng L (2016) Effects of axial spacing between counter rotors on performance and on flow field of a counter rotating fan. Int J Simul Syst Sci Technol 17(9):111–116
13. Yii R et al (2015) Effects of nozzle shape on the flow characteristics of premix injector using computational fluid dynamics (CFD). Appl Mech Mater 773:773–774
14. Xu C, Bil C, Cheung S (2014) Fluid dynamics analysis of a counter rotating ducted propeller. In: 29th congress of the international council of the aeronautical sciences, ICAS 2014

15. Manshoor B, Zaman I, Khalid A, Ghazali MF, Khandelwal MK (2020) Effect of streamwise spacing on the sound generated by flow through two square cylinders in tandem arrangement. Int J Adv Trends Comput Sci Eng 9(11):534–541
16. Bertram V (2012) Practical ship hydrodynamics
17. Manas M, Pradeep A (2020) Stall inception mechanisms in a contra-rotating fan operating at different speed combinations. Proc Inst Mech Engineers Part A J Power Energy 234(8):1041–1052

Scouring Around Rigs-to-Reefs Jacket Platform with Different Sitting Configurations on Seabed

Mohd Asamudin A. Rahman, Mohd Hairil Mohd, Ahmad Fitriadhy, Muhammad Nadzrin Nazri, Erwan Hafizi Kasiman, and Fatin Alias

Abstract The decommissioning of offshore platforms is becoming more popular due to the lack of productive fields and the unpredictability of oil prices. The Rigs-to-Reef (RTR) program, which is part of the oil and gas company's decommissioning method, involves turning jacket platforms into artificial reefs. This program creates shelter for marine life and subsequently increases marine production. However, the sediment scours around the jacket platform are one of the problems to be concerned about because it will result in RTR program failure. In this study, thc scabed profile around the jacket platform was investigated using Computational Fluid Dynamic (CFD). The simulation of the seabed scour and the deposition around, were conducted using different arrangements of jacket platform. The results showed that the flow velocity decreased when the flow was obstructed in front of the jacket platform. A horseshoe vortex occurred, then developed in front of the cylinder. The flow accelerated on the opposite side of the jacket structure, causing significant scour around the structure as well as bed shear stress. Finally, the vertical jacket platforms showed a significantly lower scour and deposition than the horizontal position.

Keywords Scouring · Jacket platform · Rigs-to-reefs · Fluid-soil interaction · Computational Fluid Dynamics (CFD)

Supplementary Information The online version contains supplementary material available at https://doi.org/10.1007/978-981-19-1577-2_37.

M. A. A. Rahman (✉) · M. H. Mohd · A. Fitriadhy · M. N. Nazri · F. Alias
Maritime Technology and Naval Architecture Programme, Faculty of Ocean Engineering Technology and Informatics, Universiti Malaysia Terengganu, 21030 Kuala Nerus, Terengganu, Malaysia
e-mail: mohdasamudin@umt.edu.my

E. H. Kasiman
School of Civil Engineering, Faculty of Engineering, Universiti Teknologi Malaysia, 81310 Skudai, Johor, Malaysia

© The Author(s), under exclusive license to Springer Nature Singapore Pte Ltd. 2023

M. H. A. Hassan et al. (eds.), *Technological Advancement in Instrumentation & Human Engineering*, Lecture Notes in Electrical Engineering 882,
https://doi.org/10.1007/978-981-19-1577-2_37

1 Introduction

Decommissioning is understood as the last phase of offshore oil and gas platform operations and to restore the field to its pre-lease condition [1]. Decommissioned platforms are expected to be removed entirely to reduce the environmental and safety risks associated with leaving unused platform structures as they are. Besides, it can also decrease the chance of disagreement between many parties regarding the environmental issues. In the past few decades, the oil and gas company had been trying to reuse and recycle the platform structure through decommissioning activities. The Rigs-to-Reef (RTR) program is one of the oil and gas company's decommissioning methods, which involves converting the platform into an artificial reef. This approach is believed to improve marine resources, which have been declining since the 1950s.

In Malaysia, around 300 fixed offshore platforms are placed in shallow water and almost 60% are close to the end of their manufacturing life. This means that Malaysia has entered the decommissioning phase for most of its oil and gas platforms of five to ten years. Baram-8 or Kenyalang Reef implemented in the South China Sea was the first major RTR program of its kind [2]. Decommissioning research carried out by Awang [3] revealed the existence of invertebrate populations of Polychaeta, Crustacea and Mollusca on the artificial reef (Fig. 1). Furthermore, in 2013, the marine study revealed that the submerged Baram-8 platform hosts populations.

Recently, Petronas has renewed interest in the decommissioning that converts jacket platforms to artificial reefs. Jacket platforms are categorized as fixed platforms and the majority of the offshore platforms are piled-jacket structures [4]. The jacket platform that undergoes decommissioning through partial removal option will be placed on the surface of the seabed as a reef. After a period of time on the seabed, there are significant changes in the morphology of the seabed that can be observed due to the movement of the seabed soil. Sediment transport is a set of processes that mediates between the flow of water and the channel boundary [5]. As the sediment is assumed cohesionless, forces will act opposite on the grain, the weight of grain

(a) (b)

Fig. 1 **a** Baram-8 platform was converted into artificial reef structures; **b** the condition of the reef after 9 years of deployment [3] will hold it on the seabed and fluid lift and drag forces will roll, lift and slide grain along the seabed.

Dynamic responses and flow characteristics around the jacket platform affect the potential growth of the coral and fish. The surface pressure and wake region of the RTR play essential roles in determining the success of the program. A higher efficiency index was achieved for the current direction normal to the flow due to the higher frontal area exposed to the incoming flow [6, 7]. Different types of structures will eventually be exposed to a different jacket structure efficiency to be considered an artificial reef [8].

Sediment transport that occurs around jacket structure may lead to an unstable position of the jacket structure itself. Local scour development might become the reason for the jacket structure to become unstable or even collapse [9]. Additionally, deposition around the structure increases the failure of the reef due to the possibility of the structure buried under the seabed. This would eventually reduce the surface area for coral to grow and fish to shelter around the reef. One of the governing factors leading to this is the arrangement and position of the jacket structure as the artificial reef.

In the last few decades, scour around the structure on the seabed were widely studied using experimental and numerical methods [10–16]. Baglio et al. [17] proposed the experimental method to understand the characteristics of scouring of a single pile in oscillatory flow using the stereo vision approach. A study by Zhang et al. [10] shows that the dimensionless pile spacing affects the flow characteristics around the structure. It also has been established that the scour depth is higher when the member of the structure is side by side and decreases when placed in tandem [5].

To date, scouring around jacket platforms has still not yet been comprehensively studied. There has been no research directly investigating the effects of different arrangements on the scouring and deposition of the partial removal of the jacket platform using Computational Fluid Dynamics (CFD). In this situation, numerical investigation of local scours around the jacket platform on seabed needs to be performed further to understand the phenomenon of seabed changes around jacket structure.

In this paper, the effects of four different arrangements of jacket structure on the growing local scour depth, variations of the scour pattern due to jacket structure, and scouring process were investigated. This study sets out to explore the suitable artificial reef deployment position to reduce the seabed's scouring and deposition. The findings should make an important contribution to the field of artificial reef in terms of morphological changes of seabed in determining the success of the reefs.

2 Methodology

2.1 Development of Scour

Study on scour depth due to morphological changes on seabed were conducted by various researchers such as Pourshahbaz et al. [18], Kurdistani & Pagliara [19], May et al. [20] and Baglio et al. [17, 21], to name a few. Validation among the current model and the earlier simulation results of the sediment scour model will be presented to determine the local scour under complicated form. To fully understand the scouring processes over a pile, flow characteristics, velocity profile, scour morphological process, scour depth, and deposition height are examined. The flow blockage and run-up can all be seen in front of a pile, resulting in both front and rear seabed elevation changes, where the flow velocity decreases significantly [17]. The boundary layer, downflow, and pressure gradient contribute to the formation of a horseshoe vortex. Furthermore, the horseshoe vortex and flow acceleration at the pile's sides create high bed shear stress, which is the principal cause of local scour surrounding the pile [10].

2.2 Sediment Transport Principles

Basically, sediment transport is a movement of particles by the fluid. The greater the velocity of the flow, the more sediment will be conveyed [22]. Sediment transport elucidates the ability to entrain and transport sediment based on gravitational forces acting on particles on seabed and drag forces will suspend inflow and shove it along downstream. The sediment transport is assumed as different properties of multiple non-cohesive sediment species. This measures the sediment's motion by estimating the sediment's erosion, advection and deposition. There have been four mechanisms for sediment transport that can explain all the processes of transporting sediments.

i. *Entrainment*

   The grains at the top of the packed bed will be removed and taken up into suspension as the bed shear stress exceeds the critical shear stress.

ii. *Suspended load transport*

   Grain is conveyed by the flowing current within a particular height above the packed bed once it has been entrained.

iii. *Deposition*

   Gravity, buoyancy, and friction cause the suspended grains to settle on the sediment bed.

iv. *Bed-load transport*

   In relation to the shear stress of fluid flow, grains will roll, hop, and slide across the packed bed surface.

2.3 Entrainment and Deposition

Entrainment and deposition take place at the same time to ensure that the net rate of change between packed and suspended sediment is accurate. The sediment entrainment lift velocity is determined as indicated in Eq. (1) [13].

$$u_{lift,i} = \alpha_i n_s d_*^{0.3}(\theta_i -)^{1.5}\sqrt{\frac{\|g\|d_i(\rho_i - \rho_f)}{\rho_f}\theta_{cr,i}} \tag{1}$$

Where:

α_i = entrainment parameter (suggested is 0.018)
n_s = outward pointing normal to the packed bed interface
$d_*^{0.3}$ = dimensionless diameter of sediment
θ_i = Shields parameter that is dependent on local bed shear stress
$\theta_{cr,\,i}$ = critical Shields parameter
$||g||$ = acceleration of gravity ($-9.8\ m/s^2$)
d_i = diameter sediment
ρ_i = density of sediment species i
ρ_f = density of the fluid

Local Shields parameter written as Eq. (2) [11].

$$\theta_i = \frac{\tau}{||g||d_i(\rho_i - \rho_f)} \tag{2}$$

τ = local shear stress. Local shear stress was calculated using the law of wall and the quadratic law of bottom shear stress, while bed surface roughness was considered.

The dimensionless critical Shields [23] parameter calculated in Eq. (3).

$$\theta_{cr,\,i} = \frac{0.3}{1 + 1.2d_*} + 0.055\left[1 - \exp(-0.02d_*)\right] \tag{3}$$

The process of sediment grains settling out of suspension on a packed bed or settling in bed-load transport is referred to as deposition. The settling velocity of sediment [23] is calculated in Eq. (4).

$$u_{setting,\,i} = \frac{v_f}{d_i}[\left(10.36^2 + 1.049d_*^3\right)^{0.5} - 10.36 \tag{4}$$

Where:

v_f = kinematic viscosity of fluid.

2.4 Suspended Load Transport

The amount of material transported downstream within the channel by water flow is known as suspended load transport. The water flow creates upward turbulence that holds particles on the seabed. The transport equation is used to compute the suspended load transport as in Eq. (5) [15].

$$\frac{\partial C_{s,i}}{\partial t} + \nabla.(u_{s,i}C_{s,i}) = \nabla.\nabla(D_f C_{s,i}) \tag{5}$$

Where:

$C_{s,i}$ = suspended sediment mass concentration of species i
$u_{s,i}$ = sediment velocity of species i
D_f = diffusivity

Each sediment in suspension, flows at its own velocity, resulting in distinct inertia and drag forces for each grain of varying mass density and size. The velocity difference between the suspended grains and the fluid-sediment mixture is referred as the grains' settling velocity. The sediment velocity of species i was calculated as shown in Eq. (6).

$$u_{s,i} = \bar{u}u_{setting,i}C_{s,i} \tag{6}$$

Where:

$\bar{u}$ = velocity of fluid-sediment mixture
$C_{s,i}$ = suspended sediment volume concentration

$\bar{u}$ was obtained by continuity and Navier–Stokes equations closed with RNG $k-\varepsilon$ turbulence model [16].

2.5 Bed-Load Transport

Bed-load transfer is a sediment transport mechanism due to rolling or bouncing over a packed bed of sediment surface. The bed-load transport velocity [10] is calculated with Eq. (7).

$$u_{bedkoad,\,i} = \frac{q_{b,i}}{\delta_i C_{b,i} f_b} \tag{7}$$

Where:

f_b = critical packing of sediment
$C_{b,i}$ = volume fraction of species i

$q_{b,i}$ = volumetric bed-load transport rate
δ_i = bed-load thickness

$q_{b,i}$ and δ_i be calculated by Eq. (8) and Eq. (9) [11]:

$$\delta_i = 0.3d_*^{0.7}\left(\frac{\theta_i}{\theta_{cr,i}} - 1\right)^{0.5} d_i \quad (8)$$

$$q_{b,i} = 8\left[\|g\|\left(\frac{\rho_i - \rho_f}{\rho_f}\right)d_i^3\right]^{0.5} \quad (9)$$

2.6 *Validation of the Present Model*

The simulation of a single pile was conducted to validate the present numerical method. A commercial CFD software, Flow3D was used in this study. The parameters of the packed sediment and test conditions for the single pile were tabulated in Table 1.

The comparison of bed elevation contours at t = 500 s around single pile between Zhang et al. [10] and the current simulation is shown in Fig. 2(a). The highest local scour depth occurs at a 45° angle to the flow direction in front of the pile, according to the current model, which is in excellent agreement with the literature. For both results, sediment deposition was seen behind the pile. The structure and scale of scour depth for both numerical models are relatively identical. The maximum scour depth around a single pile is compared at the time evolution of t = 500 s in Fig. 2(b). The maximum scour depth at 0.68 m and at t = 500 s were reported in the literature as compared to 0.75 m in this simulation. However, it is apparent from the plot that the trend of the scour over time is in agreement with the literature. The case study of the jacket platform model has considered the partial removal options. Table 2 shows the characteristic of packed sediment of the present numerical model. Four reef arrangements were studied, as shown in Fig. 3. The aim of the proposed arrangement was to study the effects of spacing and the position of the cut-leg of the jacket on the scouring phenomenon.

The numerical model's boundary conditions are shown in Fig. 4. The mesh close to the platform were refined using nested block mesh to capture the seabed changes in detail (Fig. 5). The grid size and the number of elements for the scouring simulation are included in Table 2. At the inlet boundary, the current velocity is specified. The symmetry boundary was defined on both sides and on the top domain to ensure no disturbance from the wall effects in the numerical model. Finally, the outflow was assigned at the end of the domain for continuous flow. Table 3 summarizes the boundary conditions used in the numerical simulation for different reef arrangements.

Table 1 Characteristics of packed sediment and test conditions for the single pile simulation

Parameter	Packed sediment		
	Symbol	Value	Unit
Domain height	h	7.0	m
Flow velocity	U	2.0	m/s
Grain size–fine gravel	d	5.0	mm
Grain size–medium gravel	d	10.0	mm
Sediment density	ρ	2650	kg/m^3
Shields parameter	–	0.048	Dimensionless
Bed-load coefficient	–	8.0	Dimensionless
Entrainment coefficient	–	0.018	Dimensionless
Height of packed sediment	h	3.0	m
Length of packed sediment	l	40.0	m
Width of packed sediment	w	18.0	m
Pile diameter	D	1.5	m
Pile height	h_{pile}	4.0	m
Fluid elevation	–	3.0	m
Grid domain	–	0.2	m
Grid nested	–	0.1	m

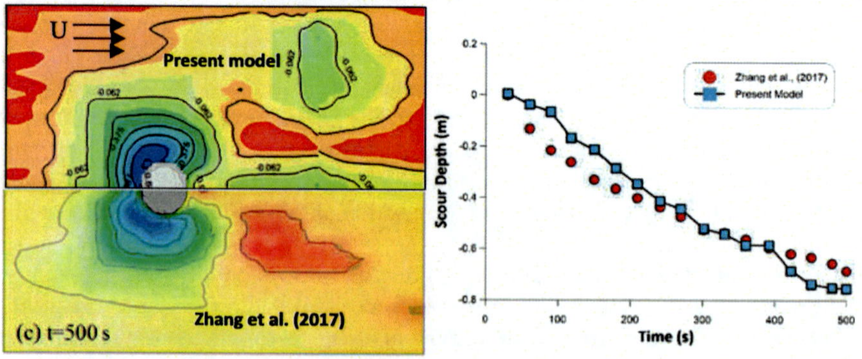

(a) (b)

Fig. 2 **a** Bed elevation contours around a single pile; **b** the scour depth changes for a single pile over time between present simulation and simulation result by Zhang et al. [10]

Table 2 Characteristics of packed sediment and mesh for the jacket platform model

Parameter	Packed sediment		
	Symbol	Value	Unit
Height of packed sediment	*h*	3.0	m
Length of packed sediment	*l*	68.0	m
Width of packed sediment	*w*	32.0	m
Fluid elevation	–	8.0	m
Grid domain	–	0.4	m
Grid nested	–	0.2	m
Global domain mesh element		81 6000	
Nested mesh element		3 360 000	
Total mesh element		4 176 000	

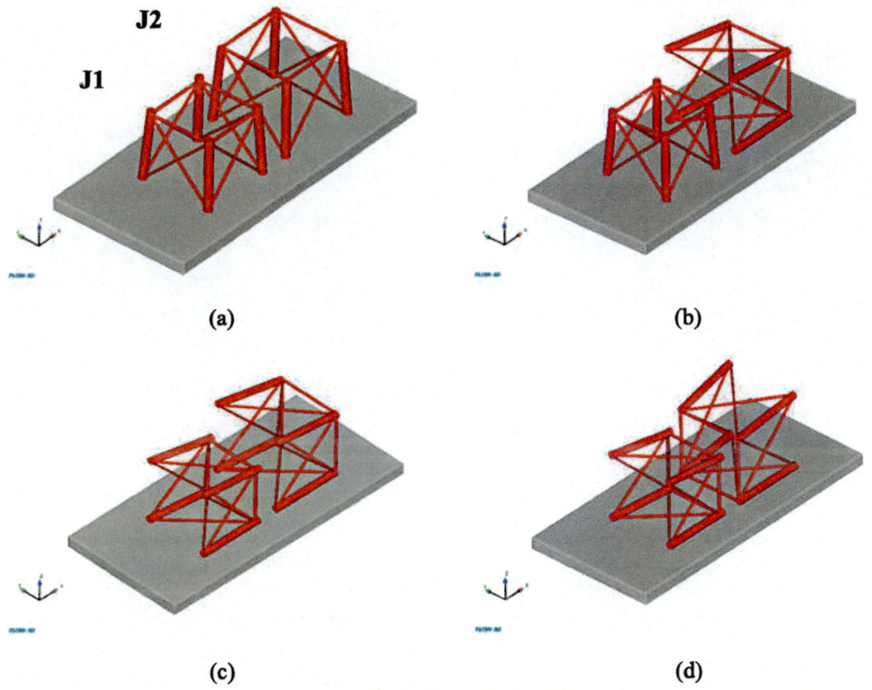

Fig. 3 The Jacket Platform Model Partial Removal Reefing Options **a** R1; **b** R2; **c**; **d** R4. front jacket structure was identified as J1 and jacket structure at the back as J2

3 Results and Discussion

Figure 6 illustrates the three-dimensional height elevation change pattern around the models. For R1 (Fig. 6a), the scour pattern was clearly visible at the front and at the

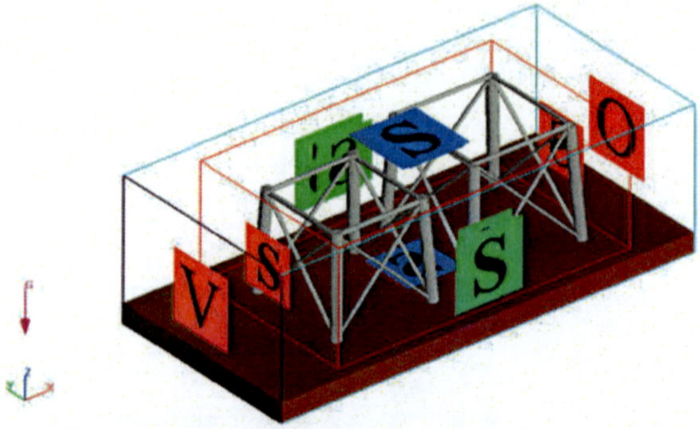

Fig. 4 Boundary conditions setting for the simulation domain

Fig. 5 Mesh elements for the simulation domain. Nested mesh is defined close to the structure

Table 3 Boundary conditions of the simulation domain

Plane axis	Function	Boundary conditions	Notation
X min	Inlet	Velocity vector	V
X max	Outlet	Outflow	O
Y min	Water domain	Symmetry	S
Y max	Water domain	Symmetry	S
Z min	Seabed	Wall	W
Z max	Water domain	Symmetry	S

side of J1, whereas the sediment deposition was observed around the back leg of J2. The maximum scour depth for J1 extended and grew gradually (Fig. 7a). After 300 s, the expending scour depth slowed down until the flow velocity became weaker around J1 which reduced the shear stress and caused the scour depth to be in an equilibrium state. The maximum deposition height was captured at J2 as the sediment transport from J1 moved further towards J2. The deposition height developed gradually. The

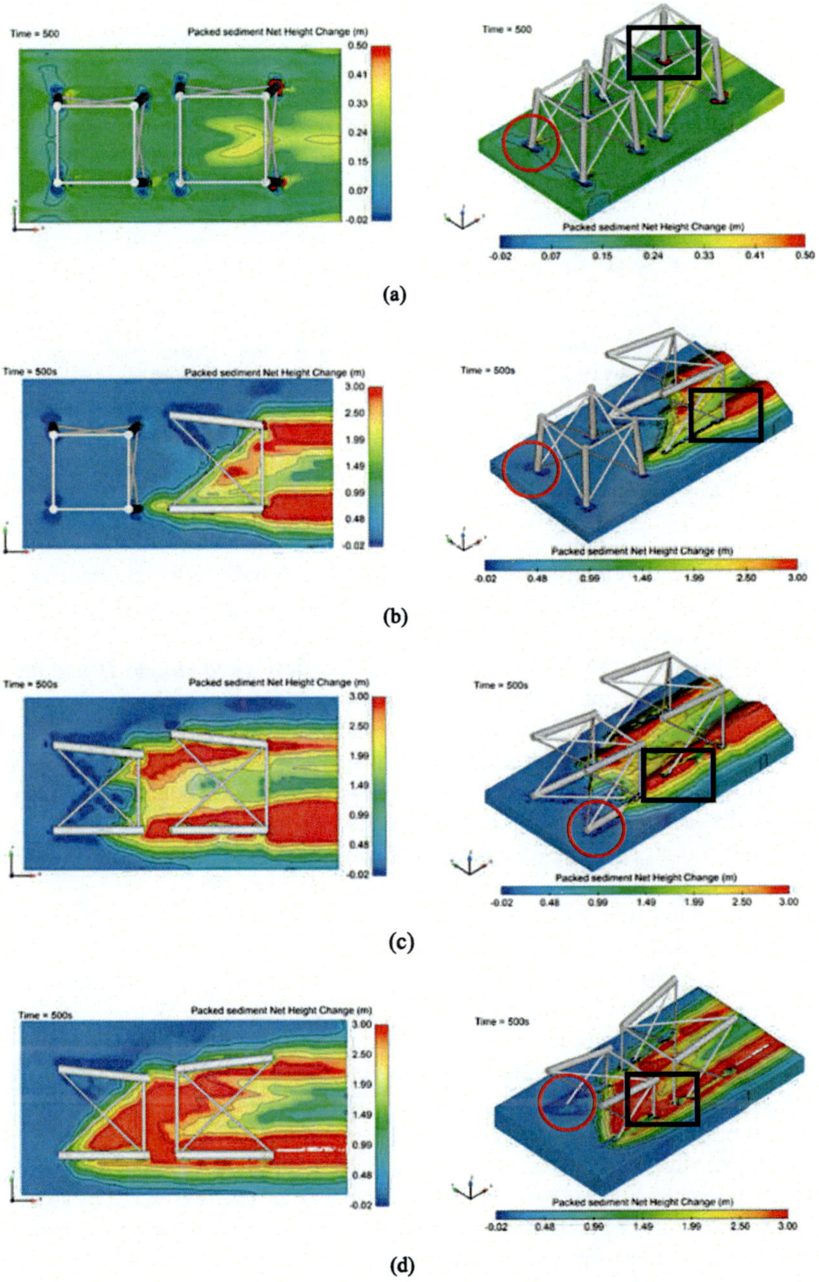

Fig. 6 Illustration of packed sediment net height change contours at t = 500 s and the maximum scour depth and deposition height for **a** R1 **b** R2 **c** R3 and **d** R4. circle area shows maximum scouring and rectangular area shows the maximum deposition location

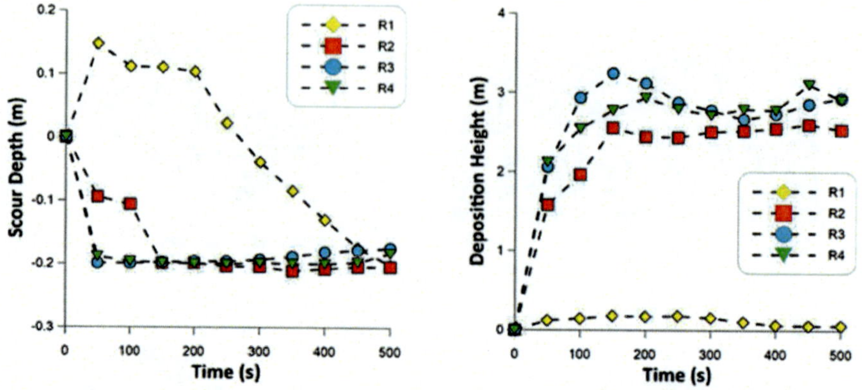

(a) (b)

Fig. 7 The time evolution of **a** scour depth; **b** deposition height for different

maximum deposition height is 0.17 m at 150 s and deposition height continued to develop until it decreased at 350 s and equilibrium was achieved at 400 s with 0.06 m height.

For R2 (Fig. 6b), it shows the scour pattern that developed around J1 and on one side of J2. Meanwhile, the sediment deposition can be obviously observed around J2 which covered most of the J2 members. The scour depth continued to develop after the 50 s and the scour pattern expanded gradually at the front and at the side of J1 until 500 s (Fig. 7a). Around J1, the scour pattern developed deliberately from the 150 s, as the scouring was still occurring but with a low scour rate. The maximum scour depth was recorded at −0.2 m at 150 s and continued to be in equilibrium until 500 s.

Sediment deposition around J2 continued to expand after the 50 s as the flow velocity decreased and led to low shear stress in which sediment deposition moved further towards around J2. The maximum deposition height at J2 was recorded at 2.55 m mainly due to the huge movement of the sediment from J1.

For R3 (Fig. 6c), the scour depth continued to develop from initial simulation up to the t = 50 s and the scour pattern expanded gradually in front and at the side of J1. The equilibrium maximum scour depth was observed at −0.2 m during the first 50 s (Fig. 7a). A significant amount of sediment deposition was observed around J2 and sediment deposition continued to develop gradually until 200 s with the maximum deposition height of 3.2 m. It can be seen that both J2 legs were covered by sediment due to the movement of the seabed sediment. The position of the jacket platform was clearly affecting the morphology of the sediment transport.

Lastly, in R4 (Fig. 6d), the scour pattern developed only around one leg of J1. Most of the jacket legs were covered by the sediment for both J1 and J2. A similar scour pattern was observed as R3 case, where the scour developed for 50 s and achieved its equilibrium at −0.2 m (Fig. 7a). The deposition pattern showed that it developed around the J1 and spread in the structure's middle forward to brace. Basically, structure sediment deposition occurred gradually by time for the entire area around bracing and main leg. The maximum deposition height was observed around J2 with 2 m height at 100 s. The sediment deposition continued to develop gradually until 500 s with 3 m maximum height (Fig. 7b). The deposition that occurred for this model is probably due to the effect of parallel leg structure and location of the structure braces sitting configurations.

Figure 8 shows the time evolution of the scour and deposition rate for all models. It can be seen from the plot that the scour develops gradually for 50 s and starts to

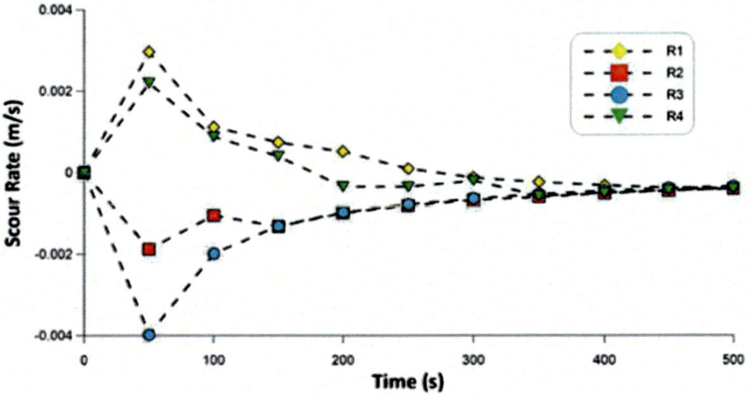

(a)

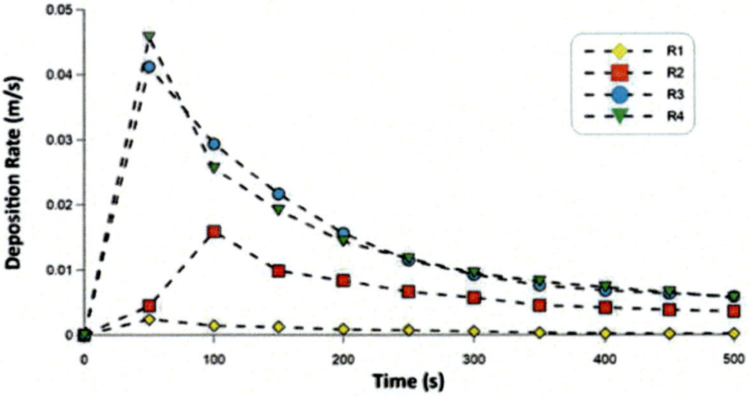

(b)

Fig. 8 The time evolution of **a** scour rate; **b** deposition rate at different sitting configurations decrease as the time increases. The maximum scour rate was recorded at −0.004 m/s for R3 model. The scour patterns were different for all models considered due to different setting configurations of the structured jacket and the design structure that involved structure bracing. The complex structure member would definitely increase the scour activity due to the flow characteristics around the members. Meanwhile, the deposition rate gradually increased for R2, R3, and R4 over the 50 to 100 s. The deposition rate decreased deliberately after 100 s of the simulation. As for R1, low deposition activity was captured around the structure. This was due to the jacket platforms that were placed in a vertical position as compared to the other arrangements. A smaller contact area between the jacket legs and the seabed causes the scour and deposition activity to be reduced significantly.

4 Conclusions

This study sets out to assess the feasibility of the jacket platform as an artificial reef in the RTR program. Four different arrangements of the artificial reef on the seabed changes were studied. Different scour and deposition patterns were observed for all models. This investigation shows that the position of jacket structure plays an important role in ensuring the scouring and deposition that occurs around the reef is smaller. Scouring is caused by the horseshoe vortex and the acceleration of flow fluids, which creates significant bed shear stress. Meanwhile, sediment deposition is primarily caused by the wake vortex, which combines the gravity and bed friction to move the sediment over time. The vertical structure decreases the scour and deposition significantly due to less contact with the seabed. Meanwhile, the structure placed horizontally on the seabed increases the surface contact with the seabed and subsequently increases the potential of scour and deposition, resulting in failure of the reef.

Acknowledgements This research was supported by Ministry of Higher Education of Malaysia (Award No. FRGS/1/2018/TK01/UMT/02/1, Vot No. 59503)

References

1. Bull AS, Love MS (2019) Worldwide oil and gas platform decommissioning: a review of practices and reefing options. Ocean Coast Manag 168:274–306
2. Zawawi NAWA, Liew MS, Na KL (2012) Decommissioning of offshore platform: a sustainable framework. CHUSER 2012 2012 IEEE Colloq Humanit Sci Eng Res 26–31
3. Awang D (2016) Oil rig as artificial reef: example of Baram 8
4. White AT, Loke CM, De Silva MWRN, Guarin FY (1990) Artificial reefs for marine habitat enhancement in Southeast Asia
5. Kim HS, Nabi M, Kimura I, Shimizu Y (2014) Numerical investigation of local scour at two adjacent cylinders. Adv Water Resour 70:131–147

6. Rahman MAA, Nazri MN, Ahmad MF, Kasiman EH, Musa MA, Alias F, Mohd MH (2020) A fundamental cfd investigation of offshore structures for artificial coral reef development. CFD Lett 12:110–125
7. Rahman MAA, Nazri MN, Alias F, Fitriadhy A, Mohd MH (2021) Computational fluid dynamics analysis of rigs-to-reefs (R2R) jacket structures. CFD Lett 13:72–83
8. Tawekal RL, Velas JD, Tawekal JR (2019) Comparative study of braced monopod and tripod jacket offshore platforms 1:14–22
9. Roulund A, Sumer BM, Fredsøe J, Michelsen J (2005) Numerical and experimental investigation of flow and scour around a circular pile. J Fluid Mech 534:351–401
10. Zhang Q, Zhou XL, Wang JH (2017) Numerical investigation of local scour around three adjacent piles with different arrangements under current. Ocean Eng 142:625–638
11. Liang SD, Zhang YL, Yang J (2015) An experimental study on pile scour mitigating measures under waves and currents. Sci China Technol Sci 58:1031–1045
12. Khosronejad A, Rennie CD, Salehi Neyshabouri SAA, Townsend RD (2007) 3D numerical modeling of flow and sediment transport in laboratory channel bends. J Hydraul Eng 133:1123–1134
13. Dey S, Helkjær A, Mutlu Sumer B, Fredsøe J (2011) Scour at vertical piles in sand-clay mixtures under waves. J Waterw Port Coast Ocean Eng 137:324–331
14. Gamvroudis C, Nikolaidis NP, Tzoraki O, Papadoulakis V, Karalemas N (2015) Water and sediment transport modeling of a large temporary river basin in Greece. Sci Total Environ 508:354–365
15. Bridge JS, Dominic DF (1984) Bed load grain velocities and sediment transport rates. Water Resour Res 20:476–490
16. Ong MC, Myrhaug D, Fu P (2016) Scour around vertical piles due to random waves alone and random waves plus currents on mild slopes. Ocean Syst Eng 6:161–189
17. Baglio S, Faraci C, Foti E, Musumeci R (2001) Measurements of the 3-D scour process around a pile in an oscillating flow through a stereo vision approach. Meas J Int Meas Confed 30:145–160
18. Pourshahbaz H, Abbasi S, Pandey M, Pu JH, Taghvaei P, Tofangdar N (2020) Morphology and hydrodynamics numerical simulation around groynes. ISH J Hydraul Eng 00:1–9
19. Kurdistani SM, Pagliara S (2017) Experimental study on cross-vane scour morphology in curved horizontal channels. J Irrig Drain Eng 143:04017013
20. May CL, Pryor B, Lisle TE, Lang M (2009) Coupling hydrodynamic modeling and empirical measures of bed mobility to predict the risk of scour and fill of salmon redds in a large regulated river. Water Resour Res 45
21. Baglio S, Faraci C, Foti E, Musumeci R (2000) Stereo vision for noninvasive dynamic measurements of the scour process around a circular cylinder in an oscillating flow. Ocean Conf Rec 2:987–992
22. Zhang Q, Zhou XL, Wang JH, Li WL (2019) Dynamic interaction between two parallel submarine pipelines considered vortex-induced vibration and local scour. Mar Georesources Geotechnol 37:609–621
23. Khosronejad A, Kang S, Sotiropoulos F (2012) Experimental and computational investigation of local scour around bridge piers. Adv Water Resour 37:73–85

Conceptualizing an Industry 4.0's Predictive Maintenance System in a Medical Devices Manufacturing Enterprise

Christian Stark and Jeng Feng Chin

Abstract The research presents a case study from a medical devices manufacturing enterprise planning for preventive maintenance from the perspective of Industry 4.0. The research aims to generate a preliminary study in the enterprise onto maintenance and use the information later to plan for a predictive maintenance system. The preliminary study focused on 12 Computer Numerical Control (CNC) 5-axis milling machines, which run the most critical processes of the enterprise. A total of 82 breakdowns of these machines were detected and investigated for over 1.5 years. They were categorized and clustered on the basis of the suitable dimensions (frequency, duration, and organization financial loss). The findings reveal that 18 types of breakdowns constituted over 85% of the total breakdown. In total, 80% of the downtimes were not over 10 h. December was observed having the highest financial loss attributed to downtime. A causality analysis was performed, and the causes (parameters) were placed in three categories to underline the degree of real-time monitoring difficulty. The management of the enterprise deliberated on the results and conceived action plans, which involve development of a computerized maintenance system and vendor collaborations. On the basis of the concept of Health and Usage Monitoring Systems (HUMS), a conceptual predictive maintenance system is presented to provide a predictive breakdown and system modelling. The case study shows the enterprise's endeavor for predictive maintenance planning. In terms of research and practical contribution, this research helps reduce the gap in literature and application by demonstrating an industry-based preliminary study onto the most common machine (e.g. CNC) in the case study company from the maintenance perspective.

Keywords Big data · CNC milling · HUMS · Machine learning · Maintenance · Industry 4.0 · IoT · Predictive

C. Stark · J. F. Chin (✉)
School of Mechanical Engineering, Universiti Sains Malaysia (USM), 14300 Nibong Tebal, Penang, Malaysia
e-mail: chinjengfeng@usm.my

© The Author(s), under exclusive license to Springer Nature Singapore Pte Ltd. 2023

M. H. A. Hassan et al. (eds.), *Technological Advancement in Instrumentation & Human Engineering*, Lecture Notes in Electrical Engineering 882,
https://doi.org/10.1007/978-981-19-1577-2_38

1 Introduction

Industry 4.0 closely relates to the interaction between machines through networks. Maintenance management is considered in Industry 4.0, given that it provides technical and economic advantages to an organization [1]. Any maintenance program should include "intelligent" behavior of machines, such as the capability to predict and monitor its performance deviations and breakdown mode. This idea motivates predictive maintenance to help not only direct avoidance of breakdown and production loss [2], but also effective management of spare parts. The new generation of maintenance relies on data collection from equipment or machine and therefore requires computer-integrated manufacturing (CIM). Fundamentally, CIM refers to the use of computer-controlled machines and automation systems, such as Computer Numerical Control (CNC) machine aided with new capabilities of monitoring and controlling machine tools and collaborating sensors in manufacturing products [3]. CIM aims to provide an error-free and automated manufacturing process. In the long run, CIM is known as "workerless" production where the Industry 4.0 transforms regular machines into self-learning and self-aware machines for great flexibility and high-quality production. Computerized maintenance management has advantages of improved resource control, cost management, scheduling, integration, and reduction of breakdowns [4]. Additionally, when the maintenance is data driven, the estimation of variables, such as risk and breakdown occurrence, are no longer subjective and rely on the experience and knowledge of employees [2].

Although new technologies are emerging and machine manufacturers are integrating advanced data collection and performance monitoring mechanisms in their latest equipment, most manufacturing organizations still depend on various existing equipment from different generations or technologies. Fusko et al. [5] commented that the implementation of predictive maintenances demands a well-structured conception and the use of emergent ICTs. The already available data must be harvested with a data collection system [6] and put to good use. Additionally, the data collection system would generate a large amount of potentially overwhelming information for any system to study them all, especially in real time. Labib et al. [4] pointed out that many computerized maintenance systems by design are greedy for data input, but are not sufficiently using these data for decision support. According to Di Bona et al. [7], the main limitations of the Industry 4.0's maintenance strategy design are the complexity and reliability of the relevant technology and advanced decision-making algorithms. Cheng et al. [8] claimed that integrating a building information modeling (BIM) and Internet of Things (IoT) into facility maintenance management (FMM) remains in its infancy stage. Consequently, the planning must study and analyze the required data selection and collection mechanism prior to predictive maintenance. Interestingly, the methodology on the maintenance planning is somewhat lacking in literature. Even for those somewhat describing the methodology, they are highly selective on several areas. For example, Labib et al. [4] proposed the decision-making grid (DMG) on the frequency and downtime for identifying and setting a plan to improve the worst performed machines on the basis of multiple criteria.

In the paper, data about the behavior of CNC 5-axis milling machines in a medical devices manufacturing enterprise in Malaysia are analyzed and interpreted through a preliminary study. Condition-based preventive maintenance [9, 10] signifies the continuous monitoring of the current state of the machine through sensors or something similar. When something suspicious (reading out of range) transpires, the machine alerts the operator to address the problem before the machine breaks down [11]. On the basis of the exercise, a predictive maintenance system is proposed to the management. The system is based on learning from incidents, whereas the ultimate predictive maintenance is rather based on condition monitoring and deviation warnings prior to incidents. The methodology first carries out a preliminary study on 12 CNC 5-axis milling machines in terms of frequency, duration, and cost against the predetermined breakdown types. Suitable graphs, including a bar chart, a bubble chart, and a causality diagram are generated to present the findings. The availability of insights into the preliminary study is essential to convince the management to consider investing in predictive maintenance. The preliminary study also provides the basis to develop the conceptual predictive maintenance system. The work represents industrial effort to embark on Industry Revolution 4.0. Unlike academic research with rigorous investigation, the industry often adopts a small-scale preliminary study as "proof-of-concept" before committing more resources to any project.

Section 2 provides the literature review. Section 3 presents the preliminary data analysis and the findings. Section 4 describes the conceptual predictive maintenance system. Section 5 concludes.

2 Literature Review

Fundamentally, maintenance management aims for "total asset life cycle optimization" [12]. Specifically, continuous effort is made to maximize the availability and reliability of these assets (equipment) to produce products in the desired quantity, quality, and cycle time. Over-maintenance and under-maintenance should be avoided in the maintenance management [4].

Many studies attempted to capture the complexity of maintenance. Filz et al. [2] defined maintenance as comprising four different layers: operations, activities, strategies, and concepts. The maintenance activity is a specific operation containing either service, inspection, or repair. Next, the coordination of activities forms different maintenance strategies, either corrective or preventive. Concepts sit at the highest layer comprising a collection of strategies, which are selected on the basis of the overarching trade-offs and influences on the maintenance goals. Established maintenance concepts include total productive maintenance, condition-based maintenance, reliability centered maintenance (RCM), and risk-based maintenance (RBM). Pintelon et al. [12] casted the maintenance as a blend of elements, such as management, technology, operations, and logistics support. The management is responsible for maintenance decisions, including strategies of utilizing other elements. Technology is the maintained asset, and operations is the maintenance service intervention. Finally, the logistics supports include planning, coordinating, and delivering the resources required in the maintenance.

Recent maintenance concepts and applications are briefly reviewed. The trend of new maintenance strategies is increasingly proactive and digitalized: they monitor symptoms and causes of breakdowns in a digitalized environment and collectively assume the anticipation actions, which cover diagnosis, prognosis, and decision making [13]. Collected from sensors, instrument inspections and interfaces with production control systems, data include usage of the asset, maintenance history, condition of asset, history of assets, and environmental data [7]. Kumar et al. [14] incorporated advanced analytics as the major component in smart maintenance or prescriptive maintenance. Cheng et al. [8] developed a data-driven predictive maintenance planning framework integrated with Internet-of-Things (IoT) technologies for facility maintenance. The framework comprises information and application layers. The first layer handles data integration and the second layer is maintenance prediction and planning. Filz et al. [2] presented a data-driven Breakdown Mode and Effect Analysis (FMEA) methodology by using deep-learning models on historical and operational data to support the maintenance planning. Wolfartsberger et al. [15] described a predictive maintenance concept that integrates mixed reality into future learning factories. In the concept, predictive maintenance would arrange a technician for a specific maintenance task and links the technician to a remote expert. Fusko et al. [3] championed the idea of Maintenance 4.0, which combines predictive maintenance, industrial IoT, and edge computing. Methods are defined to connect, collect, analyze, and take advantage of data sources for digital asset management.

As maintenance becomes proactive and digitalized, the analytic process, such as data mining, plays a critical role. Data mining extracts a large amount of data from different sources, which could be stored in separate locations and transforms them into a comprehensible structure with the development of ICT [16]. Common data mining steps are data collection, data pre-processing, data transformation, data reduction, feature and knowledge extraction, and visualization [17]. Generally, manufacturing industries use data mining instruments to enhance the intelligence and efficiency of design, production, service process, and supply chain. Data mining is mainly motivated by the profitability of data collection and processing, such as to discover useful patterns significant to prominent task and output results [18]. In [19], data mining on the machine helps recognize unknown patterns within data. A cluster analysis using k-means algorithm is applied to predict the capability of a specific process with a particular set of parameters. Accorsi et al. [20] introduced data mining models for the condition-based maintenance of a high-speed, complex packaging machine. In the grouping of machine data, clustering techniques, association rules, and classification models were exploited to detect anomalies in the machine's activity. Vazan et al. [21] used neural network (NN) to predict the manufacturing process behavior according to the production data provided by an automotive industry component supplier.

3 Preliminary Data Analysis

The study was carried out in a multinational medical manufacturing enterprise located in Malaysia. The available data for breakdown, duration, and reason of the 12 5-axis CNC milling machines in the enterprise were tracked for a period of 1.5 years. The data collected from these machines include the breakdown details, the usage of spare parts, and the service times. Reasons for some occurrences (3.4%) cannot be determined. To gain an overview of the current situation, the data were analyzed in the following aspects: frequency, loss, and duration.

3.1 Frequency

Figure 1 presents the frequency of various breakdown causes. Some causes are breakdown occurred more often in the studied period and so on. Amongst the 82 breakdowns, only the most frequent 18 types were considered (constituted over 85% of the total breakdown). The problems caused by the magazine and coolant are more prominent than others. Evaluation is required to study the impact of these breakdowns, given that frequency alone has little foundation to decide the criticality of a breakdown.

3.2 Downtime Duration

The distribution of the downtime in percentage of incidents is as shown in Fig. 2. Few cases of breakdowns lasted for up to three weeks. Most incidents were resolved within 10 h.

3.3 Organizational Loss

The loss that was caused by the downtimes were investigated, and the result is presented in Table 1. The management has decided an initial estimate of a loss amounted to US $50 for every hour a machine is not able to stay productive. Moreover, the value must be adjusted in accordance with the machine rate and the potential profit

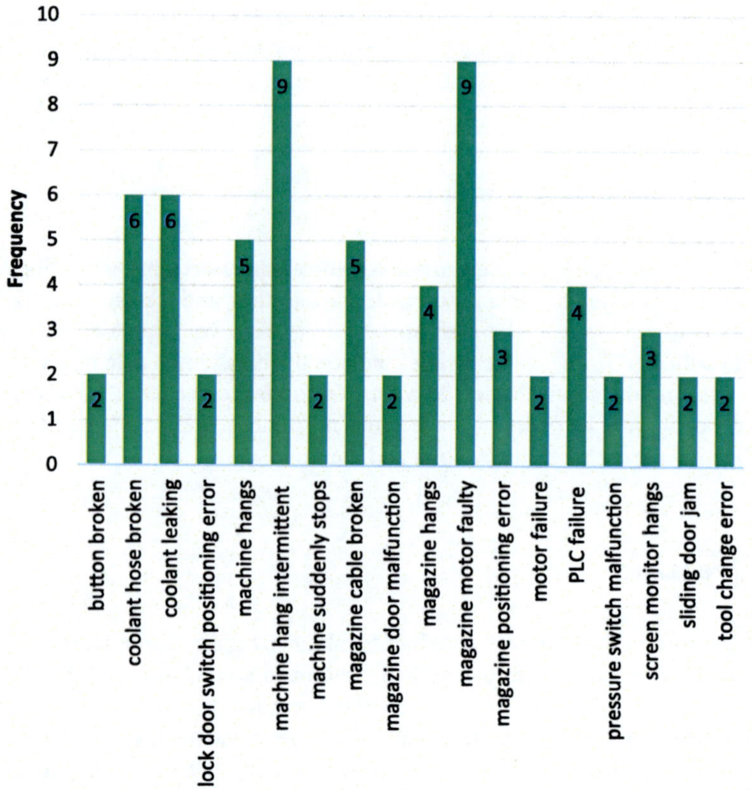

Fig. 1 Frequency of various breakdown causes

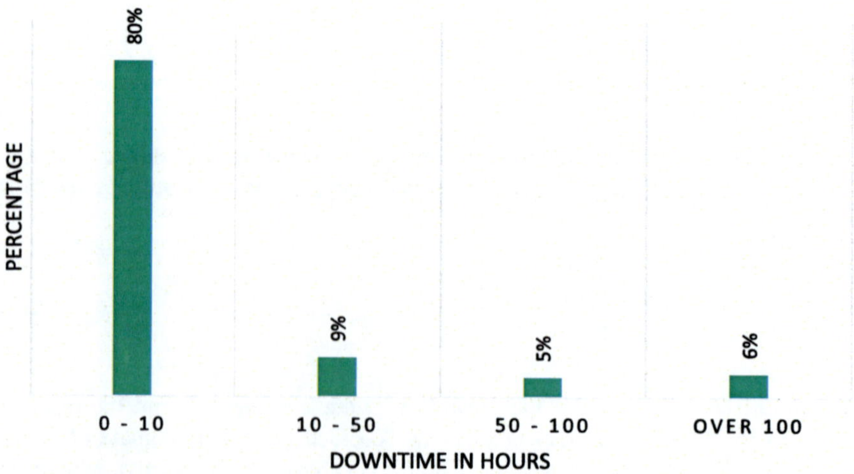

Fig. 2 Percentage distribution of downtime durations

Table 1 Loss per month depending on the downtime duration

	Observation Year 1		Observation Year 2		Observation Year 3	
Month	Total downtime	Total loss	Total downtime	Total loss	Total downtime	Total loss
January			23:30	US$587.50	84:30:00	$2,112.50
February			11:30	US$287.50		
March			104:10:00	US$2,604.25		
April			16:30	US$412.50		
May			97:40:00	US$2,441.75		
June	16:30	$412.50	7:50	US$196.00		
July			175:46:00	US$4,394.25		
August			84:18:00	US$2,107.50		
September	6:00	$150.00	116:14:00	US$2,906.00		
October	10:00	$250.00	373:27:00	US$9,336.25		
November	31:55:00	$798.00	15:30	US$387.50		
December	8:30	$212.50	1136:30:00	US$28,412.50		
Total	72:55:00	$1,823.00	2162:55:00	$54,073.50	84:30:00	$2,112.50

loss within an individual case. The loss is dependent on and directly proportional to the downtime duration.

No significant raw material costs and penalties occurred. Further losses have not been integrated in this study:

(1) Material losses: Although no handling of complex parts in exotic material was used in the study, scraping material damaged by a machine defect could be significant for many companies. Such costs have not been measured during the study.

(2) Late delivery costs: Machine downtime may delay production resulting in higher shipping costs (for example sea instead of air mode) and/or late penalties by clients depending on contractual terms.

3.4 Clustering

To derive a relationship, the next step is to form clusters of the most common causes for breakdowns in relation to the machine parts: spindle, axis, magazine, coolant, software, electric, and others. Spindle, axis, and magazine are mechanical breakdowns. The "electric" cluster includes the screen, fan, and other broken electric components. "Software" generally covers every machine stop with unknown reasons but an error code in the controller is given. Several breakdown types have no common root cause and were considered "others." In the bubble chart presented in Fig. 3, the x-axis shows the average downtime in hours, and the y-axis shows the count (the frequency

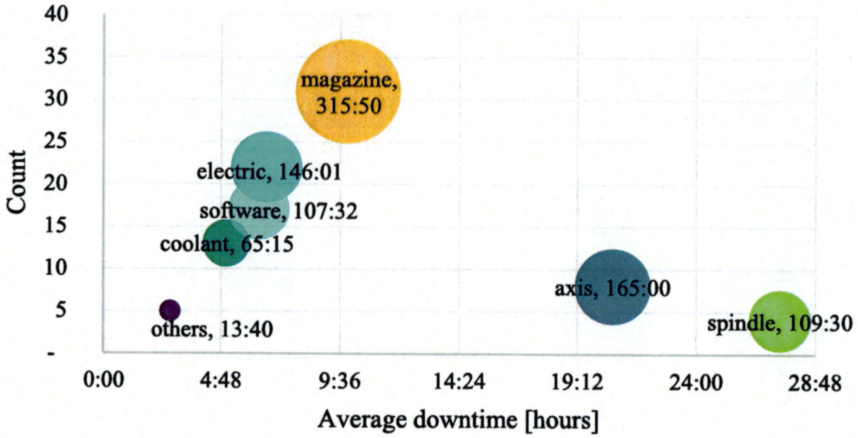

Fig. 3 Matrix of downtime clusters

of a breakdown). The size of the bubbles equates to the total downtime of the representing breakdowns. During the test period, six downtimes occurred, which lasted over 100 h. Those specific downtimes were also impacted, given that the downtime duration fell into a public holiday in which anyway a plant maintenance shutdown was scheduled. Therefore, those downtimes were not considered for the evaluation.

The information acts as the basis for the course of action. Naturally, the breakdown types that are either the ones with the longest downtime or the ones that happen the most frequent are prioritized. The focus is on the mechanical issues whereby the magazine was defined as the priority given its highest frequency and highest total downtime. Accordingly, the "axis" cluster is No. 2 and finally the "spindle" is No. 3.

As mentioned, the magazine-related breakdowns are the most critical and therefore warrant a more detailed analysis. The exact causes for the different breakdowns are presented in Fig. 4. The main breakdown cause is related to the magazine motor when it stops turning and often needing of replacement.

A root cause analysis was then prepared after a brainstorming on the possible contributing causes, as shown in Fig. 5. Four main categories of root causes were identified: environment, organization, and machine. "Environment" discusses the factors in the surroundings of the machine that could affect its performance. Everything concerning the planning and managing of the machining process is connected to the category "organization." "Material" includes every quality aspect of raw materials, including its handling and quality. Finally, "Machine" relates to machine hardware and software. Focus is placed on several findings from the brainstorming. First, root causes that originate from the machine's surroundings could be high humidity either climate related or due to the coolant mist that sprays during machine operation. Another point could be the room's temperature where the machine sits. Second, administrational-wise it could be possible that the same order always causes problems or rather the same article number. Importantly, a correlation may exist between the

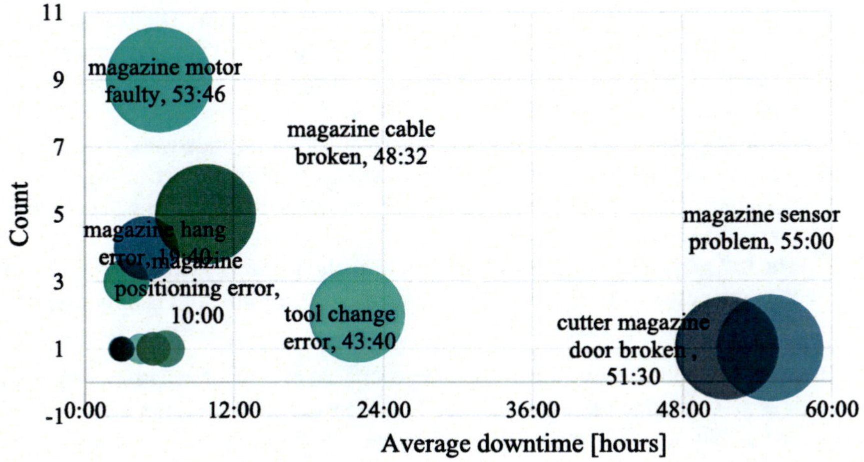

Fig. 4 Matrix of magazine breakdown reasons

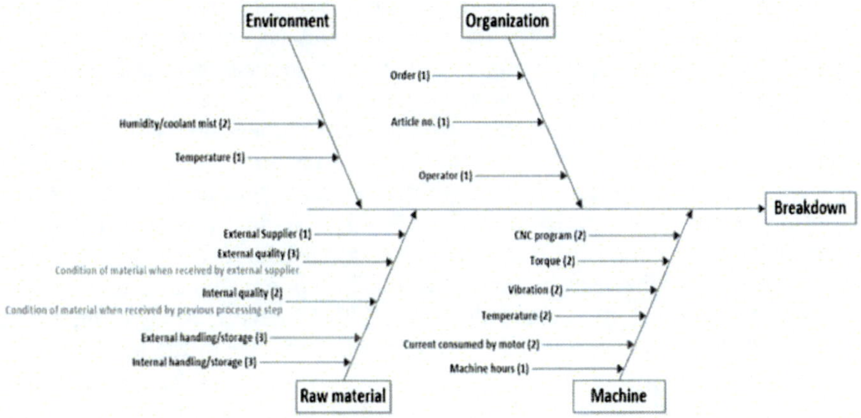

Fig. 5 Fishbone (Ishikawa) diagram for breakdown causes

breakdown of a machine and the operator who was handling it—human error. Third, for the "machine" category, the reasons are more focused on physical aspects that could wear a machine down, e.g., the operational duration and machine conditions (temperature and vibration that degrade tools or parts). This leads to the investigation of two factors which could cause signs of wear and tear: the torque and the electricity that are consumed by the motor.

The next step is to decide how these causes can be surveilled and documented best. The causes are divided into three categories in terms of the difficulty of data collection. These categories are shown in Fig. 5 through the numbers in parenthesis after the causes. They are explained as follows:

(1) The first parameter concerns more on the short-term factors. They can easily be measured and read out in real time. Everything marked with (1) is constituted by easy access to the information needed. The data are already accessible and there exists no need to take any more actions.
(2) If the required information could be measured but no means exists to do so, we are talking about middle-term parameters. Additional measuring systems must be installed to gather information. Factors that fall into this category were marked with a (2).
(3) The last parameter represents the most difficulty regarding data accessibility. It falls under the category long term, given that the information is only available from external sources.

The data described above represent a large amount of raw information, which are nowadays associated to the term Big Data [19]. The more information collected, the better the ability to refine the model and analyze the behavior of a system and the factors of breakdowns with accuracy.

These clusters are not only useful for analyzing but also for future documentation. The cluster definition allows a general classification for the technical support. During a breakdown, the breakdown type can be quickly related to one of these clusters in the maintenance database. The information facilitates comparison and acts as reference for information processing. Eventually, countermeasures can be developed and standardized on the basis of individual clusters.

The next steps involve communicating the results of the analysis to the machine manufacturer to develop a collaborative solution to further improve downtimes. As such, the data model may provide input to optimize the machine's design and/or revise maintenance and spare part management policy. During this analysis, a monitoring system with sensors or similar means was developed to address those factors. Meanwhile, various machine manufacturers offer upgrade options, and third-party service providers are offering stand-alone solutions of "IoT" devices to cover the topics highlighted in Fig. 5 under Machine and Environment. One of the most common measurable factors is vibration [20], which can provide a considerable amount of information on the machine's mechanical behaviors, cycles, and exceptional events. Such abnormalities of vibration can be linked to various tolerances to be set for each parameter.

4 Conceptual Machine Predictive Maintenance System

Predictive maintenance requires sensors to be installed to the equipment and machine learning algorithms to analyze the data stream from these sensors in real time to estimate the equipment performance over time [15]. On this premise, the concept of HUMS [22] is introduced, which comprises life analysis of big data to provide feedback to the user whenever certain deviations become abnormal. Succinctly, HUMS is a sensor-based real-time diagnostics system that collects data from criterial points

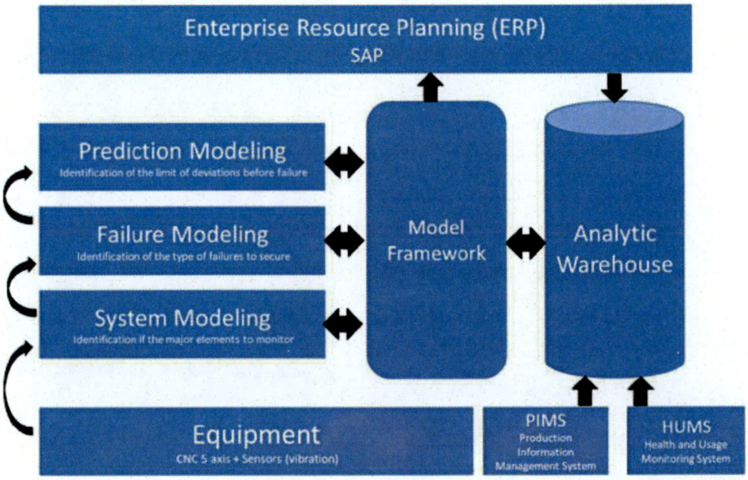

Fig. 6 Overview of the predictive maintenance system

of the mechanical part of the equipment or machine and processes the data using a predefined algorithm [23]. The overview of the system is briefly provided in Fig. 6. Equipment of interest, such as the CNC 5-axis milling machines will be installed with suitable sensors on the basis of the results of the previous study and discussion with the equipment vendor. Causes Categories 1 and 2 in machines are prioritized (Fig. 5), which include vibration, torque, temperature, and current consumed by the motor and machine hour.

HUMS will be established, which comprises two structured databases: equipment database and breakdown/maintenance database. The key relational tables in both databases are shown in Fig. 7. The advantage of relational tables is the inter-dependencies of tables are made explicit, allowing the predicting model (e.g., machine learning) to be performed effectively by the analytic warehouse. Each table would have multiple attributes, and such attributes presented in Fig. 7 are not exhaustive. The user may expand the attributes in a table when such need arises. An important attribute among these tables is the timestamp (date/time) of the event (e.g., breakdown). The attribute provides temporal information to the analytic warehouse to study the chronology of events and hence useful to determine their causality. Equipment database contains six relational tables, and the main goals are to register the sensors to the equipment, store sensor parameters, and record the readings at a specific interval as data stream. Breakdown/maintenance database has four relational tables, and its role is to track breakdown and maintenance activity and spare part usage. Meanwhile, the analytic warehouse also retrieves information from the production information management system (PIMS) and enterprise resource planning system (ERP). Integrating these two systems (on top of HUMS) permits a high level of machine learning to be performed.

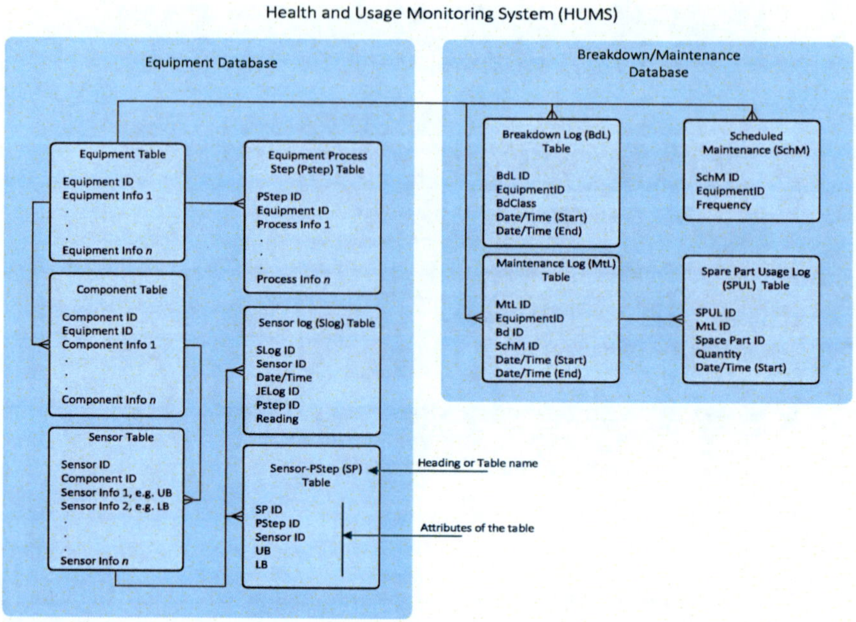

Fig. 7 Databases and relational tables in health and usage monitoring system

The collected information will be fed into the analytic warehouse to build preprogrammed analysis models. Three main models can be constructed under the model framework. System modelling identifies what the major elements are and variables to monitor. Based on the previous study, the focus will be placed on vibration (at machine level), the trends of frequency, and downtime based on breakdown causes or machines. Breakdown modelling identifies the type of breakdowns to be focused on. Statistical analysis can be used to establish the relationship between attributes. Breakdown causes will be clustered on the basis of frequency, duration, or cost of spare parts. Pareto chart can be used for breakdown prioritization. Prediction modelling identifies the limit of deviations before breakdown. Given the sufficient historical data, classification algorithm can be deployed to identify a situation or an event that leads to breakdown. Regression model can be deployed to predict the trend of a variable of interest, e.g., vibration or breakdown frequency.

Usually, model construction must be guided by experts, especially the results that link to management decision making, such as maintenance strategy planning. Nevertheless, advanced technologies that are often computationally intensive are available to facilitate autonomous modelling. These technologies would be useful to fine tune a pre-constructed model, especially when the accuracy of the existing model deteriorates due to dataset shift ("drift"). These technologies would be running either online or offline.

5 Conclusions

Industry 4.0 promotes predictive ability in the manufacturing system. The preliminary study represents industrial effort to embark on Industry Revolution 4.0 on the front of maintenance. The preliminary study focused on 12 CNC 5-axis milling machines to help gain insights into such ability in maintenance. In the course of the breakdown analysis, vital information was gathered. The clusters magazine, axis, spindle, electric, software, coolant, and others determined a basic framework for the preliminary analysis in this research. The identification of the key causes that induce a breakdown could be used to monitor the machines more precisely and therefore more expediently. This compilation and analysis of information is the fundamental basis to define further steps for the predictive maintenance system. Depending on the results of the internal and external approaches, the next steps were delivered to further inclusion of technologies into the existing range of machines for improved monitoring of deviations. Through that information, the endeavor was made to develop a conceptual predictive maintenance framework for holistically surveilling the machine and for performance improvement. In terms of research and practical contribution, this study helps reduce the research gap and application by demonstrating an industry-based preliminary study onto the most common machine in the case study company from the maintenance perspective and later generates conceptualizing the predictive maintenance framework from the inputs derived from the preliminary study.

References

1. Mosyurchak A, Veselkov V, Turygin A, Hammer M (2017) Prognosis of behaviour of machine tool spindles, their diagnostics and maintenance. MM Sci J (December) 2100–2104
2. Filz M, Langner JEB, Herrmann, C, Thiede S (2021) Data-driven failure mode and effect analysis (FMEA) to enhance maintenance planning. Comput Ind 129:1–17
3. Mourtzis D, Milas N, Athinaios N (2018) Towards machine shop 4.0: a general machine model for CNC machine-tools through OPC-UA. Procedia CIRP 78:301–306
4. Labib A, Kobbacy K, Murthy D (2008) Computerised Maintenance Management Systems. In: Complex System Maintenance Handbook. Springer Series in Reliability Engineering, pp 417–435. Springer, London. https://doi.org/10.1007/978-1-84800-011-7_17
5. Fusko M, Bučkova M (2019) Gola a key concepts of maintenance in industry 4.0 concepts. In: book: InvEnt 2019: Industrial Engineering – Invention for Enterprise Publisher: Wydawnictwo Akademii Techniczno-Humnistycznej w Bielsku-Białej, pp 44–47
6. Guo Y, Sun Y, Wu K (2020) Research and development of monitoring system and data monitoring system and data acquisition of CNC machine tool in intelligent manufacturing. Int J Adv Robot Syst (March-April):1–12
7. Poór P, Basl J (2020) Machinery maintenance model for evaluating and increasing maintenance, repairs and operations within industry 4.0 concept. IOP Conf Ser Mater Sci Eng 947:1–7
8. Cheng JCP, Chen WW, Chen K, Wang, Q (2020) Data-driven predictive maintenance planning framework for MEP components based on BIM and IoT using machine learning algorithms. Autom Constr 112:1–21
9. Zargarbashi SHH, Angeles J (2015) Identification of error sources in a five-axis machine tool using FFT analysis. Int J Adv Manuf Technol 76:1353–1363

10. Bink R, Zschech P (2018) Predictive maintenance in der industriellen praxis – entwicklung eines prognoseansatzes unter eingeschränkter informationslage. HMD Praxis der Wirtschaftsinformatik 55:552–565
11. Lödding H, Riedel R, Thoben K, Von Cieminski G, Kiritsis D (Eds.) (2017) Advances in Production Management Systems–The Path to Intelligent, Collaborative and Sustainable Manufacturing. Springer International Publishing, Heidelberg. https://doi.org/10.1007/978-3-319-66926-7
12. Pintelon L, Parodi-Herz A (2008) Maintenance: An Evolutionary Perspective. In: Complex System Maintenance Handbook. Springer Series in Reliability Engineering, pp 21–48. Springer, London. https://doi.org/10.1007/978-1-84800-011-7_2
13. Fusko M, Rakyta M, Krajčovič M, Dulina L, Gašo M, Grznár P (2018) Basics of designing maintenance processes in industry 4.0. MM Sci J 14:2252–2259
14. Kumar R, Singh SP, Lamba K (2018) Sustainable robust layout using big data approach: a key towards industry 4.0. J Clean Prod 204(3):643–659
15. Wolfartsberger J, Zenisek J, Wild N (2020) Data-driven maintenance: combining predictive maintenance and mixed reality-supported remote assistance. Procedia Manuf 45:307–312
16. Wang RL et al (2018) Review on mining data from multiple data sources. Patterns Recognit Lett 109:120–128
17. Gandhi K, Schmidt B, Ng, AHC (2018) Towards data mining-based decision support in manufacturing maintenance. Procedia CIRP 72:261–265
18. Gullo F (2015) From patterns in data to knowledge discovery: what data mining can do. Phys Procedia 6:18–22
19. Denkena B, Schmidt J, Kruger M (2014) Data mining approach for knowledge-based process planning. Procedia Technol 15:406–415
20. Accorsi R, Manzini R, Pascarella P, Patella M, Sassi S (2017) Data mining and machine learning for condition-based maintenance. Procedia Manuf 11:1153–1161
21. Vazan P, Janikova D, Tanuska P, Kebisek M, Cervenanska Z (2017) Using data mining methods for manufacturing process control. IFAC PapersOnLine 50(1):6178–6183
22. Dicicco M (March 2016) Machines that tell you when they're sick. https://www.nasa.gov/offices/oct/feature/machines-that-tell-you-when-theyre-sick. Accessed 22 July 2021
23. Zaman T, Bayoumi AE (2014) Analysis of health and usage monitoring system (HUMS) users' perspective towards mission benefits using regression analysis. In: Proceedings of the AHS 70th Annual Forum, Montréal, pp. 1621–1626, the American Helicopter Society International, Québec, Canada

Optimization of Temperature Rise in Turning Using Single Objective Genetic Algorithm

Mimi Muzlina Mukri and Nor Atiqah binti Zolpakar

Abstract Temperature rise is an essential element that must be consider during machining process which will contribute to the satisfactory end products. All the factors such as cutting environments, methods and work material must be emphasize since it will influence the outcome. The optimization of the machining process was carried out in this research to optimize the machining process by minimizing temperature rise for turning machining. The parameters involve during this optimization are cutting speed, feed rate, depth of cut and nose radius by using genetic algorithm optimization. This study was separated into three sections, each of which was optimized to see what effect each parameter had on temperature rise. The minimum temperature attained was 23.07 °C, while the cutting variables for cutting speed, feed rate, and depth of cut were 81.22 m/min, 0.08 mm/rev, and 0.12 mm, respectively.

Keywords Optimization · Genetic Algorithm · Temperature rise · Machining

1 Introduction

The geometry or surface characteristics of a product are created using machining techniques. Aerospace, biological, electrical, agricultural, communications, and car sectors are all in critical need of high-precision miniature parts [1]. Turning is a machining or material removal technique for creating rotational components by removing excess material. Temperature is a crucial variable in every industrial activity that must be considered throughout the metal cutting process. Temperature is an essential consideration during machining since it affects the machine's efficiency. Several factors involved in the metal removal process, such as material surface quality, cutting tool life, work and tool material composition, are all controlled by

Supplementary Information The online version contains supplementary material available at https://doi.org/10.1007/978-981-19-1577-2_39.

M. M. Mukri · N. A. binti Zolpakar (✉)
Faculty of Mechanical & Automotive Engineering Technology, Universiti Malaysia Pahang, 26600 Pekan, Pahang, Malaysia
e-mail: noratiqahz@ump.edu.my

© The Author(s), under exclusive license to Springer Nature Singapore Pte Ltd. 2023
M. H. A. Hassan et al. (eds.), *Technological Advancement in Instrumentation & Human Engineering*, Lecture Notes in Electrical Engineering 882,
https://doi.org/10.1007/978-981-19-1577-2_39

Fig. 1 Heat sources during machining

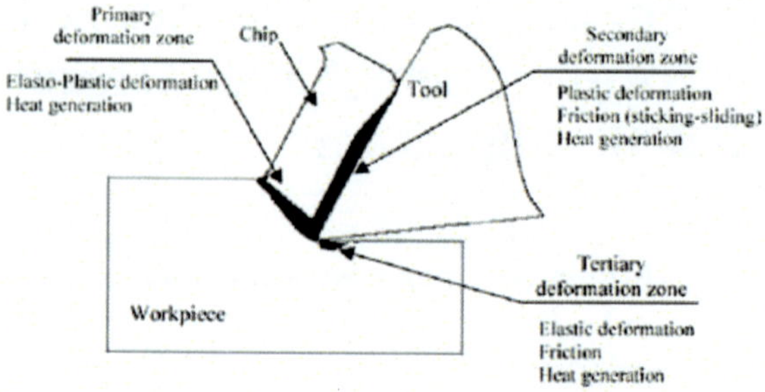

temperature [2]. At machined surface temperatures, the heat energy from the cutting operation will cause the chosen shape to deteriorate. When it comes to cutting cylindrical workpiece, this is certainly relevant [3]. Figure 1 shows the heat generated during machining process.

Cutting speed, depth of cut, feed rate, nose radius, tool geometry, coolant type and more factors that associated with machining all have an effect on dispersion of heat to the cutting tool, generated chip, and workpiece material [1, 2]. Parameters of cutting such as velocity of cutting, rate of feed, and the depth of cutting affect temperature, which impacts strength of the material and hardness, according to Motorcu et al. and Okokpujie [3, 4]. The temperature in the area of the cutting zone is controlled by the contact length between tool and chip, cutting forces, and material friction between tool and work all influence the material in the cutting zone [5]. Meanwhile, Gosai [6] discovered that the increment of the temperature in the tool clearance face or the tool rake face for cutting will reduce the tool's life. The effect of the temperature rises to the machining process also studied by Akhil et al. [7] discovered generated heat is unsteadily escalate in the midst of chip, cutting tool and blank workpiece. The assigning model is determined by the configuration, scale and process thermal conductivity. The tool will consume about 10–20% of the total heat generated, while the rest being dispersed through the medium such workpiece and chips. The control in heat produced is essential for the tool's efficiency and the condition of the completed product since the influence of the cutting temperature, specifically during it maximum state, sometimes will damaging to both the tool and the workpiece. As a result, understanding of temperature of cutting during machining is required to optimize machining and tool life in order to decrease temperature increase during the metal cutting process [7]. Previous researchers did several optimization studies to reduce temperature increase throughout the machining process. Das et al. [8] optimizing the cutting parameters such as cutting speed, depth of cut and feed in dry turning of AISI D2 steel to obtain the lowest tool wear, the lowest surface temperature of workpiece and the highest rate of material removal. Saravanan [9] focus on optimizing the machining parameters for turning cylindrical stock into continued finished profiles with respect to the minimum production cost by applying genetic algorithm (GA) and simulated annealing (SA). Using the information presented previously by Gopal [10] the cutting speed, feed rate, depth of cut and nose radius all will affect the value of temperature rise throughout the turning operation. From the past research, it can be conclude that heat generated during metal removal is the factor that will deteriorate both workpiece and tool in many way such as dimensional accuracy, crater and flank wear. The goal of this study was to use genetic algorithm optimization to reduce temperature rise in the time of machining by optimizing cutting speed, feed rate and depth of cut. Mathematical model that used significant parameters in this paper derived from the past research by Gopal [10]. The model is adopt as the objective function for utilizing a genetic algorithm in optimizing process parameters during turning process. This approach aids in obtaining the optimum tool geometry and cutting conditions feasible for turning.

2 Genetic Algorithm

According to Kalita [11], the past researchers used different approach in optimizing parameters of the machine in order to get the best results such as particle swarm optimization (PSO), artificial bee colony, genetic algorithm and ant colony optimization. However, the most widely used optimization approach is the Genetic Algorithm (GA) [12]. The theory of natural evolution by Charles Darwin shaped GA. It was a natural selection process in which the fittest individuals were picked for reproduction in order to produce their heiress for the next generation. It was a heuristic search approach that is used in computing to get the precise or approximate answers to optimization and search issues [13]. According to the theory of fittest survival, the fittest members in the population owns the best chance of reproducing and surviving to the next generation, therefore enhancing future generations. Nonetheless, inferior individuals also have meager opportunity to survive and reproduce. GA is population-based search capability purposely to figure out multiple type of problems by go into all regions of the stated space and range [12].

2.1 Single Objective Genetic Algorithm Optimization

The aim of the Genetic Algorithm is to discover the optimum value for the decision parameters and constraints in order to achieve the best value for the objective function. In relation to the constraint parameters, a single objective genetic algorithm was designed to minimize or maximize the function [14]. The optimization scheme is applied in MATLAB toolbox. The focus of this research is to reduce temperature rise in turning machine during machining. The temperature rise objective function is made up of parameters that will be optimized throughout the optimization process. Cutting speed, feed rate, depth of cut, and nose radius are the parameters that will be optimized in this study. The objective function that used to represent the temperature rise in turning operation stated in Eq. 1. Equation 1 is based on the study done by Mahesh and this equation is chosen to be optimized because Eq. 1 is validated using Central Composite Design by using Response Surface Methodology (RSM) of Design of Experiments and it can be represent as follow [13, 15]:

$$\text{Temperature Rise} = 125.64792 - 1.85694x_1 - 19.62963x_2 + 36.41667x_3 - 46.70833x_4 - 0.018519x_1.x_2 - 0.258333x_1.x_3 + 0.433333x_1.x_4 - 22.91667x_2.x_3 + 24.30556x_2.x_4 - 28.43750x_3.x_4 + 0.008824x_1^2 + 33.69342x_2^2 + 41.90802389x_3^2 + 12.44792x_3^2 5.26042x_4^2 \quad (1)$$

Within range of parameters,

$$75\,\text{m/min} \leq \textbf{cutting speed},\ \mathbf{x_1} \leq 120\,\text{m/min}$$
$$0.18\,\text{mm/rev} \leq \textbf{feed rate},\ \mathbf{x_2} \leq 0.36\,\text{mm/rev}$$
$$0.20\,\text{mm} \leq \textbf{depth of cut},\ \mathbf{x_3} \leq 0.60\,\text{mm}$$
$$0.40\,\text{mm} \leq \textbf{nose radius},\ \mathbf{x_4} \leq 0.80\,\text{mm}$$

The range of the cutting parameters has been chosen by range is already tested by Gopal [10]. The range of parameccccccters must be selected based on the practical values that used in the real machining process. By optimization using GA, The algorithm begins by generating a random sample of solutions with the chosen number of population. The number of population will actuate the number of solution in the search space. If the population size is too small, the search region may not be fully explored [11, 16]. This population is evaluated and fitness values are appoint. In this case, the fitness values is represent by the value of the temperature rise. Through Genetic Algorithm operators which are selection, crossover and mutation, the population evolves to create better solutions called offspring. A sequence of interchanges of chromosomes that have been employed as representations of the optimized variables completes the generations of selection, crossover and mutation outcomes. Best fitness and individuals were created by the dominant solution which is a result of the comparison of solutions. Optimized cutting parameters are obtained when the objective function has been optimizing. Figure 2 design the flowchart of the GA for present work and Table 1 lists the parameters set in GA.

Table 1 shows the GA operators setup in the MATLAB Toolbox. The selection of the operator's setup is based on the trial and error since GA optimization is based on the probability concept. Thus, the GA operators that listed in the Table 1 are the operators that produced minimum temperature rise during the turning machining process.

3 Results and Discussion

In this work, the optimization was done by applying an optimization technique which is genetic algorithm along with different machine circumstances (cutting speed, feed rate, depth of cut, nose radius). The present study was separated in three optimization

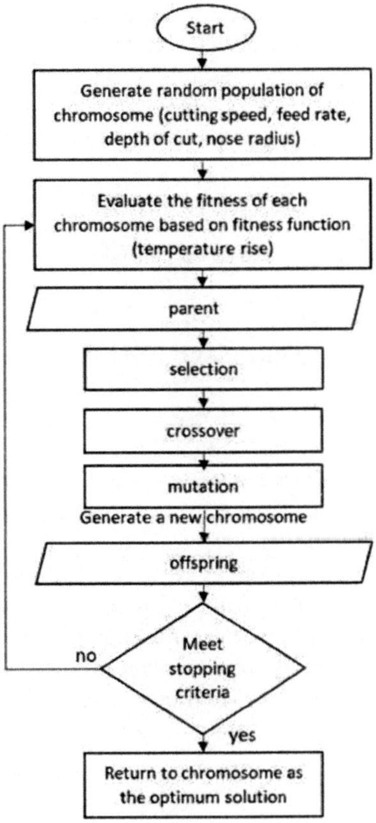

Fig. 2 Flowchart of temperature rise using GA

Table 1. Parameters of Genetic Algorithm

Parameters	Values
Population size	100
Scalling function	Rank
Function	Tournament
Mutation function	Gaussian
Mutation rate	0.1
Crossover function	Scattered
Crossover rate	1.0
Generations	300

phases in order to search the impact of each parameter on temperature rise. The first variables to be optimized were cutting speed and feed rate. The algorithm's second optimization focused on cutting speed, feed rate, and depth of cut while the third optimization added nose radius as an optimal parameter. Cutting speed, feed rate, depth of cut and nose radius restrictions are all being optimized to achieve the lowest temperature increase, providing for a better knowledge of how machining factors affect temperature rise. In order to study the cause and effect of the machining parameters on temperature rise during turning, the direct interaction impacts of the machining parameters on temperature increase during turning were plotted.

3.1 Optimizing Cutting Speed and Feed Rate

In this study, the response of cutting speed and feed rate on temperature rise were investigate and sketched while depth of cut and nose radius remain constant at value of 0.592 and 0.18 mm in order to get only the aftereffect of this two control variable on temperature rise. Figure 3 shows the fitness values versus generation that generated from GA optimization. When the termination requirements were fulfilled, the algorithm converged after 51 generations, as shown in Fig. 3. The best and mean values of the objective function for each generation indicated by the point on the figure. The result also shows that as the generation increases, the objective function and mean values are improving until the algorithm terminated.

As shown in Figs. 4 and 5, this two factor which are cutting speed and feed rate have a major effect on temperature rise. Each figures show that elevating both of these factors increase the temperature rise. The temperature is 34.33 °C when the cutting speed is 76.55 m/min, and it is 30.93 °C when the cutting speed is 94.57 m/min. When

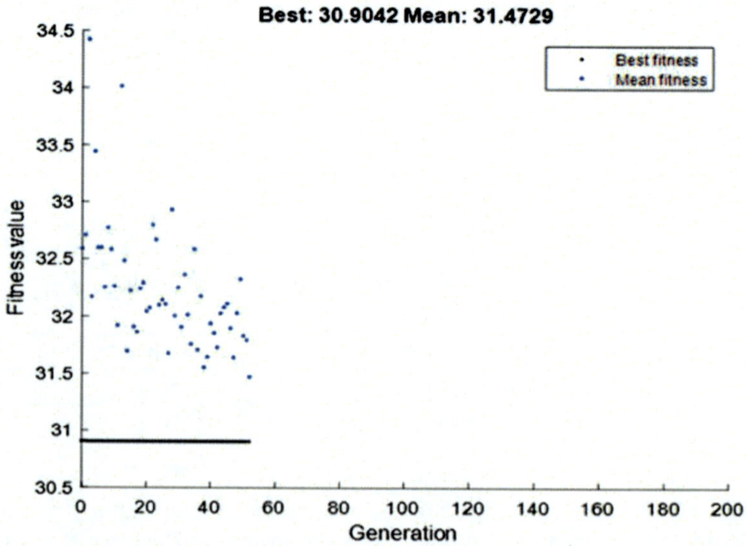

Fig. 3 The performance of fitness value versus generation for cutting speed and feed rate optimization

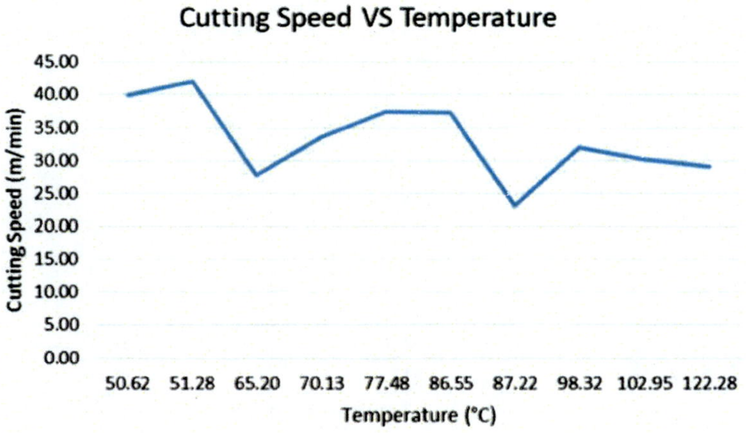

Fig. 4 Interaction effect between cutting speed and temperature for 2 parameters optimization

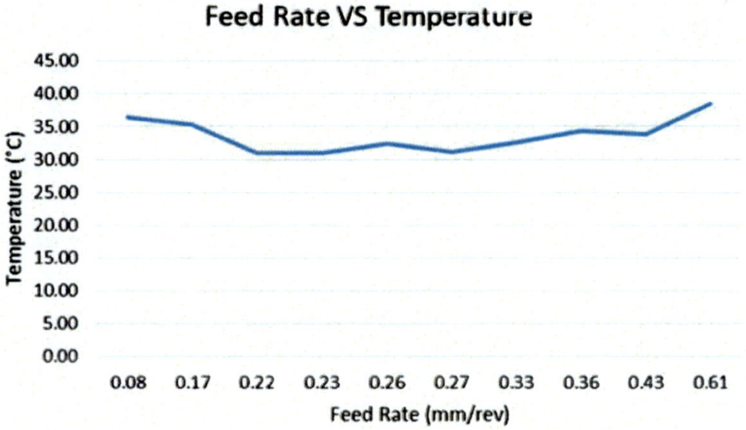

Fig. 5 Interaction effect between feed rate and temperature for 2 parameters optimization

the feed rate is 0.08, the temperature is 36.45 °C, and when the feed rate is 0.23 mm, the temperature is 30.90 °C. The maximum temperature become 38.58 °C when the cutting speed and feed rate are 112.47 m/min and 0.61 mm/rev. This happened because when the cutting speed and feed rate increase, the energy lost during the cutting process due to plastic deformation was converted into heat, heating up in the cutting region. Heat generation is inextricably connected to plastic deformation and friction. Friction increases as cutting speed increases, resulting in a rise in cutting zone temperature. As the feed rate rises, the chip section increases gradually, resulting in increased friction and higher temperatures.

Since GA has the ability to find the minimum temperature rise by optimizing more than one parameters, the results listed in Table 2 shows that the minimum temperature

Table 2. Optimize process predicted by GA

No	Cutting speed (m/min)	Feed rate (mm/rev)	Temperature (°C)
1	94.57	0.23	30.90
2	76.55	0.36	34.33
3	116.25	0.17	35.25
4	106.82	0.33	32.53
5	96.22	0.22	30.93
6	107.58	0.26	32.43
7	90.73	0.27	31.09
8	112.47	0.61	38.58
9	81.28	0.43	33.89
10	117.54	0.08	36.45

rise achieved at 30.90 °C with 94.57 m/min for cutting speed and 0.23 mm/rev for feed rate. The significant of this result is that to obtain minimum temperature rise it is not compulsory to set the both parameters (cutting speed and feed rate) to the lowest values, instead it should be set to the optimum values while to maintain as minimum temperature rise and also able to achieve other important criteria in machining such as production rate and surface roughness.

3.2 *Optimizing Cutting Speed, Feed Rate and Depth of Cut*

Further down, the effect of associated process factors on temperature rise is investigated. In this setup, the value of nose radius is constant in 0.18 mm. Figure 6 shows the fitness values versus generation that generated from GA optimization. Based on Fig. 6, the algorithm converged at 60 generations when the stopping criteria was met. The point on the graph represents the best and mean values of the objective function for each generation. The result also indicates that as the number of generations increase, the objective function and mean values improve until the algorithm terminates.

Figures 7, 8, and 9 show the relationship between temperature, cutting speed, feed rate and depth of cut. The cutting speed was determined to have a significant influence the temperature rise throughout turning process. Figure 7 indicates that increasing the cutting speed resulted in a temperature increase and an ideal cutting speed of 87 m/min. This is because the rate at which energy is absorbed by plastic deformation and friction increases as the cutting speed increases. As a result, the rate of heat generation in the cutting zone increases, producing a high cutting temperature. As the cutting speed increases, the cutting temperature rises, producing automatic dispersion between the tool and the workpiece material, propagating tool wear. Figures 8 and 9 show that the feed rate and depth of cut have a substantial impact on the temperature

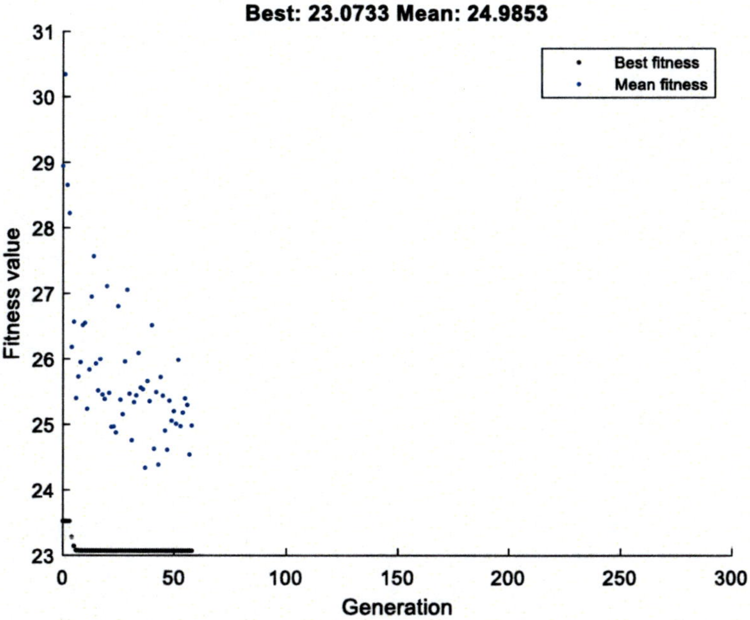

Fig. 6 The performance of fitness value versus generation for cutting speed, feed rate, and depth of cut optimization

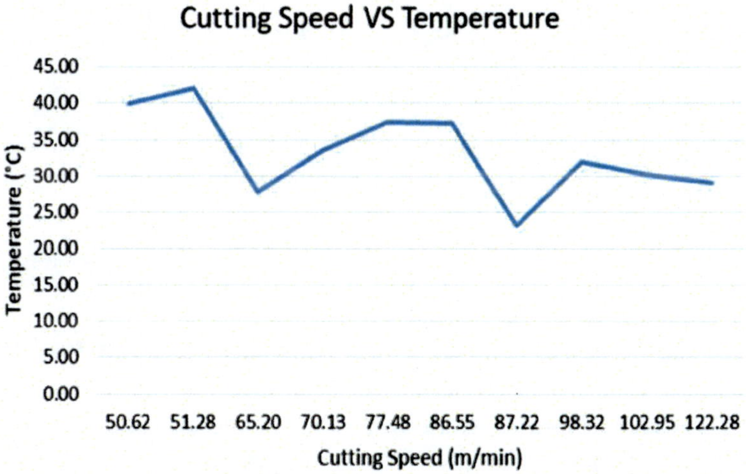

Fig. 7 Interaction effect between cutting speed and temperature for three parameters optimization

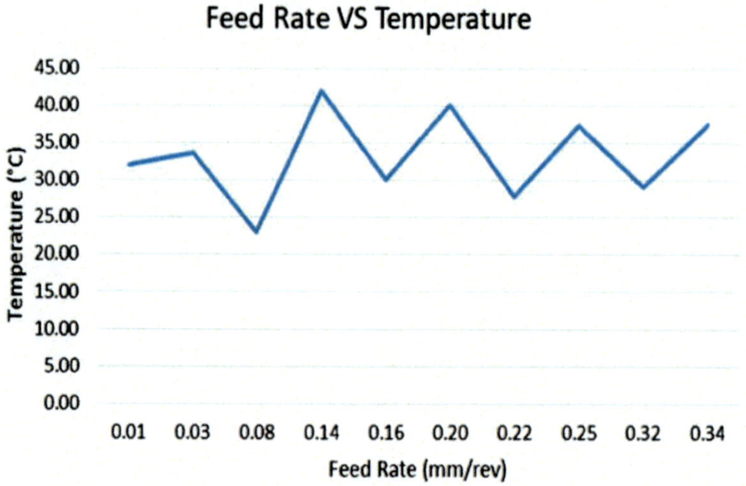

Fig. 8 Interaction effect between feed rate and temperature for three parameters optimization

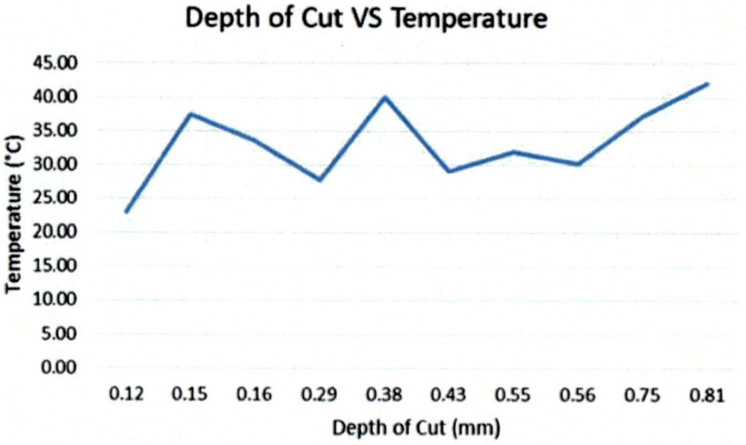

Fig. 9 Interaction effect between depth of cut and temperature for three parameters optimization

rise during machining process. When the feed rate is increased, the amount of chip and friction energy spent on the tool face increases, resulting in a rise in cutting zone temperature [12, 17]. Figure 9 illustrates that the temperature increases as the depth of cut increases. The amount of workpiece materials to be removed increases when the depth of cut is increased, resulting in an increase in cutting temperature.

When cutting at a lower depth, the material of a workpiece sticks to the tool's flank less than when cutting at a deeper depth. The temperature rises as the workpiece material adheres to the in tool flank. The section of chip and friction power spent on

Table 3. Three parameters optimization result using GA

No	Cutting speed (m/min)	Feed rate (mm/rev)	Depth of cut (mm)	Temperature (°C)
1	87.22	0.08	0.12	23.07
2	50.62	0.20	0.38	39.98
3	65.20	0.22	0.29	27.80
4	70.13	0.03	0.16	33.59
5	122.28	0.32	0.43	29.07
6	77.48	0.34	0.15	37.48
7	51.28	0.14	0.81	42.04
8	86.55	0.25	0.75	37.28
9	102.95	0.16	0.56	30.15
10	98.32	0.01	0.55	32.00

the tool face will rise during the increasing in feed rate which will in increase cutting zone area. As GA may optimize several parameters to obtain the lowest temperature increase, the results in Table 3 show that the lowest temperature rise was achieved at 23.07 °C with 87.22 m/min for cutting speed, 0.08 mm/rev for feed rate and 0.12 mm depth of cut.

3.3 *Optimizing Cutting Speed, Feed Rate, Depth of Cut and Nose Radius*

The impact of cutting speed, feed rate, depth of cut, and nose radius on temperature rise were investigated and presented in the third optimization. Figure 10 shows the fitness values versus generation that generated from GA optimization. When the stopping requirements were fulfilled, the algorithm converged after 255 generations. The best and mean values of the objective function for each generation are indicated by the point on the figure. The result also demonstrates that the objectives function and mean values improve as the generation increases until the algorithm terminates. Based on this result, it shows that when number of parameters that needed to be optimized is increasing, the algorithm required higher number of generations to converge into final solution. When number of parameters that need to be optimized in increasing, the search space in the algorithm become wider. Thus, the GA operators that responsible to do exploration and exploitation in the search which are crossover and mutation took longer generation to find the optimum solution.

The energy dissipated in metal cutting is transformed into heat at the cutting edge of the tool and this heating exploit is directly or indirectly responsible for many of the machining-related economic and technological obstacles [13, 18]. The area of contact allowed for conduction between the tool and the workpiece is less when using a tool with a small tool nose radius than when using a tool with a larger tool

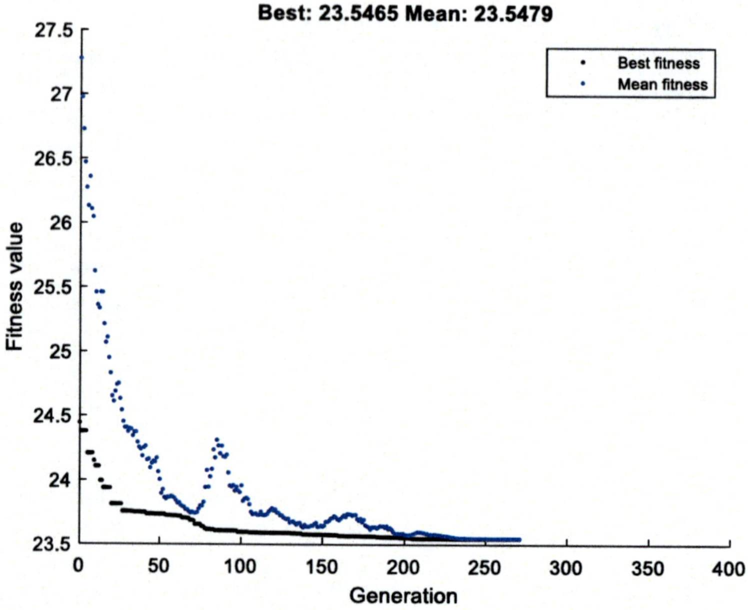

Fig. 10 The performance of fitness value versus generation for cutting speed, feed rate, depth of cut, and nose radius optimization

nose radius. Local temperature increase at the cutting edge is aided by the decreased heat conduction field. However, as demonstrated in Table 4, the nose radius has no influence on the temperature rise during the turning phase. It also implies that the optimization procedure does not have to incorporate all parameters. According to the results in Fig. 10 and Table 4, adding nose radius to the algorithm increased

Table 4. Four parameters optimization result using GA

No	Cutting speed (m/min)	Feed rate (mm/rev)	Depth of cut (mm)	Nose radius (mm)	Temperature (°C)
1	90.7498	0.1800	0.2000	0.8000	23.5465
2	90.7499	0.1800	0.2001	0.7999	23.5472
3	90.7505	0.1800	0.2006	0.8000	23.5481
4	90.7501	0.1801	0.2001	0.8000	23.5472
5	90.7506	0.1800	0.2000	0.7999	23.5471
6	90.7505	0.1802	0.2001	0.7999	23.5485
7	90.7504	0.1800	0.2006	0.8000	23.5481
8	90.7506	0.1800	0.2000	0.7999	23.5471
9	90.7507	0.1800	0.2001	0.7999	23.5472
10	90.7500	0.1801	0.2003	0.8000	23.5477

the number of generations necessary to converge to a final solution by 86% to the optimization with only three parameters, but the optimal result was still not attained. This optimization yielded a minimum temperature of 23.5465 °C, which is higher than the 23.07 °C achieved in the three-parameter optimization. When the nose radius is 0.8000 mm, the maximum temperature is 23.5485 °C, and when the nose radius is 0.7999 mm, the maximum temperature is 23.5485 °C.

4 Conclusions

The research reported in this study has considerably improved the minimal temperature rise during turning machining using GA optimization. To achieve the lowest temperature rise during turning machining, an optimization approach based on GA optimization was utilized to determine the best cutting speed, feed rate, and depth of cut. The GA recommends a minimum temperature of 23.0733 °C, with cutting variable values for cutting speed, feed rate, and depth of cut is 81.22 m/min, 0.08 mm/rev, and 0.12 mm, respectively, based on this research. Out of the three optimization approaches, the run solver view results from MATLAB software show that the second optimization generates the best individual performance of variables. It has decreased by 3.62% as compared to the previous optimization, which had a temperature of 23.94 °C [13, 15]. The results of this study also show that the nose radius has the least impact on getting the lowest temperature rise throughout the machining processes.

Acknowledgements The authors wish to thank Universiti Malaysia Pahang (UMP) for the use of research facility and grant RDU1803144 for providing financial support to complete the study.

References

1. Thakare A, Nordgren A (2015) Experimental study and modeling of steady state temperature distributions in coated cemented carbide tools in turning. Procedia CIRP 31:234–239. https://doi.org/10.1016/j.procir.2015.03.024
2. Chinchanikar S, Choudhury SK (2014) Evaluation of chip-tool interface temperature: effect of tool coating and cutting parameters during turning hardened AISI 4340 steel. Procedia Mater Sci 6:996–1005. https://doi.org/10.1016/j.mspro.2014.07.170
3. Motorcu AR, Isik Y, Kus A, Cakir MC (2016) Analysis of the cutting temperature and surface roughness during the orthogonal machining of AISI 4140 alloy steel via the Taguchi method. Mater Technol 50(3):343–351. https://doi.org/10.17222/mit.2015.021
4. Okokpujie IP, Okonkwo UC (2015) Effects of cutting parameters on surface roughness during end milling of aluminium under minimum quantity lubrication (MQL). Int J Sci Res 4(5):2937–2942
5. Ejieji T, Adedayo SM, Bello OW, Abdulkareem S (2018) Effect of machining variables and coolant application on HSS tool temperature during turning on a CNC lathe. IOP Conf Ser Mater Sci Eng 413(1). https://doi.org/10.1088/1757-899X/413/1/012004

6. Gosai M, Bhavsar SN (2016) Experimental study on temperature measurement in turning operation of hardened steel (EN36). Procedia Technol 23:311–318. https://doi.org/10.1016/j.protcy.2016.03.032
7. Akhil CS, Ananthavishnu MH, Akhil CK, Afeez PM, Akhilesh R, Rajan R (2016) Measurement of cutting temperature during machining. J Mech Civ Eng 13(2):102–116. https://doi.org/10.9790/1684-130201102116
8. Das SR (2012) Optimization of cutting parameters on tool wear, workpiece surface temperature and material removal rate during turning of AISI D2 steel Orissa Engineering College Orissa Engineering College 1(5):1–10
9. Saravanan R, Asokan P, Vijayakumar K (2003) Machining parameters optimisation for turning cylindrical stock into a continuous finished profile using genetic algorithm (GA) and simulated annealing (SA). Int J Adv Manuf Technol 21(1):1–9. https://doi.org/10.1007/s001700300000
10. Gopal M (2020) Optimization of machining parameters on temperature rise in CNC turning process of aluminium 6061 using RSM and genetic algorithm. Int J Mod Manuf Technol 12(1):36–43
11. Kalita K, Shivakoti I, Kumar R (2017) Optimizing process parameters for laser beam micro-marking using a genetic algorithm and particle swarm optimization. Mater Manuf Process 32(10):1101–1108. https://doi.org/10.1080/10426914.2017.1303156
12. Chandrasekaran M, Muralidhar M, Krishna CM, Dixit US (2010) Application of soft computing techniques in machining performance prediction and optimization: A literature review. Int J Adv Manuf Technol 46(5–8):445–464. https://doi.org/10.1007/s00170-009-2104-x
13. Rudrapati R, Pal PK, Bandyopadhyay A (2015) Modeling and optimization of machining parameters in cylindrical grinding process. Int J Adv Manuf Technol 82:2167–2182. https://doi.org/10.1007/s00170-015-7500-9
14. Singh D, Venkateswara Rao P (2007) Optimization of tool geometry and cutting parameters for hard turning. Mater Manuf Process 22(1):15–21. https://doi.org/10.1080/10426910601015816
15. Mahesh G, Muthu S, Devadasan SR (2015) Prediction of surface roughness of end milling operation using genetic algorithm. Int J Adv Manuf Technol 77(1):369–381. https://doi.org/10.1007/s00170-014-6425-z
16. Zolpakar NA, Mohd-Ghazali N, Ahmad R (2016) Experimental investigations of the performance of a standing wave thermoacoustic refrigerator based on multi-objective genetic algorithm optimized parameters. Appl Therm Eng 100:296–303. https://doi.org/10.1016/j.applthermaleng.2016.02.028
17. Sulaiman S, Roshan A, Borazjani S (2014) Effect of cutting parameters on tool-chip interface temperature in an orthogonal turning process. Adv Mater Res 903:21–26. https://doi.org/10.4028/www.scientific.net/AMR.903.21
18. Kaushik VS, Subramanian M, Sakthivel M (2018) Optimization of processes parameters on temperature rise in CNC end milling of Al 7068 using hybrid techniques. Mater Today Proc 5(2):7037–7046. https://doi.org/10.1016/j.matpr.2017.11.367

Alternative Railway Tools and Sustainability in RAMS: A Review

M. A. Muhammed Nor, A. F. Yusop, M. A. Hamidi, M. N. Omar, N. A. Abdul Hamid, and W. M. Wan Mohamed

Abstract RAMS is a tool and methodology that combines reliability engineering, availability, maintainability, and safety in a way that is tailored to the system's goals. A comprehensive view on RAMS's components and theory behind the underlying mathematical model is not to be found in journal publication. This paper would also discuss several benefits and sustainability of RAMS. Life Cycle Cost (LCC) would also being introduce as a complementary discipline in term of costing that normally regarded parallel to RAMS. There are a series of methods that being utilized at every discipline of the RAMS component such as Fault Tree Analysis (FTA), Failure Mode Effect Critical Analysis (FMECA), Reliability Block Diagram and many more. Some commonly used methods would be highlighted in this paper. RAMS application and implementation will aid asset owners, contractors, and operators in efficiently procuring, developing, and operating their assets. However, further research and analysis is needed in the railway industry to build a viable framework for project and operation implementation using both tools.

Keywords RAMS tools · LCC · Railway sustainability

1 Introduction

RAMS is an integrated discipline that includes measures and characteristics for system reliability, availability, maintainability, and safety that are tailored to operational and project objectives. It is commonly used in the operation and engineering

M. A. Muhammed Nor (✉) · A. F. Yusop · M. A. Hamidi · M. N. Omar
Faculty of Mechanical and Automotif Engineering, Universiti Malaysia Pahang, 26600 Pekan, Pahang, Malaysia
e-mail: mazri.mnor@gmail.com

N. A. Abdul Hamid
Faculty of Technology Management and Business, Universiti Tun Hussien Onn Malaysia, 86400 Parit Raja, Batu Pahat, Johor, Malaysia

W. M. Wan Mohamed
Malaysia Institute of Transport (MITRANS), Universiti Teknologi Mara, 40450 Shah Alam, Selangor, Malaysia

© The Author(s), under exclusive license to Springer Nature Singapore Pte Ltd. 2023

M. H. A. Hassan et al. (eds.), *Technological Advancement in Instrumentation & Human Engineering*, Lecture Notes in Electrical Engineering 882,
https://doi.org/10.1007/978-981-19-1577-2_40

of railway system [1] s. RAMS, according to Alstom, must be followed internationally by rolling stock suppliers for the development of Mass Rail Transit Systems (MRTS) [2]. It has become a widely discussed and researched discipline in recent years due to its robustness and flexibility in achieving defined objectives such as service time, safety, and cost limitation. Railway assets are designed for long-term use with high reliability and availability without compromising safety; therefore, asset maintenance must be carried out optimally and cost-effectively. This management tool has the potential to significantly improve the effectiveness and economic competitiveness of railway transportation in comparison to other modes of transportation. Modern railway systems are complex, incorporate multiple technologies, and operate in an environment where it is difficult to pinpoint precise system responses and behaviors. The use of today's technology, such as computers, microprocessors, interconnected communication, and information technologies, in combination with historically developed electromechanical components, has dramatically increased the complexity of railway systems [3]. A primary goal for completing RAMS-related duties today is to obtain a safe, highly dependable, and available railway system, as well as an innovative and sustainable railway system. RAMS activities are also critical in this context for extending the lifespan of railway system. Railway RAMS-related standards require that railway manufacturers and operators to install a RAMS management system and verify compliance with specific safety and RAM requirements. While the standards only provide a general framework for RAMS activities, real-world implementation is still being researched [4].

This paper provides a general analysis of RAMS's impact. The primary goal of this work is to inform engineers, industrialists, and researchers who are interested in RAMS as a promising technology in railway systems. In scientific indexes, many literatures, including the most recent articles, are reviewed from highly rated journals.

As technology advances, the environment changes, and consumer needs change, railway system designers and operators are constantly upgrading their various operational tasks. A secure and dependable network with sufficient capacity and availability is required. A railway system's goal is to achieve a specific level of rail traffic in each time frame while remaining safe and within budget. The Railway RAMS procedure determines the system's confidence in achieving this goal. Railway RAMS has a significant impact on customer service quality [5]. Table 1 shows component of RAMS:

Common tools used to assess RAMS are Fault Tree Analysis (FTA) [7], Failure Mode Effect Analysis (FMEA) [8], Reliability Block Diagram [9] and many more.

2 Railway Reliability

The reliability of a product is closely connected to its quality. This criterion is one of the most important considerations during the various stages of product design, testing, and operation [10]. Reliability is a function of time, and it decreases as the

Table 1 RAMS component

Reliability	**Definition**: Ability of an item to perform a required function under given condition for a given period [6]
Availability	**Definition**: the ability of a product to be in a state to perform a required function under given conditions at a given instant of time or over a given time interval, assuming that the required external resources are provided [6]
Maintainability	**Definition**: the probability that a given active maintenance action, for an item under given conditions of use, can be carried out within a stated time interval when the maintenance is performed under stated conditions and using stated procedures and resources [6]
Safety	**Definition**: the state of technical system freedom from unacceptable risk of harm [5]
	Basic Item: Risk Assessment and Hazard Control

period of operation lengthens. Because the cost of procuring each railway asset is very high, a high reliability product or system over a long period of time is required.

Aside from technical definition, railway reliability can also be defined in terms of train operationality. Vromans, M. (2005) investigated the reliability of a railway using train punctuality [11]. A train is considered reliable when it can run properly all of the time, allowing goods and other services to be delivered on time.

According to Durivage, M. A., reliability is "probability that an item will perform a required function without failure under stated conditions for a specific period of time [12]." According to this interpretation, studying the reliability of a system or component entails investigating its failure behavior. The failure event would be collected, tabulated, and plotted stochastically to understand and compute the relevant information.

To investigate the probability of failure, a population of products or systems must be observed over time. Gerokostopoulos et al. (2015) proposed an estimation approach as well as a risk control approach for calculating sample size for a reliability study [13]. A Probability Density Function (PDF) of a failure event could be developed with an adequate number of samples. A PDF's data will contain the Time To Failure (TTF) of the samples, which is the time it takes for an individual sample to fail. The collected data is known as life data, and it is used to calculate the product's lifespan. In the study of Life Data Analysis (LDA), it is known that there are a few failure distributions that most likely fit the collected data, which are exponential, lognormal, and Weibull Distributions [14].

A histogram of failure numbers versus observed time is plotted, and the line of best fit is calculated. This TTF Distribution, f(x), will be used to calculate the Reliability, R(t). A cumulative probability function (CDF) is found by integrating the plotted TTF Distribution. In this case, CDF is also known as the probability of failure, F(t). R(t) is defined as the complement of F(t) [15, 16].

From PDF, mean time to failure, MTTF could be calculated. Hence, failure rate, λ could be computed with the formula below:

$$mean,\ MTTF = \frac{\int_a^b f(t)dt}{b-a} \tag{1}$$

$$\lambda = \frac{1}{MTTF} \tag{2}$$

Because of its constant failure rate, an exponential pdf is used to predict reliability when there is no historical data on the operation of a system [17].

The failure rate is typically high at the beginning and end of a system's operation. When all the components start to work together, some minor adjustments may be required. As shown in the Bathtub graph, after a long period of operation, the components begin to wear out and contribute to a higher failure rate [18, 19]. The useful operational period with a low and steady failure rate is defined as the constant failure rate after the adjustment and wear out period [20]. The PDF for exponential distributed failure is:

$$f(t) = \lambda e^{-\lambda t} \tag{3}$$

$$F(t) = \int f(t)dt = 1 - e^{-\lambda t} \tag{4}$$

$$R(t) = 1 - F(t) = e^{-\lambda t} \tag{5}$$

Component reliability can also be calculated using historical data. If the component is highly reliable, a long period of operation is required to collect and capture adequate data on failure events. The failure probability density function could be plotted, and the best fit distribution found. With such a case study, the true reliability of the system or component could be discovered, and the value could be compared to the theoretical and predicted value.

Rail systems are made up of several subsystems, each with its own set of functionalities and reliability characteristics [21]. For example, an electro-pneumatic brake control system includes an electric, pneumatic, braking system, compressor, and other components [22]. To calculate the reliability of a system, all the subsystems that are interconnected must be determined and arranged in an orderly manner. Subsystem or component configurations are typically in series, parallel, or complex. The inter-reliability of the system could be calculated based on the configuration. This method is known as a reliability block diagram (RBD) [23].

2.1 Reliability Methodology

To successfully finding reliability using RBD, the boundary of the system under review needs to be defined. All the subsystems need to be arranged accordingly. FTA is a widely used method to be used together with RBD to determine the reliability and risk analysis of a system [24]. The arrangement of subsystem will be determined, either series, parallel or mixed combination. The calculation is straightforward and could be done using software such as Reliasoft (Fig. 1).

Figure 2(a) shows the series combination, 2(b) shows parallel and 2(c) shows mixed combination. Reliability for series combination $R_T = R_1 * R_2 * \cdots * R_n$ and $R_T = \frac{1}{R_1 * R_2 * \cdots * R_n}$ for parallel arrangement. To compute the mixed combination, the system needs to be divided into smaller series or parallel subsystems and using the two previous described equation to find the system's reliability.

3 Railway Availability

The availability of the required and relevant systems is critical for a railway asset manager's train operation. To keep the line running smoothly, the number of inoperable rolling stock must be kept to a minimum. Furthermore, the electrical section of

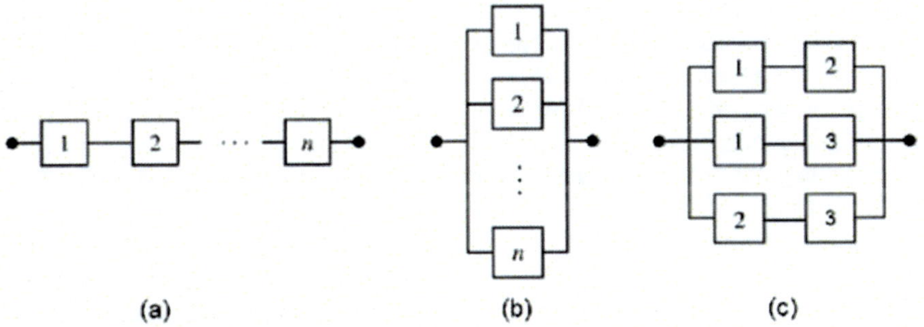

Fig. 1 Possible arrangement of subsystems

Fig. 2 Uptime and Downtime of Railway Operation

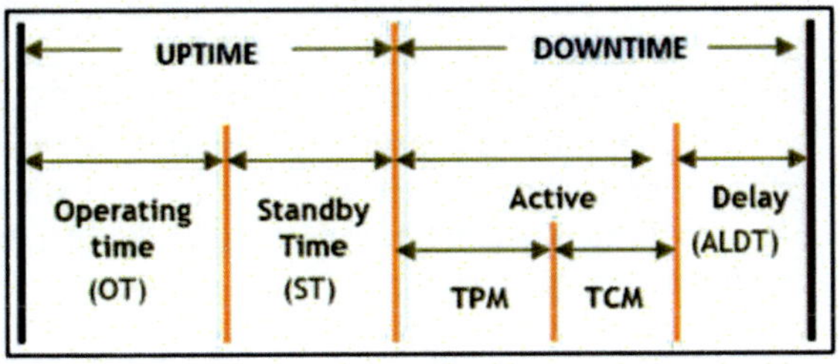

the rolling stock, as well as power traction, must be available and operable. Availability is closely tied to reliability and availability [25, 26]. Availability is defined as the sum of the total time the system is working properly or Uptime and the total time the system is not working properly or Downtime.

Where TPM is total preventive maintenance, TCM is total corrective maintenance and ALDT is Administrative and Logistic delay time. Figure 2.0 shows that availability is a combination of dependability (Uptime) and maintainability (Downtime) [27].

The time recorded when the train is in operation and in standby mode is referred to as uptime. Down time, on the other hand, is time recorded as maintenance time plus all administrative and logistical delays incurred to complete the maintenance schedule. Varies to the asset manager's maintenance philosophy, the downtime could be much longer or shorter. This would almost certainly have an impact on a system's availability. Thus, the expanded availability, A formula is as follow,

$$A = \frac{OT + ST}{OT + ST + TPM + TCM + ALDT} \quad (6)$$

4 Railway Maintainability

The main objective of railway track maintenance and renewal is to ensure safety and meet quality standards [28]. The availability of a system or service is heavily influenced by an asset's maintainability. The best maintained system is one that can always be relied on and is available when its service is required. To accomplish this, maintenance must be performed as much as needed but as infrequently as possible. The maintainability philosophy and methodology need to be adopted wisely by Train Operator Company (TOC).

There are several philosophies in planning and scheduling maintenance. Afzali et al. (2019) proposed a new model for reliability-centered maintenance (RCM) of electrical power distribution [29]. A reliability team thoroughly evaluates each critical component in this approach, and all failure modes are identified. The maintenance requirements will then be identified, and a preventive maintenance (PM) schedule will be created.

Su et al. (2019) are investigating another approach known as condition-based maintenance (CBM) for railway track maintenance in the Netherlands [30]. This is not significantly different from RBM. While RBM determines the PM through failure analyses, CBM considers the machine's condition as well. Maintenance is performed only when it is necessary, based on continuous observation of the system or item conditions. Although monitoring required a small number of skilled workers at regular intervals, this method provides efficient use of the asset's useful life [31].

4.1 Maintainability Methodology

There are several methodologies available to achieve the asset management philosophy. There are two main methods used in common practice: Corrective Maintenance (CM) and Preventive Maintenance (PM) [32]. The CM is tasked with restoring a faulty system to its pre-crash state. While PM is being completed on a set schedule or at a predetermined time. It could also be carried out based on the health of a specific system component. PM includes routine inspection and checking. To ensure high operational availability of the railway, PM is typically performed when the opportunity arises, such as late at night when the train is shut down. Opportunity maintenance is another name for this kind of maintenance [33].

Another two methods that gaining more interest in maintenance strategy are Proactive Maintenance (PaM) [34] and Predictive Maintenance (PdM) [35]. PaM focuses on resolving problems before they become failures. PdM, on the other hand, is a process that analyses and monitors machine performance and operational parameters to detect and diagnose developing problems before they cause failure and significant damage [36]. PdM can be performed using techniques such as oil analysis, mechanical ultrasound, vibration analysis, and wear particle analysis [37].

Total continuous monitoring is now possible thanks to advancements in information technology. The asset or plant under maintenance would only be monitored by sensors, which would record a wide range of data from the physical movements of a structure or piece of equipment, such as temperature, vibrations, and conductivity, among other things. The Internet of Things (IoT) is an important part of the process because it allows multiple systems to work together to translate and analyze recorded data to forecast when maintenance should be performed [38] Furthermore, new machine-learning technologies have the potential to increase the accuracy of predictive algorithms over time, resulting in even better performance [39].

Maintaining assets would necessarily require a maintenance team repairing or replacing faulty components or systems. The component's reliability would normally decrease once it was repaired or replaced [40]. Repair is classified into three levels: perfect repair, minimal repair, and imperfect repair. Perfect repair restores the component to its original state; minimal repair restores the component to its pre-maintenance state; and imperfect repair restores the component to its pre-maintenance state [41].

Discussing all the philosophy and methodology of maintenance's possibilities, it appears that the asset could always be in the best condition. The reality is that maintenance costs a lot of money. For example, Prasarana Malaysia Berhad, the operator of the Rapid Rail network, spends RM350 million per year on maintenance cost [42]. Around 70% of the maintenance cost, which includes employee wages and other corporate and management costs, will be used for technical maintenance. Manual inspection and monitoring will then consume approximately 30% of the technical maintenance cost [43]. The goal of all asset managers is to have a reliable, highly available, and safe operation at the lowest possible cost. This requires the best and most efficient maintenance scheduling and activities. Maintenance cost analysis could be used to calculate and determine the rate and scheduling of maintenance [44].

5 Railway Safety

A robust design with high reliability and accessibility to maintenance, combined with good management, would result in a high safety standard for railway operation. EN 50,129, as a guide for primarily electronic systems such as signaling, communication, and processing systems, discusses and assesses RAMS safety in railway [45]. EN 50,126-2 is another standard that describes railway safety requirements [6]. This standard, which supplements EN 50,129 on safety procedures, discusses other railway applications such as command, control, and signaling, rolling stock, and fixed installation. Wang et al. (2021) investigates the method for safety analysis using cusp catastrophe model. This method describes the ever-changing process of railway system safety and considers the emergent property of safety [46]. Liu et al. (2020) proposes a comprehensive model by combining commonly used practice in safety analysis such as Analytic Hierarchy Process (AHP) and Maximum Entropy Method (MEM) [47].

5.1 Safety Methodology

The main subtopics in Safety Analysis are Risk Assessment and Hazard Control. The EN 50,126-2 proposes Hourglass Model to describes activities conducted in safety analysis.

Hazard identification, consequence analysis, and risk acceptance are all components of risk assessment. It defines high-level system safety requirements, more specifically safety requirements for the system under consideration from the perspective of the railway duty holder and operator. It considers operational safety, previous rail application experience, and regulatory requirements.

Activities such as Causal Analysis, Hazard Identification (refinement), Common Cause Analysis, and Show of Compliance, on the other hand, must be carried out in accordance with Hazard Control standards. Hazard Identification in Risk Assessment focuses on high-level hazards derived from system functions (black boxes) and related system operation, as well as the system's environment, whereas Hazard Identification in Hazard Control focuses on the event's cause. Because there could be several causes for a hazard to occur, an iterated hazard identification process would normally be carried out in Hazard Control. The Bowtie Model is a method that is frequently used as a methodology for safety analysis [48].

FTA, FMEA and event tree analysis (ETA) are common techniques used for system reliability and safety [49]. FTA is a powerful diagnostic technique that uses logical and functional links between components, processes, and subsystems to identify the underlying causes of potential risks. A fault tree (FT) is a model that logically and graphically shows the various combinations of likely events in a system, both faulty and normal, that result in unexpected events or states. FTs can be used to determine the source of potential hazards. Faults can be caused by hardware failure, software error, or human error. Traditional FTA involves events and gates and is based on Boolean algebra. Logic modelling is a visual representation of basic event relationships.

6 Sustainability of RAMS

RAMS is an extensive tool that covers the entire life cycle of a system, project, or component. This effective tool is widely used in the design and operation of Critical Infrastructure (CI) such as petroleum platforms, servers, critical public construction and building, and many others [50]. To ensure its sustainability, RAMS has specific requirements for suppliers, project executors, tendering departments, project owners, and rail operators. It would ensure the operation's dependability and safety. Aside from that, the costs incurred from project concept to project design phase and final decommissioning phase could be calculated and determined [51]. RAMS, as previously stated, is becoming a common term among railway engineering practitioners. This is due to the railway's history, which dates back more than 200 years, with the first railway track being built in 1825 [52]. As a result, considering the age of railway engineering, the implementation of RAMS in railway is relatively new.

RAMS is a tool for understanding how a system works, particularly a complex and multi-interface system like a railway. Predicting vulnerabilities, weaknesses, and potential failures and explaining how they affected the system's quality and performance is a significant management task. It also used a strategy to improve the operation's quality to achieve optimal long-term availability and choose the best maintenance solutions [53]. An asset manager can use it to predict and plan appropriate maintenance strategies and technologies to be integrated into the existing system. The manager could take appropriate action at the lowest possible cost to ensure that the service is not disrupted.

The railway-based vehicle transportation system is both mission and safety critical [54] As a result, to ensure the safety and reliability of a railway system, all potential hazards that affect system components could be detected, analyzed, and controlled via RAMS. Safety and availability are higher level RAMS characteristics that can only be attained by meeting all reliability and maintainability requirements as well as controlling ongoing, long-term operation and maintenance activities. In short, RAMS provides a wide range of methodologies and approaches to engineers, system designers, clients, and system operators to address relevant concerns during the construction and operation of a specific system.

7 Life Cycle Cost Analysis

While RAMS is taking care of technical parameters in determining reliability, availability, maintainability, and safety along railway's life cycle, LCCA determines whether the investment in meeting the RAMS requirements is leveraged or is there any other meaningful alternative [55]. Like RAMS, LCCA could be performed from the conceptual to operation and decommissioning phase. Any parameter and risk identified would have cost attribute, thus alternative or other option need to be defined in RAMS [56]. Both LCCA and RAMS are needed to make a right decision in railway project or operation.

Liden et al. (2016) investigate availability of train operation and maintainability (RAMS parameter) versus cost impact (LCCA) [57]. A model for assessing and dimensioning such maintenance windows is described in this study, which considers marginal effects on both maintenance costs and predicted train traffic demand. Senaratne et al. (2020) assessed the alternative for railway support material using LCC analysis [58]. Banar et al. (2015) evaluated the investment benefit of Turkey railway system for the passenger [59].

Generally, there are several stages in performing LCCA. The first stage involves the definition of objectives, assumption development, gathering all source materials and preparation of input data. In the second stage, RAMS parameter is prepared and analyze for all the proposed variants. Later, the LCC model is developed based on RAMS parameter and assumption in the first stage. In the fourth stage, the model is being analyze and all the calculation take place. The result of the analysis is being reviewed in the fifth stage. The model and calculation will then be verified in the sixth and final stage that required continuous monitoring and reassessment with real operational data.

8 Conclusion

RAMS is a major theme in system design, project execution, and asset management. As stated in System Life Cycle, it controls the product or system life cycle from the requirement and design phase to the decommissioning phase. RAMS introduces a wide range of interdisciplinary engineering, technical, logistic, and cost-effective methods, making it the best candidate for complex system management tools.

Because the use of RAMS in century-old railway infrastructure is not as widespread as in new industries such as petroleum, chemicals, and aviation, it is now becoming a new standard for the procurement and construction of new railway assets. As the application of RAMS is still in its early stages, it provides an excellent opportunity for engineers, consultants, and scientists to build and develop a better, standardized, and widely applicable RAMS methodology for the railway industry.

There are considerable limitations in implementing RAMS in railway industry. From authority, supply chain, management, sand also technical knowledge as well as player awareness on RAMS importance need a paradigm shift to fully implement RAMS. A workable policy needs to be drafted by the government that to be followed by all railway stakeholders. Absence of such a policy would affect the RAMS implementation throughout the lifecycle of railway. Often during developing RAMS, significant information from supplier is missing. Many suppliers do not reveal their design and technical specification to the degree that RAMS engineer could use for the calculation. A widely enforced standard needs to be developed to make sure suppliers meet the requirement for information relegation to relevant parties throughout railway construction and operation. These are a few future works that need to be addressed and researched as a first step into implementing RAMS.

Acknowledgements The authors would like to thank the Ministry of Higher Education for providing financial support under Konsortium Kecemerlangan Penyelidikan (KKP) No. JPT (BPKI) 1000/016/018/25(53) or RDU210702 (Railway Reliability, Availability, Maintainability and Safety Excellence Consortium) and University Malaysia Pahang for laboratory facilities.

References

1. Park MG (2014) Integration of RAMS management into railways system engineering. http://etheses.bham.ac.uk/4750/
2. Sudhir K (2013) Report of the sub-committee on rolling stock for metro railways
3. Mahboob Q, Zio E (2018) Handbook of RAMS in Railway Systems: Theory and Practice. Taylor & Francis, Abingdon-on-Thames
4. Al-Douri A, Kazantzi V, Currie-Gregg N, El-Halwagi MM (2021) Integrating uncertainty quantification in reliability, availability, and maintainability (RAM) analysis in the conceptual and preliminary stages of chemical process design. Chem Eng Res Des 167:281–291. https://doi.org/10.1016/j.cherd.2021.01.014
5. Patra AP (2009) Maintenance Decision Support Models for Railway Infrastructure using RAMS & LCC Analyses
6. BS EN 50126-2:2017, 50126 BSI Standards Publication Railway Applications–The Specification and Demonstration of Reliability, Availability , Maintainability and Safety (RAMS) National foreword, no. Part 1 (2017). file:///C:/Users/mlincoln/OneDrive/Mark/Msc materials/Project PRCP/Notes and Refs/BS EN 50126-2-2017–[2019-08-28–06-07-25 PM].pdf
7. Márquez FPG, Pérez JMP, Marugán AP, Papaelias M (2016) Identification of critical components of wind turbines using FTA over the time. Renew Energy 87:869–883. https://doi.org/10.1016/j.renene.2015.09.038
8. Dinmohammadi F, Alkali B, Shafiee M, Bérenguer C, Labib A (2016) Risk evaluation of railway rolling stock failures using FMECA technique: a case study of passenger door system. Urban Rail Transit 2(3–4):128–145. https://doi.org/10.1007/s40864-016-0043-z
9. Jagtap HP, Bewoor AK, Kumar R, Ahmadi MH, El Haj Assad M, Sharifpur M (2021) RAM analysis and availability optimization of thermal power plant water circulation system using PSO. Energy Rep 7: 1133–1153. https://doi.org/10.1016/j.egyr.2020.12.025
10. Kumar U, Ahmadi A, Verma AK, Varde P (2016) Current Trends in Reliability, Availability, Maintainability and Safety: An Industri Perspective. Springer International Publishing Switzerland, Heidelberg. https://doi.org/10.1007/978-3-319-23597-4
11. Vromans MJCM (2005) Reliability of railway systems
12. Durivage MA (2017) The Certified Reliability Engineer Handbook. Quality Press, Milwaukee

13. Gerokostopoulos A, Guo H, Pohl E (2015) Determining the right sample size for your test: theory and application. In: Annual RELIABILITY and MAINTAINABILITY Symposium, pp 1–22. https://www.weibull.com/pubs/2015_RAMS_right_sample_size.pdf
14. HBM Prenscia (2021): Life Data Analysis (Weibull Analysis), HOTTINGER BRUEL & KJAER INC. https://www.weibull.com/basics/lifedata.htm
15. Gao K, Liu G (2021) Novel nonlinear time-varying fatigue reliability analysis based on the probability density evolution method. Int J Fatigue 149:106257. https://doi.org/10.1016/J.IJFATIGUE.2021.106257
16. Kwon K, Frangopol DM (2010) Bridge fatigue reliability assessment using probability density functions of equivalent stress range based on field monitoring data. Int J Fatigue 32(8):1221–1232. https://doi.org/10.1016/J.IJFATIGUE.2010.01.002
17. Tian Y, Wang Y, Jin X, Huang Z (2020) Identifying explicit expression of response probability density of nonlinear stochastic system: information-theoretic method. Mech Syst Signal Process 143:106839. https://doi.org/10.1016/J.YMSSP.2020.106839
18. Sarhan A (2019) Bayesian analysis of the discrete two-parameter bathtub hazard distribution. J Math Sci Model 2(3):183–192. https://doi.org/10.33187/jmsm.570880
19. Haselgruber N, Capser SP, Vignati GI (2021) Early life reliability growth testing with non-constant failure intensity. Procedia Comput Sci 180:608–617. https://doi.org/10.1016/j.procs.2021.01.283
20. Wang S, Tomovic M, Liu H (2016) Comprehensive reliability design of aircraft hydraulic system. Commerc Aircr Hydraul Syst 115–169. https://doi.org/10.1016/B978-0-12-419972-9.00003-6
21. Intergovernmental Organisation for International Carriage by Rail, Status (2012): IN FORCE APTU Uniform Rules (Appendix F to COTIF 1999) Uniform Technical Prescriptions (UTP) relating to the Subsystem Rolling Stock FREIGHT WAGONS, vol 06, no December 2010, pp 2011–2013
22. Günay M, Korkmaz ME, Özmen R (2020) An investigation on braking systems used in railway vehicles. Eng Sci Technol Int J 23(2):421–431. https://doi.org/10.1016/J.JESTCH.2020.01.009
23. Kim MC (2011) Reliability block diagram with general gates and its application to system reliability analysis. Ann Nucl Energy 38(11):2456–2461. https://doi.org/10.1016/J.ANUCENE.2011.07.013
24. Jia L, Ren Y, Yang D, Feng Q, Sun B, Qian C (2019) Reliability analysis of dynamic reliability block diagram based on dynamic uncertain causality graph. J Loss Prev Process Ind 62:103947. https://doi.org/10.1016/J.JLP.2019.103947
25. Barone G, Frangopol DM (2014) Life-cycle maintenance of deteriorating structures by multi-objective optimization involving reliability, risk, availability, hazard and cost. Struct Saf 48:40–50. https://doi.org/10.1016/j.strusafe.2014.02.002
26. Agrawal AK, Murthy VMSR, Chattopadhyaya S (2019) Investigations into reliability, maintainability and availability of tunnel boring machine operating in mixed ground condition using Markov chains. Eng Fail Anal 105:477–489. https://doi.org/10.1016/j.engfailanal.2019.07.013
27. Cevasco D, Koukoura S, Kolios AJ (2021) Reliability, availability, maintainability data review for the identification of trends in offshore wind energy applications. Renew Sustain Energy Rev 136:110414. https://doi.org/10.1016/J.RSER.2020.110414
28. Sedghi M, Kauppila O, Bergquist B, Vanhatalo E, Kulahci M (2021) A taxonomy of railway track maintenance planning and scheduling: a review and research trends. Reliab Eng Syst Saf 215:107827. https://doi.org/10.1016/J.RESS.2021.107827
29. Afzali P, Keynia F, Rashidinejad M (2019) A new model for reliability-centered maintenance prioritisation of distribution feeders. Energy 171:701–709. https://doi.org/10.1016/J.ENERGY.2019.01.040
30. Su Z, Jamshidi A, Núñez A, Baldi S, de Schutter B (2019) Integrated condition-based track maintenance planning and crew scheduling of railway networks. Transp Res Part C Emerg Technol 105:359–384. https://doi.org/10.1016/J.TRC.2019.05.045
31. Consilvio A, di Febbraro A, Meo R, Sacco N (2019) Risk-based optimal scheduling of maintenance activities in a railway network. EURO J Transp Logist 8(5):435–465. https://doi.org/10.1007/S13676-018-0117-Z

32. Syamsundar A, Naikan VNA, Wu S (2021) Estimating maintenance effectiveness of a repairable system under time-based preventive maintenance. Comput Ind Eng 156:107278. https://doi.org/10.1016/j.cie.2021.107278
33. Rezvanizaniani SM, Barabady J, Valibeigloo M, Asghari M, Kumar U (2009) Reliability analysis of the rolling stock industry: a case study. Int J Perform Eng 5(2):167–175
34. Özgür-Ünlüakın D, Türkali B (2021) Evaluation of proactive maintenance policies on a stochastically dependent hidden multi-component system using DBNs. Reliab Eng Syst Saf 211:107559. https://doi.org/10.1016/J.RESS.2021.107559
35. Fitouri C, Fnaiech N, Varnier C, Fnaiech F, Zerhouni N (2016) A decison-making approach for job shop scheduling with job depending degradation and predictive maintenance. IFAC-PapersOnLine 49(12):1490–1495. https://doi.org/10.1016/J.IFACOL.2016.07.782
36. Meyer VFA (2019) Challenges and reliability of predictive maintenance. Fac Commun Environ (March): 16. https://doi.org/10.13140/RG.2.2.35379.89129.
37. Mobley RK (2002) An introduction to predictive maintenance, Second. New York
38. Gbadamosi AQ et al (2021) IoT for predictive assets monitoring and maintenance: an implementation strategy for the UK rail industry. Autom Constr 122:103486. https://doi.org/10.1016/J.AUTCON.2020.103486
39. Coleman C, Damofaran S, Deuel E (2017) Predictive maintenance and the smart factory, Deloitte, p 8
40. Yang DY, Wu CH (2021) Evaluation of the availability and reliability of a standby repairable system incorporating imperfect switchovers and working breakdowns. Reliab Eng Syst Saf 207:107366. https://doi.org/10.1016/J.RESS.2020.107366
41. Zhang F, Shen J, Ma Y (2020) Optimal maintenance policy considering imperfect repairs and non-constant probabilities of inspection errors. Reliab Eng Syst Saf 193:106615. https://doi.org/10.1016/J.RESS.2019.106615
42. R. The M. R. IZZAT (2018): Prasarana aims more local parts in rail biz, The Malaysian Reserve. https://themalaysianreserve.com/2018/04/11/prasarana-aims-more-local-parts-in-rail-biz/
43. Pamela L (2011) The high maintenance cost of railway track equipment imposes new solutions. https://www.railwaypro.com/wp/the-high-maintenance-cost-of-railway-track-equipment-imposes-new-solutions/
44. Durazo-Cardenas I et al (2018) An autonomous system for maintenance scheduling data-rich complex infrastructure: fusing the railways' condition, planning and cost. Transp Res Part C Emerg Technol 89:234–253. https://doi.org/10.1016/J.TRC.2018.02.010
45. BS EN 50129 (2019): 2018 BSI Standards Publication Railway applications–Communication, signalling and processing systems–Safety related electronic systems for signalling, no April
46. Wang Y, Liu SYM, Cho L, Lee K, Tam H (2021) A method of railway system safety analysis based on cusp catastrophe model. Accid Anal Prev 151:105935. https://doi.org/10.1016/J.AAP.2020.105935
47. Liu C, Yang S, Cui Y, Yang Y (2020) An improved risk assessment method based on a comprehensive weighting algorithm in railway signaling safety analysis. Saf Sci 128:104768. https://doi.org/10.1016/J.SSCI.2020.104768
48. Ehlers UC, Ryeng EO, McCormack E, Khan F, Ehlers S (2017) Assessing the safety effects of cooperative intelligent transport systems: a bowtie analysis approach. Accid Anal Prev 99:125–141. https://doi.org/10.1016/j.aap.2016.11.014
49. Peng Z, Lu Y, Miller A, Johnson C, Zhao T (2016) Risk assessment of railway transportation systems using timed fault trees. Qual Reliab Eng Int 32(1):181–194. https://doi.org/10.1002/qre.1738
50. Pirbhulal S, Gkioulos V, Katsikas S (2021) A systematic literature review on RAMS analysis for critical infrastructures protection. Int J Crit Infrastruct Prot 33: 100427. https://doi.org/10.1016/j.ijcip.2021.100427.
51. Belany P, Hrabovsky P, Kolkova Z (2021) Combination of lighting retrofit and life cycle cost analysis for energy efficiency improvement in buildings. Energy Rep 7:2470–2483. https://doi.org/10.1016/j.egyr.2021.04.044
52. Acharya R (2000) Indian railways: where the commuter is king. Jpn Railw Transp Rev 25:34–45

53. Raoul R (2018) RAMS: Retter Results in Less Time. https://www.stork.com/en/about-us/blog/rams-better-results-in-less-time
54. Janušová L, Čičmancová S (2016) Improving safety of transportation by using intelligent transport systems. Procedia Eng 134:14–22. https://doi.org/10.1016/j.proeng.2016.01.031
55. Ripke B, Nawabi W, Lamper A, Juntti U, Paragreen J (2006) Guideline for LCC and RAMS analysis. Deliverable D6.5.4 from Project No. TIP5-CT-2006-031415, p 115
56. Satora M, Szkoda M (2018) Method for effectiveness assessment of rolling stock investments using LCC (Life Cycle Cost) analysis. Transp Econ Logist 80(December):217–224. https://doi.org/10.26881/etil.2018.80.22
57. Lidén T, Joborn M (2016) Dimensioning windows for railway infrastructure maintenance: cost efficiency versus traffic impact. J Rail Transp Plan Manag 6(1):32–47. https://doi.org/10.1016/J.JRTPM.2016.03.002
58. Senaratne S, Mirza O, Dekruif T, Camille C (2020) Life cycle cost analysis of alternative railway track support material: a case study of the Sydney harbour bridge. J Clean Prod 276:124258. https://doi.org/10.1016/J.JCLEPRO.2020.124258
59. Banar M, Özdemir A (2015) An evaluation of railway passenger transport in Turkey using life cycle assessment and life cycle cost methods. Transp Res Part D Transp Environ 41:88–105. https://doi.org/10.1016/J.TRD.2015.09.017

Brief Review on Recent Advancement of Computational Analysis on Hemodynamics in Peripheral Artery Disease

U. Z. Shahrulakmar, M. N. Omar, and N. H. Johari

Abstract Peripheral Artery Disease (PAD) occurs due to presence of atherosclerotic plaque in blood vessels which reduce the blood flow to extremities at lower part of the body. Failure in treating this impairment will lead to serious vascular disorders such as ulceration and gangrene of feet. Computational Fluid Dynamics (CFD) and Fluid-Structure Interaction (FSI) can be considered as non-invasive atherosclerosis diagnosis method to evaluate the blood flow parameters and wall structural interaction. Previous studies illustrated that uncertainty of stenosis area majorly relies on the assumption of the model geometry to evaluate the pathology of peripheral artery. However, the interaction between mechanical and flow condition that occurs in the peripheral artery using independent computational model of cardiovascular system is still poorly resolved. This study aimed to systematically review the recent progress on the implementation of CFD and FSI in PAD from 2017 to 2021. The findings were classified into geometry, viscosity models, analysis method, wall characteristic and validation, accordingly. This study may provide a systematic information to future research in producing a better computational analysis.

Keywords Peripheral Arterial Disease · Computational Fluid Dynamics · Fluid Structure Interaction

1 Introduction

Peripheral Arterial Disease (PAD) is a form of cardiovascular disease (CVD) that affects the blood vessels in the lower part of the body. Until the year 2020, it has

Supplementary Information The online version contains supplementary material available at https://doi.org/10.1007/978-981-19-1577-2_41.

U. Z. Shahrulakmar · M. N. Omar · N. H. Johari (✉)
Faculty of Mechanical and Automotive Engineering Technology, Universiti Malaysia Pahang, 26600 Pekan, Pahang, Malaysia
e-mail: nhadi@ump.edu.my

Centre for Human Technology, Universiti Malaysia Pahang, 26600 Pekan, Pahang, Malaysia

© The Author(s), under exclusive license to Springer Nature Singapore Pte Ltd. 2023

M. H. A. Hassan et al. (eds.), *Technological Advancement in Instrumentation & Human Engineering*, Lecture Notes in Electrical Engineering 882,
https://doi.org/10.1007/978-981-19-1577-2_41

already affected almost 200 million people globally [1, 2]. The PAD prevalence cases recorded an increase globally in all ages in the past 15 years [3]. This shows that the seriousness of PAD should be highlighted and given critical attention. The recorded CVD deaths were 85% due to heart and stroke complications. PAD can increase the risk of heart attack, transient ischemic attack and stroke. Among the risk factors for PAD are diabetes, smoking, hypertension, hyperlipaemia, heart disease and stroke [4].

Most of PAD patients are asymptomatic, but many experiences intermittent claudication such as pain or cramping, leg discomforts and numbness when walking or climbing stairs. If left untreated, PAD could lead to impairment of leg ulceration, gangrene and eventually amputation that will affect the patient's lifestyle. The impairment occurs due to occlusion of blood supply, in which atherosclerosis is the main aetiology (Fig. 1). Atherosclerosis is the formation of substance called plaque in the artery wall due to deposition of fats, cholesterol and other substances. Specifically it is prone to occur at the profunda femoris artery which is the branching of femoral artery, part of external iliac arteries [5].

Low wall shear stress (WSS) at certain regions in the arteries promotes the accumulation of the substances by regulating less atheroprotective genes and elevate the atherogenic ones that will lead to plaque formation [1]. The plaque includes dead foam cells, macrophages, smooth muscle cells and extracellular matrix [6]. The plaque will cause the affected regions to narrow its inner diameter and reduce the arterial compliance (increased arterial stiffness) [7]. The narrowing occurrence also disturbs the local normal blood flow—in the cases of high grade stenosis, the local blood flow could transit from laminar to turbulent in the downstream region [8]. Sudden high WSS on the endothelium layer could induce the plaque to rupture and

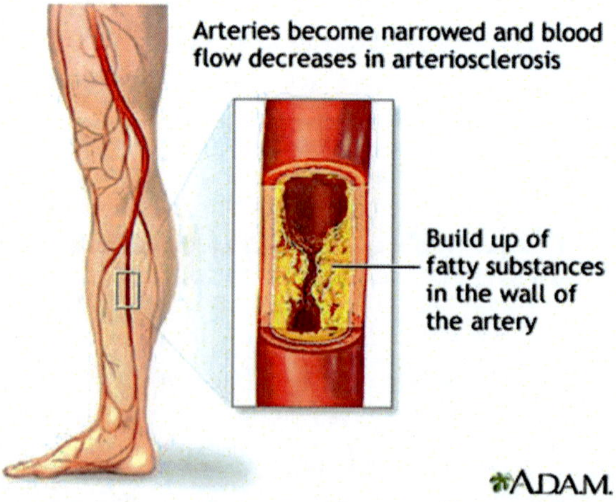

Fig. 1 Atherosclerosis in peripheral arterial disease (PAD). (Courtesy of U.S. National Library of Medicine (NLM))

Table 1 Endovascular therapy and non-invasive physiological test for peripheral arterial disease [5, 17]

Invasive treatment	Non-invasive vascular test
1. Percutaneous Transluminal Angioplasty 2. Vein Graft Bypass 3. Drug Eluting Balloon 4. Drug Eluting Stent 5. Bare Metal Stent	1. Echocardiography 2. Computed Tomography 3. Magnetic Resonance Imaging 4. Computational Analysis 5. ABI/TBI 6. Duplex Ultrasound

lead to thrombosis (formation of blood clot) that may cause shortage of blood supply to the lower parts of peripheral arteries.

Apart from prescribing common medications like anti-platelet agents such as aspirin or clopidogrel and statins [9], severe PAD patients with advanced plaque sizes may be advised to go for endovascular treatment. Surgeons may refer the patients for either invasive treatment like angioplasty and stenting or open surgical atherectomy intervention [10]. There are two types of endovascular treatment i.e. invasive and non-invasive techniques (Table 1). Stent implantation is usually performed to widen the affected arteries from narrowing again. The stenting and angioplasty techniques are recommended for short and long/medium lesion, respectively [2]. The implementation of drug-coated balloons and bioresorbable stents is one of the advancement techniques in Percutaneous Transluminal Angioplasty (PTA) to completely resolve the constriction by "leaving nothing behind" strategy [11]. Artery stiffness may get affected after the stent implantation with the decrement on its axial elongation but this does not pose effect on the average arterial curvature [12]. Correlatedly, these implantations also restrict the axial shortening of the artery and experienced low flow velocity during leg flexion [13]. Even though the angioplasty or the usage of statin lower the risk of the stent implantation on the patient, the long-term primary femoropopliteal patency after endovascular treatment remains low with rates around 25% [14]. The development of restenosis may strongly happen after 6–12 months follow-up on the treated area [15, 16].

Computer analysis (CA) is another strategy for early diagnosis to evaluate the blood flow disturbance in arteries [17]. In early 1970s, research on hemodynamic factors that influence the development of artery wall has been studied by using CA as problem solving method for CVDs issues [18, 19]. Simulation on blood flow and interaction of arterial wall can be computed through CA by using Computational Fluid Dynamics (CFD). It consists of prominent steps such as Problem Identification, Pre-processing, Determination of material properties, Boundary conditions and Turbulence models, Optimization of Governing Equations and Blood Flow Mechanics, and lastly followed by Post-processing.

CFD is a significant tool in science and biomedical field to illustrate the flow behaviour in vitro experiments that relate with vascular hemodynamic. It has been carried out by many researchers throughout the past two decades. The CFD simulations are widely conducted by using a set of vascular remodelling mechanism leading to atherosclerosis development and possible rupture on the diseased location in line with morphologic factors and hemodynamic forces. Besides, CFD also has been a prime tool in examining different endovascular treatment modalities and validating evolution of endovascular devices [2, 14, 16, 20]. Advancement on understanding of vascular pathologies such as atherosclerosis [1, 21] and the flow perturbations at bifurcations found to have large gap of knowledge existence. In the field of PADs research, the element of pathobiological processes is limitedly discovered due to the complexity of the blood flow [5, 21] and the wall structure deformation in correlation with the artery behaviour [11–13, 22–25].

However this assessment alone could not illustrate the real rigid wall characterization where most of recent CFD studies used the idealized geometry, CTA/MRA reconstruction method to analyze the arterial wall compliance on the local blood hemodynamic which possess direct impact on blood behavior [11–13]. These limitations can be overcome by using Fluid-Structure Interaction (FSI) method, which is the combination of CFD with Finite Element Analysis (FEA). Numerical method between fluid domain and structure domain allow the investigation of the fluid behavior, structural behavior and how they interact with each other [18, 19, 26]. It could also reduce the degree of uncertainty on development of stenosis regions. Moreover, integrating all the factors with real patient-specific datasets will further help the clinician in delivering suitable treatment plan for PAD patients.

Most of the papers report that their working models have yet to achieve the expected results in term of its precision [13, 21, 22]. Limitations in selecting factors such as types of geometry, viscosity model whether it's Newtonian or Non-Newtonian, type of flow, boundary conditions at the inlet and outlet of the artery, and the solver used to solve its governing equations still need to be taken into consideration in their works from time to time. Thus, unveiling and improving those assessments is vital and need to be approached before proceeding the investigation in order to understand the behavior of the PADs. In the next section, the description on selecting parameters will be explained further.

This paper presents a systematic review of important recent research articles from 2017 to 2021 in the investigation of blood hemodynamic in PAD using computer simulation analysis. The goal of this study is to illustrate the methodologies used in the computational modelling and analysis of this phenomena. The systematic review may provide a better strategy to future research in producing higher impact computational analysis.

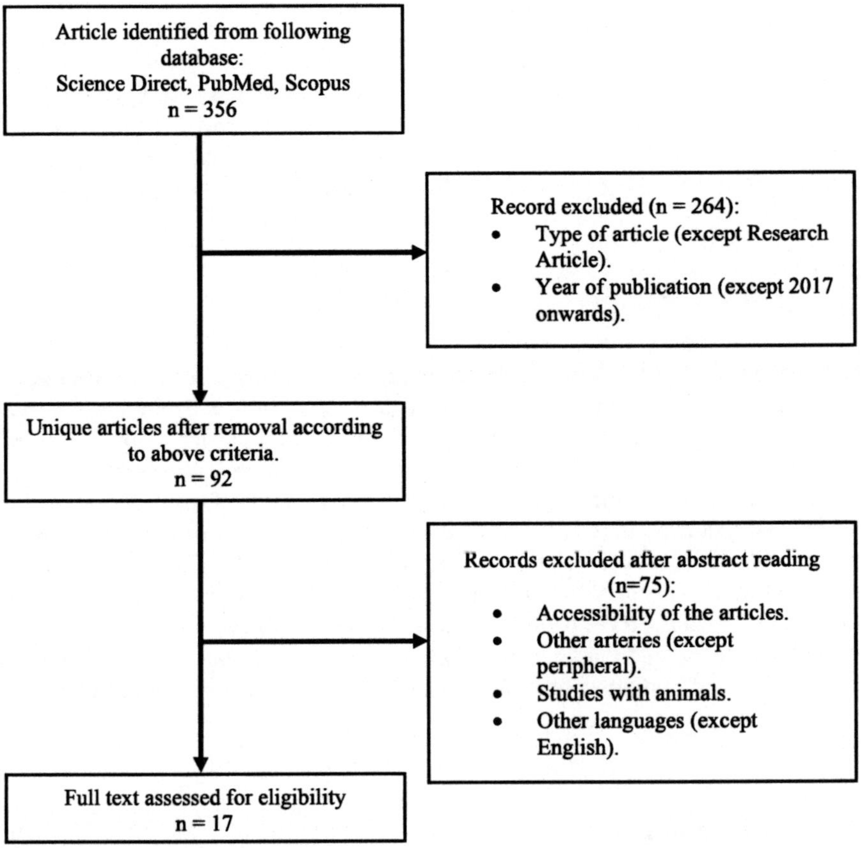

Fig. 2 Flowchart of the systematic review

2 Methods

This systematic review was performed according to an agreed predefined protocol. The review was conducted and presented according to the statement standards of the Preferred Reporting Items for Systematic Reviews and Meta-Analyses (PRISMA). PRISMA was initiated to assist authors to provide a wide range of systematic reviews on advantages and disadvantages of health care intervention [Liberati et al. 2009; Moher et al. 2009].

2.1 Search Strategy

The search of the literature was carried out until 30th of June 2021, with the use of the keywords "Peripheral Arterial Disease", "Atherosclerosis", "Computational Fluid Dynamics", "Fluid Structure Interaction" with the Boolean operators AND, OR and NOT. The search databases include Scopus, PubMed and Science Direct.

2.2 Study Selection

In line with PRISMA guidelines, search references were obtained and organized to their significance of study. Firstly, the classification of articles from database was done by screening according to type of article and limiting the year of publication from 2017 onwards. Then, the remaining papers were reviewed for their quality according to their relevancy of topics through manual reading of abstracts and conclusions. The exclusion criteria were limited to (1) accessibility of the articles in the database, (2) types of arteries that do not exist from hip to toe (lower limb extremities), (3) study on animals and (4) full texts in other languages besides English. At the end of screening, there were 17 articles eligible to fully assess for its novelty of the content. Some of the articles from similar author also found to have insufficient information and findings where meta-analysis approach was conducted to resolve these issues.

3 Results and Discussion

Elimination of duplicates and exclusion through the abstract reading, a total of 17 papers are eligible for this work out of 92 research papers. This prominent analysis was performed to classify the articles primarily into two categories, CFD and FSI approaches. CFD simulations only priorities on the behavior of the fluid domain (blood) while FSI study emphasizes on behavior of both fluid and structure domain (blood and artery) during numerical analysis of blood flow.

3.1 Study Characteristics

Following the screening method for this work through systematic review according to PRISMA guidelines as shown in Fig. 2, possible significant article papers were identified. Of the included articles; 15/17 (88%) corresponded to only CFD works, while remaining 2/17 (12%) demonstrated FSI works. Thus, the review findings were summarized by several classifications in Table 2. Classification was done to highlight

Table 2 Summary of CFD and FSI studies in peripheral arteries disease

Article	Geometry	Image source	Analysis method	Viscosity model	Wall characteristic	Validation method
CFD studies						
Pengcheng Xu et al. [23]	Patient-specific (stenosis, bending, branching)	MRA (MIMICS)	CFD (Scientific report of comparison between CFD data and UDV result)	Newtonian	Rigid	Ultrasound Doppler
Huseyin et al. [21]	Patient-specific (stenosed)	CTA (MIMICS)	CFD (Analysis of skin response to identify the location of stenosis)	–	Single layer structured (FEA)	–
	Idealised	–			Rigid	–
Anna Corti et al. [1]	Idealised (healthy SFA)	–	CFD-FLUENT	Non-Newtonian (Carreau)	Rigid (based on ABM Framework)	–
Monika Colombo et al. [2]	Patient-specific (stented)	CTA (CAD)	CFD-FLUENT	Non-Newtonian (Carreau)	Rigid	Rigid 3D printed phantoms
Aneesha Gogineni et al. [5]	Idealised (bifurcation)	–	CFD-FLUENT	Newtonian	Rigid	–
Can Gokgol et al. [13]	Patient-specific (stented, kinking)	CTA (MIMICS)	CFD-CFX	Newtonian	Rigid	–
Anastasia Desyatova et al. [24]	Patient-specific (aging, FPA)	CTA	CFD-CFX	Newtonian	FEA (Abaqus)	–
Francesca Donadoni et al. [27]	Patient-specific (stenosed)	USA	CFD-CFX	Non-Newtonian (Carreau-Yasuda)	Compliant	Ultrasound Doppler
Kaiyun Gu et al. [28]	Idealised (peripheral ECMO)	–	CFD based on Finite Element Method	Newtonian	Rigid	–
John Gounley et al. [29]	Patient-specific (stenosed)	CTA, MRA	HARVEY	Newtonian	Rigid	Experiment validation

(continued)

Table 2 (continued)

Article	Geometry	Image source	Analysis method	Viscosity model	Wall characteristic	Validation method
A Ferrarini et al. [30]	Patient Specific (FPA)	CTA	CFD-FLUENT	Newtonian, non-Newtonian (Carreau)	Rigid	–
Francesca Donadoni et al. [31]	Patient-specific (vein graft failure)	CTA (Scan IP)	CFD-CFX	Newtonian	Rigid	Clinical data
N.B Wood et al. [32]	Patient-specific (curvature, tortuosity)	MRA (MIMICS) USA (HDI Lab, MATLAB)	CFD-CFX	Newtonian	Rigid	–
Xuanyu Li et al. [33]	Patient-specific (tortuosity)	CTA (MIMICS)	CFD-CFX	Newtonian	Rigid	–
Patrick M. McGah et al. [34]	Patient-specific (curvature)	USA (MATLAB)	CFD-FLUENT	–	Rigid	–
FSI studies						
Danyang Wang et al. [35]	Patient-specific (stenosed)	CTA (MIMICS)	CFD-FLUENT, One-Way	Non-Newtonian (Carreau)	Nonlinear, homogenous, and hyperelastic material	–
Monica Colombo et al. [36]	Idealised (FPA)	–	LS-DYNA (ALE)	Newtonian	Rigid	–

CAD—Computer Aided Design; MIMICS—Materialise Interactive Medical Image Control System; Harvey®—The Cardiopulmonary Patient Simulator; FEA—Finite Element Analysis; FPA—Femoralpopliteal Artery; ECMO—Extracorporeal Membrane Oxygenation; ALE—Arbitrary Lagrangian Eulerian the important parameters explored in pre-processing and post- processing simulation method for PAD studies in respective papers. For pre-processing parameters, the types of geometry models approached were described first, followed by source of the image, the analysis method utilized and lastly explanation on the types of viscosity models and wall characteristics that were taken into account during simulation. Consequently, the post-processing parameter was indicated in the 'Validation method' category. This category represents the method used in the study to analyze the validity and the quality of obtained simulation results. Validation methods may vary by different approaches in accordance to author's preference and consideration.

3.2 Geometry Construction

The selection of PAD geometrical configuration is crucial in the initial stage of simulation to obtain qualitative result in peripheral blood flow analysis. There are two types of geometries commonly used: idealised and patient-specific model. Out of 17 articles (CFD and FSI), patient-specific (12/17) had been widely used compared to idealised model (4/17). In addition, comparison study (1/17) had also been conducted with both idealised and patient-specific model. Pengcheng Xu et al. [23] discussed that patient specific model represents more detailed structure for stenosis, branching, bifurcation and indirectly impacts the Wall Shear Stress (WSS) distribution. This structure also contributes advancement for the hemodynamic study which is influenced by changes on a certain lesion of the diseased geometry. Nevertheless, another scenario illustrated by Monica Colombo et al. [36] investigated the impact of leg movement by mimicking the movement of hip rotation, knee flexion and complete movement of walking on FPA hemodynamic with idealised model. The author believes inclusion of patient-specific model will be challenging in this study because of its complexity. Besides, CTA and MRA also neglect the leg bending during imaging process as per the clinical standard imaging protocol. Hence, involvement of idealised model by taking into account relevant boundary conditions in accordance to patient specific data also distinctly indicate key factors which influence the phenomenon. Monica Colombo et al. [36] and A. Ferrarini et al. [30] strongly support that patient specific boundary condition is much reliable in leg flexion investigation and directly assure absence of unrealistic results. Apart from that, Huseyin et al. [21] investigated the identification of stenosis location by sound vibration response toward the skin along the inner arterial wall of peripheral arteries. The hyper-elastic of skin, fat, artery and muscle define the relation between stress and strain by using hyper-elastic Ogden and Mooney-Rivlin approach. Meanwhile for the viscoelastic soft body tissue behavior is modelled using Maxwell approach. The outcome of the study shows sum of amplitude attains the maximum value near the stenosis location.

From the obtained research articles, Computed Tomography Angiography (CTA) image source is found to be commonly employed for the patient-specific models followed by Magnetic Resonance Angiography (MRA), whereas Ultrasound Angiography (USA) is least likely utilised among the reviewed studies. However, several

Table 3 The comparison of stenosis diagnosis method [17, 39]

Method	Hemodynamic data	Anatomic measurements	Time consumption	Resolution	Risk
Computed Tomography Angiography (CTA)	No	Yes	Low	High	Radiation exposure
Magnetic resonance Angiography (MRA)	Yes	Yes	High	Low	–
Ultrasound Angiography (USA)	No	Yes	High	High (have depth limit)	Imaging modality (operator and patient dependent)

studies done by John Gounley et al. [29], N.B Wood et al. [32] have used more than one image source. Table 3 describes the common type of imaging procedure in clinical diagnosis. Generally, CTA is a method to generate high contrast and resolution internal body X-ray images along with injection of contrast agent [19]. Pengcheng Xu et al. [23] stated the presence of stenosis may confound by the silhouette of curvature in CTA images. Unlike CTA technique, MRA scanning produces detailed images with strong magnetic field and blood acts as its own contrast agent. MRA approach is also beneficial by creating 4D flow pattern in real time PC-MRA [19]. However, it is a comparatively more sophisticated technique and has a higher time requirement for acquisition and interpretation of the images. Apart from CTA and MRA, USA is categorised as another commonly used source image in non-invasive technique. USA images are generated by sound wave frequency to scan part of the body. Comparison between MRA and USA data was conducted by N.B Wood et al. [32] and his paper stated that USA is excellent in terms of measurement accuracy of measured circular diameter of SFA than by MRA technique. HDI Lab 1.9 (ATL-Phillips) and custom-written Matlab 7.0.4 software were used to conduct the analysis (The Mathworks, Natick, MA). The results of MRA in the USA were compared in terms of diameter measurement accuracy using MRA images with an L12–5 linear array transducer and an ATL HDI 5000 ultrasound machine (ATL Phillips, Bothell, WA). Meanwhile, D. Lopes et al. [37] also reported that USA known to be low-cost technique but the accuracy of the method is still highly influenced by the skill and expertise of the operator. Furthermore, imaging of unusual anatomical circumstances, such as extensively stenosed carotid vessels, is limited in the USA Adla & Adlova [38], resulting in the lowest prevalence rate.

In general, geometry construction can be categorized into two categories; idealised model and patient-specific model. Realistic geometry for patient specific model obtained by MRA and CTA techniques. However, these geometries are more crucial to be reconstructed as required model for the computational analysis. Through this approach, artery behaviour and mechanical structure will not be neglected.

3.3 Viscosity Models

Blood is a complex fluid of erythrocytes, leukocytes, and thrombocytes suspended in plasma as mentioned by Ameenuddin et al. [40]. Haematocrit is the volume fraction in which erythrocytes make up roughly 45 percent of blood volume. Although plasma is known to be a Newtonian fluid, the entire fluid possesses non-Newtonian properties, especially in narrow passages like capillaries. The shear-thinning behavior of blood was first described by Charm and Kurland et al. [41], who demonstrated that viscosity decreases as shear and strain rates increase. In present work, 10 out of 17 research articles assumed blood flow as Newtonian fluid, whereas 5 papers represent non-Newtonian viscosity model. Likewise, one paper indicates the comparison study with both Newtonian and non-Newtonian viscosity models while remaining papers do not highlight any assumption on type of blood flow implied. Carreau (4/5studies) and Carreau-Yasuda (1/5 studies) are two non-Newtonian viscosity models that are featured. Table 4 shows the equations that describe these models as well as the values that are typically employed to imitate blood rheology.

Majority of arterial simulation investigations simulate blood flow as a Newtonian model. This Newtonian blood assumption is a major flaw in blood flow modelling approximations [42]. Even though plasma is a Newtonian fluid entirely, however, small channels like capillaries exhibit non-Newtonian characteristic due to viscosity changes of the blood flow in the arterial surrounding [43]. Several studies Guerciotti and Vergara et al. [43], Kumar et al. [44], Lee et al. [45], Urevc et al. [46] have found considerable discrepancies between Newtonian and non-Newtonian outcomes, particularly in local hemodynamics due to lower velocities at stenosed locations. This

Table 4 Viscosity models frequently used to mimic the blood rheology [37]

Model	Equation	Parameters
Newtonian	$\mu = \mu_\infty$	$\mu_\infty = 0.0035$ pa.s
Power Law	$\mu = k(y')^{n-1}$	$k = 0.017$ $n = 0.708$
Carreau–Yasuda	$\mu = \mu_\infty + (\mu - \mu_\infty)[(1 + \lambda y')^2]^{\frac{n-1}{q}}$	$\mu_0 = 0.056$ pa.s $\lambda = 1.902$ s $\mu_\infty = 0.0345$ pa.s $n = 0.22q$ $q = 1.25$
Casson	$\mu = \left\{\sqrt{\frac{\tau_0}{\lvert y' \rvert}}[1 - e^{(-my')}] + \sqrt{\mu_0}\right\}^2$	$\mu_0 = 0.004$ pa.s $\tau_0 = 0.004$ pa m = 100

contribute to lower shear rates at the downstream region of the stenosis area subsequently increase in its viscosity. These findings suggest that in peripheral blood flow simulations, it may be appropriate to simulate blood as a non-Newtonian fluid. Further, there are few kind of non-Newtonian models being used to predict shear thinning of blood [47]. Firstly, Power law model creates attentive interest as generalized non-Newtonian flow behavior. The power-law method is the most widely adopted rheology to construct empirical relationship because of its simplicity. The unbounded power-law model, on the other hand, has two drawbacks: first, it fails to reflect the linear shear-strain relationship at very low and very high shear rates, which are common in most real systems, and second, the viscosity curve becomes singular in the vanishing shear limit. Besides, the Casson model, known to be another mathematical modelling of blood flow specifically evaluate in narrow arteries at low shear rates only. Subsequently, Carreau or Carreau-Yasuda model a four-parameter model that is applicable to a wide range of shear rates, proposed when there is a substantial variance from the other models. In addition, this model was constructed to act as a Newtonian fluid at high and low shear rates, and as a Power law fluid in the intermediate shear phase [37].

As a summary, consideration of viscosity types, Newtonian or non- Newtonian model is essential in blood flow analysis. Naturally, plasma exists as non-Newtonian fluid. Studies have investigated the effect of both condition in blood and the outcome shows significant difference. Basically, Newtonian model is a simplest and practical way with consistent viscosity consideration, however it does not perfectly fit by blood in nature. On the other side, non-Newtonian consideration is more prominent due to inclusion of viscosity changes according to blood flow with the arterial surrounding. Nonetheless, the viscosity model consideration hugely depends on the case that need to be deal by the researcher.

3.4 Software Used

There are 17 studies in Table 2 that employed CFD method in PAD investigations. CFD includes a number of applications, including commercial softwares like ANSYS CFD, Comsol-Multiphysics, Autodesk CFD, STAR-CCM, SolidWorks Flow Simulation, and Open Source codes like Visual-CFD, simFlow, OpenFoam, SimScale, HARVEY, etc. The most often used CFD commercial packages for modelling blood flow are ANSYS Inc.'s Fluent and CFX. Thirteen (13) articles in Table 2 employed ANSYS-CFX and ANSYS Fluent approach, and one (1) used HARVEY [29] hemodynamic program that includes the Lattice Boltzmann Method (LBM) to resolve the unsteady Navier–Stokes equations. This is attributable to the scaling's effectiveness and easiness to which boundary conditions in complex structures are addressed. However, the simulation still considers the vessel wall as a rigid wall. The simulation was run on Vulcan, an IBM Blue Gene/Q supercomputer with 24, 576 nodes.

On the other hand, Fluent and CFX are two commercialised solver in ANSYS which widely being used for its own advantages. Firstly, Fluent is highly capable of running main elements such as pre-processor, solver and post-processor under one single window. In terms of meshing, the grid quality in CFX is much more permissive where it focuses on cell-vertex node that divides each element into sub volumes. Besides, convergence is a key concern in CFD since iterative methods are used to solve the governing equations. Different studies employ different criteria of convergence, but it's usual to look at residuals that should drop below a certain predefined threshold, along with specific local and global flow characteristics that should stabilize [19]. By examining the data at various monitor points, both Fluent and CFX are able to assess the convergence's validity. In such cases, CFX excelled Fluent in terms of visibility and ease of use. In most situations, this is advantageous; nonetheless, each iteration takes longer, making it harder to identify whether the solution is converging. Meanwhile, Fluent needs more number of iterations, but this one is computationally efficient, and you can discern whether the solution is converging or diverging during the early phases of a simulation. In addition, CFX can only run 3D simulation while Fluent can run both 2D and 3D simulations where Fluent is able to resolve simulations for axisymmetric model [48].

Overall, Fluent and CFX are competent solver that can solve most flow analyses according to the need of the user. For inexperienced CFD users, there is a thorough description on how CFD software would aid users to tackle the Navier–Stokes equations rather than depending on commercialised CFD codes [49]. The robust CFD simulation has the capability to be a useful tool in treatment decision-making systems [23].

3.5 Rigid Wall Assumptions Toward Their Works

Numerous studies Pengcheng Xu et al. [23], Can Gokgol et al. [13], Monica Colombo et al. [36], Francesca Donadoni et al. [27], Anastasia Desyatova et al. [24], Xuanyu Li et al. [33] have stated that the implementation of an FSI method is essential for future requirements of their work. According to Pengcheng Xu et al. [23], a change in WSS distribution can readily lead to rupture and the formation of calcified plaque. This indicates that hardened plaque is the primary cause of atherosclerosis. The specific interaction between geometric features and WSS change is still unknown. Due to rigid wall presumption used in the framework, the displacement of the artery wall caused by blood is neglected.

Simulating a rigid wall also restricts study of the model's structural element. As a result, choosing particular mechanisms and stimuli for the infected region may reduce the accuracy of the outcome as the disease progresses. For the purpose of simplicity and computational efficiency, Anastasia Desyatova et al. [24] conducted the analysis by separating the solid and fluid studies. The effects of fluid–structure interaction are minor in terms of qualitative results, but quantitative implications might be considerable. Another study carried out to determine the distribution and severity of hemodynamic disturbances by a combination of fluid–structure interaction and numerous different geometries of the SFA (including not only the magnitude and distribution of vascular). Hence, through the use of fluid structure interaction to quantify the movement of the artery wall when subjected to hemodynamic stresses, might hugely impact the findings of the study.

FSI approach is still futuristic and most researchers are not paying keen interest on these approaches toward the lower limb study. Limited study has been explored by using other arteries and they are majorly focused on carotid and coronary arteries. Only 1/18 FSI studies by Danyang Wang et al. [35], was done in stenosed femoral artery for the prediction of the hemodynamic performance with multiple plaques (calcified and lipid plaques) using Mooney-Rivlin material property.

Despite geometries from patient specific models, the presence of limitation is still undeniable due to the complex structure of arterial wall which consists of smooth muscle cells, elastin and collagen fibrils [35]. Moreover, FSI still needs to comply with several boundary conditions which restraint the accuracy of output similar to realistic arterial model [30, 33, 36].

3.6 Validation and Verification

The verification and validation of CFD hemodynamic research is a major concern. The term "verification" refers to a review of the computational model's numerical precision. The presence of an analytical solution for the model is typically required for this process. On the contrary, validation is the act of determining and measuring the degree to which a computer model accurately represents the real world. When creating novel hemodynamic models, it is evident that verification must proceed prior validation. If the discrepancies between the simulation's findings and actual data are minimal, and the uncertainties in the experimental data are similarly modest, the simulation is considered verified. Roache et al. [50] and Babuska & Oden [51] provided a distinct line between verification and validation, describing verification as "solving the equations correctly," and validation as "solving the proper equations."

Similarly, it is explained that "There can be no validation without experimental data," based on the current ASME Standard for Verification and Validation in Computational Fluid Dynamics and Heat Transfer, and indeed the data can be generated through in vivo or in vitro studies. As per Prantil et al. [52] in their textbook "Lying by Approximation: The truth about Finite Element Analysis," modelling validation incorporating theoretical or experimental data is required to obtain accurate simulation results.

Some researchers Schumann et al. [12], Can Gokgol et al. [11], Anastasia Desyatova et al. [24], Patrick M. Mc Gah et al. [34] suggested follow up on clinical datasets or experimental data in the scenario of leg flexion induced by stent implantation to evaluate the conformity of their findings. Complex hemodynamic characteristics, as well as their significance in influencing long-term morphologic changes and, more importantly, clinical consequences, may be shown using computational simulations combined with longitudinal patient scans. Additionally, applying existing in vivo data to validate numerical models is a common approach among researchers. The benefits of published work are that it is easy to get and affordable, require less time to conduct particular trials, does not require ethical approval, and can be compared against a variety of databases [17]. The simulation result could be compared to the actual in vivo data either indirectly or directly. The indirect comparison technique is either (a) the CFD user alters the initial boundary conditions and/or geometry to correspond with published clinical data, or (b) the CFD user compares the published research to the initial simulation results [53].

On contrary, the verification procedure can be divided into two methods. The first method examines the sensitivity of the output to the elements on meshed domain's variable quality level. This mesh sensitivity test examines the feasibility of the number of mesh elements capturing the flow pattern by focusing on the magnitudes of residual error. While the second technique focuses on and evaluates governing equation solver problems. Alternately, direct comparative approach (c) has a good level of validity where it creates a 3D structure and pathway for the CFD simulation based on the patient's specific data and boundary conditions [54]. Then, the response is compared to the patient's initial data. The shortcoming of this strategy is that it requires time to gather patient imaging data from CTA or MRA, extract and reconstruct data for modelling, secure ethics clearance, and patient enrolment and consent [17].

4 Conclusion

Peripheral Arterial Disease (PAD) is a type of cardiovascular disease (CVD) that impacts the lower body by restricting the lumen patency of blood arteries. Recent technology allows the use of computational software to imitate the conditions at the affected area. However, it is still a challenging task to obtain an accurate representation of the conditions due to the behavior of blood and the characteristics of arteries wall. This paper summarizes the parameters of geometrical construction, viscosity models, analysis methods, and wall characteristics taken into consideration by the researchers to identify and simulate the blood flood flow in the stenosis area. This review also presents the methods used to validate the identification and simulation.

Overall, this study illustrates the current efforts in the computational analysis of PAD. The findings can help researchers to develop a better computational analysis that can assists for early diagnosis of PAD diseases, and appropriate therapeutic approach.

Acknowledgements The authors would like to be obliged to Universiti Malaysia Pahang for providing laboratory facilities and financial assistance under the grant no. RDU200746 and PGRS210346.

References

1. Corti A, Chiastra C, Colombo M, Garbey M, Migliavacca F, Casarin S (2020) A fully coupled computational fluid dynamics – agent-based model of atherosclerotic plaque development: multiscale modeling framework and parameter sensitivity analysis. Comput Biol Med 118:103623. https://doi.org/10.1016/j.compbiomed.2020.103623
2. Colombo M et al (2020) Computing patient-specific hemodynamics in stented femoral artery models obtained from computed tomography using a validated 3D reconstruction method. Med Eng Phys 75:23–35. https://doi.org/10.1016/j.medengphy.2019.10.005
3. Fowkes FGR, Aboyans V, Fowkes FJI, McDermott MM, Sampson UKA, Criqui MH (2017) Peripheral artery disease: epidemiology and global perspectives. Nat Rev Cardiol 14(3):156–170. https://doi.org/10.1038/nrcardio.2016.179
4. Morley RL, Sharma A, Horsch AD, Hinchliffe RJ (2018) Peripheral artery disease. BMJ 360:1–8. https://doi.org/10.1136/bmj.j5842
5. Gogineni A, Ravigururajan TS (2017) Flow dynamics and wall shear stresses in a bifurcated femoral artery. J Biomed Eng Med Devices 02(02):2017. https://doi.org/10.4172/2475-7586.1000130
6. Sun N, Torii R, Wood NB, Hughes AD, Thom SAM, Xu XY (2009) Computational modeling of LDL and albumin transport in an in vivo CT image-based human right coronary artery. J Biomech Eng 131(2). https://doi.org/10.1115/1.3005161
7. Hewlin RL, Kizito JP (2018) Development of an experimental and digital cardiovascular arterial model for transient hemodynamic and postural change studies: 'a preliminary framework analysis'. Cardiovasc Eng Technol 9(1). https://doi.org/10.1007/s13239-017-0332-z
8. Johari NH et al (2019) Disturbed flow in a stenosed carotid artery bifurcation: comparison of RANS-based transitional model and LES with experimental measurements. Int J Appl Mech 11(4):1–21. https://doi.org/10.1142/S1758825119500327
9. Saratzis A et al (2019) Observational study of the medical management of patients with peripheral artery disease. Br J Surg 106(9):1168–1177. https://doi.org/10.1002/bjs.11214
10. Cristalli G, Mills DCB (1993) Identification of a receptor for ADP on blood platelets by photoaffinity labelling. Biochem J 291(3):875–881. https://doi.org/10.1042/bj2910875
11. Gökgöl C, Schumann S, Diehm N, Zheng G, Büchler P (2017) In vivo quantification of the deformations of the femoropopliteal segment. J Endovasc Ther 24(1):27–34. https://doi.org/10.1177/1526602816677530
12. Schumann S, Gökgöl C, Diehm N, Büchler P, Zheng G (2017) Effect of stent implantation on the deformations of the superficial femoral artery and popliteal artery. In vivo three-dimensional deformational analysis from two-dimensional radiographs. J Vasc Interv Radiol 28(1):142–146. https://doi.org/10.1016/j.jvir.2016.04.023
13. Gökgöl C, Diehm N, Räber L, Büchler P (2019) Prediction of restenosis based on hemodynamical markers in revascularized femoro-popliteal arteries during leg flexion. Biomech Model Mechanobiol 18(6):1883–1893. https://doi.org/10.1007/s10237-019-01183-9
14. Miller AJ, Takahashi EA, Harmsen WS, Mara KC, Misra S (2017) Treatment of superficial femoral artery restenosis. J Vasc Interv Radiol 28(12):1681–1686. https://doi.org/10.1016/j.jvir.2017.07.032
15. El Sayed H, Kerensky R, Stecher M, Mohanty P, Davies M (2016) A randomized phase II study of Xilonix, a targeted therapy against interleukin 1α, for the prevention of superficial femoral artery restenosis after percutaneous revascularization. J Vasc Surg 63(1):133-141.e1. https://doi.org/10.1016/j.jvs.2015.08.069
16. Kawarada O et al (2020) Peak systolic velocity ratio derived from quantitative vessel analysis for restenosis after femoropopliteal intervention: a multidisciplinary review from Endovascular Asia. Cardiovasc Interv Ther 35(1):52–61. https://doi.org/10.1007/s12928-019-00602-z
17. Zakikhani P, Ho R, Wang W, Li Z (2019) Biomechanical assessment of aortic valve stenosis: advantages and limitations. Med Nov Technol Devices 2:100009. https://doi.org/10.1016/j.medntd.2019.100009

18. Saqr KM et al (2020) What does computational fluid dynamics tell us about intracranial aneurysms? A meta-analysis and critical review. J Cereb Blood Flow Metab 40(5):1021–1039. https://doi.org/10.1177/0271678X19854640
19. Caballero AD, Laín S (2013) A review on computational fluid dynamics modelling in human thoracic aorta. Cardiovasc Eng Technol 4(2):103–130. https://doi.org/10.1007/s13239-013-0146-6
20. Razavi MK, Flanigan DPT, White SM, Rice TB (2021) A real-time blood flow measurement device for patients with peripheral artery disease. J Vasc Interv Radiol 32(3):453–458. https://doi.org/10.1016/j.jvir.2020.09.006
21. Salman HE, Yazicioglu Y (2019) Computational analysis for non-invasive detection of stenosis in peripheral arteries. Med Eng Phys 70:39–50. https://doi.org/10.1016/j.medengphy.2019.06.007
22. Carallo C, Tripolino C, de Franceschi MS, Irace C, Xu XY, Gnasso A (2016) Carotid endothelial shear stress reduction with aging is associated with plaque development in twelve years. Atherosclerosis 251:63–69. https://doi.org/10.1016/j.atherosclerosis.2016.05.048
23. Xu P, et al (2016) Patient-specific structural effects on hemodynamics in the ischemic lower limb artery. Sci Rep 6. https://doi.org/10.1038/srep39225
24. Desyatova A, Mactaggart J, Romarowski R, Poulson W, Conti M, Kamenskiy A (2018) Effect of aging on mechanical stresses, deformations, and hemodynamics in human femoropopliteal artery due to limb flexion. Biomech Model Mechanobiol 17(1):181–189. https://doi.org/10.1007/s10237-017-0953-z
25. Johari NH, Hamady M, Xu XY (2020) A computational study of the effect of stent design on local hemodynamic factors at the carotid artery bifurcation. Artery Res 26(3):161–169. https://doi.org/10.2991/artres.k.200603.001
26. Maleckis K et al (2018) Nitinol stents in the femoropopliteal artery: a mechanical perspective on material, design, and performance. Ann Biomed Eng 46(5):684–704. https://doi.org/10.1007/s10439-018-1990-1
27. Donadoni F, Bonfanti M, Pichardo-Almarza C, Homer-Vanniasinkam S, Dardik A, Díaz-Zuccarini V (2019) An in silico study of the influence of vessel wall deformation on neointimal hyperplasia progression in peripheral bypass grafts. Med Eng Phys 74:137–145. https://doi.org/10.1016/j.medengphy.2019.09.011
28. Gu K, Zhang Z, Gao B, Chang Y, Wan F (2018) Hemodynamic effects of perfusion level of peripheral ECMO on cardiovascular system. Biomed Eng Online 17(1). https://doi.org/10.1186/s12938-018-0493-5
29. Gounley J et al (2019) Computing the ankle-brachial index with parallel computational fluid dynamics. J Biomech 82:28–37. https://doi.org/10.1016/j.jbiomech.2018.10.007
30. Ferrarini A et al (2021) Impact of leg bending in the patient-specific computational fluid dynamics of popliteal stenting. Acta Mech Sin Xuebao 37(2):279–291. https://doi.org/10.1007/s10409-021-01066-2
31. Donadoni F, Pichardo-Almarza C, Bartlett M, Dardik A, Homer-Vanniasinkam S, Díaz-Zuccarini V (2017) Patient-specific, multi-scale modeling of neointimal hyperplasia in vein grafts. Front Physiol 8. https://doi.org/10.3389/fphys.2017.00226
32. Wood NB et al (2006) Curvature and tortuosity of the superficial femoral artery: a possible risk factor for peripheral arterial disease Downloaded from. J Appl Physiol 101:1412–1418. https://doi.org/10.1152/japplphysiol.00051.2006.-Ath
33. Li X, Liu X, Li X, Xu L, Chen X, Liang F (2019) Tortuosity of the superficial femoral artery and its influence on blood flow patterns and risk of atherosclerosis. Biomech Model Mechanobiol 18(4):883–896. https://doi.org/10.1007/s10237-019-01118-4
34. McGah PM, Leotta DF, Beach KW, Zierler RE, Riley JJ, Aliseda A (2012) Hemodynamic conditions in a failing peripheral artery bypass graft. J Vasc Surg 56(2):403–409. https://doi.org/10.1016/j.jvs.2012.01.045
35. Wang D, Serracino-Inglott F, Feng J (2021) Numerical simulations of patient-specific models with multiple plaques in human peripheral artery: a fluid-structure interaction analysis. Biomech Model Mechanobiol 20(1):255–265. https://doi.org/10.1007/s10237-020-01381-w

36. Colombo M et al (2020) Impact of lower limb movement on the hemodynamics of femoropopliteal arteries: a computational study. Med Eng Phys 81:105–117. https://doi.org/10.1016/j.medengphy.2020.05.004
37. Lopes D, Puga H, Teixeira J, Lima R (2020) Blood flow simulations in patient-specific geometries of the carotid artery: a systematic review. J Biomech 111. https://doi.org/10.1016/j.jbiomech.2020.110019
38. Adla T, Adlova R (2014) Multimodality imaging of carotid stenosis. Int J Angiol 24(3):179–184. https://doi.org/10.1055/s-0035-1556056
39. Wallyn J, Anton N, Akram S, Vandamme TF (2019) Biomedical imaging: principles, technologies, clinical aspects, contrast agents, limitations and future trends in nanomedicines. Pharm Res 36(6). https://doi.org/10.1007/s11095-019-2608-5
40. Ameenuddin M, Anand M, Massoudi M (2019) Effects of shear-dependent viscosity and hematocrit on blood flow. Appl Math Comput 356:299–311. https://doi.org/10.1016/j.amc.2019.03.028
41. Topps J, Elliott RC (1965) © 1965 nature publishing group. Nat Publ Gr 205(5007):498–499
42. Roy M, Sikarwar BS, Bhandwal M, Ranjan P (2017) Modelling of blood flow in stenosed arteries. Proc Comput Sci 115:821–830. https://doi.org/10.1016/j.procs.2017.09.164
43. Guerciotti B, Vergara C Newtonian and non-Newtonian blood rheologies in stenotic vessels, pp 169–183. https://doi.org/10.1007/978-3-319-59548-1
44. Kumar N, Khader A, Pai R, Kyriacou P, Khan S, Koteshwara P (2019) Computational fluid dynamic study on effect of Carreau-Yasuda and Newtonian blood viscosity models on hemodynamic parameters. J Comput Methods Sci Eng 19(2):465–477. https://doi.org/10.3233/JCM-181004
45. Lee SH, Han KS, Hur N, Cho YI, Jeong SK (2019) The effect of patient-specific non-Newtonian blood viscosity on arterial hemodynamics predictions. J Mech Med Biol 19(8):1–15. https://doi.org/10.1142/S0219519419400542
46. Urevc J, Zun I, Brumen M, Stok B (2017) Modeling the effect of red blood cells deformability on blood flow conditions in human carotid artery bifurcation. J Biomech Eng 139(1):1–11. https://doi.org/10.1115/1.4035122
47. Pang B, Wang S, Chen W, Hassan M, Lu H (2020) Effects of flow behavior index and consistency coefficient on hydrodynamics of power-law fluids and particles in fluidized beds. Powder Technol 366:249–260. https://doi.org/10.1016/j.powtec.2020.01.061
48. Acharya R (2016) Investigation of differences in Ansys solvers CFX and fluent, pp 1–48
49. Maulana S (2016) Pemanfaatan computational fluid dynamics (CFD) Dalama Strategi Penelitian Simulasi Model Pada Teknologi Penghawaan Ruang, vol 2, no 2
50. Roache PJ (1998) Verification and validation in computational science and engineering. Comput Sci Eng 107–240. http://www.hermosa-pub.com/hermosa%0Ahttp://scholar.google.com/scholar?hl=en&btnG=Search&q=intitle:Verification+and+validation+in+computational+science+and+engineering#0
51. Babuska I, Oden JT (2004) Verification and validation in computational engineering and science: basic concepts. Comput Methods Appl Mech Eng 193(36–38):4057–4066. https://doi.org/10.1016/j.cma.2004.03.002
52. Prantil VC, Papadopoulos C, Gessler PD (2013) Lying by approximation. Lying by Approx Truth about Finite Elem Anal
53. Mao W, Caballero A, McKay R, Primiano C, Sun W (2017) Fully-coupled fluid-structure interaction simulation of the aortic and mitral valves in a realistic 3D left ventricle model. PLoS ONE 12(9):1–21. https://doi.org/10.1371/journal.pone.0184729
54. Coleman HW (2009) Standard for verification and validation in computational fluid dynamics and heat transfer: ASME V&V 20. Am Soc Mech Eng 1–26

Using Microsoft Project as Planning, Monitoring and Controlling Tool for Project Success – A Case Study of Construction Projects in Malaysia

Kanesan Muthusamy, Nagesparan Ainarappan, Elango Natarajan, and Batumalay Kaliannan

Abstract The growth of performance has become more critical to the success of engineering projects. Different companies follow different project management methodologies based on their experience and client requirements. However, some common steps are followed such as project planning, project execution, progress tracking, and project closure. With the increasing complexity of cities projects and the scarcity of resources also the expertise, there is a need for better, more sophisticated tools for project planning and management. The objective of this quantitative base research is to evaluate the effectiveness of using Microsoft Project as planning, monitoring, and controlling tools in achieving engineering project success. Questionnaire methodology was adopted. This research study concludes that it is very important for all stakeholders to have an effective management and an effective project team towards the overall project success.

Keywords Project management · Planning tools · Microsoft project

1 Introduction

In [1] assumed that cities development to be characterized and perceived as project(s). The life cycle of a project is a series of phases that a project passes through from its inception to its completion or closure. According to PMBOK® Guide, areas of expertise of knowledge and skills for effective project monitoring and controlling are application on area of knowledge, standard and regulation, understanding the project environment, general management knowledge and skills and interpersonal skills. Scheduling software has now been widely developed and implemented.

K. Muthusamy (✉) · N. Ainarappan · E. Natarajan
Faculty of Engineering and Technology, UCSI University, 56000 Kuala Lumpur, Malaysia
e-mail: kanesan@ucsiuniversity.edu.my

B. Kaliannan
Faculty of Business and Management, Asia Metropolitan University, 81750 Bandar Seri Alam, Malaysia

© The Author(s), under exclusive license to Springer Nature Singapore Pte Ltd. 2023

M. H. A. Hassan et al. (eds.), *Technological Advancement in Instrumentation & Human Engineering*, Lecture Notes in Electrical Engineering 882,
https://doi.org/10.1007/978-981-19-1577-2_42

Cities project success relies on careful preparation, tracking, control, and monitoring. Therefore, keeping a project on track may also reflect the projects "Triple Constraints". It is about balancing the time (duration of the project as per agreed during tender), scope (or certain may refer to the quality) and the cost (contract sum amount). With the increasing complexity of cities projects and the scarcity of resources also the expertise, there is a need for better, more sophisticated tools for project planning and management. The need for a more efficient system that can facilitate better coordination and communication between project team members and communicate the same ideas project ideas to all stakeholders involved in the life cycle of the cities projects [2]. In [3] mentioned that the blame for project delays were laid mostly contractor-related which cause a time delay; contractor's problem in planning, inefficient site management, company finances and payment for the work's completion, subcontractor problem (late contract award), insufficient material (late submit, late approval, shortage of supply), lack of labor, equipment problems, lack of parties, mistakes, and accidents during construction progress due to all these problems. According to [4], construction projects has been considered as one of the most demanding job domains whereby accidents may happen while expediating to complete the delayed tasks which can add further delays to the project.

2 Literature Review

Kerzner [5] described that efficient scheduling software requires more than good planning, it requires that relevant information be obtained, analyzed, and reviewed in a timely manner. This may also be considered as pre-analysis of identifying potential risks that could damage the progress of the project. A realistic work program is a must, so that the project progress run smoothly, it considers as many variables as it can be. With the help of realistic work program, Project Managers should be able of making some of the adjustment when there are any unforeseen circumstances. If the contractor is not working based on the work program, there will be risks that might occur. According to Lock [6], that the large projects involve numerous differentiated activities that must focus on one final target. Microsoft Project able to improve the project planning and scheduling more efficiently by identifying dependencies of a task, able to allocate the resources, and setting up the deadlines. The data updating and tracking can be done on daily basis or weekly basis depends on when the reporting need to be done or when is the monthly meeting. With the Microsoft Project, the Project Leader or Project Manager can easily delegate project tasks to members and find who is currently available. Thus, it provides intuitive toolbars and menus that help users learn project management fundamentals. However, planning a project is a mentally demanding and complicated process, having to be done under constantly changing conditions, without complete information being always available. The resources need to be assigned manually to different tasks based on their experience to optimize resource usage, shorten the cycle time, and control the evolutionary nature of project development. In [7] suggested that both planner (Microsoft

Project expert) and the Project Manager, should be required to carry out construction planning. In [8] reported that more than 40% of failed projects were unsuccessful because of ineffective planning of human resources and project tasks in China.

However,

a) In [9] stated that Microsoft Project is the modern tool of Project Management that aid to overcome the obstacles faced owing to traditional way of Planning and Management. It helps for the optimum and effective organization of activities which helps to give the vision to complete the project in planned duration and within economy.
b) In [10] stated that the influence of project management software tools on project success are used all over the world. Microsoft Project is utilized in 48% of cases, Primavera Project Planner in 14% of the, and there is 9% of Microsoft Excel usage.
c) In [11] stated that a realistic work program must be prepared by the contractor in such a way that all the planning and scheduling are done ahead of time and keep on updating without00 any delay.

Monitoring is a process of evaluating and comparing objectives, as well as the project's "strategic fit" with enterprise purpose. Monitoring may also call as observation. Daily monitoring means to uncover issues, technological problems interface mismatch, communication misunderstanding etc.

Controlling means seeing that everything occurs in conformity with establish rule and expressed command. Some would define control related to the analysis of the impacts of any schedule deviations and evaluations of what remedial action, if any should be taken. To be an effective project control tool, it is important that the schedule should be an accurate reflection of the job constructed.

The **project team member** is known to influence the task productivity. There is increasing evidence that for large or complex projects, the use of an autonomous, full-time, and empowered core project team can contribute to the delivery of projects on time and within budget [12]. The contractor can improve their performance by hiring more experienced and knowledgeable team members.

Project does therefore find success if:

- It achieves all or much of what it said it will, regardless of timeline or budget results.
- It achieves as it has promised, on schedule and / or under the approved budget.
- It delivers what it said it would, on schedule within the agreed budget and in accordance with planned quality level.
- It delivers all agreed goals within the scope, schedule, budget, quality, or outcomes (i.e., goals to be achieved or strategic positions to be attained).
- The commodity created by the project provides tremendous new value for the company upon completion of the project (Fig. 1).

The research variables are identified as Independent Variable (IV) and Dependent Variable (DV) as following:

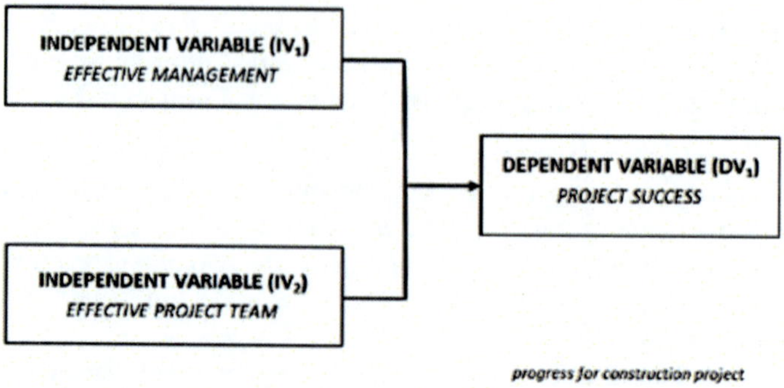

Fig. 1 Conceptual model and framework of research study

IV_1 or X_1: Effective Management.
IV_2 or X_2: Effective Project Team.
DV or Y: Project Success.

3 Research Methodology

According to Kothari [13], the research design is the conceptual structure within which the research is conducted. It provides a framework or plan of action for the researcher to solve the research problem [14]. This research study aims to examine the link between effective management and the project team towards the project success. The research methodology for this study is divided into three major stages as show in Fig. 2.

Data collection or sampling is a process of identifying the source of the data and understand the limitation of the research. However, the questionnaire template needs

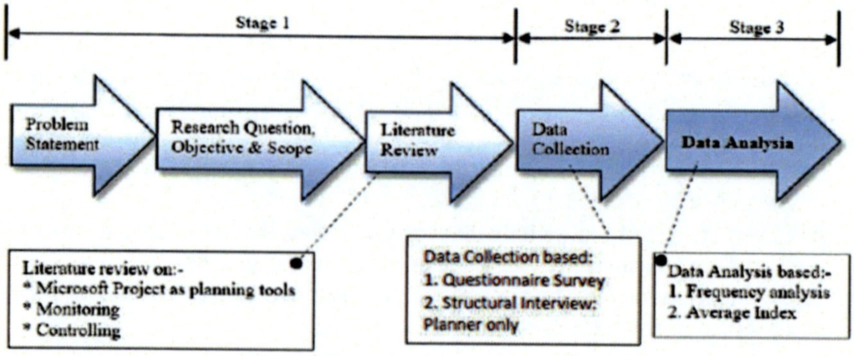

Fig. 2 Research methodology process

Table 1 Projects selected

No.	Cities project no.	Main contractor (peoples)	Consultant (peoples)	Client (peoples)
1	Project 1	15	15	10
2	Project 2	15	10	10
	Total respondents	30	25	20

to be short and simple. In this research, a set of questionnaires are prepared divided into seven (7) sections.

- Section 1: To identify the respondent particular
- Section 2: To identify the knowledge of Microsoft Project usage as a Planning, Monitoring and Controlling tools for a project
- Section 3: To identify the level of competency of a user using Microsoft Project as a Planning, Monitoring and Controlling tools for a project
- Section 4: To identify the factor that obstructs Main Contractor of having fully usage of the Microsoft Project
- Section 5: To identify the top management involvement in Project Planning, Monitoring and Controlling process to a Project
- Section 6: To identify the benefits of Microsoft Project
- Section 7: To identify the level of understanding of project success

In quantitative data analysis, the raw numbers will be turn into meaningful data through the application of rational and critical thinking. The objective of measurement is to numerically represent degree of attributes. A Likert Scale which is five-point or seven-point was employed. The choices of range from Strongly Agree to Strongly Disagree, so that the researcher can get a holistic view of people's opinion. Likert Scale is a type of non-comparative scale. It's an orderly scale from which the respondents choose the best option that suits their opinion. It is a powerful tool to measure the extent to which they agree or disagree with a statement.

The research samples are however limited to two cities projects from the Klang Valley region in Malaysia as listed in Table 1. Samples are randomly selected amongst all the staff from all departments within the cities project.

4 Data Analysis

All the data gathered are analyzed using Statistical Package for Social Sciences *(SPSS)*, Version 22. In summary there are 75 sets of questionnaires distributed to all respondents. Each questionnaire consists of 48 questions addressing the independent and dependent variables, i.e., effective management, effective project team and project success. From the 75 respondents, 73 (97.33%) of them responded but only 68 responses (90.67%) are usable for further analysis and the balance of 7 responses (9.33%) were discarded due to incomplete and/or error. The high response rate of

90.67% is good because it increases the probability of the samples in representing the population and to avoid nonresponsive biasness.

Respondent Particulars (S1)

Majority of the respondents are male (45 respondents, 66.18%) compared to female (23 respondents, 33.82%). The nature of site work, under the sun is more suitable for male than females. Most of respondents are 6–10 years of working experienced (24 respondents, 35.29%). Only 8 respondents (11.76%) have more than 20 years of working experience. This shows that majority of site staff are middle level where their working experience is more than 5 years. 48 respondents (70.59%) are bachelor's degree holders.

To Identify the Knowledge of Microsoft Project Usage as a Planning, Monitoring and Controlling Tools for a Project (S2)

The highest percentage shows that most of the respondent's poor about *"How good are you in using the tools/features provided in the Microsoft Project"* (48.5%) and "*Will you able to do Project Management Plan for this project"* (45.6%). However, the respondents also good about "*Did you know, Microsoft Excel can be used for monitoring tools"* (38.2%) and "*Are you able to read a Project Master Plan"* (35.3%). The highest mean showed respondents good about "*Did you know, Microsoft Excel can be used for monitoring tools"* (with a mean of 3.147, SD = 0.981). While the lowest mean showed the respondents poor about "*How good are you in using the tools/features provided in the Microsoft Project"* (with a mean of 1.955, SD = 1.138). The overall mean knowledge of Microsoft Project is 2.399 and standard deviation is 0.959. These shows that the respondents were fair about knowledge of Microsoft Project.

To Identify the Level of Competency of a User Using Microsoft Project as a Planning, Monitoring and Controlling Tools for a Project (S3)

Most of the respondent's poor about *"Can you generate Microsoft for the purpose of monitoring and controlling independently"* (48.5%) and "*What is your experience of using Microsoft Project to prepare the work program in the initial stage"* (45.6%). The respondents also choose fair about "*Did you know, what is the project slippage meaning"* (38.2%). However, the respondents choose good about "*Do you know how Microsoft Project help does to monitor a project"* (35.3%). The highest mean showed respondents fair about "*Do you know how monitor a project"* (with a mean of 2.602, SD = 1.446). While the lowest mean showed the respondents poor about *"Can you generate Microsoft for the purpose of monitoring and controlling independently"* (with a mean of 1.882, SD = 1.015). The overall mean competency of a user using Microsoft Project is 2.18 and standard deviation is 0.936. These shows that the respondents were **fair** about competency of a user using Microsoft Project.

To Identify the Factor that Obstructs Main Contractor of having Fully Usage of Microsoft Project (S4)

The biggest percentage shows that most of the respondents strongly agree about *"Project Management Plan is the responsibility of Planner"* (48.5%) and "*The usage of Microsoft project is only for the Planner"* (42.6%). The respondents also choose extremely "agree" about "*No support from management (such as insufficient workers at site) planning preparation is for the documentation @ audit only"* and "*Management decided to supply insufficient machineries, even though theoretically proven by daily production rate will slow the progress*" (42.6%). However, the respondents choose "disagree" about "*Software provided to the planner is illegal software (not all features are active and not all function is available)"* (35.3%) and "*Work program prepared by Planner, but the information is not share to the project team"* (30.9%). The highest mean showed respondents moderately agree about "*No support from management (such as insufficient workers at site) planning preparation is for the documentation @ audit only"* (with a mean of 3.735, SD = 1.399). While the lowest mean showed the respondents slightly agree about *"Work program prepared by PLANNER, but the planning is not realistic and unachievable"* " (with a mean of 2.867, SD = 1.359). The overall mean factors that obstruct the usage of Microsoft Project is 3.346 and standard deviation is 1.119.

To Identify the top Management Involvement in Project Planning, Monitoring and Controlling Process to a Project (S5)

Most of the respondents strongly agree about *"The usage of Microsoft Project is only for documentation only, reporting basis and for the evaluate the monthly claim"* (39.7%). The respondents also choose moderately to agree about "*Management willing to support all the sources needed to ensure the project using the Microsoft Project*" (38.2%). The highest mean showed respondents moderately agree about "*The Microsoft Project is for the project monitoring only"* (with a mean of 3.397, SD = 1.477). While the lowest mean showed the respondents also moderately agree about *"The management priority is the cash flow* (with a mean of 3.044, SD = 1.568). The overall mean the top management involvement in Project Planning, Monitoring and Controlling process to a project is 3.156 and standard deviation is 1.203. These shown that the respondents were moderately agree about the top management involvement in Project Planning, Monitoring and Controlling process to a project.

To Identify the Benefits of Microsoft Project (S6)

The highest percentage shows that most of the respondents extremely agree about *"Microsoft Project able to generate various of table comparison, so the project controlling will be more efficient"* and "*The calculation of a head or delay impact in days' unit can be done*" (55.9%). The respondents also choose "strongly agree" about "*Microsoft project give a clear view to understand a project*" and "*Easy to visualize the flow and network for each activity"* (48.5%). The highest mean showed respondents strongly agree about "*Microsoft Project able to generate various of table comparison, so the project controlling will be more efficient"* (with a mean of 4.47, SD = 0.657). While the lowest mean showed the respondents also strongly agree about *"Relation between activities is understood by networking* (with a mean of 4.044, SD = 0.818). The overall mean the benefits of Microsoft Project is 4.323 and standard deviation is 0.628. These shows that the respondents were strongly agree about the benefits of Microsoft Project.

To Identify the Level of Understanding of Project Success (S7)

Most of the respondents extremely agree about *"Project is acceptable by stakeholder"* (54.4%). It was followed by *"Quality element cannot be compromise"* and *"If a project able to manage the triple element constraint (scope, time, cost) it also considers as project success"* (52.9%). The respondents also choose "strongly agree" about "*Project have specific start date and finish date*" and "*Project duration is fixed based on Letter Award received*" (48.5%). The highest mean showed respondents strongly agree about "*If a project able to manage the triple element constraint (scope, time, cost) it also considers as project success*" (with a mean of 4.455, SD = 0.656). While the lowest mean showed the respondents also strongly agree about *"Project consider success, if the contractor able to deliver the project on time"* (with a mean of 4.044, SD = 0.854). The overall mean the understanding of project success is 4.296 and standard deviation is 0.615. These shows that the respondents were strongly agree about the understanding of project success.

Unstandardized Coefficients information is used in this regression analysis to arrange the multi linear regression equation. The regression plot is shown in Fig. 3.

$$Y = 1.950 + 0.244X_1 + 0.343X_2 + e$$

where:
Y = Project success (dependent variables).
a = constant.
X_1 = Effective management (Independent Variables, IV_1).
X_2 = Effective Project Team (independent Variables, IV_2).
e = residual error (random error term).

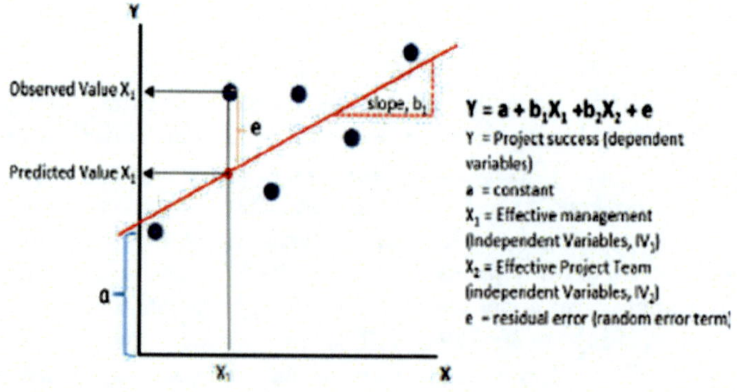

Fig. 3 Regression plot

5 Discussion

From the analysis, it can be established that a positive linear relationship between project success (dependent variable, DV) with both effective management (independent variable, IV_1) and the effective project team (independent variable, IV_2). However, to identify the level of relationship between both IV_1 and IV_2 towards the DV, the Coefficient of Determination of R^2 is calculated. Based on the calculated amount, R value equals to 0.508. Therefore, the R^2 are 0.258, which means 25.80% of factor contributes to project success can be explained by the combined factors of effective management and effective of project team. The remaining of 74.2% variances towards project success are derived from other factors which are not being investigated in this research study. According to De Wit [15], measuring success is complex because it depends on the stakeholders' points of view, and it is time dependent. The definition of project success is different for each stakeholder, but it is based on the basic concept of overall achievement of project goals and expectations. The researcher does agree with Baccarini [16] in which, the criteria for measuring project success must be set out at the beginning of the project. Finally, no matter what, the contracting company (main contractor) is the next particularly important participant or stakeholder who is the main party responsible for the successful project delivery.

Research Question No. 1: What are the benefits from using Microsoft Project as project planning, monitoring, and controlling tools for a project?

Main function of any planning tools software is to offer help and enhance the quality of output with less effort than having it done manually. In addition, as the project increase in size, the construction's activities become more adamant. The Microsoft Project software tools aid the project management team with planning, monitoring, and controlling the project, including scheduling, cost estimating, communication collaboration, risk analysis, configuration management, and many mores. In Fig. 4, results analyzed the overall mean is 4.32 and standard deviation is 0.628. These shown that the respondents were strongly up to extremely agreed about the benefit of having Microsoft Project Software (MPS) for a project. Concerning the question, 91.1% of respondents agreed that by having the MPS helps them to understand the overall scope of a project, and with the features available makes them easy to visualize the network for each activity. By having MPS, the work activity can be planned to move concurrently (parallel) and still the "critical path" can be calculated and be able to identify the total float of each activity compared to have it manually calculated. MPS helps to produce compatible report with the supporting of various table comparison, and most importantly it helps the person preparing it finish on time. 54.4% agreed that, with MPS the project manager will be able to assign resources (money, manpower, material, and machineries) to all activity and rearrange the work according to priority and the suitable the project condition. 55.9% of respondent agreed that MPS usage ease the calculation of the project achievement (actual progress). Lastly, another important aspect is that the MPS usage helps to support the application of "Extension of Time".

Statement	3 (%)	4 (%)	5 (%)	Mean
Microsoft Project give a clear view to understand a project	8.80	48.5	42.6	4.34
Easy to visualize the flow and network for each activity	16.2	48.5	35.3	4.19
Relation between activities is understood by networking	30.9	33.8	35.3	4.04
Using Microsoft Project, certain activity can be scheduled to move parallel, thus this can help reduce some time	16.2	33.8	50	4.34
Microsoft Project able to identify the critical path (total float =0)	5.90	42.6	51.5	4.46
Resources allocation (assigned resources) can be done for each activity; thus this will help to plan the work effectively	19.1	26.5	54.4	4.35
Microsoft Project able to generate various of table comparison, so the project controlling will be more efficient	8.80	35.3	55.9	4.47
The calculation of a head or delay impact in days unit can be done	16.2	27.9	55.9	4.4
Overall				**4.32**

Fig. 4 Microsoft project usage benefits

Research Question No. 2: What are the barriers and limitation of Microsoft Project software for contractor and the client?

There is growing recognition that different types of projects required different approaches to their management, requiring management procedures tailored to the needs of the project [17] and project managers selected with appropriate competencies [18]. Some suggests that, in choosing a project management methodology, the project sponsor or project manager should identify the relevant success factors to increase the chance of achieving those success criteria, and then select a project management methodology that delivers those success factors. Further to this statement, 48.50% of respondent choose to agree that the Project Management Plan is the responsibility of Planner. Only 33.80% vote for this task is the responsibility of Project Manager. It seems to be, for these projects, the workload and total responsibility of the overall project planning for the project has been shifted from Project Manager to the Planner. Support with 32.4% of the respondent agreed that the work planning prepared is solely because of the contract requirement. 42.6% respondents agreed that the usage of MPS is only for the Planner. Most of the respondents is honestly rating for either "poor" or "fair". However, 20.6% of respondent able to read the MPS only. Unfortunately, 48.5% of respondents did not use the MPS.

According to Kumar [19], the early development of strategies, philosophies and methodologies of project implementation are the most important factor in achieving success. He suggested that by gathering sufficient site information and being aware of project consideration and constrains; it is possible to tailor strategies and methodologies which are specific to a certain situation. 33.8% of respondents agreed that if there are lots of additional Variation Order (VO) for a project may also affected the baseline planning and early estimation of the daily production rate achievements. The VO need to be carried out through the process but the time impact subject to the approval of the client. The project team is facing unprecedented changes when 36.8% of them agreed that any increased of the quantity may also jeopardize the earlier baseline planning. In summary, not only the automated initial plans, but also any adjustments to the scope of specifications or external events would cause revised plans to suit the new assumptions. These modifications cannot be easily accomplished by manual methods; this is where the MPS or any planning tools are a necessity for a construction project.

Research Question No. 3: What is the commitment from top management (from the view of contractor side) towards supporting the usage of Microsoft Project as a planning, monitoring, and controlling tools?

From Fig. 5, it is observed that, 32.4% of respondents agreed that top management interest of knowing the project progress, and 29.4% says the management interest in project monitoring only. Cash flow is important to any Contractor and the project owner because it give an important evaluation in cost performance at any stage of a project. 27.9% of respondents agreed with this statement. However, 39.7% respondents agreed that the usage of MPS is for the purpose of documentation only. This is the major obstruction to MPS usage if the top management is not supportive and not comply to all the planning requirements.

Statement	1 (%)	2 (%)	3 (%)	4 (%)	5 (%)	Mean
The management interest of knowing the project progress only	27.9	10.3	2.9	32.4	26.5	3.19
The management priority is the cash flow only	25.0	16.2	16.2	14.7	27.9	3.04
The Microsoft Project is for project monitoring only	17.6	13.2	10.3	29.4	29.4	3.39
The purchase of Microsoft software solely because it is compulsory according to the COC	17.6	13.2	32.4	17.6	19.1	3.07
The usage of Microsoft Project is only for documentation only, reporting basis and for the evaluate the monthly claim	8.8	32.4	8.8	39.7	10.3	3.1
Management willing to support all the resources needed to ensure the project using the Microsoft Project	8.8	20.6	38.2	13.3	19.1	3.13
Overall						**3.16**

Fig. 5 Commitment from top management

6 Conclusion

In conclusion, an effective project management requires planning with a determination to complete the project; careful selection of a professional project manager, allocating sufficient time when defining the project, proper planning of activities, ensuring the correct and sufficient flows of information, adjusting activities to accommodate regular changes, accommodating employees' personal goals with performance and rewards, and making a fresh start when mistakes in implementation. Having MPS allows a project manager to identically plan resources needed. Resources are commonly inclusive the 4 M, i.e., methods, materials, machineries, and manpower. Mostly, projects are delayed due to improper methods, material shortage, lack of manpower, late deployment, or late rental of machineries, and all this due to inadequate planning.

Acknowledgements The author would like to express utmost gratitude towards the financial support from UCSI University. The author would also like to extend gratitude to Ms Raja Noraida Raja Aziz for her support and assistance in completing this research paper.

References

1. Schipper RPJR, Silvius AJG (2018) Characteristics of smart sustainable city development: implications for project management. Smart Cities 1:75–97
2. Muthusamy K, Chew L (2020) critical success factors for the implementation of building information modeling (BIM) among construction industry development board (CIDB) G7 contractors in the Klang Valley, Malaysia. In: 2020 IEEE European Technology and Engineering Management Summit (E-TEMS), pp 1–6
3. Munns AK, Bjeirmi BF (1996) The role of project management in achieving project success. Int J Proj Manag 14(2):81–87
4. Muthusamy K, Gunasegaran HR, Natarajan E, Renganathan K (2021) Analysis of potential project work accidents: a case study of a construction project in Malaysia. In: 2021 IEEE European Technology and Engineering Management Summit (E-TEMS), pp 21–26
5. Kerzner H (2013) Project Management: A System Approach to Planning, Scheduling and Controlling, 11th edn. Wiley, Hoboken, NJ
6. Lock D (2003) Project Management, Eight Edition. Gower Publishing, Aldershot
7. Laufer A, Tucker RL (1988) Competence and timing dilemma in construction planning. Constr Manag Econ 6:339–355
8. Ding RG, Jing XH (2003) Five principles of project management in software companies. Proj Manag Technol 1
9. Wale PM, Jain ND, Godhani NR, Beniwal SR, Mir AA (2015) Planning and scheduling of project using Microsoft project (case study of a building in India). IOSR J Mech Civ Eng 12(3):57–63
10. Omazić MA, Baljkas S (2005) Projektni menadžment, Sinergija, Zagreb
11. Hasmori MF et al (2018) Significant factors of construction delays among contractors in Klang Valley and its mitigation. Int J Integr Eng 10(2):32–36
12. Michael B, Renee D, James H (2002) skunkworks approach to project management. J Manag Eng 18(1)
13. Kothari CR (2013) Research Methodology: Methods and Techniques. New Age Int., New Delhi
14. Dahlan AN (2015) Business Research Methods. Open University Malaysia, Kuala Lumpur
15. De Wit A (1998) Measurement of project success. Int J Proj Manag 6(3):164–170
16. Baccarini D (1999) The logical framework method for defining project success. Proj Manag J 30(4):25–32
17. Crawford L, Hobbs B, Turner JR (2015) Project Categorization Systems. Project Management Institute, Newton Square, PA, USA
18. Turner JR, Muller R (2006) Choosing Appropriate Project Managers: Matching their Leadership style to the Type of Project. Project Management Institute, Newton
19. Kumar D (1989) Developing strategies and philosophies early for successful project implementation. Proj Manag 7(3):164–171

Numerical Studies for Small-Scale Solar Chimney Power Plants with Various Geometric Configurations

Mohd Noor Asril Saadun, Nor Azwadi Che Sidik, Teng Meng Xian, and Mohd Afzanizam Mohd Rosli

Abstract With annual increases in energy demand, solar chimney power plants (SCPP) are regarded as one of the renewable energy power plants. It can potentially be an alternative renewable energy source such as photovoltaic (PV) cells and wind when properly developed. This study primarily focuses on determining the optimal size of a SCPP by examining various geometrical characteristics of the SCPP. The approach for this investigation included utilizing Computational Fluid Dynamics (CFD) to simulate the flow within the SCPP. Three essential parameters were examined in this investigation: the chimney's height, its diameter, and the shape of the SCPP's solar collector. The model was developed using 3D visualization tools, and the results were validated to demonstrate that it fulfils the degree needed for the model's intended purpose or application. In addition, the simulation was used to investigate the temperature and air velocity of the SCPP. It was discovered that increasing the chimney height, decreasing the diameter of the chimney, and utilising a square collector will help improving the SCPP's performance. By expanding the chimney height, the temperature and air velocity will be higher, but the air velocity decreases after exceeding 4.0 m of chimney height. The optimal diameter shows at 16 cm when both temperature and air velocity are at their maximum values. In contrast, the air velocity generated by the square collector is more than the air velocity produced by the circular collector. In order to improve simulation results, it is recommended to cross-reference with experimental data specifically.

Keywords Solar updraft tower · Solar chimney · Renewable energy

---

M. N. A. Saadun · N. A. Che Sidik (✉)
Malaysia – Japan International Institute of Technology (MJIIT), University Teknologi Malaysia, Jalan Sultan Yahya Petra, 54100 Kuala Lumpur, Malaysia
e-mail: azwadi@utm.my

T. M. Xian · M. A. Mohd Rosli
Fakulti Kejuruteraan Mekanikal, Universiti Teknikal Malaysia Melaka, Hang Tuah Jaya, 76100 Durian Tunggal, Melaka, Malaysia

M. N. A. Saadun
Fakulti Teknologi Kejuruteraan Mekanikal dan Pembuatan, Universiti Teknikal Malaysia Melaka, Hang Tuah Jaya, 76100 Durian Tunggal, Melaka, Malaysia

© The Author(s), under exclusive license to Springer Nature Singapore Pte Ltd. 2023
M. H. A. Hassan et al. (eds.), *Technological Advancement in Instrumentation & Human Engineering*, Lecture Notes in Electrical Engineering 882,
https://doi.org/10.1007/978-981-19-1577-2_43

1 Introduction

Solar chimney power plant (SCPP) is a type of renewable energy power plant for generating electricity by utilizing the surrounding warm air. A simple SCPP consists of three main components, a solar collector, chimney and wind turbine. All those main components will work together in conjunction with each other to produce air flow to power the turbine. It is revealed that the various geometrical factors had an advantageous effect on the system performance and the system performance efficiency [1]. The solar collector will be made from transparent material such as glass or semi-transparent plastic to act as a greenhouse that traps sun's energy under the solar collector and heat up the air inside the collector. Next, the chimney will serve as the thermal engine of the plant which creates suction that draws in surrounding air into the collector further increases the temperature of air while also decreasing the air density which increase the velocity of air. Finally, the turbine will act as the energy conversion for the SCPP which converts the kinetic energy of air into mechanical energy and to electrical energy which enables collection and utilization.

Therefore, the SCPP operates the best at places with lots of sun and on cheap flat land such as desserts or low nutrients soil. Besides that, because of SCPP is fueled by hot air, it is able to operate in the day using the sun to heat the surrounding air and also in the night where the ground releases heat absorbed by the sunlight during the day. Therefore, this give SCPP an distinct advantage over the conventional renewable energy such as Photovoltaic (PV) cell or the wind turbine as both only generates energy when certain requirement is met whereas SCPP are able to operate during the day and night. Besides that, there exist studies and prototypes of the SCPP and it is proven that this method of harvesting energy using the sun is viable as the efficiency of the SCPP will keep increases when the size and height of the solar collector and chimney increases. The actual results demonstrate that the device can produce a driving force simply by observing the turbine spin. This driving force is directly proportional to changes in temperature and pressure, but the humidity has an indirect influence on these two critical system variables [2]. However, with this advantage of increasing efficiency there is also the disadvantages of increasing cost when the size of the SCPP increases. As a result, it is best to build a SCPP with its optimum parameters to prevent high costs while maintaining the highest efficiency of SCPP.

Numerous scholars have led to the development of the solar chimney, either experimental or numerical simulation. The precision of simulation software about the distribution behaviour of the flow in the SUT system may boost the CFD application in this field. Bernandes et al. [3] applied CFD and numerical analysis to simulate the flow through the solar chimney of a radial solar heater with natural convection. They explored the various connection shapes at the collector base to estimate the thermos-hydrodynamic activity. They studied the different connection shapes at the collector base to assess the thermos-hydrodynamic activity. To validate the experimental data of the scaled model, commercial CFD software, such as ANSYS-CFX, was used to forecast the performance of a large-scale solar chimney. Due of the apparent consistency between numerical data and experimental, Kirstein and Backstrom examined the flow inside a collector and chimney during the transitional phase using a numerical method [4].

Tingzhen et al. [5] carried out the simulation using code and then validated the CFD code with a Manzanares SUT tri-bladed turbine. The quantitative model was validated using both numerical and experimental methods, as well as the CFD code. In order to provide a reference for the design of large-scale SUT systems five bladed turbines were tested with MW type SUT. Another way for numerical simulations was to build a mathematical model of the SUT based on the Navier–Stokes, continuity and energy equations. Sangi et al. [6] carried out a research on the Manzanares SUT using the commercial CFD software FLUENT. The results revealed a high agreement between the numerical model, experimental data mathematical model.

Pasumarthi et al. [7, 8] designed and constructed a convergent SUT, using a basic analytical model to demonstrate that the convergent SUT was useful in increasing the plant power production with a turbine installed in the system. In a specific section, Koonsrisuk and Chitsomboon used CFX to evaluate the effect of tower area modifications on an SUT [9]. Because of the changing tower areas, they found that using a diverging tower can enhance mass flow rate and kinetic energy. The kinetic energy of a convergent tower is equivalent to that of a constant area tower since the velocity is higher but the mass flow rate is lower at the tower outlet. They also believed that the highest kinetic energy is generated near the tower's inlet. The findings demonstrates that conceptually divergent chimneys are up to 18 times performing better than the same-height cylindrical chimney [10].

Ming et al. [11] conducted a recent analysis on the effect of ambient crosswind on SUT performance using CFD. According to their findings, the performance of the SUT was affected both positively and negatively by the ambient crosswind. This phenomenon resulted in changes in power output due to changes in the flow field and mass flow rate. When the flow field worsened due to a light ambient crosswind, the SUT output power decreased. When the ambient crosswind was sufficiently strong, the output power rose as the mass flow rate increased. This increase in mass flow rate was generated by a wind suction effect on the chimney's top caused by the high velocity wind (Bernoulli principle). Ming et al. [12] performed another numerical analysis using an obstruction a few metres away from the collector intake aperture to avoid the unfavourable effect of high ambient crosswind. They said that the barriers served to mitigate the harmful impacts of the strong ambient crosswinds.

In order to increase the SCPP efficient while limiting the height of the chimney, different shape of chimney will be utilized. For example, there is a study conducted by Haythem et al. [13] on design SCPP with hyperbolic chimney shape as shown in Fig. 1. The study claimed that instead of increasing the height of chimney, the shape of hyperbolic chimney is proposed and according to simulation the efficiency of the power generated is increase by 295%.

The SCPP is often constructed with a circular solar collector. However, a recent study conducted by Ramin et al. [14] in the performance of SCPP by changing the shape of the solar collector to a square collector has improved the performance of SCPP, as illustrated in Fig. 2. Compared to the circular solar chimney, the square

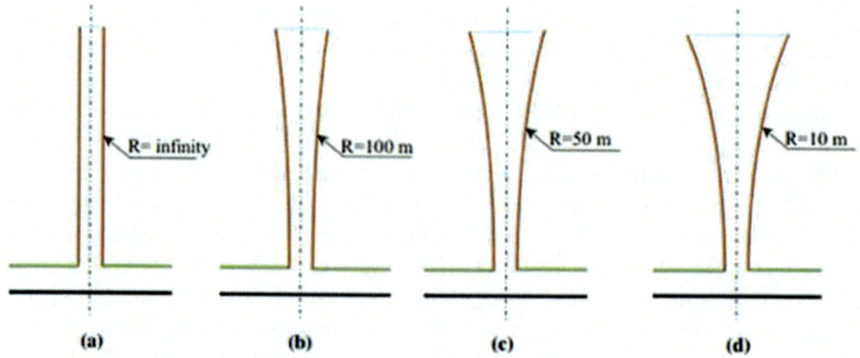

Fig. 1 Different configurations design of solar chimney [13]

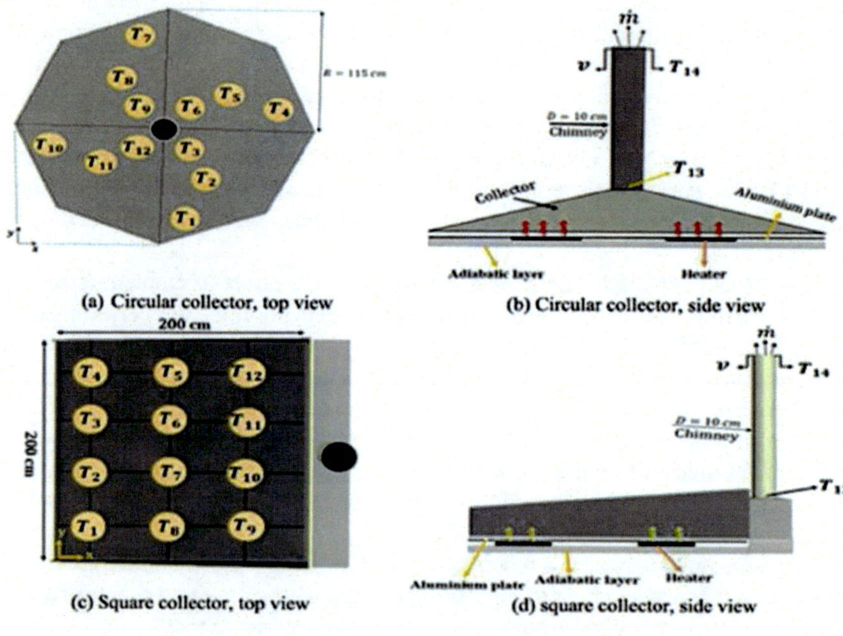

Fig. 2 Schematic comparison between the two different shapes of solar collector [14]

collector excels the circular collector in terms of air flow velocity, Nusselt Number, and thermal efficiency, with values of 245, 1225, and 169%, respectively. In a study to create the ideal geometry for the solar chimney using simulations, Saadun et al. [15] have found that 0 degrees of path collector and 0.05 m for collector entry gap contributed to boost the system power generation. According to Walid et al. [16], geometric calculation shows that the square collector had an increase of inlet area by 11.4% compared to a circular collector. When compared to a circular collector, the square collector produces more power because to the increased intake area, which results in a 7.6% increase in mass flow rate. Increasing radius of collectors above a particular value for a certain collector height does not improve the SCPP performance especially because it entails additional investment costs [17].

Fluri et al. [18] conducted a comparison of various turbine configurations in 2008 and discovered that the difference in performance of various configurations is minor, with roughly 1.6% differences. This is mainly because the turbine size and area do not change, hence the power generated remains constant. Majority of earlier research did not conduct experiments to validate their data, but instead validated it by comparing the simulation performance to previous SCPP prototype experiments. As a result, multiple studies will have comparable SCPP parameters in their design. For example, both the SCPP models of Lu et al. [19] and Ali et al. [20] contain the same parameters since they validate their data using the Manzanares SCPP, mirroring the model in the process. Osama et al. [21] stated that the best optimization of diameter and height of SCPP's tower are 0.3 and 3.4 m respectively. Further increase in height or diameter pass the critical point will loss efficiency.

To attract future investors, the SCPP construction must be as efficient as possible. As a result, in order to acquire the optimum efficiency of the SCPP, research must be conducted to determine what geometric parameters will effect the efficiency of the SCPP and what is the ideal configuration for the tower's construction. Some studies have shown the beneficial effects of numerical on SCPP but others showed a deterioration in ideal geometric setup. This study therefore set out to assess the effect on the determining the optimal size of a SCPP by examining various geometrical characteristics of the SCPP. The approach to empirical research adopted for this study was utilizing Computational Fluid Dynamics (CFD) to simulate the flow within the SCPP. Three essential parameters were examined in this investigation including the chimney's height, its diameter, and the shape of the SCPP's solar collector.

2 Methodology

2.1 Geometrical Setup

Ansys Fluent was used to construct a CFD model for a parametric investigation of a solar updraft tower's structure. For various configurations, the current results and prior studies will be compared to determine the best velocity and temperature profile for each configuration. Because the generator is a pressure turbine, the output is proportional to the velocity of the air passing through the turbine, whereas the greenhouse effect required a much more consistent temperature. A validated model was conducted to evaluate the effects of various geometry configurations on the original shape. The geometry and configuration used in this study correspond to Osama Nsaif [21] simulation and results, where the model was duplicated using the

Table 1 Solar chimney configuration used by Osama Nsaif [21]

Case	Chimney tower parameters		Collector parameters	
	H_{ch} (cm)	D_{ch} (cm)	R_c (cm)	H_c (cm)
1 (Verification Case)	300	20	150	6

same parameters. The parameters employed by Osama in their analysis are shown in Table 1.

2.2 Boundary Condition

The boundary conditions for this study are shown in Table 2. The pressure inlet and outlet are treated as the domain's inlet and outlet, respectively. Aside from that, the height of the chimney wall is set to be adiabatic with zero heat flux. ANSYS Geometry Module is used to create the geometry of the SCPP. Figure 3 depicts the shape of the SCPP and its parameters that will be used in this study.

It was assumed that the air flow into the update tower was turbulent and that the ANSYS Fluent software simulated using the k-epsilon turbulence model. It was assumed that the air flow into the update tower was turbulent and that the ANSYS Fluent software simulated the k-epsilon turbulence model. The standard k-model is used because it is a part of the turbulence families of the Reynolds-averaged Stokes (RANS), which model all causes of turbulence. In addition, the traditional k model is a two-equation model that resolves two transport equations with turbulent kinetic energy k variables, which determine the turbulence energy and the turbulent dissipation rates that define the turbulent dissipation rate of kinetic energy. The following equations represent the turbulent kinetic energy k and the turbulent kinetic energy dissipation rate k:

Turbulent kinetic energy k,

Table 2 Description of the boundary conditions

Boundary condition	Type	Parameters
Collector inlet	Pressure inlet	$T_0 = 302$ K
Chimney outlet	Pressure outlet	$P_0 = 0$ Pa
Chimney wall	Heat flux (Adiabatic Wall)	$Q = 0$ W/m^2
Solar collector	Semi-transparent wall (Mixed)	$h = 5$ W/m^{2}K
Ground	Heat flux (Opaque Wall)	$Q = 765$ W/m^2

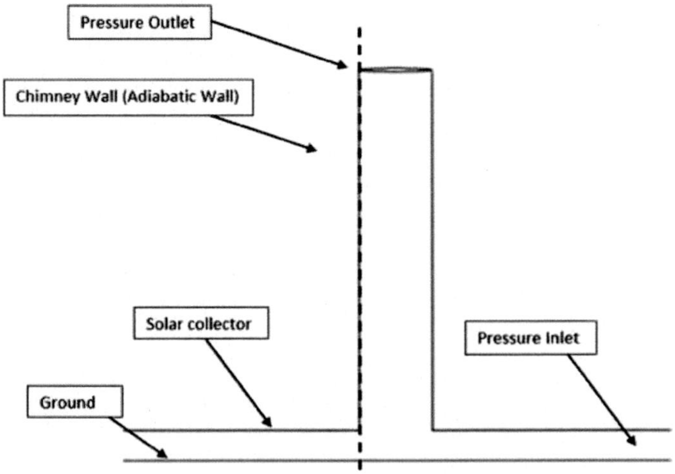

Fig. 3 The illustration of boundary conditions of the SCPP setup

$$\frac{d}{dt}(pk) + \frac{d(pkuj)}{dxi} = \frac{d}{dxj}\left(\left(\mu + \frac{\mu t}{\sigma k}\right)\frac{\partial k}{\partial xj}\right) + G_k + G_b - Y_M - \rho_\varepsilon + S_k \quad (1)$$

Turbulent kinetic energy ε,

$$\frac{d}{dt}(pk) + \frac{d(pkuj)}{dxi} = \frac{d}{dxj}\left(\left(\mu + \frac{\mu t}{\sigma k}\right)\frac{\partial k}{\partial xj}\right) + C_1\frac{\varepsilon}{k}(P_k + C_3 P_b) + C_2 p\frac{^2}{k} + S_k \quad (2)$$

Launder and Sharma [22] establish the k-turbulence model constants as C1 = 1.44, C2 = 1.92, = 1.3, and k = 1. The Launder and Sharma model is known as the standard k-model, and it is still used to accurately estimate flow, heat, and mass transfer in a domain.

2.3 *Grid Independency Test*

The grid independence test is a procedure for determining the optimum grid condition with the smallest number of grids that produces minimal changes in numerical output. Aside from that, grid independence testing saves a lot of time because increasing the number of mesh elements does not change the simulation's result but increases the simulation's processing time. The element size is employed in this study to generate the amount of mesh elements on the chimney; the smaller the element size, the more mesh elements will be matched to the model used as shown in Table 3.

Table 3 The number of mesh elements and the output velocity of the chimney

Element size	Number of mesh elements	Output velocity
30	15,197	1.9443
25	16,293	1.9450
11	17,754	1.9462
8	19,402	1.9528
7	20,560	1.9658
5	32,772	1.9751
3	93,589	1.9778

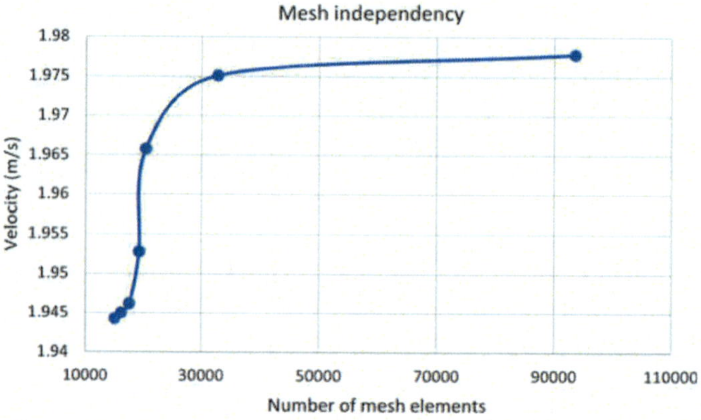

Fig. 4 Number of mesh elements in relation to the output velocity of the chimney

As demonstrated in Fig. 4, the velocity difference of element sizes from 8 to 5 cm are substantial and unsuitable for deployment, but from 5 to 3 cm, the changes in velocity are minimal even though the number of mesh elements increases dramatically. As a result, 5 cm element size is chosen as the starting point of the converging graph and it takes significantly less time to simulate.

3 Results

3.1 Model Validation

To validate the numerical results, velocity distribution under the collector was compared with the previous simulation data by Osama Nsaif [21]. Therefore, a comparison of velocity at different points of the solar collector radius is used to validate the data obtained from the simulation. The numerical results demonstrate that at 0.1 m of collector radius, the velocity flow through the collector is 1.6 m/s. The

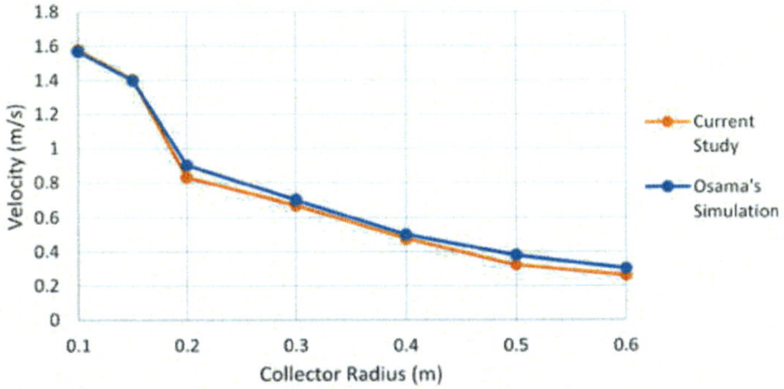

Fig. 5 Simulated results from this experiment vs the theoretical data derived from Osama Naif's study

previously data by Osama and the current studies shows a good quantitative agreement. The slight deviation, as indicated in Fig. 5, was caused by mesh generation and geometry inaccuracy, which may have resulted in some data variation from the current results. The CFD modelling can then be utilised using the software FLUENT to determine the best shape for designing a Solar Updraft Tower for the system's performance and efficiency.

Furthermore, in order to calculate the % error, the greatest discrepancy between the experimental and theoretical data values is selected. Thus, the percentage error obtained is 7.78% or less, which is minimal, and therefore this concludes that the data received from this experiment is similar to the data obtained from Osama Nsaif's study.

3.2 *Influence of Chimney Height*

In this study, chimney heights ranging from 2 to 5 m with 0.5 m increments are examined. Figure 6 presents the temperature and velocity at the turbine with varying heights of the chimney, while the collector radius remains unchanged at 3 m and the diameter of the chimney at 0.2 m. The velocity produced will increases as the chimney tower height increases. However, the increase in velocity will peak between 3.5 and 4 m of chimney tower height and will decrease dramatically when the height is increased higher. As a result, the optimum height of the chimney tower is estimated to be 4.3 m, as this is the point at which the temperature of the SCPP is high while the velocity is acceptable.

Figure 7 and Fig. 8 illustrate the temperature and velocity distributions, respectively. At the starting point, where the chimney is set at 2 m, the temperature distribution is considered to remain constant, particularly in the collector's bottom part. According to these findings, the chimney has no effect on the temperature along the chimney. The results appear to make sense and to be compatible with our expectations

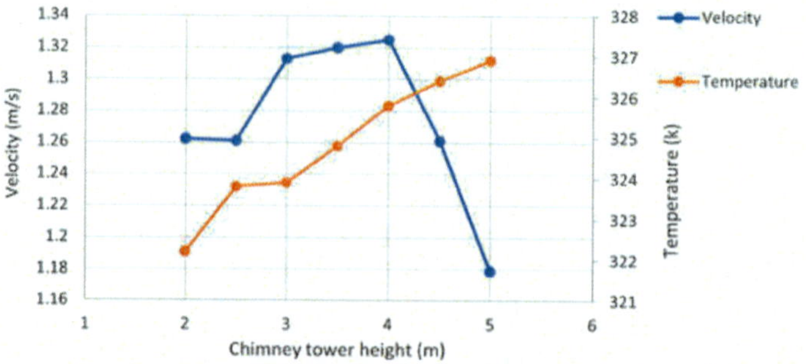

Fig. 6 The graph of velocity and temperature at the turbine with varying chimney heights

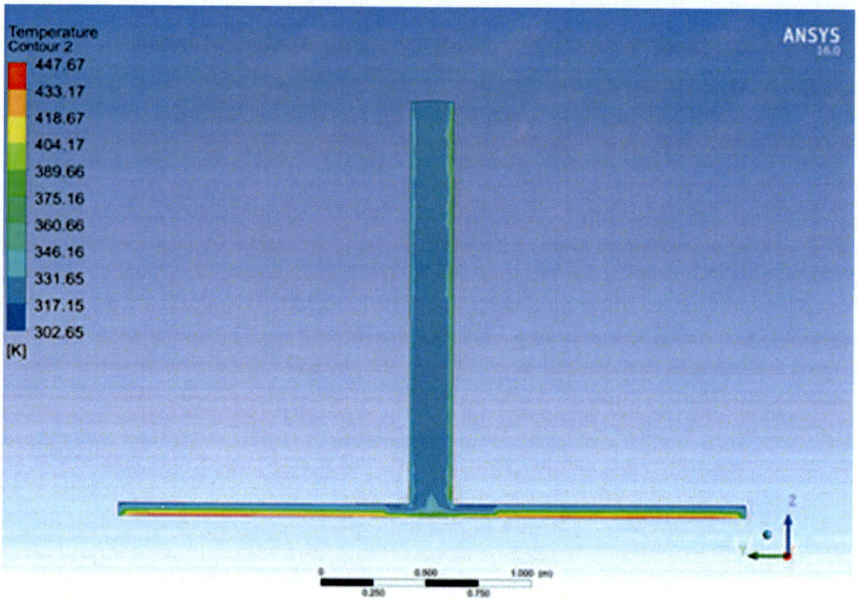

Fig. 7 The temperature distribution of SCPP when Chimney height at 2.0 m

where the velocity distribution clearly occur at the chimney inlet. The results indicate that the hot air is moving upward, which can be interpreted in part by the greenhouse effect. Increasing velocity and temperature cause a drop in pressure values due to the ideal gas equation of state. As a result, it is possible to conclude that greatest velocity will assure maximum power production of the system.

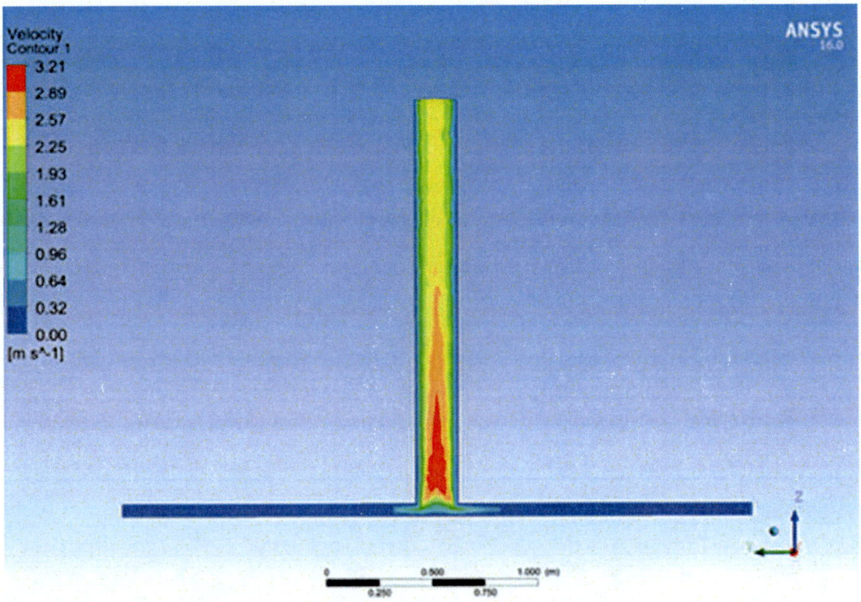

Fig. 8 The velocity distribution of SCPP when Chimney height at 2.0 m

3.3 Influence of Chimney Diameter

The diameter of the solar chimneys examined in this study ranged from 15 to 45 cm. Figure 8 illustrates the temperature and velocity at the turbine with varying chimney diameters while fixing the collector radius at 3 m and the tower height at 3.5 m.

In general, there appeared to be a trend when the diameter of the chimney tower of the SCPP is increased, the air temperature inside the SCPP at the turbine section will drops. This is because as the diameter of the chimney tower increases, the volume of air surrounding the turbine increases accordingly. A greater volume of air must be heated by the same amount of collector surface, resulting in a lowered temperature of air [16]. As illustrated in Fig. 8, the turbine's speed will decrease as the tower's diameter grows. Due to the high velocity and temperature, the intersection of 16 cm diameter velocity and temperature is believed to be the optimal location for the chimney tower's diameter measurement.

The results obtained from the simulation analysis of temperature and velocity distribution with different diameter are shown in Fig. 9 and Fig. 10. For a 20 cm diameter chimney, the regions having the higher velocity and temperature are represented in contour form. The velocity was kept at a constant level along the height of the chimney, where the hot air appears to push upward. After 20 cm, output power begins to decline, and this has a negative impact on power and efficiency since the air temperature and pressure differential reduces significantly after that point. As a

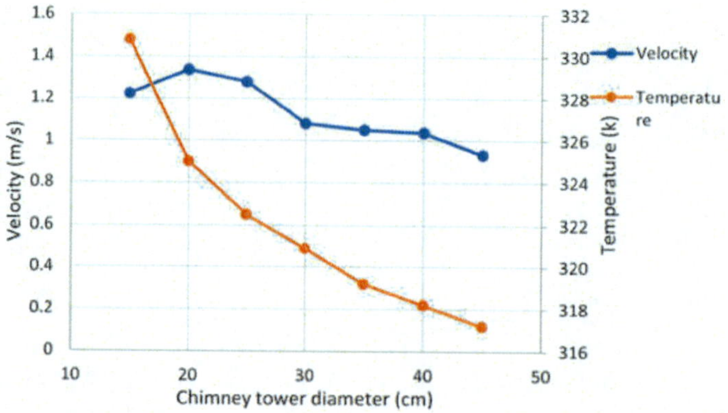

Fig. 9 The graph of velocity and temperature at the turbine with different diameter of chimney

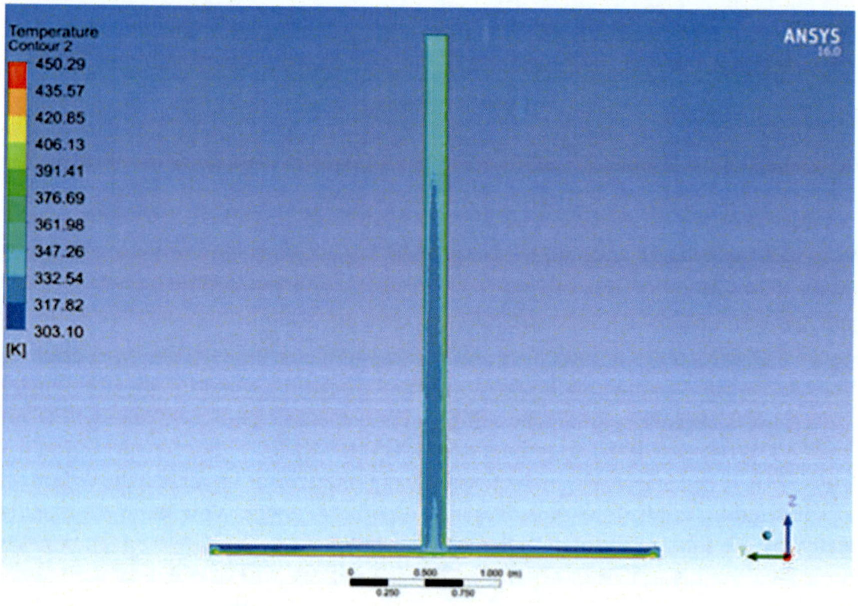

Fig. 10 The Temperature contour of SCPP when Chimney diameter at 20 cm

result, the increase in the diameter of the chimney is restricted by the power factor, which is critical for selecting the appropriate diameter for the tower. It is apparent that the diameter of the chimney tower has an important influence on temperature distribution, particularly at the entrance of the chimney tower, where a turbine will be installed.

3.4 Influence of Shape

Ramin et al. [14] proposed an alternative to the circular sun collector shown in Fig. 11 by using a square solar collector with the solar chimney positioned on the ends side of the square and raising the height of the solar collector. According to the findings of the study, changing the square solar collector produces a higher velocity at the turbine. However, since the chimney is near the back end of the square, the airflow from behind the chimney could be wasted. Therefore, in this study, the chimney is repositioned to the center of the solar chimney, allowing the air to travel in all directions, as illustrated in Fig. 12.

A varied chimney height is used in this study to compare the circular and square collector SCPPs. Each of SCPP's collectors, whether circular or square, have the same surface area. On the other hand, the peak velocity of square collectors can be achieved at shorter chimney tower heights than circular collectors, as demonstrated in Fig. 14. Fig. 15 depicts the increasing temperature of the intake regions between square and circular collectors, with the circular shape outperforming the square collector in terms of heat performance.

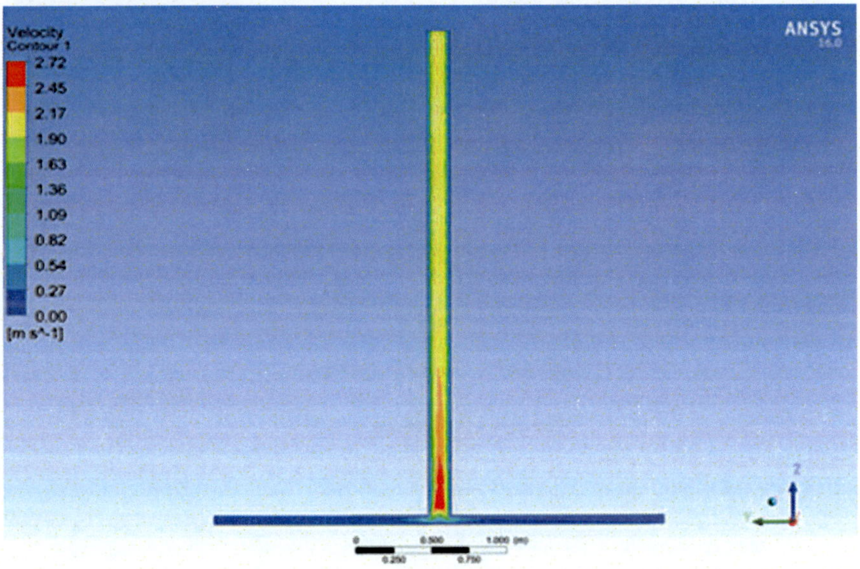

Fig. 11 The velocity distribution of SCPP when Chimney diameter at 20 cm

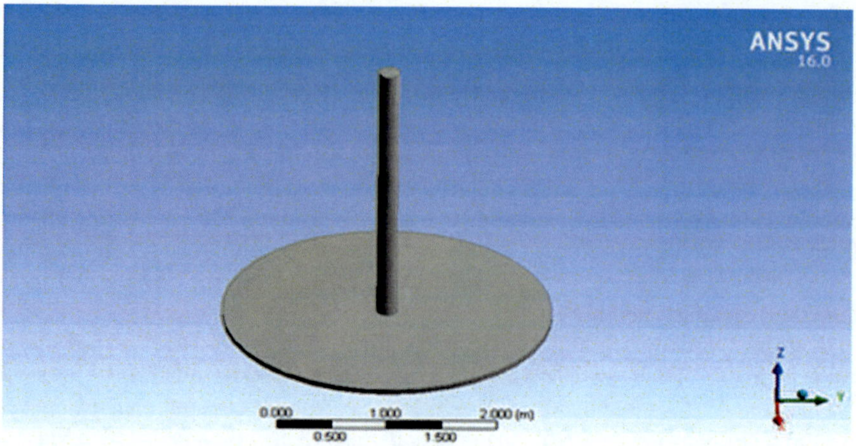

Fig. 12 Top view of the circular collector SCPP

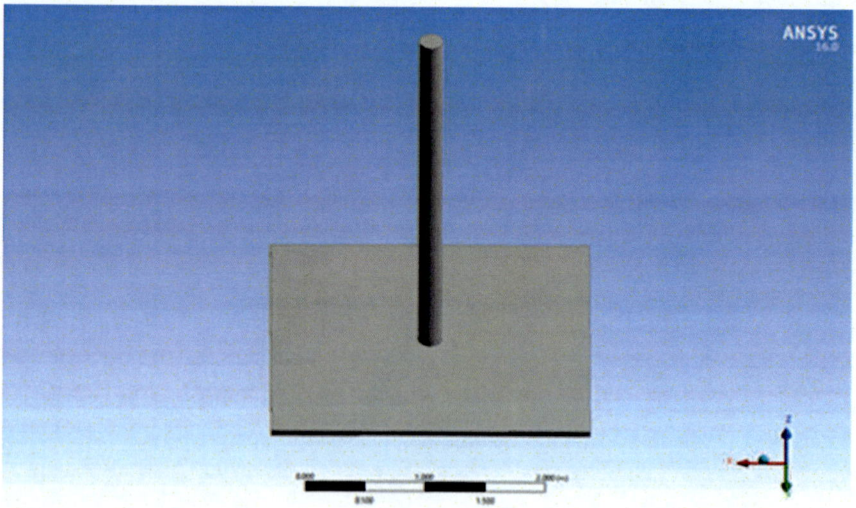

Fig. 13 Top view of the square collector SCPP

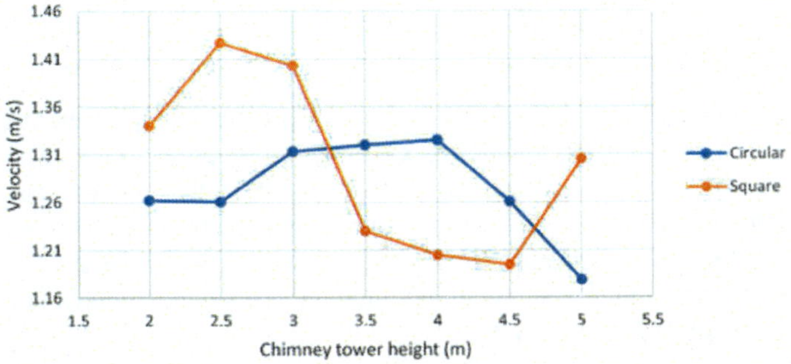

Fig. 14 The graph of velocity at the turbine with different height of chimney of circular and square collector SCPP

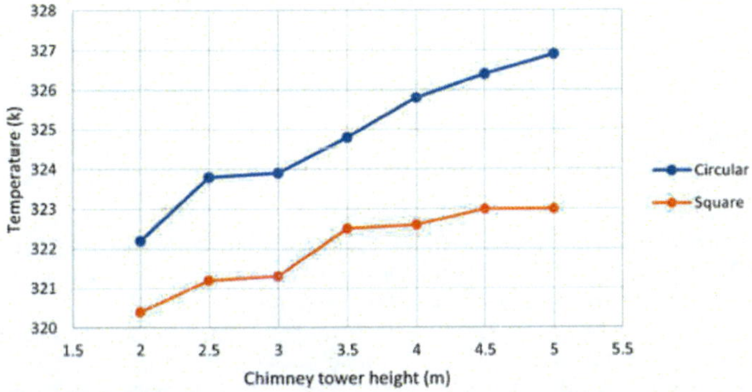

Fig. 15 The graph of average temperature of the SCPP with different height of chimney of circular and square collector SCPP

4 Conclusions

This study performed CFD simulation to determine the impacts of various parameters on the SCPP to identify the optimal configuration while designing a SCPP. The key findings of this study are organized into points below.

- For the chimney height, it is determined that the best height is 4.3 m, as this is the point at which both temperature and velocity are the greatest at 326.1 K and 1.293 m/s, respectively. The temperature and velocity increases as the chimney height increase, but as the chimney height exceeds 4.0 m, the velocity begins to decrease.
- The optimum diameter is 16 cm, where both temperature and velocity are at their peak value of 328 K and 1.28 m/s, respectively. The temperature and velocity begin to drop as the diameter of the chimney increases.
- When compared to circular collectors, square collectors are supposed to be the best built for the shape of the solar collector. The velocity produced by the square

collector is 1.427 m/s, which is higher than the velocity produced by the circular collector, which is 1.32 m/s.

The production of secondary flows such as free convection flow and reverse flow may be caused by the lower temperature of the square collector. As a recommendation, the results obtained can be improved by cross-referencing the data obtained to an experimental data. This is because Ansys CFD fluent excels at modelling and forecasting a flow in a controlled environment; however, such a controlled environment does not exist in the real world, and there are always extra elements that can disrupt and affect SCPP performance.

Acknowledgements This research was funded by Takasago Thermal Engineering Co. Ltd. Grant (R.K.130000.7343.4B422, R.K.1300007343.4B472). The authors also want to thank Universiti Teknologi Malaysia and Universiti Teknikal Malaysia Melaka as well as the Ministry of Higher Education Malaysia for giving a scholarship for this study.

References

1. Too JHY, Azwadi CSN (2016) Numerical analysis for optimizing solar updraft tower design using computational fluid dynamics (CFD). J Adv Res Fluid Mech Therm Sci 22(1):8–36
2. Saadun MNA, Sidik NAC (2020) Experimental study on the performance of small solar updraft tower in the climate region. Int J Automot Mech Eng 17(4):8372–8383
3. dos Santos Bernardes MA, Valle RM, Cortez MFB (1999) Numerical analysis of natural laminar convection in a radial solar heater. Int J Therm Sci 38: 42–50
4. Kirstein CF, von Backström TW (2006) Flow through a solar chimney power plant collector-to-chimney transition section. J. Solar Energy Eng 128(3):312–17
5. Tingzhen M, Wei L, Guoling X, Yanbin X, Xuhu G, Yuan P (2008) Numerical simulation of the solar chimney power plant systems coupled with turbine. Renew Energy 33(5): 897–905
6. Sangi R, Amidour M, Hosseinizadeh B (2011) Modelling and numerical simulation of solar chimney power plants. Sol Energy 85:829–838
7. Pasumarthi N, Sherif SA (1998) Experimental and theoretical performance of a demonstration solar chimney model—part i mathematical model development. Int J Energy Res 22:277–288
8. Pasumarthi N, Sherif SA (1998) Experimental and theoretical performance of a demonstration solar chimney model—part ii experimental and theoretical results and economic analysis. Int J Energy Res 22:443–461
9. Koonsrisuk A, Chitsomboon T (2013) Effects of flow area changes on the potential of solar chimney power plants. Energy 51:400–406
10. Jawad A, Misaran MS, Rahman MM, Ismail MA (2021) Experimental investigation on the effect of divergent tower solar chimney on the theoretical power potential. J Adv Res Fluid Mech Therm Sci. 81(1): 140–149
11. Ming T, Wang X, De Richter RK, Liu W, Wu T, Pan Y (2012) Numerical analysis on the influence of ambient crosswind on the performance of solar updraft power plant system. Renew Sustain Energy Rev 16(8): 5567–83
12. Ming T, Gui J, de Richter RK, Pan Y, Xu G (2013) Numerical analysis on the solar updraft power plant system with a blockage. Sol Energy 98: 58–69
13. Haythem N, Zied D, Ahmad A, Hedi K (2019) Numerical and experimental study of the aerothermal characteristics in solar chimney power plant with hyperbolic chimney shape. Arabian J Sci Eng. 44(9):7491–7504

14. Ramin M, Zahra B, Sajad G, Sohel M (2020) Geometry modification of solar collector to improve performance of solar chimneys. Renew Energy 162(2020):160–170
15. Saadun MNA, Sidik NAC, Muhammadu M (2018) Design and optimising of geometric for solar updraft tower using computational fluid dynamics (CFD). J Adv Res Fluid Mech Therm Sci 51(1):8–18
16. Walid M, Elmagid A, Istvan K, Lidiko M, Esmail FC, Tarek M (2019) Studying the collector performance of updraft solar chimney power plant. J Eng Technol 11(1):65–82
17. Daimallah A, Lebbi M, Lounici MS, Boutina L (2020) Effect of thermal collector height and radius on hydrodynamic flow control in small solar chimney. J Adv Res Fluid Mech Therm Sci 71(2): 10–25
18. Fluri TP, Von Backström TW (2008) Performance analysis of the power conversion unit of a solar chimney power plant. Sol Energy 82(11):999–1008
19. Lu Z, Pengzhan D, Zihan L, Ning Q, Ling D, Bo Q, Yue Y (2020) Numerical analysis of wind supercharging solar chimney power plant combined with seawater desalination and gas waste heat. ELESEVIER
20. Habibollahzade A, Houshfar E, Ashjaee M, Ekradi K (2020) Continuous power generation through a novel solar/geothermal chimney system: Technical/cost analyses and multi-objective particle swarm optimization. J Cleaner Prod :124666
21. Osama N, Ece A, Ekin O (2020) Numerical investigation on the performance of a small-scale solar chimney power plant for different geometrical parameters. J Cleaner Prod 276:22908
22. Launder BE, Sharma BI (1974) Application of the energy-dissipation model of turbulence to the calculation of flow near a spinning disc. Lett Heat Mass Transf 1(1974):131–138

Design and Simulation of Diffuser Augmented Wind Turbine (DAWT) for Urban Areas

Delan S. Bacus and Cresencio P. Genobiagon Jr.

Abstract This paper is concerned with developing a shrouded design for a diffuser-augmented wind turbine (DAWT) suitable for use in urban areas, such as on the rooftops of buildings or houses. This study would specifically investigate the effects of the developed designs on wind pressure and velocity conditions at the diffusers' inlet and outlet areas. Using the SOLIDWORKS application, two design concepts were created: single-layered and double-layered. The first concept used a single cone-shaped structured diffuser, whereas the second concept used a cone-shaped structure but with a smaller cone-shaped structure placed inside the first structure. To determine the pressure and velocity variations along the diffuser's inlet and outlet areas, these concepts were simulated using the ANSYS Fluent® application. The results strongly favored the double shrouded design of Concept 2 in both velocity and pressure simulations, with a velocity increment of 102% and a drag force of −0.161 N. The actual test results also show promising data that would make the prototype feasible for use in an urban area, with an average efficiency of 40% and a high efficiency compared to other urban-built wind turbines.

Keywords Diffuser Augmented Wind Turbine · Simulation · Urban area

1 Introduction

Wind energy is one of the cleanest and most sustainable ways to generate electricity. Reasons why many researchers are trying to develop and innovate a range of different designs of a wind turbine to make it more convenient, efficient, and effective [1].

Supplementary Information The online version contains supplementary material available at https://doi.org/10.1007/978-981-19-1577-2_44.

D. S. Bacus (✉) · C. P. Genobiagon Jr. (✉)
Master of Engineering-Mechanical Engineering Program, Professionals School University of Mindanao, Matina, Davao, Philippines
e-mail: dbacus@umindanao.edu.ph

C. P. Genobiagon Jr.
e-mail: cresencio_genobiagon@umindanao.edu.ph

© The Author(s), under exclusive license to Springer Nature Singapore Pte Ltd. 2023

M. H. A. Hassan et al. (eds.), *Technological Advancement in Instrumentation & Human Engineering*, Lecture Notes in Electrical Engineering 882,
https://doi.org/10.1007/978-981-19-1577-2_44

Developing efficient and cost-effective wind turbines for the urban is a new application that can further reduce dependency on fossil fuels, thus reducing greenhouse gas emissions. In Southeast Asia, the Philippines ranks number 2 for having a high electricity tariff as the end of July 2018, posting an average industry rate of P5.84 per kWh. The country also accounts for having outstanding commercial and household rates of P7.49 and P8.90 per kWh, respectively [2]. Thus, paying electrical bills in a simple household in an urban area proves to be burdensome. A case study conducted by the GreenPeace organization in the Visayas estimated that premature deaths are attributable to operating coal-fired power plants and the projected premature deaths from the joined emissions of existing and new plants [3]. These financial, environmental and health risk suggests exploring and developing a much cleaner and cost-friendly alternative to the recent power generation machinery in the country today. To address these challenges in power production, this study deals with designing a shroud for a diffuser augmented wind turbine suitable in urban areas. However, a challenge for urban wind energy is the lower mean wind speeds in the urban environment. This is due to an increased surface roughness of the free stream winds and reduced installation heights of small wind turbines. Next, the incoming wind will have higher turbulence intensity [4]. The researchers have chosen to use a diffuser augmented type wind turbine to respond to these challenges, commonly known as D.A.W.T.

D.A.W.T. is best described as a wind turbine modified with a cone-shaped wind diffuser. In other words, the rotor blades of the wind turbine are set inside a diffuser. The addition of a diffuser to a wind turbine has been shown to increase power output for a given wind speed [5] and to maintain power production capabilities in turbulent conditions, making a DAWT more suitable for small-scale energy production in an urban area than a traditional Horizontal Axis Wind Turbine [6]. A comparative study showed that reducing power output in the urban built environment would be well compensated by augmenting the turbine with a diffuser [7].

Matsushima, Takagi, and Muroyama [8] studied the improved effects of a diffuser on the output power of a wind turbine. They found out that the length and expansion angle greatly influence the wind speed inside the turbine. Similar findings were also observed in the study conducted by Abe and Ohya [5]; the results showed that the performance of the diffusers significantly depends on the expansion angle of the diffuser. Kannan et al. [8] investigated an optimal design for a diffuser by simulating different geometry of the diffusers. The result showed that an optimal diffuser with a 61.25% velocity increase has dimensions of 16° diffuser opening angle coupled with a 0.5 m diffuser splitter and a 4° splitter opening angle.

Previous studies focus more on the effects on the wind characteristics brought about by the changes in the diffuser characteristics. Although these studies investigated these in both field and wind tunnel tests, implementation in an account of actual conditions on an urban built environment is uncommon; thus, performance data in actual life application of these turbines is scarce [10].

This study primarily aims to design a shroud for a diffuser augmented wind turbine that will suit the conditions in an urban area, particularly the urban settings in Davao city. To achieve this objective, the study would design an optimal diffuser ideal for urban environments. Like the previous studies, it also investigates the pressure and velocity variations that would develop along the inlet and outlet area of the design concepts using simulation programs. This study also aims to fabricate a prototype for the simulation's most optimal design and then test and calculate the maximum power output it generates.

This research will create a wind turbine system that will help Filipino households in urban areas, such as subdivisions, produce their own electricity, thereby bypassing the structure of the Philippine energy industry and lowering costs by avoiding the charges imposed by the various sectors of this industry. At the same time, this research may reduce carbon emissions produced by thermal plants. Thus, reducing pollution and the health risk it may impose on the people. It would help ordinary Filipino people financially in producing energy, but it would also help the environment be much safer and more sustainable.

This would focus on developing a design for a shroud of a diffuser augmented wind turbine regarding the diffuser's geometry only. Excluded from this study are factors that could potentially affect the concepts like the rotor blade and electrical system. This study is limited to determining the variations of wind pressure and velocity along the inlet and outlet peripheries of the design concepts. Davao City's urban area is the research locale.

2 Materials and Methods

This study is an applied type of research where the academe's results from various theoretical research are used to give a probable solution to multiple problems. This study provides a solution for residential households to minimize dependence on the grid in producing electricity using clean energy. It is achieved by developing a design for a diffuser augmented wind turbine system through improving and adapting past pieces of literature.

2.1 Conceptual Framework

Figure 1 represents the conceptual framework of the study, where it shows an input-process-output type of concept. It suggests that the input of this study is the gathering of all the data needed and calculations considering the related literature of past reviews. After gathering the data, it would go through the processes involving the design and simulation process, fabrication, evaluation, and testing of the study. The success of these reveals the final output.

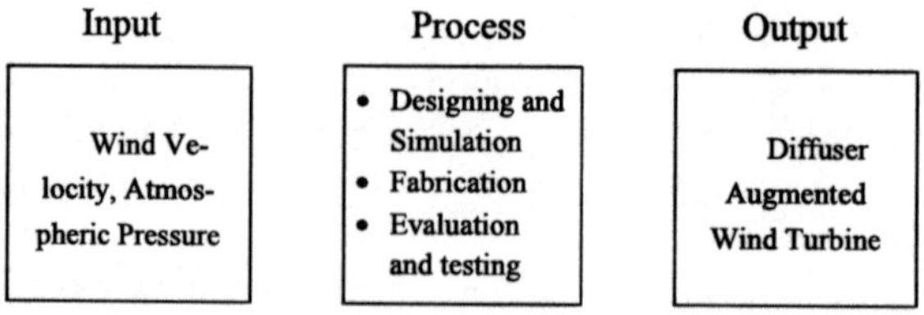

Fig. 1 Conceptual framework

2.2 *Material and Resources*

To accomplish this study, various equipment is needed. An ABS filament will serve as the material for the 3D printing of the diffuser. Ceramic Bearing was used to maintain proper alignment of turning parts and enable high-speed rotation. A Tachometer is a device used to measure the rotational speed of the turbine. A spring balance is a device used to measure the force induced by the rotating shafts.

2.3 *Design Conceptualization*

This would compare two design concepts that would achieve the goal of the study. The design concepts were drawn using SOLIDWORKS application in adherence to the previous scholarly works and related literature. Two designs concepts were developed, namely the single-layered and the double layered concepts. The first concept utilized a single cone-shaped structured diffuser. The second concept also utilized a cone-shaped structure; however, it differed by placing a smaller cone-shaped structure inside the first structure.

2.4 *Flow Simulations*

This utilized ANSYS Fluent®, Flow Simulation application in investigating the pressure and velocity variation needed in this study. ANSYS Flow Simulation is an intuitive Computational Fluid Dynamics (CFD) solution embedded within ANSYS Fluent® that enables the researcher to simulate liquid and gas flows quickly and easily through and around the designs [11]. Using steady-state 3D conservation equations of mass and momentum for incompressible turbulent flow [17], the computational analysis was carried out, with the governing equations being as follows:

$$\frac{\partial u_i}{\partial x_i} = 0 \tag{1}$$

$$\frac{\partial \rho u_i u_j}{\partial x_j} = -\frac{\partial p}{\partial x_j} + \rho g_i + \frac{\partial}{\partial x_j}\left[(\mu + \mu t)\left(\frac{\partial \mu_i}{\partial x_j} + \frac{\partial \mu_i}{\partial \mu_j}\right)\right] \tag{2}$$

The symbol represents the velocity component in the *i* direction, while the pressure, density, laminar viscosity, turbulent viscosity, and gravitational force are represented by the symbols p, ρ, μ, μt and ρg_i, respectively [18–21].

The simulation was set to be an incompressible flow, and the designs were assumed to be adiabatic. Initial parameters were set to have a 1.5 m/s velocity, the average wind velocity in Davao city. Placing pressures are at atmospheric pressures of 101.325 kPa. The simulation also used a single reference frame (SRF) model to simulate the incompressible and steady-state flow field.

2.5 Fabrication

The chosen optimal design was then fabricated using a 3D printer. The diffuser was divided into six parts and assembled using an epoxy supported by a 3D printing pen. The material used for printing was ABS plastic. Ceramic Bearing will maintain proper alignment of turning parts and enable high-speed rotation of the blade

2.6 Testing and Evaluation

The prototype was then tested to determine the power output and its coefficient of performance. Subjecting the rotor shaft of the turbine with a radius of 4 mm to a belt pulley system determines the power output. Whereby a belt is subjected to the rotating shaft and two spring balance is attached to its end. Using a tachometer determines the angular speed *(f)*. The power transmitted *(P_{out})* by the shaft, as stated by [12], would be then computed using the formula presented in Eq. (3)

$$P_{out} = 2\pi f T \tag{3}$$

whereby torque *(T)* is determined by net belt pull *(F_1 and F_2)* of the spring balance and radius as presented in Eq. (4)

$$Torque = (F_2 - F_1)x\,radius \tag{4}$$

The coefficient of performance *(Cp)* as discussed in [9] is determined using the formula as presented in Eq. (3)

$$Cp = \frac{P_{out}}{0.5x\rho_{air}xAxv^3} \tag{5}$$

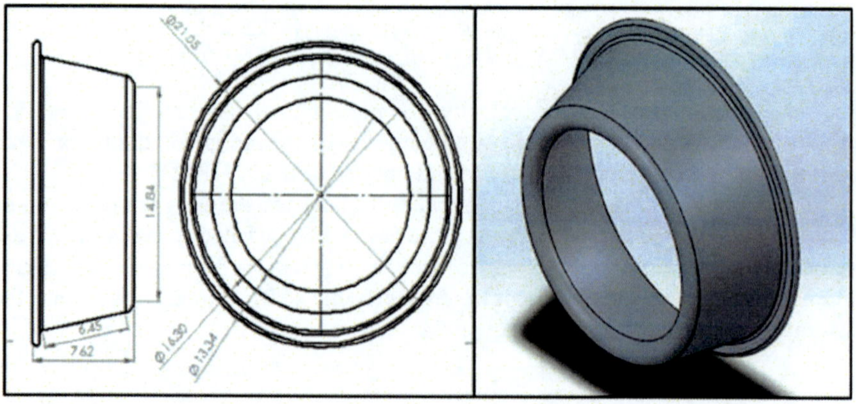

Fig. 2 Design Concept 1

Equation (5) sets the density (ρ_{air}) as 1.2 kg/m^3 and velocity (v) as 1.5 m/s following the environmental conditions of the research locale. The swept area (A) is computed using the diameter of the diffuser rather than the blade diameter [13].

3 Results and Discussions

3.1 Design Concepts

Concept 1, as shown in Fig. 2, is a single shroud diffuser with dimensions at the left and the 3D visualization at the right. A 12° angle from the research of Kannan et al. [8] becomes the basis for the design angle.

Concept 2 as shown in Fig. 3 is a double shrouded diffuser with dimension presented in the left and the 3D visualization at the right. This design concept utilized two cone-shaped diffusers by surrounding a diffuser by a larger one.

3.2 Pressure and Velocity Simulations

Figure 4 presents the velocity cut plot simulation of the two concepts. The cut plot simulation revealed that Concept 2 has a higher velocity of 3.04 m/s at the turbine location than Concept 1 of 2.3 m/s at the same spot. It is an increase of 102% by Concept 2. The results contradict the study conducted by Kannan et al. [8], which favors a single shroud compared to a double shroud.

Figure 5 presents the pressure cut plot simulation between the two concepts. Based on the simulation, it is evident that there is a bigger pressure difference in Concept

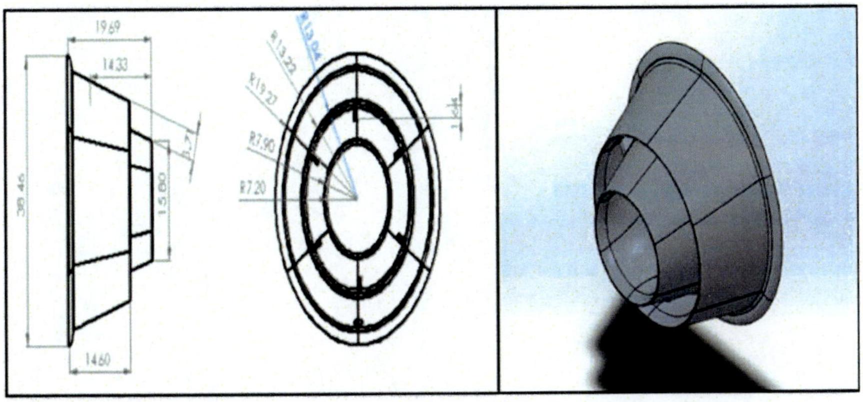

Fig. 3 Design Concept 2

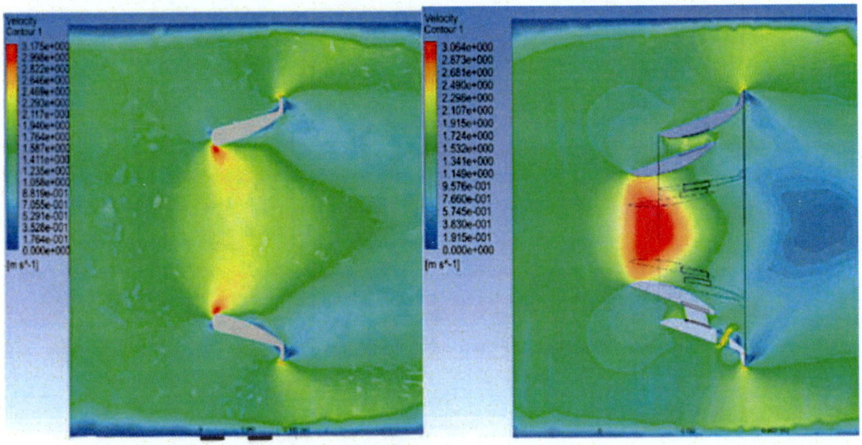

Fig. 4 Velocity cut plot of single and double shroud diffuser

2 than in Concept 1. The simulation also suggests a more considerable pressure variation from the double shrouded diffuser than the single shroud. The drag force generated during the simulation was the single shroud–0.0361 N while the double shroud has–0.161 N, which means that the double shroud created a considerable vacuum pressure at the outlet that affects the increase of wind velocity and pressure.

3.3 Design Fabricated

Figure 6 presents the fabricated design. The prototype uses ABS plastic as the filament for printing in the 3D printer. It has a red color due to the material that was used. The

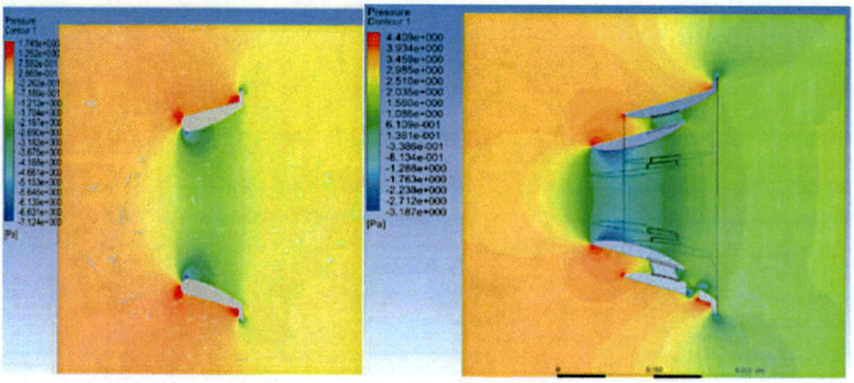

Fig. 5 Pressure cut plot of single and double shroud diffuser

Fig. 6 Double shroud diffuser augmented wind turbine

diffuser is made up of six parts, and epoxy was used to assemble them. However, using 3D printing has inevitable setbacks in the accuracy of the sizes.

3.4 Testing and Evaluation

Table 1 presents the actual operational data gathered from the prototype. The data suggest that the turbine augmented by a double shrouded has an average power output of 0.0309 watts with an efficiency of 40%. This efficiency is within the range for typical horizontal axis wind turbines as presented in [14], where current typical HAWTs are within 30 to 50% efficiency and below the Beltz's limit [15]. However, these ranges are for a typical wind turbine where urban built conditions are not

Table 1 Actual data from the prototype double shroud diffuser augmented wind turbine

F1 (Newton)	F2 (Newton)	Torque (Joules)	RPM	Power (Watts)	Cp
0.981	0.4905	0.001962	149.25	0.0307	0.392
0.981	0.4905	0.001962	151.89	0.0312	0.399
0.981	0.4905	0.001962	151.51	0.0311	0.398
0.981	0.4905	0.001962	150.37	0.0309	0.395
0.981	0.4905	0.001962	150.01	0.0308	0.394
Average				0.0309	0.396

recognized. In another study, efficiency ranges from 18 to 41% when comparing the current available urban-built wind turbines [16]. Thus, having a coefficient of performance of about 40% is very high compared to the current available urban wind turbines. It is almost on par with a swift turbine, as stated in [16]

4 Conclusions and Future Works

This study compared two design concepts for a diffuser augmented turbine that would be more optimal to be used in an urban area, using ANSYS Fluent® Simulation. The results favored the double shrouded design of concept 2 in both velocity and pressure variations of the simulations, with a velocity increment of 102% and a drag force of −0.161 N. The actual test also shows promising data that would make the prototype feasible to be used in an urban area that yielded an average of 40% efficiency. It has a high efficiency compared to other urban-built wind turbines. It means that a double shroud diffuser is more optimal for diffuser geometry to be used in an urban area than a single shroud, having a very high efficiency compared to current available urban wind turbines. Using 3D printing to fabricate the device makes it more economical than the present available urban wind turbines.

The research further recommends including an electrical system for this device and the blade design as part of the optimization analysis for the turbine. A supplementary study on the effects of losses to the turbine is also needed.

References

1. Jaber S (2014) Environmental impacts of wind energy. J Clean Energy Technol 57:251–254
2. Rivera D (2018) Philippine Electricity Rates Still Highest in Southeast Asia, 24 August 2018. www.philstar.com. Accessed 20 November 2018
3. Coal: A Public Health Crisis. Internet. https://www.greenpeace.org/seasia/ph/PageFiles/718084/Coal_A_Public_Health_Crisis.pdf. 29 November 2003
4. Stathopoulos T et al (2018) Urban wind energy: some views on potential and challenges. J Wind Eng Ind Aerodyn 179:146–157

5. Ohya Y, Karasudani T, Sakurai A, Abe K-I, Inoue M (2008) Development of a shrouded wind turbine with a flanged diffuser. J Wind Eng Ind Aerodyn 96:524–539
6. Kesby J, Bradney D, Clausen P (2016) Determining diffuser augmented wind turbine performance using a combined CFD/BEM method. J Phys Conf Ser 753:082033
7. Kc A, Whale J, Urmee T (2018) Urban wind conditions and small wind turbines in the built environment: a review. Renew Energy 13:268–283
8. Matsushima T, Takagi S, Muroyama S (2005) Characteristics of a highly efficient propeller type small wind turbine with a diffuser. Renew Energy 31:1343–1354
9. Kannan TS, Mutasher SA, Lau YH (2013) Design and flow velocity simulation of diffuser augmented wind turbine using CFD. J Eng Sci Technol 8:372–384
10. Dilimulati A, Stathopoulos T, Paraschivoiu M (2018) Wind turbine designs for urban applications: a case study of shrouded diffuser casing for turbines. J Wind Eng Ind Aerodyn 175:179–192
11. The Solidworks website (2002). https://www.solidworks.com/product/solidworks-flow-simulation
12. Beer FP, Johnston ER, Dewolf JT, Mazurek DF (2012) Mechanics of Materials, 6th edn. McGraw-Hill, New York
13. Heikal HA, Abu-Elyazeed OS, Nawar MA, Attai YA, Mohamed MM (2018) On the actual power coefficient by theoretical developing of the diffuser flange of wind lens turbine. Renew Energy 125:295–305
14. Schubel PJ, Crossley RJ (2012) Wind turbine blade design. Energies 5:3425–3449
15. Bukala J, Damaziak K, Kroszczynski K, Krzeszowiec M (2015) Investigation of parameters influencing the efficiency of small wind turbines. J Wind Eng Ind Aerodyn 146:29–38
16. Ani SO, Polinder H, Ferreira JA (2011) Energy yield of small wind turbines in low wind speed areas. In: 3rd IEEE international conference on adaptive science and technology
17. Yang AS, Su YM, Wen CY, Juan YH, Wang WS, Cheng CH (2016) Estimation of wind power generation in dense urban area. Appl Energy 171:213–230. https://doi.org/10.1016/j.apenergy.2016.03.007
18. ANSYS I (2013) ANSYS ICEM CFD 15.0 User's Manual
19. Doormaal JPV, Raithby GD (1984) Enhancements of the simple method for predicting incompressible fluid flows. Numer Heat Transfer 7:147–163
20. Jang DS, Jetli R, Acharya S (1986) Comparison of the PISO, SIMPLER, and SIMPLEC algorithms for the treatment of the pressure-velocity coupling in steady flow problems. Numer Heat Transfer 10:209–228
21. Ramponi R, Blocken B, de Coo LB, Janssen WD (2015) CFD simulation of outdoor ventilation of generic urban configurations with different urban densities and equal and unequal street widths. Build Environ 92:152–166

Modeling the Effect of Different Locations of Carotid Atherosclerosis on Hemodynamics Parameters

A. Fahmi Huwaidi M. Noor and Nasrul Hadi Johari

Abstract Carotid atherosclerosis is a potentially fatal diseases caused by plaques buildup in arteries that deliver blood to the brain. Over the years, this pathological condition particularly the stenosis size severity has been studied, and computational fluid dynamics has played an essential role in investigating the blood flow behavior. However, the study on the effect of stenosis location especially at the area of bifurcation is still lacking. This study aims to investigate which stenosis location would cause higher risk factor and high flow disturbance to the blood flow. The area susceptible for atherosclerosis is usually at the carotid bifurcation where common carotid artery (CCA) is bifurcated into internal carotid artery (ICA) and external carotid artery (ECA). The computational simulations were performed using idealized geometry of carotid artery with different locations of stenoses. Type I, II and III stenoses were grouped according to the most common type and location stenosis. The results show that the Type I geometry with stenosis extended toward the ICA had higher possibility for the atherosclerosis plaque to grow. Velocity profiles and low wall shear stress contours predicted more complex helical and recirculation blood flow at post-stenotic region of Type I as compared to the other two. The findings indicate that atherosclerosis plaque in the ICA could provide higher risk to the patient and immediate medical treatment shall be required.

Keywords Atherosclerosis · Computational fluid dynamics · Location of stenosis

Supplementary Information The online version contains supplementary material available at https://doi.org/10.1007/978-981-19-1577-2_45.

A. F. H. M. Noor · N. H. Johari (✉)
Faculty of Mechanical and Automotive Engineering Technology, Universiti Malaysia Pahang, 26600 Pekan, Pahang, Malaysia
e-mail: nhadi@ump.edu.my

Centre for Human Technology, Universiti Malaysia Pahang, 26600 Pekan, Pahang, Malaysia

© The Author(s), under exclusive license to Springer Nature Singapore Pte Ltd. 2023
M. H. A. Hassan et al. (eds.), *Technological Advancement in Instrumentation & Human Engineering*, Lecture Notes in Electrical Engineering 882,
https://doi.org/10.1007/978-981-19-1577-2_45

1 Introduction

Cardiovascular disease (CVD) was reported by the World Health Organization (WHO) as the number one cause of death globally for the past 15 years (WHO, 2016). Atherosclerosis is a major contributor to CVD that causes thickening and hardening of vessel walls due to growth of plaques. The atherosclerosis plaque accumulates and grows mostly in the tunica intima, the innermost layer of arteries which has direct contact with blood. The plaque includes the dead foam cells, macrophages, smooth muscle cells and extracellular matrix [3]. The growing plaque will gradually reduce the lumen's diameter to become narrow and eventually increase the stiffness of the arterial wall that affects its compliance characteristic. The inflammatory process may trigger the blood contain with high level low of density lipoprotein (LDL) as well as abnormal wall shear stress [1, 2].

Computational modeling has been widely utilized to analyze the blood flow behavior in stenotic arteries, resulting insights into the cardiovascular disease condition, its progression and therapeutic optimization [5, 6]. The simulations with the assistance of medical imaging techniques for the geometry and realistic boundary conditions are done either using computational fluid dynamics (CFD) or fluid structure interaction (FSI). CFD focuses on hemodynamics parameters in geometries with rigid wall, and only fluid domain is considered. If the wall is responding to the blood flow which usually referred as complaint, both fluid lumen and solid wall domains are often referred to FSI. CFD has been applied in many cardiovascular computational simulations especially in investigating the critical role of hemodynamics to predict area that is prone to stenosis.

Previous studies have investigated the effect of severity size of stenosis to the blood flow [7], the influence of bifurcation angle [8], and turbulence model for stenosis cases. DiCarlo et al. [7] has analyzed the effect of different stenosis sizes i.e. 30, 50 and 70% on the hemodynamics parameters. The study reported that only the 70% stenosis size affects the blood flow to become abnormal at the downstream of the stenosis. Hence, a proper turbulent model is required to visualize the flow vorticity and eddy viscosity at that area. Johari et al. [9] has compared the effectiveness of two promising turbulence models i.e. Reynold-averaged Navier-Stokes (RANS) shear stress transport- transitional model (SST-Tran) and large eddy simulation (LES). Both SST-Tran and LES models were compared with particle image velocimetry (PIV) measurement to evaluate the velocity profiles in a patient-specific carotid bifurcation model with 80% stenosis size. Both SST-Tran and LES predicted the experimental velocity profiles reasonably well, with LES being slightly superior especially at the post-stenotic area. However, SST-Tran also managed to capture important flow features as observed in the experiment although slightly differed distal to the stenosis. Other studies like in [10, 11] also reported the influence of stenosis sizes in idealized straight carotid and patient-specific carotid geometry [9] that extend the flow separation and recirculation zone distal to stenosis.

Hemodynamic parameters i.e. blood pressure, time-averaged wall shear stress, relative residence time, and oscillating shear index are important in the onset and development of atherosclerosis [14, 15] and have been extensively studied. Abnormal flow condition such as flow separation and flow recirculation in areas with low or high shear stress have shown to play a role in the formation or development of atherosclerosis. The plaque vulnerability correlated with shear stress distribution and regime. The low velocity and recirculation flow occurring near the carotid bifurcation and carotid sinus also contribute progression of atherosclerosis [16].

Based on the previous carotid stenosis studies, the severity of stenosis size is reported to cause different levels of disturbed flows especially in the downstream of the stenosis. However, the study on the effect of location of the severe stenosis is still lacking. Hence, the present study aimed to investigate the influence of stenosis locations [17] that contribute to higher risk of flow disturbances. The decrease of velocity field and flow separation zone will result in low wall shear stress that could potentially initiate progression of atherosclerosis [18].

2 Methodology

2.1 *Geometry Reconstruction and Mesh Generation*

The study had reconstructed three idealized carotid artery bifurcations with different types of stenosis based on a recent clinical report by Lu et al. [17]. The geometries have a 70% stenosis size at the common carotid artery (CCA) to internal carotid artery (ICA), mostly in the carotid bulb but in different shape and location (Fig. 1). The type I, II and III are based on atherosclerotic plaques in the carotid arteries measured by magnetic resonance images (MRI). For type I, the arc-length of plaque is extended to the upstream, type II is when the plaque is equal in upstream and downstream whilst type III is the plaque is extended to the downstream. Due to the lack of statistical dimensions of the patients' carotid geometries in [17], an idealized geometry of carotid bifurcation was used for the fluid domain (Fig. 2). Then, the shape

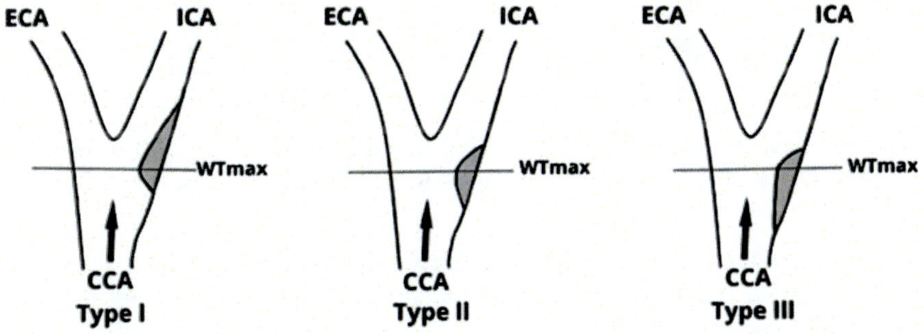

Fig. 1 Different types of atherosclerosis plaque shapes at the area of CCA-ICA. WTmax is the maximum wall thickness measured [17]

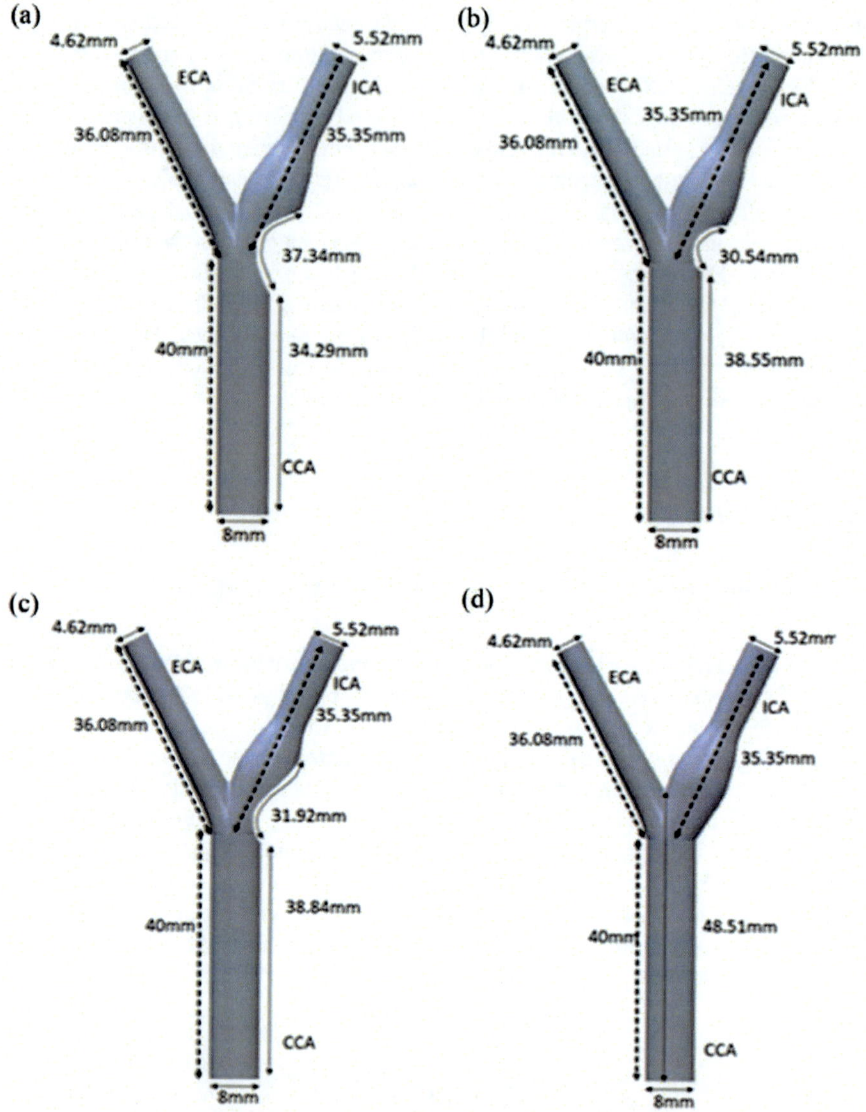

Fig. 2 Idealised geometry model **a** Type I (upstream stenosis), **b** Type II (equal upstream and downstream), **c** Type III (downstream stenosis) and **d** healthy carotid artery

of atherosclerotic plaques was reconstructed and attached to the idealized geometry according to the type I-III. A healthy carotid bifurcation was also reconstructed for the control geometry.

Table 1 Mesh element numbers of different geometries

Geometry	Mesh element 1	Mesh element 2	Mesh element 3
Healthy artery	109,678	400,705	896,171
Upstream stenosis (type I)	204,316	550,394	1,039,971
Middle stenosis (type II)	121,622	426,678	896,171
Downstream stenosis (type III)	89,841	317,747	653,033

The stenosed carotid geometries were exported to ANSYS WORKBENCH for mesh generation. Unstructured tetrahedral elements were adopted for all geometries with local refinement around the area of stenosis. Grid independence was tested for simulations using the SST-Tran turbulence models in stenotic geometries by comparing post-stenotic velocity profiles obtained with different mesh sizes (Table 1). The difference in terms of peak velocity was more than 5% for mesh element 1 (course mesh) compared to the medium mesh element size i.e. mesh element 2. However, the percentage difference between mesh element 2 and 3 was approximately 4 to 5%. Hence, mesh element 2 was used in the study.

2.2 *Flow Simulations*

The RANS-based SST-Tran model [19, 20] was adopted in the simulations. The SST-Tran model combines $k-\varepsilon$ and $k-\omega$ models where the $k-\varepsilon$ is dedicated to resolve flow in the inner region of boundary layer whilst the $k-\omega$ is for the outer region and free shear flows. The model has additional of a transitional model comprising two additional formulated transport equations for intermittency and the transition onset criterion in terms of the momentum thickness Reynolds number [21]. The correlation between the $k-\varepsilon$ and $k-\omega$ together with the transition model has been successfully employed in cardiovascular flow applications with promising results [9, 11, 22]. For the healthy model, laminar model was adopted in the simulation.

The blood flow is assumed to be steady and incompressible Newtonian flow with blood density of 1060 kg/m3 and viscosity of 0.0035 Pa.s. The fully developed flow profile was specified at the CCA inlet with typical Reynolds number of Re = 1940. At the ECA and ICA outlets, a relative 0 Pa of gauge pressure was applied. The walls were assumed to be rigid with no-slip conditions. All simulations to solve the governing equations were completed using a finite volume based CFD code ANSYS Fluent.

The pre-processor, solver, and post-processor are all part of the computational fluid dynamics method. The contribution of the development model from geometry and mesh recognition is included in the first technique. Next, the actual computations are performed by the solver, and in this solving phase, computational simulation is required using ANSYS software. Finally, in the post-processing phase, the collected findings are displayed and evaluated.

3 Results and Discussion

3.1 *Flow Patterns (3D Velocity Streamlines, 2D Velocity Streamlines, Velocity Vectors)*

Figure 3 illustrates the velocity streamline in the three models of stenotic carotid artery with different locations of stenosis patches, and a healthy carotid model. The healthy model was observed with low velocity profiles in the carotid bulb due to localized expansion of lumen area where flow is recirculating. Figure 3 for a, b and c show area for recirculating flow will initiate high risk of progressive plaque on blue arrow. The presence of stenosis had dramatically altered the flow pattern, particularly distal to stenosis with higher velocity at the stenosis throat, flow separation immediately after stenosis and strong helical flow structure distal the stenosis (Fig. 4). Similar observation was also recorded in the previous study by [9–13].

Generally, flow in Type I–III geometries have slightly similar flow pattern but differed in the proximal and distal regions of artery. Type 1 shows a complex helical post-stenotic flow with low velocity streamlines before reattached to the axial flow stream. This indicates the area of that prone to the stenosis to grow more and extend [13]. Meanwhile velocity streamlines in Type II and III are comparable with the large area of low velocity flow recirculates at distal to stenosis.

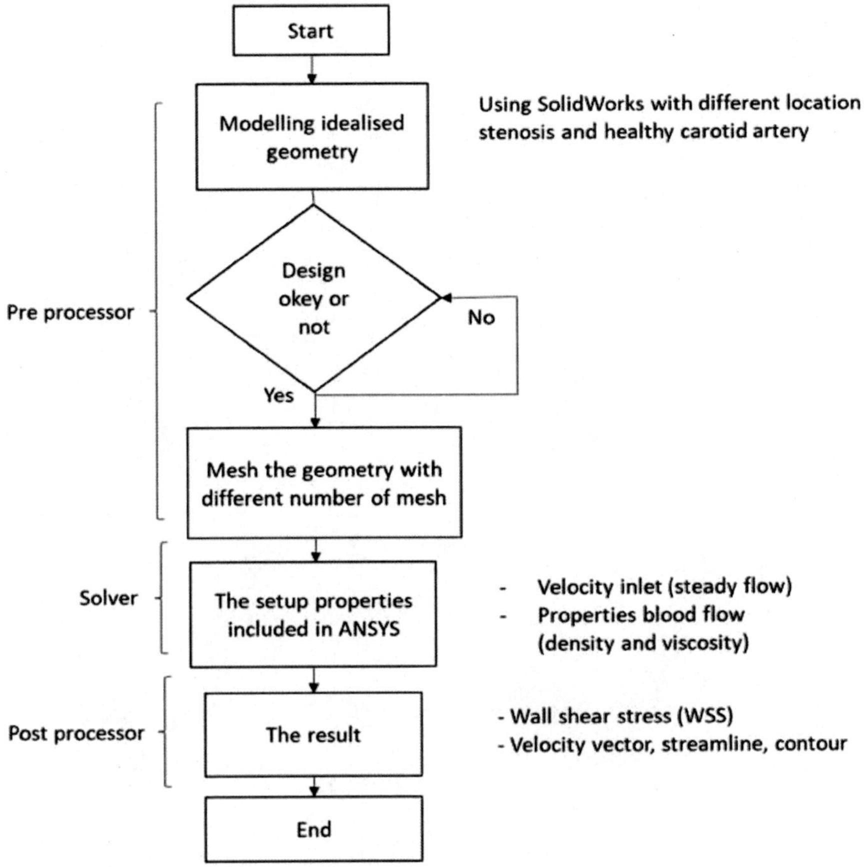

Fig. 3 Flow chart process of computational study

To further examine the different of velocity profiles, Fig. 5 shows velocity vectors in all carotid models. The arrows represent the flow direction, while the color shown the velocity magnitude. It is clear that 70% stenosis has altered the flow with different locations of recirculation zones as compared to the healthy model. However, detail comparison between Type I-III is not really obvious, probably due to the similar size of stenosis. On the other hand, high-density of velocity vector in Type I is recorded immediately after the stenosis and closer to the outer wall.

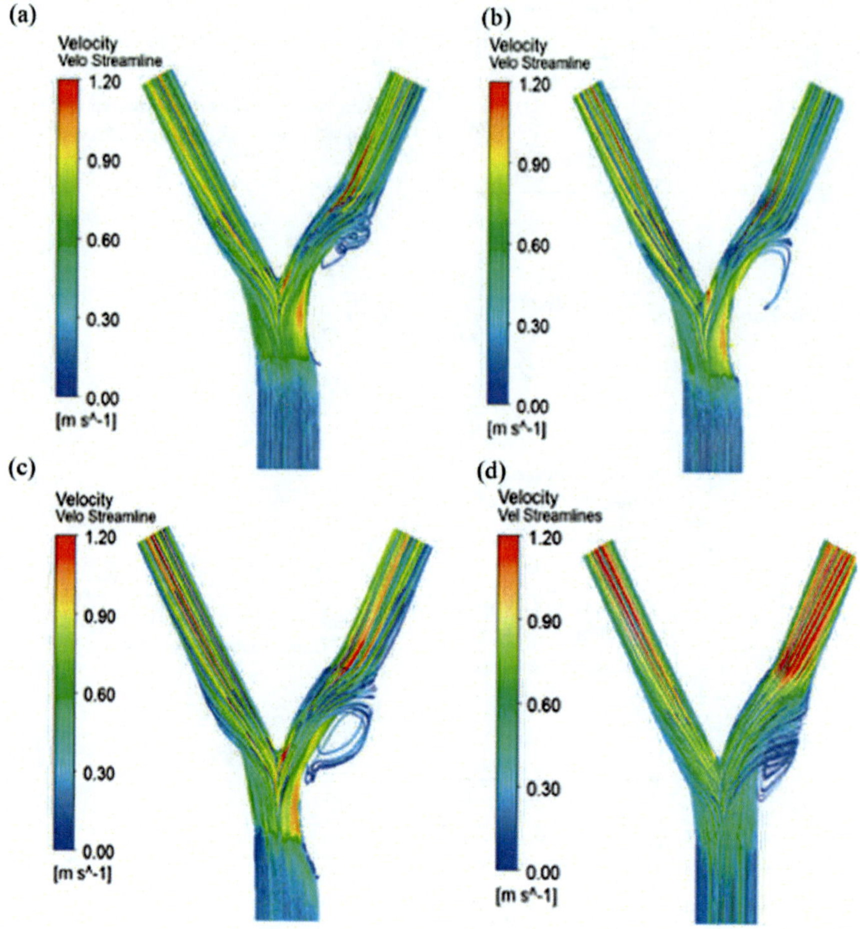

Fig. 4 3D velocity streamline in **a** Type I, **b** Type II, **c** Type III and **d** Healthy model

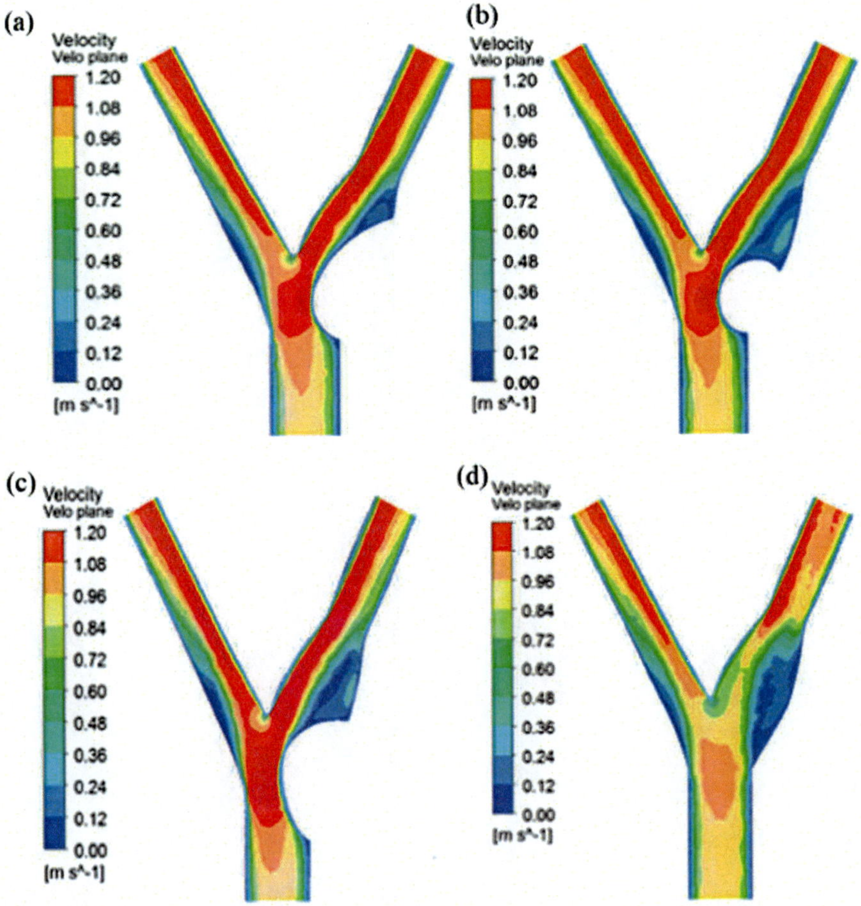

Fig. 5 Velocity contours at mid-plane in **a** Type I, **b** Type II, **c** Type III and **d** Healthy model

3.2 Wall Shear Stress

Wall shear stress was evaluated to identify regions of adverse hemodynamic conditions that might lead to plaque development. As shown in Fig. 6, the maximum WSS contour was set to be 10 Pa for a consistent comparison between different models, and to focus more WSS in the carotid bulb and in the ICA. The WSS in the healthy carotid model was relatively low. As expected, very low WSS values observed in the carotid bulb is due to localized flow recirculation. Compared to the healthy model, WSS in the stenotic carotid models exhibited more spatial variation. The 70% stenosis caused high WSS at the throat, and in the distal region along to the inner wall impinged by high velocity flow. Wall shear stress (WSS) directly proportional to velocity gradient can compare with Figs. 4 and 6. Value of WSS will

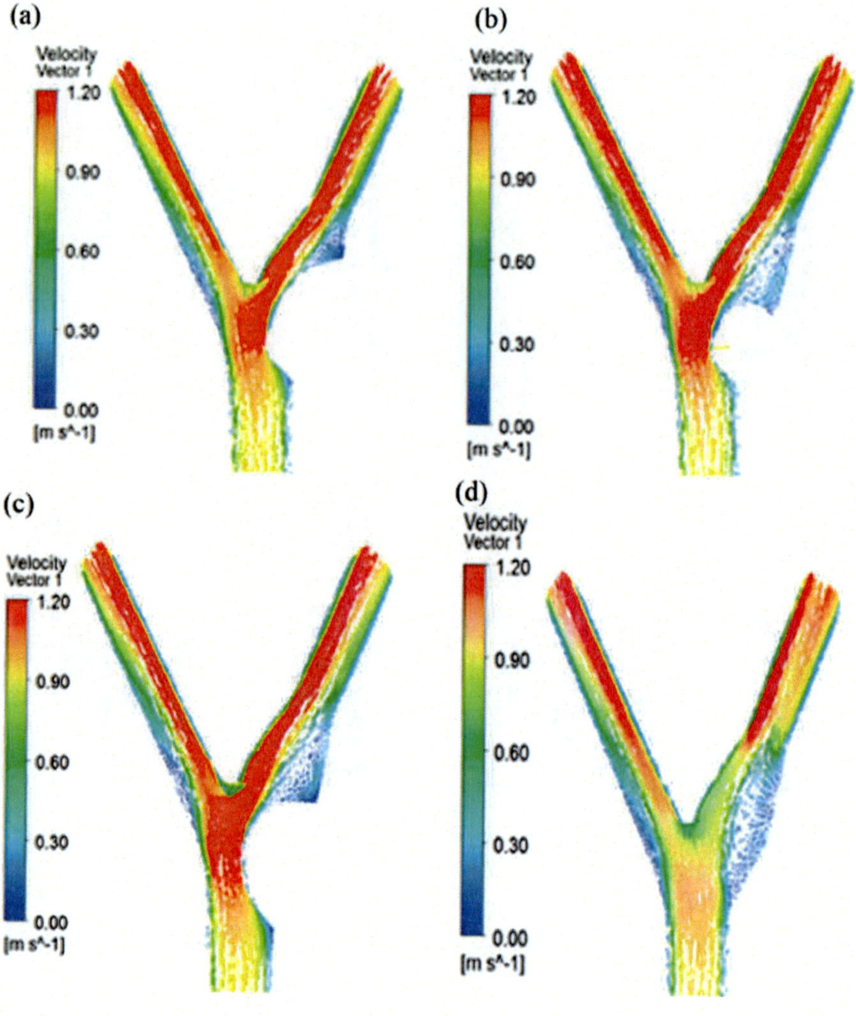

Fig. 6 Velocity vectors in **a** Type I, **b** Type II, **c** Type III and **d** Healthy model

increase proportional with velocity. Similar observation was also recorded earlier by [9, 13]. Regions of high WSS are more prone to produce matrix degradation and weakening of the plaque cap, which can lead to rupture [23, 24]. High WSS had been possible causative factor on development of plaque (Fig. 7).

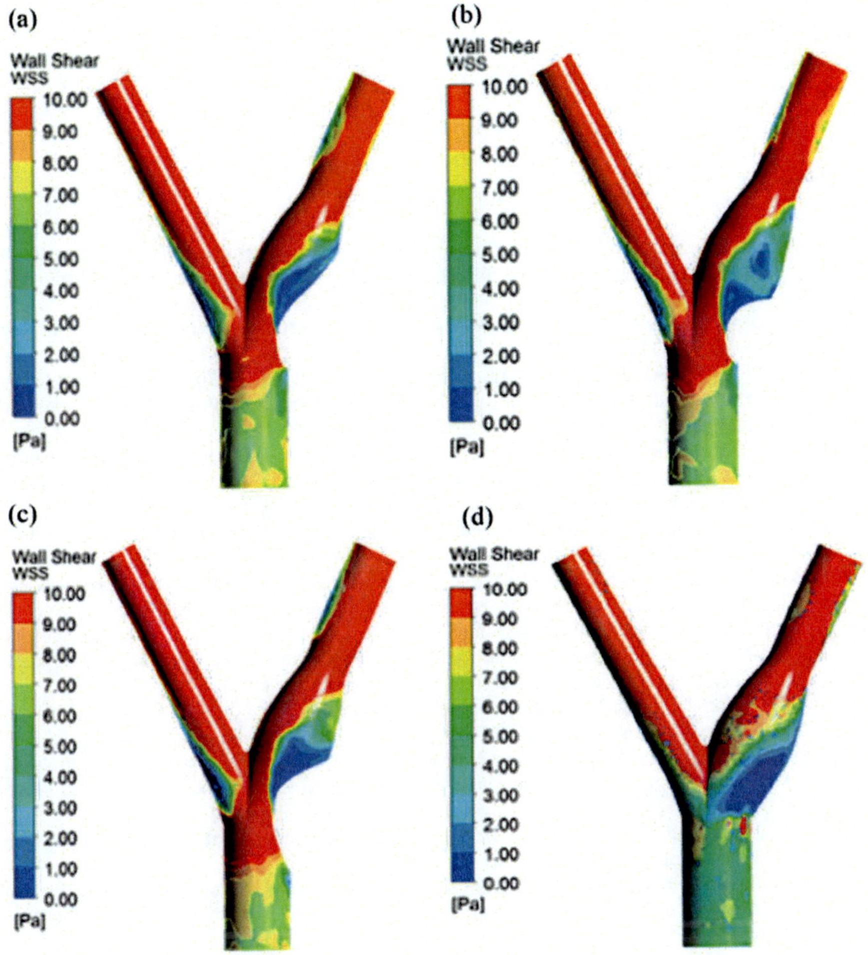

Fig. 7 Wall shear stress in **a** Type I, (b) Type II, (c) Type III and **d** Healthy model

4 Conclusion

Carotid artery bifurcation is one the most prone arteries to atherosclerosis which could lead to ischemic stroke disease. In this paper, a preliminary study on the velocity profiles and wall shear stress analysis were carried out on an idealized carotid artery bifurcation with different locations of stenoses. The location of stenosis extended toward the end of ICA (Type I) produced higher risk compared to other stenosis locations. This indicates an urgency to the patient to get immediate treatment, especially the blood flow from the ICA to the brain which is very crucial. Further extended study with variant inlet velocity using transient flow profiles and realistic boundary conditions will be explored to support the findings in this paper.

Acknowledgements This work has been supported by a Research Grant RDU190343 and PGRS210343 from Universiti Malaysia Pahang.

References

1. Sakellarios A et al (2017) Prediction of atherosclerotic disease progression using LDL transport modelling: a serial computed tomographic coronary angiographic study. Eur Heart J Cardiovasc Imag 18(1):11–18
2. Tarbell JM (2003) Mass transport in arteries and the localization of atherosclerosis. Ann Rev Biomed Eng 5:79–118
3. Sun N, Torii R, Wood NB, Hughes AD, Thom SAM, Xu XY (2009) Computational modeling of LDL and albumin transport in an in vivo CT image-based human right coronary artery. J Biomech Eng 131(2):1–9
4. Ho WH, Tshimanga IJ, Ngoepe MN, Jermy MC, Geoghegan PH (2020) Evaluation of a desktop 3D printed rigid refractive-indexed-matched flow phantom for PIV measurements on Cerebral Aneurysms. Cardiovasc Eng Technol 11(1):14–23
5. Carvalho V et al (2021) In vitro biomodels in stenotic arteries to perform blood analogues flow visualizations and measurements: a review. Open Biomed Eng J 14(1):87–102
6. Lopes D, Puga H, Teixeira J, Lima R (2020) Blood flow simulations in patient-specific geometries of the carotid artery: a systematic review. J Biomech 111:110019
7. DiCarlo AL, Holdsworth DW, Poepping TL (2019) Study of the effect of stenosis severity and non-Newtonian viscosity on multidirectional wall shear stress and flow disturbances in the carotid artery using particle image velocimetry. Med Eng Phys 65:8–23
8. Hong H, Yeom E, Ji HS, Kim HD, Kim KC (2017) Characteristics of pulsatile flows in curved stenosed channels. PLoS ONE 12(10):1–19
9. Johari NH et al (2019) Disturbed flow in a stenosed carotid artery bifurcation: comparison of RANS-Based transitional model and les with experimental measurements. Int J Appl Mech 11(4):1–21
10. Banks J, Bressloff NW (2007) Turbulence modeling in three-dimensional stenosed arterial bifurcations. J Biomech Eng 129(1):40–50
11. Tan FPP, Wood NB, Tabor G, Xu XY (2011) Comparison of les of steady transitional flow in an idealized stenosed axisymmetric artery model with a RANS transitional model. J Biomech Eng 133(5):1–12
12. Lancellotti RM, Vergara C, Valdettaro L, Bose S, Quarteroni A (2017) Large eddy simulations for blood dynamics in realistic stenotic carotids. Int. J. Numer. Method. Biomed. Eng. 33(11):e2868
13. Li ZY, Tan FPP, Soloperto G, Wood NB, Xu XY, Gillard JH (2015) Flow pattern analysis in a highly stenotic patient-specific carotid bifurcation model using a turbulence model. Comput Methods Biomech Biomed Eng 18(10):1099–1107
14. Moradicheghamahi J, Jahangiri M, Mousaviraad M, Sadeghi MR (2020) Computational studies of comparative and cumulative effects of turbulence, fluid–structure interactions, and uniform magnetic fields on pulsatile non-Newtonian flow in a patient-specific carotid artery. J Braz Soc Mech Sci Eng 42(10):1–22
15. Ahmadpour-B M, Nooraeen A, Tafazzoli-Shadpour M, Taghizadeh H (2021) Contribution of atherosclerotic plaque location and severity to the near-wall hemodynamics of the carotid bifurcation: an experimental study and FSI modeling. Biomech Model Mechanobiol 20(3):1069–1085
16. Gharahi H et al (2017) Measured with magnetic resonance imaging 8(1):40–60
17. Lu M, Cui Y, Peng P, Qiao H, Cai J, Zhao X (2019) Shape and location of carotid atherosclerotic plaque and intraplaque hemorrhage: a high-resolution magnetic resonance imaging study. J Atheroscler Thromb 26(8):720–727

18. Nagargoje M, Gupta R (2020) Effect of sinus size and position on hemodynamics during pulsatile flow in a carotid artery bifurcation. Comput Methods Programs Biomed 192:105440
19. Menter FR (1994) Two-equation eddy-viscosity turbulence models for engineering applications. AIAA J 32(8):1598–1605
20. Langtry RB, Menter FR (2009) Correlation-based transition modeling for unstructured parallelized computational fluid dynamics codes. AIAA J 47(12):2894–2906
21. Menter FR, Langtry RB, Likki SR, Suzen YB, Huang PG, Völker S (2006) A correlation-based transition model using local variables - part I: model formulation. J Turbomach 128(3):413–422
22. Johari NH, Hamady M, Xu XY (2020) A computational study of the effect of stent design on local hemodynamic factors at the carotid artery bifurcation. Artery Res 26:161–169
23. Malek AM, Alper SL, Izumo S (1999) Hemodynamics shear stress and its role in atherosclerosis. JAMA 282(21):2035–2042
24. Slager CJ et al (2005) The role of shear stress in the destabilization of vulnerable plaques and related therapeutic implications. Nat Clin Pract Cardiovasc Med 2(9):456–464

Preliminary Study of Offshore Wind Turbine Foundation in Malaysia

M. A. N. Mutasim, M. H. Mansor, and H. A. Salaam

Abstract One of the most promising prospects being actively researched and implemented is the offshore wind turbine. The offshore wind turbine has proven to have great potential when looking to supply high amounts of clean and renewable energy. However, here in Malaysia, wind turbines productions are limited. In this study, the aim is to determine a potential installation of wind turbine. Wind data are determined, and a model of wind turbine are then subjected to the loading to analyze its movement and foundation in these different conditions. In this paper, a design study is presented which aims to encounter the problem involving the offshore wind turbine structure and analyze the structures load transferred to the surrounding soil. This paper also reviews previous research work and then presents a simple dynamic system of calculations along with free body diagrams to help define this problem. The model is developed and simulated with SIMULINK to obtain time domain force distribution. The results are then applied into the equations of motion to analyse the relationship between the turbine and soil foundation dynamics. The conclusion of this design analysis will help designers determine the best design for offshore wind turbine foundations and allow offshore wind turbines to be implemented in Malaysian region.

Keywords Offshore wind turbine · Foundations · Dynamics system modeling

Nomenclature

α	*Angular velocity*
D	*Rotor Diameter*

M. A. N. Mutasim (✉) · M. H. Mansor
Faculty of Mechanical and Automotive Engineering Technology (FTKMA), Universiti Malaysia Pahang, 26600 Pekan, Pahang Darul Makmur, Malaysia
e-mail: ammar@ump.edu.my

H. A. Salaam
College of Engineering, Universiti Malaysia Pahang, Lebuhraya Tun Razak, 26300 Gambang, Pahang Darul Makmur, Malaysia

© The Author(s), under exclusive license to Springer Nature Singapore Pte Ltd. 2023

M. H. A. Hassan et al. (eds.), *Technological Advancement in Instrumentation & Human Engineering*, Lecture Notes in Electrical Engineering 882,
https://doi.org/10.1007/978-981-19-1577-2_46

F_{k1} *Soil Spring force*
F_{k2} *Soil Spring force*
F_{b1} *Damping force*
F_{b2} *Damping force*
H *Tower height, hub to soil level*
l_w *Length of hub to soil base*
l_1 *Length of soil base to midpoint of soil 1*
l_2 *Length of soil base to midpoint of soil 2*
M *Moment on structure foundation*
ρ *Density*
θ *Angle of rotation*
R *Radius of soil foundation*
T_F *Force acting on the hub*
G *Shear modulus*
γ *Dimensionless coefficient of damping*
υ *Poisson ratio*
w *Width of the tower*
ω *Angular velocity*
z *Depth of soil level*

1 Introduction

According to National Energy Balance 2018 [1], Malaysia electricity is primarily generated by coal and coke with the biggest growth 7.3% from the previous year. It is due to the higher demand from the power sector that contribute to the significant growth [1]. In country throughout the world, like United States, China, Russia, Australia, India, and also ASEAN country including Indonesia, Singapore, Thailand, Philippines and Malaysia, are still heavily relying on fossil fuel as the main source for electricity production, with more than 79% in their energy mix on average [2]. This use of non-renewable energy impacted the environment through air and thermal pollutants. From 1990 until 2020, Malaysian electricity demand has increased up to 20% [3]. With increase in oil prices, due to its limitation, electricity expenditures will keep growing. To ensure sustainability, the Malaysian government proposed a renewable energy implementation. The eights Malaysian National renewable energy policy plan to achieve 20% renewable energy by 2025 [4]. Currently, the implementation of renewable energy is mainly focused on solar and hydro power [1, 2]. The wind power on the other hand, which can generate more than 50% compared to other renewable energy, is very much restricted. Currently, there are only 2 wind turbines installed in Malaysian onshore [5]. Unlike onshore wind, the offshore wind supplies more power compared to the former [6, 7].

Offshore wind turbines are significantly better than onshore wind turbines. Since it is built further from human confines, the noise, visual and space limitations are negligible [6, 7]. Aside from logistical problem for cost and transportation, in open space, the design of wind turbines can extend to longer diameter blades and larger generators [6]. However, the challenges for offshore turbines are more focused on maintaining the wind turbines structural integrity and reliability due to higher wind loads. To keep the costs of the offshore wind turbine on par with the onshore wind turbine, suggestions to reduce the overall weight of the tower structure was brought into focus with an ability to respond to wind induced forces [7]. The operating frequency of offshore wind turbines are typically higher than the structures natural frequency [8]. A slight change in dimensions may cause changes in structural stiffness that can results in different frequency response. Ideally, to make it more realistic, a sufficient spring and damping provision of the entire structure is required to respond to the forces acting on the wind turbine.

The structure of the wind turbine is difficult to present as a mass spring system due to its structural design. One way to overcome this deficit is, to present the wind turbine soil foundation as the mass spring systems and thus producing a capability to estimate the natural frequency of the wind turbine system [9]. This, however, requires a proper analysis of the basic dynamic system of the wind turbine structure and its soil foundation. Since the maximum force acting at the tip and the hub of the wind turbine rotor will lead to a higher inertial moment in the base of the soil foundation, analysis between wind loads and moment of inertia of the soil in time domain is critical. The dynamic behavior of the soil structure is almost too complex to analyze; thus, it may require calculating the parameters of spring and damping coefficient with the individual properties of soil layers.

Based on previously existing methods for assessing the dynamic system of wind turbine foundations, one of the most important analysis is the Soil-Structure Interaction (SSI). Among the earliest of these studies on the SSI are done by Novak and Hifnawy [10] which focused on the dynamic wind loading in restricted buildings in seismic zones. Others are, failure analysis with flexible soil foundations where researched to determine the natural frequency from turbine operation by Hamaydeh and Hussian [11] and a sensitivity analysis of the natural frequency for the multiple wind turbine support structures by Zaaijer [12]. To this date, SSI has increasingly become part of the design codes and guideline for foundation design with dynamics system of structural building design.

In other studies, attempts to evaluate natural frequency for support structure was done using finite-element method (FEA). Among those are FEA with radiating and transmitting boundary conditions by Higdon [13] and Krenk [14] using the boundary-element method, the domain-transformation method, analytical and semi-analytical approaches. These are based on a transfer matrix for a layered half-space, originally proposed by Thomson [15], and Damgaard et al. [16] to analyzed a rigid surface foundation on a horizontal viscoelastic layer which proved that the impedance in terms of its magnitude and phase angle followed a second order polynomial without any resonance peaks. However, the presence of stratified soil showed a drastic change in the frequency response with local tips and dips. Similar conclusions were drawn by Emperador and Domínguez [17], and Liingaard et al. [18], who used boundary

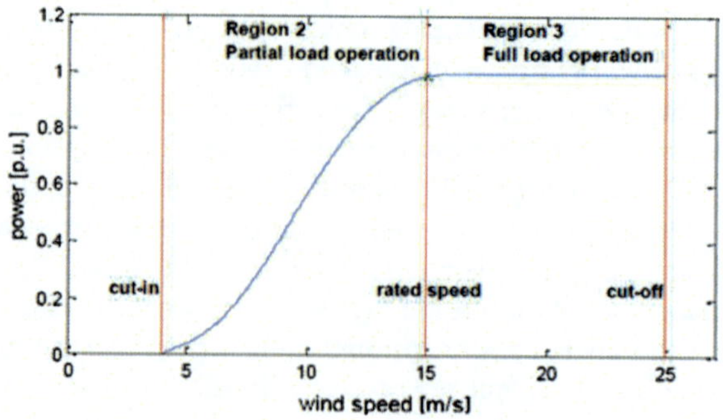

Fig. 1 Wind power curve [20]

element and coupled boundary-element/finite-element models to study the soil structure foundation. In this paper, the model of the soil structure damper and spring coefficient is determined from suction caissons for offshore wind turbine foundations by Houlsby et al. [19]. The damping coeffiecient is derived from three possible causes which are the viscous material damping, and the plastic dissipation of energy in the soil and the radiation damping. The impedence, for this case, is calculated by the imaginary part and the real part of the stiffness and the inertial effects as well as factors related to shear wave velocity, v_s.

For simplicity of the design analysis, wind loads will be presented with variable wind speed operation in Fig. 1 below, where there will be three regions to consider. Since Malaysia has monsoon season, from May to September and from November to march, wind speeds can reach up to 15 m/s. These sites are mostly in the east coast of peninsular Malaysia and Sabah coast. Region 1 is where the systems are operated below the rated wind speed. In region 2, the wind speed is in partial load operation, this is where design of the soil foundation is mostly needed to adapt. This analysis of wind and structure of the wind turbine will be subjected to these 2 regions.

In this paper, a simple foundation design using monopile will be presented with a generic wind turbine of 2 MW. The scope of the study will be presented using a design from [20] with a tubular structure extended to seabed in a 20 m water depth. The wind turbine tower dimensions is based on Jonkman [21] with 90 m of length and 0.02 m thickness. The tower top diameter and base diameter are set as constant for simple analysis purpose. The main goal of this paper is to determine the correlation between the flexible soil foundation and wind loads acting on the wind turbine and subsequently will illustrate the best soil foundation for the preferred three regions, producing a method to evaluate turbine and soil movements with varying wind loading.

2 Mathematical Modelling

The framework presented in this paper will discuss a simple presentable approach of wind-structure-soil interaction of the given wind turbine design. The approach will present the construction of free body diagram which lays out the interactions of the mentioned three components and the derivation of equations of motion of this system will be derived for the produced FBD relationships. The system will then be modeled based on a Multi Degree of Freedom (MDOF) method. The developed equations of motion will be validated with data of soil properties from previous researchers and an analysis using SIMULINK will be presented to obtain the acceleration, velocity and displacement of soil foundation and turbine movement with the range velocity in time domain.

2.1 Wind Turbine Design

The proposed method used to investigate the dynamic model of the offshore wind turbine foundation will be carried out by a simple static design calculation using free body diagram (FBD). The offshore wind turbine structure is shown in Fig. 2. The inputs from the turbine design are defined as a generic wind turbine having a Rotor Dimeter (D) of 100 m, a hub to the soil surface height of 90 m, and width or tower diameter of w = 3 m [22]. The wind turbine is assumed to be buried in the multiple layers of soil at its foundation which serves to counter the forces applied to the structure from wind loading. These different soil layers are used as this foundation to the wind turbine structure. Each soil layer has a constant depth of z. The foundation is in an elliptical shape of diameter of r and r/2. For simplicity this model considers the turbines vertical structure as rigid whereas in actual, the structure will have some dynamic movement due to wind loading. The turbines moment will be measured with degrees of rotation from the vertical, with the model allowing for a variable placement of the structures rotational point. Simulations, however, will be placing the point of rotation at the bottom of all available soil layers as seen in Fig. 4. The following presented assumptions cover the basic set of governing rules which our model followed when determining the system relationships.

The static design calculation is presented with assumptions:

1. The tower is rigid body with constant geometry dimension
2. Seabed is assumed to have n layers of soil
3. System is model with 2 DOF
4. The Moment is acting on the base of the soil.
5. No net vertical movement

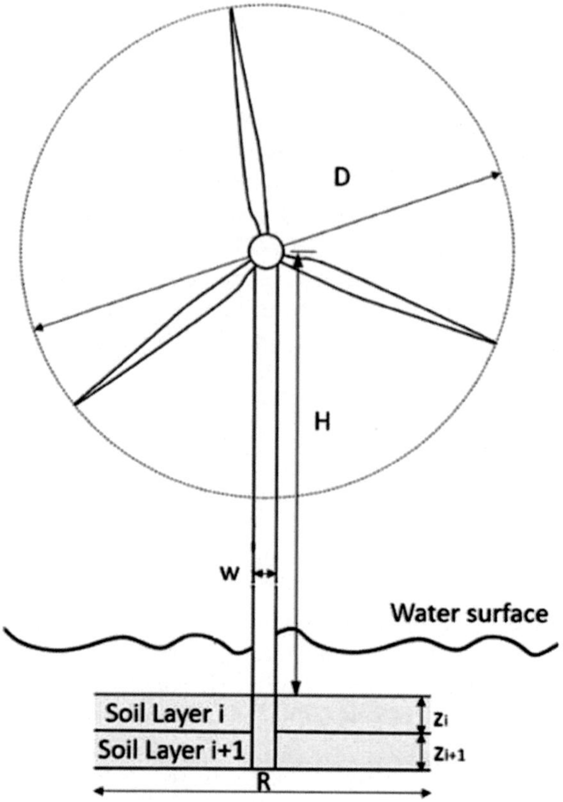

Fig. 2 Offshore wind turbine structure

2.2 Wind Structure Interaction

Wind loads are generally unpredictable and forces acting towards the structure fluctuates with time. The ideal approach to demonstrate a design is to maintain the turbine at a nearly constant rotor speed. This can be presented with a generalized wind turbine power curve where region 2 acts as partial load operation and region 3 acts as full load operation. A steady wind loading can cause vibrations from vortex shedding phenomenon while an unsteady wind loads with extreme winds may impact in higher stress to the structure resulting in a different range of behaviors.

In this paper, the Bernoulli equation is used to determine the pressure on the structure from wind loads classified as the thrust load, T_f and is calculated using Eq. (1) presented below where A_{rotor} is the swept area of the rotor with D is the diameter; ρ_{air} is the air density; $U_{upstream}$ and $U_{downstream}$ are the upstream velocity and downstream wind velocities experienced by the system respectively shown in Eq. (1) below. The air density is calculated at 760 mm of mercury and at temperature of 15 °C throughout this model but can be changed for varying turbine environment characteristics.

$$T_f = 0.5 \times \rho_{air} \times A_{rotor}\left(U_{upstream}^2 - U_{downstream}^2\right) \tag{1}$$

where

$$A_{rotor} = \pi\left(\frac{D}{2}\right)^2$$

2.3 Structure to Soil Interaction

The overall height of monopile design of the wind turbine embedded into the soils from hub height to the base is 110 m. The rotor-nacelle mass is presented as 130,000 kg and the diameter of the tubular is presented as 3 m. The base radius of the foundation is measured as 15 m with approximately ten times the width of the monopile. Each soil layer are having a different rotational stiffness coefficient, k and rotational damping coefficient, b given by Eq. (2) and Eq. (3) by [23] seen in Fig. 3.

$$k = \frac{8GR^3}{3(1 - \upsilon)} \tag{2}$$

$$b = \frac{\gamma Rk}{\upsilon_s} \tag{3}$$

The ν_s, which is the shear wave velocity is calculated using Eq. (4) below, where ρ is the soil density.

$$\upsilon_s = \sqrt{\frac{G}{\rho_{soil}}} \tag{4}$$

The inertial force, I_0 acting on the tower at point O can be calculated with Eq. (5) below with the moment of inertial acting on the mass of the tower, mass of the rotor and mass of the generator.

$$I_o = \frac{1}{3}m_{tower} \times l_w^2 + \frac{2}{5}m_{rotor}\left(\frac{D}{2}\right)^2 + \frac{1}{3}m_{generator}\left(l_w + \left(\frac{D}{2}\right)\right)^2 \tag{5}$$

Here the system can be modeled as a spring mass damper system seen in Fig. 4 placing a spring and damper to represent the soil characteristics of each layer.

The presentable Free body diagram of the soil-structure interaction system is shown below. Utilizing the FBD, the equation of motion can be presented.

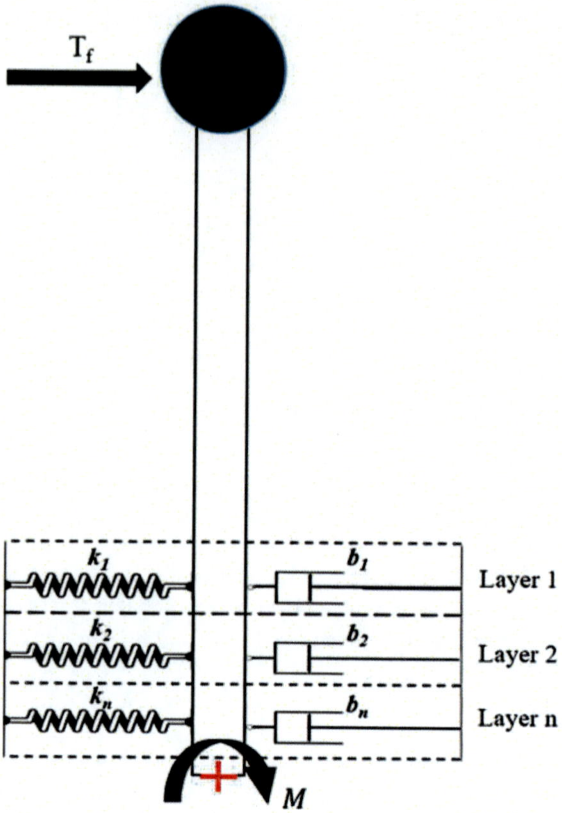

Fig. 3 Mass damper system

The equation of moment at the base of the soil at point O is expressed in τ_o as the summation of torque at the base of the soil; $\ddot{\theta}$ is the angular acceleration of the tower; τ_f is the torque of the force at the hub; τ_{k1}, τ_{k2} is the torque stiffness of the soil and τ_{b1} and τ_{b2} is the torque damping stiffness of the soil.

$$I_o\ddot{\theta} = T_f l_w cos\theta - sin\theta cos\theta\left(k_1 l_1^2 + k_2 l_2^2\right) - \dot{\theta}cos^2(\theta) \times \left(b_1 l_1^2 + b_2 l_2^2\right) \quad (6)$$

The above Eq. 6 represents one of the outputs of the system, which allows for the analysis of the movement of the tower under various wind loading conditions. This system was then modeled using SIMULINK to represent the motion of the turbine and foundation system allowing for the given variables to be easily changed and variations compared and quantified to aid in the better understanding of the overall dynamic system.

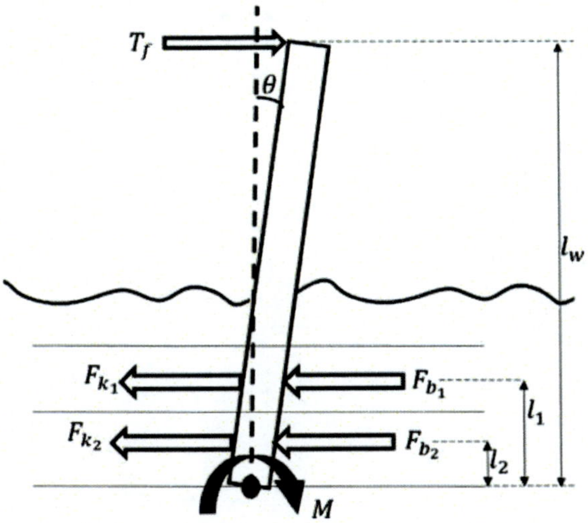

Fig. 4 Free body diagram of the system

3 Results and Discussions

The validation involves the comparison of results from the previous work of [8] using this research with soil properties applied to the system to allow for a more qualitative verification. Due to assumptions made earlier the results are slightly different. Looking at the displacement vs time plot this can be compared to the torsional displacement motion of the previous study. The plot of the previous study is show in Fig. 6. It the results found in this study follows similar trend as in the previous study. As the wind load is applied to the wind turbine the wind turbine has a torsional displacement and over time the movement is dampened out. As the wind load increases the time to dampened out the system is increased as well (Fig. 5).

The model was simulated using the soil properties and varying wind loading inputs of Malaysia wind speed and coefficient of the soil properties for each layer shown in Table 1. Soil 1, correspond to medium dense sand overlaying a thin layer

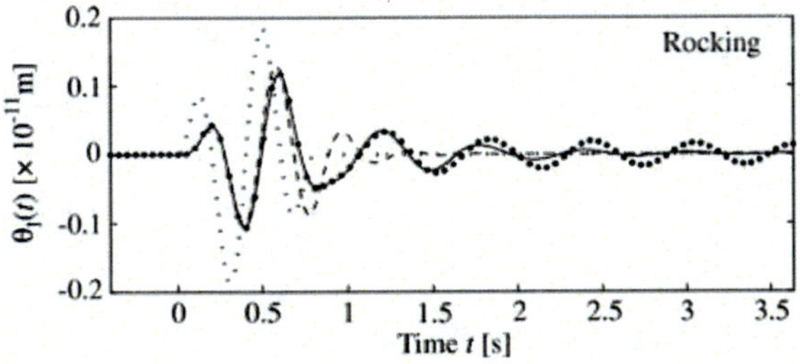

Fig. 5 Qualitative validation with [24]

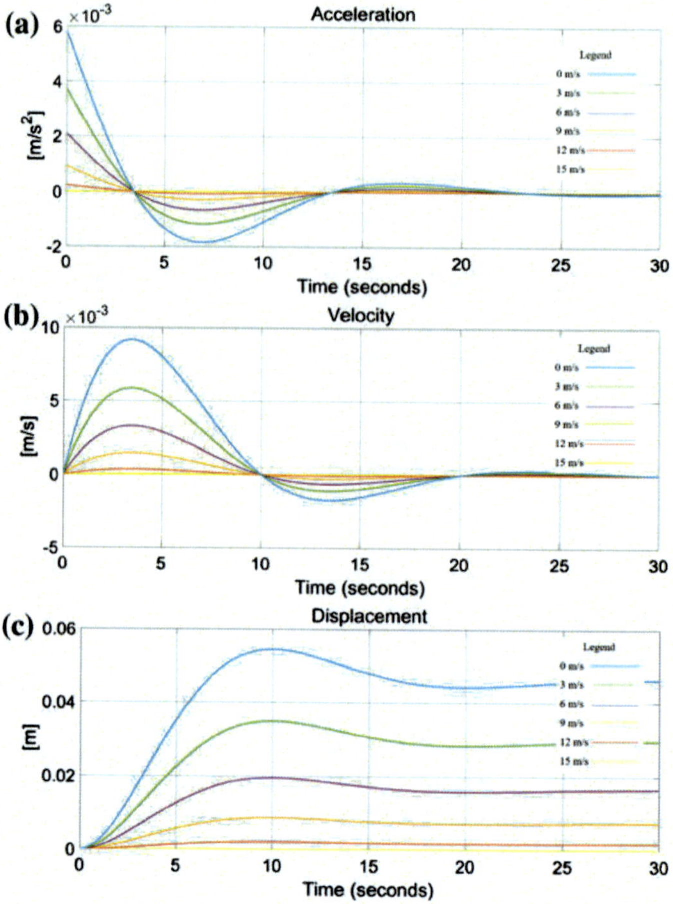

Fig. 6 Simulation result for region 1 wind loads in time domain **a** Acceleration **b** Velocity **c** Displacement

Table 1 Soil properties [16]

Soil	Z[m]	*G*[MPa]	$\upsilon[-]$	$\gamma[-]$	ρ_{soil}[kg/m^3]
Layer1	10	75	0.35	0.04	1700
Layer2	10	134	0.45	0.03	2000

of highly consolidated drained firm clay and a half-space of brittle rock, the five non-zero dynamic stiffness coefficients. Layer 2 refer to less granular type of soft clay. The dynamic vibration response of the investigated wind turbine is analyzed for the soil conditions given in Table 1. For each stratum, the following two load cases are considered with a simulation length of 100 s.

(i) Load case 1: Below cut in operating turbine with very low wind speed of 0 to 5 m/s.
(ii) Load case 2: Normal operating turbine with normal rated wind loads between 6 to 15 m/s.

In Figs. 6, 7, it can be seen of clear implication of pattern of induced wind loads against rotational stiffness. This increase in stiffness of the system may be attributed due to the compaction of the granular material under the cyclic action. The overall damping of the system has been observed to increase three times fold as the induced wind loads at 19 to 25 m/s is subjected. This measure of damping is due to the components from the structure and the soil, but the increase is more likely due to the soil–structure interaction. The overall behaviour of the pattern for any given wind

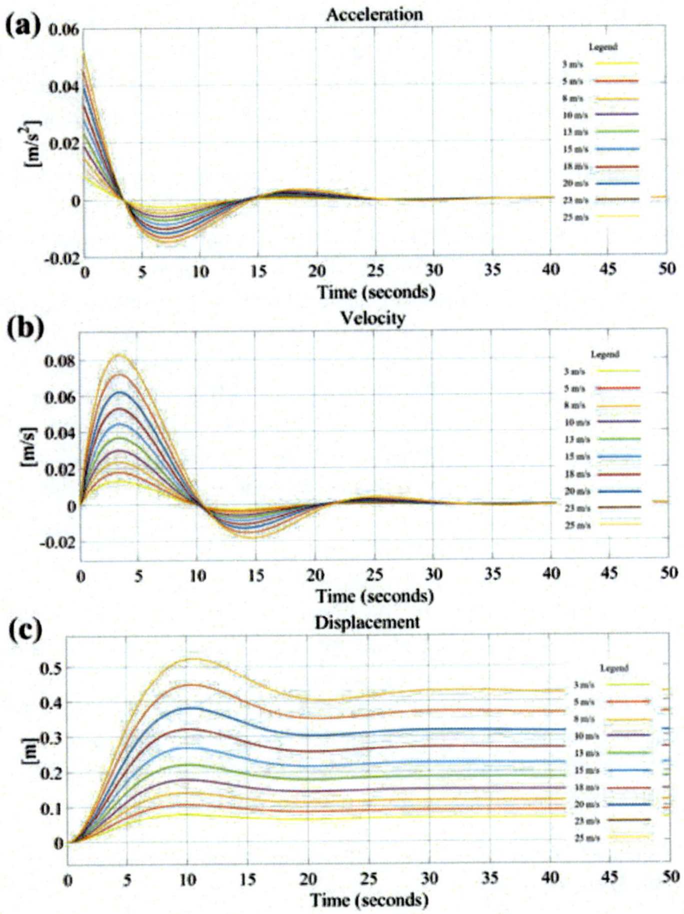

Fig. 7 Simulation result for region 2 wind loads in time domain **a** Acceleration **b** Velocity **c** Displacement

loading shows that the system dampens out and becomes stable over time. For region 1 it takes approximately 25 s for the system to become stable, and the maximum displacement is 0.06 m which is very small. For region 2 it takes approximately 30 s for the system to become stable, and the maximum displacement is 0.5 m. Lastly for lower region 3 winds the system becomes stable after approximately 60 s. For higher region 3 winds closer to 25 m/s the system does not become stable. The displacement for the higher region 3 winds is close to 2.5 m. This is a result of a futter aerodynamics and can cause a fatigue in the structure. Operating at this condition is not optimal.

4 Conclusions

A wind forced vibration analysis of an offshore wind turbine has been carried out including the effects of soil flexibility underneath the foundation. A MDOF wind turbine model was developed using simple equations of motion by derivation from FBD. The soil-foundation interaction was modeled by spring impedance functions obtained from a previous literature study and the damping parameter based on three possibilities which is the viscous material damping, the plastic dissipation of energy in the soil and the radiation damping where considered. The equations of motion were solved in the time domain to allow for a comparison of length of time each scenario takes to achieve steady state. Two soil profiles were examined in this study, both having a uniform soil profile. The input forces acting on the hub diameter of the wind turbine shown to have an impact on the displacement, velocity and the acceleration of the tower. As for future work, an optimization approach is needed to solve the possible impedance and the damping coefficient with soil properties for variational wind loads. The future presentation can ensure that by optimizing the soil foundation, wind turbine life span can be increased. Even by reducing the weight of the material and making the tower slenderer and more flexible to allow minimizing the overall cost of the wind turbine system. This system could also be adapted in future work to accommodate structure flexure and movement allowing to produce an even more accurate system model. Likewise, this model could also be applied to the turbines system of equations when considering power generation allowing for the velocity changes of the movement of the system to be accounted for when looking the regions wind velocity allowing for a more accurate power generation analysis to be conducted. Researchers have also begun optimizing the given system of equations allowing for the elimination of the assumption that the foundation is pivoting about its base, which is projected to be useful when foundation and soil layers allow for the movement of the foundation around an offset point.

Acknowledgements The support of the UMP Internal Research Funding (Grant No. RDU 1803127) is greatly acknowledged.

References

1. National Energy Balance 2018; Suruhanjaya Tenaga. ISSN No. 0128-6323, 1 June 2020. https://meih.st.gov.my/. ST(P)
2. BP (2013) Statistical review of world energy. http://large.stanford.edu/courses/2013/ph240/lim1/docs/bpreview.pdf. Accessed 14 Apr 2021
3. National Energy Balance 2010; Suruhanjaya Tenaga. https://meih.st.gov.my/. Accessed 13 May 2021
4. National Renewable Energy Policy and Action Plan, Overarching Policy, Suruhanjaya Tenaga (2009) Issued by: Minister of Energy, Green Technology and Water
5. Saberi Z, Fudholi A, Sopian K (2019) Potential evaluation of wind energy in Kuala Terengganu, Malaysia through Weibull distribution method. In: International conference on sustainable energy and green technology 2018, IOP conference series: earth and environmental science, vol 268, p 012074. IOP Publishing. https://doi.org/10.1088/1755-1315/268/1/012074
6. Adedipe O, Abolarin MS, Mamman RO (2018) A review of onshore and offshore wind energy potential in Nigeria. IOP Conf Ser Mater Sci Eng 413:012039. https://doi.org/10.1088/1757-899X/413/1/012039
7. Junginger M, Hittinger E, Williams E, Wiser R (2020) Onshore wind energy. In: Junginger M, Louwen A (eds) Technological Learning in the Transition to a Low-Carbon Energy System, pp 87–102. Academic Press. ISBN 978-0-12-818762-3
8. Ellabban O, Abu-Rub H, Blaabjerg F (2014) Renewable energy resources: current status, future prospects and their enabling technology. Renew Sustain Energy Rev 39:748–764
9. Schubel PJ, Crossley RJ (2012) Wind turbine blade design. Energies 5:3425–3449
10. Novak M, El Hifnawy L (1988) Structural response to wind with soil-structure interaction. J Wind Eng Ind Aerodyn 28(1–3):329–338. https://doi.org/10.1016/0167-6105(88)90129-8
11. AlHamaydeh M, Hussain S (2011) Optimized frequency-based foundation design for wind turbine towers utilizing soil–structure interaction. J Franklin Inst 348(7):1470–1487
12. Zaaijer MB (2006) Foundation modelling to assess dynamic behaviour of offshore wind turbines. Appl Ocean Res 28(1):45–57
13. Higdon RL (1990) Radiation boundary conditions for elastic wave propagation. SIAM J Numer Anal 27(4):831–869
14. Krenk S (2002) Unified formulation of radiation conditions for the wave equation. Int J Numer Meth Eng 53(2):275–295
15. Thomson WT (1950) Transmission of elastic waves through a stratified solid medium. J Appl Phys 21(2):89–93
16. Damgaard M, Andersen LV, Ibsen LB (2015) Assessment of dynamic substructuring of a wind turbine foundation applicable for aeroelastic simulations. Wind Energy 18(8):1387–1401
17. Emperador JM, Dominguez J (1989) Dynamic response of axisymmetric embedded foundations. Earthq Eng Struct Dyn 18(8):1105–1117
18. Liingaard M, Andersen L, Ibsen LB (2007) Impedance of flexible suction caissons. Earthq Eng Struct Dyn 36(14):2249–2271
19. Houlsby GT et al (2005) Field trials of suction caissons in clay for offshore wind turbine foundations. Geotechnique 55(4):287–296
20. Hall JF et al (2011) Wind energy conversion with a variable-ratio gearbox: design and analysis. Renew Energy 36(3):1075–1080
21. Lozano-Minguez E, Kolios AJ, Brennan FP (2011) Multi-criteria assessment of offshore wind turbine support structures. Renew Energy 36(11):2831–2837
22. Jonkman JM (2009) Dynamics of offshore floating wind turbines-model development and verification. Wind Energy 12(5):459–492
23. Kansal HK, Singh S, Kumar P (2005) Parametric optimization of powder mixed electrical discharge machining by response surface methodology. J Mater Process Technol 169:427–436
24. Damgaard M, Andersen LV, Ibsen LB (2014) Computationally efficient modelling of dynamic soil-structure interaction of offshore wind turbines on gravity footings. Renew Energy 68:289–303

Study on the Recognition of Driver's Starting Intentions Based on Fuzzy Inference-SVM Cascade Algorithm

Hongtao Hao and Chao Zhang

Abstract Research of the identification method on driver's starting intention is very necessary. It can provide basis and support for subsequent vehicle clutch control, optimized shift curve and driving style recognition, and also contribute to vehicle assisted driving and intelligent driving. This paper presents a fuzzy inference-support vector machine (SVM) cascade algorithm to recognize the driver's starting intentions, which can make up the low accuracy of fuzzy inference algorithm and overcome the difficulty to identify large samples of SVM algorithm. The proposed recognition method of driver's starting intentions includes two-layer: the first layer is fuzzy inference layer while the second layer is SVM layer. At the same time, the fuzzy inference-SVM cascade algorithm is trained and tested with the sample data acquired from the actual vehicle. The experimental results show that the cascade algorithm has high recognition accuracy and moderate recognition time. Accordingly, the fuzzy inference-SVM cascade algorithm is an effective way to recognize driver's starting intentions.

Keywords Fuzzy inference · Support vector machine · Cascade algorithm · Starting intention

1 Introduction

Research of the recognition methods on driver's starting intentions is able to provide the basis and support for subsequent automobile clutch control optimization, shifting curve optimization, and driving style recognition, and also assists in auxiliary driving and intelligent driving of cars [1, 2]. Scholars worldwide have studied recognition and classification of driver's intentions. Li et al. proposed a novel method to identify the driver's starting intention based on an artificial error back-propagation neural

H. Hao (✉)
School of Mechanical Engineering, Ningxia University, Yinchuan 750021, China
e-mail: haoht_03@126.com

C. Zhang
Yinchuan University of Science and Technology, Yinchuan 750021, China

© The Author(s), under exclusive license to Springer Nature Singapore Pte Ltd. 2023
M. H. A. Hassan et al. (eds.), *Technological Advancement in Instrumentation & Human Engineering*, Lecture Notes in Electrical Engineering 882,
https://doi.org/10.1007/978-981-19-1577-2_47

network [3]. Lei et al. [4] proposed a method to recognize the driving intention based on an improved Gustafson-Kessel clustering analysis, which can be used to adapt gearshift decision. Ikenishi et al. [5] proposed an estimation algorithm of driver's longitudinal intention by using the brain current distribution. An inference model based on a dynamic Bayesian network which considers the influence of driver's past behavior on the current behavior was proposed by Li et al. [6] and it can be used to identify the driver's intention for the prediction of the driving risk in advance. Weidl et al. [7] optimized a Bayesian recognition network to distinguish driver maneuver in order to reduce the average computation time. Wang et al. used Markov chain to estimate driving conditions and classify driver's style [8].

Recognition of a driver's starting intention is a typical empirical model, which makes it very difficult to describe with an accurate mathematical model. In handling the empirical models, fuzzy inference has certain advantages in fault tolerance, adaptability, and robustness, but fuzzification and fuzzy rule design are excessively dependent on operator's experience. When recognizing driver's starting intentions, SVM is a small-sample learning method capable of handling nonlinear sample data classification problem, but encounters some difficulties in handling large sample data. Based on the above analysis, a new recognition method of driver's starting intentions based on fuzzy inference-SVM cascade algorithm is presented. The proposed recognition method of driver's starting intentions includes two-layer: the first layer is fuzzy inference layer while the second layer is SVM layer. In the first layer, the model deals with recognizable intentions and exports the easily confounding samples to the second layer for further recognition, hence this algorithm overcomes drawbacks of a single algorithm.

2 Data Acquisition

Test vehicle is Ford Escape compact SUV, its engine model is CAF479WQ1, and the real time acquisition scene is shown in Fig. 1. In the test system, accelerator opening sensor sends its signal to electronic control unit (ECU) and USBCAN-OBD tool is used to acquire test data. During real vehicle data acquisition, a total of 60 sets of data are acquired (35 data points per set, totally 2100 data points). These 60 sets of samples include the accelerator pedal opening degree varying with time corresponding to quick start, moderate start, and slow start intentions, respectively.

Fig. 1 Real vehicle acquisition scene

3 Fuzzy Inference Recognition Method

3.1 Fundamentals of Fuzzy Inference

Driver's starting intention recognition belongs to a typical empirical model, and fuzzy inference decision has an advantage to handle empirical models. Fuzzy inference process of the driver's starting intentions recognition is shown in Fig. 2, where α, $\dot{\alpha}$ and $\ddot{\alpha}$ indicate accelerator pedal opening degree, its first derivative and second derivative, respectively. Depending on vehicle driving practice, driver's starting intentions are divided into three typical intentions: quick start (*Q*), moderate start (*M*), and slow start (*S*).

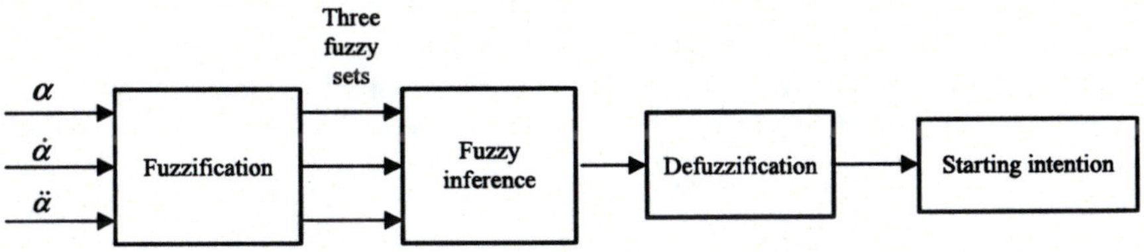

Fig. 2 Block diagram of fuzzy inference for starting intention recognition

3.2 *Fuzzy Inference of Driver's Starting Intention*

Fuzzy Subsets and Membership Function Selection

For accelerator pedal opening degree $\alpha\alpha$, domain of discourse is defined as [0, 100], fuzzy domain is defined as $E = \{0, 1, 2, 3, 4, 5, 6, 7\}$, measured factor K_e is set to 0.07, and corresponding fuzzy subsets is defined as {*VS, S, M, B, VB*}, among which, *VS* denotes a very small accelerator pedal opening degree, *S* denotes a small accelerator pedal opening degree, *M* denotes a moderate accelerator pedal opening degree, *B* denotes a great accelerator pedal opening degree, *VB* denotes a great accelerator pedal opening degree. Fuzzy subsets of α, $\dot{\alpha}$, $\ddot{\alpha}$, and starting intentions are shown in Fig. 3.

For the first derivative $\dot{\alpha}$ of accelerator pedal opening degree, domain of discourse is defined as [0, 68], fuzzy domain is defined as $E = \{0, 1, 2, 3, 4, 5, 6, 7\}$, measured factor K_e is set to 0.102, and corresponding fuzzy subset is defined as {*VS, S, M, B, VB*} where *VS* denotes a very small $\dot{\alpha}$, *S* denotes a small $\dot{\alpha}$, *M* denotes a moderate $\dot{\alpha}$, *B* denotes a great $\dot{\alpha}$, *VB* denotes a very great $\dot{\alpha}$.

For the second derivative $\ddot{\alpha}$ of accelerator pedal opening degree, domain of discourse is defined as [−125, 300], fuzzy domain is defined as $E = \{-7, -6, -5, -4, -3, -2, -1, 0, 1, 2, 3, 4, 5, 6, 7\}$, measured factor is set to 0.033, and corresponding fuzzy subset is defined as {*NB, NS, Z, PS, PB*} where *NB* denotes a very great $\ddot{\alpha}$ in negative direction, *NS* denotes a very small $\ddot{\alpha}$ in negative direction, *Z* denotes that $\ddot{\alpha}$ is 0, *PS* denotes that $\ddot{\alpha}$ is very small in positive direction, and *PB* denotes that $\ddot{\alpha}$ is very great in positive direction.

Intention means output starting intention. Domain of discourse is defined as [0, 100], fuzzy domain is defined as $E = \{0, 1, 2, 3, 4, 5, 6, 7, 8, 9, 10\}$, and corresponding fuzzy subset is defined as {*S, M, Q*},where *S* denotes slow start, *M* denotes moderate start, and *Q* denotes quick start.

Fuzzy Inference

In the current study, 125 IF-THEN rules have been used to predict driver's starting intention. The rules processed are shown in Table 1. Fuzzy inference process is realized by Matlab/Simulink. "Three-inputs-and-one-output" pattern is adopted in the inference process, where "three inputs" are accelerator pedal opening degree, first derivative of accelerator pedal opening degree and second derivative of accelerator pedal opening degree, while "one output" is driver's starting intention. Mamdani fuzzy inference is used to compute the mapping from the input values to the output values.

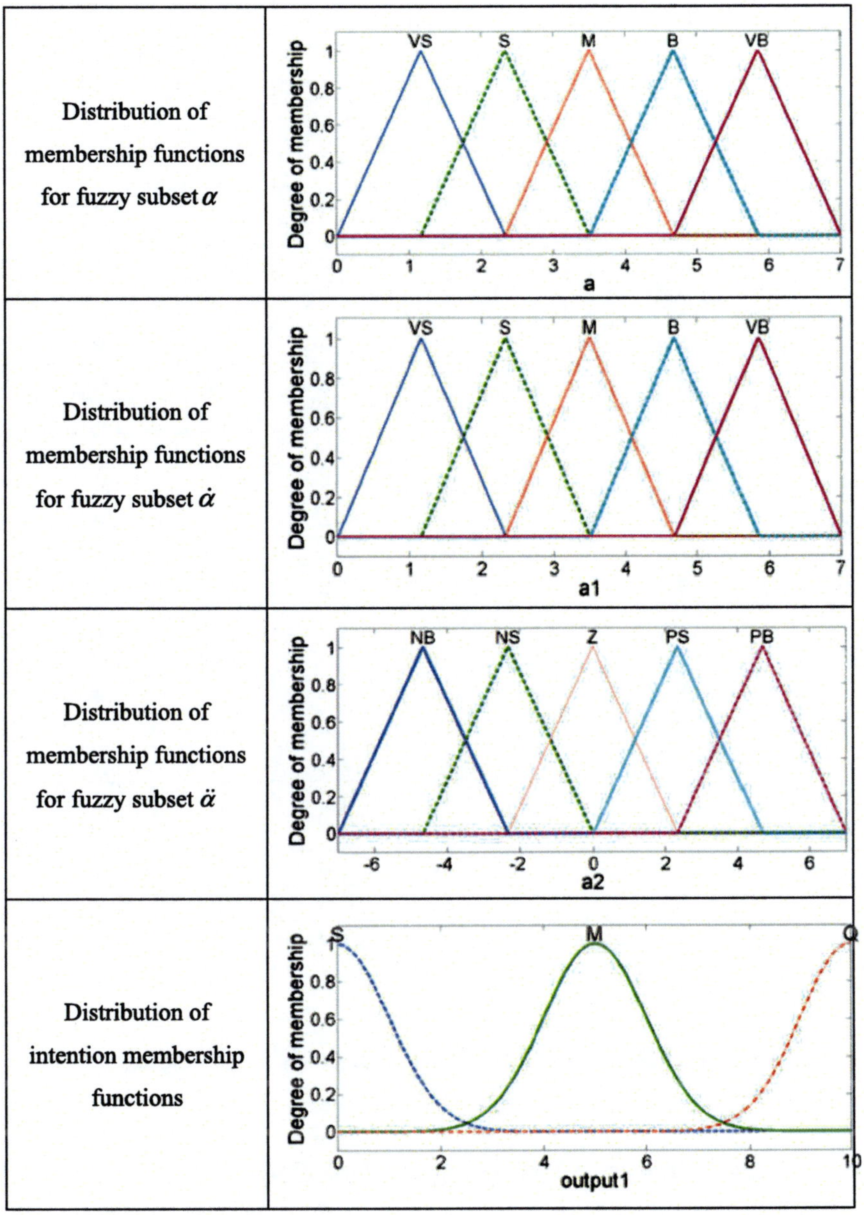

Fig. 3 Distribution of fuzzy subsets of α, $\dot{\alpha}$, $\ddot{\alpha}$ and intention

Table 1 Fuzzy rules for starting intention recognition

Intention	$\ddot{\alpha}$					
α	$\dot{\alpha}$	*NB*	*NS*	*Z*	*PS*	*PB*
VS	*VS*	*S*	*S*	*S*	*S*	*S*
	S	*S*	*S*	*S*	*S*	*S*
	M	*S*	*S*	*S*	*S*	*M*
	B	*S*	*S*	*S*	*M*	*M*
	VB	*S*	*S*	*S*	*M*	*Q*
S	*VS*	*S*	*S*	*S*	*S*	*S*
	S	*S*	*S*	*S*	*S*	*M*
	M	*S*	*S*	*S*	*M*	*M*
	B	*S*	*S*	*S*	*M*	*Q*
	VB	*S*	*S*	*M*	*M*	*Q*
M	*VS*	*S*	*S*	*S*	*S*	*S*
	S	*S*	*S*	*S*	*S*	*M*
	M	*S*	*S*	*S*	*M*	*Q*
	B	*S*	*S*	*M*	*M*	*Q*
	VB	*S*	*M*	*M*	*Q*	*Q*
B	*VS*	*S*	*S*	*S*	*M*	*M*
	S	*S*	*S*	*S*	*M*	*M*
	M	*S*	*S*	*M*	*M*	*Q*
	B	*S*	*S*	*M*	*Q*	*Q*
	VB	*S*	*M*	*Q*	*Q*	*Q*
VB	*VS*	*S*	*S*	*S*	*M*	*M*
	S	*S*	*S*	*M*	*M*	*Q*
	M	*S*	*S*	*M*	*Q*	*Q*
	B	*S*	*M*	*M*	*Q*	*Q*
	VB	*M*	*M*	*Q*	*Q*	*Q*

4 Recognition Based on Fuzzy Inference–SVM Cascade Algorithm

4.1 *Fuzzy Inference-SVM Cascade Algorithm*

Flowchart of the fuzzy inference-SVM cascade algorithm developed in this paper is shown in Fig. 4. Sample data of driver's starting intention to be identified are imported into the first-layer fuzzy inference model for likelihood classification to eliminate small-probability categories and recognize intentions not prone to confounding. Then, confounding intentions with low recognition rate in the first layer are served

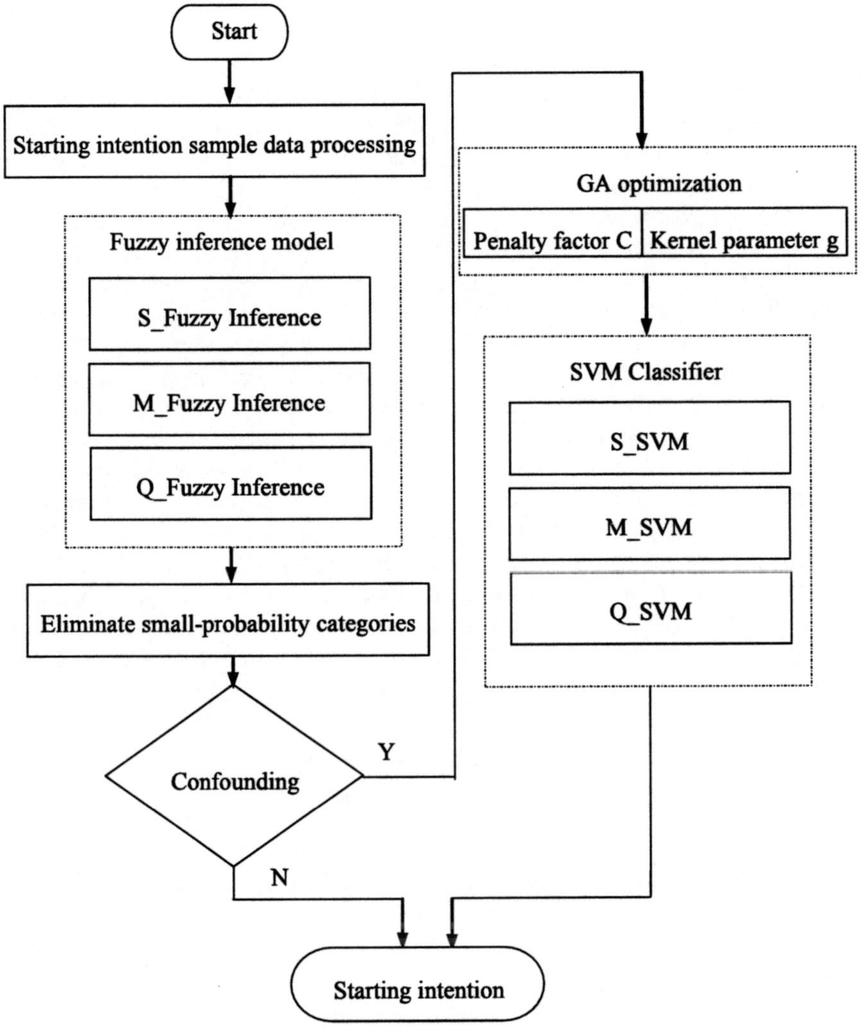

Fig. 4 Flowchart of recognizing starting intentions by fuzzy inference-SVM cascade algorithm

as candidate set and are input to SVM classifier of the second layer (a kernel function has been selected for SVM classifier, and penalty factor *C* and kernel function g have been optimized by GA), and then the intentions to be identified are classified and recognized to improve driver's starting intention recognition accuracy.

4.2 *Testing Cascade Algorithm*

Firstly, 60 sets of samples (35 data points per set, 2100 data points) acquired from the real vehicle are introduced into the SVM classifier, 1050 data are used for training and the remaining are used for testing. Secondly, the remaining 30 sets acquired from the real vehicle are imported to the first-layer, where fuzzy inference algorithm is used to recognize driver's starting intention. Recognition errors occurred in 4 sets (140 data points), among which, slow starting intentions in 2 sets are recognized as moderate start, moderate starting intention in 1 set is recognized as quick start, and quick starting intention in the other set is recognized as moderate. Accuracy of recognizing driver's starting intention by fuzzy inference algorithm alone is 86.66%. In the second-layer SVM, GA is used to optimize parameter fitness variability, as shown in Fig. 5. It shows that the optimal penalty factor C is 1.4288 and kernel parameter g is 75.8563 with parameter optimization accuracy being 94.2857%. Four sets of confounding samples (140 data points) are imported to the second-layer SVM classifiers. The comparison between the actual classification results and the predicted classification results is shown in Fig. 6. For test samples 1–70, driver's starting intention is classified as type 1 tag, manifested as slow start; for test samples 71–105, driver's starting intention is classified as type 2 tag, manifested as moderate start; and for test samples 106–140, driver's starting intention is classified as type 3 tag, manifested as quick start. Experimental results show that the recognition accuracy of starting intention achieves 97.8571% (137/140).

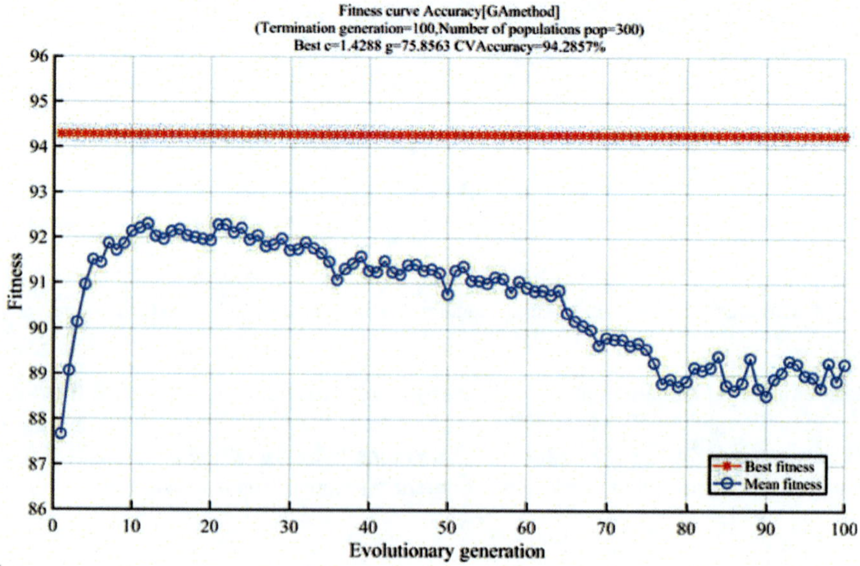

Fig. 5 Optimization of parameter fitness variability

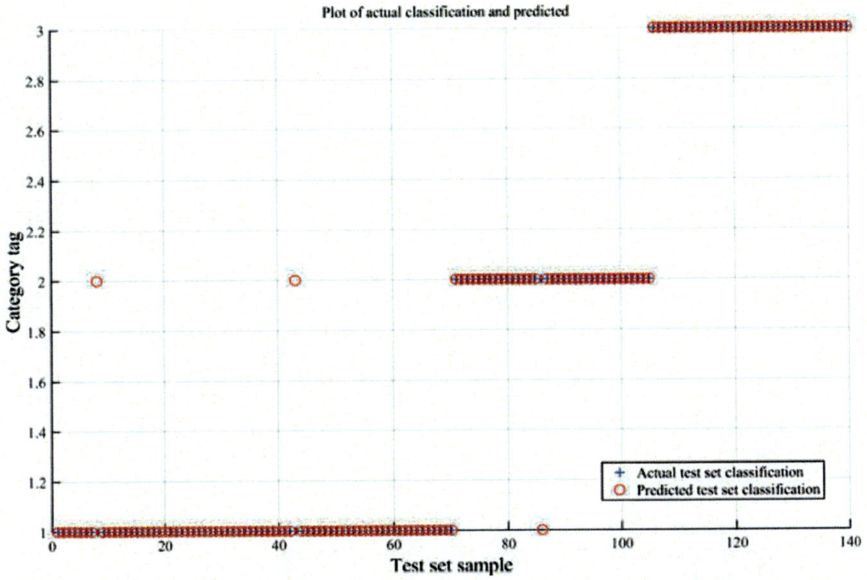

Fig. 6 Actual classification versus predicted classification of the second-layer SVM of the cascade algorithm

Table 2 Comparison of three algorithms in recognition accuracy and recognition time

Algorithm	Accuracy (%)	Mean time (s)
Fuzzy inference	86.66	0.167
SVM	90.85	0.201
Fuzzy inference–SVM cascade	97.86	0.183

Recognition results of the sample data acquired from the real vehicle by fuzzy inference algorithm, SVM algorithm, and cascade algorithm are compared in Table 2. The recognition accuracy by the fuzzy inference-SVM cascade algorithm is the highest among the three methods, and the recognition time is moderate. The fuzzy inference-SVM cascade recognition algorithm combines the advantages of fuzzy inference algorithm and SVM algorithm, and overcomes the low precision disadvantage of fuzzy inference algorithm and the difficulty for SVM to cope with large sample recognition during driver's starting intention recognition.

5 Conclusions

By comparing advantages and disadvantages of fuzzy inference and SVM, the fuzzy inference–SVM cascade algorithm is proposed, and implementation process of this algorithm is described. Then, the fuzzy inference-SVM cascade algorithm is trained and tested with the sample data acquired from the real vehicle. The test results show that the fuzzy inference-SVM cascade algorithm is an effective way to recognize driver's starting intention. In the fuzzy inference-SVM cascade algorithm, the first-layer fuzzy inference model eliminates the intentions of easy recognition, and the second-layer SVM is used to handle the confounding samples, which not only alleviates the burden on second-layer SVM classifiers but also improves recognition accuracy.

Acknowledgements This research is supported by Natural Science Foundation of China (Grant No. 51665048).

References

1. Gruosso G, Bascetta L (2018) A model-based approach for the analysis and simulation of a hybrid bus in an urban context. Int J Veh Perform 4(3):222–236
2. Zhao H, Zhang G, Wang B et al (2020) An energy-saving strategy for steering-motors of steer-by-wire vehicles. Int J Veh Perform 6(2):234–262
3. Liang L, Zaobei Z, Xiangyu W et al (2016) Identification of a driver's starting intention based on an artificial neural network for vehicles equipped with an automated manual transmission. Proc Inst Mech Eng Part D J Automob Eng 230(10):1417–1429
4. Lei Y, Zhang Y, Fu Y et al (2018) Research on adaptive gearshift decision method based on driving intention recognition. Adv Mech Eng 10(10):1–12
5. Ikenishi T, Kamada T (2015) Estimation of driver's longitudinal intention for the preceding car by brain current distribution estimation method. In: 18 international conferences on intelligent transportation systems, pp 1311–1316. IEEE, Las Palmas, Spain
6. Li F, Wang W, Feng G et al (2014) Driving intention inference based on dynamic Bayesian networks. Adv Intell Syst Comput 279(2014):1109–1119
7. Weidl G, Madsen AL, Kasper D et al (2014) Optimizing Bayesian networks for recognition of driving maneuvers to meet the automotive requirements. In: IEEE international symposium on intelligent control, pp 1626–1631, Antibes/Nice, France
8. Wang R, Lukic SM (2011) Review of driving conditions prediction and driving style recognition based control algorithms for hybrid electric vehicles. In: Vehicle power and propulsion conference (VPPC), pp 1–7. IEEE, Chicago

Signal Processing

A Review on the Application of Fiber Bragg Grating Sensors in Bolted Joints Health Monitoring

M. S. N. A. Adhreena and Z. M. Hafizi

Abstract Bolted joint is a critical component used in all engineering structures and industrial piping systems, including the petrochemical industry. Monitoring bolted joints structures are vital to ensure structural safety. In recent years, fiber Bragg grating (FBG) sensors are becoming more appealing for sensing technology in structural health monitoring due to their promising features. In contrast to conventional electrical sensors, FBG sensors have a relatively high potential to withstand harsh conditions and are suitable for long-term applications due to their miniature size, ability to multiplex, immune toward electromagnetic interferences (EMI), higher resolution and fast response. This article aims to provide a comprehensive overview of the experimental techniques applied for bolted joints health monitoring. Their relevant contributions for monitoring the degree of bolt looseness, bolt clamping force, bolt shear force and bolted flanges gap are also classified.

Keywords Fiber Bragg grating · Bolted joints monitoring · Structural health monitoring

1 Introduction

Bolted joints offer significant benefits over welding connections, including excellent fatigue resistance, convenient installation on-site, and rapid construction [1]. Due to the fact that the joints are demountable, this would be the point where the leakage is inevitable. It is crucial to identify this leakage point at an early design stage and if possible, strictly controlled them during the assembly or operational stage of the pipeline systems. The loosening of bolts may be induced and accelerated

Supplementary Information The online version contains supplementary material available at https://doi.org/10.1007/978-981-19-1577-2_48.

M. S. N. A. Adhreena · Z. M. Hafizi (✉)
Advanced Structural Integrity and Vibration Research (ASIVR), Faculty of Mechanical and Automotive Engineering Technology, Universiti Malaysia Pahang (UMP), 26600 Pekan, Pahang, Malaysia
e-mail: hafizi@ump.edu.my

© The Author(s), under exclusive license to Springer Nature Singapore Pte Ltd. 2023

M. H. A. Hassan et al. (eds.), *Technological Advancement in Instrumentation & Human Engineering*, Lecture Notes in Electrical Engineering 882,
https://doi.org/10.1007/978-981-19-1577-2_48

by severe work circumstances such as cyclic loads [2, 3], mechanical overloads [4, 5], corrosion [6] and also poor joint installation [7–9]. During service, these bolted joints may be subjected to external loads such as axial, bending and torsional stresses. These circumstances may result in accidents and casualties if these conditions are not adequately identified and addressed at an early stage.

Several works have been done in the past few years to anticipate the interaction of bolted connections and monitoring their health status. Prior works implemented the techniques of visual observation done by an expert inspector and image processing approaches by acquiring a pictured image from a digital camera. Visual inspection technique refers to the traditional process of an in-situ inspection done by experienced personnel to search and sense anomalous conditions on the structures [10]. For a large-scale structure, this method is deemed inefficient and inaccurate since the accuracy relies on the inspector's ability and expertise to decide the bolt loosening. On the other hand, the image processing technique necessitates a set of images from the bolted joints periodically [11]. By using some algorithms, bolt loosening can be identified by measuring the changes in rotation angles of nuts [12]. Both visual inspection and image processing techniques are relatively simple, but the methods are unreliable for real-time monitoring since they have significant measurement errors and could not be utilized to detect bolt looseness at an early stage [13].

In contrast to these methods, the emerging structural health monitoring (SHM) applications that employ integrated sensors have garnered more attention due to the capability to perform real-time or remote online monitoring [14, 15]. Strain gauge, piezoelectric and fiber Bragg grating (FBG) sensors are the most frequently reported sensors used for bolted joints monitoring. Among them, FBG sensors have sparked considerable attention due to their compact size and can easily be embedded inside a structure and provide information about the internal strains of the host without significantly affecting its structural integrity. To the best of our knowledge, only a few recent articles have highlighted the application of FBG sensors in bolted joints detection technology. This article aims to provide an overview of FBG technology in bolted joints health monitoring in terms of its principles and relevant contributions.

2 Fiber Bragg Grating Sensors

2.1 Introduction to FBG Sensor

The emergence of fiber Bragg grating (FBG) sensors revolutionized the growing demand of mainstream industrial facilities, predominantly in telecommunication and sensing technology, where cost-effective production, safety and efficiency are required [16]. The use of FBG as a sensing element can be applied to measure various physical quantities, including strain, temperature, pressure, force, acceleration and flow rate. FBG sensors can effectively overcome the limitation of the conventional electrical sensors by a multiplexing capability that combining several sensors on a

Table 1 Advantages and limitations of conventional electrical sensors and FBG sensors

Sensors	Advantages	Disadvantages	References
Conventional electrical sensor	1) Cost-effective 2) Small in size 3) Ease in manufacturing 4) High accuracy 5) Available in abundance	1) Not reliable for prolonged measurement in harsh environments 2) Short survival rate 3) Lower range capabilities 4) Time-consuming installation 5) Easily affected to EMI 6) Complex wiring	[21–24]
FBG sensor	1) High strength and fatigue limit 2) Lightweight and compactness 3) Long-distance transmission 4) Multiplexing ability 5) Immune toward EMI 6) Lower aging ratc 7) Higher resolution and sensitivity 8) Ability to measure different magnitudes	1) Higher market price 2) Thermal sensitive 3) Difficult to modulate wavelength shift	

single cable. Table 1 lists a comparison of advantages and disadvantages for both conventional electrical sensors and FBG sensors.

2.2 *Working Principle of FBG Sensor*

FBG sensor is generally formed by creating a periodic structure of refractive index inside the fiber core, which is done by exposing the inside core of fiber to an intense ultraviolet technology [17]. This process will create a tiny and fixed periodic distribution called Bragg grating. This grating section will reflect only the specific wavelengths of light while all the others will be transmitted along the fiber. The first invention of FBG was developed in 1978 by Hill et al. [18]. After a decade of research, Meltz et al. [19] discovered external imprinting of the FBG fabrication process, which leads to tremendous success and received widespread adoption among other researchers.

For many engineering structures, strain is dominated as the critical parameter of interest. In the case of strain sensing structural monitoring, the load is being transferred by shear forces directly from the host materials towards a specific position of the fiber grating section. This mechanism causes the length of the grating section will be significantly changed and also resulting in the refractive index of the fiber core

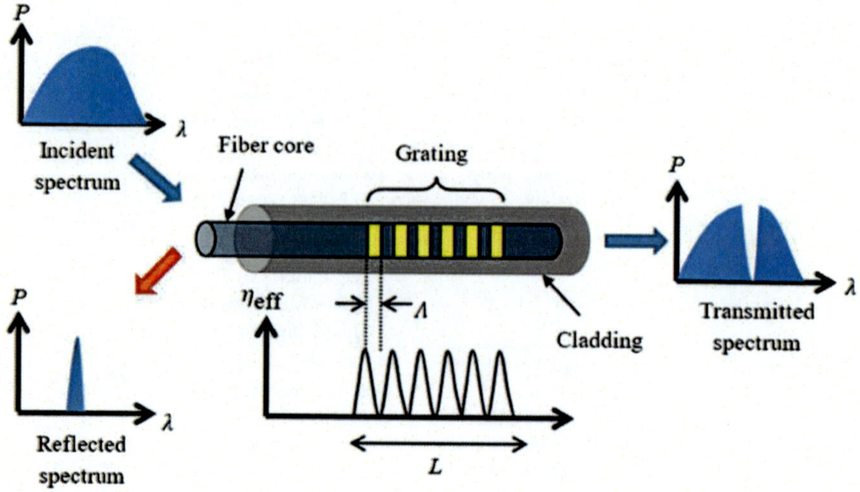

Fig. 1 Schematic illustration of FBG structure and its concept

varying over time. When a broadband light source is transmitted through the FBG sensor, only a narrow spectrum of the incoming light signal will be partially reflected back by the grating fringe of the sensor while the rest will be passing through the sensor. These are resulting in a missing transmission signal [20]. Figure 1 depicts the schematic illustration of FBG working operation.

The reflected signal will be centered around on the Bragg wavelength, λ_B and it depends on the spatial grating period of the fiber core, Λ and also the effective refractive index, n_{eff} as following this Eq. (1):

$$\lambda_B = 2n_{eff}\Lambda \tag{1}$$

The back-reflected Bragg wavelength will be shifted accordingly to any mechanical or thermal disruption imposed on the FBG sensor [25]. The sensitivity for both strain and temperature of Bragg wavelength can be presented as:

$$(\Delta\lambda_B)/\lambda_B = (1 - \rho_e)\,\varepsilon + (\alpha - \xi)\,\Delta T \tag{2}$$

The sensitivities of FBG sensor for strain and temperature recorded at a wavelength of 1550 nm are approximately 1.2 pm/με and 13.7 pm/°C, respectively. For the vast majority of applications where only strain parameters are required, the cross-sensitivity of FBG sensor to thermal effects must be considered and compensated with an appropriate implemented technique. Otherwise, the increase in temperature may go undetected and cause an error in measurements [24, 26].

3 Bolted Joints Monitoring Application

3.1 *Degree of Bolt Loosening*

Bolt looseness is one of the most reported failures in bolted structures. Common methods to detect bolt looseness are based on ultrasonic [27], electromechanical impedance [28, 29] and vibration [30–32]. However, the results of these methods were limited since they were only able to detect a severe bolt loosening and could not provide precise information on the extent of damage [33]. Shin et al. [34] had proposed a fiber optic-based sensor for simultaneous multi-bolt monitoring involving angle determination, event detection and localization of multi-bolt loosening. Similarly, to address the issue of bolt loosening, Liu et al. [35] proposed a technique based on operational strain derived from FBG sensor, which utilized the Structural Damage Indicator (SDI) to quantify the relative degree of bolt looseness. Based on the results, the SDI can precisely locate and detect the loose bolts. However, in this work, they only considered the adjacent bolts and did not focus on the non-adjacent bolts to validate the proposed method.

3.2 *Bolt Preload or Clamping Force*

It had been reported that almost 90% of the bolted joints failure are caused by improper bolt tightening [36]. Thus, controlling the bolt tension throughout the assembly process and maintaining their clamping force during operation become significant concerns. The small transverse dimensions of FBG sensors make them ideal for measuring the bolt strain with high precision. Hence, the diameter of the drilled hole along the bolt shank for sensor embedment can be significantly reduced than is needed for electrical resistance strain gauge sensors, and its associated effect on the bolt structural integrity can be substantially decreased. In recent years, the FBG sensors had been successfully embedded into the bolt structure, allowing the clamping force to be determined by measuring strain in bolts. The first concept of utilizing embedded FBG sensors into a bolt was patented by Hay [37]. This FBG-based smart bolt design was subsequently used by Pran et al. to monitor and detect creep in a bolted glass fiber reinforced polymer (GRP) sample [38]. They conducted a long-term investigation of the tension level in bolted GRP. After a year, they noticed a significant result in data which are strongly influenced by the composite's thermal expansion and also the insufficient adherence of the acrylate coating to the glass fiber had a substantial impact.

However, the technique of permanently embedded the sensors into the host structure and leaving them to remain in place after the joint assembly makes such designs are not commercially appealing for non-destructive evaluation (NDE) and SHM practices. By modifying the design, Khomenko et al. [37] introduced a detachable and reusable embedded FBG sensor known as a bolt tension monitor device for

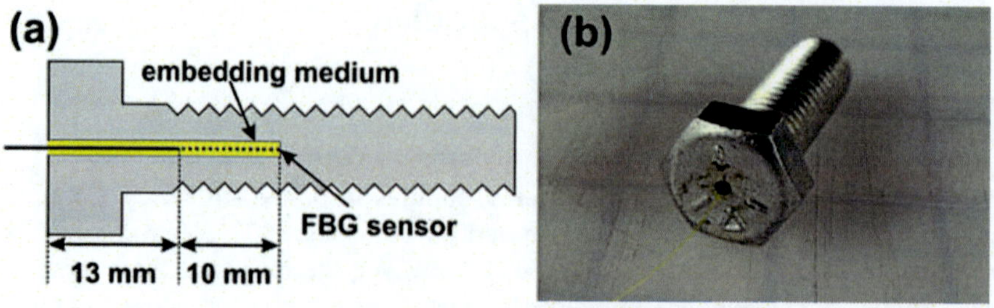

Fig. 2 **a** Cross-sectional diagram of bolt tension monitor and **b** prototype utilizing thermoplastic for sensor embedding [37]

precise bolt tightening measurement. The cross-section of the device is illustrated in Fig. 2(a), while Fig. 2(b) presents a thermoplastic prototype that is being used for FBG sensor embedding. A hole with a diameter of 1.8 mm was drilled along the bolt axis to provide quick and easy sensor installation. It was validated that the embedment medium of thermoplastic was satisfactory for monitoring the bolt clamping force. Furthermore, this device successfully showed that no residual strains were generated during tightening in either the bolt or the embedment medium.

This successful development also attracted many other researchers to further investigate the capability of FBG-based smart bolt [39–41]. At some point, the distributed characteristics of the smart bolts are eventually restricted to the through-hole host structures. To address the shortcoming of the smart bolt, a design of smart washer was introduced as illustrated in Fig. 3 [42, 43]. This smart washer was fabricated by encircling and embedding the FBG sensor into a pre-machined groove along the circumferential surface of the washer. This simple yet effective design can monitor the bolt preload by directly measuring the circumferential strain change.

With no incorporation of a temperature-compensated scheme, questions have been raised as the distributed features of these smart bolt and smart washer are impeded to monitor the bolt clamping force in high temperatures. In many cases, FBG sensors are frequently embedded into the host structures by using epoxy, which is indeed a straightforward procedure at room temperature. However, most epoxy materials severely deteriorate when exposed to a high temperature above 400 °C. Recently, Tu

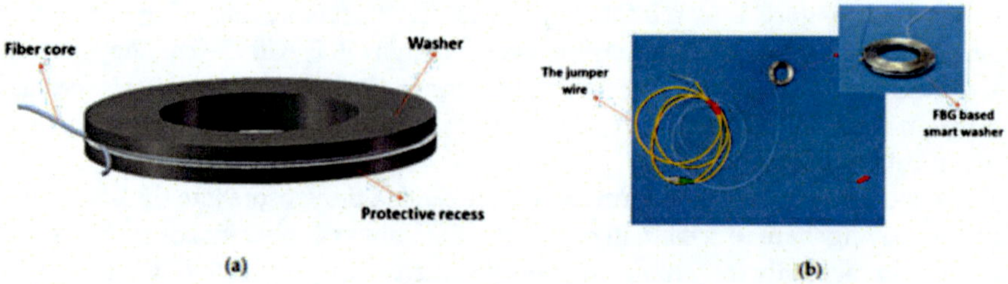

Fig. 3 **a** Illustrated design of smart washer and **b** photo of FBG-based smart washer [42].

et al. [44] provided an alternative solution by proposing the use of metal-packaged regenerated fiber Bragg grating (RFBG) strain sensors for monitoring bolt clamping force in high temperature environments. To validate the operational ability of this design, the prototype was heated up to 500 °C at an interval of 50 °C. The method of using the metal-packaged RFBG sensor showed excellent linear response, stability and repeatability to the axial load applied on the bolt simultaneously when exposed to constant high temperature. For future improvement, one should include testing at higher temperature exposure for a longer period of time and the enhancement of alternative mounting techniques to suit diverse applications.

3.3 *Bolt Shear Forces*

In most cases, high-strength bolted joints often depend on the friction between the contact surfaces to provide support for the host structures. To ensure and sustain the structural integrity, there are stringent conditions for the allowable maximum load imposed on the bolted connections [45]. Although certain restrictions on the bolt load are required, bolted connections are often inadvertently subjected to higher shear forces during field deployment. Large shear loads when combined with excessive axial loads may impair the capacity of these joints, thus its structural integrity will degrade and be susceptible to failure [46, 47]. Therefore, it is essential to monitor both axial and shear forces to avoid overloading the bolted structures. Many studies have recently focused on monitoring the shear loads during operation [48–50].

The method of monitoring shear force based on FBG sensors was proposed by Suresh et al. [51–53]. The basic design comprises layers of carbon composite material (CCM) and embedded FBG sensors. A deformable layer of silicon rubber between the CCM layers will transmit the shear forces to the FBG sensors, which then altering the Bragg wavelength of the FBG sensors. Ren et al. [54] presented a concept of a smart bolt utilizing four FBG sensors (as illustrated in Fig. 4) to monitor axial and shear loads simultaneously.

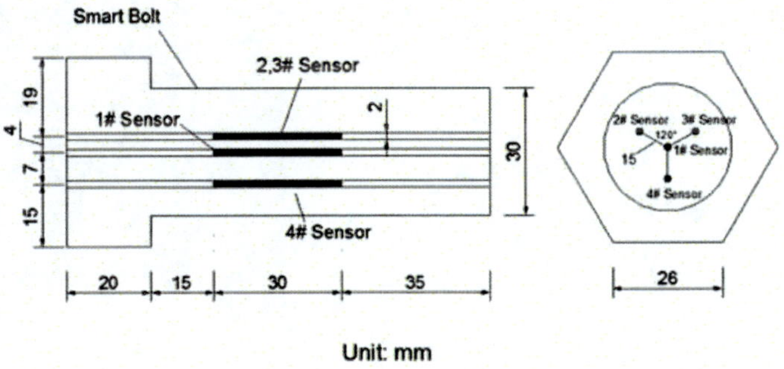

Fig. 4 The layout illustration of embedded FBG sensors in the smart bolt [54]

One FBG sensor was embedded along the bolt axis to measure the axial force, while another three FBG sensors were distributed along the central of bolt axis to measure the shear force directly. Based on the acquired data, this integrated FBG-based smart bolt can be a reliable tool for bolted joints health monitoring applications.

3.4 Gap Openings Between the Flanges

The flange leakage is majorly influenced by the gap between the flanges. A frequent gap opening of bolted flanges may result in cumulative fatigue damage within the bolted region. In most instances, the stress-induced due to thermal gradients also affects the fatigue life of the flange joint, particularly when they are going through sudden increases in temperature. The flange joint's thermal gradient is influenced by several factors, including flange dimension, the gap between the flanges, and thermal boundary conditions on both the inner and outer sides of the joint. To the best of our knowledge, no prior studies are addressing the detection for flange gap developments by using the FBG sensor. In the current study, considerable attention is given to detecting bolted joints health conditions by embedding FBG sensors through bolts or special-engraved washers. Kim et al. [1] focused on the study that included the development of gaps between the flanges under tensile loading. Twelve flanges specimens were examined under monotonic tensile load. To measure the flange gap development, a digital caliper was being utilized as shown in Fig. 5.

The same practice was also adopted by Bausman et al. [55] which using a digital caliper to indirectly measure an elastic and plastic flange rotation. This indirect flange rotation measurement was done by measuring the height variation of the flange joints. Then, the gap measurement was taken when the load was utterly removed from the flanges as shown in Fig. 6. Although these methods of utilizing digital caliper might seem reliable, this technique is often impractical and unattainable since the bolt

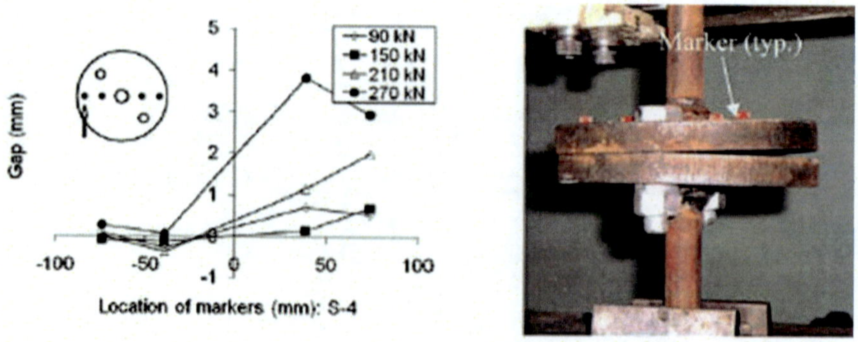

(a) (b)

Fig. 5 Typical gap development: **a** along the flange section; **b** test picture [1]

Table 2 Summary of sensor's characteristic in bolted joints evaluation

Features	Sensors		
	Strain gauge sensor	Piezoelectric sensor	FBG sensor
Type of method	Direct	Indirect	Indirect
Source of energy	Electrical	Electrical	Optical
Nature of signal	No signal	Frequency	Wavelength
Signal processing	No	Yes	Yes
Equipment cost	Low	High	General
Real-time monitoring	No	No	Yes
Sensor accuracy	Low	Good	Excellent
Immunity toward EMI	No	Yes	Yes
Suitable for harsh environment	No	No	Yes
References	[57–59]	[60, 61]	[42, 43, 54]

Fig. 6 Experimental validation of flange assembly [55]

elongation is so small, making it difficult to measure it precisely. Moreover, using a digital caliper to measure gap opening and bolt elongation is inappropriate for in-situ monitoring [48].

For a pipeline bolted flange, the water may flow out from the flange joint through the holes of the bolts due to prying force exerted at the flanges. On the other hand, due to the larger gap opening between the flanges plate, the pressure contains inside the joint structure does not encounter any hindrances and leads to greater flow out of the structure. La Sarandra et al. [56] executed monotonic and cyclic tests to examine the performance of bolted flange joints. To determine the axial and stiffness of the bolted flange, eight displacement transducers were installed in different points to measure the flange gap. The experimental findings showed that the bolted joints were performed well under cyclic loading which virtually no degradation was observed.

4 Conclusions

Bolted joints are an essential connection which often utilized in many engineering structures. A proper and correct bolt tightening force is a critical element in connection safety. However, due to the interaction effect, it is difficult to uniformly tighten a set of bolts, and bolts that have previously been tightened will ineluctably loosen under high-stress conditions. Therefore, effectively measuring the bolt tension or preload is a great importance to monitor these structures. Up to date, numerous sensors have been developed as a source to monitor the bolted joints structures. However, certain limitations of sensors could restrict their application in SHM practices. Table 2 summarized the characteristics of sensors which widely utilized for bolted joints detection technology.

Among other conventional sensors, FBG sensing technology provides an attractive alternative for SHM monitoring. Compared to them, FBG sensors have the advantages of being lightweight, smaller in size, ease in the implementation, minimally invasive, high accuracy, well-correlated, multiplexing ability and also long-term stability.

Several research had been conducted to demonstrate their ability to monitor bolted joints structures. Being smaller in size, FBG sensors can be easily embedded into the bolt without significantly degrading their host's structural integrity. The proposed monitoring techniques based on FBG sensors demonstrated a successful implementation, thus solving and minimizing the failure of engineering structures. However, considerable issues are still worthy of further investigation and study for the final successful engineering application, considering reliable adhesive material, design optimization, and the establishment of a remote sensing network.

Acknowledgements The authors would like to thank the Faculty of Mechanical & Automotive Engineering Technology, Universiti Malaysia Pahang (http://www.ump.edu.my/) for providing the laboratory facilities. The authors would also like to acknowledge the Institute of Postgraduate Studies (IPS), Universiti Malaysia Pahang for funding through the Master Research Scheme (MRS) scholarship. Finally, special thanks to the UMP Research and Innovation Department for providing the internal research grant under grant no. RDU1903120; and Postgraduate Research Grants Scheme (PGRS) under grant no. PGRS2003145.

References

1. Kim YJ, Madugula MKS (2010) Behavior of bolted circular flange connections subject to tensile loading. Int J Steel Struct 10(1):65–71
2. Ataei A et al (2019) Cyclic behaviour of bolt and screw shear connectors in steel-timber composite (STC) beams. J Constr Steel Res 161:328–340
3. Giannopoulos IK et al (2017) Effects of bolt torque tightening on the strength and fatigue life of airframe FRP laminate bolted joints. Compos B Eng 125:19–26
4. Aziz N et al (2014) Rock bolt corrosion–an experimental study. Min Technol 123(2):69–77
5. Peng J et al (2019) Flexural behavior of corroded HPS beams. Eng Struct 195:274–287
6. Mohammadi M, Salimi HR (2007) Failure analysis of a gas turbine marriage bolt. J Fail Anal Prev 7(2):81–86
7. Noda N-A et al (2016) Effect of pitch difference between the bolt–nut connections upon the anti-loosening performance and fatigue life. Mater Des 96:476–489
8. Yin H et al (2016) A smart washer for bolt looseness monitoring based on piezoelectric active sensing method. Appl Sci 6:320
9. Yuan R et al (2019) Percussion-based bolt looseness monitoring using intrinsic multiscale entropy analysis and BP neural network. Smart Mater Struct 28(12):125001
10. Park J, Kim T, Kim J (2015) Image-based bolt-loosening detection technique of bolt joint in steel bridges. In: 6th international conference on advances in experimental structural engineering. University of Illinois, Urbana-Champaign
11. Ho H-N et al (2013) An efficient image-based damage detection for cable surface in cable-stayed bridges. NDT E Int 58:18–23
12. Yamaguchi T, Hashimoto S (2010) Fast crack detection method for large-size concrete surface images using percolation-based image processing. Mach Vis Appl 21(5):797–809
13. Ho H-N, et al (2012) A synchronized multipoint vision-based system for displacement measurement of civil infrastructures. Sci World J 2012
14. Chen B et al (2018) Pipeline two-dimensional impact location determination using time of arrival with instant phase (TOAIP) with piezoceramic transducer array. Smart Mater Struct 27(10):105003
15. Liao W-I, Chiu C-K (2019) Seismic health monitoring of a space reinforced concrete frame structure using piezoceramic-based sensors. J Aerosp Eng 32(3):04019015
16. Mihailov SJ (2012) Fiber Bragg grating sensors for harsh environments. Sensors (Basel, Switzerland) 12(2):1898–1918
17. Yau M, et al (2011) Using Fiber Bragg Grating (FBG) sensors to measure vertical displacements of bridges-a preliminary study. In: Proceedings of the first international conference on engineering, designing and developing the built environment for sustainable wellbeing. Queensland University of Technology
18. Hill K et al (1978) Photosensitivity in optical fiber waveguides: application to reflection filter fabrication. Appl Phys Lett 32(10):647–649
19. Meltz G, Morey WW, Glenn W (1989) Formation of Bragg gratings in optical fibers by a transverse holographic method. Opt Lett 14(15):823–825
20. Luyckx G et al (2011) Strain measurements of composite laminates with embedded fibre bragg gratings: criticism and opportunities for research. Sensors (Basel, Switzerland) 11(1):384–408
21. Sun L et al (2019) The strain transfer mechanism of fiber Bragg grating sensor for extra large strain monitoring. Sensors (Basel, Switzerland) 19(8):1851

22. Leng J, Asundi A (2003) Structural health monitoring of smart composite materials by using EFPI and FBG sensors. Sens Actuators A 103(3):330–340
23. Baxter S et al (2010) Validation of a novel fiber optic strain gauge in a cryogenic and high magnetic field environment. Cryogenics 50(10):700–707
24. Roriz P et al (2014) From conventional sensors to fibre optic sensors for strain and force measurements in biomechanics applications: a review. J Biomech 47(6):1251–1261
25. Gao X et al (2020) A dual-parameter fiber sensor based on few-mode fiber and fiber Bragg grating for strain and temperature sensing. Opt Commun 454:124441
26. Zhu L et al (2018) Metallic-packaging fiber Bragg grating sensor based on ultrasonic welding for strain-insensitive temperature measurement. Rev Sci Instrum 89:045005
27. Jhang K-Y et al (2006) Estimation of clamping force in high-tension bolts through ultrasonic velocity measurement. Ultrasonics 44:e1339–e1342
28. Annamdas VGM, Yang Y, Soh CK (2007) Influence of loading on the electromechanical admittance of piezoceramic transducers. Smart Mater Struct 16(5):1888
29. Lim HJ et al (2011) Impedance based damage detection under varying temperature and loading conditions. NDT and E Int 44(8):740–750
30. Qu W et al (2017) Using sub-harmonic resonance to detect bolted joint looseness. J Vibr Meas Diagn 37:279–283
31. Zhang M et al (2017) Application of subharmonic resonance for the detection of bolted joint looseness. Nonlinear Dyn 88(3):1643–1653
32. Gou B et al (2015) Bolt tightening force detection using outlier analysis of structural natural frequencies. J Vibr Shock 34(23):77–82
33. Park G, et al. (2010) Monitoring of bolted joints using piezoelectric active-sensing for aerospace applications. Los Alamos National Lab. (LANL), Los Alamos, NM (United States)
34. Shin H, Park CY (2014) A novel fiber optic bolt loosening monitoring sensor system for aircraft bolt joints. J Intell Mater Syst Struct 25:647–653
35. Liu M, et al (2019) Looseness detection utilizing FBG based operational modal strain for a bolted cantilever plate. In: 2019 prognostics and system health management conference (PHM-Qingdao)
36. Huang YH et al (2009) Real-time monitoring of clamping force of a bolted joint by use of automatic digital image correlation. Opt Laser Technol 41(4):408–414
37. Khomenko A et al (2015) Bolt tension monitoring with reusable fiber Bragg-grating sensors. J Strain Anal Eng Des 51:101–108
38. Pran K, Farsund O, Wang G (2002) Fibre Bragg grating smart bolt monitoring creep in bolted GRP composite. In: 2002 15th optical fiber sensors conference technical digest. OFS 2002 (Cat. No. 02EX533). IEEE
39. Zhao Y et al (2018) Study on the optimal groove shape and glue material for fiber Bragg grating measuring bolts. Sensors 18(6):1799
40. You R, Ren L, Song G (2020) A novel comparative study of European, Chinese and American codes on bolt tightening sequence using smart bolts. Int J Steel Struct 20
41. Zhang L, et al (2020) A loose self-checking method based on FBG for axle counting sensor. In: IOP conference series: materials science and engineering. IOP Publishing
42. Chen D et al (2018) A fiber Bragg grating (FBG)-enabled smart washer for bolt pre-load measurement: design, analysis, calibration, and experimental validation. Sensors 18:2586
43. Yeager M, Whitaker A, Todd M (2017) A method for monitoring bolt torque in a composite connection using an embedded fiber Bragg grating sensor. J Intell Mater Syst Struct 29:1045389X1770521
44. Tu Y, Huang Y-K, Tu S-T (2019) Real-time monitoring of bolt clamping force at high temperatures using metal-packaged regenerated fiber Bragg grating sensors. Int J Press Vessels Pip 172:119–126
45. Li J, Hao H (2016) Health monitoring of joint conditions in steel truss bridges with relative displacement sensors. Measurement 88:360–371
46. Ibrahim R, Pettit C (2005) Uncertainties and dynamic problems of bolted joints and other fasteners. J Sound Vib 279(3–5):857–936

47. Chang F-K, Chang K-Y (1987) Post-failure analysis of bolted composite joints in tension or shear-out mode failure. J Compos Mater 21(9):809–833
48. Wang T et al (2013) Review of bolted connection monitoring. Int J Distrib Sens Netw 2013:1–8
49. Caccese V, Mewer R, Vel SS (2004) Detection of bolt load loss in hybrid composite/metal bolted connections. Eng Struct 26(7):895–906
50. Ritdumrongkul S et al (2003) Quantitative health monitoring of bolted joints using a piezoceramic actuator–sensor. Smart Mater Struct 13(1):20
51. Tjin S, Suresh R, Ngo N (2004) Fiber Bragg grating based shear-force sensor: modeling and testing. J Lightwave Technol 22(7):1728
52. Suresh R, Tjin S, Ngo N (2004) Shear force sensing by strain transformation using non-rectilinearly embedded fiber Bragg grating. Sens Actuators A 116(1):107–118
53. Suresh R, Tjin SC, Ngo NQ (2005) Application of a new fiber Bragg grating based shear force sensor for monitoring civil structural components. Smart Mater Struct 14(5):982
54. Ren L et al (2018) A smart "shear sensing" bolt based on FBG sensors. Measurement 122:240–246
55. Bausman AR, Waterland AF III (2014) A survey method to optimize bolted flanged joint assembly torques for ASME B16.5 flanges. In: ASME 2014 pressure vessels and piping conference
56. La Salandra V, et al (2016) Cyclic response of enhanced bolted flange joints for piping systems. In: ASME 2016 pressure vessels and piping conference
57. Huo L et al (2017) Smart washer - a piezoceramic-based transducer to monitor looseness of bolted connection. Smart Mater Struct 26:025033
58. Huo L et al (2017) Impedance based bolt pre-load monitoring using piezoceramic smart washer. Smart Mater Struct 26(5):057004
59. Jiang T et al (2018) Monitoring of bolt looseness-induced damage in steel truss arch structure using piezoceramic transducers. IEEE Sens J 18(16):6677–6685
60. Argatov I, Sevostianov I (2010) Health monitoring of bolted joints via electrical conductivity measurements. Int J Eng Sci 48(10):874–887
61. Liu X-C et al (2019) Tension–bend–shear capacity of bolted-flange connection for square steel tube column. Eng Struct 201:109798

High-Speed Camera Analysis of Tool Eccentricity During Friction Stir Welding of Thick Plate Aluminium Alloys

L. H. Ahmad Shah, S. Walbridge, and A. Gerlich

Abstract This study examines the change in tool eccentric motion during friction stir welding of aluminium alloys. 9.5 mm-thick AA5052-H32 and AA6061-T6511 aluminium alloys were used for this study. A specially designed tool was utilized to form the artificial eccentric motion, while a typical tool was used for the aligned setup. High-speed camera was employed to measure the eccentric motion during jog state and welding process. Similar observations were recorded for both materials. The aligned setup shows an increase in the eccentric motion up to 113%. However, the eccentric motion was dampened up to 84% in the artificial eccentricity setups. It can be concluded that varying base materials does not affect the change in tool eccentricity during welding process. However, the changes can be attributed to the initial eccentric setup.

Keywords Friction stir welding · Tool eccentricity · Aluminium alloy · High-speed camera

1 Introduction

Friction stir welding (FSW) is a solid-state welding process that uses a specially designed tool consisting of a shoulder and pin to generate heat through friction between the tool and the workpiece caused by rapid rotation. Once the workpiece is sufficiently plastically deformed, the tool is then traversed linearly in order to achieve consolidation between the workpieces [1]. FSW has been an active field of research

Supplementary Information The online version contains supplementary material available at https://doi.org/10.1007/978-981-19-1577-2_49.

L. H. A. Shah (✉)
Faculty of Mechanical and Automotive Engineering Technology, Universiti Malaysia Pahang, 26600 Pekan, Pahang, Malaysia
e-mail: luqmanhakim@ump.edu.my

S. Walbridge · A. Gerlich
Faculty of Engineering, University of Waterloo, 200 University Avenue West, Waterloo, ON N2L 3G1, Canada

© The Author(s), under exclusive license to Springer Nature Singapore Pte Ltd. 2023

M. H. A. Hassan et al. (eds.), *Technological Advancement in Instrumentation & Human Engineering*, Lecture Notes in Electrical Engineering 882,
https://doi.org/10.1007/978-981-19-1577-2_49

since its invention due to its numerous advantages and applications in various fields [2–4].

Tool eccentricity is the wobbling motion of the rotating tool caused by a slight misalignment between the spindle's rotational axis and the tool's rotational axis. Depending on the severity of the eccentric motion, such undesired motion could cause unnecessary additional load on the tool, leading to accelerated tool wear or even premature tool fracture [5]. Therefore, accurate measurement of this phenomenon could facilitate the understanding of the detrimental effect of eccentric motion on FSW tool. Shah et al. [5] have classified tool eccentricity into two categories, namely inherent tool eccentricity, i.e., eccentricity that is caused by collet thread wear or improper fit-up, and artificial tool eccentricity, i.e., eccentricity that is present through modifications of the tool design. Usually, both eccentricities are identified by whether the value is less than 0.1 mm (inherent eccentricity) or equal to/larger than 0.1 mm (artificial eccentricity).

While many studies have been conducted using inherent tool eccentricity [6–10] and artificial eccentricity [11–18], measuring the change in eccentric motion of the tool during the welding process have been very challenging. Yan et al. [8] employed a high resolution linear variable displacement transducer (LVDT) to monitor the displacement of the tool during the welding process and reported that tool eccentricity was suppressed by 50% during FSW. On the other hand, Gratecap et al. [6] and Shah et al. [19] utilized a high-speed camera to observe the tool motion during the welding process. While the former did not mention any quantitative values, the later noted that the change in eccentricity varies according to the FSW machine, the tool design and initial eccentricity values [19]. However, to the best of the authors' knowledge, no study so far has looked into how eccentric motion can be affected by varying the base material to be welded. Therefore, this study aims to look into the effect on tool eccentricity by varying aluminium alloy base materials. Two alloys, namely AA5052 and AA6061 are of particular interest in this study as they represent two of the most used aluminium series around the world.

2 Materials and Method

9.5 mm-thick AA5052-H32 and AA6061-T6511 aluminium alloys were used for this study. The nominal chemical compositions of both base materials are tabulated in Table 1. For the welding process, an H13 steel tool with an 8° tapered, M6-threaded and tri-flats design was utilized. The dimensions include a 15 mm shoulder diameter,

Table 1 Chemical composition of AA5052 and AA6061 alloy (in weight percentage, wt%)

Alloy	Al	Mg	Si	Mn	Cr	Cu	Fe
5052	Balance	2.8	0.25	0.1	0.35	0.1	0.4
6061	Balance	0.8	0.53	0.08	0.06	0.2	0.2

Table 2 Welding parameters used for eccentric change analysis

Welding parameter	Value
Tool eccentricity (mm)	0 (aligned), 0.2, 0.4
Rotational speed (rpm)	1120
Weld speed (mm/min)	63
Tool tilt angle (°)	2.5

6 mm pin root diameter and 9.3 mm pin length. The tools were modified at the shoulder in order to achieve 0.2 mm and 0.4 mm artificial eccentricity, as explained in [16]. The welding parameters are shown in Table 2. Separate welds were made for both base materials using the bead-on-plate approach, i.e., there was no faying surface present, but full tool penetration was achieved.

Accurate tool eccentricity observations were conducted using a Photron FASTCAM Mini UX50 type 160 km high-speed camera (HSC) which was mounted on a 7.5 HP Jafo Universal Milling Machine stage using a simple fixture as shown in Fig. 1. A light source was fixed at the back of the spindle to provide a much sharper contrast of the tool outline for accurate measurements. The tool eccentricity changes before welding (jog, i.e., free run) and during the welding process was recorded using a 500 frames-per-setup (fps) HSC setup.

First, the spindle was rotated at 1120 rpm without any welding done, i.e., jogging state. For comparison, the rotation of the spindle was significantly reduced to 56 rpm in order to measure the eccentricity during the jog using a dial gauge with the tolerance of ±0.013 mm (0.0005 in). Next the welding process was recorded using the HSC. The recorded image data was finally analyzed using Fiji image analysis software. The eccentric values were measured based on its ratio to the diameter of the tool shoulder (Ø15 mm) in the HSC images taken. Examples of the images taken during jog and welding process can be seen in Fig. 2. Each welding parameter conditions were repeated three times to ensure consistency.

Fig. 1 High-speed camera observation setup

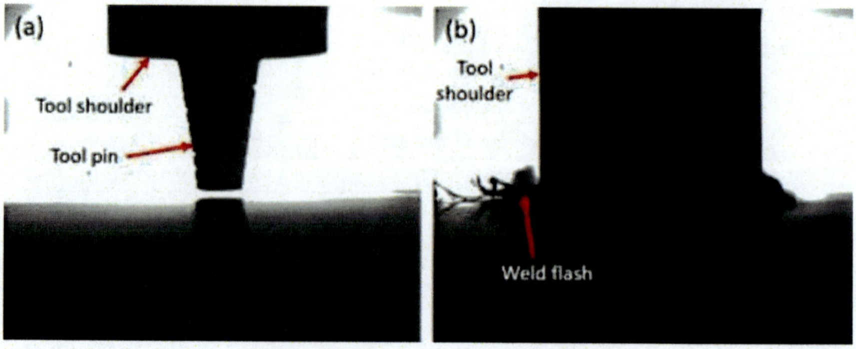

Fig. 2 High-speed camera images during **a** jog and **b** welding process

3 Results and Discussion

Before the welding processes, the validity of the eccentricity values measured from the HSC images at 1120 rpm rotational speed were compared with the tool jog measurements at 56 rpm rotational speed. Both measurements are tabulated in Table 3 during the setup of both 5052 and 6061 samples. The dial gauge values were converted from inches. In both setups, the dial gauge and HSC values showed very good agreement between each other, indicating the reliability of the measurements in the HSC setup. Note that, for the aligned setup, an inherent eccentricity is expected within the range of <0.05 mm, as observed in both sample setups.

The results of the tool eccentricity change for both 5052 and 6061 samples are shown in Fig. 3. Regardless of the base material, a similar trend can be observed. During the welding process of the aligned setup (<0.05 mm eccentricity), a slight increase of the tool eccentricity was recorded, as compared to its jog stage counterpart. However, when artificial eccentricity of 0.2 and 0.4 mm were introduced, tool eccentricity showed a decrease in value. The percentage change in eccentricity between jog and weld is presented in Table 4.

Table 3 Eccentricity during jog stage using dial gauge (at 56 rpm) and HSC (at 1120 rpm)

Eccentricity (mm)	Dial gauge values (mm)	HSC values (mm)
5052 sample setups		
0 (aligned setup)	0.046 ± 0.013	0.046 ± 0.007
0.2	0.216 ± 0.013	0.224 ± 0.012
0.4	0.406 ± 0.013	0.407 ± 0.007
6061 sample setups		
0 (aligned setup)	0.046 ± 0.013	0.046 ± 0.008
0.2	0.234 ± 0.013	0.257 ± 0.006
0.4	0.381 ± 0.013	0.406 ± 0.015

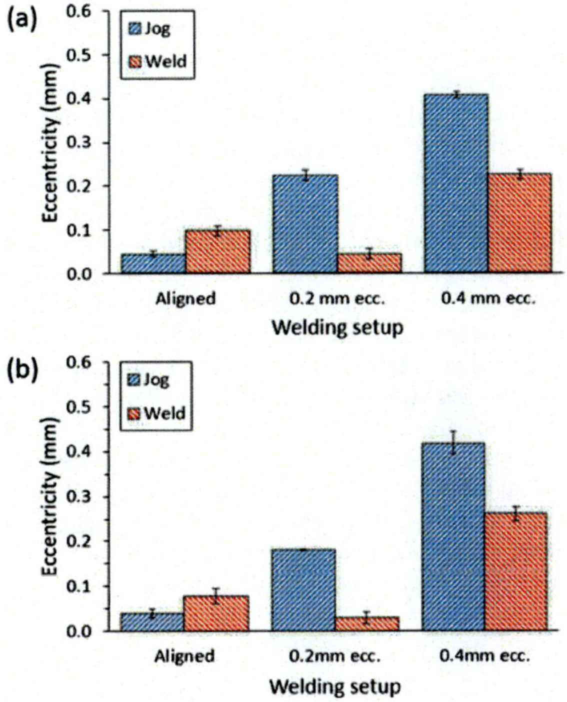

Fig. 3 Tool eccentricity change with varying eccentric setup for **a** 5052 and **b** 6061 samples

Table 4 Percentage change in eccentricity between jog and weld

Base metal	Aligned (%)	0.2 mm eccentricity (%)	0.4 mm eccentricity (%)
5052	113	−80	−44
6061	100	−84	−38

From the table, it can be concluded that, regardless of the initial hardness of AA5052 alloy (68 HV) and AA6061 alloy (108 HV) samples, all eccentric setups show almost identical trends when varying tool eccentricity. Previous study by the authors [19] on the forces acting on the tool indicated that the initial plunging stage exerted the largest force on the tool. However, given the plastic state of the material because of the elevated temperature during the plunging stage, the force exerted on the tool during the traverse stage is minimal. In a separate analysis, the authors have shown that the flow stress decreases significantly when the temperature exceeds 80% of the melting temperature of the base material, which is typically the case in FSW processes [16]. Table 4 provides additional evidence that minimal resistance is expected from both base materials during the traverse stage where severe plastic deformation is taking place [20].

Given the minimal resistance during the traverse stage, the question therefore arises as to why a decrease in eccentricity can be seen. In general, the findings from this study supports our previous claim that the reason for the eccentricity change can be mainly attributed to the asymmetric (tapered, threaded, tri-flats) tool design and the rigidity of the components of the spindle system [19], which are both constant parameters in this study. While the tool design is critical to enhance material mixing [21, 22], the asymmetric pin exerts an uneven force distribution around the pin, aggravating the eccentric motion of the aligned setup. A similar aligned setup in a previous study using a tapered and threaded pin with no tri-flats showed that the eccentricity decreases, since a relatively even force distribution is imposed during welding [19]. On the other hand, the 0.2 and 0.4 mm eccentricity samples show dampening effect during the welding process. This is expected, since the eccentricity of the system is suppressed by the solid base metal and rigid clamping system [8].

While the eccentricity does show a decrease in value for 0.2 and 0.4 mm eccentric samples during welding, the previous study also pointed out that the tool underwent severe cyclic bending forces throughout the welding process [19], which may accelerate tool fracture. This study has further proven that, regardless of the aluminium alloy base metal, the trend is similar and relatively large eccentricity ($\geq$0.2 mm) should ideally be avoided and checked using a dial gauge prior to welding process.

4 Conclusion

Friction stir welding of AA5052 and AA6061 aluminium alloys with varying tool eccentricity setup were successfully conducted. The findings from this study are highlighted below:

- The dial gauge and high-speed camera measurements show good agreement between each other, indicating reliability of the high-speed camera setup.
- For both base materials, the aligned setup shows an increase in the eccentric motion during the welding process compared to the jogging stage.
- However, the eccentric motion was dampened in the artificial eccentricity setups for both base materials.
- Varying base materials does not affect the change in tool eccentricity during welding process. However, the changes can be attributed to the initial eccentric setup.

Acknowledgements The authors would like to acknowledge the financial support by Universiti Malaysia Pahang under Internal Research grant RDU210319, Ministry of Higher Education Malaysia and the National Science and Engineering Research Council of Canada (NSERC).

References

1. Lohwasser D, Chen Z (2010) Friction stir welding: from basics to applications. Woodhead Publishing Limited, Cambridge
2. Shah LH, Othman NH, Gerlich A (2018) Review of research progress on aluminium–magnesium dissimilar friction stir welding. Sci Technol Weld Join 23:256–270. https://doi.org/10.1080/13621718.2017.1370193
3. Çam G, Mistikoglu S (2014) Recent developments in friction stir welding of al-Alloys. J Mater Eng Perform 23:1936–1953. https://doi.org/10.1007/s11665-014-0968-x
4. Gibson BT, Lammlein DH, Prater TJ, Longhurst WR, Cox CD, Ballun MC, Dharmaraj KJ, Cook GE, Strauss AM (2014) Friction stir welding: process, automation, and control. J Manuf Process 16:56–73. https://doi.org/10.1016/j.jmapro.2013.04.002
5. Shah LH, Walbridge S, Gerlich A (2019) Tool eccentricity in friction stir welding: a comprehensive review. Sci Technol Weld Join 24:566–578. https://doi.org/10.1080/13621718.2019.1573010
6. Gratecap F, Girard M, Marya S, Racineux G (2012) Exploring material flow in friction stir welding: tool eccentricity and formation of banded structures. Int J Mater Form 5:99–107. https://doi.org/10.1007/s12289-010-1008-5
7. Zaeh MF, Gebhard P (2010) Dynamical behaviour of machine tools during friction stir welding. Prod Eng Res Dev 4:615–624. https://doi.org/10.1007/s11740-010-0273-y
8. Yan JH, Sutton MA, Reynolds AP (2007) Processing and banding in AA2524 and AA2024 friction stir welding. Sci Technol Weld Join 12:390–401. https://doi.org/10.1179/174329307X213639
9. Fonda RW, Bingert JF (2007) Texture variations in an aluminum friction stir weld. Scr Mater 57:1052–1055. https://doi.org/10.1016/j.scriptamat.2007.06.068
10. Guo N, Wang MR, Meng Q, Zhou L, Tang DY (2015) Effect of tool eccentricity on surface periodic banded structures in friction stir welding. IOP Conf Ser Mater Sci Eng 103. https://doi.org/10.1088/1757899X/103/1/012020
11. Chen Y, Wang H, Wang X, Ding H, Zhao J, Zhang F, Ren Z (2019) Influence of tool pin eccentricity on microstructural evolution and mechanical properties of friction stir processed Al-5052 alloy. Mater Sci Eng A 739:272–276. https://doi.org/10.1016/j.msea.2018.10.057
12. Amini S, Amiri MR, Barani A (2015) Investigation of the effect of tool geometry on friction stir welding of 5083-O aluminum alloy. Int J Adv Manuf Technol 76:255–261
13. Mao Y, Ke L, Liu F, Liu Q, Huang C, Xing L (2014) Effect of tool pin eccentricity on microstructure and mechanical properties in friction stir welded 7075 aluminum alloy thick plate. Mater Des 62:334–343. https://doi.org/10.1016/j.matdes.2014.05.038
14. Essa ARS, Ahmed MMZ, Mohamed AKYA, El-Nikhaily AE (2016) An analytical model of heat generation for eccentric cylindrical pin in friction stir welding. J Mater Res Technol 5:234–240. https://doi.org/10.1016/j.jmrt.2015.11.009
15. Burek R, Wydrzyński D, Andres J, Wrońska A(2017) Effect of tool eccentricity on microstructure and properties of FSW joints made of Al 7075 T6 alloy. Adv Sci Technol Res J 11:333–338. https://doi.org/10.12913/22998624/80938
16. Ahmad Shah LH, Midawi ARH, Walbridge S, Gerlich A (2020) Influence of tool eccentricity on the material flow and microstructural properties of AA6061 aluminum alloy friction stir welds. J Alloys Compd 826. https://doi.org/10.1016/j.jallcom.2020.154219
17. Shah LH, Guo S, Walbridge S, Gerlich A (2018) Effect of tool eccentricity on the properties of friction stir welded AA6061 aluminum alloys. Manuf Lett 15:14–17. https://doi.org/10.1016/j.mfglet.2017.12.019
18. Shah LH, Huda N, Esmaeili S, Walbridge S, Gerlich AP (2020) Structural morphology of Al-Mg-Si alloy friction stir welds through tool eccentricity. Mater Lett 275:128098. https://doi.org/10.1016/j.matlet.2020.128098
19. Shah LH, Fleury A, St-George L, Walbridge S, Gerlich AP (2020) Evolution of process parameters in friction stir welding of AA6061 aluminum alloy by varying tool eccentricity. Int J Adv Manuf Technol 109:1601–1612

20. Schneider J, Nunes AC (2004) Characterization of plastic flow and resulting microtextures in a friction stir weld. Metall Mater Trans B 35:777–783. https://doi.org/10.1007/s11663-004-0018-4
21. Zhang YN, Cao X, Larose S, Wanjara P (2012) Review of tools for friction stir welding and processing. Can Metall Q. https://doi.org/10.1179/1879139512Y.0000000015
22. Rai R, De A, Bhadeshia HKDH, DebRoy T (2011) Review: friction stir welding tools. Sci Technol Weld Join 16:325–342. https://doi.org/10.1179/1362171811Y.0000000023

Lamb Wave Actuation Techniques for SHM System-A Review

N. Ismail, Z. M. Hafizi, Kok-Sing Lim, and Harith Ahmad

Abstract Lamb waves-based structural health monitoring (SHM) system has gained significant consideration due to its excellent performance in both precision and versatile accessibility. This paper specifically reviews the active actuation techniques of Lamb waves in the SHM system for diagnostic analysis. It is axiomatic that appropriate signal actuation is of vital necessity and important for a reliable SHM system. Multimodal and dispersive characteristics of the Lamb waves are the main factors for the variety of applied actuation techniques. Multiple Lamb modes synchronously exist, and their dispersive properties are not identical which lead to complex superimposed signals. Various techniques for single mode generation have been proposed to reduce the complexity of the measured signals. Generally, the applied techniques can be clustered to frequency tuning and actuator configuration. The findings showed that the applied techniques were proven to be effective in producing pure mode for further analysis. All these studies have revealed the evolution of Lamb waves-based SHM system for controlling and monitoring the condition of the structure.

Keywords Structural health monitoring · Lamb waves · Actuation techniques

1 Introduction

Among the SHM based techniques, Lamb Waves (LW) method is proven to be a promising and widely employed techniques in many areas. LW were used as a means for the internal and surface damage detection due to their special characteristics that are able to propagate throughout the entire thickness of the plate [1]. Thus, the valuable information related with structural integrity can be extracted from the

N. Ismail (✉) · Z. M. Hafizi
Advances Structural Integrity and Vibration Research (ASIVR), Faculty of Mechanical and Automotive Engineering Technology (FTKMA), Universiti Malaysia Pahang (UMP), 26600 Pekan, Pahang, Malaysia
e-mail: nurazima@ump.edu.my

K.-S. Lim · H. Ahmad
Photonics Research Centre, Universiti Malaya, 50603 Kuala Lumpur, Malaysia

© The Author(s), under exclusive license to Springer Nature Singapore Pte Ltd. 2023

M. H. A. Hassan et al. (eds.), *Technological Advancement in Instrumentation & Human Engineering*, Lecture Notes in Electrical Engineering 882,
https://doi.org/10.1007/978-981-19-1577-2_50

detected waves. Its ability to perform inspection for a large structure from limited access helps this technique to grow rapidly [2].

The active LW technique for SHM involves the utilisation of wave actuator and the sensor. The actuator generates the LW on the monitored structures and the sensor receives the generated LW. Before any meaningful analysis is performed, an appropriate actuation technique is necessary for a reliable signal acquisition by the sensor. Considering the complexity of the dispersive and multimode LW, it is desirable to generate a single or pure LW mode to simplify the signal analysis.

To achieve this objective, some factors need to be considered during the LW actuation such as the type of actuator used, the configuration of the actuator during the testing and the input actuation parameter. In the literatures, various different actuation methods for single mode generation have been suggested. This article summaries the actuation method available in literatures up to the authors knowledge. This information is important, prior to designing a proper LW-based SHM system set up.

1.1 Dispersive and Multimode of the Lamb Waves

Information of dispersive and multimode characteristics can be extracted from dispersion curve as shown in Fig. 1(a). This curve can be generated by Eqs. 1 and 2, describes about the motion of LW propagation which can be in symmetrical (S) or anti-symmetrical (A) mode, respectively.

$$\frac{\tan(qh)}{\tan(ph)} = -\frac{4k^2qp}{\left(k^2 - q^2\right)^2} \tag{1}$$

$$\frac{\tan(qh)}{\tan(ph)} = -\frac{\left(k^2 - q^2\right)^2}{4k^2qp} \tag{2}$$

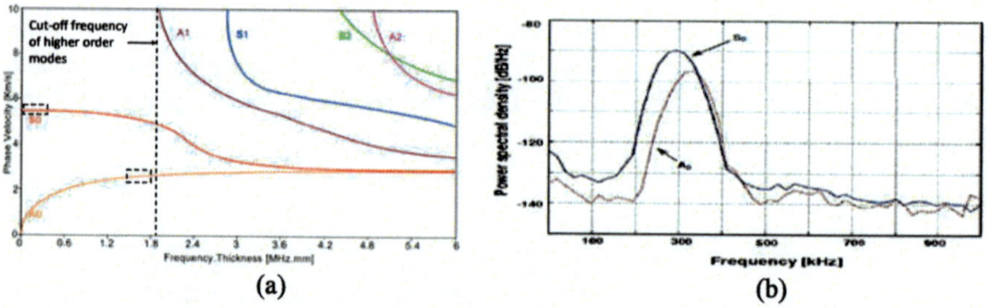

Fig. 1 **a** The dispersion curve; **b** Frequency domain features of the fundamental symmetric and anti-symmetric at original excitation frequency of 300 kHz [3]

where $p^2 = (\omega^2/c_L{}^2) - k^2$, $q^2 = (\omega^2/c_T{}^2) - k^2$, $k = \omega/c_p$. k, h, c_L, c_T, c_p and ω are the wavenumber, plate thickness, velocities of longitudinal and transverse modes, phase velocity and wave circular frequency, respectively. This dispersion phenomena leads to the spreading of wave-packet in space and time as it transverses through a monitored structure. This property affects the spatial and temporal resolutions, thus can potentially compromise the receiver detectability.

Consider a plate structure, actuated at a center frequency of 300 kHz. The result of frequency domain analysis using the measured LW signal displayed the peak occurring at 293 kHz for symmetric and 332 kHz for an anti-symmetric mode as presented in Fig. 1(b). This indicates the dispersion phenomenon since the measured center frequency is not identical to the original center frequency.

Meanwhile, the multimode property refers to the existence of at least two modes that coexist at any given frequency. At low frequency, minimum two modes (fundamental symmetric and anti-symmetrical mode) can propagate. As the frequency increases, more modes are possible and the interpretation of signals tends to become more complicated [4]. This characteristic leads to the complexity of the response signals which are difficult to be analysed and interpreted. By considering these problems, many studies were focused on devising their experimental set up to generate single Lamb mode to ease the post-processing analysis. Different researchers performed different methods based on the applicability and suitability of their applications. The next section provides the comprehensive description about the approaches that have been reported by the researchers in SHM system using LW approach.

2 Generation of Single Lamb Mode

2.1 *Frequency Tuning*

In frequency tuning approach, the frequency of the input wave contributes most to the actuation of the desired modes. Most of the researchers called this procedure the mode tuning method which aims to identify the frequency points where single-mode LW can be tuned into. In a simple way, researchers referred to the theoretical dispersion curve to determine the appropriate frequency point for single-mode generation. In other words, knowledge of the dispersion curve is required before performing the mode tuning. The selected frequency is based on minimum number of modes that exist at the given excitation frequency; and, in the less dispersive frequency region, as highlighted by the dotted rectangular in Fig. 1(a). The higher steepness slope of the curve indicates higher dispersion. Thus, multimode and dispersive problems can be avoided by choosing the suitable actuating frequency which is in a less-dispersive region at low frequency range, which is lower than cut-off higher frequency value.

In the meantime, several studies had derived the LW mode tuning curves analytically. For instance [5] derived the curve indicates the frequencies that exist when the A_0 mode or the S_0 mode can be either suppressed or enhanced. Verification of these tuning curves was extensively performed using experimental tests, and the results exhibited an agreement with each other. The findings was implemented for monitoring of a large area using Piezoelectric Wafer Active Sensors (PWAS).

Meanwhile, different types of actuators applied different strategies for frequency tuning techniques. The excitation of pure mode using electromagnetic acoustic transducers (EMATs) is determined by the meander-line coil spacing interval L, the exciting-current frequency f, and the thickness of specimen d for single mode generation. G. Zhai et al. [6] have established the excitation equation for choosing a reasonable operating point to generate a single mode of LW using EMATs. Z. Guo et al. [7] had selected the L size to be twice the interested mode wavelength λ. The authors were successfully generated the pure A_0 mode to study its interaction with the crack-like notch. Successful single-mode generation using EMAT also can be found in [8] and [9]. Some innovations to the existing EMAT have also been invented for single-mode generation.

For flexible interdigital polyvinylidene fluoride (PVDF) transducers, an individual modes can be excited by adjusting the electrode finger spacing. The basic principle is based on the wavelength of mode of interest is equal to the PVDF comb transducer's finger spacing [10]. Meanwhile, interdigital transducers (IDT) were designed with apodisation to enhance A_0 mode and attenuate S_0 mode dedicated for the CFRP laminates [11]. Modification of the frequency function response was used to determine the frequency range that generated the lowest order of symmetric and anti-symmetric modes. T. Liu et al. [12] developed a numerical simulation tool to investigate the acoustic wave field excited by IDT in composite plates. The developed tool could perform the mode isolation based on certain factors such as the apodisation, number of array elements, thickness, and width of the elements. The thickness and materials of possible backings should also be considered. For this purpose, frequency tuning was applied by choosing half wavelength of the targeted mode as the spatial distance between two adjacent elements at the centre frequency of the excitation pulse.

J. Li and J. L. Rose [13] built a multi-channel time-delay system to conduct the mode tuning procedure to solve the problems related with the conventional transducer array for the generation of individual wave modes. In principle, the wavelength of the modes of interest should be identical to the element spacing and all elements should work in-phase in order to perform frequency tuning. Automatic time delay algorithm was an added value for the real practice implementations.

2.2 *Actuator Configuration: Double Side Excitation*

Other than frequency tuning, many research works have manipulated the actuator configuration for the single mode generation. Utilisation of the double side actuation has shown its outstanding ability for pure modes generation. Pure symmetric and anti-symmetric modes were excited by in-phase and out-of-phase voltages application. Two actuators were utilised to implement this phase cancellation method. X. Lin and

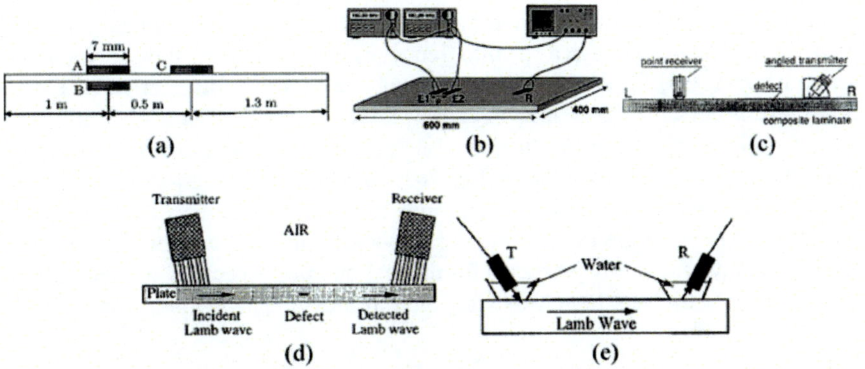

Fig. 2 **a** Double side actuation [15]; **b** opposite side actuation [17]; **c** angled actuator [18]; **d** angled air-coupled transducer [19]; **e** angled fluid-coupled transducer [20]

F. G. Yuan [14] and H.-Y. Zhang and J.-B. Yu [15] have explained this principle in detail in their reports. A pair of the PZT actuators was symmetrically attached on the upper and lower surfaces (double side actuation) of a structure as shown in Fig. 2(a). Some researchers applied this method using EMAT [16] and macrofibre composite (MFC) actuators and it showed that the technique is capable to generate a single wave mode.

2.3 Actuator Configuration: Opposite Side Excitation

On the other hand, N. Mori et al. [21] and Y.-H. Kim et al. [22] performed the excitation strategies using a pair of transducers which were surface-bonded opposite to each other (opposite side actuation) to generate a single-LW mode. The method is known as dual-element transducer setup as shown in Fig. 2(b). In this method, two actuators were placed opposite of each other at an optimised distance and were excited in-phase to dominantly generate the desired modes. The distance between two PZT elements is the main contributory aspect for the excitation of modes of interest. Despite the promising result from the method in the single mode generation, the requirements for such an experimental system should be taken into consideration. The system demands double transducer numbers and higher hardware requirement if compared with the single side excitation method.

2.4 Angled Actuator

Another common method in generation of pure Lamb mode is by the angle adjustment of the actuator. The principle used in this technique is based on Snell's law. Depending on the elasticity and acoustic properties of the medium, the desired phase velocity of the excited Lamb mode at a specific frequency can be controlled by adjusting the incident angle. [18, 23] adjusted the angle probe to excite the selected mode for the delamination detection on the carbon/epoxy prepreg laminates. General experimental setup for this approach can be illustrated by Fig. 2(c). This method can be performed in air [19] and fluid [20] as shown in Fig. 2(d) and (e). However, air- or fluid-coupled transducers suffers from a low efficiency due to the large differences in mechanical impedances between the air or fluid and the objects under detection. This method also leads to the flexibility problems for the real practice especially when accessing in-service systems even though showing effectiveness in single mode generation.

2.5 *Actuator Embedment*

Actuator embedment strategies are other approach that was introduced for the single mode excitation. [24] have demonstrated the use of embedded microstructured optical fibre (MOFBG) based sensors to selectively generate two fundamental LW modes. The excitation of the interested modes was based on the sensor orientation. Its selective sensitivity behaviour to axial and transverse strains allows the generation of preferential modes. The experimental arrangement for this technique can be referred to Fig. 2(f). B. Hailu et al. [25] using embedded IDT for the single mode excitation. The position of the transducer embedment in the plate thickness defined the generated modes. Symmetric mode was excited if the transducer is embedded exactly at the centre of a plate. However, when the transducer position deviates from the centre, symmetric and antisymmetric will simultaneously generated.

2.6 *Other Method*

Zi li et al. [26] proposed a new way of a selective generation of Lamb waves using laser scanning method. The essential procedure of this method is to vary the scanning speed of the beam to match the phase velocity of a specified mode (Table 1).

3 Conclusions

LW method have shown a great potential as a SHM tool due to their advantages. Offering cost-effective evaluation, covering relatively long distances, being highly susceptible to even barely visible internal damage, have made them widely used for structural health monitoring. However, the acquired LW signal is a complex product of its its dispersive and multimodal characteristics. This can affect the accuracy of the signal interpretation. Hence, various different approaches were introduced

Table 1 The summarised information about the excitation approaches for the pure mode generation

Method	The procedure	Limitations
Frequency tuning	The frequency is tuned accordingly for the single mode generation	1. Time consuming and a tuning frequency may not be available for real practice
Double side actuation	A pair of actuators was surface mounted or embedded on the top and bottom of the structure	1. Demanding double transducer numbers, which is increases the hardware requirement 2. High consistency in the location of transducer pair is required
Opposite side actuation	A pair of the actuator was attached in the opposite position to each other on the structure surface	
Angle of incidence: 1. Direct 2. Air-coupled 3. Fluid-coupled	The generation of the selected modes can be obtained by adjusting the angle of incidence of the actuator	1. Time consuming and technically challenging for in-service assessment
		2.Suffering from a low efficiency due to the large differences in mechanical impedances between the air or fluid and the objects under detection
Actuator embedment	The actuator was embedded at certain thickness of the structure for the separation of the different modes	Symmetric and antisymmetric will simultaneously generated when the transducer position deviates from its original position

by researchers to improve the LW signal interpretation. One of the approaches is by controlling the Lamb waves excitation. This strategy was applied in order to overcome the effect of multimodal and dispersive problem of the Lamb wave. Some flexible and practical excitation strategies were implemented with the purpose to generate selective pure modes. The mode of interest was decided based on the sensitivity and suitability of the application. Such selection is proven to ease the following post-processing analysis and provide reliable information during the inspection.

Acknowledgements The authors would like to thank the Ministry of Higher Education for providing financial support under Fundamental Research Grant Scheme (FRGS) No. FRGS/1/2019/TK03/UMP/02/7 (University reference RDU1901116) and Universiti Malaysia Pahang for laboratory facilities as well as additional financial support under Internal Research grant PGRS180376.

References

1. Gorgin R, Luo Y, Wu Z (2020) Environmental and operational conditions effects on Lamb wave based structural health monitoring systems: a review. Ultrasonics 105:106114. https://doi.org/10.1016/j.ultras.2020.106114

2. Qing X, Li W, Yishou Wang H, Sun H (2019) Piezoelectric transducer-based structural health monitoring for aircraft applications. Sensors 19(3):545. https://doi.org/10.3390/s19030545
3. Su Z, Ye L, Lu Y (2006) Guided Lamb waves for identification of damage in composite structures: a review. J Sound Vib 295(3–5):753–780. https://doi.org/10.1016/j.jsv.2006.01.020
4. Rathod VT, Raju G, Udpa L, Udpa S, Deng Y (2020) Multimode guided wave extraction capabilities using embedded thin film sensors in a composite laminated beam. Sens Actuators A Phys 309:112040. https://doi.org/10.1016/j.sna.2020.112040
5. Santoni GB, Yu L, Xu B, Giurgiutiu V (2007) Lamb wave-mode tuning of piezoelectric wafer active sensors for structural health monitoring. J Vib Acoust 129(6):11
6. Zhai G, Jiang T, Kang L, Wang S (2010) Minimizing influence of multi-modes and dispersion of electromagnetic ultrasonic lamb waves. IEEE Trans Ultrason Ferroelectr Freq Control 57(12):2725–2733. https://doi.org/10.1109/TUFFC.2010.1746
7. Guo Z, Achenbach JD, Krishnaswamy S (1997) EMAT generation and laser detection of single lamb wave modes. Ultrasonics 35(6):423–429. https://doi.org/10.1016/S0041-624X(97)00024-3
8. Liu Z, Hu Y, Xie M, Wu B, He C (2018) Development of omnidirectional A0 mode EMAT employing a concentric permanent magnet pairs with opposite polarity for plate inspection. NDT E Int 94:13–21. https://doi.org/10.1016/j.ndteint.2017.11.001
9. Huang X, Saniie J, Bakhtiari S, Heifetz A (2018) Applying EMAT for ultrasonic communication through steel plates and pipes. In: 2018 IEEE international conference on electro/information technology (EIT), pp 379–383. https://doi.org/10.1109/EIT.2018.8500148
10. Zhu Y, Zeng X, Deng M, Han K, Gao D (2018) Detection of nonlinear Lamb wave using a PVDF comb transducer. NDT E Int 93:110–116. https://doi.org/10.1016/j.ndteint.2017.09.012
11. Schmidt D, Sinapius M, Wierach P (2013) Design of mode selective actuators for Lamb wave excitation in composite plates. CEAS Aeronaut J. https://doi.org/10.1007/s13272-012-0059-3
12. Liu T, Veidt M, Kitipornchai S (2002) Single mode Lamb waves in composite laminated plates generated by piezoelectric transducers. Compos Struct 58(3):381–396. https://doi.org/10.1016/S0263-8223(02)00191-5
13. Li J, Rose JL (2001) Implementing guided wave mode control by use of a phased transducer array. IEEE Trans Ultrason Ferroelectr Freq Control 48(3):761–768. https://doi.org/10.1109/58.920708
14. Lin X, Yuan FG (2001) Diagnostic Lamb waves in an integrated piezoelectric sensor/actuator plate: analytical and experimental studies. Smart Mater Struct 10(5):907. http://stacks.iop.org/0964-1726/10/i=5/a=307
15. Zhang H-Y, Yu J-B (2011) Piezoelectric transducer parameter selection for exciting a single mode from multiple modes of Lamb waves. Chinese Phys B 20(9):94301. http://stacks.iop.org/1674-1056/20/i=9/a=094301.
16. Kubrusly AC, Freitas MA, von der Weid JP, Dixon S (2018) Mode selectivity of SH guided waves by dual excitation and reception applied to mode conversion analysis. IEEE Trans Ultrason Ferroelectr Freq Control 65(7):1239–1249. https://doi.org/10.1109/TUFFC.2018.2835299
17. Grondel S, Paget C, Delebarre C, Assaad J, Levin K (2002) Design of optimal configuration for generating A0 Lamb mode in a composite plate using piezoceramic transducers. J Acoust Soc Am 112(1):84–90. https://doi.org/10.1121/1.1481062
18. Guo N, Cawley P (1993) Lamb wave propagation in composite laminates and its relationship with acousto-ultrasonics. NDT E Int 26(2):75–84. https://doi.org/10.1016/0963-8695(93)90257-U
19. Castaings M, Cawley P (1996) The generation, propagation, and detection of Lamb waves in plates using air-coupled ultrasonic transducers. J Acoust Soc Am 100(5):3070–3077. https://doi.org/10.1121/1.417193
20. Ghosh T, Kundu T, Karpur P (1998) Efficient use of Lamb modes for detecting defects in large plates. Ultrasonics 36(7):791–801. https://doi.org/10.1016/S0041-624X(98)00012-2
21. Mori N, Biwa S, Kusaka T (2019) Damage localization method for plates based on the time reversal of the mode-converted Lamb waves. Ultrasonics 91:19–29. https://doi.org/10.1016/j.ultras.2018.07.007

22. Kim Y-H, Kim D-H, Han J-H, Kim C-G (2007) Damage assessment in layered composites using spectral analysis and Lamb wave. Compos Part B Eng 38(7):800–809. https://doi.org/10.1016/j.compositesb.2006.12.010
23. Tan KS, Guo N, Wong BS, Tui CG (1995) Experimental evaluation of delaminations in composite plates by the use of Lamb waves. Compos Sci Technol 53(1):77–84. https://doi.org/10.1016/0266-3538(94)00076-X
24. De Pauw B, Goossens S, Geernaert T, Habas D, Thienpont H, Berghmans F (2017) Fibre bragg gratings in embedded microstructured optical fibres allow distinguishing between symmetric and anti-symmetric lamb waves in carbon fibre reinforced composites. Sensors (Basel) 17(9):1948. https://doi.org/10.3390/s17091948
25. Hailu B, Gachacan A, Hayward G, McNab A (1999) Embedded piezoelectric transducers for structural health monitoring. In: 1999 IEEE ultrasonics symposium. proceedings. International symposium (Cat. No. 99CH37027), vol 1, pp 735–738. https://doi.org/10.1109/ULTSYM.1999.849506
26. Li Z, Lomonosov AM, Ni C, Han B, Shen Z (2018) Selective generation of Lamb modes by a moving continuous-wave laser. Opt Lett 43(1):78–81. https://doi.org/10.1364/OL.43.000078

The Development of Euler Solver Based on Flux Vector Splitting and Modified TVD Schemes

Iman Fitri Ismail, Bambang Basuno, Akmal Nizam Mohammed, Farzad Ismail, and Nurul Farhana Mohd Yusof

Abstract The study of shock waves is usually conducted using Computational Fluid Dynamics (CFD) methods because they are inexpensive and practical. Plenty of numerical schemes can be employed in predicting and analysing the discontinuities flow phenomena that may present in the high-speed flow. These schemes have been integrated into software packages such as Open-source Field Operation and Manipulation (OpenFOAM) and ANSYS, which incorporate a graphical user interface (GUI) that complicates user interaction and are computationally demanding. The previous implementation of the Flux Vector Splitting (FVS) schemes is limited to a typical wave propagating flow only. Hence, the objective of this research is to develop a first-order Flux Vector Splitting (FVS) schemes and extend the solution capabilities to flow with two shocks such as the Woodward-Collela blast wave problem. Three FVS schemes have been selected, Steger-Warming, Van Leer, and Liou-Steffen scheme. Execution of these schemes over the test cases has shown different flow patterns according to each test case's extremities. The ability of the schemes in capturing shock is also analysed. The results over the one-dimensional test cases have shown that the Steger-Warming scheme produced the largest dissipation. In contrast, the Liou-Steffen scheme has shown itself as the most accurate, with very few overshoots and undershoots. The fourth-order Runge-Kutta (RK) method augmented with second-order Time Variation Diminishing (TVD) has been previously implemented but limited to internal flow within a body surface. This research

I. F. Ismail (✉) · B. Basuno
Faculty of Mechanical and Manufacturing Engineering, Universiti Tun Hussein Onn Malaysia, 86400 Parit Raja, Batu Pahat, Johor, Malaysia
e-mail: ifitri.ismail@gmail.com

A. N. Mohammed
Flow Analysis Simulation and Turbulence Research Group, Universiti Tun Hussein Onn Malaysia, 86400 Parit Raja, Batu Pahat, Johor, Malaysia

F. Ismail
School of Aerospace Engineering, Universiti Sains Malaysia, 14300 Nibong Tebal, Pulau Pinang, Malaysia

N. F. Mohd Yusof
School of Mechanical Engineering, Universiti Sains Malaysia, 14300 Nibong Tebal, Pulau Pinang, Malaysia

© The Author(s), under exclusive license to Springer Nature Singapore Pte Ltd. 2023

M. H. A. Hassan et al. (eds.), *Technological Advancement in Instrumentation & Human Engineering*, Lecture Notes in Electrical Engineering 882,
https://doi.org/10.1007/978-981-19-1577-2_51

aims to extend the capabilities of the method for previously untested geometries such as flow over a wedge and a blunt-body. The results obtained from the scheme exhibits oblique shock waves at the compression corners and expansion waves at expansion corners. The presence of the waves is apparent at Mach number $M\infty = 2.0$ for the inlet, wedge, and bump channel geometries, while bow shocks can be simulated using the blunt-body geometry. The results obtained for the one-dimensional and two-dimensional problems confirm the shock-capturing ability of these schemes.

Keywords Flux vector splitting · TVD · Euler · Shockwave · Steger-Warming · Van Leer · Liou-Steffen

1 Introduction

With the recent developments of the aerospace field in mind, it is understood that most of the engineering applications of Computational Fluid Dynamics (CFD) pertaining to the field concern the Mach number that is over than $Ma = 0.3$ [1]. The variation of density within the flow changes even more drastically as the Mach number's value progresses higher. The changes from incompressible to compressible state introduced interesting physical phenomena such as shockwave and Mach. Not just that, the larger the Reynolds number of a flow is, the thinner the boundary layers, which are relative to its streamlined shaped size, which gives the flow an inviscid behaviour.

Compressible flows can be solved via finite difference methods (FDM), which includes the central schemes, upwind scheme, Navier-Stokes equations via Pressure-Implicit with Splitting of Operators (PISO) schemes [2], and flow field-dependent variation method. Due to discontinuities within a compressible supersonic flow, shock-capturing methods are needed to address these flow discontinuities better. This is an unsuitable task for central difference schemes as they primarily address problems with continuous and smooth variations of flow field variables.

First-order upwind schemes were implemented to compensate for the excessive damping with shock discontinuities. One of the first pioneering works in the first-order upwind scheme is the Godunov approach. Godunov proposed piecing together solutions to the exact Euler equations. Overshoots and undershoots still occur at the discontinuities within the first-order upwind scheme, which can be mitigated by incorporating the low or high-resolution second-order upwind scheme. Godunov proposed the solution of a general flow field by piecing together local regions to synthesize the general flow field [3].

The inviscid conservation form of the Euler equation has nonlinear flux vectors containing homogeneous functions of degree one. The name of the scheme is coined after the ability of the flux vector to be split. Hence, the flux vector splitting (FVS). One of the first pioneers for FVS methods is Steger and Warming, with the publication in 1979 regarding the use of FVS for inviscid gas dynamic problems [4]. The basic strategy of applying the FVS method is by splitting the flux and eigenvalues into positive and negative components.

Various engineering problems deal with the inviscid and compressible flow, especially at high speed and Mach number past through a streamlined body. Such applications include flow past through compressor and turbine in the gas turbine engine or flow past through aircraft at transonic or supersonic flight. At such a high Mach number, shockwave and discontinuities flow phenomena may exist in the flow field. The presence of such flow phenomena depends on the body's geometry that is immersed within the flow field, speed, and the direction of the flow with respect to the body.

The Flux Vector Splitting (FVS) methods have been extensively studied over the years. It is widely understood in the computational physics field that the interest in these particular schemes continuously fades with the advent of newer, stabler, and computationally inexpensive solvers. However, improving the shock-capturing method is beneficial to the community, even at its most fundamental level.

Therefore, this research is initiated not to develop a solver that defeats the capability of currently available CFD codes but to explore the possibility of building a better mathematical and programming approach for the development of a more robust CFD solver. Additionally, currently available open-source programs are usually limited to one type of flow condition, such as one-dimensional shock tube or two-dimensional airfoil analysis. Integration of FVS into CFD software packages such as ANSYS and Open-Source Field Operation and Manipulation (OpenFOAM) requires intensive training for the user to adjust the settings for the geometry, mesh, and initial flow parameters. The programs also added a graphical user interface (GUI) that is not tailored for low-end personal computers. Therefore, it is believed that the benefits of recent technological incorporation into CFD should outweigh the hardware requirements they demand.

There is a lack of dedicated standalone programs that can generate a solution for compressible and inviscid Euler equations based on the users' limited inputs. This research concerns the development of programming codes as tools to mitigate such problems. The program must be intuitive, generate mesh automatically based on the user's preference, and approximate the compressible flow with or without shockwave. The program should also generate different flow solutions based on the selection of FVS methods such as Steger-Warming, Van Leer, Liou-Steffen, and TVD schemes.

2 Literature Review

Navier-Stokes equation can be simplified into different forms according to the compressibility and viscosity of the flow. This research concerns the behaviours of the compressible inviscid flow. Therefore, the parameters of the flow governed by a reduced form of the Navier-Stokes equation will be used. The Euler equation governs the inviscid compressible flow as the flow moves in the region that is far from the solid surface. Therefore, the effect of its viscosity is small and negligible [5].

Configuration of flow for a non-viscous, non-heat conducting fluid is generally defined using Euler equations. Prandtl determined in 1904 that the viscous effects are confined to a thin boundary layer close to the solid boundary surface at a large Reynolds number. Hence, the interaction in the flow between its inviscid portion and boundary layer is negligible. If the flow's interaction is not negligible, separate sets of equations for the two regions can be used and solved iteratively [6].

Heat conduction and shear stresses are neglected as valid approximations regarding flows at high Reynolds numbers and outside the viscous regions that are developing near the solid surface. A drastic change in the form of the mathematical formulation was introduced by the approximation since the Navier-Stokes equation in its primitive form still considers the flow's viscosity [7].

The slip condition is required for the Euler equation at the solid surface, while the Navier-Stokes equation requires the opposite. This is due to viscous effects confined to the solid surface region for flows with a high Reynolds number. The region is known as the boundary layer. Outside the boundary layer region, zero shear stresses are zero due to negligible velocity gradients, and this region is referred to as the inviscid region. The formulation used to solve the inviscid flow region is governed by the equations known as the Euler equations for the case two-dimensional flow written in vector notation as given in Eq. (1)

$$\frac{\partial Q}{\partial t} + \frac{\partial E}{\partial x} + \frac{\partial F}{\partial y} = 0 \tag{1}$$

The conserved variable vector Q, the flux vector E, and F are defined respectively as:

$$Q = \begin{bmatrix} \rho \\ \rho u \\ \rho v \\ \rho e_t \end{bmatrix}; \quad E = \begin{bmatrix} \rho u \\ \rho u^2 + p \\ \rho u v \\ (\rho e_t + p)u \end{bmatrix} \quad \text{and } F = \begin{bmatrix} \rho v \\ \rho u v \\ \rho v^2 + p \\ (\rho e_t + p)u \end{bmatrix} \tag{2}$$

The notation of variables that appears in the above equation are defined as commonly used in CFD, where ρ is the air density, u and v are the component velocities in x and y direction respectively, p is the static pressure and e_t is total energy. Strictly speaking, Eq. (1) consists of four equations with 5 unknowns. Hence additional equation is needed in order the system Eq. (1) is solvable and gives a unique solution. The additional relationship may be in the form of the relation between total energy e_t, p, ρ and velocity as:

$$e_t = \frac{p}{\rho(\gamma - 1)} + \frac{1}{2}\left(u^2 + v^2\right) \tag{3}$$

The quantity γ which is known as a non-dimensional quantity represents the ratio between the heat coefficient at constant pressure and the heat coefficient at a constant volume. At standard atmosphere, this quantity can be set to equal 1.4.

The Euler equation as presented by Eq. (1) represents the Navier Stokes equation without viscosity, hence a lot of features of the solution of the Navier Stokes still represent the solution of the Euler equation such as a contact discontinuity, vortex flow phenomena dan shock wave. Maintaining the governing equation for the Euler equation as an unsteady equation although the flow problem is a steady flow problem, is due to the fact to keep the Euler equation belong to class a hyperbolic partial differential equation and so the steady flow problem can be treated as a solution of time marching goes to a steady-state solution [8].

2.1 Flux Vector Splitting Schemes

Euler equations can be solved via three significant categories of solution schemes, the first one being central schemes, first-order upwind schemes, and second-order upwind schemes [9]. Central-difference schemes are used to solve the compressible Euler Equation frequently demand dissipation scheme for numerical stability. The benefit of Upwind schemes is that they contain dissipation needed to manage the numerical instabilities.

Central schemes do not resolve shock discontinuities correctly by causing excessive damping. Therefore, the application of first-order upwind schemes usually exhibits trends of overshoots and undershoots that can be observed at the discontinuities. These can be mitigated by using the high or low resolution of second-order upwind schemes. Two groups made up the first-order upwind schemes, which are flux vector splitting and Godunov schemes.

One of the methods in flux-vector splitting schemes separates the vectors of convective fluxes into two parts. Then, they are discretised by upwind biased differences. First of flux-vector splitting schemes were pioneered by Steger-Warming [4] and Van Leer [10]. Another class decomposes the flux vector into convective and pressure parts, this was proposed in the Advection Upstream Splitting Method (AUSM) by [11] and Convective Upwind Split Pressure (CUSP) scheme [12].

Similar approaches were proposed in the low-diffusion flux-splitting scheme (LDFSS) [13] and Rossow's Mach number-based advection pressure splitting (MAPS) scheme [14]. Flux-vector splitting can be used for real gas flows with ease compared to the flux-difference splitting of TVD schemes.

AUSM scheme especially has been revamped over the years with the incorporation of flux difference splitting (FDS). The scheme that has been developed suggested that the generated solution will be much more robust against carbuncle instability. The scheme is able to resolve discontinuities and dampen numerical overshoots which made it reliable for a wider range of Mach numbers [15, 16].

Woodward Collela blast wave problem, is difficult to capture on a typical Riemann solver due to the presence of strong shocks and interactions within the flow, along with contact discontinuities and rarefactions [17]. Hence, this research aims to extend the FVS schemes by allocating three pressurised regions of different magnitude that will propel shocks towards the center while launching strong rarefactions at the walls.

2.2 TVD Schemes

Chung proclaimed that Total Variation Diminishing (TVD) might be the most monumental development in CFD fields [9]. The following requirements of physical properties reinforce the importance of these schemes. First, the scheme must not admit association entropy decrease with expansion shocks. Second, the monotonicity condition must be implemented to prevent oscillation within the solution. Finally, the total variation of the admissible solution cannot increase in time. It can be deduced from the requirement of physical properties that monotone schemes are TVD, and the monotonicity is preserved in TVD schemes [18].

Limiters are added to the TVD scheme in order to alter the properties of the numerical solutions. For example, combining the explicit Euler time-marching method with the linear convection equation will cause the extrema in the solution to be maintained. The solution will also have non-decreasing minima and, non-increasing maxima. Hence, the solution will not contain spurious oscillations in the region of discontinuities [19]. To reduce the dispersion error within the solution domain, the Runge-Kutta (RK) scheme employed in this study will be augmented with Davis-Yee symmetric TVD.

Second-order implicit-explicit TVD schemes have been extended for the low Mach regime. The solutions that were proposed in the work, suggested that it is possible for the construction of accurate implicit TVD methods [20]. Another research suggested that incorporating alternating limiters to the TVD scheme allows it to balance numerical compression and diffusion. Hence, the numerical errors were reduced up to 60% based on the benchmarked test performed. TDVal, the scheme proposed by the researcher is combined with splitting methods to capture advection in multidimensional cases [21]. The TVD method employed within this paper incorporates a limiter and is extended into multiple previously untested geometries.

3 Methodology

3.1 One-Dimensional FVS Methods

In approximating the spatial term of one-dimensional Euler equation, Steger-Warming FVS is used, the flux vector E is split and separated into positive and negative components, which is written as

$$S\frac{\Delta Q}{\Delta t} + \frac{\partial}{\partial x}(E^{+} + E^{-})^{n} - H^{n} = 0 \tag{4}$$

where $\Delta Q = Q^{n+1} - Q^{n}$. A First-order or second-order one-sided approximation can be used to replace the spatial derivative. Backward difference approximation is used in calculating the positive term, while forward difference approximation is used for the negative term. Hence, first-order approximation in space is

$$\Delta Q = -\frac{1}{S}\left(\frac{\Delta t}{\Delta x}\right)E_i^+ - E_{i-1}^+ + E_{i+1}^{-E_i^-} + \frac{1}{S}(\Delta t)H_i \quad (5)$$

For supersonic flow, the flux vector E for Steger-Warming is given by.

$$E = S\frac{\rho}{2\gamma}\begin{bmatrix} 2\gamma u + a - u \\ 2(\gamma-1)u^2 + (u+a)^2 \\ (\gamma-1)u^3 + \frac{(u+a)^3}{2} + \frac{(3-\gamma)(u+a)a^2}{2(\gamma-1)} \end{bmatrix} \quad (6)$$

The flux vector for the van Leer method is expressed as

$$E = S\begin{bmatrix} \rho a M \\ \rho a^2\left(M^2 + \frac{1}{\gamma}\right) \\ \rho a^3 M\left(\frac{1}{2}M^2 + \frac{1}{\gamma-1}\right) \end{bmatrix} \quad (7)$$

There are restrictions on the flux vectors which have been imposed by Van Leer. Therefore, E^+ and $\frac{\partial E^+}{\partial M}$ must disappear as M decreases towards $M = -1$. A similar restriction is imposed on E^- and $\frac{\partial E^-}{\partial M}$ as they will disappear as M is increased towards $M = 1$. The fifth restriction requires that the factor of $(1+M)^2$ and $(1-M)^2$ are included in E^+ and E^-. Therefore, the flux vectors of E^+ and E^- are written as

$$E^+ = E_1^+\begin{bmatrix} 1 \\ a\left(\frac{\gamma-1}{\gamma}M + \frac{2}{\gamma}\right) \\ a^2\frac{\gamma^2}{2(\gamma^2-1)}\left(\frac{\gamma-1}{\gamma}M + \frac{2}{\gamma}\right)^2 \end{bmatrix} \quad (8)$$

And

$$E^- = E_1^-\begin{bmatrix} 1 \\ -a\left(-\frac{\gamma-1}{\gamma}M + \frac{2}{\gamma}\right) \\ -a^2\frac{\gamma^2}{2(\gamma^2-1)}\left(-\frac{\gamma-1}{\gamma}M + \frac{2}{\gamma}\right)^2 \end{bmatrix} \quad (9)$$

Contrary to Steger-Warming and Van Leer method, the splitting of Mach number and pressure following van Leer is

$$M^\pm = \begin{Bmatrix} \pm\frac{1}{4}(M\pm 1)^2 \text{if}|M| \le 1 \\ \frac{1}{2}(M \pm |M|)\text{if}|M| > 1 \end{Bmatrix} \quad (10)$$

$$p^\pm = \begin{Bmatrix} \pm\frac{1}{2}p(1\pm M)^2 \text{if}|M| \le 1 \\ \frac{1}{2}p\frac{(M\pm|M|)}{M}\text{if}|M| > 1 \end{Bmatrix} \quad (11)$$

3.2 Modified Fourth-Order Runge-Kutta Scheme

The averaging step in Runge-Kutta is eliminated to reduce the storage requirement of the scheme. The modified formulation of a fourth-order RK for one-dimensional flow is

$$u_i^{(1)} = u_i^n \tag{12}$$

$$u_i^{(2)} = u_i^n - \frac{\Delta t}{4}\left(\frac{\partial E}{\partial x}\right)_i^{(1)} \tag{13}$$

$$u_i^{(3)} = u_i^n - \frac{\Delta t}{3}\left(\frac{\partial E}{\partial x}\right)_i^{(2)} \tag{14}$$

$$u_i^{(4)} = u_i^n - \frac{\Delta t}{2}\left(\frac{\partial E}{\partial x}\right)_i^{(3)} \tag{15}$$

$$u_i^{(n+1)} = u_i^n - \Delta t\left(\frac{\partial E}{\partial x}\right)_i^{(4)} \tag{16}$$

Second-order accuracy of the central-difference approximation is applied to the convective term $\partial E/\partial x$ again. The modified fourth-order RK scheme introduced from Eq. 10 to Eq. 14 can be extended to a two-dimensional Euler equation to generate the following finite difference equation

$$\overline{Q}_{ij}^{(1)} = \overline{Q}_{ij}^n \tag{17}$$

$$\overline{Q}_{i,j}^{(2)} = \overline{Q}_{i,j}^n - \frac{\Delta\tau}{4}\left[\left(\frac{\partial\overline{E}}{\partial\xi}\right)_{i,j}^{(1)} + \left(\frac{\partial\overline{F}}{\partial\eta}\right)_{i,j}^{(1)} + \alpha\overline{H}_{i,j}^{(1)}\right] \tag{18}$$

$$\overline{Q}_{i,j}^{(3)} = \overline{Q}_{i,j}^n - \frac{\Delta\tau}{3}\left[\left(\frac{\partial\overline{E}}{\partial\xi}\right)_{i,j}^{(2)} + \left(\frac{\partial\overline{F}}{\partial\eta}\right)_{i,j}^{(2)} + \alpha\overline{H}_{i,j}^{(2)}\right] \tag{19}$$

$$\overline{Q}_{i,j}^{(4)} = \overline{Q}_{i,j}^n - \frac{\Delta\tau}{2}\left[\left(\frac{\partial\overline{E}}{\partial\xi}\right)_{i,j}^{(3)} + \left(\frac{\partial\overline{F}}{\partial\eta}\right)_{i,j}^{(3)} + \alpha\overline{H}_{i,j}^{(3)}\right] \tag{20}$$

$$\overline{Q}_{i,j}^{n+1} = \overline{Q}_{i,j}^n - \Delta\tau\left[\left(\frac{\partial\overline{E}}{\partial\xi}\right)_{i,j}^{(4)} + \left(\frac{\partial\overline{F}}{\partial\eta}\right)_{i,j}^{(4)} + \alpha\overline{H}_{i,j}^{(4)}\right] \tag{21}$$

In this research, dispersion error is reduced by the addition of Total Variant Diminishing (TVD) schemes such as the Davis-Yee Symmetric TVD. It is implemented after the computation of Eq. 19, and the updated value of $\overline{Q}$ is as follows.

$$\begin{aligned}\overline{Q}_{i,j}^{n+1} = \overline{Q}_{i,j}^{n+1} &- \tfrac{1}{2}\tfrac{\Delta\tau}{\Delta\xi}\Big[(X_A)^n_{i+\frac{1}{2}j}(\Phi_\xi)^n_{i+\frac{1}{2}j} - (X_A)^n_{i-\frac{1}{2}j}(\Phi_\xi)^n_{i-\frac{1}{2}j}\Big] \\ &-\tfrac{1}{2}\tfrac{\Delta\tau}{\Delta\eta}\Big[(X_B)^n_{i,j+\frac{1}{2}}(\Phi_\eta)^n_{i,j+\frac{1}{2}} - (X_B)^n_{i,j-\frac{1}{2}}(\Phi_\eta)^n_{i,j-\frac{1}{2}}\Big]\end{aligned} \tag{22}$$

Several limiters that can be used for Davis-Yee Symmetric TVD and the general expression of the flux limiter was elaborated by Hoffman [22], the following limiter is used.

$$(G_\eta)_{i,j+\frac{1}{2}} = minmod\left\{2(\delta_\eta)_{i,j-\frac{1}{2}}, 2(\delta_\eta)_{i,j+\frac{1}{2}}, 2(\delta_\eta)_{i,j+\frac{3}{2}}, \frac{1}{2}\Big[(\delta_\eta)_{i,j-\frac{1}{2}} + (\delta_\eta)_{i,j+\frac{3}{2}}\Big]\right\} \tag{23}$$

4 Numerical Results

4.1 *One-Dimensional Shock Tube Problem*

The first test case's role is to test the FVS schemes' ability to address the shockwaves within the shock tube flow involving sonic points in rarefaction. When the virtual diaphragm of the shock tube is ruptured, expansion waves propagate into high-pressure gas while the contact discontinuity and shockwave propagate into the region of low-pressure.

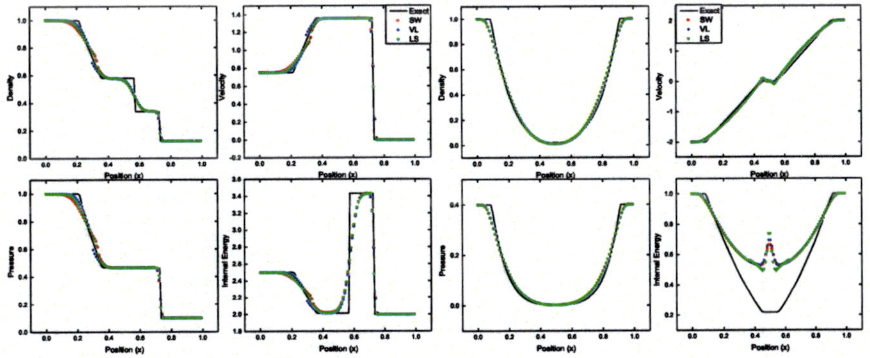

Fig. 1 Comparison of FVS schemes against exact solution applied to **a** Problem 1 for $x_o = 0.3$ at time $t = 0.2$; **b** Problem 2 for $x_0 = 0.5$ at time $t = 0.15$

As observed in Fig. 1(a), the plots for Problem 1 exhibit the FVS scheme trend in the region of the shockwaves. The overall distribution of the density, velocity, pressure and internal energy for the adopted FVS schemes are consistent with the exact solution. Steger-Warming and Van Leer produce noticeable dissipation at the region of the discontinuities. The dissipative nature of these FVS schemes is due to their first-order algorithms. As observed in Fig. 1(a), these numerical schemes possess better shockwave resolution than contact discontinuities. The second-order scheme can capture regions of discontinuities much better than the first-order counterparts [23].

Problem 3 is a test case designed to test the FVS schemes severely since its solution consists of three strong discontinuities that travel to the shock tube's right; while the left shock travels slowly [24]. As observed in Fig. 2(a), the computed distributions between the schemes of FVS are almost identical. Again, the van Leer scheme gives more accurate results than the FVS scheme of Steger-Warming.

While Steger-Warming and van Leer's results show a repeating undershoot reading especially on the density after the shock, Liou-Steffen produced a noticeable overshoot. FVS results that have been produced confirm the known defects and weaknesses of FVS schemes, especially in dealing with slowly moving shocks, as significant differences occur around the foot of the discontinuities [25].

Problem 4 concerns a blast wave in which rarefaction reflected from the right wall interacts with the contact discontinuity. Multiple interactions of discontinuities and strong nonlinear waves are involved in this problem. Hence, this problem tax the Euler FVS schemes' ability since it is difficult to calculate using a uniform Eulerian mesh [26].

Even though the FVS schemes' behaviour seems to agree with the exact solution, Liou-Steffen's scheme failed in this problem. Steger-Warming generated more considerable dissipation from the exact solution than Van Leer when it comes to Internal Energy distribution.

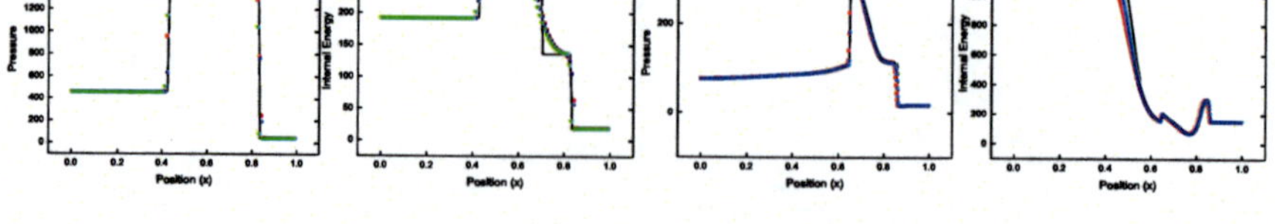

Fig. 2 Comparison of FVS schemes against exact solution applied to **a** Problem 3 for $x_o = 0.4$ at time $t = 0.035$; **b** Problem 4, with $x_0 = 0.1$ and $x_1 = 0.9$ at time $t = 0.038$

4.2 Two-Dimensional Axisymmetric Problem

The credibility of the modified four-order Runge-Kutta has already been validated against analytical results for Mach number $M_\infty = 2.0$ by Hoffman and Chiang, as shown in Fig. 3a. Here, the comparison in terms of pressure distribution between Hoffmann-Chiang's and the presently developed codes are shown in Fig. 3b and Fig. 4a. From these two figures, it is clear that the presently developed codes are able to produce the result as Homman-Chiang. Using the same grid generation, the calculation result at Mach Number $M_\infty = 1.5$ and $M_\infty = 2.0$ are shown in Figs. 4, 5, 6, 7.

Hoffman has detailed the capabilities of modified fourth-order Runge-Kutta augmented with Davis-Yee regarding its robustness and accuracy [27]. The free-stream pressure of 100 kPa and the free-stream temperature of 300 K were implemented in all four geometries. Davis-Yee symmetric TVD model with limiters is added as a dissipation mechanism to the numerical scheme of Runge-Kutta. In all

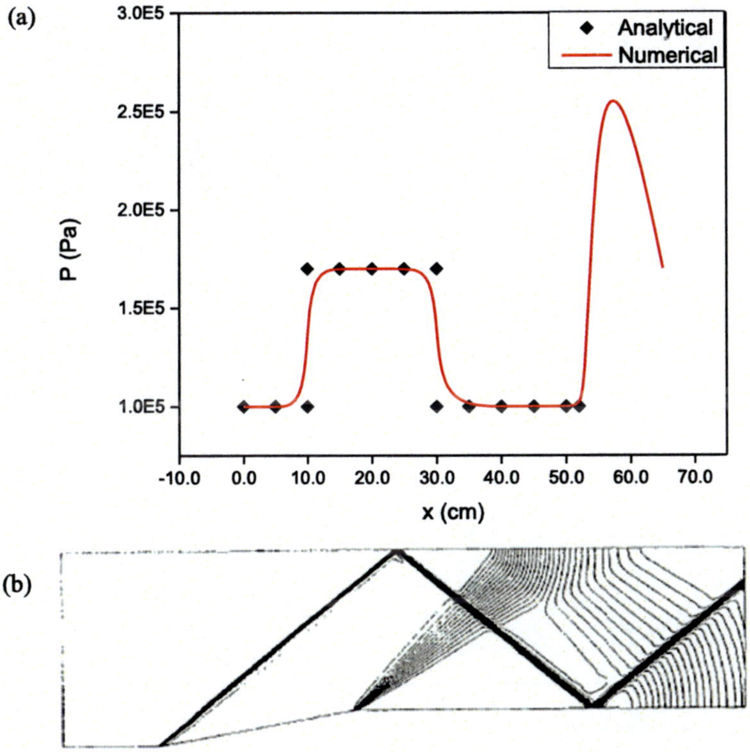

Fig. 3 **a** Validation of modified four-order RK (Numerical) against analytical method for lower surface pressure distribution at $M_\infty = 2.0$; **b** Pressure pattern developed by the modified fourth-order RK

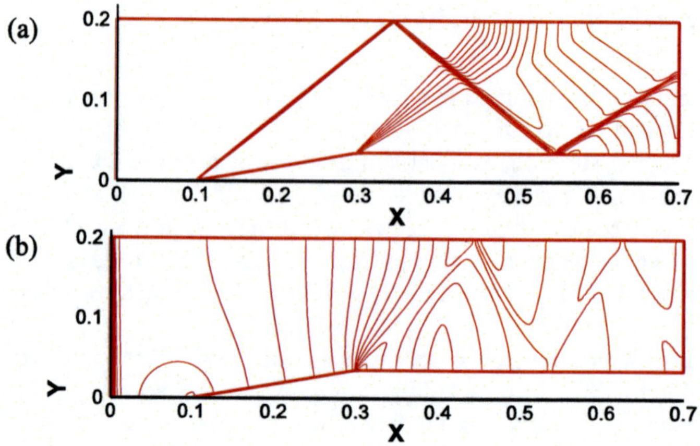

Fig. 4 Pressure contours of internal inlet flow for **a** $M_\infty = 2.0$; **b** $M_\infty = 1.5$

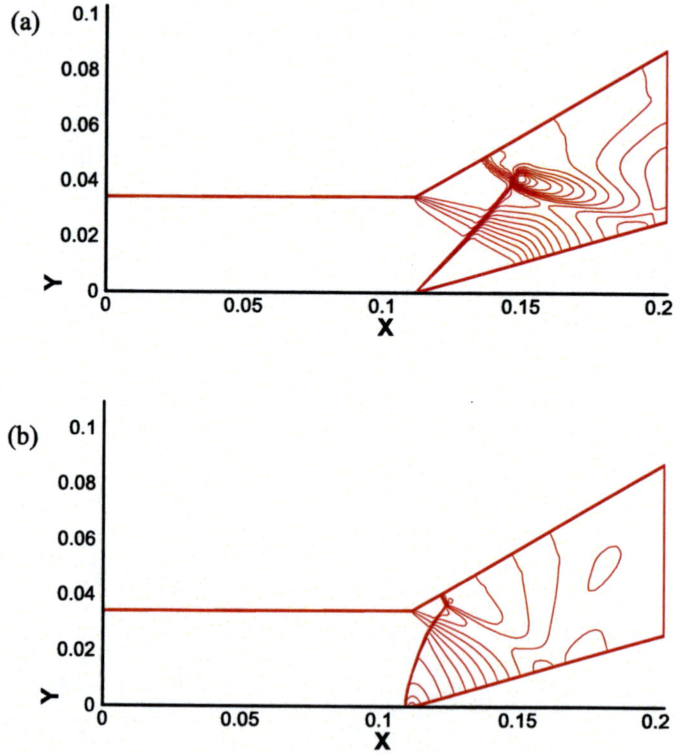

Fig. 5 Pressure contours of wedge flow for **a** $M_\infty = 2.0$; **b** $M_\infty = 1.5$

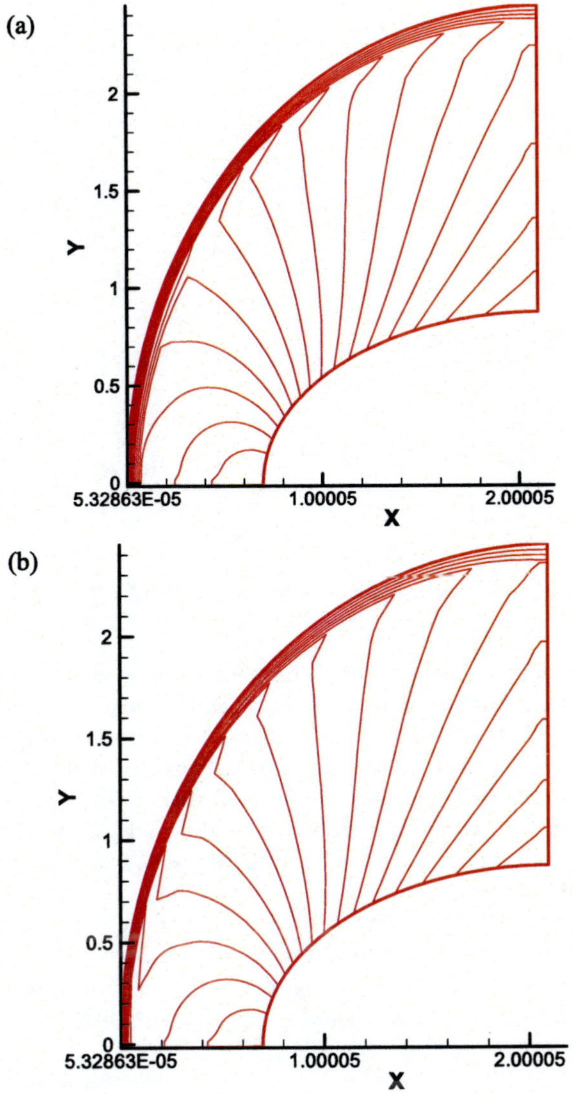

Fig. 6 Pressure contours of external blunt-body flow for **a** $M_\infty = 2.0$; **b** $M_\infty = 1.5$

of the geometries, grid clustering is conducted at the compression and expansion corners in which one might expect the location of the shock. Therefore, the pressure contours illustrated the formation of an expansion wave and oblique shock. At both the upper and lower surfaces, the surfaces' reflected oblique shock was accompanied by an expansion wave.

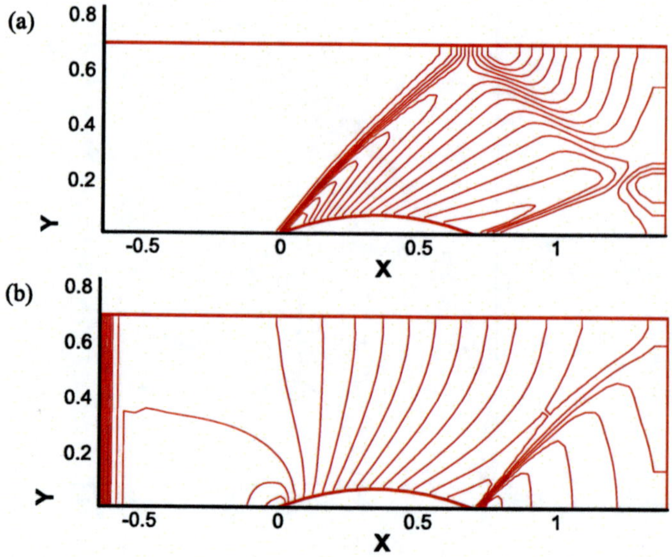

Fig. 7 Pressure contours of internal bump channel flow for **a** $M_{\infty} = 2.0$; **b** $M_{\infty} = 1.5$

Close observation of the pressure distribution at the inlet and wedge shows that for the flow of Mach number $M_{\infty} = 1.5$, the reflection and interaction of these waves exist in a weaker form in which the oblique shock is less sharp (Fig. 4(b) and 5(b)) compared to a Mach number $M_{\infty} = 2.0$. Bow shocks can be observed on the blunt-body problem caused by the geometric induced changes in pressure, temperature, and density when supersonic flow imparts a body with an angle of deviation [28]. Reducing the Mach number of the flow segregates the bow shocks further from each other, as seen in Fig. 6.

Bump in a channel is a specifically unique geometry, especially for its ability to induce bow shocks along with oblique shock and expansion waves. In Fig. 7(a), the oblique shock is shown at the beginning of the compression corner. But when the Mach number is $M_{\infty} = 1.5$, bow shocks and expansion waves can be seen along with the bump channel flow, as seen in Fig. 7(b).

5 Conclusion

The validation of the results obtained from FVS schemes against its exact solution confirms the ability of the Steger-Warming, van Leer, and Liou-Steffen schemes to address shock waves within the solution shock tube problems. The degree of smearing is very apparent with the Steger-Warming scheme, followed by the van Leer scheme. These two schemes particularly agree with each other based on the observed overshoots on the region after the shocks and their smearing pattern. The reason for these trends is the ability of these FVS schemes that cause diffusion at contact discontinuities. In general, the Liou-Steffen scheme produces overall good numerical accuracy compared to the previous two methods except for the Woodward-Collela blast wave problem. The low-pressure condition at the diaphragm caused the numerical solution to fail. In the future implementation of the method, a higher-pressure fix is needed in order to allow the scheme to converge. Therefore, the first two objectives of this study are accomplished after the solutions of the FVS schemes were compared against the exact solution.

The next phase of the study involves the development of the programming codes, and the focus shifted to the alteration of the boundary condition to complement different geometries that are studied. The fourth-order Runge-Kutta was modified and augmented with Davis-Yee symmetric TVD limiter to reduce dissipation within the solution results. Obviously, the oblique shockwave and expansion wave are much more apparent at high Mach number as observed in Chapter 4. These waves' patterns are observed for the internal flow of the inlet, wedge, and bump channel. The blunt-body geometry is able to simulate the bow shocks due to the flow being supersonic, and the segregation of the bow shocks increases as the Mach number decreases. The presence of bow shocks and expansion waves can be observed for the flow in a bump channel. Therefore, the aim of using the scheme as a tool to observe these flow phenomena is concluded based on the observations that can be made within the pressure and density distribution obtained from different geometry.

Acknowledgements The authors would like to express gratitude to the Ministry of Higher Education (MOHE) and Universiti Tun Hussein Onn Malaysia for supporting this research under Fundamental Research Grant Scheme Vot. No. (FRGS/1/2020/STG06/UTHM/02/1).

References

1. Anderson JD (2017) Fundamentals of Aerodynamics, 6th edn. McGraw-Hill Education, New York
2. Burgreen G (1987) Studies of pressure-velocity coupling schemes for analysis of incompressible and compressible flows. NASA STI/Recon Technical Report N
3. Godunov SK (1959) A difference method for numerical calculation of discontinuous solutions of the equations of hydrodynamics [English title page], Mat. Sb
4. Steger JL, Warming RF (1981) Flux vector splitting of the inviscid gasdynamic equations with application to finite-difference methods. J Comput Phys. https://doi.org/10.1016/0021-9991(81)90210-2
5. Ferziger JH, Perić M (2002) Computational methods for fluid dynamics
6. Anderson D, Tannehill JC, Pletcher RH (2016) Computational fluid mechanics and heat transfer, 3rd edn
7. Hirsch C (2007) Numerical computation of internal and external flows: the fundamentals of computational fluid dynamics
8. Computational fluid dynamics for engineers: from panel to Navier-Stokes methods with computer programs (2006). Choice Rev. Online. https://doi.org/10.5860/choice.43-3431
9. Chung TJ (2010) Computational fluid dynamics, 2nd edn.
10. van Leer B (1982) flux-vector splitting for the Euler equations

11. Liou MS, Steffen CJ (1993) A new flux splitting scheme. J Comput Phys. https://doi.org/10.1006/jcph.1993.1122
12. Jameson A (1995) Positive schemes and shock modelling for compressible flows. Int J Numer Methods Fluids 20(8–9):743–776. https://doi.org/10.1002/fld.1650200805
13. Edwards JR (1997) A low-diffusion flux-splitting scheme for Navier-Stokes calculations. Comput Fluids 26(6):635–659. https://doi.org/10.1016/S0045-7930(97)00014-5
14. Rossow CC (2000) A Flux-Splitting scheme for compressible and incompressible flows. J Comput Phys 164(1):104–122. https://doi.org/10.1006/jcph.2000.6586
15. Chen S-S, Cai F-J, Xue H-C, Wang N, Yan C (2020) An improved AUSM-family scheme with robustness and accuracy for all Mach number flows. Appl Math Model 77:1065–1081. https://doi.org/10.1016/j.apm.2019.09.005
16. Yoo YL, Sung HG (2021) A hybrid AUSM scheme (HAUS) for multi-phase flows with all Mach numbers. Comput Fluids 227:105050. https://doi.org/10.1016/j.compfluid.2021.105050
17. Shen H, Parsani M (2021) A rotated characteristic decomposition technique for high-order reconstructions in multi-dimensions. J Sci Comput 88(3):01602. https://doi.org/10.1007/s10915-021-01602-z
18. Harten A (1983) High resolution schemes for hyperbolic conservation laws. J Comput Phys 135(2):260–278. https://doi.org/10.1016/0021-9991(83)90136-5
19. Pulliam TH, Zingg DW (2014) Fundamental algorithms in computational fluid dynamics. Springer, Cham https://doi.org/10.1007/978-3-319-05053-9
20. Dimarco G, Loubère R, Michel-Dansac V, Vignal M-H (2018) Second-order implicit-explicit total variation diminishing schemes for the Euler system in the low Mach regime. J Comput Phys 372:178–201. https://doi.org/10.1016/j.jcp.2018.06.022
21. Lin L, Liu Z (2019) TVDal: Total variation diminishing scheme with alternating limiters to balance numerical compression and diffusion. Ocean Model 134:42–50. https://doi.org/10.1016/j.ocemod.2019.01.002
22. Hoffman KA, Chiang ST (2000) Computational fluid dynamics, 4th edn. vol 2, Engineering Education System, Wichita
23. Ning J (2020) Comparison and analysis of different numerical schemes in sod's one-dimensional shock tube problems. J Phys Conf Ser 1550(3):032049. https://doi.org/10.1088/1742-6596/1550/3/032049
24. Sun M, Takayama K (2003) An artificially upstream flux vector splitting scheme for the Euler equations. J Comput Phys 189(1):305–329. https://doi.org/10.1016/S0021-9991(03)00212-2
25. Lyra PRM, Morgan K (2000) A review and comparative study of upwind biased schemes for compressible flow computation. Part I: 1—D first—order schemes. Arch Comput Methods Eng 7(1):19–55. https://doi.org/10.1007/BF02736185
26. Woodward P, Colella P (1984) The numerical simulation of two-dimensional fluid flow with strong shocks. J Comput Phys 54(1):115–173. https://doi.org/10.1016/0021-9991(84)90142-6
27. Harada S, Hoffmann KA, Augustinus J (1998) Development of a modified runge-kutta scheme with TVD limiters for the ideal two-dimensional MHD equations. https://doi.org/10.2514/6.1998-981
28. Landau LD, Lifshitz EM (1987) Fluid mechanics, 2nd edn.

Applicability of Peak Detection Methods for Composite Fatigue FBG Wavelength

M. Loman, M. H. Zohari, and F. Lamin

Abstract This study compares the accuracy and effectiveness of existing peak detection methods on the FBG spectrum in sensing the fatigue strain in a composite. It aims to give a solution to measure the FBG spectrum and further processing it, including selecting the FBG peak wavelength. The FBG sensor was bonded to the fatigue test specimen and the spectrums were acquired during the fatigue test were progressing. Three different peak detection algorithms were utilised, which include maximum, polynomial and centroid algorithms. Results showed that the centroid has the lowest precision compared to the polynomial and maximum algorithms. For fatigue wavelength data, it was found out that the maximum algorithm and third-order polynomial fitting are applicable as the peak seeking method for fatigue strain data from FBG sensor.

Keywords Composite · Fatigue · FBG · Peak detection · Structural health monitoring

1 Introduction

Fibre Bragg grating (FBG) has gained attention in the structural health monitoring area by its abundance of specialities such as lightweight and small [1]. It is also has multiplexing behaviour besides does not change the structure where it is bonded to. Additionally, these sensors are compatible with communications systems and allow for remote sensing [2]. The FBGs can be adapted as sensing elements to measure temperature, pressure, strain, vibration, inclination, load, and displacement [3]. Fibre Bragg grating is a periodic structure embedded in the core of an optical fibre with varying refractive indices [4]. A narrow wavelength band is reflected when a broad

M. Loman (✉) · M. H. Zohari
Faculty of Mechanical and Automotive Engineering Technology (FTKMA), Universiti Malaysia Pahang, 26600 Pekan, Pahang, Malaysia
e-mail: miminorazean@ump.edu.my

F. Lamin
Vehicle Safety and Biomechanics Research Centre, Malaysian Institute of Road Safety Research, Lot 125-135, Jalan TKS 1, Taman Kajang Sentral, 43000 Kajang, Selangor, Malaysia

© The Author(s), under exclusive license to Springer Nature Singapore Pte Ltd. 2023
M. H. A. Hassan et al. (eds.), *Technological Advancement in Instrumentation & Human Engineering*, Lecture Notes in Electrical Engineering 882,
https://doi.org/10.1007/978-981-19-1577-2_52

spectrum of wavelengths passes through the FBG, while all other wavelengths are transmitted [5, 6]. The wavelength at maximum reflectivity is referred to as the Bragg wavelength, λ_B and is determined by the condition as expressed in Eq. (1).

$$\lambda_B = 2n_{eff}\Lambda \tag{1}$$

where Λ is the period of the index of refraction variation and n_{eff} is the optical fibre's effective refractive index [7].

In the FBG interrogation system, firstly, it will measure the spectrum, identify and calculate the Bragg wavelength and convert the wavelength into a few other parameters [8]. Usually, the wavelength measurement is not very straightforward; thus, the general principle is to convert the wavelength shift to easily measured parameters, such as amplitude, phase, or frequency [8]. The central wavelength for each spectrum is also commonly known as peak wavelength or maximum central wavelength [9]. This parameter is used as a representation to infer the physical changes on the structures measured. A few peak detection algorithms have been introduced and applied in vibration structures [10]. Maximum and centroid are categorised under the direct method, while polynomial is under the least square method [11]. Stress, strain, impact and any physical differences to the structure's original condition may lead to shifts of the FBG peak wavelength from its initial wavelength before the structure is loaded. These shifts are monitored as the change of peak wavelength of a data series.

During FBG development and investigation, optical spectrum analysers (OSA) are often used to monitor grating transmission or reflection spectra [8]. However, optical spectrum analysers are not attractive in practical applications due to their slow scanning speed and limited resolution capability [8]. Measurement techniques such as the tunable-laser and OSA systems measure the FBG's reflectance as a function of wavelength, which leaves the problem of determining the peak, centre, or centroid wavelength of the FBG from the measured data [9]. For that reason, the peak detection algorithm is essential to analyse further the FBG spectrum obtained from the composite fatigue test. Since each algorithm has its own benefit and drawbacks, its utilisation and performances depend on the types of spectrum measured [11]. The fatigue sensing task is a significant challenge, especially on a composite structure due to its brittle behaviour.

The purpose of this study is to verify the applicability of maximum amplitude, polynomial and centroid algorithms in processing the FBG data focusing on strain from fatigue test on a composite structure. Since different systems and their current health conditions transmit unique signal ranges, it is crucial to determine the most appropriate peak detection algorithm for a specific health condition monitoring.

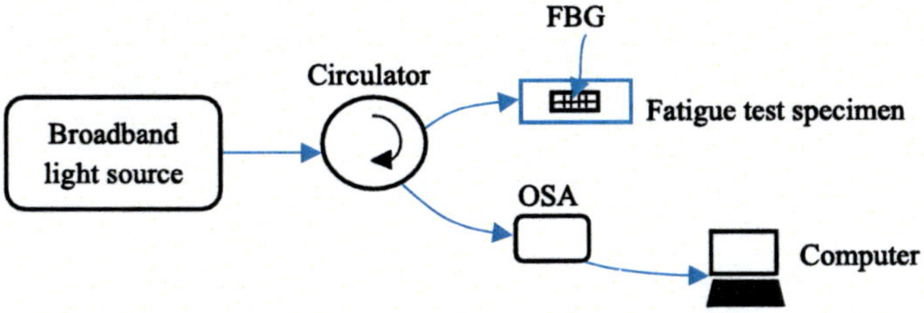

Fig. 1 Schematic diagram of the experimental setup

2 Methodology

2.1 Experimental Setup

The schematic diagram of the experimental setup is shown in Fig. 1. A single FBG was connected to the optical spectrometer analyser (OSA) and NI-9235 via a coupler. Next, FBG was attached to the composite fatigue test specimen and connected to the FBG interrogation system in Fig. 1. During the fatigue test, FBG spectrums were simultaneously acquired and directly stored in the computer.

2.2 Peak Detection Methods

The massive amount of data acquired from experimental work need to be appropriately segmented before it is converted and further analysed. Moreover, the converted data was monitored adequately via the FBG peak wavelength recognition method. In order to track wavelength shifts in FBG response spectrums from the fatigue test, a few types of peak tracking algorithms have been utilised. These algorithms are practical to track the peak position for the FBG wavelength shifts counting [12]. This study compares the output peak wavelength obtained from the centroid, maximum amplitude and polynomial algorithms. These algorithms have been used in a few previous studies [12–18] in heat transfer, structural health monitoring, and many more.

The centroid algorithm is calculated using the relationship in Eq. (2) and represents the geometric centroid of a spectrum.

$$\lambda_B = \frac{\sum_{i=1}^{N} \lambda_i A_i}{\sum_{i=1}^{N} A_i} \tag{2}$$

In Eq. (2), the Bragg wavelength, λ_B is calculated from the summation of ith point wavelength, λ_i and reflectivity amplitude, A_i respectively, with N is the number of FBG vector lengths. The maximum algorithm is based on the search for the highest

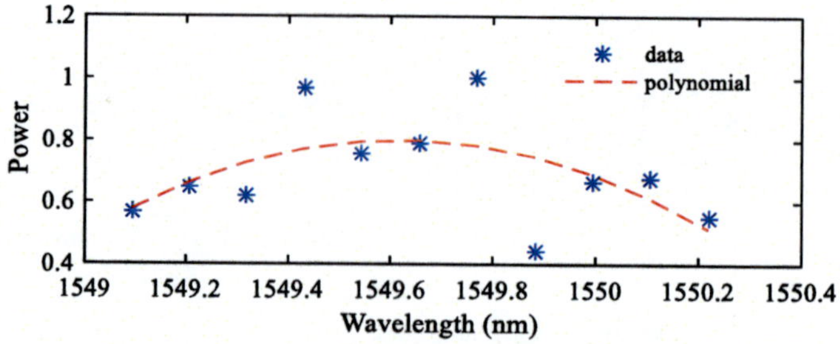

Fig. 2 Example of polynomial fitting for the fatigue test spectrum from FBG

amplitude of the input power [12]. For the maximum peak algorithm, it is the search for the wavelength with the highest amplitude.

The second, third and fourth-order polynomials were calculated in seeking the peak using the polynomial algorithm. The analysis then compares the polynomial order with the optimum consistency with the previous peak detection methods. In the polynomial fitting, the amplitudes of the spectrum are normalised between 0 and 1. The example of the polynomial fitting is shown in Fig. 2.

3 Results and Discussion

Three types of peak detection methods have been evaluated in this study purposely to compare their effectiveness in analysing the fatigue strain peak wavelength on composite structures. The methods include maximum peak amplitude, centroid and polynomial least-square fit. As previously described in the methodology section, these methods are the most commonly utilised to track the FBG peak wavelength during FBG sensing.

As for the polynomial method, this study compared peaks obtained by several polynomial orders. The visual impact of different polynomial orders does not show unless a very high polynomial order is utilised. The absolute error was calculated relative to the absolute relative approximate error of the lower polynomial order as in Table 1. The third and fourth-order polynomial suit this data more as reflected by the minimum error calculated, less than 0.05%. Dyer et al. [9] also reported a comparable finding in detecting peak wavelength for several typical wavelength measurement systems. It includes tunable laser, interferometric, and OSA, used in this study. Therefore, this study further extends the previous finding on the applicability of the polynomial orders, particularly for fatigue sensing on the composite structure.

Figure 3 shows the comparison of the peak wavelength acquired by each polynomial order from second to fourth. The fourth-order seems the most preferred for further analysis as it offers an optimum fit between the second and third-order. Furthermore, the third-order polynomial is simpler than the fourth-order polynomial as it consists of lesser coefficients and shows an almost similar fit curve as the

Table 1 Comparison of FBG peak wavelength by different polynomial orders

	Peak wavelength (nm)				
Polynomial 2nd order	1547.75	1547.85	1547.92	1548.04	1548.1
Polynomial 3rd order	1547.81	1547.91	1547.94	1548.06	1548.16
Absolute relative approximate error (%)	0.016	0.021	0.021	0.028	0.032
Polynomial 4th order	1547.85	1547.95	1548.01	1548.12	1548.22
Absolute relative approximate error (%)	0.0026	0.0026	0.0045	0.0039	0.0039

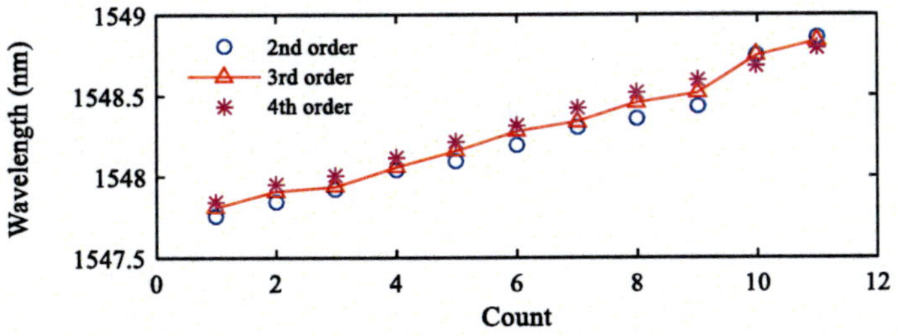

Fig. 3 Peak wavelength comparison gained from a different order of polynomials

fourth-order polynomial. The higher order of the polynomial is not preferred as it is very likely distorted by noise interference. Based on these arguments, the third-order polynomial was selected, and the following discussion focuses on its applicability relative to other peak detection methods.

Figure 4 compares the three peak detection methods utilised in this study, with eleven data selected at the early stage of the fatigue test progression. In this stage, an upward trend is expected due to the introduction of a maximum loading threshold prior to the cyclic phase. The centroid method shows the most deviated peak wavelength compared to polynomial and maximum amplitude methods. It also means that

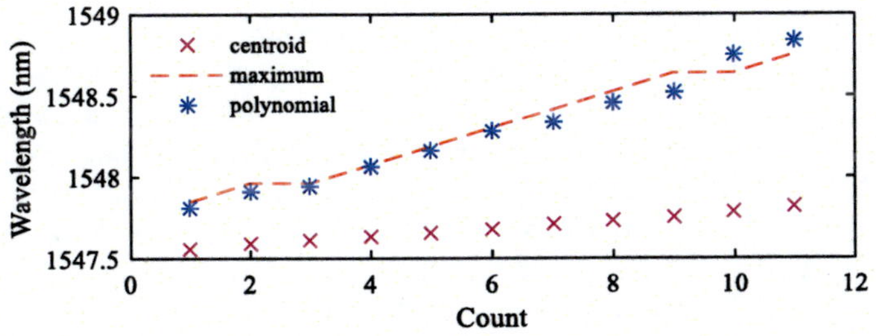

Fig. 4 Peak wavelength comparison obtained from different methods during the early stage of fatigue test

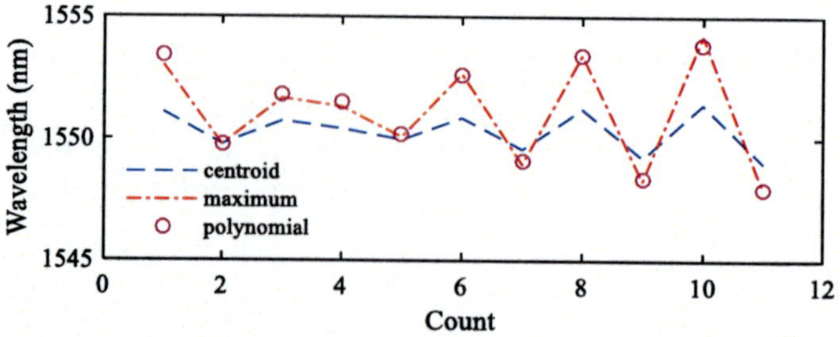

Fig. 5 Peak wavelength comparison obtained from different methods during the middle stage of fatigue test

the centroid algorithm has the worst accuracy and is incomparable to the polynomial and maximum amplitude algorithms. This result is consistent with Dyer et al. [9].

Figure 5 shows peak wavelength comparison obtained from different methods during the middle stage of the fatigue test. The cyclic trend is expected due to the physical condition of the composite structure under compression and tension loading. At this stage, the crack propagation was still developing and producing inconsistent FBG peak wavelength. Regardless of the fatigue loading stage, the analysis shows a consistent result. A noticeable deviation of the centroid curve of FBG peak wavelength compared to the maximum and polynomial curves was observed. Both maximum and polynomial algorithms demonstrated a comparable result. This finding evidenced that the centroid algorithm does not show good precision even though the early stage of the fatigue test spectrum was more stable than the middle stage.

4 Conclusions

This paper attempted to compare the most applicable peak tracking method for processing fatigue data obtained from the FBG sensor on the composite structure. The presented analysis found that the centroid algorithm is not suitable for the composite structures fatigue sensing. The precision is low and not effective as a peak detection method for the tested FBG fatigue data. The other two ways, the maximum algorithm and third-order polynomial fitting, are more compatible with the fatigue wavelength acquired by the FBG. Therefore, these methods' applicability was verified as the peak seeking method for fatigue strain data from the FBG sensor. Determining the most suitable peak detection algorithm is the key to successful composite structure health monitoring. This effort is crucial, especially when the monitored structure is highly exposed to cyclic loading that may lead to abrupt failure.

Acknowledgements The authors would like to thanks Universiti Malaysia Pahang for providing laboratory facilities and financial assistance under the grant no. RDU190353.

References

1. Nicolas MJ, Sullivan RW, Lance Richards W (2016) Large scale applications using FBG sensors: determination of in-flight loads and shape of a composite aircraft wing. Aerospace 3:18
2. Campanella CM, Cuccovillo A, Campanella C, Yurt A, Passaro VMN (2018) Fibre Bragg grating based strain sensors: review of technology and applications. Sensors 18:3115
3. Vorathin E, Hafizi ZM, Ismail N, Loman M (2020) Review of high sensitivity fibre-optic pressure sensors for low pressure sensing. Opt Laser Technol 121:105841
4. Sheverev V, Stepaniuk V, Narang AS (2019) Principles of drag force flow sensor. In: Handbook of pharmaceutical wet granulation, pp 317–349
5. Peters KJ, Inaudi D (2014) Fiber optic sensors for assessing and monitoring civil infrastructures. Sens Technol Civ Infrastruct 1:121–158
6. Tosi D, Poegge S, Iordachita I, Schena E (2018) Fiber optic sensors for biomedical applications. In: Opto-mechanical fiber optic sensors, pp 301–333
7. Tosi D (2017) Review and analysis of peak tracking techniques for fiber Bragg grating sensors. Sensors 17:2368
8. Ganziy D (2017) Technology for polymer optical fiber Bragg grating fabrication and interrogation. Ph.d. thesis
9. Dyer SD, Williams PA, Espejo RJ, Kofler JD, Etzel SM (2005) Fundamental limits in fiber Bragg grating peak wavelength measurements. In: 17th international conference on optical fibre sensors, Marc Voet, Reinhardt Willsch
10. Wang J, Huang T, Duan F, Cheng Q, Zhang F, Qu X (2020) Fast Peak-tracking method for FBG reflection spectrum and nonlinear error compensation. Opt Lett 45(2):451
11. Huang T, Wang J, Duan F, Jianga J, Fua X, Xu X (2019) A simple algorithm for the implementation of second-order-polynomial based peak-tracking methods. Opt Fiber Technol 47:192–196
12. Negri L, Nied A, Kalinowski H, Paterno A (2011) Benchmark for peak detection algorithms in fiber Bragg grating interrogation and a new neural network for its performance improvement. Sensors 11:3466–3482
13. Xu O et al (2020) A multi-peak detection algorithm for Fiber Bragg Grating sensing systems. Opt Fiber Technol 58:102311
14. Dinardo G, Fabbiano L, Vacca G (2015) An innovative algorithm for fiber Bragg grating sensors interrogation. In: XXI IMEKO world congress measurement in research and industry, 30 August–4 September 2015, Prague, Czech Republic
15. Tosi D (2012) Performance analysis of peak tracking techniques for fiber Bragg grating. J Microwaves Optoelectron Electromagn Appl 11(2):252
16. Paterno AS et al (2016) FBG interrogation and the benchmark for algorithms in the processing of experimental data. In: SBMO/IEEE MTT-S international microwave and optoelectronics conference proceedings
17. Bodendorfer T, Muller MS, Hirth F, Koch AW (2009) Comparison of different peak detection algorithms with regards to spectrometic fiber Bragg grating interrogation systems. In: ISOT 2009 - International symposium on optomechatronic technologies
18. Ganziy D, Jespersen O, Rose B, Bang O (2015) An efficient and fast detection algorithm for multimode FBG sensing. In: Proceedings of SPIE, vol 9634, p 963445. © 2015 SPIE

Controllability Analysis of Convergence on Water Distribution System Architecture with Fault Using Cellular Automata Sequence

Nurul Hannah Mohd Yusof, Nurul Adilla Mohd Subha, Norikhwan Hamzah, Fatimah Sham Ismail, Mohd Ariffanan Mohd Basri, and Anita Ahmad

Abstract This paper develops a consensus-based control protocol for a resilient Water Distribution System (WDS) intending to allow continuous water supply on the WDS facilities in the presence of faulty event(s). The consensus is based on logical inputs which iterate by utilizing the cellular automata (CA) concept. The simulation is performed on MATLAB based on the WDS leakage problem. As a result, the consensus is reached within a five-cycle time, which indicates the triggered value in two agents' readings. The controllability of the system is analyzed using the existencc of the Hamiltonian cycle (HC) and full rank matrix which appears in the reachability matrix The approach suggested in this paper is expected to be scalable and can be applied to any huge scale of WDS layouts.

Keywords Convergence · Logical control · Consensus-based control

1 Introduction

The main objective of a consensus protocol application within a Networked Multi-Agent System (NMAS) is to achieve an agreement between all agents within the system through the exchange of local information via network [1]. This type of protocol has been used in various types of systems such as intelligent transportations [2] and water distribution systems (WDS) [3].

Logical consensus is a part of NMAS which is developed by introducing the system's communication and visibility capability which represented its network topology. The purpose of logical consensus is to facilitate their decisions with a set of logical values. In this consensus, to reach convergence on a distributed system, the

N. H. M. Yusof (✉) · N. A. M. Subha · F. S. Ismail · M. A. M. Basri · A. Ahmad
School of Electrical Engineering, Faculty of Engineering, Universiti Teknologi Malaysia, 81310 Johor Bahru, Johor, Malaysia
e-mail: nhannahmyusof@gmail.com

N. Hamzah
School of Mechanical Engineering, Faculty of Engineering, Universiti Teknologi Malaysia, 81310 Johor Bahru, Johor, Malaysia

© The Author(s), under exclusive license to Springer Nature Singapore Pte Ltd. 2023
M. H. A. Hassan et al. (eds.), *Technological Advancement in Instrumentation & Human Engineering*, Lecture Notes in Electrical Engineering 882,
https://doi.org/10.1007/978-981-19-1577-2_53

agents exchange information, which is not represented by real numbers but instead by logical values. The problem is formulated based on the combination of communication matrix, visibility matrix, reachability, and a decision system of a function. The study of logical consensus is applied to several practical approaches; Intrusion Detection System (IDS) [4] and urban area surveillance system [5], and distributed multi-robot IDS [6].

In control theory, other than stability, controllability and observability are also crucial to be analyzed and discussed. Controllability is also known as reachability, there exist control signals which allow the system to reach any state in a finite amount of time. On the other hand, observability implies that all critical states can be known by measuring the output of the systems, where on a side note, it is impractical to know every state. Hence, only critical states are significant to be known. In this study, controllability analysis is considered because it is important to highlight when a controllable system is moving towards an uncontrollable one which can jeopardize the stability of the system in practice [7].

In the control system, by referring to Fig. 1, the controller combines the input and output to obtain the actuator commands. The system includes the actuators and sensors as a plant, where the actuators impart the control forces and energy, and the sensors measure the states and produce the output. The controllability within the system and the actuators imply that there is a need for the actuators to be able to influence the system. On the other hand, the observability within the system and the sensors let the system acknowledged any changes.

Dridi proved the controllability of logical CA by utilizing the Markov chain method in [8]. However, the Markov chain is only suitable for a probabilistic CA. Motivated by studies of Dridi, this study considers two significant notions of CA; the existence of Hamiltonian cycle (HC) and full rank of the controllability element.

A Hamiltonian graph is defined as a graph that contains an HC named after William Rowan Hamilton in 1857, who invented the Icosian puzzle game to solve the problem of symmetries of an icosahedron [8]. A Hamiltonian path exhibits a path that visits every vertex exactly once without any repeats but does not necessarily start and end at the same vertex, which means the starting vertex does not equal the destination vertex. In contrast, an HC is a closed circuit that visits every vertex exactly once without any repeats but must start and end at the same vertex, which means the

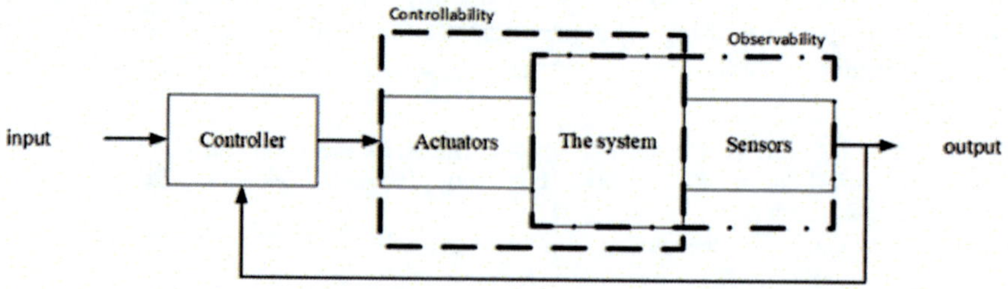

Fig. 1 Relationship of controllability and observability in control system design

starting vertex is similar to the destination vertex. The presence of HC with a closed-circuit indicates that the CA is regionally controllable which means all information of agents is reachable throughout the entire system or on any sub-region of the system.

Another convenient test for controllability is the testing of the rank [7]. This method provides a simpler approach than the use of eigenvalue or eigenvector decompositions. A pair of reachability matrices is introduced; communication matrix and visibility matrix, to test for the rank. The system is said to be controllable if there exists a full rank matrix within the combination of these matrices.

In this work, we propose a consensus-based control algorithm for WDS which the system is based on a logical or binary framework for detection that integrates opening or closing of control elements to allow the affected area to keep receiving supply through alternative routes available. Application of an efficient control system through data analytics gathered from sensors allow for detection and recovery from water leakages which not only help to minimize water wastage but also to restrict disruption to the users. In this study, an automated intelligent control is needed to provide a decision-making facility that provides a sustainable and resilient infrastructure. Two significant notions of CA are considered to investigate the controllability of the system. The controllability is analyzed using the existence of HC and the full rank of the controllability element. The importance of controllability analysis is to provide validation on the logical convergence for the system.

2 Research Methodology

Figure 2 depicted the flow of the research from the initial WDS study until the controllability analysis performed throughout this paper.

First, CA is a dynamic model with discrete space and time. The important entities of CA are lattice, state, neighborhood, and transition function.

Definition 1 A CA consists of a 4-tuple $\mathcal{A} = (\mathcal{L}, \mathcal{S}, \mathcal{N}, f)$, where

1. $\mathcal{L}$ is represented as a lattice (cellular space) that is a discrete cells grid on $\mathbb{R}^d$, where d is the dimension of the lattice. $c \in \mathcal{L}$ for every cell, c in the lattice which is described by its position.
2. $\mathcal{S}$ is a set of states that consists of all the possible states that each cell might hold at every time step. The cell state is denoted as $s_t(c) \in \{1, 0\}$.
3. $\mathcal{N}$ indicates the neighborhood that maps every cell and its neighbor into a neighborhood which is defined as

$$c \to \mathcal{N}(c) = \{c_1^*, c_2^*, \ldots, c_k^*\}, \tag{1}$$

where k is the size of the neighborhood, c_i^* is a cell for $i = 1, 2, \ldots, k$, and r is the neighborhood radius. The radius of the neighborhood can be evolved around (left, right, above, and below) each cell by a cell or more.

Fig. 2 Research flowchart

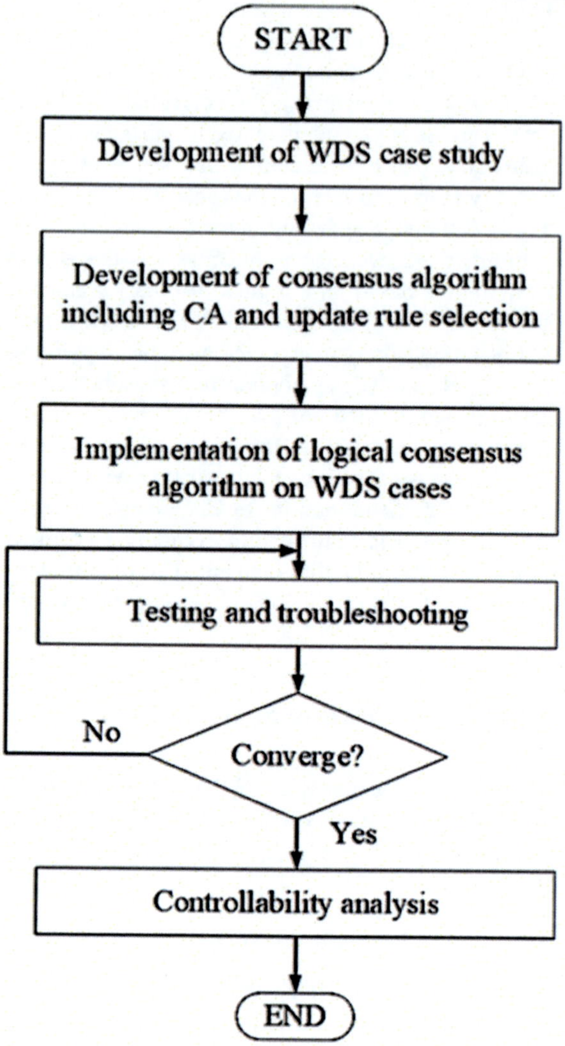

4. f represents a transition function (rule) that gives the state of $s_{t+1}(c)$ at the time step, $t + 1$ depends on the neighborhood's state, $s_t(\mathcal{N}(c))$ at a t time step. The transition function is written as

$$s_t(\mathcal{N}(c)) \rightarrow s_{t+1}(c) = f(s_t(\mathcal{N}(c))), \tag{2}$$

As a summary, the definition of CA basic composition is illustrated in Fig. 3. In this study, to carry out the logical consensus for the system, all the agents' information is formulated using the CA model. A basic combined configuration is used to provide

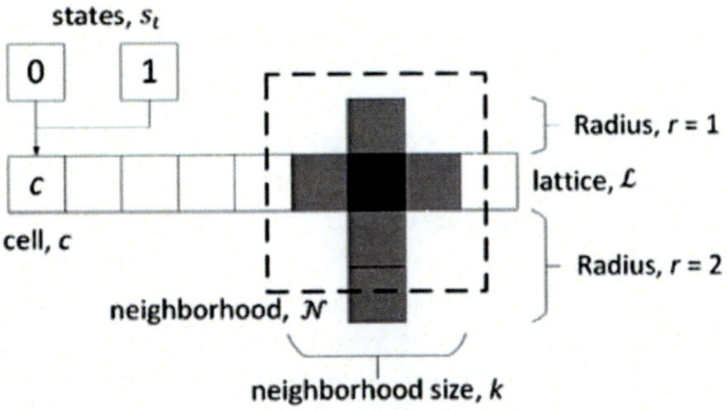

Fig. 3 Lattice in CA

a fundamental framework [9]. The configuration is signified by four water storage tanks, with interconnected pipelines between each other.

The pipelines are described by the dashed line as in Fig. 4(a). The isolation valves in between pipelines are denoted as *p, q, r,* and *s*. The variables include flow rate and water level. The manipulated variable is represented by the valve and the pump voltage is assumed to be constant. A set of possible *m* events are constructed after the configuration is determined to signify the state of fault conditions, and the $\mathcal{f}$decision according to each event is defined. Each agent has a binary vector state,

$$X_i = \left(X_{i,1}, \ldots, X_{i,m}\right) \tag{3}$$

where the state logical inputs, $X_{i,j} \in \{1, 0\}$, where *i* is the local agent, while *j* is the neighboring agent to the *i* agent. All cells that contain the states of every agent are arranged according to sequential order regardless of their geographical locations. The data of every agent are arranged in the form of lattice or array. Before planning for *m* events and ι decision, initially, each agent must acknowledge the input binary state that will lead to an output decision vector,

Fig. 4 **a** The basic combined configuration of WDS **b** the network communication topology

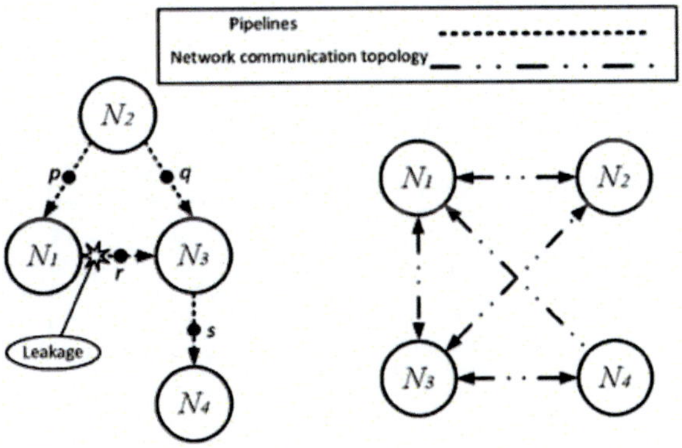

$$Y_i = (y_{i,1}, \ldots, y_{i,f}) \tag{4}$$

The isolation threshold, f_{th} is introduced to specify each event. The status of the flowrate, $f_{i,j}$ is observed, where $f_{i,j}$ indicates the flow rate of water flows from i to j. By referring to the layout in Fig. 4(a), the events, m, and the decisions, f are adopted with the f_{th} value is set to 7 ℓ/s. Based on the user preference, this value can be varied according to the type of hardware or WDS configuration.

Four events are constructed with four possible fault conditions as in Fig. 5. Fault events are derived from the states of any abnormal detection on one or more sensor(s) in the WDS, in which the flowrate measurement falls behind the pre-defined threshold. For instance, event 3 indicates the occurrence of leakage or burst along the pipeline between N_1 and N_3, which means the flowrate, $f_{1,3}$ is less than 7, thus decision 3 is considered, where valve r needs to be closed to isolate the fault and reduce the water wastage.

Due diversity of geographical locations that the agents, the generic agent i has the possibility of being able or not to see the input event, u_j. Therefore, a visibility matrix, $V \in \mathbb{B}^{n \times m}$ is systematically introduced according to the agents' sequence, where n is the number of agents and m represents the chosen event. $V(i, j) = 1$, if and only if the agent that is placed in a cell, c_i can consent the input event, u_j. In other words, agent i is said to have a direct reachable from the j-th input.

As a pair to visibility matrix, V, later, to gain the reachability throughout the system, communication matrix, $C \in \mathbb{B}^{n \times n}$ is conveniently introduced. Each agent in the cell, c_j can receive data from the agent, c_i with input event, u_j. Communication matrix, C is developed using the vertex-edge incidence matrix, where the agents become the vertex on the column and the transmission lines connected to the vertex

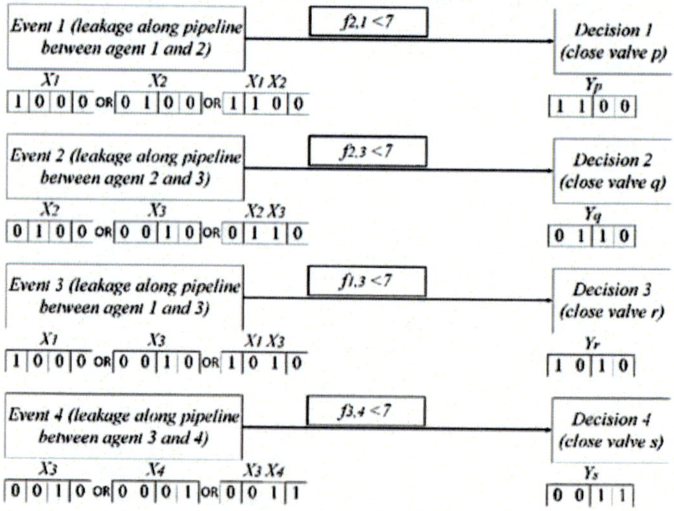

Fig. 5 Composition of input and output for each agent

Table 1 Communication matrix composition

		vertex			
		1	2	3	4
edge	1	0	1	0	0
	2	0	0	0	0
	3	1	1	0	0
	4	0	0	1	0

are the edge on the row as tabulated in Table 1. The transmission lines moving away from the *j*-th vertex will be '1'.

To check for the consensus feasibility, the reachability matrix, R_j is constructed based on C and V_j as follows. A detailed study of the consensus feasibility can be found in [4].

$$R_j = \left[V_j C V_j C^2 V_j \ldots C^{n-1} V_j\right] \tag{5}$$

Theorem 1 A pair of C and V_j is considered as completely reachable if and only if the reachability matrix, $R_j(C, V_j)$ spans for the entire graph.

The sequence $\{C^k V_j\}$ for the k-steps, $k = 0, \ldots, n-1$ meaning that all the agents have received the input value,u_j exactly after the $k-1$ step.

$$I_j = \sum_{i=0}^{i=k} C^k V_j \tag{6}$$

$$R_j = \left[I_1 \; I_2 \; I_3 \; I_4\right] \tag{7}$$

The span of the graph indicates the input event, u_j can be propagated throughout the node-set that contains the agents. The consensus is said to be feasible if, and only if the extend of $R_j(C, V_j)$ is the whole graph for the node-set, $N_{\mathcal{R}} = N$ means that all elements on I_j in the most right column is '1'.

A simple one-dimensional cellular automaton (CA) consists of cells that use adjoining neighbors' information which is located on the left and right sides. The next state of each cell is computed at the same time and continued at every discrete time step. Each cell is indexed as, $(c_i)i \in \{1, 2, \ldots, \mathcal{N}_{\mathcal{L}}\}$ and the state of all cells in the lattice at time t is described as an integer, C_t. The transition function for the entire lattice is given by,

$$(I - T)s_d = D \tag{8}$$

where the s_d is the desired convergent state and D is the constant. The matrix T must be an upper-diagonal or lower-diagonal matrix.

The transition function for the overall array is denoted as $f(s_t(\mathcal{N}(c_i)))$. If the automata always converge to a single global pattern regardless of its starting state, the final pattern must be independent of its initial pattern with the presence of sufficient iterations. Hence, the equation will be,

$$\begin{bmatrix} 1 & 0 & 0 & 0 \\ -x & 1 & 0 & 0 \\ 0 & -x & 1 & 0 \\ 0 & 0 & -x & 1 \end{bmatrix} \begin{bmatrix} c_{1,t+n} \\ c_{2,t+n} \\ c_{3,t+n} \\ c_{4,t+n} \end{bmatrix} = \begin{bmatrix} d \\ d \\ d \\ d \end{bmatrix} \quad (9)$$

The simultaneous equation will produce,

$$\begin{aligned} c_{1,t+n} &= d \\ -xc_{1,t+n} + c_{2,t+n} &= d \\ -xc_{2,t+n} + c_{3,t+n} &= d \\ -xc_{3,t+n} + c_{4,t+n} &= d \end{aligned} \quad (10)$$

3 Simulation Results

By referring to the layout in Fig. 4(a), event 3 is considered with the occurrence of leakage or burst along the pipeline between N_1 and N_3, which means the flowrate,$f_{1,3}$ is less than 7, thus decision 3 is considered and valve r need to be closed to isolate the fault and reduce wastage. In between the agents, there are valves, p, q, r, and s. The CA of the WDS, $\mathcal{A} = (\mathcal{L}, \mathcal{S}, \mathcal{N}, f)$ is considered with lattice $\mathcal{L} = \{c_1, \ldots, c_{\mathcal{N}_\mathcal{L}}\}$, where $\mathcal{N}_\mathcal{L} = 4$ is presented as the number of agents available in a WDS. Every cell, c_i which represents an agent, can take a state value of binary in the set of state $\mathcal{S} = \{0, 1\} \in \mathbb{B}$. Let the desired state, $s_d = 1010$. The convergence is succeeded when $s_T = s_d$, where s_T is the final state of all agents.

The pair matrices, C and V_j are given as

$$C = \begin{bmatrix} 0\,1\,1\,0 \\ 1\,0\,1\,0 \\ 1\,1\,0\,1 \\ 0\,0\,1\,0 \end{bmatrix} \text{ and } V_j = \begin{bmatrix} 1 \\ 0 \\ 1 \\ 0 \end{bmatrix} \quad (11)$$

According to the physical configuration, these matrices are developed to avoid any unnecessary changes in the hardware. Without losing the advantage in exchanging information, the decision-making in opening and closing valves within the pipelines is involved. Based on Eq. (5), matrix R_j is calculated as

$$R_j = \begin{bmatrix} 1\,1\,1\,1 \\ 0\,1\,1\,1 \\ 1\,1\,1\,1 \\ 0\,1\,1\,1 \end{bmatrix} \quad (12)$$

The R_j described the final spanning matrix $I_4 = \begin{bmatrix} 1 & 1 & 1 & 1 \end{bmatrix}^T$, thus, the consensus is feasible and iteration can be done. At each iteration, states of agent i are obtained as follows:

$$
\begin{array}{llll}
t = 0: & & X_1 = X_2 = X_3 = X_4 = \begin{bmatrix} 0\,0\,0\,0 \end{bmatrix} & \\
t = 1: & X_1 = \begin{bmatrix} 1\,0\,0\,0 \end{bmatrix} & X_3 = \begin{bmatrix} 0\,0\,1\,0 \end{bmatrix} & X_2 = X_4 = \begin{bmatrix} 0\,0\,0\,0 \end{bmatrix} \\
t = 2: & X_1 = \begin{bmatrix} 1\,0\,1\,1 \end{bmatrix} & X_3 = \begin{bmatrix} 1\,1\,1\,0 \end{bmatrix} & X_2 = X_4 = \begin{bmatrix} 1\,1\,1\,1 \end{bmatrix} \\
t = 3: & X_1 = \begin{bmatrix} 1\,0\,1\,0 \end{bmatrix} & X_3 = \begin{bmatrix} 1\,0\,0\,0 \end{bmatrix} & X_2 = X_4 = \begin{bmatrix} 1\,0\,0\,0 \end{bmatrix} \\
t = 4: & X_1 = \begin{bmatrix} 1\,0\,1\,0 \end{bmatrix} & X_3 = \begin{bmatrix} 1\,0\,1\,1 \end{bmatrix} & X_2 = X_4 = \begin{bmatrix} 1\,0\,1\,1 \end{bmatrix} \\
t = 5: & X_1 = \begin{bmatrix} 1\,0\,1\,0 \end{bmatrix} & X_3 = \begin{bmatrix} 1\,0\,1\,0 \end{bmatrix} & X_2 = X_4 = \begin{bmatrix} 1\,0\,1\,0 \end{bmatrix}
\end{array}
\tag{13}
$$

Figure 6 shows that every cell represents the state of the sensor for every agent. The results are arranged as in the 1-D cells of the lattice. The black highlighted indicate logic '1' and vice versa.

As an initialization, at $t = 0$, $X_1 = X_2 = X_3 = X_4 = \begin{bmatrix} 0\,0\,0\,0 \end{bmatrix}$ indicating all valves are in the open condition to allow water to be transported throughout the system since no fault is detected. At $t = 1$, an event of leakage has occurred within a pipeline between N_1 and N_3, the detected sensor's state turns to the value '1' respectively. Then, at $t = 2$, based on the simultaneous in (10), the states of all agents are updated and the iteration is done until $s_T = s_d$. At the final iteration, $t = 5$, the convergence value is reached, where all of the agents achieved the consensus on the activated sensors and appeared with a global decision to close the valve r.

The information is disseminated to all agents to reach for consensus with the presence of a spanning matrix that holds for the entire graph. The 1-D pattern converges in five cycles. In addition, when the consensus is reached, all agents are capable to

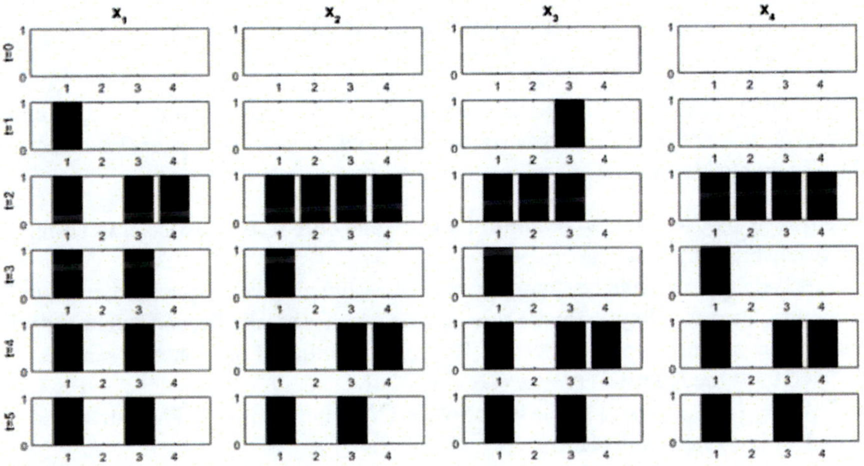

Fig. 6 The convergence states for all agents using 1-D pattern

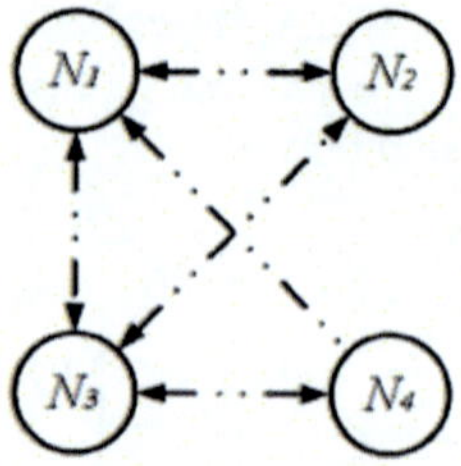

Fig. 7 Network communication topology of the WDS

consider contingency planning and execute local decisions for every individual agent correspondingly.

4 Reachability Analysis

In this section, the concept of reachability analysis is introduced. Reachability links to the ability to reach the desired state. A system is considered reachable if the system can achieve a specified final state regardless of its initial state. Most of the time, reachability can be referred to as controllability. There are a few criteria to check on the controllability such as HC and full rank existence. For a more complex system, a regional sub-region is introduced and analysis can be done by parts. The regional controllability will be analyzed by using HC and full rank existence.

Definition 2 A Hamiltonian cycle is a circuit that visits every vertex once with no repeats along the path. To consider an HC as a circuit, the cycle must be started end ended at the same vertex, which means, the starting point vertex is equal to the destination vertex.

Theorem 2 A CA is said to be controllable if, and only if, there exists an HC inside the graph on the communication matrix, C.

Definition 3 The reachability matrix, R_j has full rank n means that the final column on the R_j is a non-zero matrix, where n is the total number of vertices involved on the communication matrix, C.

Theorem 3 A CA is said to be controllable if, and only if, there exists n full rank of the reachability matrix, R_j.

Precisely, the communication matrix, C is designed to comply with the network communication topology, to make sure that the graph is reachable, and a unique decision of convergence can be presented. The controllability of the system is commonly referred to as the reachability pair (C, V_j).

Let $\mathcal{L} = \{c_1, \ldots, c_{\mathcal{N}_\mathcal{L}}\}$ be the lattice of the CA, where $\mathcal{N}_\mathcal{L} = 4$. The pair matrices, C and V_j are given as in (12).

By referring to **Definition 2**, the network communication topology in Fig. 7 contains an HC, $1 \rightarrow 2 \rightarrow 3 \rightarrow 4 \rightarrow 1$ with 1 as the starting and ending vertex.

The availability of HC on the communication graph denotes that the whole system is reachable as referred to in **Theorem 2**, which means the input signal can be propagated to all agents. With the pair matrices presented in (11), the reachable matrix, R_j is obtained as in (12). By referring to **Theorem 3**, it is convenient to observe that R_j has full rank (rank 4), which means all states are reachable using both inputs on V_j.

5 Conclusion

In this study, logical consensus control is adopted to facilitate decision-making for a WDS. The convergence and iteration of the algorithm are constructed based on the 1-D CA framework. The consensus on the WDS is reached in five cycles. The existence of HC and rank testing are determined to validate the reachability within the network communication topology of the system. The communication matrix can be manipulated to find the optimum network topology for a specific system configuration based on the user or designer preference thus the fault impact can be minimized. The present method is adaptable and scalable with the advantage that the WDS can be extended into any arbitrary physical configuration.

References

1. Gulzar MM, Rizvi STH, Javed MY, Munir U, Asif H (2018) Multi-agent cooperative control consensus : a comparative review. Electronics 7(2):22
2. Cocîrlea D, Dobre C, Hîrţan LA, Purnichescu-Purtan R (2020) Blockchain in intelligent transportation systems. Electron 9(10):1–24
3. Fagiolini A, Housh M, Ostfeld A, Bicchi A (2014) Distributed estimation and control of water distribution networks by logical consensus. In: International Symposium Control. Communication Signal Processing, pp 239–242 (2014)
4. Fagiolini A, Bicchi A (2013) On the robust synthesis of logical consensus algorithms for distributed intrusion detection. Automatica 49(8):2339–2350
5. Martini S, Fagiolini A, Bicchi A, Dini G (2009) Safety and security in networked robotic systems via logical consensus. In: 6th international conference informatics control automation robotics, vol 1
6. Martini S et al (2015) Distributed motion misbehavior detection in teams of heterogeneous aerial robots. Rob Auton Syst 74:30–39
7. Datta BN (2004) Controllability, observability, and distance to uncontrollability. In: Numerical methods for linear control systems. Academic Press, pp 159–199
8. Dridi S (2019) Recent advances in regional controllability of cellular automata.
9. Trifunovic N (2006) Introduction to urban water distribution. Taylor & Francis/Balkema
10. Murty PSR (2017) Incidence matrices. In: Power Systems Analysis, vol 6, pp 19–33 Elsevier Ltd.

Fault Diagnosis of Spark Plug in a Spark Ignition Engine by Using Wavelet Power Spectrum

A. A. Azrin, I. M. Yusri, A. Aziz, M. F. Jamlos, and R. Mamat

Abstract In a Spark Ignition (SI) engine, the study of significant parameters that influence spark plug deterioration is very critical. It allows the detection of faulty characteristics/behavior and displays the types of spark plug faults, allowing for a more thorough understanding of the health status of spark plug. Faulty behavior of spark plug was identified based on the breakdown voltage picked up in the secondary ignition signal, by analyzing the statistical features of the data gathered using global Wavelet Power Spectra (WPS). The overall power spectra of five spark plugs was studied, and the results revealed a high degree of precision.

Keywords Spark plug · Fault diagnosis · Wavelet

1 Introduction

The major cause of performance degradation and increased fuel consumption in current SI engines is a problem in the ignition system. The most extensively utilized approach to mitigate the risk of engine breakdown is predictive maintenance techniques. Since SI engine must have spark plugs, it is critical to develop a condition monitoring method that will avoid malfunction during operation. Fault identification is important to avert engine problem, which could result in significant vehicle damage. Fault diagnosis is a crucial step in spark plug condition monitoring that prevents significant damage. Wavelet analysis is commonly employed in engine

A. A. Azrin (✉) · I. M. Yusri · A. Aziz
Faculty of Mechanical and Automotive Engineering Technology, Universiti Malaysia Pahang, 26600 Pekan, Pahang, Malaysia
e-mail: azrynabdulazyz@gmail.com

I. M. Yusri · A. Aziz
Center for Automotive Engineering, Universiti Malaysia Pahang, 26600 Pekan, Pahang, Malaysia

M. F. Jamlos · R. Mamat
College of Engineering, Universiti Malaysia Pahang, Lebuhraya Tun Razak, 26300 Gambang, Kuantan Pahang, Malaysia

© The Author(s), under exclusive license to Springer Nature Singapore Pte Ltd. 2023
M. H. A. Hassan et al. (eds.), *Technological Advancement in Instrumentation & Human Engineering*, Lecture Notes in Electrical Engineering 882,
https://doi.org/10.1007/978-981-19-1577-2_54

fault detection and diagnosis as condition monitoring as well as for other mechanical components including gears and bearings [1].

Analog signals are analysed using a variety of techniques, including time, frequency, and time–frequency domain techniques [2, 3]. The frequency domain assesses the condition based on the frequency content of the signal using the Fast Fourier Transform (FFT) of the time domain signal, power spectrum, kepstrum, and other techniques. Fourier Transform (FT) depicts a signal as a series of exponents with an the time length, therefore it is excellent for recognising harmonic signals [4]. Despite of its frequency resolution and constant, it is ineffective for evaluating transitory signal components [5]. Knocking and misfire are caused by a failure in a spark plug, which might result in a non-stationary signal due to transitory signal alteration [6]. Because these non-stationary components typically contain a wealth of information about spark plug problems, it is critical to examine them. The signal's non-stationary nature necessitates the employment of time-frequency approaches, which allow researchers to examine the frequency content of the signal over time. Non-stationary signal analysis is not suited for FFT-based condition monitoring approaches [7]. The analysis of complex machines' non-stationary signals necessitates innovative approaches that go beyond the standard Fourier approach. There are many different temporal variant approaches; [8].

The Wavelet Transform (WT) is able to analyse both time frequency domains which is very useful to determine faults in spark plug by extracting the transient signal properties of dynamic analogue signals. When compared to the FFT spectrum, Wavelet analysis provides more information about the issue [9]. Dynamic analogue signal recorded is filtered into a number of frequency bands in WT, then later separated into energy assessed, time and their frequency contents. The wavelet is changed in time to pick the segment of the signal to be studied. The wavelet is then compressed or expanded to target on a specific range or number of oscillations during the analysis. The wavelet focuses on low-frequency signal components when it enlarges. When the wavelet is compressed, it focuses on high-frequency signal components [10]. The WT accomplishes a weighted set of scaled waves from a time scale decomposition of the signal.

In order to extract the signal's fault features more effectively, the right wavelet base function should be chosen [11]. The selection of the appropriate wavelet basis function that is most similar to the fault feature to be identified and the analysis of the generated wavelet coefficients to extract the fault-related features are the main areas of investigation related to the use of wavelet analysis in fault detection and diagnosis [12]. A number of wavelet-based features are proposed for high-sensitivity spark plug fault detection. To improve the fault detection, Morlet wavelet and impulse wavelet are used to extract the features of spark plug faults and these wavelet parameters are set based on the maximum kurtosis [13].

Various engine condition monitoring techniques employ analog monitoring. Despite this, there is no widespread adoption of online condition monitoring systems in practise. The Morlet wavelet is used for signal processing since the Morlet wavelet does give smoothing but is not a highly flexible tool in signal analysis [14].

In this study, the spark plug fault recognition was investigated using the WT in this study. Since its theoretical development in 1984 by Grossman and Morlet, the WT has gained a lot of attention in signal processing [15]. Its popularity as a substitute for the FT in retaining local, non-periodic, multi-scaled phenomena is quickly growing. It has advantages over traditional spectral analysis in that it may analyse many scales of temporal variability and does not require a stationary series. As a result, it's suitable for analysing irregularly distributed events and time series with non-stationary power at a variety of frequencies. The purpose of this research is to investigate the reliability of using wavelet analysis to help identify faulty behavior of spark plug in a SI engine.

2 Materials and Methods

2.1 Experimental Setup

A series of tests were conducted on NGK BPR6EF conventional copper spark plug with Cadic KD-9007A Ignition Coil, which is driven by trigger signal from micro-controller Arduino Uno with the frequency equivalent to 1000 rpm until 5000 rpm. The microcontroller controls the frequency and duty cycle of the ignition system through the "sketch" program which is uploaded using Arduino IDE platform. The element of Internet of Things (IoT) is implemented on the microcontroller with the help of ESP8266 NodeMCU WiFi module. FLUKE Secondary Ignition Pickup and RS Pro RSDS 1102CML+ Digital Storage Oscilloscope were used to record the breakdown voltage signal of the spark plug (Fig. 1).

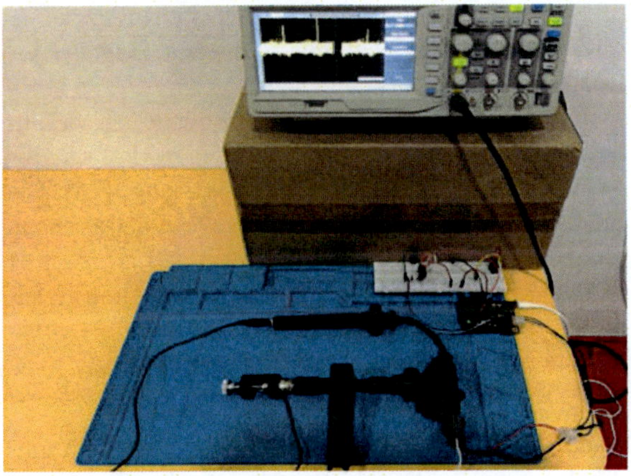

Fig. 1 The experimental setup for data collection

Table 1 Table of experimental parameters

Engine speed (RPM)	Mileage (km)
1000 (baseline)	0 (baseline)
2000	7500
3000	15,000
4000	22,500
5000	30,000

2.2 *Selected Data*

A single experiment covers from 0 to 30,000 km mileage. The ignition coil and spark plug type were held constant throughout the data collection. Table 1 shows the key operating parameter for the present experiments.

Tests were conducted with different engine speeds to determine their influences on the breakdown voltage signal and overall spark plug behaviour. Spark plugs of the similar spark plug type but with different engine speed applied were tested. The test involved operating the ignition coil with (a) base-line, and (b) adjusted frequency of ignition system.

2.3 *Wavelet Transform*

The original signals are subjected to mathematical changes in order to extract previously unknown information. There are various transformations that can be used, the most popular of which being the FT.

By simultaneously decomposing or changing a one-dimensional time series into a diffuse two-dimensional time-frequency image, wavelet analysis maintains time and frequency localization in a signal analysis. The amplitude of all "periodic" signals within the series can then be determined, as well as how that amplitude changes over time. Wavelet analysis uses fundamental waves, or mother wavelets as they are called in the literature, to achieve this change [16]. The Paul and Gaussian derivatives (DOG) are two examples of mother wavelets, as shown in Fig. 2 with their respective orders (m), and the Morlet wavelet (Fig. 3), which is defined in practise as the product of a complex exponential wave and a Gaussian envelope:

$$\psi_0(\eta) = \pi^{-\frac{1}{4}} e^{-i\omega_0 n} e^{\left(-\frac{\eta^2}{2}\right)} \tag{1}$$

where $\psi_0(\eta)$ is the wavelet value at dimensionless time η and ω_0 is the dimensionless frequency (equal to 6 in this study) (Fig. 3). A new time series of the projection amplitude, where 0 is the dimensionless frequency (in this case equal to 6) and 0 is the wavelet value at dimensionless time (Fig. 3). By moving this wavelet along the time series, you can create a new time series of the projection amplitude versus time.

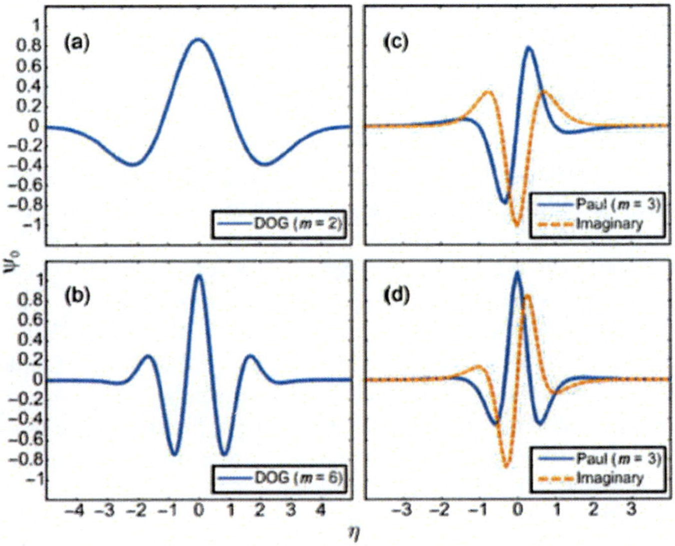

Fig. 2 **a** derivative of Gaussian (DOG) with $m = 2$; **b** DOG with $m = 6$; **c** Paul with $m = 3$; and **d** Paul with $m = 4$

Fig. 3 Example of the Morlet mother wavelet including the imaginary part. Main plot: Morlet mother wavelet with $\omega_0 = 6$. Inset: Morlet mother wavelet with $\omega_0 = 2$

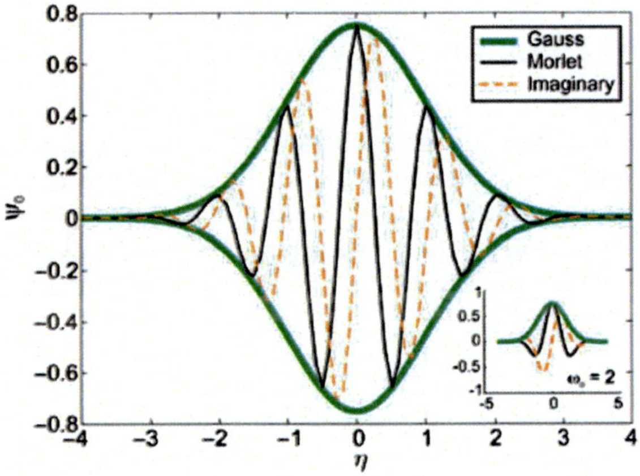

Finally, the “scaling” of the wavelet can be changed by adjusting its width. Technical data and related physical properties describing all parameters used in the different mother wavelets can be found in the corresponding literature [16]. The projection of the amplitude against time can be created by moving this wavelet along the time series. Finally, the “scaling” of the wavelet can be adjusted by varying its width. The relevant literature provides technical details and associated physical properties detailing all the parameters used in the various mother wavelets [16].

3 Results and Discussion

3.1 Wavelet Power Spectrum

Because the current data is disseminated with number of cycles, the wavelet analysis settings are set as follows: time interval dt = 1 cycle, start scale $s_0 = 2$ cycles, scale width j = 0.1, resulting in 10 sub-octaves per octave and seven powers (4, 8, 16, 32, 64, 128 and 256). The results for other spark plug are identical to those shown in Fig. 4, which displays the example of spark plug #1, while Fig. 4(a) shows the raw breakdown voltage data. Figure 4(b) illustrates the WPS (absolute value squared) for the average breakdown voltage reading in Fig. 4(a), which is recorded from 0 to 30,000 km. The Morlet wavelet was chosen because, when compared to other mother wavelets, it provide more accurate frequency information. The concentration of power can be clearly seen in the frequency or time domain, i.e., an annual frequency along the entire time series (kilometres travelled), as shown in Fig. 4(b). The healthy and faulty behaviour of the spark plug is represented by the variance of the power in the range of 2 to 8 cycles, as shown in Fig. 4(c); i.e., when the power decreases significantly in this range, it is a healthy behaviour, and when the power reaches its maximum, it is a faulty behaviour.

3.2 Scale-Average Time Series

A time series of the average variance in a given band is the scale-average wavelet power. It is the 2- to 8- cycles band in the case of Fig. 4(d). Within the same time series, this can be utilised to investigate modulation of one frequency by another. Figure 4(d) represents the average of Fig. 4(b) across all scales between 2 and 8 cycles, giving a measure of average monthly variance versus time.

3.3 Global Wavelet Power Spectrum

Integrating the power across the time spectrum assuming red noise (represented by the dashed lines) confirms the trends in the time series (Fig. 4(c)). White noise or red noise can be used to model various geophysical time series. The univariate lag-1 autoregressive process is a fundamental model for red noise [16]. The lag-1 (1) is the correlation between the time series and itself, but shifted (or lagged) by one time unit. This would be a one month shift in the current situation.

The lag-1 is a statistic that indicates how long an anomaly lasts from one month to the next. If lag-1 is 0.4, the actual lag-1 should be calculated as α = (α_1 + 〚α_2〛^(1/2))/2, where α_1 is the autocorrelation lag-1 and α_2 is the autocorrelation lag-2, which is equal to lag-1 but delayed by two cycles instead of one. When α_1 is

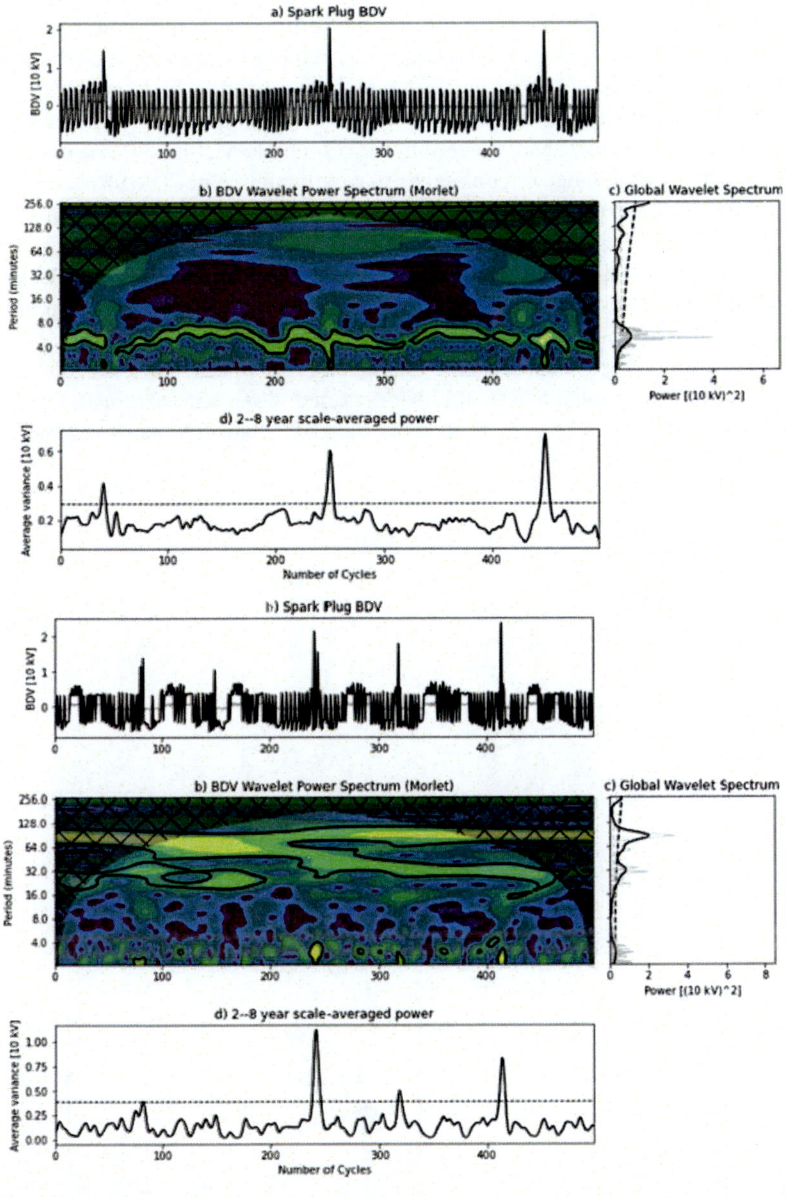

Fig. 4 The comparison between the graphs of spark plug #1 and #5 with 1000 RPM and 5000 RPM for 30,000 km, respectively. **a** Raw spark breakdown voltage data for 30,000 km. **b** The WPS using the Morlet nut wavelet. The thick contour encloses regions with greater than 95% confidence for a red noise process and an α-coefficient of 0.776. **c** The global WPS. The dashed line represents the 5% significance level, using a background spectrum with red noise. **d** Scale average of wavelet power over the band from 2 to 8 cycles. The dashed line also represents the 95% confidence level, assuming red noise

0.4, it is recommended to model the series as white noise ($\alpha = 0$). The null hypothesis for the WPS is described as assuming that the time series has a mean power spectrum; with some percentage confidence, it can then be assumed to be a true feature. For the sake of definition, "significant at the 5% level" is identical to "95% confidence level" and refers to a test against a certain background level, while "95% confidence interval" refers to the range of confidence in a particular value. The 95% confidence interval means that 5% of the wavelet power should be higher than this value [16].

The global wavelet spectra estimate the time series true power in an impartial as well as consistent manner. In Figs. 4(c) for spark plug #1 for example, one peak rises above the others, indicating that there is one main frequency; however, for spark plug #5 there are two or more peaks instead of one, regardless of the frequency. In Fig. 5, the peaks are highlighted. Following this analogy to other selected datasets, two primary patterns of global WPS could be identified: three with one main frequency referred to as Pattern A and two with more than one main frequency referred to as Pattern B (Fig. 6).

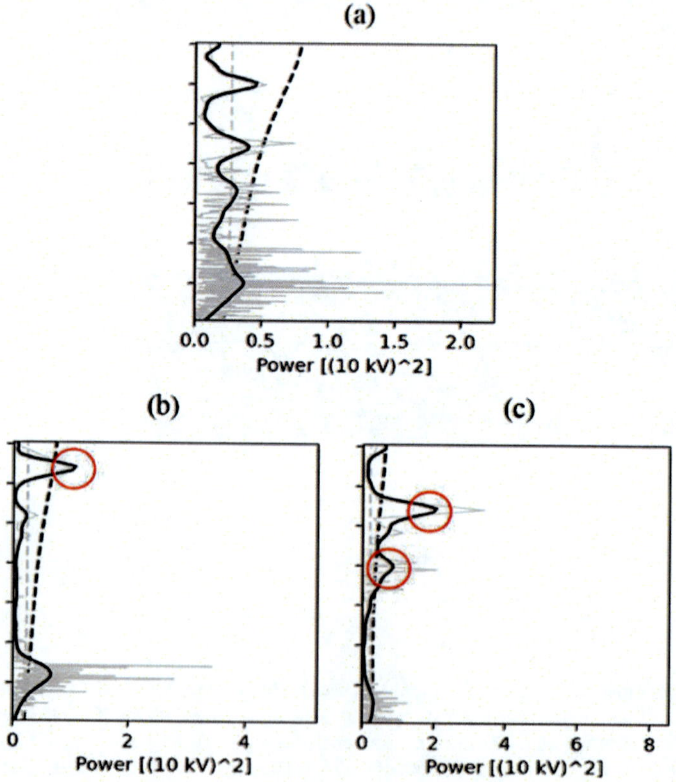

Fig. 5 Global wavelet power spectra: **a** baseline, **b** classified as Pattern A with one main frequency, and **c** classified as Pattern B with more than one main frequency

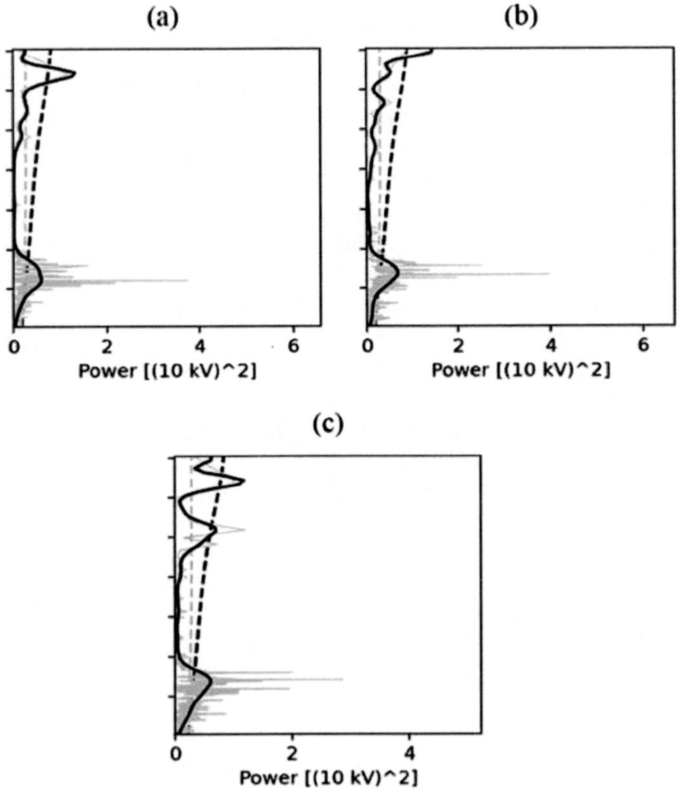

Fig. 6 Global wavelet power spectra: **a** and **b** classified as frequency Pattern A with one main frequency at the 2- to 8- cycles band; and **c** classified as frequency Pattern B

As a result of this, two separate spark plug condition (Pattern A and Pattern B) were discovered. Pattern A indicates that the categorised time series has just one main frequency, which represents the stable breakdown voltage signal and so the shape of its global wavelet spectrum differs from Pattern B. Pattern B has more than one main frequency which indicates the irregularity of the breakdown voltage signal of spark plug.

4 Conclusion

Based on the methodology used and the results obtained, it could be concluded that WT is an effective tool for the detection of spark plug faults in SI engines. Despite ongoing research on the usefulness of spark plug condition monitoring in spark plug fault detection, a complete picture of the situation is still far away. In this study, the efficiency of condition monitoring, represented by WT analysis, in detecting spark plug faults was investigated at different engine speeds. Data from five spark plugs were analysed, and results across the performance spectrum showed high accuracy. However, other engine speeds that were not considered in this study are of great importance, which represent changes in the effectiveness of spark plug fault detection. Although the calculated global wavelet power spectra exhibited this main frequency, they showed peculiar patterns that could be used to characterise the state of the spark plug. Thus, two states of the spark plug were identified: healthy state with frequency pattern A, faulty state with frequency pattern B. The global wavelet spectra provided an unbiased and consistent estimate of the true power spectrum of the breakdown voltage signal time series and could be used to characterise the time series variability of the spark plug breakdown voltage signal in a simple and reliable method. However, the global wavelet spectra could be combined with other methods to improve the diagnosis of damaged spark plugs, such as multivariate statistical analysis or machine learning classification. More research with a larger number of datasets and more important operating parameters to describe spark plug condition is recommended.

Acknowledgements The authors would like to express their gratitude to the sponsorship by Universiti Malaysia Pahang under research university grants RDU1903101 and PGRS2003142 for laboratory facilities and financial aid.

Conflicts of Interest The authors declare that they have no conflicts of interest to report regarding the present study.

References

1. Heydarzadeh M, Kia SH, Nourani M, Henao H, Capolino G-A (2016) Gear fault diagnosis using discrete wavelet transform and deep neural networks. In: IECON 2016 - 42nd annual conference of the IEEE industrial electronics society, pp 1494–1500, October 2016. https://doi.org/10.1109/IECON.2016.7793549
2. Bakhtadze N, Sakrutina E (2016) Applying the multi-scale wavelet-transform to the identification of non-linear time-varying plants. IFAC PapersOnLine 49(12):1927–1932. https://doi.org/10.1016/j.ifacol.2016.07.912
3. Abuhamdia T, Taheri S (2017) Wavelets as a tool for systems analysis and control. J Vib Control 23(9):1377–1416. https://doi.org/10.1177/1077546315620923
4. Vernekar K, Kumar H, Gangadharan KV (2014) Gear fault detection using vibration analysis and continuous wavelet transform. Procedia Mater Sci 5:1846–1852. https://doi.org/10.1016/j.mspro.2014.07.492
5. Sharma S, Tiwari SK, Singh S (2019) Diagnosis of gear tooth fault in a bevel gearbox using discrete wavelet transform and autoregressive modeling. Life Cycle Reliab Saf Eng 8(1):21–32. https://doi.org/10.1007/s41872-018-0061-9
6. Hartono D, Halim D, Roberts GW (2019) Gear fault diagnosis using the general linear chirplet transform with vibration and acoustic measurements. J Low Freq Noise Vib Act Control 38(1):36–52. https://doi.org/10.1177/1461348418811717
7. Haddad S, Karel J, Peeters R, Westra R, Serdijn W (2004) Complex wavelet transform for analog signal processing

8. Wang Y, Yan Y, Wang Q (2016) Wavelet-based feature extraction in fault diagnosis for biquad high-pass filter circuit. Math Probl Eng 2016:5682847. https://doi.org/10.1155/2016/5682847
9. Cohen MX (2018) A better way to define and describe Morlet wavelets for time-frequency analysis. bioRxiv. https://doi.org/10.1101/397182
10. Czajkowski KM, Pastuszczak A, Kotyński R (2018) Single-pixel imaging with Morlet wavelet correlated random patterns. Sci Rep 8(1):466. https://doi.org/10.1038/s41598-017-18968-6
11. Tiscareno MS, Hedman MM (2018) A review of Morlet wavelet analysis of radial profiles of Saturn's rings. Philos Trans R Soc A Math Phys Eng Sci 376(2126):20180046. https://doi.org/10.1098/rsta.2018.0046
12. Boako G, Alagidede P (2017) Co-movement of Africa's equity markets: regional and global analysis in the frequency–time domains. Phys A Stat Mech Appl 468:359–380. https://doi.org/10.1016/j.physa.2016.10.088
13. Yi H, Ouyang P, Yu T, Zhang T (2019) An algorithm for Morlet wavelet transform based on generalized discrete Fourier transform. Int J Wavelets Multiresolution Inf Process 17(05):1950030. https://doi.org/10.1142/S0219691319500309
14. Husaini TEP, Ali N (2018) The Morlet wavelet transform for reducing fatigue testing time of an automotive suspension signal. In: AIP conference proceedings, vol 1983, no. 1, p. 30003, July 2018. https://doi.org/10.1063/1.5046238
15. Goupillaud P, Grossmann A, Morlet J (1984) Cycle-octave and related transforms in seismic signal analysis. Geoexploration 23(1):85–102. https://doi.org/10.1016/0016-7142(84)90025-5
16. Torrence C, Compo GP (1998) A practical guide to wavelet analysis. Bull Am Meteorol Soc 79(1):61–78. https://doi.org/10.1175/1520-0477(1998)079%3c0061:APGTWA%3e2.0.CO;2

Image Reconstruction Technique Using Radon Transform

Teh Chia Ai, Wan Zailah binti Wan Said, Norsuzlin Mohd Sahar, and Mohammad Tariqul Islam

Abstract Image reconstruction is a very important processing step in the analysis of medical images. This paper describes the framework of image reconstruction developed using filtered backprojection algorithm based on Radon transform in MATLAB. Acquisition of image data and image pre-processing were conducted to format the images for further analysis of the images. The sample image of CT slice was converted to sinogram by Radon transform and the reconstructed image was generated using filtered backprojection. The simulated results are analyzed and evaluated using image quality assessment methods. The internal structure and main features are shown in the reconstructed image. The image has a Structural Similarity Index (SSIM) value of 0.9839 and a mean-squared error (MSE) value of 0.0002. The reconstruction is precise within the framework of suggested image reconstruction system as the reconstructed image has a high similarity with the original image. The effects of number of projections, varying filters and varying cut-off frequency in the image reconstruction system were studied and analyzed. Further studies can aim at using more advanced filters and interpolation as well as implement other methods or techniques on the presented process.

Keywords Image reconstruction · Radon transform · Filtered backprojection

T. C. Ai · W. Z. binti Wan Said (✉)
Department of Mechanical and Mechatronic Engineering, Faculty of Engineering, Technology and Built Environment, UCSI University, 56000 Kuala Lumpur, Malaysia
e-mail: zailah@ucsiuniversity.edu.my

N. M. Sahar
Space Science Centre, Institute of Climate Change, Universiti Kebangsaan Malaysia, UKM, Bangi, Selangor, Malaysia

M. T. Islam
Department of Electrical, Electronic and Systems Engineering, Universiti Kebangsaan Malaysia, UKM, Bangi, Selangor, Malaysia

© The Author(s), under exclusive license to Springer Nature Singapore Pte Ltd. 2023
M. H. A. Hassan et al. (eds.), *Technological Advancement in Instrumentation & Human Engineering*, Lecture Notes in Electrical Engineering 882,
https://doi.org/10.1007/978-981-19-1577-2_55

1 Introduction

The study of medical imaging has resulted in the development of techniques that are essential in the practice of medicine to produce a readable and functional image. [1]. Image reconstruction techniques have been created to increase the quality of images for better visual perception, understanding, and analysis [2]. Many forms of medical imaging require image reconstruction [3]. The image reconstruction process' aim is to convert raw data into images that can be evaluated in a clinical setting. The most widely used medical imaging modalities for the studying of complex structures and tissues of human body are computed tomography (CT), magnetic resonance imaging (MRI), positron electron tomography (PET) and single photon emission computed tomography (SPECT) [3, 4, 5]. The diseases can be recognised and detected by using the medical imaging modalities.

The image reconstruction aims to create a cross section image of an object from a series of projections. The Radon transform (RT) is commonly used in image reconstruction in these tomography techniques. The Radon transform is a two-dimensional data processing mapping that produces one-dimensional projections [6]. The Radon transform, as well as its inverse, are used to reconstruct tomographic images from measured projection or scattering results [7].

The aim of this project is to create a comprehensive reference source for image reconstruction and analysis techniques that have been accelerated using high-performance computing. The proposed approach for image reconstruction using Radon transform is expected to perform by building and designing a GUI with App Designer in MATLAB. The performance of the image reconstruction system is analysed and evaluated using image quality assessment techniques.

2 Mathematical Theories of Radon Transform

The Radon Transform is a type of linear transform that is used in projection-based image analysis [8]. The line-integral model reflects an idealisation of transmission and emission tomography systems, which obtain measurements that are close to blurred line integrals [9]. Figure 1 shows the geometry of the line integral associated with the 2D Radon transform [10]. The line integral through f(x, y) along the line L(r, φ) is denoted by p(r). The Radon transform allows to calculate the total density of a function f along a given line L. An angle φ from the x-axis and a distance r from the origin to determine this line L. The resulting projection is a line integral, which is the sum of the intensities of the pixels in each direction [11]:

$$p(r, \varphi) = \int_{-\infty}^{\infty} \int_{-\infty}^{\infty} f(x, y)\delta(x \cos \varphi + y \sin \varphi - \xi)dxdy \tag{1}$$

As a result, a new image R (ρ, φ) is generated:

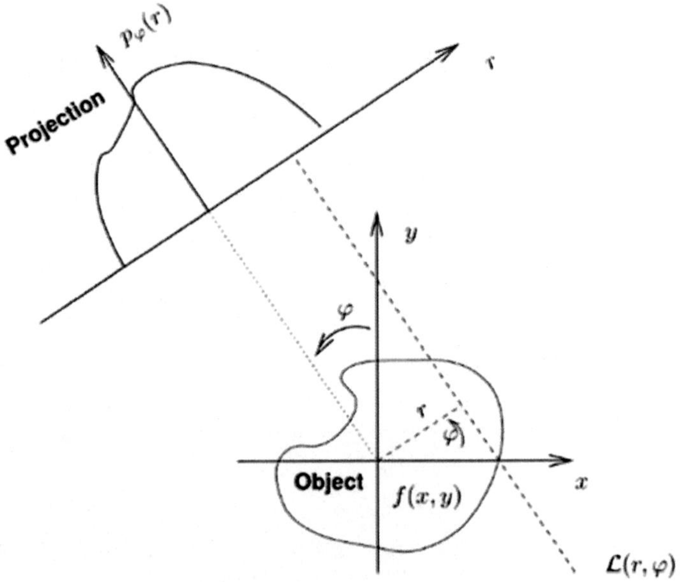

Fig. 1 The geometry of the line integral associated with the 2D Radon transform

$$R(p, \varphi) = \int_L f(x, y)dL \tag{2}$$

3 Methodology

The image processing steps of the project were conducted to reconstruct image using MATLAB. GUI of this project is designed and created to display the result images of each of image processing steps. The proposed block diagram of image reconstruction system is shown in figure below. The first process of the project is image data acquisition, and then image pre-processing including colour conversion of image, image resizing and image denoising. The last process is the image reconstruction as shown in Fig. 2.

3.1 Acquisition of Image Data

Sample images of CT slice of brain, neck, abdomen, and chest are obtained from online. The sample images are saved in the directory in the MATLAB defined path. Image Processing Toolbox in MATLAB is used to display, analyse, and process

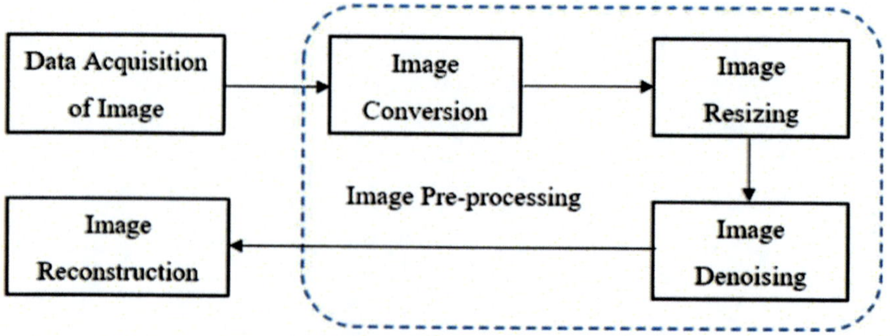

Fig. 2 The overall progress of this study

Fig. 3 The CT slice of abdomen

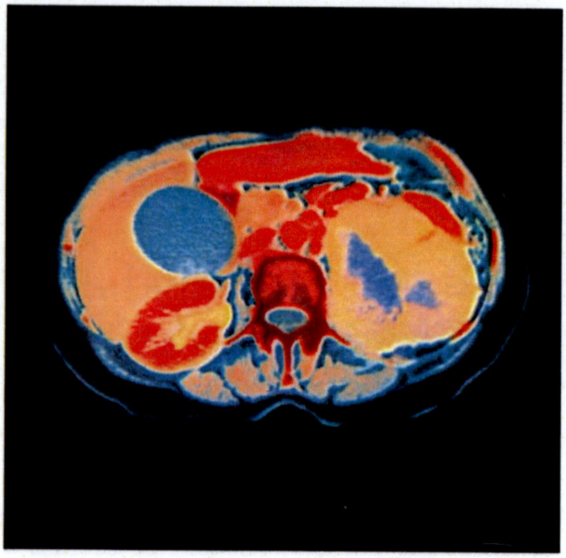

images in a variety of data kinds. The sample image of CT slice of abdomen is chosen to carry out the image reconstruction system as shown in Fig. 3. After the image is loaded into the MATLAB framework, the data is acquired from the image and the data input of the image is stored into MATLAB workspace.

3.2 *Image Pre-processing*

The original image is required to carry out image pre-processing procedures such as image colour conversion, image resizing and image denoising. The original RGB image is converted to grayscale image by function 'rgb2gray(original image) in

MATLAB. The grayscale image is then resized to 256 × 256 pixels. Function 'imresize' is used to resize the image to specifying rows and columns. Image resizing is needed during the pre-processing phase to establish a base size for all images and fed into the algorithm. After image resizing, image denoising is applied to remove unwanted noise or distortions from the image by smoothing it [12]. Image denoising is carried out using the spatial domain filtering method on the input image. Image denoising is achieved by applying various types of spatial filters to two noise models which are Gaussian noise and salt and pepper noise [13]. Noisy images are made by inserting Gaussian noise and salt and pepper noise separately in the initial noise free image. The denoising filtering used on the corrupted images is median filter, Gaussian filter, bilateral filter and wiener filter. Quantitative performance measures signal-topeak ratio (SNR) is used to measure the performance of denoising algorithms and the visual quality of denoised images [14]. After the performance measurement are done on denoised images with different filters, the best filter is chosen to use on the original image and the denoised image will be used in image reconstruction system.

3.3 *Image Reconstruction*

The filtered backprojection algorithm based on Radon transform is used in the image reconstruction system as shown in Fig. 4. The input image undergoes the whole process to reconstruct an image.

First step of the image reconstruction process is the acquisition of sinogram from the 2D image 256 × 256 pixels. The data are acquired from the image into sinogram space by using forward projection. Radon transformation is applied on the image over the number of angles. The projection data are formed to generate the sinogram. The output matrix from the Radon transform is transformed using the Fast Fourier Transform (FFT) [5]. The sinogram are then filtered using a filter in the frequency

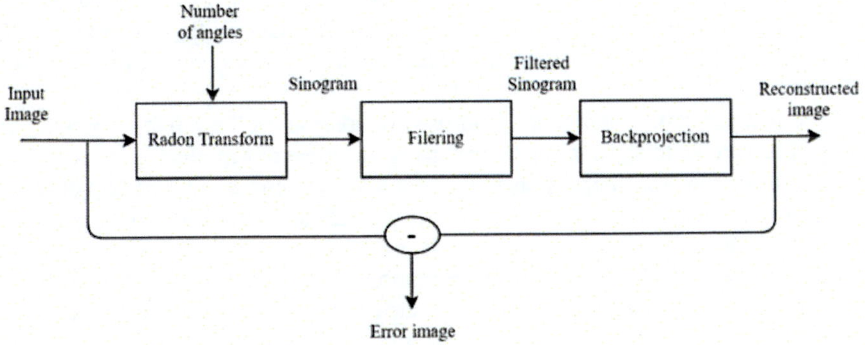

Fig. 4 The block diagram of image reconstruction system

domain. A high pass filter is implemented on the sinogram data in the frequency domain to overcome the effect of blurred image [15]. A useful and required ramp filter can be implemented easily because it is simply a frequency domain multiplication feature that can be quickly transformed using the Fourier transform.

The image is applied by the inverse Radon transform in order to reconstruct it from the sinogram [6]. The IRT algorithm was then used to reconstruct projections of the object slice-by-slice using the sinogram generated [16]. Backprojection produces sinogram data predictions based on the slice. The number of projections used in reconstruction is the number of columns of sinogram. Each projection profile in the sinogram, which was generated and stored when the image was acquired, can be back-projected at the angle at which it was acquired [17]. The number of projections is set at 180 that applied on the image to reconstruct and the backprojection of the sinogram data is set at 30°. Information needs to be backprojected from all 180 projections of acquisition data to accurately reconstruct an image. The projections of each degree sinogram is accumulated to generate the reconstructed image.

3.4 Image Quality Assessment

The image quality assessment is conducted by comparing the original image and reconstructed image. The image quality evaluation metric used are Structural Similarity Index (SSIM) and mean-squared error (MSE). The structured similarity indexing method (SSIM) calculates the normalised mean structural similarity between two images. SSIM metric extracts three main features from an image which are luminance, contrast and structure. The comparison of the two images is based on these three characteristics [18]. Mean-squared error is an image quality assessment technique which focus on calculating between a reference image and a test image. MSE quantifying the difference in the values of each of the corresponding pixels between the sample and reference images [19]. The original image is assumed to be the reference image and it has a SSIM value of 1 and a MSE value of 0. The general value range of SSIM value is zero to one. Both original image and reconstructed image must have the same data type to calculate the mean SSIM value and MSE value. The original image which has a data type of unit 8 is converted to data type of double.

The image quality metrics are also used to analyse on the effect of number of projections on the image quality. In evaluating the effect of sampling, Ram Lak filter is used to compare the images under the same condition. In testing the effect of filters on image reconstruction system, 180 of projections and 0.5 of cut off frequency are applied on the original image. Eight different filters that are used to evaluate and analyze are Ram-Lak, Shepp-Logan, Hamming, Hann, Bartlett and Bartlett-Han. The effect of cut-off frequency of filters on the quality of reconstructed image is also analysed.

3.5 Image Reconstruction GUI

The image reconstruction system using Radon transform GUI (Graphical User Interface) is built by using the App Designer in MATLAB. This GUI is created and designed to provide and present a concise summary of the image output once each of the project's primary image processing phases has been completed including image pre-processing, Radon transformation and image reconstruction. Relevant display functions and visual components were required to design in the GUI to get a precise layout. The four processed images that needed to show in GUI are original image, pre-processed image, sinogram and reconstructed image. The number of projections can be adjusted and the type of filters can be selected in the GUI. SSIM and MSE are quality assessment technique required in the GUI to evaluate the quality of reconstructed image. The GUI designed using MATLAB App Designer is shown in Fig. 5.

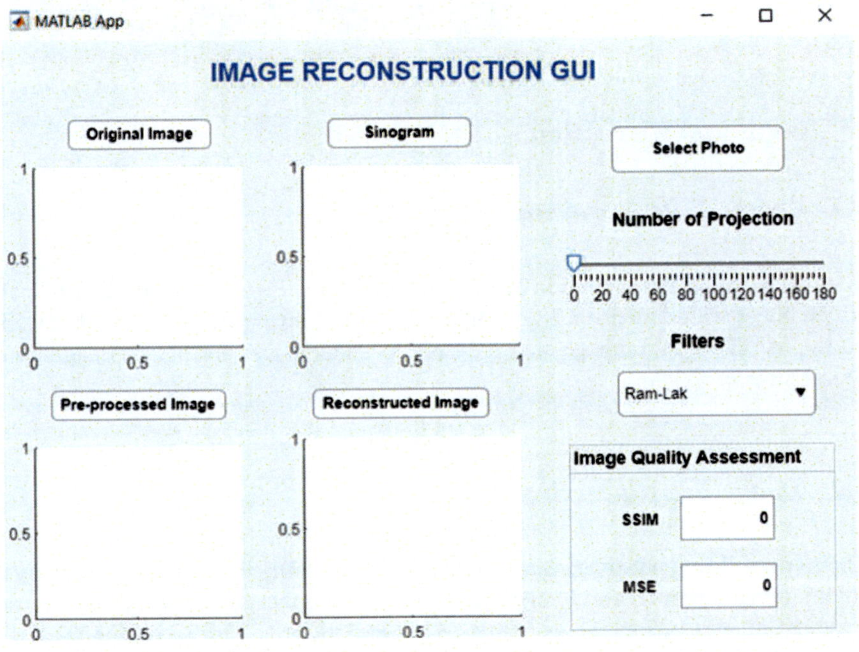

Fig. 5 Image Reconstruction GUI

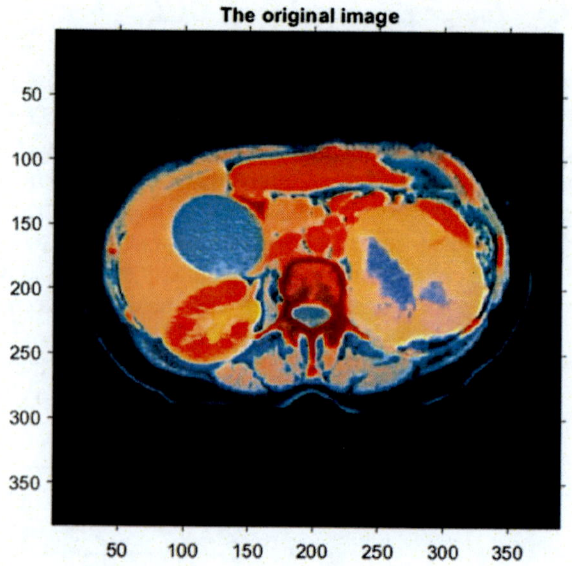

Fig. 6 The original image with scale

4 Result and Discussion

4.1 Image Data Acquisition

The image data was acquired in MATLAB workspace. Figure 6 shows the original scaled image with dimension 383 × 390 × 3 in. After image was read and displayed on the MATLAB, the image was prepared to undergo the process of image pre-processing.

4.2 Image Pre-processing

The steps of image pre-processing in the project including the conversion of image colour, image resizing, and image denoising. The original image was converted to grayscale image of CT slice of abdomen and resized to 256 × 256 pixels in MATLAB. The result image after conducting colour conversion and image resizing is in grey colour and has a size of 256 × 256 pixels as shown in Fig. 7.

The noise models of Gaussian noise and salt and pepper noise have added to the separate images to form noise images. The image denoising filtering was then applied on the noisy image to eliminate noises. SNR is used as representative quantitative measurements to evaluate the performance metrics of image denoising methods. After comparing and analysing the denoised images with different filters, it was found that the Median filtering method outperform the other filtering methods. Therefore,

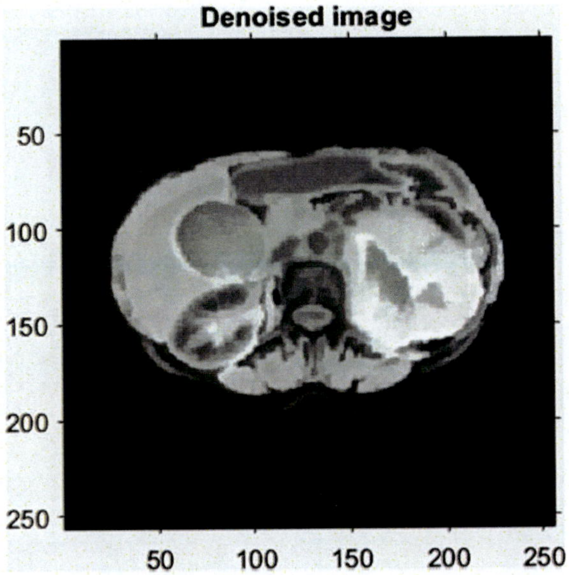

Fig. 7 The pre-processed image

Median filter was used to applied on the image with salt and pepper noise. The internal structure and main features are clearly shown in the denoised image as shown in Fig. 7.

4.3 Image Reconstruction

The sinogram was generated by using Radon transformation. Sinogram was required to create before the backprojection takes place. Figure 8 shows the non-filtered sinogram generated from the original image. The data of image was acquired into the sinogram space.

The simple backprojection generated a blurred image due to the oversampling at the center and less sampling at the edge. Appropriate filters can be used to reduce the blurring of the image and increase the signal-to-peak ratio. Filtered backprojection is a technique used to correct the blurring encountered in the simple backprojection. The ramp filter was created and applied on the sinogram. The Fourier transformation is computed to generate the filtered sinogram as shown in Fig. 9.

After the ramp filter was applied on the sinogram data, the backprojection using inverse Radon transform was done to reconstruct image. The sinogram data was backprojected at 1° and the backprojection summation the reconstructed image with 180 projections. Maximum projection was chosen to reconstruct the image because more projections produce an accurate reconstruction.

Figure 10 is the visualisation of original image and reconstructed image. The reconstructed image is shown to be a little blurry compared to the original image

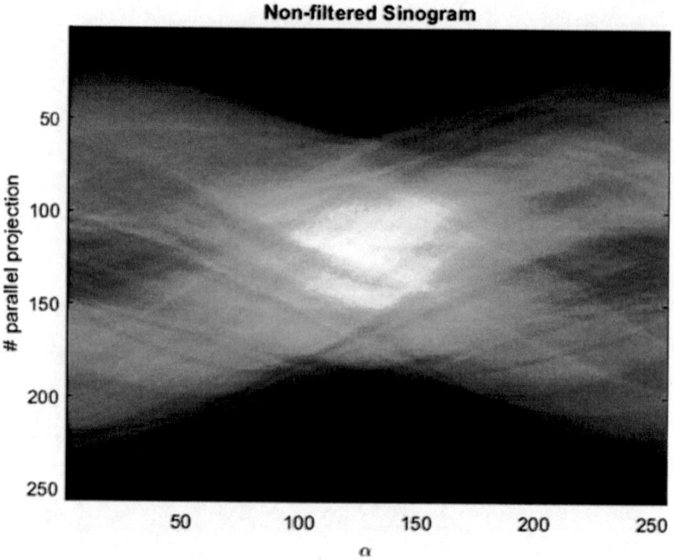

Fig. 8 The non-filtered sinogram

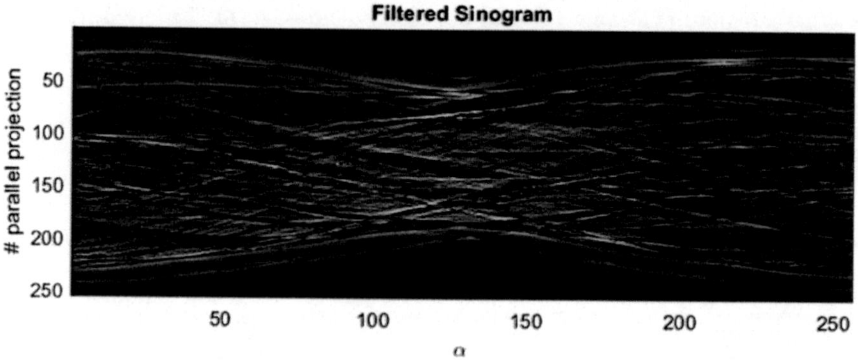

Fig. 9 The filtered sinogram

due to the processes of image transmission and image pre-processing. They have affected the quality of the reconstructed image as some information might have lost during the processes. However, the reconstruction is considered precise within the framework of image reconstruction system as the internal structure and the important features of the original image has shown in the reconstructed image.

During the image acquisition, pre-processing, transmission, and reconstruction of images, a wide range of distortions can occur, all of which can lead to a loss of visual quality. The information provided by the features of the image was distorted after the completion of image processing. The quality of the reconstructed image

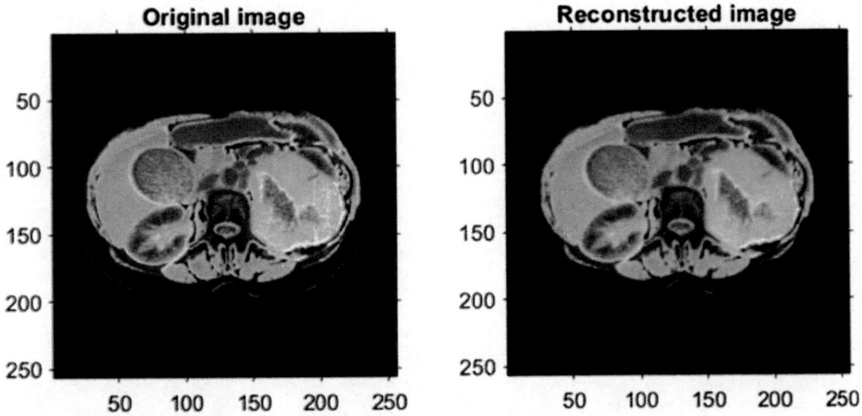

Fig. 10 Comparison of original image and reconstructed image

is comparing it with the original image by using image quality metrics. SSIM and MSE were used in evaluating and assessing the reconstructed image. The SSIM value of reconstructed image is 0.97966 and MSE obtained is 0.0003. This is shown that the reconstructed image and the original image has a high similarity after image pre-processing and image reconstruction were conducted.

4.4 Effect of Number of Projections

The number of projections and the sampling interval of the acquired projection have a substantial effect on the reconstructed image quality. Ramlak filter was used to apply on the sinogram of different number of projections in the image reconstruction system. Therefore, comparison on the quality of reconstructed image can be made under the same conditions. Several projection numbers were applied to analyse if there was a change in the quality of the reconstructed image. Six sampling data was tested by changing the number of projections including 5, 10, 20, 40, 100 and 180 of projections. Figure 11 shows the effect of number of projections on quality of image. It is shown that as the number of projections increases, the clearer the image becomes. Image with 180 projections applied on it has the highest similarity to the original image.

There are 6 of number of projections have been taken to evaluate the quality of the image. The SSIM values and MSE values obtained from the comparison between original image and reconstructed image were tabulated as shown in Table 1. The SSIM value increases and MSE value decreases as the number of projection increases. The results show that when more projections are applied, the quality of the reconstructed image is improved. This is because greater projection for a single point means a total

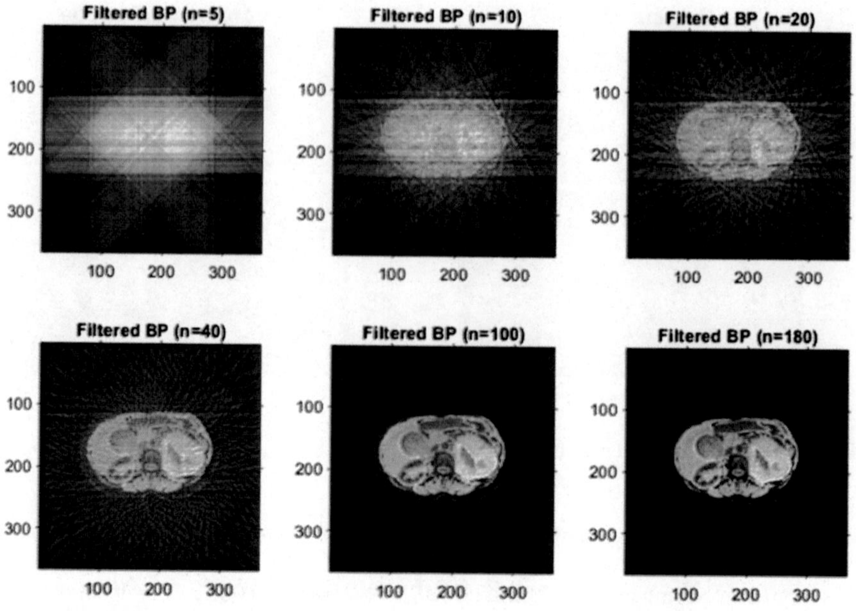

Fig. 11 The effect of number of projections on quality of reconstructed image

Table 1 The effect of number of projections

Number of projections	Image quality assessment techniques	
	SSIM	MSE
5	0.03814	1.2831
10	0.08264	0.4911
20	0.15089	0.1834
40	0.3239	0.0391
100	0.6183	0.0086
180	0.9871	0.0002

of backprojection is closer to the whole integral. As a result, additional information can be reconstructed.

4.5 Effect of Varying Filters

The effect of varying filters on the quality of reconstructed image is studied. Eight filters were applied on the sinogram to compare the original image and reconstructed image. Three traditional backprojection filters which are Ram-Lak, Shepp-Logan

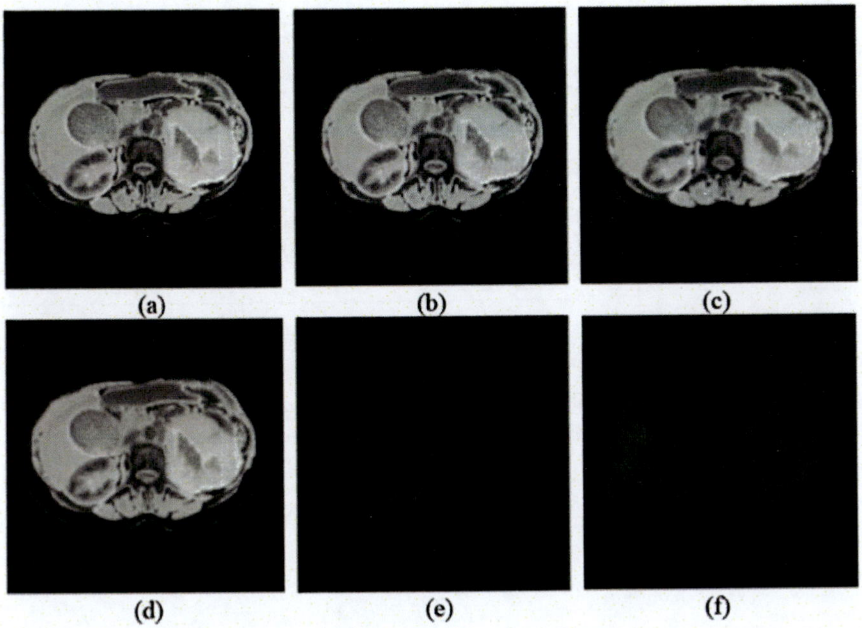

Fig. 12 The reconstructed image with different filters. **a** Reconstructed image with Ram-Lak filter, **b** Reconstructed image with Shepp-Logan filter, **c** Reconstructed image with Hamming filter, **d** Reconstructed image with Hann filter, **e** Reconstructed image with Barlett filter, **f** Reconstructed image with Barlett-Han filter

and Hamming filters and another three high-pass filters which are Hann, Bartlett and Bartlett-Han filters were used. In testing the effect of filters on the quality of reconstructed image under the same condition, 180 of projections was applied on the image and 0.5 of cut off frequency was used on all the filters.

After the filtered backprojection was conducted, visual inspection and contrast measurement were done on all the result images. Figure 12 shows the reconstructed image with different filters. The results show that backprojection with Ram-Lak filter and Shepp-logan filters generated images that has high similarity to the original image. In terms of quality of reconstructed image, all filters that applied on the sinogram produced images that look very similar to the original image except Barlett filter and Barlett-Hann filter. They kept the sharp edge of the images, which were lost when reconstructing without any filters. The structures and important features of the reconstructed images were visible. Barlett filtered backprojection had reconstructed an image that is darker in colour and the structure of the image are hardly can be seen.

SSIM and MSE values were calculated to evaluate and examine the effect of filters on the image reconstruction system. The values of quality metrics obtained were tabulated as shown in Table 2. Ram-Lak filter has the highest SSIM value and the lowest MSE value. It is shown that backprojection using Ram-Lak filter produce

Table 2 The effect of different filters

Type of filters	Image quality assessment techniques	
	SSIM	MSE
Ram-Lak	0.9839	0.0002
Shepp-Logan	0.98966	0.0003
Hamming	0.95834	0.0006
Hann	0.9553	0.0006
Barlett	0.25439	0.1171
Barlett-Hann	0.60328	0.0637

the highest similarity of reconstructed image compared to the original image. As shown in the table, Barlett filter has the smallest SSIM value and biggest MSE value among the six filters. It is proved that Barlett filter generate the poorest quality of image.

4.6 *Effect of Varying Filters*

The cut-off frequency controls how the filter affects image noise as well as resolution. A high cut-off frequency improves spatial resolution, allowing for more information to be seen. However, a high cut-off frequency remains the image noisy. A low cut-off frequency boosts smoothing but reduces image contrast in the final reconstruction. Therefore, the quality of reconstructed image is evaluated by varying the cut-off frequency of Ram-Lak filter. The reconstructed images are compared using cut-off frequency 0.2, 0.4, 0.6, 0.8, 1.0 and 1.2. Figure 13 show the reconstructed images with different cut-off frequency of Ram-Lak filter. By visual inspection, reconstructed image with cut off frequency of 0.8 has the highest similarity to the original image. The sharp edges and important features of the images are shown very clearly in the figure. The image with cut-off frequency of 1.0 and 1.2 looked very bright and the internal structures and features of the image are not visible.

The images of different cut-off frequency of the same filter were compared quantitatively using methods SSIM and MSE and the values obtained were tabulated as shown in Table 3. It is shown that the image with cut-off frequency of 0.6 has the highest SSIM value while the image with cut-off frequency of 1.2 has the lowest SSIM value. The image with 0.8 cut-off frequency of filter has the lowest MSE value whereas the images with 1.0 and 1.2 cut-off frequency has the highest MSE value. Both images with cut-off frequency of 1.0 and 1.2 have lower SSIM values and higher MSE values indicating they have the worst image quality.

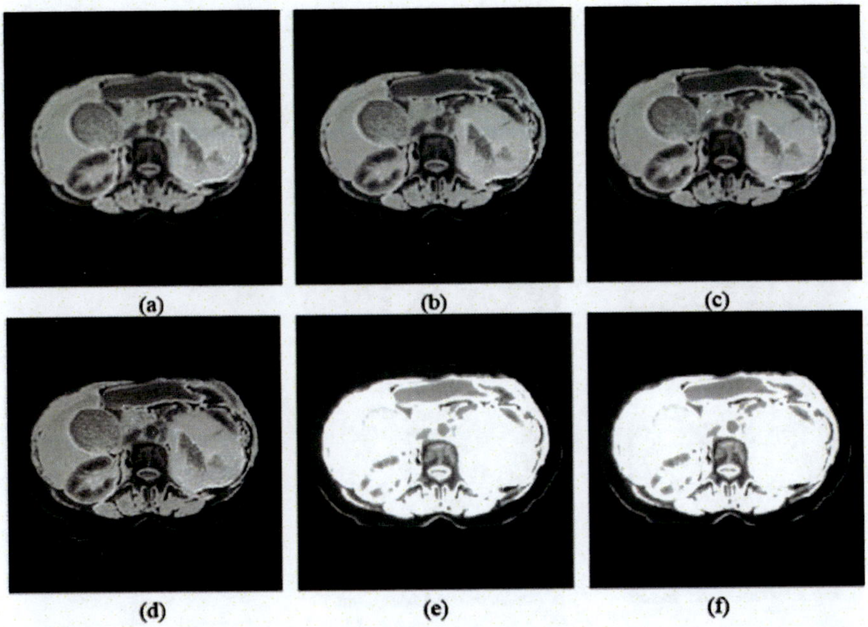

Fig. 13 The reconstructed image with varying cut-off frequencies. **a** cut-off frequency = 0.2, **b** cut-off frequency = 0.4, **c** cut-off frequency = 0.6, **d** cut-off frequency = 0.8, **e** cut-off frequency = 1.0, **f** cut-off frequency = 1.2

Table 3 The effect of varying cut-off frequency

Cut-off frequency	Image quality assessment techniques	
	SSIM	MSE
0.2	0.92882	0.0008
0.4	0.98235	0.0002
0.6	0.98399	0.0002
0.8	0.97525	0.0001
1.0	0.77626	0.1189
1.2	0.77512	0.1189

4.7 *Image Reconstruction GUI*

The Image Reconstruction GUI was designed and created to display the result image of each image processing step. The functionality and flexibility of GUI was tested with two different sample images of CT slice of medical images. One image tested is the CT slice of abdomen and another is the CT slice of brain. The images were reconstructed as expected as shown in Figs. 14, 15 and 16. The original image, pre-processed image, sinogram and reconstructed image are displayed on the GUI after

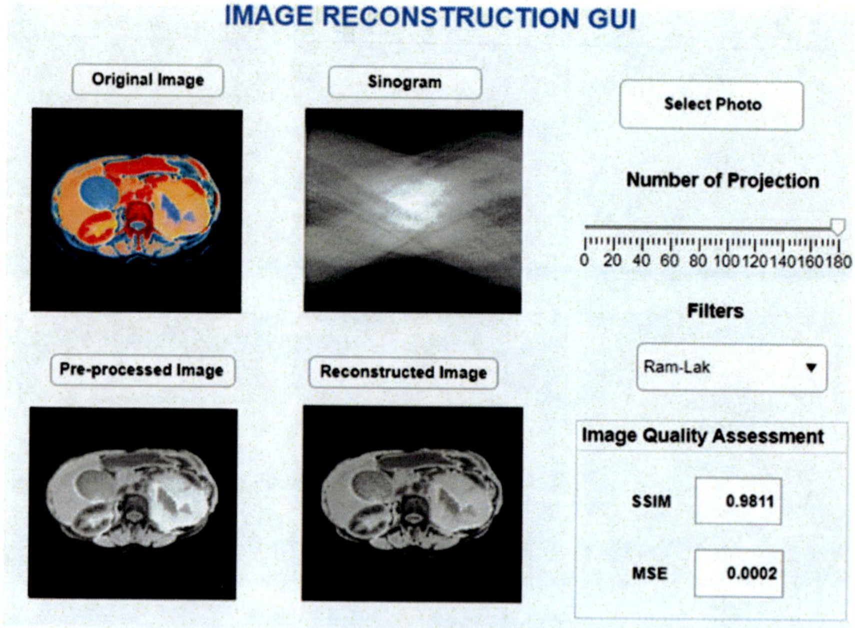

Fig. 14 The Image Reconstruction GUI (CT slice of abdomen, n = 180)

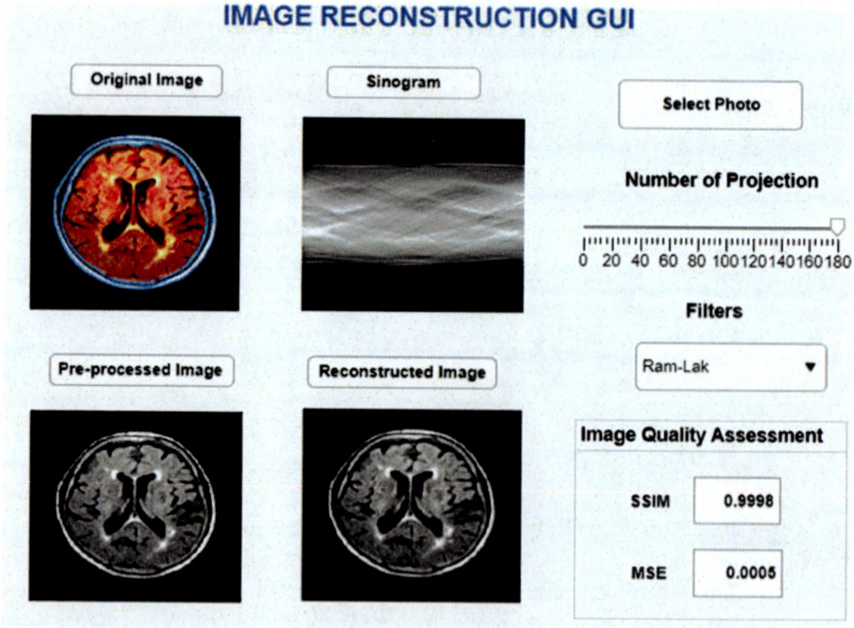

Fig. 15 The Image Reconstruction GUI (CT slice of brain, n = 180)

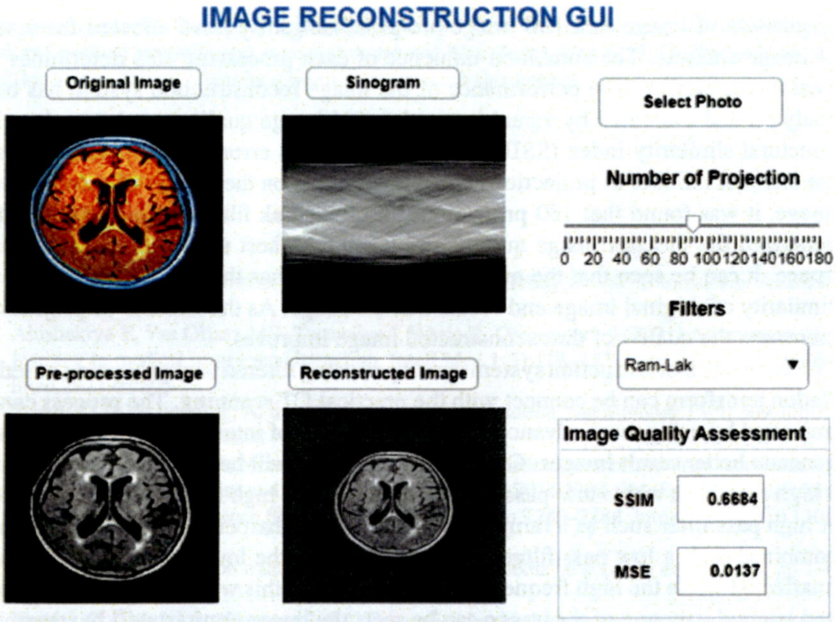

Fig. 16 The Image Reconstruction GUI (CT slice of brain, n = 90)

the button were pressed to callback each of the UI (User Interface) components' function. 180 projections are adjusted to applied on the image and Ram-Lak filter was selected to evaluate the performance of the image reconstruction GUI developed in MATLAB. The number of projections was also tested on the GUI. Figure 15 shown the result of image reconstruction GUI with sample image of CT slice of brain by using 180 projections while Fig. 16 shows 90 projections were applied on the image reconstruction system in GUI. It is shown that the larger the number of projections applied on the original image, the more accurate the reconstructed image becomes.

5 Conclusion and Recommendations

All of the projects' goals were achieved and all three of the project's objectives were met as well, therefore this project is considered as completed successfully. The framework of the image reconstruction system was developed using MATLAB. Multiple steps are involved in the image reconstruction process. The image reconstruction process using filtered backprojection algorithm based on Radon transform was conducted successfully. There are several image processing processes in this approach, each of which has an impact on image quality. It is essential to determine the impact of each step on image quality by thorough study. This is because